Advances in Experimental Medicine and Biology

Volume 1217

AF847521

Series Editors

More information about this series at http://www.springer.com/series/5584

Yi Sun • Wenyi Wei • Jianping Jin
Editors

Cullin-RING Ligases and Protein Neddylation

Biology and Therapeutics

Springer

Editors
Yi Sun
Radiation Oncology
University of Michigan
Ann Arbor, Michigan, USA

Wenyi Wei
Pathology
Beth Israel Deaconess Medical Center,
Harvard Medical School
Boston, Massachusetts, USA

Jianping Jin
Life Science Institute
Zhejiang University
Hangzhou, China

ISSN 0065-2598 ISSN 2214-8019 (electronic)
Advances in Experimental Medicine and Biology
ISBN 978-981-15-1024-3 ISBN 978-981-15-1025-0 (eBook)
https://doi.org/10.1007/978-981-15-1025-0

This Springer imprint is published by the registered company Springer Nature Singapore Pte Ltd.
The registered company address is: 152 Beach Road, #21-01/04 Gateway East, Singapore 189721, Singapore

Yi Sun would like to dedicate this book to his dear father Qian Sun and dear mother Ruiying Zhang, particularly to his mother, who died on October 22, 2019 from recurrent esophageal squamous cell carcinoma 4.5 years post radiation therapy; and to his dear wife, Hua Li, for her 34 years of companionship, heartfelt love and continuous support.

Contents

Introduction

1

Yi Sun

Abstract

Protein ubiquitylation is one of the most important posttranslational protein modifications, catalyzed by an enzyme cascade of E1/E2/E3. Neddylation, like ubiquitylation, is also catalyzed by an E1/E2/E3 enzyme cascade to covalently attach the ubiquitin-like molecule NEDD8 to a lysine residue of a substrate, not for degradation, but for modulation of substrate activity. The best known neddylation substrates are cullin family members, the scaffold component of the cullin-RING ligase (CRL), which is the largest family of E3 ligases that catalyze the ubiquitylation of ~20% of cellular proteins doomed for the degradation by proteasome system. The activity of CRLs requires cullin neddylation, which facilitates the adaptation of CRLs to an open conformation for easy substrate access. Since the substrates of CRLs are various key molecules that regulate a variety of cellular functions, CRLs and their neddylation activation, therefore, play the essential roles in many biological processes. Indeed, CRLs are abnormally regulated in many human diseases and serve as the therapeutic targets at least for cancer. This book will summarize most recent progress in this field with three sections, covering (1) structure and regulation of CRL and neddylation, (2) biological function of each CRLs, and (3) drug discovery efforts to target CRL/neddylation for cancer therapy.

Keywords

Cullin-RING ligase · Neddylation · RBX1/2 · UBE2F/2M · Ubiquitylation

Abbreviations

APC/C	Anaphase-promoting complex/cyclosome
CRL	Cullin-RING ligase
CSN	COP9 signalosome
DND	Deneddylase
DUBs	Deubiquitylases
HECT	Homologous to E6-AP C-terminus
NAE	NEDD8-activating enzyme
NEDD8	Neural precursor cell expressed developmentally downregulated protein 8
RBX1	RING-box protein 1
RBX2	RING-box protein 2
RING	Really interesting new gene
SAG	Sensitive to apoptosis gene
UAE	Ubiquitin-activating enzyme
UBL	Ubiquitin-like

Y. Sun (✉)
Division of Radiation and Cancer Biology, Department of Radiation Oncology, University of Michigan, Ann Arbor, MI, USA
e-mail: sunyi@umich.edu

Y. Sun et al. (eds.), *Cullin-RING Ligases and Protein Neddylation*, Advances in Experimental Medicine and Biology 1217, https://doi.org/10.1007/978-981-15-1025-0_1

1.1 Protein Ubiquitylation

Protein ubiquitylation is one of the most important posttranslational protein modifications which plays fundamental role in the maintenance of cell homeostasis and for normal cell physiology and many cellular functions (Ciechanover 1998; Hershko and Ciechanover 1998). Protein ubiquitylation is catalyzed by three sequential enzymatic reactions: First, the C-terminal Gly residue of ubiquitin is activated in the presence of ATP by an E1 activating enzyme to form a thioester linkage between ubiquitin and a Cys residue of E1 active center. Second, activated ubiquitin is transferred to an active site Cys residue of an E2 conjugating enzyme also via a thioester linkage. Finally, ubiquitin is linked by its C-terminus in an amide isopeptide linkage to an ε-amino group of the Lys residues of a substrate protein, catalyzed by an E3 ubiquitin ligase (Hershko and Ciechanover 1998). One run of this reaction causes substrate monoubiquitylation, which normally alters protein localization or function. Multiple runs of this reaction lead to formation of polyubiquitylated substrates, which are recognized and then degraded by proteasome system or possess altered function in a manner dependent of the linkage between ubiquitins. Ubiquitin is removed from ubiquitylated substrates and recycled by deubiquitylase (DUB) (Fig. 1.1, top panel). The UPS is abnormally regulated in many human diseases, particularly neurodegenerative diseases and cancers (Ciechanover 2003; Ciechanover and Iwai 2004). The UPS as an attractive anticancer target was fully demonstrated by the fact that bortezomib (Velcade), a first-in-class proteasome inhibitor, has been used to treat patients with multiple myeloma and relapsed mantle cell lymphoma since 2003 (Orlowski and Kuhn 2008; Speedy approvals for new cancer treatments 2003).

In mammalian cells, there are 2 types of E1s, 38 of E2s, more than 600 of E3s (Zhao and Sun 2013), and 106 DUBs (Harrigan et al. 2018). Structurally, the E3 ligases are categorized into four major types: N-end rule E3s; HECT (homologous to E6-AP C-terminus) domain-containing E3s; RING (really interesting new gene) finger-containing E3s; and APC/C (anaphase-promoting complex/cyclosome) E3 (Hershko and Ciechanover 1998; Zhao and Sun 2013; Zheng and Shabek 2017; Zhang et al. 2014). The HECT E3s contain a conserved cysteine residue that forms a thioester intermediate with ubiquitin from ubiquitin-loaded E2 and pass ubiquitin onto a substrate. In contrast, the RING-finger containing E3s promote the direct transfer of ubiquitin from the ubiquitin-loaded E2 to a substrate (Hershko and Ciechanover 1998; Lipkowitz and Weissman 2011). The RING-finger E3s have two subclasses: single peptide E3s, containing both the RING-finger domain and substrate-binding motif (e.g., MDM2) (Haupt et al. 1997; Kubbutat et al. 1997), and the multicomponent-containing E3s, in which the RING-finger and substrate-binding units are assembled together from different individual E3s (such as CRLs) (Sun 2003; Willems et al. 2004).

1.2 Protein Neddylation

Protein neddylation is a process of tagging NEDD8 onto a substrate protein, not for degradation, but for modulation of protein activity and function. NEDD8 is one of the most studied UBL (ubiquitin-like) proteins and is 60% identical and 80% homologous to ubiquitin (Kamitani et al. 1997). Unlike ubiquitin, NEDD8 is first synthesized as a precursor with five additional residues downstream from Gly76, which are cleaved by C-terminal hydrolases, including UCH-L3 (Johnston et al. 1997; Larsen et al. 1996) and NEDP1/DEN1/SENP8 (Gan-Erdene et al. 2003; Mendoza et al. 2003; Wu et al. 2003). After this processing, NEDD8, like ubiquitin, is attached covalently to its substrates by an isopeptide linkage between the C-terminal Gly76 of NEDD8 and a lysine residue of the target protein, catalyzed by an E1 NEDD8-activating enzyme (NAE), an E2 NEDD8-conjugating enzyme, and an E3 ligase, and ultimately conjugates NEDD8 to its specific substrates. Neddylated substrates can be

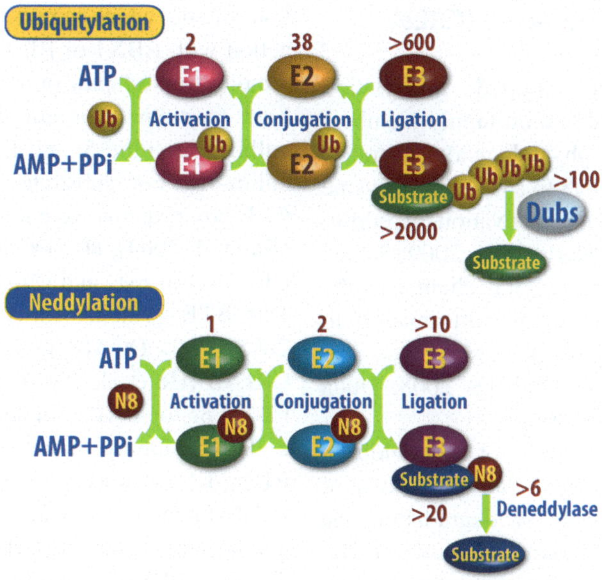

Fig. 1.1 Protein modifications by ubiquitylation and neddylation. Protein ubiquitylation is catalyzed by E1 ubiquitin-activating enzyme, E2 ubiquitin-conjugating enzyme, and E3 ubiquitin ligase. Ubiquitylated substrates are deubiquitylated by deubiquitylase (DUB). In mammals, there are 2 E1s, 38 E2s, more than 600 E3s, over hundreds and thousands of substrates, and 100 DUBs. Likewise, protein neddylation is catalyzed by E1 NEDD8-activating enzyme, E2 NEDD8-conjugating enzyme, and E3 NEDD8 ligase. Neddylated substrates are deneddylated by deneddylase (DND). In mammals, there are 1 E1, 2 E2s, over 10 E3s, about 20 substrates, and a little over 6 DNDs

deneddylated by a DND, such as CSN, and the free NEDD8 is recycled (Fig. 1.1, bottom panel).

In mammalian cells, there are a single E1 (NAE), two E2s (UBE2M, also known as UBC12, and UBE2F), and a dozen or so E3s (Fig. 1.2). NAE is a heterodimer, consisting of a regulatory subunit NAE1/APPBP1 and a catalytic subunit UBA3/NAEβ (Bohnsack and Haas 2003), and is structurally and biochemically similar to ubiquitin-activating enzyme (UAE). NAE1 and UBA3 are homologous to the amino and carboxyl regions of UAE, respectively. UBE2M preferentially promotes neddylation of RBX1-associated cullins (CUL1–3, CUL4A, and CUL4B), whereas UBE2F promotes neddylation of RBX2-associated CUL5 (Huang et al. 2009). The majority of NEDD8 E3 ligases contain a RING domain, including c-CBL (Oved et al. 2006; Zuo et al. 2013), FBXO11 (Abida et al. 2007), IAPs (Broemer et al. 2010; Nagano et al. 2012), MDM2 (Xirodimas et al. 2004), RBX1 and RBX2 (Huang et al. 2009; Kamura et al. 1999; Duan et al. 1999), RNF111 (Ma et al. 2013), TFB3 (Rabut et al. 2011), and TRIM40 (Noguchi et al. 2011). DCN1 and its family members (Kurz et al. 2005, 2008), however, do not contain a RING domain for its catalytic activity, but cooperate with RBX1 to facilitate cullin neddylation (Meyer-Schaller et al. 2009). Furthermore, DCN1 directly interacts with UBE2M E2 on a surface that overlaps with the E1-binding site (Kurz et al. 2008). More strikingly, crystal structure analysis revealed that the N-terminal methionine of UBE2M is acetylation, which along with few downstream amino acid residues is inserted into a hydrophobic pocket of DCN1 to promote cullin neddylation (Scott et al. 2011). Disruption of this UBE2M-DCN1 binding by a small molecule (DI591), discovered via structure-based design, selectively inhibited cullin-3 neddylation, leading to the accumulation of its substrate NRF2 (Zhou et al. 2017).

1.3 Cullin-RING Ligases (CRLs)

Among a dozen of reported neddylation substrates (Fig. 1.2), only cullin family members are considered as the physiological substrates. The attachment of NEDD8 to a C-terminal lysine residue of cullins leads to activation of cullin-RING ligases (CRLs) (Merlet et al. 2009; Sakata et al. 2007), owing to structural change in the CRLs complex with an open conformation to facilitate the access of the substrates for ubiquitylation (Saha and Deshaies 2008; Duda et al. 2008; Zheng et al. 2002).

CRL is the largest family of E3 ubiquitin ligase, responsible for ubiquitylation of ~20% of cellular proteins doomed for degradation via ubiquitin-proteasome system (Soucy et al. 2009). CRL is a multiunit E3, consisting of four components, including a cullin, an adaptor protein, a substrate-recognizing receptor, and one RING protein. Structurally, cullins act as scaffold proteins with the N-terminus binding to an adaptor protein, a substrate receptor protein, and the C-terminus binding to a RING component protein (Fig. 1.3) (Zhao and Sun 2013). In the human genome, eight cullin family members (CUL1, CUL2, CUL3, CUL4A, CUL4B, CUL5, CUL7, and CUL9) were identified with an evolutionarily conserved cullin homology domain at the C-terminus, which is responsible for cullin interaction with RBX1 or RBX2 (Zheng et al. 2002). Among four adaptor proteins, SKP1 is shown to bind to cullin-1 or cullin-7, Elongin B/C for cullin-2 or cullin-5, and DDB1 for cullin-4A or cullin-4B. For substrate recognition receptors, 69 F-box proteins were identified for CRL1/SCF (Jin et al. 2004), 80 SOCS proteins for CRL2 or CRL5 (Linossi and Nicholson 2012), about 180 BTB proteins for CRL3 (Stogios et al. 2005), 90 DCAF proteins for CRL4A or CRL4B (He et al. 2006), and 1 adaptor protein, FBXW8, for CRL7 (Scheufele et al. 2014) and 2 RING family members RBX1/ROC1 and RBX2/ROC2 (also known as SAG) with a RING-finger domain at the C-terminus to bind 2 zinc atoms via the $C_3H_2C_3$ motif (Fig. 1.3). Thus, CRLs consist of more than 400 components, which form 8 cullin-based RING-dependent E3 ubiquitin ligases, and are responsible for ubiquitylation and degradation of thousands of substrates (Deshaies and Joazeiro 2009; Sarikas et al. 2011).

In general, most of short-lived proteins are the substrates of CRLs, particularly for CRL1, including cell cycle regulators, signal molecules, transcription factors, and proteins regulating translation and apoptosis. The major substrate of CRL2 is HIF1α, which plays a critical role in

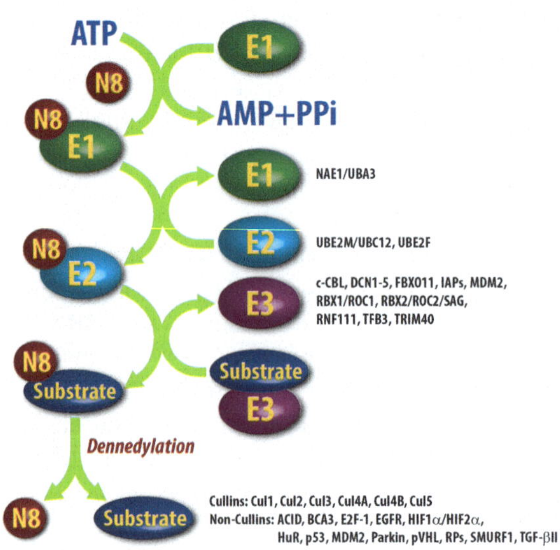

Fig. 1.2 Protein neddylation by three catalytic enzymes and neddylation substrates. Protein neddylation is catalyzed by E1 NEDD8-activating enzyme, which is a heterodimer of NAE1/APPBP1 (regulatory subunit) and UBA3/NAEβ (catalytic subunit); E2 NEDD8-conjugating enzymes with two family members: UBE2M/UBC12 and UBE2F; and E3 NEDD8 ligases. The physiological substrates of neddylation are cullin family members

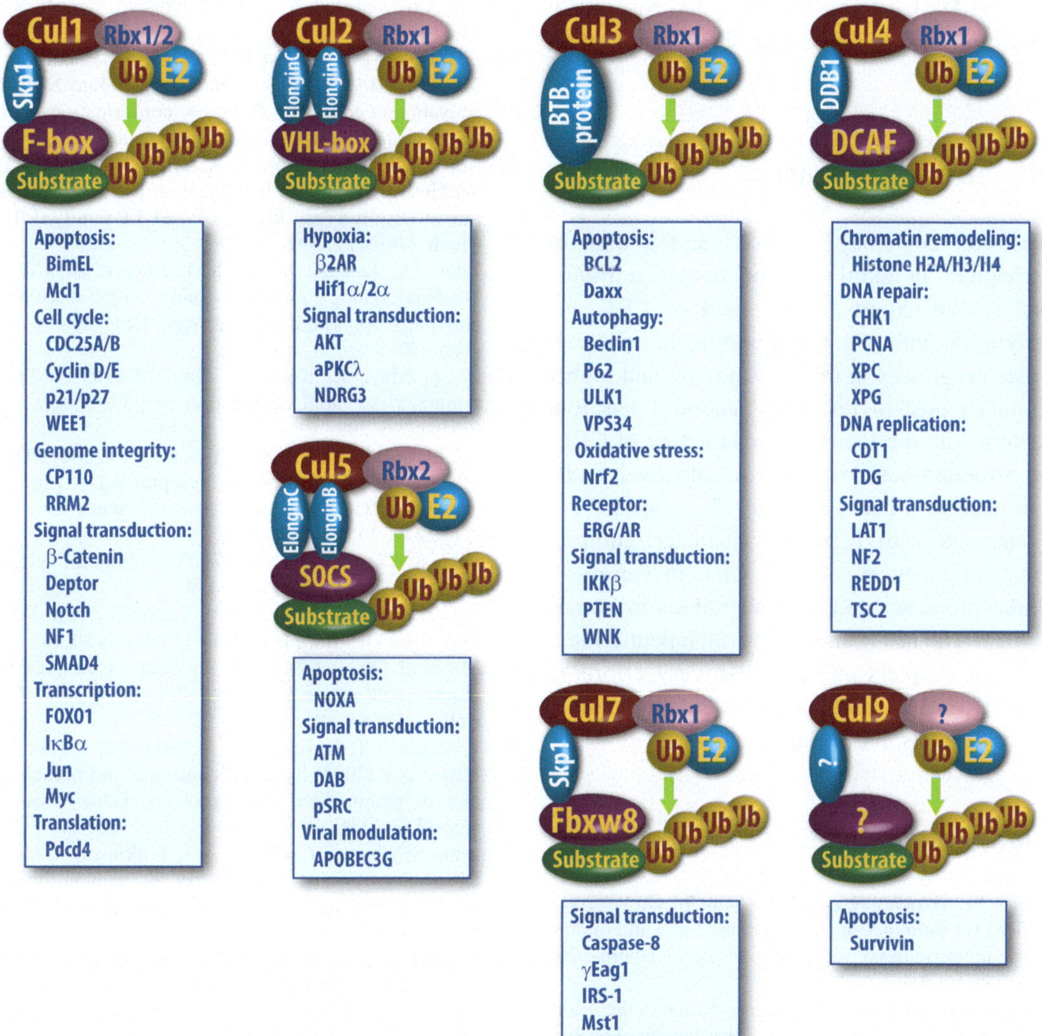

Fig. 1.3 Family of cullin-RING ligases: components and substrates. Cullin-RING ligases (CRLs) consist of four components: a scaffold cullin, an adaptor protein, a substrate-recognizing subunit, and a RING protein. Eight members of cullin family were identified including cullin-1, cullin-2, cullin-3, cullin-4A, cullin-4B, cullin-5, cullin-7, and cullin-9. The RING protein RBX1 prefers to bind to cullins 1–5 and 7, whereas RBX2 prefers to bind to cullin-5 as well as cullin-1. CRLs are the largest family of E3 ligases, which catalyze the ubiquitylation of ~20% cellular proteins doomed for proteasome degradation

regulation of hypoxia response, whereas the major substrate of CRL3 is NRF2, an antioxidant transcription factor responsible for oxidative stress. The CRL4 substrates play the essential role in DNA replication and DNA damage response, while the CRL5 substrates are mainly involved in regulation of apoptosis and viral infection. The substrates for CRL7 and CRL7 are rarely studied with very limited substrates identified (Fig. 1.3). Given the substrates of CRLs are various key molecules that regulate a variety of cellular functions, it is not surprising that CRLs and their neddylation activation play the essential roles in many biological processes. Indeed, CRLs are abnormally regulated in many human diseases and served as the therapeutic targets at least for cancer (Zhao and Sun 2013; Skaar et al. 2014; Zhao et al. 2014; Zhou et al. 2013, 2019).

1.4 Book Contents

This book contains 3 sections and a total of 20 chapters, in addition to this overall introduction, by the experts in the field of CRL and neddylation with summarization of the most up-to-date progresses in this fast-moving and highly significant area of research. Section 1 has four chapters with the focus on the structure of CRLs and structure-based regulation of neddylation modification. The Section 2 consists of 12 chapters with focus on characterization of biological functions of each CRL in variety of cellular processes and their alterations in human diseases. The last section is the therapeutic focus with four chapters on drug discovery efforts to target CRLs and neddylation for the treatment of human diseases, particularly cancer.

References

Abida WM, Nikolaev A, Zhao W, Zhang W, Gu W (2007) FBXO11 promotes the neddylation of p53 and inhibits its transcriptional activity. J Biol Chem 282 (3):1797–1804

Bohnsack RN, Haas AL (2003) Conservation in the mechanism of Nedd8 activation by the human AppBp1-Uba3 heterodimer. J Biol Chem 278(29):26823–26830

Broemer M, Tenev T, Rigbolt KTG, Hempel S, Blagoev B, Silke J, Ditzel M, Meier P (2010) Systematic in vivo RNAi analysis identifies IAPs as NEDD8-E3 ligases. Mol Cell 40(5):810–822

Ciechanover A (1998) The ubiquitin-proteasome pathway: on protein death and cell life. EMBO J 17 (24):7151–7160

Ciechanover A (2003) The ubiquitin proteolytic system and pathogenesis of human diseases: a novel platform for mechanism-based drug targeting. Biochem Soc Trans 31(2):474–481

Ciechanover A, Iwai K (2004) The ubiquitin system: from basic mechanisms to the patient bed. IUBMB Life 56 (4):193–201

Deshaies RJ, Joazeiro CA (2009) RING domain E3 ubiquitin ligases. Annu Rev Biochem 78:399–434

Duan H, Wang Y, Aviram M, Swaroop M, Loo JA, Bian J, Tian Y, Mueller T, Bisgaier CL, Sun Y (1999) SAG, a novel zinc RING finger protein that protects cells from apoptosis induced by redox agents. Mol Cell Biol 19:3145–3155

Duda DM, Borg LA, Scott DC, Hunt HW, Hammel M, Schulman BA (2008) Structural insights into NEDD8 activation of cullin-RING ligases: conformational control of conjugation. Cell 134(6):995–1006

Gan-Erdene T, Nagamalleswari K, Yin L, Wu K, Pan ZQ, Wilkinson KD (2003) Identification and characterization of DEN1, a deneddylase of the ULP family. J Biol Chem 278(31):28892–28900

Harrigan JA, Jacq X, Martin NM, Jackson SP (2018) Deubiquitylating enzymes and drug discovery: emerging opportunities. Nat Rev Drug Discov 17 (1):57–78

Haupt Y, Maya R, Kazaz A, Oren M (1997) Mdm2 promotes the rapid degradation of p53. Nature 387 (6630):296–299

He YJ, McCall CM, Hu J, Zeng Y, Xiong Y (2006) DDB1 functions as a linker to recruit receptor WD40 proteins to CUL4-ROC1 ubiquitin ligases. Genes Dev 20 (21):2949–2954

Hershko A, Ciechanover A (1998) The ubiquitin system. Annu Rev Biochem 67:425–479

Huang DT, Ayrault O, Hunt HW, Taherbhoy AM, Duda DM, Scott DC, Borg LA, Neale G, Murray PJ, Roussel MF et al (2009) E2-RING expansion of the NEDD8 cascade confers specificity to cullin modification. Mol Cell 33(4):483–495

Jin J, Cardozo T, Lovering RC, Elledge SJ, Pagano M, Harper JW (2004) Systematic analysis and nomenclature of mammalian F-box proteins. Genes Dev 18 (21):2573–2580

Johnston SC, Larsen CN, Cook WJ, Wilkinson KD, Hill CP (1997) Crystal structure of a deubiquitinating enzyme (human UCH-L3) at 1.8 A resolution. EMBO J 16(13):3787–3796

Kamitani T, Kito K, Nguyen HP, Yeh ETH (1997) Characterization of NEDD8, a developmentally down-regulated ubiquitin-like protein. J Biol Chem 272 (45):28557–28562

Kamura T, Conrad MN, Yan Q, Conaway RC, Conaway JW (1999) The Rbx1 subunit of SCF and VHL E3 ubiquitin ligase activates Rub1 modification of cullins Cdc53 and Cul2. Genes Dev 13(22):2928–2933

Kubbutat MH, Jones SN, Vousden KH (1997) Regulation of p53 stability by Mdm2. Nature 387(6630):299–303

Kurz T, Ozlu N, Rudolf F, O'Rourke SM, Luke B, Hofmann K, Hyman AA, Bowerman B, Peter M (2005) The conserved protein DCN-1/Dcn1p is required for cullin neddylation in C. elegans and S. cerevisiae. Nature 435(7046):1257–1261

Kurz T, Chou Y-C, Willems AR, Meyer-Schaller N, Hecht M-L, Tyers M, Peter M, Sicheri F (2008) Dcn1 functions as a scaffold-type E3 ligase for Cullin neddylation. Mol Cell 29(1):23–35

Larsen CN, Price JS, Wilkinson KD (1996) Substrate binding and catalysis by ubiquitin C-terminal hydrolases: identification of two active site residues. Biochemistry 35(21):6735–6744

Linossi EM, Nicholson SE (2012) The SOCS box-adapting proteins for ubiquitination and proteasomal degradation. IUBMB Life 64(4):316–323
Lipkowitz S, Weissman AM (2011) RINGs of good and evil: RING finger ubiquitin ligases at the crossroads of tumour suppression and oncogenesis. Nat Rev Cancer 11(9):629–643
Ma T, Chen Y, Zhang F, Yang C-Y, Wang S, Yu X (2013) RNF111-dependent neddylation activates DNA damage-induced ubiquitination. Mol Cell 49 (5):897–907
Mendoza HM, Shen LN, Botting C, Lewis A, Chen J, Ink B, Hay RT (2003) NEDP1, a highly conserved cysteine protease that deNEDDylates cullins. J Biol Chem 278(28):25637–25643
Merlet J, Burger J, Gomes JE, Pintard L (2009) Regulation of cullin-RING E3 ubiquitin-ligases by neddylation and dimerization. Cell Mol Life Sci 66(11–12):1924–1938
Meyer-Schaller N, Chou Y-C, Sumara I, Martin DDO, Kurz T, Katheder N, Hofmann K, Berthiaume LG, Sicheri F, Peter M (2009) The human Dcn1-like protein DCNL3 promotes Cul3 neddylation at membranes. Proc Natl Acad Sci 106(30):12365–12370
Nagano T, Hashimoto T, Nakashima A, Kikkawa U, Kamada S (2012) X-linked inhibitor of apoptosis protein mediates neddylation by itself but does not function as a NEDD8–E3 ligase for caspase-7. Febs Lett 586(11):1612–1616
Noguchi K, Okumura F, Takahashi N, Kataoka A, Kamiyama T, Todo S, Hatakeyama S (2011) TRIM40 promotes neddylation of IKKγ and is downregulated in gastrointestinal cancers. Carcinogenesis 32(7):995–1004
Orlowski RZ, Kuhn DJ (2008) Proteasome inhibitors in cancer therapy: lessons from the first decade. Clin Cancer Res 14(6):1649–1657
Oved S, Mosesson Y, Zwang Y, Santonico E, Shtiegman K, Marmor MD, Kochupurakkal BS, Katz M, Lavi S, Cesareni G et al (2006) Conjugation to Nedd8 instigates ubiquitylation and down-regulation of activated receptor tyrosine kinases. J Biol Chem 281(31):21640–21651
Rabut G, Le Dez G, Verma R, Makhnevych T, Knebel A, Kurz T, Boone C, Deshaies Raymond J, Peter M (2011) The TFIIH subunit Tfb3 regulates Cullin neddylation. Mol Cell 43(3):488–495
Saha A, Deshaies RJ (2008) Multimodal activation of the ubiquitin ligase SCF by Nedd8 conjugation. Mol Cell 32(1):21–31
Sakata E, Yamaguchi Y, Miyauchi Y, Iwai K, Chiba T, Saeki Y, Matsuda N, Tanaka K, Kato K (2007) Direct interactions between NEDD8 and ubiquitin E2 conjugating enzymes upregulate cullin-based E3 ligase activity. Nat Struct Mol Biol 14(2):167–168
Sarikas A, Hartmann T, Pan ZQ (2011) The cullin protein family. Genome Biol 12(4):220
Scheufele F, Wolf B, Kruse M, Hartmann T, Lempart J, Muhlich S, Pfeiffer AFH, Field LJ, Charron MJ, Pan ZQ et al (2014) Evidence for a regulatory role of Cullin-RING E3 ubiquitin ligase 7 in insulin signaling. Cell Signal 26(2):233–239
Scott DC, Monda JK, Bennett EJ, Harper JW, Schulman BA (2011) N-terminal acetylation acts as an avidity enhancer within an interconnected multiprotein complex. Science 334(6056):674–678
Skaar JR, Pagan JK, Pagano M (2014) SCF ubiquitin ligase-targeted therapies. Nat Rev Drug Discov 13 (12):889–903
Soucy TA, Smith PG, Milhollen MA, Berger AJ, Gavin JM, Adhikari S, Brownell JE, Burke KE, Cardin DP, Critchley S et al (2009) An inhibitor of NEDD8-activating enzyme as a new approach to treat cancer. Nature 458(7239):732–736
Speedy approvals for new cancer treatments (2003) FDA Consum 37(4):36
Stogios PJ, Downs GS, Jauhal JJ, Nandra SK, Prive GG (2005) Sequence and structural analysis of BTB domain proteins. Genome Biol 6(10):R82
Sun Y (2003) Targeting E3 ubiquitin ligases for cancer therapy. Cancer Biol Therapy 2:623–629
Willems AR, Schwab M, Tyers M (2004) A hitchhiker's guide to the cullin ubiquitin ligases: SCF and its kin. Biochim Biophys Acta 1695(1–3):133–170
Wu K, Yamoah K, Dolios G, Gan-Erdene T, Tan P, Chen A, Lee CG, Wei N, Wilkinson KD, Wang R et al (2003) DEN1 is a dual function protease capable of processing the C terminus of Nedd8 and deconjugating hyper-neddylated CUL1. J Biol Chem 278(31):28882–28891
Xirodimas DP, Saville MK, Bourdon J-C, Hay RT, Lane DP (2004) Mdm2-mediated NEDD8 conjugation of p53 inhibits its transcriptional activity. Cell 118 (1):83–97
Zhang J, Wan L, Dai X, Sun Y, Wei W (2014) Functional characterization of anaphase promoting complex/ cyclosome (APC/C) E3 ubiquitin ligases in tumorigenesis. Biochim Biophys Acta 1845 (2):277–293
Zhao Y, Sun Y (2013) Cullin-RING ligases as attractive anti-cancer targets. Curr Pharm Des 19(18):3215–3225
Zhao Y, Morgan MA, Sun Y (2014) Targeting neddylation pathways to inactivate Cullin-RING ligases for anti-cancer therapy. Antioxid Redox Signal 21 (17):2383–2400
Zheng N, Shabek N (2017) Ubiquitin ligases: structure, function, and regulation. Annu Rev Biochem 86:129–157
Zheng N, Schulman BA, Song LZ, Miller JJ, Jeffrey PD, Wang P, Chu C, Koepp DM, Elledge SJ, Pagano M et al (2002) Structure of the Cul1-Rbx1-Skp1-F boxSkp2 SCF ubiquitin ligase complex. Nature 416 (6882):703–709

Zhou H, Lu J, Liu L, Bernard D, Yang CY, Fernandez-Salas E, Chinnaswamy K, Layton S, Stuckey J, Yu Q et al (2017) A potent small-molecule inhibitor of the DCN1-UBC12 interaction that selectively blocks cullin 3 neddylation. Nat Commun 8 (1):1150

Zhou L, Jiang Y, Luo Q, Li L, Jia L (2019) Neddylation: a novel modulator of the tumor microenvironment. Mol Cancer 18(1):77

Zhou W, Wei W, Sun Y (2013) Genetically engineered mouse models for functional studies of SKP1-CUL1-F-box-protein (SCF) E3 ubiquitin ligases. Cell Res 23 (5):599–619

Zuo W, Huang F, Chiang YJ, Li M, Du J, Ding Y, Zhang T, Lee Hyuk W, Jeong Lak S, Chen Y et al (2013) c-Cbl-Mediated neddylation antagonizes ubiquitination and degradation of the TGF-β Type II receptor. Mol Cell 49(3):499–510

Structural Biology of CRL Ubiquitin Ligases

2

Domnița-Valeria Rusnac and Ning Zheng

Abstract

Cullin-RING ubiquitin ligases (CRLs) represent the largest superfamily of multi-subunit E3s conserved in all eukaryotes. Soon after the discovery of these important ubiquitin ligase machineries, structural studies have made tremendous contributions to our understanding of their functions. Identification of the key components of CRLs by early studies raised immediate questions as to how these multi-subunit complexes assemble to promote the polyubiquitination of substrates. Specifically, how do the CRL subunits interact with each other to form a versatile E3 platform? How do they recognize specific substrates? How are the CRL-substrate interactions regulated in response to upstream signals? How are the CRL E3s themselves activated and deactivated, and how are substrate receptor subunits of CRLs exchanged in the cell? Even though we might not yet have complete answers to these questions, extensive structural analyses of CRL complexes in the past two decades have begun to unveil the themes and variations of CRL biology. In this chapter we will discuss both classic and emerging structures that help elucidate the overall architecture of CRLs, their substrate recognition modes, and regulatory mechanism of CRLs by NEDD8 modification.

Keywords

Cullin-RING ubiquitin ligases · Structural biology · Ubiquitination · Neddylation · DCN1

Abbreviations

APOBEC3	Apolipoprotein B mRNA editing enzyme catalytic polypeptide-like 3
ARIH1	Ariadne-1 homolog
ASK1	Apoptosis signal-regulating kinase 1
AUX/IAA	Auxin/indole-3-acetic acid
BPA	Beta-propeller A
BPB	Beta-propeller B
BPC	Beta-propeller C
BTB	Broad-complex, tramtrack, and bric-a-brac
CAND1	Cullin-associated NEDD8-dissociated protein 1
CBF-β	Core-binding factor beta
CDC34	Cell division cycle 34/ubiquitin-conjugating enzyme E3 R1

D.-V. Rusnac
Department of Pharmacology, University of Washington, Seattle, WA, USA

N. Zheng (✉)
Department of Pharmacology, University of Washington, Seattle, WA, USA

Howard Hughes Medical Institute, University of Washington, Seattle, WA, USA
e-mail: nzheng@uw.edu

Y. Sun et al. (eds.), *Cullin-RING Ligases and Protein Neddylation*, Advances in Experimental Medicine and Biology 1217, https://doi.org/10.1007/978-981-15-1025-0_2

CDK2	Cyclin-dependent kinase 2
CK1α	Casein kinase 1
CKS1	Cyclin-dependent kinases regulatory subunit 1
COI1	Coronatine-insensitive protein 1
COP9	Constitutive photomorphogenesis 9
CRBN	Cereblon
CRL	Cullin-RING ubiquitin ligases
CRY1/2	Cryptochrome circadian regulator 1/2
CSN	COP9 signalosome
CTD	C-terminal domain
CUL	Cullin
DCAFs	DDB1-CUL4A-associated factors
DCN1	Defective in cullin neddylation protein 1-like protein 1
DDB1/2	DNA damage-binding protein 1/2
EB	Elongin B
EC	Elongin C
FBXL	F-box and leucine-rich repeat proteins
FBXO	F-box and other repeat-containing proteins
FBXW	F-box and WD repeat-containing proteins
GLMN	Glomulin
GSPT1	G1 to S phase transition 1
HHARI/ARIH1	Ariadne-1 homolog
HIF1α	Hypoxia-inducible factor 1 subunit alpha
IKZF1/3	IKAROS family zinc finger 1/3
IP6	Inositol hexakisphosphate
JA	Jasmonic acid
JAZ	Jasmonate ZIM domain
KEAP1	Kelch-like ECH-associated protein 1
KLHDC2	Kelch domain-containing protein 2
LRR	Leucine-rich repeat
MATH	Meprin and TRAF-C homology
MEIS2	Myeloid ecotropic viral integration site 1 homolog 2
NEDD8	Neural precursor cell-expressed developmentally downregulated protein 8
NRF2	Nuclear factor erythroid 2-related factor 2
NTD	N-terminal domain
ODD	Oxygen-dependent degradation
PTM	Post-translational modification
RBX1/2	RING box protein 1/2
RING	Really interesting new genes
RING-IBR-RING RBR	RING – in between ring – RING
SALL4	Sal-like protein 4
SCF	SKP1, CUL1, and F-box proteins
SELK	Selenoprotein K
SELS	Selenoprotein S
SKP1/2	S-phase kinase-associated protein 1/2
SPOP	Speckle-type POZ protein
STAT	Signal transducer and activator of transcription
SV5-V	Simian virus 5 V protein
TIR1	Transport inhibitor response 1
UBC12/UBE2M	Ubiquitin-conjugating enzyme E2 M
USP1	Ubiquitin specific peptidase 1
VHL	von Hippel-Lindau tumor suppressor
Vif	Viral infectivity factor
WHB	Winged-helix B
Wnt	Wingless
β-TrCP	Beta-transducin repeat-containing Protein

2.1 Overall Architecture of CRLs

2.1.1 Introduction

CRLs are modular E3 ligases, which utilize interchangeable substrate receptors to recruit a variety of substrates onto a common catalytic scaffold (Fig. 2.1a). All CRLs contain a scaffolding cullin protein, namely, CUL1, CUL2, CUL3, CUL4A, CUL4B, or CUL5. Most of these cullin proteins bind a specific adaptor polypeptide through their N-terminal regions, which helps engage interchangeable substrate receptors to dock their cognate substrates to the E3. The ubiquitination reaction involves the transfer of ubiquitin from an E2 ubiquitin-conjugating enzyme to the

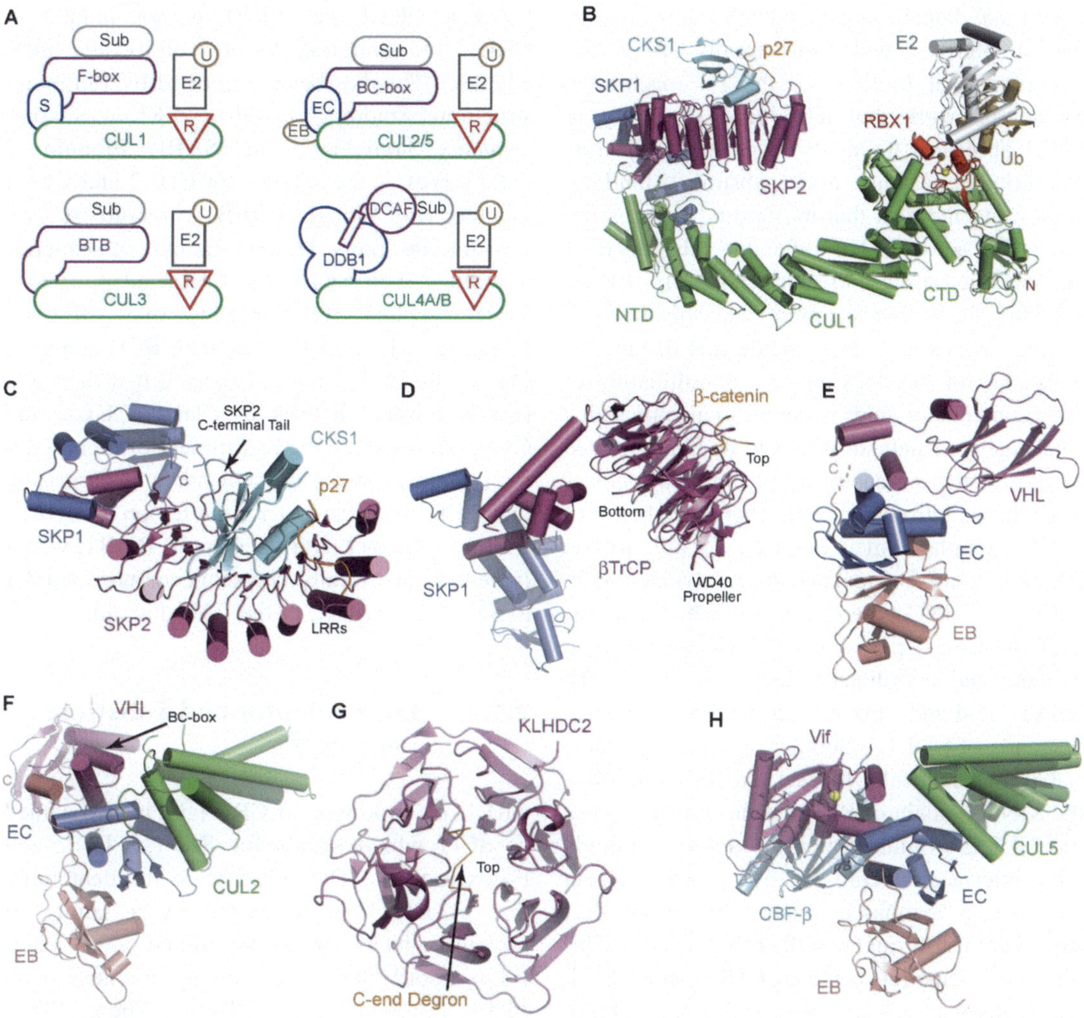

Fig. 2.1 Overall architecture of CRLs and structural components of CRL1 and CRL2/5. (**a**) Schematic drawing of all CRLs. S, SKP1; R, RBX1; EB, Elongin B; EC, Elongin C; U, ubiquitin; Sub, substrate. (**b**) A model of CRL1(SCF)$^{SKP2\text{-}CKS1}$ in complex with substrate p27 and a ubiquitin-charged E2. (**c**) An orthogonal view of the SKP1-SKP2-CKS1-p27 complex (PDB: 1LDK, 2AST). (**d**) Crystal structure of the SKP1 in complex with the WD40-repeat domain containing F-box protein, β-TrCP, which interacts with the degron of the substrate protein, β-catenin (PDB: 1P22). (**e**) Crystal structure of the BC-box protein, VHL, in complex with Elongin B and Elongin C (PDB: 1VCB). (**f**) A different view of VHL-EB-EC complex bound to CUL2 with the stabilized C-terminus of EB labeled with "C" (PDB: 4WQO). (**g**) The substrate-binding kelch repeat domain of the BC-box protein KLHDC2 (PDB: 6DO3). (**h**) A crystal structure of the HIV accessory protein Vif in complex with EB, EC, CUL5, and a cellular factor, CBF-β (PDB: 4N9F)

substrate. The cullin scaffold facilitates this process by using its C-terminus to house the catalytic subunit, RBX1 or RBX2, which directly interacts with the ubiquitin-conjugated E2 enzyme.

The first glimpse of the architecture of a CRL complex came from the X-ray crystal structure of human CUL1-RBX1 (Fig. 2.1b) (Zheng et al. 2002b). In the structure, CUL1 adopts a highly elongated shape with two functional domains. The stalk-like N-terminal domain of CUL1 is made of helical repeats that interact with the CRL1 adaptor protein, SKP1. The C-terminal domain of the scaffold has a more globular fold, which harbors RBX1. The N-terminal and

C-terminal domains are coupled through a hydrophobic interface, which structurally affixes the two functional portions of the E3 machinery. Mutational work that introduced flexibility to CUL1 abolished its ability to promote substrate ubiquitination, but not substrate binding, reinforcing the idea that its rigidity is necessary for its E3 activity. The inflexible CUL1-RBX1 ubiquitin ligase scaffold spans more than 100 Å, allowing the E3 to accommodate substrates of various shapes and sizes. While this distance is beneficial for the docking and ubiquitination of large substrate proteins, it seems to also present a challenge for small substrates to be ubiquitinated by the E2 that is tens of ångström away, unless flexibility is introduced somewhere (see below).

The globular domain of CUL1 not only fosters RBX1, but also intercalates the catalytic subunit. RBX1 is a RING-type zinc finger, which consists of a N-terminal β-strand and a C-terminal core domain that coordinates three zinc ions. The RBX1 β-strand inserts itself between five β-strands of CUL1's globular domain to create a stable intermolecular β-sheet. At the same time, the RING domain of RBX1 is hosted by the rest of CUL1 C-terminal region through a seemingly loose interface. Overall, the two proteins appear to exist in a "symbiotic" relationship, where they stay together throughout their life cycles. This bimodal interaction between CUL1 and RBX1, which involves an anchored end and a relaxed interface, allows for the RING domain of RBX1 to pop out upon the post-translational modification (PTM) of CUL1 by NEDD8, a ubiquitin-like molecule. Such an intermolecular topological change provides flexibility to CRL1 that helps reposition the ubiquitin-conjugated E2 to potentially approach the substrate. Further discussion about the regulation of CRLs by neddylation will be in Sect. 2.3.1.

CUL1 uses two α-helices to bind SKP1. The majority of the residues clustered at this interface are strictly conserved between CUL1 orthologues. Interestingly, these residues are not conserved across CUL1 paralogues, i.e., CUL1–CUL5. But they are conserved within orthologues of each cullin family member (Zheng et al. 2002b), suggesting that the binding mode between CUL1 and SKP1 is not unique to CRL1, but common to all cullins and their adaptors. This has been confirmed by follow-up structural studies of other CRLs. Sharing sequence homology with SKP1, Elongin C (EC) serves as the adaptor for CUL2 and CUL5 together with Elongin B (EB) to recruit BC-box substrate receptors. Despite the lack of sequence homology with SKP1, the CUL3 adaptors/substrate receptors, BTB proteins, share structure homology with SKP1 within their BTB domains. On one hand, it is not unexpected that the same two N-terminal helices of CUL1, CUL2, and CUL3 are used for the recruitment of their adaptor proteins. On the other hand, it is surprising that CUL4 uses the exact same region as CUL1–CUL3 to interact with its adaptor, DDB1, which is neither structurally nor sequence-wise similar to SKP1, Elongin C, or the BTB domain.

2.1.2 CRL1 Adaptor and Substrate Receptors

CRL1, the prototype of CRL E3s, is also named as SCF, which stands for SKP1, CUL1, and F-box proteins. The substrate recruitment function of CRL1 is conducted by a family of proteins, which share an N-terminal ~40 residues F-box motif that constitutively interacts with SKP1 (Skowyra et al. 1997). These F-box proteins feature characteristic C-terminal protein-protein interaction domains, which are responsible for binding specific substrates. Based on their predicted tertiary structures, the 69 human F-box proteins have been classified in three groups: FBXL for leucine-rich repeat (LRR)-containing proteins, FBXW for the ones containing WD40 repeats, and FBXO for F-box protein with other folds (Jin et al. 2004). The diverse protein-protein interaction domains used by F-box proteins enable them to recognize different substrates with high specificity and optimally orient these substrates to receive ubiquitin from the RBX1-bound E2 enzyme. In most, if not all, structures of F-box proteins in complex with SKP1, the N-terminal domain of the substrate receptors appears to be structurally coupled to the SKP1-F-box module, which displays little structural variation. Together with the stable SKP1-CUL1 interface, the entire CUL1-SKP1-F-box protein complex has been postulated to play a role in spatially positioning the bound substrate relative to the catalytic end of the E3 platform (Zheng et al. 2002b).

The first structure of an F-box protein, SKP2, was determined almost 20 years ago, which revealed the binding mode of the F-box motif to SKP1 (Fig. 2.1b, c) (Schulman et al. 2000). CRL1^{SKP2} is a key regulator of mammalian cell cycle progression. The overall structure of the SKP1-SKP2 complex resembles a sickle. The handle of the sickle is made of the SKP1-F-box structural module, while the variable linker and leucine-rich repeats of SKP2 constitute the blade. SKP1 and the F-box motif of SKP2 interact via an extensive interdigitated interface that consists of four alternating layers from each protein. Although part of the interface is mediated by residues that are not strictly conserved among all F-box motifs, it is generally believed, and subsequently validated, that all F-box proteins engage SKP1 in a similar binding mode.

The C-terminal domain of SKP2 is composed of ten LRRs, each of which is made of an α-helix and a β-strand. These LRRs pack in tandem and give rise to an arc-shaped structure, which is characterized by a concave side formed by parallel β-strands and a convex surface presented by α-helices. In many known non-E3 LRR proteins, the concave surface is involved in protein-protein interaction. The FBXLs, including SKP2, are no exception. Remarkably, the FBXL subfamily of F-box proteins have evolved different numbers of LRRs, which give rise to curved structures with different diameters, arc lengths, and pitches. These features presumably allow the FBXLs to hold diverse substrates. Interestingly, SKP2 also features a long C-terminal tail, which wraps back into the concave surface of the LRR domain and provides an additional structural element for substrate recruitment.

Unlike many F-box proteins, which directly recruit CRL1 substrates, SKP2 requires yet another adaptor protein, CKS1, to bind and ubiquitinate its substrate p27^{Kip1}. The crystal structure of the SKP1-SKP2-CKS1-p27 complex shows that CKS1 is anchored to the concave surface of SKP2 LRR domain and is supported underneath by the SKP2 C-terminal tail (Fig. 2.1c) (Hao et al. 2005). The residues that are involved in the CKS1-SKP2 interaction are conserved in animal orthologues on both sides of the interface, underlying the functional importance of CKS1. Furthermore, CKS1 has been found to form a stable complex with CDK2-cyclin A, which might contribute to the binding of p27^{Kip1} to the CRL1 machinery.

The crystal structure of SKP1-β-TrCP offered the first sight of an FBXW-type F-box protein, which plays a major role in the Wnt signaling pathway by catalyzing the polyubiquitination and degradation of β-catenin (Fig. 2.1d) (Wu et al. 2003). The SKP1-β-TrCP complex adopts a bi-lobal structure, displaying the substrate at the opposite side of SKP1. The F-box motif of β-TrCP interacts with SKP1 in a similar fashion as SKP2. The substrate-binding function of the F-box protein is performed by its C-terminal WD40-repeat domain, whose name is derived from the ~40 amino acids sequence repeat that contains structurally essential tryptophan (W) and aspartic acid (D) amino acids. WD40 repeats are known to fold into a β-propeller structure, which is usually made of seven β-sheets arranged in a circular manner around a central narrow channel. In β-TrCP and other FBXW proteins with known structures, the amino acids located on the "top" surface are involved in substrate recognition. Through a rigid linker helix connecting to the SKP1-F-box module, β-TrCP is thought to position its substrate toward the E2 for ubiquitin transfer.

2.1.3 CRL2 and CRL5

CRL2 is organized in a similar manner as CRL1, with CUL2 serving as the scaffolding protein, EB-EC assisting as adaptor proteins, and members from the BC-box family of proteins acting as substrate receptors (Mahrour et al. 2008). CUL2 shares 38% of sequence identity with CUL1.

As predicted by homology models and subsequently confirmed by crystal structure, CUL2 adopts a similar elongated structure as CUL1 (Cardote et al. 2017). As the adaptor protein of CUL2, EC shares about 30% sequence identity with SKP1 and has a BTB core fold. Unlike SKP1, EC does not act alone to bridge the substrate receptor BC-box proteins to CUL2 and mandates its association with the ubiquitin-like molecule EB. It is not clear yet if EB with the ubiquitin-like fold plays any particular role in CRL2 besides stabilizing EC. In comparison with the F-box proteins, the BC-box proteins contain structural elements that do not only interact with the adaptor proteins, but also make direct contacts with their cognate cullin scaffolds. Besides a short BC-box motif, some BC-box proteins feature a CUL2-box motif that specifically recognizes CUL2. Interestingly, these two motifs are always consecutively localized in the BC-box protein sequences but could be found at different positions relative to the substrate-binding domain in the polypeptide. The best studied member from this family of proteins is the VHL tumor suppressor protein that promotes the ubiquitination of HIF1α under normoxia (Kaelin 2005). When its substrate-binding domain or BC-box motif is mutated, VHL fails to target HIF1α for ubiquitination and degradation, which leads to von Hippel-Lindau disorder, a hereditary predisposition to develop tumors in a variety of organs.

The structure of VHL-EB-EC unveils the portion of VHL that is involved in substrate recognition as a beta-sandwich (Fig. 2.1e) (Stebbins et al. 1999). The BC-box and CUL2-binding motifs of VHL together form a three α-helices structural module that is very similar to the structure of the F-box motif. Moreover, EC can be superimposed almost perfectly on part of SKP1. As predicted, the other adaptor protein, EB, adopts a ubiquitin-like α-/β-roll structure. In this crystal structure, the C-terminal tail of EB interacts with EC and points toward VHL with the last 20 amino acids disordered. Snapshots of VHL-EB-EC in complex with the first N-terminal repeat of CUL2 or the full-length protein reveal that the C-terminal tail of EB adopts an ordered structure interacting with VHL (Nguyen et al. 2015; Cardote et al. 2017). In addition, an internal loop of EC that is disordered in the VHL-EB-EC complex becomes structured upon its interaction with CUL2 (Fig. 2.1f). Remarkably, the CUL2-box of VHL makes polar interactions with part of the cullin scaffold, which hints at how different BC-box proteins could differentiate between CUL2 and CUL5, even if they use the same adaptor proteins.

A myriad of BC-box proteins with various folds have been identified by biochemical and bioinformatical approaches (Mahrour et al. 2008). A variety of substrate-binding domains have been mapped for the ones that bind CUL2, such as leucine-rich repeats, ankyrin repeats, tetratricopeptide repeats, armadillo repeats, kelch repeats, or SWIM zinc fingers. Recently, the structure of a kelch repeat BC-box protein, KLHDC2, has been determined (Fig. 2.1g) (Rusnac et al. 2018). It adopts a β-propeller fold composed of six β-sheets that are arranged in a circular fashion around a central axis. As observed for most kelch repeat domains, the "top" surface of KLHDC2 sports long variable loops that give rise to a deep binding pocket, which recognizes the degron of its substrates. The "bottom" part of the repeat is predicted to neighbor the BC-box and CUL2-box motifs, which have been removed for ease of crystallization.

Similar to CUL2, CUL5 employs the same adaptor proteins EB and EC to associate with a distinct subfamily of the BC-box proteins (Mahrour et al. 2008). The substrate receptors that bind CUL5 can have one of the following folds: Src homology 2 phosphotyrosine-binding domain, ankyrin repeats, SP1a and ryanodine receptor domain, WD40 repeats, Rab-like GTPase domain, protein-L-isoaspartate carboxymethyltransferase, and transcription factor SII-like domain. Unexpectedly, not only cellular BC-box proteins have been found to bind CUL5-EB-EC. A very interesting study has demonstrated that HIV-1 virion is able to hijack CRL5 via an accessory protein, Vif, to promote the degradation of APOBEC3, the host restriction factor that blocks the replication of the virus (Yu et al. 2003). Recent studies showed that this function of Vif entails yet another host protein, the transcription factor CBF-β that plays a role in APOBEC3 expression (Kim et al. 2013). The structure of the Vif-CBF-β-$CUL5_N$-EB-EC complex reveals an overall U shape, with Vif-CBF-β and $CUL5_N$ making up the arms and EB-EC sticking out at the bottom (Fig. 2.1h) (Guo et al. 2014). Akin to cellular BC-box proteins, Vif uses its BC-box and CUL5-box to bridge EC and CUL5, mimicking the positioning of SKP1 and SKP2 in the context of CRL1. This structure provides further evidence supporting the notion that there is a conserved assembly mode for most CRLs.

2.1.4 CRL3

CUL1, CUL2, CUL4, and CUL5 all use adaptor proteins that recruit specific substrate receptors. CUL3, on the other hand, recruits BTB domain-containing proteins that combine the adaptor and substrate receptor functions into a single polypeptide (Pintard et al. 2003; Xu et al. 2003; Geyer et al. 2003). Without sharing sequence homology to SKP1 or EC, the BTB domain adopts a structural fold analogous to these two CRL adaptors and anchors to CUL3 in a similar manner. In addition, the BTB domain is known to dimerize, which facilitates the formation of homodimeric CRL3 complexes with two copies of each components of the E3 machinery. This property is reminiscent of select F-box proteins, such as FBXW7 and β-TrCP, which also contain a dimerization domain on the N-terminal side of the F-box motif. Similar to the CRL1 and CRL2/5 substrate receptors, the BTB family of proteins features different protein-protein interaction domains to recruit substrates. These include, but are not limited to, the well-characterized MATH and kelch repeats domains that are present in SPOP and KEAP1, respectively.

KEAP1 is one of the best studied CUL3 substrate-specific adaptors, which controls the degradation of NRF2 transcription factor involved in oxidative stress response pathway (Yamamoto et al. 2018). KEAP1 consists of four functional domains: BTB, intervening region (IVR), double glycine repeats (DGR, aka kelch repeats), and C-terminal region (CTR). While the DGR and CTR give rise to the kelch repeat β-propeller, the IVR domain contains two reactive cysteine residues that can be modified by toxic electrophiles and subsequently alter the spatial configuration of other domains of the BTB protein. Despite extensive studies, it is still unclear how these chemical-induced structural changes affect the overall architecture of the KEAP1 dimer, thereby affecting NRF2 polyubiquitination and degradation.

SPOP is a BTB-domain protein that is frequently mutated in human cancer. The dimerization and substrate-binding mode of SPOP have been determined in its near full-length form (Fig. 2.2a) (Zhuang et al. 2009). The MATH domain of SPOP forms an antiparallel β-sandwich, with a central shallow grove that is used by substrates to dock. Dimerization of the CRL3 substrate receptor takes place through a hydrophobic interface between the BTB domains. Mutational analysis indicates that BTB domain dimerization is not necessary for CUL3 binding, which is consistent with its spatial separation from the BTB-CUL3 interface (Fig. 2.2b). Nonetheless, defects in BTB dimerization negatively impact the polyubiquitination of the SPOP substrates, suggesting that the dimeric architecture of the CRL3 complex is critical for productive substrate ubiquitination. Interestingly, the two MATH domains in the crystal structure are asymmetrically arranged with one MATH domain cradled by the grove between the BTB domains and the other one pushed away from its BTB. Based on different crystal structures obtained for the same complex, slight topological differences have been observed, suggesting that the linker between the two functional domains of SPOP confers structural flexibility to the E3 complex. This structural plasticity has not been previously observed for other CRL family members. The MATH domain of SPOP can recognize a host of substrate degron motifs. It is possible that MATH domain-containing BTB proteins, including SPOP, have evolved a mechanism of engaging substrates with high affinity by simultaneously recognizing two low-affinity degron motifs.

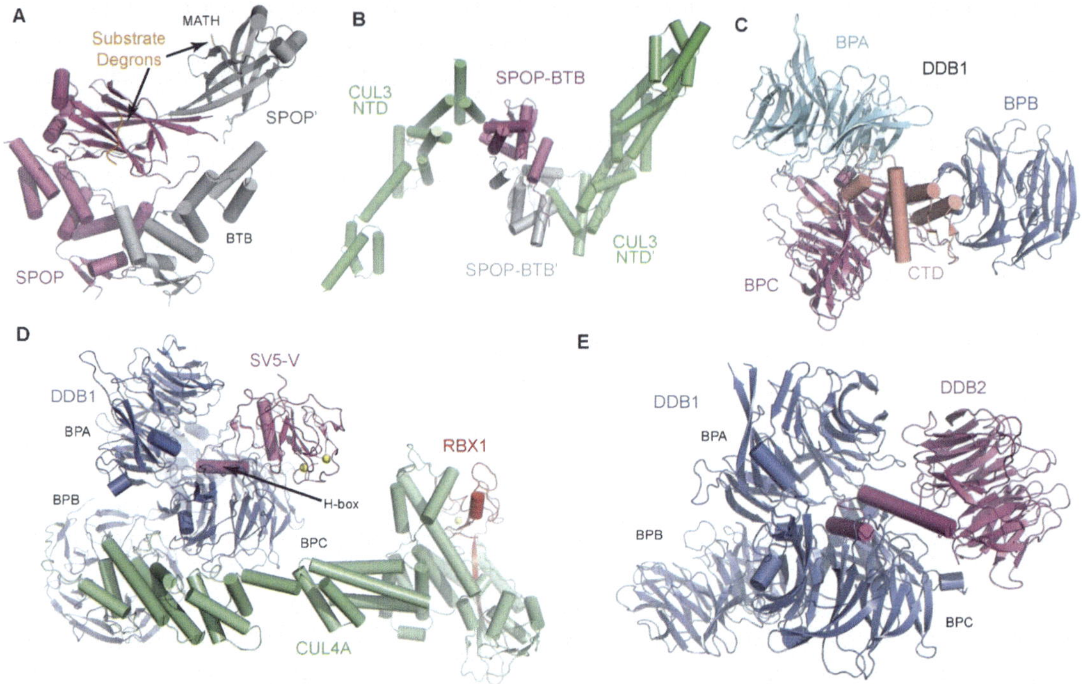

Fig. 2.2 Overall architecture and structural components of CRL3 and CRL4. (**a**) An asymmetric dimeric structure of the BTB-domain protein, SPOP, in complex with two substrate degron peptides (PDB: 3HQI). (**b**) The dimeric SPOP BTB domain in complex with CUL3-NTD (PDB: 4EOZ). (**c**) Domain architecture of DDB1 with three propellers (BPA, BPB, and BPC) and a helical CTD (PDB: 2B5M). (**d**) Complex structure of CUL4A-RBX1-DDB1-SV5-V. The H-box motif of SV5-V responsible for binding DDB1 is labeled (2HYE). (**e**) Crystal structure of DDB1 in complex with the DCAF protein, DDB2 (PDB: 3EI3)

Although the BTB domains of KEAP1 and SPOP are responsible for CUL3 binding, not all BTB domain-containing proteins function as CRL3 substrate adaptors. An interesting finding from the SPOP structure is a pair of C-terminal helices that have structural equivalence in the F-box or CUL2-box. Because the function of the BTB domain was thought to bridge CUL3 and the MATH domain of SPOP, it seemed unnecessary to have such a vestigial element. Surprisingly, this helix pair of the SPOP BTB domain has been proven to be crucial for CUL3 interaction. Because this structural element, named 3-box, is found in some, but not all, BTB proteins, it could be the structural determinant that allows a subset of BTB proteins to function as CRL3 substrate adaptors.

2.1.5 CRL4

While most CRL adaptors adopt a BTB or BTB-like fold, CRL4 employs a 127 kDa protein, DDB1, to dock substrate receptors. DDB1 is a multi-domain protein, made of three β-propellers, labeled as BPA, BPB, and BPC, and a C-terminal helical domain (CTD) (Fig. 2.2c) (Li et al. 2006). Remarkably, the three β-propellers are not folded in a linear fashion within the polypeptide. Instead, two of the β-propellers, BPA and BPB, are inserted into two internal loops of BPC. Together, the three propellers adopt a compact tri-star structure with the CTD housed in the middle. The BPB propeller binds the N-terminal domain (NTD) of CUL4 via two interfaces, one resembling SKP1 binding to CUL1 and the other involving an

N-terminal conserved sequence of CUL4 cradling DDB1 (Fig. 2.2d) (Angers et al. 2006). The BPA and BPC propellers, on the other hand, pack against each other to create an open clam-shaped structure, which is responsible for holding CRL4 substrate receptors.

To date, a number of DDB1 structures have been documented. Strikingly, the linker between the BPB domain and the BPA-BPC double propeller appears to have a large degree of plasticity, which enables the two functional modules of DDB1 to adopt different orientations relative to each other (Fig. 2.2d, e). This feature might allow CRL4 to accommodate and polyubiquitinate substrates of different shapes and sizes. In addition, the structural flexibility of DDB1 enables CUL4 to rotate up to 150° around the substrate receptor, thereby creating a ubiquitination zone that could help detect various lysines on a substrate and promote their ubiquitination. The precise role of the structural flexibility within CRL4 remains to be elucidated.

Similar to CRL5, CRL4 is also known to be hijacked by viruses. The SV5-V protein encoded by paramyxovirus has been shown to functionally mimic the CRL4 substrate receptors to mediate the polyubiquitination and degradation of the otherwise stable STAT proteins in the interferon pathway (Horvath 2004). SV5-V anchors itself to DDB1 by inserting its N-terminal helix into the opening of the BPA-BPC double propeller, interacting predominantly with the "top" surface of BPC (Fig. 2.2d) (Li et al. 2006). The C-terminal region of SV5-V folds into unique globular structure featuring a bowl-shaped depression with many conserved hydrophobic and nonpolar residues. This surface region of the viral protein is critical for the recruitment of STATs. Perhaps due to its rich structural features, DDB1 appears to be a frequent target for viral hijacking. Besides SV5-V, hepatitis B virus X protein (HBx) and woodchuck hepatitis virus X protein (WHx) have also been reported to reprogram the CRL4 adaptor to degrade host factors (Decorsière et al. 2016). Peptide motifs from these proteins have been mapped and crystalized in complex with DDB1. Despite their divergent sequences, these motifs form a common three-turn α-helix (termed H-box) that anchors to the "top" surface of BPC in a similar fashion as the SV5-V N-terminal helix (Fig. 2.2d) (Li et al. 2010).

The structural insights obtained from the DDB1-viral hijacker complexes prompted multiple proteomics studies that were aimed at identifying possible cellular CRL4 substrate receptors (Angers et al. 2006; Jin et al. 2006). Multiple DDB1-CUL4A-associated factors (DCAFs), most of which contain a WD40-repeat domain, have been classified as subunits of CRL4 E3 complexes for recruiting substrates. Interestingly, DDB2, which was originally identified together with DDB1 as a UV-damaged DNA-binding protein, anchors to DDB1, like a canonical DCAF protein, but functions to recognize DNA abduct in the nuclear excision repair pathway (Fig. 2.2e) (Fischer et al. 2011). With the structural knowledge gathered from the DDB1-hijacking viral proteins, the H-box motif has been found in a number of DCAFs and validated by crystallography. A natural question arises as to whether the H-box motif exists in other DCAFs. However, the lack of an obvious consensus sequence for the H-box motif has made it challenging to find the answer to this question.

Distinct from most DCAFs, cereblon (CRBN) does not contain a WD40-repeat domain. Instead, it consists of a seven-stranded β-sheet NTD, a bundle domain composed of seven α-helices (HBD), and an eight-stranded β-sheet CTD (Petzold et al. 2016; Matyskiela et al. 2016). Unlike DCAFs, CRBN uses its HBD with its α-helices to bind DDB1 in the cavity between BPA and BPC. Its CTD domain is involved in the recruitment of native substrates like MEIS2 via a conserved binding pocket. Among all DCAFs, CRBN stands out by being the target of thalidomide, which has a notorious history in biomedicine, but is now repurposed to treat multiple myeloma (see Sect. 2.2.4).

The combination of structural, biochemical, and proteomic approaches has helped delineate the composition and architecture of the CRL E3 superfamily in great details. Although there might be outliers unknown to us, most CRL complexes are expected to assemble following the structural principles described above. In contrast to the common architecture shared among different CRL E3s, the mechanisms by which these ubiquitin ligases recognize their specific substrates in response to different cellular cues are incredibly diverse.

2.2 Substrate Recognition by CRLs

To achieve spatially and temporally controlled degradation, a substrate needs to be recognized by its cognate E3 with high specificity. This interaction is often mediated by a short linear sequence motif, termed degron, present on the substrate. It has been challenging to identify substrates for different E3 ligases and to fully understand how their recognition is regulated. To date, some substrates have been found to have unmodified degrons that are recognized by the E3 either directly or with the help of a small molecule, while others utilize post-translationally modified degrons to engage their E3 ligases. In some emerging cases, the whole globular domain of a substrate protein contributes to the specific interaction. It is expected that more regulatory mechanisms will be revealed in the future.

2.2.1 PTM-Dependent Substrate Degron Recognition

Protein phosphorylation is one of the most common forms of post-translational modification and is involved in regulating essentially all cellular processes. Many degrons are under the control of this PTM. One classic example is the cell cycle regulatory protein, cyclin E, which has not just one, but two phosphorylated degrons. These phospho-degrons are specifically recognized by FBXW7, a substrate receptor for CRL1 that is responsible for the degradation of cyclin E (Welcker and Clurman 2008). The FBXW7 F-box protein contains a D-box dimerization domain N-terminal to its F-box motif and a C-terminal eight β-sheets WD40 propeller. The FBXW7 D-box promotes the dimerization of the F-box protein, which enables simultaneous binding of two different phosphorylated degrons that independently dock to the "top" surface of the β-propeller in an extended conformation (Fig. 2.3a).

The two cyclin E degrons share similarities but also differ in certain aspects. Both phosphorylated degrons have been co-crystalized with FBXW7 (Hao et al. 2007). The degron located at the C-terminus of cyclin E has three phosphorylated residues and is characterized as a strong degron due to its nanomolar affinity to FBXW7. This tight interaction can be easily explained by the numerous contacts that the degron makes with the substrate receptor, including both polar interactions and van der Waals packings. It is notable that two out of the three phosphate groups present on the degron are recognized by three arginine residues that are strictly conserved among FBXW7 orthologues. By contrast, the N-terminal degron of cyclin E features only one phosphorylated threonine residue and makes fewer interactions with FBXW7, which leads to weaker binding. Interestingly, the three conserved FBXW7 arginine residues are frequently mutated in human cancers, rendering the E3 incapable of promoting the degradation of cyclin E and possibly other substrates. Despite the well-elucidated mechanism of the FBXW7-cyclin E interaction, the necessity of having two degrons on cyclin E for its productive degradation remains unclear. Cell-based study suggests that the strong degron alone is sufficient for substrate degradation (Welcker and Clurman 2008).

Proline hydroxylation is another well-known, albeit less common, form of PTM that is involved in degron regulation. Hypoxia-inducible factor 1-alpha (HIF1α) contains a critical proline in its oxygen-dependent degradation domain (ODD) that gets hydroxylated in the presence of oxygen (Ivan et al. 2001). This normoxia-associated modification allows VHL, a BC-box protein, to bind, ubiquitinate, and target HIF1α for degradation. Under low oxygen conditions, HIF1α is spared from degradation and functions as a transcription factor activating angiogenic gene expression. The crystal structure of VHL-EB-EC in complex with a partial ODD peptide from HIF1α elucidated how the post-translational modification dictates

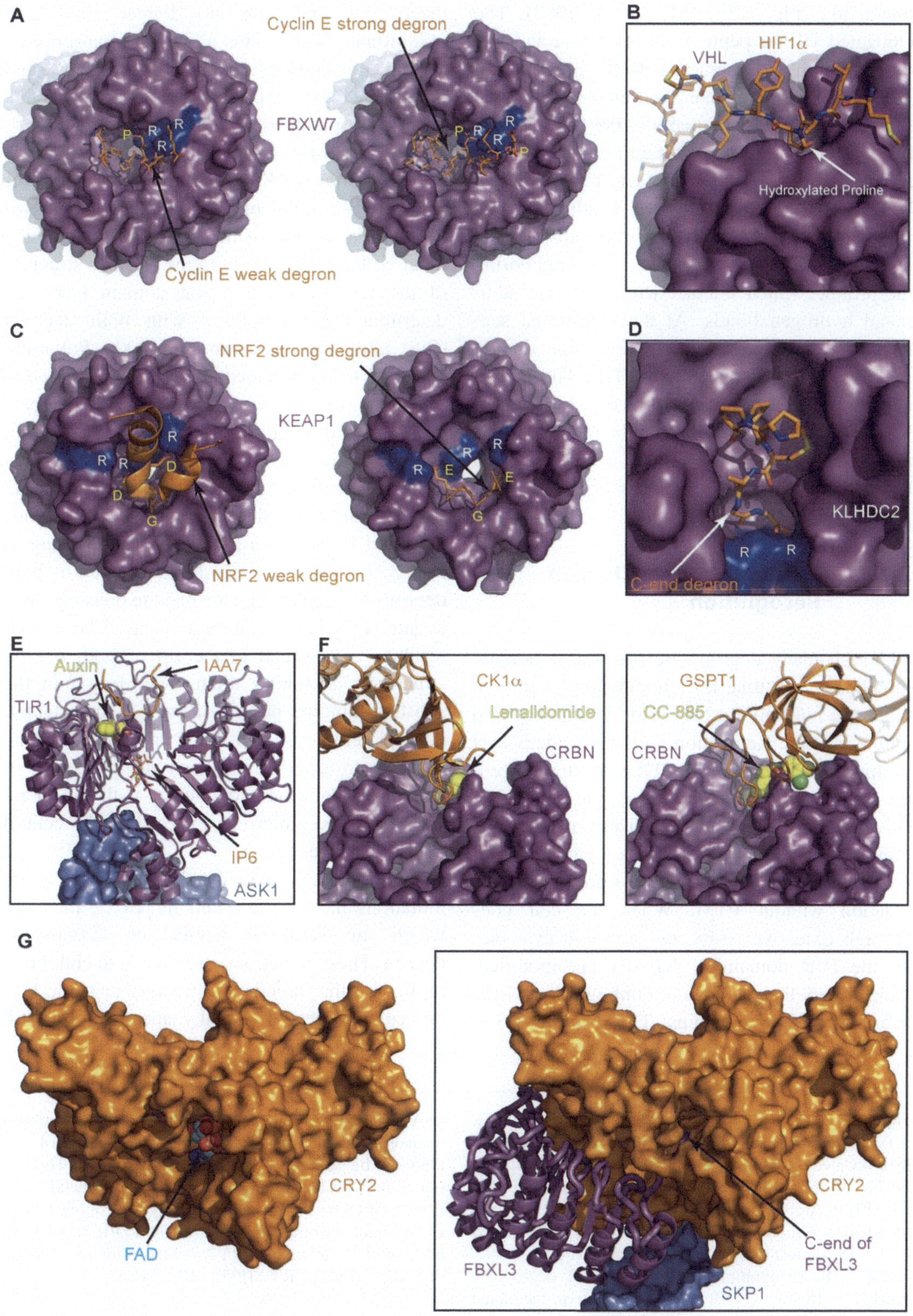

Fig. 2.3 Structural mechanisms of substrate-CRL interactions. (**a**) Binding modes of the strong and weak degrons of cyclin E to FBXW7. R, Arg; P, phosphate. Phosphate-binding arginine residues of FBXW7 are

interaction (Fig. 2.3b) (Min et al. 2002). The elongated HIF1α peptide adopts a β-strand-like conformation and interacts with the β-domain of VHL in a bipartite manner. Within the N-terminal segment of the HIF1α peptide, the hydroxyproline is deeply embedded at the interface, forming multiple van der Waals contacts and hydrogen bonds with highly conserved residues of the BC-box protein. This interaction is substantiated by the backbones of the amino acids neighboring the proline, which interact with VHL via additional hydrogen bonds. At the C-terminal segment, there are a few interactions that do not seem essential for binding to VHL. The importance of hydroxyproline interface is highlighted by the clustering of cancer inducing mutations involved in von Hippel-Lindau disorder around this site.

2.2.2 Native Substrate Degron Recognition

Although many degrons are post-translationally modified, substrate-E3 interaction can be regulated not only on the substrate side but also on the E3 side. Regulation on the E3 side is best exemplified by KEAP1, a CUL3 substrate receptor (Yamamoto et al. 2018). Under normal cellular conditions, KEAP1 seizes the native transcription factor NRF2 in the cytoplasm and mediates its constitutive ubiquitination and degradation without PTM. When the cell goes through oxidative stress, two reactive cysteines on the IVR domain of KEAP1 are modified, which alters the topological configuration of the substrate receptor domains. The resulting E3 is inactivated and can no longer bind and ubiquitinate NRF2. This allows the transcription factor to translocate into the nucleus and upregulate the expression of cytoprotective genes in response to oxidative insults.

The proper positioning of the two kelch repeat domains from the KEAP1 dimer is essential for productive binding of NRF2, which holds two degron motifs, one with a low affinity and the other with a high affinity to the E3. The structure of the KEAP1 kelch repeat domain has been determined in complexes with both degrons (Fig. 2.3c) (Padmanabhan et al. 2006; Fukutomi et al. 2014). As expected, the two degrons bind to the same "top" surface of KEAP1 kelch repeat propeller, although they seem to adopt different conformations. The weak degron contains 35 amino acids and forms a three-helix structure, whereas the strong degron is only 9 amino acids long and folds into a β-hairpin upon binding to the E3. The only common feature shared by both degrons is a central glycine residue preceded by a negatively charged amino acid. These two residues are located at the tip of both degron structures, anchoring themselves deep into the substrate-binding pocket of the E3. Importantly, KEAP1 employs multiple conserved arginine residues to stabilize the two degrons, two of which are dedicated to interact with the negatively charged amino acid preceding the central glycine residue.

Multiple cancer-related loss-of-function mutations have been found in NRF2, many of which are localized within the low-affinity degron. These mutations correlate with either disruption of the three-helix structure or steric hindrance introduced by a bulky side chain. It is still

Fig. 2.3 (continued) colored in blue (PDB: 2OVR, 2OVQ). (**b**) Recognition of HIF1α degron with a hydroxylated proline by VHL (PDB: 1LM8). (**c**) Binding modes of the strong and weak degrons of NRF2 to KEAP1. R, Arg; D, Asp; E, Glu; G, Gly. Arginine residues of KEAP1 responsible for binding negatively charged NRF2 degron residues are colored in blue (PDB: 3WN7, 1X2R). (**d**) Recognition of SELK C-end degron by KLHDC2. R: Arg (PDB: 6DO3). (**e**) Auxin-facilitated IAA7 degron recognition by TIR1. Inositol hexakisphosphate (IP6) serves as a cofactor of the ubiquitin ligase-based receptor (PDB: 2P1Q). (**f**) The recognition of CK1α by CRBN is promoted by lenalidomide (PDB: 5FQD). Structure of CRBN in complex with CC-885 and substrate protein GSPT1 (PDB: 5HXB). (**g**) FAD binding to human CRY2 and the complex structure of SKP1-FBXL3-CRY2 (PDB: 4I6G, 4I6J)

unclear how the full-length NRF2 substrate interacts with KEAP1 dimer at the structural level. Depicting how NRF2 is positioned relative to the E3 and determining which lysine residues are ubiquitinated would clarify a significant mystery in the field.

Besides internal short linear motifs, a new class of unmodified degrons has recently been identified at the C-terminus of various early terminated proteins. These so-called C-end degrons feature key glycine, arginine, or a combination of both at critical positions (Lin et al. 2015, 2018; Koren et al. 2018). One such degron was found in the early terminated selenoproteins, SELK and SELS, as well as the N-terminal fragment of USP1, which cleaves itself at an internal site. The common characteristic of these three polypeptides is a degenerate C-terminal sequence that ends in a diglycine motif. Recent studies have revealed that this diglycine-containing C-end degron is specifically recognized by the BC-box protein KLHDC2. To understand the structural mechanism by which KLHDC2 recognizes such a simple degron, the crystal structure of the kelch repeat domain of the BC-box protein in complexes with the degron peptide of each of the three polypeptides has been determined (Fig. 2.3d) (Rusnac et al. 2018). The "top" surface of the KLHDC2 β-propeller engages the degrons that adopt a common helical fold. The C-terminus of the degrons inserts itself into a deep binding pocket, while the N-terminus is more solvent exposed. The extreme C-terminal carboxyl group of the degron is recognized by two strictly conserved arginine residues and a highly conserved serine residue. The penultimate glycine adopts a phi/psi angle that is in the generously allowed region of the Ramachandran plot, suggesting that only a glycine would be accommodated at this position. Strikingly, the remainder of the peptides interacts with KLHDC2 exclusively via its backbone. The lack of interactions between the E3 and the side chains of the degrons explains how these degenerate degradation motifs can be contained in the binding pocket in an identical manner. An unexpected finding from these studies is the single-digit nanomolar affinity of these seemingly simple C-end degrons, which is in stark contrast to the N-end degrons previously documented (Choi et al. 2010; Matta-Camacho et al. 2010; Chen et al. 2017; Dong et al. 2018). Such a tight interaction suggests that the substrates of KLHDC2 are either in low abundance or highly toxic to the cell.

2.2.3 Compound-Dependent Substrate Degron Recognition

Besides PTMs, cellular signals such as hormones and secondary metabolites can directly participate in degron recognition by E3 ligases. In the green kingdom of life, two plant hormones, auxin and jasmonate (JA), have been shown to serve as molecular glue bridging the CRL1 F-box proteins TIR1 and COI1 to their specific substrates, respectively (Shabek and Zheng 2014). These are the first known cases where E3s function in hormone perception and as a result regulate gene expression. The transcriptional regulation is achieved in the auxin and JA signaling pathways through the ubiquitination and degradation of AUX/IAA and JAZ transcription repressor proteins, respectively.

The crystal structure of the F-box protein, TIR1, has been determined in complex with ASK1, one of the plant homologues for SKP1 (Fig. 2.3e) (Tan et al. 2007). The complex adopts an overall mushroom-shaped structure, where the cap is composed of the TIR1 LRR domain and the stem contains the TIR1 F-box motif and ASK1. TIR1 folds into a twisted horseshoe-shaped solenoid and provides its top surface pocket for auxin-mediated degron binding. The crystal structure of TIR1 in complex with auxin and an AUX/IAA degron reveals that the hormone fills up a gap at the protein-protein interaction interface without inducing any detectable conformational changes of the F-box protein. Its unique mechanism of action helps raise the concept of molecular glue. By serendipity, an inositol hexakisphosphate (IP6) was discovered in the middle of the TIR1 LRR domain underneath the hormone-binding pocket. Its strategic location and its conserved binding site strongly argue for a cofactor role in stabilizing hormone binding. Remarkably, JA-isoleucine, the active form of JA, acts through the same mechanism as auxin, mediating the interaction between the F-box protein COI1 and its substrate JAZ proteins (Sheard et al. 2010). Surprisingly, instead of IP6, COI1 uses a specific inositol pentakisphosphate as its cofactor, which is essential for its hormone-sensing function. The TIR1 and COI1 F-box proteins, therefore, are each regulated by two naturally occurring small molecules, a molecular glue hormonal signal and an inositol polyphosphate cofactor, which might function as a proxy signal for phosphate abundance (Wild et al. 2016).

Previously, we briefly mentioned CRBN as the target of the anticancer drug, thalidomide. Recent studies have shown that thalidomide and its derivatives, lenalidomide and pomalidomide, also work as molecular glues, rewiring CRBN to bind and degrade several clinically relevant substrates, including IKZF1, IKZF3, CK1α, and SALL4 (Krönke et al. 2014, 2015; Lu et al. 2014; Matyskiela et al. 2018; Donovan et al. 2018). These substrates do not normally interact with CRBN in the absence of the small molecules. Therefore, they have been referred to as neo-substrates. The structure of DDB-CRBN-lenalidomide-CK1α complex has elucidated the detailed mechanism of interaction among all the components (Fig. 2.3f). The small-molecule drug targets a tryptophan cage surface pocket of CRBN and is stabilized by a combination of hydrogen bonds and hydrophobic interactions. Upon binding to the E3, lenalidomide offers a novel binding surface for the neo-substrate to exploit. CK1α uses a β-hairpin loop to dock into the newly formed hydrophobic pocket, contacting both the drug and the E3 substrate receptor.

CC-885 is a lenalidomide derivative that reprograms CRBN to ubiquitinate yet another neo-substrate, GSPT1 (Matyskiela et al. 2016). This compound shares part of its structure with lenalidomide and interacts with CRBN in an almost identical manner (Fig. 2.3f). The additional moiety of CC-885 that differs from lenalidomide participates in the formation of additional hydrogen bonds with the E3 and presents a new hydrophobic interface and hydrogen bonding for GSPT1 docking. Despite adopting a completely different fold and packing against CRBN from a different direction, GSPT1 interacts with the compound-reshaped E3 pocket using a β-hairpin motif that is superimposable to the one found in CK1α. Based on structural and mutational analyses of the CRBN-substrate complexes, a key glycine residue within the β-hairpin shared among all neo-substrates has emerged. Although the target of thalidomide and its derivatives was elucidated post hoc, their mechanism of action closely resembles that of the naturally occurring plant hormones, auxin and JA. Together, these E3-reshaping molecular-glue small molecules inspire the discovery of novel compounds with therapeutic potentials through targeted protein degradation.

2.2.4 Globular Substrate Protein Recognition

Although degron-mediated substrate-E3 interaction has become a widely accepted dogma, an increasing number of studies have revealed an alternative strategy for certain CRL E3s to recognize their cognate substrates with high specificity. Perhaps the best example comes from the mammalian cryptochrome proteins, CRY1 and CRY2, which are central components of the circadian clock in mammals (Takahashi 2017). The mammalian circadian rhythm is an internal timing system that synchronizes physiological processes to the ~24 h solar day. In all mammalian cells, the circadian clock is driven by a transcription-translation negative feedback loop, in which the CRY1/2 and PERIOD proteins heterodimerize and suppress their own gene expression. Protein degradation plays an important role in oscillating the clock by periodically removing both proteins, thereby alleviating their inhibitory effects. While the PERIOD proteins are polyubiquitinated by $CRL1^{\beta\text{-TrCP}}$, which recognizes their phosphorylated degrons, the CRY1/2 protein is destabilized by $CRL1^{FBXL3}$ without an obvious degron (Shirogane et al. 2005; Busino et al. 2007; Siepka et al. 2007; Godinho et al. 2007). Similar to their orthologues in insects and plants, mammalian CRY1/2 adopts a large globular fold with a deep binding pocket for flavin adenine dinucleotide (FAD) (Xing et al. 2013). The structure of the mammalian CRY2 in complex with FAD shows a partially solvent-exposed pocket, which differs from the closed pocket seen in its plant and insect orthologues (Fig. 2.3g). In the crystal structure of the CRY2-FBXL3-SKP1 complex, the LRR domain of FBXL3 adopts an expected arc-shaped structure, whose concave surface wraps around the CRY2 globular domain, burying many residues that are only spatially but not sequence-wise connected. A surprising and crucial element of the FBXL3-CRY2 interaction involves the C-terminal tail of the F-box protein, which inserts into the FAD-binding pocket of CRY2. This unexpected interface strongly suggests that FAD might be able to compete with the ubiquitin ligase and protect CRY2 from polyubiquitination. Moreover, the surface area of CRY2 involved in binding FBXL3 overlaps with PERIOD2-binding interface, indicating that the PERIOD proteins might also play an antagonistic role in keeping the E3 ligase in check (Nangle et al. 2013; Schmalen et al. 2014). Because the cellular circadian clock can be entrained by many signals, such as metabolism and hormones, the complex binding mode of FBXL3-CRY2 might have been evolved to allow the single substrate-E3 interacting pair to be regulated through multi-faceted mechanisms. As more substrate-E3 interactions are mechanistically interrogated, it is expected that a wider variety of regulatory and structural factors will be revealed beyond the simple degron.

2.3 Regulation of CRLs by NEDD8 Modification

As the central ubiquitin ligase machineries regulating diverse cellular pathways, CRLs rely on a multitude of substrate receptors to recognize and recruit their specific substrates. How do these interchangeable substrate receptor subunits share the common cullin scaffolds without interfering with each other's function? How is the ubiquitin ligase activity of the resulting E3 complexes modulated in the cell? Since CRLs were discovered, a battery of cullin-interacting proteins has been identified as important cellular factors that coordinate CRL complex assembly and control their ubiquitin ligase functions. The structural biology approach has not only helped establish the structural framework for investigating the regulation of CRL E3s but also revealed the detailed mechanisms for several key steps.

2.3.1 NEDD8-Modified CRLs

Many cellular enzymes catalyzing a form of post-translational modification are themselves subject to the same modification. For example, protein kinases are often activated by phosphorylation. CRL E3s follow this trend with a slight variation. All cullins can be modified by the ubiquitin-like molecule, NEDD8, at a specific lysine residue in their C-terminal WHB domain, which is close to the RBX1/2 binding site (Hori et al. 1999). This form of cullin modification, often referred to as neddylation, is conserved from fungi to humans and plays a role in stimulating the E3 activity of CRLs. Although cullin neddylation is not essential in budding yeast, it has been shown to alleviate the autoinhibition of CRLs through augmenting CRL-E2 interaction, closing the gap between the CRL-bound substrate and RBX1-bound E2, and promoting the amide bond formation at the E2 active site (Saha and Deshaies 2008; Yamoah et al. 2008).

A major breakthrough in our understanding of the effect of cullin neddylation came from the crystal structure of a NEDD8-modified CUL5-CTD-RBX1 complex (Fig. 2.4a, b) (Duda et al. 2008). Upon NEDD8 conjugation, the C-terminal portion of the CUL5-CTD undergoes a large degree of rotation, which reorients the WHB domain relative to the rest of the cullin scaffold. Because the WHB domain and its preceding long α-helix are responsible for cradling and stabilizing the globular RING domain of RBX1 in the unmodified form of cullin, this neddylation-induced conformational change releases the RING domain of the catalytic subunit from the cullin CTD. Due to the stable intermolecular

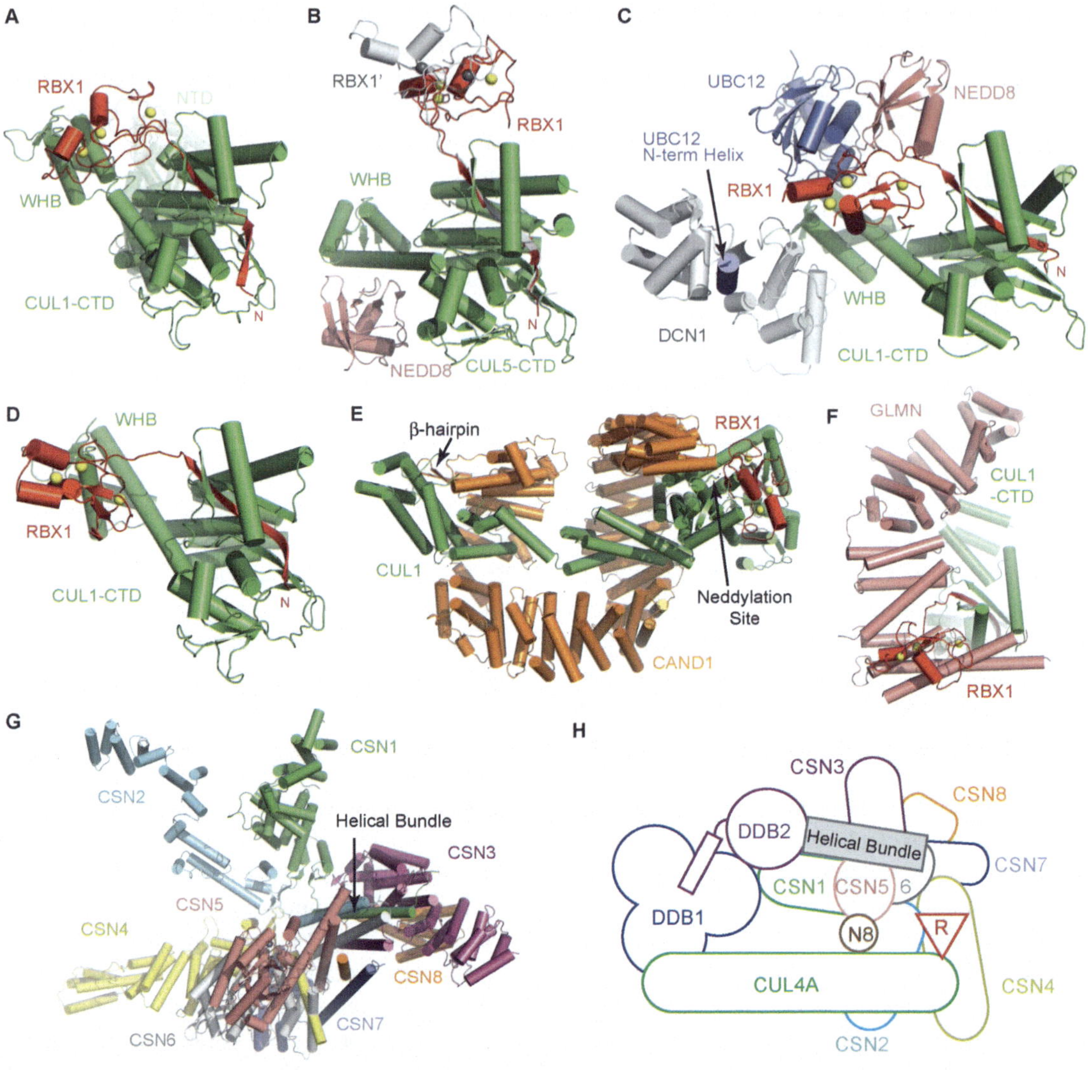

Fig. 2.4 Regulatory mechanisms of CRL assembly and function. (**a**) Interaction between CUL1-CTD and RBX1 viewed 90 degrees from Fig. 2.1b (PDB: 1LDJ). The N-terminus of RBX1 is indicated by "N." Yellow spheres represent zinc ions. (**b**) Dislodging of RBX1 RING domain from CUL5-CTD pocket upon cullin neddylation (PDB: 3DQV). Two orientations of the RBX1 RING domain captured in the crystal structure are shown. (**c**) Complex structure of CUL1-CTD-RBX1 bound to NEDD8-charged UBC12 and DCN1 (PDB: 4P5O). The linker between the N-terminal helix and catalytic domain of UBC12 is disordered. (**d**) A new position of RBX1 RING domain revealed by a CUL1-CTD-RBX1 complex structure (PDB: 3RTR). (**e**) CAND1 wraps around CUL1-RBX1, burying the cullin neddylation site lysine residue and blocking SKP1-binding site with a β-hairpin (PDB: 1U6G). (**f**) GLMN binds and blocks the E2-binding surface of RBX1 RING domain, which is flexibly tethered to the CUL1-CTD via an N-terminal β-strand (PDB: 4F52). (**g**) The overall architecture of COP9 signalosome (PDB: 4D10). (**h**) A schematic drawing of NEDD8-modified CUL4A-RBX1 in complex with DDB1-DDB2 and COP9 signalosome. R: RBX1. 6: CSN6. N8: NEDD8

β-sheet formed between the N-terminal β-strand of RBX1 and the α/β sub-domain of CUL5-CTD, RBX1 remains bound to the cullin scaffold with its RING domain gaining a significant degree of freedom to move around. This topological change of the cullin-RBX1 complex is thought to help bring the RBX1-bound E2 closer to the substrate anchored on the substrate receptor.

Just like protein ubiquitination, cullin neddylation requires the actions of the NEDD8-specific E1, E2 (UBC12), and E3 (DCN1 and DCN1 paralogues) enzymes (Liakopoulos et al. 1998; Osaka et al. 1998; Kurz et al. 2005). Similar to all ubiquitin-specific E2s, UBC12 (aka UBE2M) features a canonical E2 catalytic core domain, harboring an active site cysteine residue that can form a thioester bond with NEDD8 after it is activated by the E1 enzyme. Distinct from most ubiquitin-specific E2s, however, UBC12 contains an N-terminal extension sequence, whose extreme N-terminus has been shown to be acetylated. Remarkably, the acetylated UBC12 N-terminal extension adopts an α-helical conformation and specifically interacts with the neddylation E3 protein, DCN1 (Scott et al. 2011). With two EF hand-like sub-domains juxtaposed together, DCN1 is a compact all-helical protein with a slightly elongated shape (Fig. 2.4c). At the center of the protein is a hydrophobic pocket, which can specifically recognize the acetyl group of UBC12 N-terminus and the first methionine residue. Separate from this pocket, DCN1 also features a surface area that is able to engage cullin C-terminal WHB domain. Together, these interactions represent major interfaces through which DCN1 recruits the NEDD8-specific E2, UBC12, to catalyze the NEDD8 transfer reaction.

Strictly speaking, the NEDD8 E3 ligase function is performed by DCN1 in conjugation with RBX1, which plays a critical role in docking and activating the NEDD8-charged UBC12 catalytic core for cullin neddylation. Given the structural similarity of the catalytic domain between UBC12 and ubiquitin-specific E2s, the NEDD8 E2 is expected to interact with RBX1 RING domain in a similar fashion as ubiquitin-specific E2s to RING E3s. A simple modeling of a UBC12-RBX1 complex in the context of the unmodified cullin-RBX1 structures, however, readily reveals a long distance between the catalytic cysteine residue of UBC12 and the cullin neddylation site. For the cullin's lysine residue to attack the thioester bond formed between the UBC12 active site cysteine and the carboxyl terminus of NEDD8, the two residues have to be close to each other. This geometrical requirement strongly suggests that the RBX1 RING domain has to be reoriented before cullin neddylation can take place. Indeed, the crystal structure of an isolated CUL1-CTD-RBX1 complex revealed that the RING domain of RBX1 can be disengaged from its binding site on the CUL1-CTD in the absence of neddylation (Fig. 2.4d) (Calabrese et al. 2011). When UBC12 was modeled onto RBX1 in this structure, the gap between the UBC12 catalytic site and the cullin neddylation site is mostly closed.

The final picture of cullin neddylation has been depicted by the crystal structure of CUL1-CTD-RBX1 in complex with DCN1 and NEDD8-charged UBC12 (Fig. 2.4d) (Scott et al. 2014). In this structure, the CUL1 C-terminal WHB domain and its preceding long α-helix are shifted away from the rest of the CUL1-CTD, which allows the RING domain of RBX1 to adopt yet a new orientation. Resembling the previously reported docking model of ubiquitin-charged E2 to RING E3s, NEDD8-charged UBC12 is anchored to the RBX1 RING domain, and their compact structure is stabilized by a "linchpin" arginine residue unique to RBX1. Importantly, the NEDD8 molecule conjugated to the E2 also makes contacts with the linker sequence that connects the N-terminal β-strand of RBX1 to its RING domain, thereby optimally positioning the NEDD8 transfer module so that the catalytic site of the NEDD8 E2 is placed right next to the CUL1 neddylation site. Consistent with this notion, the interface between UBC12 catalytic domain and CUL1-WHB, which harbors the neddylation site, is kept minimal. Although DCN1 is also co-crystallized with the complex, it does not make direct interactions with the UBC12 catalytic domain (Fig. 2.4c). A flexible linker between the catalytic domain of the NEDD8 E2 and its N-terminal extension, which stably binds DCN1, is thought to accommodate the movement of the NEDD8 transfer module formed between RBX1 and NEDD8-charged UBC12 catalytic domain relative to the cullin scaffold.

2.3.2 CAND1 and Cullin Cycle

Cullin-associated and neddylation-dissociated protein 1 (CAND1) was the first cullin-binding protein identified that does not belong to the basal subunits of CRLs (Zheng et al. 2002a; Liu et al. 2002). It is a 120 kDa HEAT-repeat proteins that can form a stable complex with native, but not neddylated, cullin-RBX1 catalytic core. Interestingly, CAND1 binding seems to inhibit CUL1 from binding SKP1 and the substrate receptor F-box proteins, suggesting that CAND1 and SKP1-F-box proteins are mutually exclusive on the CRL1 scaffold. The crystal structure of a CAND1-CUL1-RBX1 complex unveiled the structural basis of all these biochemical activities of CAND1 (Fig. 2.4e) (Goldenberg et al. 2004). The 120 kDa protein adopts a superhelical structure with 27 consecutively stacked HEAT repeats that together form a long but highly sinuous fold. By curving around the entire CUL1-RBX1 structure, CAND1 grasps onto CUL1 like a two-pronged clamp. Importantly, CAND1 sports a β-hairpin projecting out of one of its HEAT repeats and reaching to the SKP1-binding site of the cullin scaffold. In doing so, CAND1 is able to compete with SKP1 for binding to the N-terminal end of CUL1. At the opposite end, the first two HEAT repeats of CAND1 closely pack against the WHB domain of CUL1, burying the neddylation site lysine residue. This suggests that CUL1 neddylation would sterically block CAND1 from binding.

Because SKP1-F-box proteins are responsible for recruiting substrates and CUL1 neddylation is thought to activate the E3 complex, the binding mode and biochemical properties of CAND1 seem to suggest that it acts as an inhibitor of CRL1. However, genetic studies indicate that CAND1 plays a positive role in regulating substrate ubiquitination and degradation by the E3 machinery. A growing body of evidence has helped raise an interesting model designating CAND1 as an exchange factor of CRL-substrate receptors (Pierce et al. 2013; Reitsma et al. 2017; Liu et al. 2018; Wu et al. 2013; Zemla et al. 2013). In this model, CAND1 can promote the disassembly of a SKP1-F-box protein complex from CUL1-RBX1, thereby allowing another SKP1-F-box protein complex to engage the cullin scaffold. While details of this model are discussed in Chap. 3 of this book, the structure of the CAND1-CUL1-RBX1 complex supports the notion that NEDD8 conjugation not only stimulates the activity of the E3 complex but also prevents CAND1 from dislodging an existing SKP1-F-box protein on the cullin scaffold.

In addition to CAND1, two other cellular factors have been documented to regulate a subset of CRLs, an α-helical protein known as Glomulin (GLMN) and a RING-IBR-RING protein, HHARI (aka ARIH1). The gene encoding GLMN is mutated in the hereditary disease glomuvenous malformations, which are characterized by venous lesions involving glomus cells. GLMN was initially identified as a protein that binds the C-terminus of CUL7, a distinct family member of CRLs (Arai et al. 2003). It was later shown to directly interact with the RBX1 RING domain and block its E3 ubiquitin ligase activity (Tron et al. 2012). The crystal structure of GLMN in complex with RBX1 bound to a fragment of CUL1-CTD revealed that GLMN contains two HEAT repeat-like sub-domains, which show structural similarity to each other (Fig. 2.4f) (Duda et al. 2012). One side of the GLMN C-terminal domain forms an extensive interface with the RING domain of RBX1, masking its E2-binding site. Although the CUL1-CTD fragment was present in the crystal and makes contact with GLMN C-terminal domain, it plays minimal role in stabilizing the complex formation. Owing to the orientational flexibility of RBX1 RING domain relative to the rest of the CRL1, GLMN binding is compatible with CRL assembly both at the RBX1-cullin interface and the cullin-adaptor-substrate receptor site. Even cullin neddylation showed no effect to GLMN-RBX1 interaction. Overall, GLMN appears to be an RBX1-specific inhibitor. Nonetheless, GLMN only binds a small subset of CRLs in human cells, suggesting that an unknown mechanism is involved in selectively controlling GLMN-RBX1 interaction in the context of the CRL functions.

RING-IBR-RING E3s represent a distinct class of ubiquitin ligases, which is a hybrid of the canonical RING-type and the HECT-type E3s (Zheng and Shabek 2017). RING-IBR-RINGSs are characterized by multiple RING domains and thioester intermediates they form with ubiquitin before the modifier is transferred to a substrate. ARH1, a member of the RING-IBR-RING E3s, has recently been identified to be preferentially associated with NEDD8-modified CRL1-CRL3 (but not CRL4) (Scott et al. 2016). Interestingly, it catalyzes mono-ubiquitination of representative substrates of these CRL E3s, which can be further polyubiquitinated by CDC34, the cognate E2 for RBX1. The precise mechanism by which ARH1 coordinates with neddylated CRLs to mediate the ubiquitin transfer reaction awaits future structural studies.

2.3.3 COP9 Signalosome-CRL Interactions

Just like protein ubiquitination, cullin neddylation is reversible. Deconjugation of NEDD8 from cullins is catalyzed by an evolutionarily conserved eight-subunit protein complex, known as the COP9 signalosome (CSN). CSN was first identified in plants based on mutants that showed a *constitutive photomorphogenesis* (*COP*) phenotype (Wei et al. 1994). These plant mutants turned out to carry mutations in eight genes, whose protein products form a stable complex with each subunit sharing sequence homology with one component of the eight-subunit lid complex of the 19S proteasome (Chamovitz et al. 1996; Wei et al. 1994). Among the eight CSN subunits, CSN5 is a zinc-containing metalloprotease that is responsible for cleaving the iso-peptide bond between cullins and NEDD8 (Cope et al. 2002). The assembly mechanism of the COP9 signalosome was first revealed in the crystal structure of the human CSN complex (Fig. 2.4g) (Lingaraju et al. 2014). Each of the CSN subunits employs one or two α-helices to build a superhelical bundle, which contributes to the stable assembly of the deneddylase complex. Meanwhile, six of the CSN subunits encircle a horseshoe-shaped ring structure with elongated α-helical PCI (proteasome lid-CSN-initiation factor 3) domains projecting away from the center. CSN5 and CSN6 share a common MPN (MPR1/PAD1) domain with a metalloprotease fold and together form a heterodimer. With their C-terminal regions integrated into the superhelical bundle, these two subunits anchor themselves onto one side of the CSN ring structure. Interestingly, the catalytic site of CSN5 in the CSN holoenzyme was found in an autoinhibited state, suggesting that CSN has to be activated upon binding to its substrate.

Recent advances in cryo-EM technology have enabled several studies that have shed light on how CSN interacts with different CRL complexes. Despite limited resolution, single particle analysis of CSN bound to neddylated CUL1-RBX1 in complexes with SKP1-SKP2-CKS1 and monomeric SKP1-FBXW7 offered the first glimpse of the CSN-CRL interaction (Enchev et al. 2012). In the structural models derived from the EM density maps, CSN2 appears to make major contacts with CUL1-CTD, whereas the distal end of the two F-box proteins is located close to CSN1 and CSN3. Structural modeling and biochemical analysis indicated that CSN competes with both substrates and E2 for binding the E3 platform, thereby raising the possibility that substrate-loaded CRL1 might protect neddylated CUL1 by preventing CSN from accessing the E3. This notion was subsequently supported by the cryo-EM structure of CSN bound to the neddylated CUL4A-RBX1-DDB1-DDB2 complex (Fig. 2.4h) (Cavadini et al. 2016). In this super-assembly, CSN2 not only interacts with CUL4A-CTD but also sandwiches RBX1 RING domain together with CSN4, thereby preventing the E3 scaffold from recruiting an E2 molecule. While CSN1 makes specific interactions with DDB1, DDB2 is snugly situated in between DDB1 and the CSN helical bundle. In comparison with the CSN holoenzyme structure, the CSN helical bundle is repositioned to accommodate the CUL4A-DDB1 substrate receptor. Interestingly, despite its topological flexibility, the CSN helical bundle cannot accommodate additional cellular factors that interact with DDB2, corroborating the idea that substrate binding to CRL4 will introduce steric hindrance for

CSN binding. Nevertheless, question remains as to whether CSN has the ability to differentiate variations of CRL-substrate receptor in size and shape from the binding of a small degron as part of a flexible substrate polypeptide.

Upon binding to CRL4, CSN undergoes multiple conformational changes to not only adapt to the landscape of its substrate but also alleviate its autoinhibition. These changes include movements of the PCI domains of CSN2 and CSN4 for clamping down to CUL4A-CTD and RBX1 and the translocation of the CSN5-CSN6 dimer to approach NEDD8. Although CRL4 is significantly different from other CRLs, similar structural changes have also been observed in the EM structure of CSN in complex with neddylated CUL1 with SKP1-SKP2-CKS1 bound. Although biochemical analyses have helped identify several structural elements that relay these structural changes to the alleviation of CSN autoinhibition, the detailed structural mechanism underlying CSN5 activation requires structural analysis at a higher resolution (Cavadini et al. 2016; Mosadeghi et al. 2016).

Acknowledgments We thank all members of the Zheng laboratory for their support and help. We apologize to all colleagues whose work could not be cited due to space constraints and the timing of manuscript preparation. Both D-V.R. and N.Z. are supported by Howard Hughes Medical Institute. N.Z. is a Howard Hughes Medical Institute Investigator.

References

Angers S, Li T, Yi X, Maccoss MJ, Moon RT, Zheng N (2006) Molecular architecture and assembly of the DDB1-CUL4A ubiquitin ligase machinery. Nature 443:590–593

Arai T, Kasper JS, Skaar JR, Ali SH, Takahashi C, Decaprio JA (2003) Targeted disruption of p185/Cul7 gene results in abnormal vascular morphogenesis. Proc Natl Acad Sci U S A 100:9855–9860

Busino L, Bassermann F, Maiolica A, Lee C, Nolan PM, Godinho SI, Draetta GF, Pagano M (2007) SCFFbxl3 controls the oscillation of the circadian clock by directing the degradation of cryptochrome proteins. Science 316:900–904

Calabrese MF, Scott DC, Duda DM, Grace CR, Kurinov I, Kriwacki RW, Schulman BA (2011) A RING E3-substrate complex poised for ubiquitin-like protein transfer: structural insights into cullin-RING ligases. Nat Struct Mol Biol 18:947–949

Cardote TAF, Gadd MS, Ciulli A (2017) Crystal structure of the Cul2-Rbx1-EloBC-VHL ubiquitin ligase complex. Structure 25:901–911.e3

Cavadini S, Fischer ES, Bunker RD, Potenza A, Lingaraju GM, Goldie KN, Mohamed WI, Faty M, Petzold G, Beckwith RE, Tichkule RB, Hassiepen U, Abdulrahman W, Pantelic RS, Matsumoto S, Sugasawa K, Stahlberg H, Thomä NH (2016) Cullin-RING ubiquitin E3 ligase regulation by the COP9 signalosome. Nature 531:598–603

Chamovitz DA, Wei N, Osterlund MT, Von Arnim AG, Staub JM, Matsui M, Deng XW (1996) The COP9 complex, a novel multisubunit nuclear regulator involved in light control of a plant developmental switch. Cell 86:115–121

Chen SJ, Wu X, Wadas B, Oh JH, Varshavsky A (2017) An N-end rule pathway that recognizes proline and destroys gluconeogenic enzymes. Science 355

Choi WS, Jeong BC, Joo YJ, Lee MR, Kim J, Eck MJ, Song HK (2010) Structural basis for the recognition of N-end rule substrates by the UBR box of ubiquitin ligases. Nat Struct Mol Biol 17:1175–1181

Cope GA, Suh GS, Aravind L, Schwarz SE, Zipursky SL, Koonin EV, Deshaies RJ (2002) Role of predicted metalloprotease motif of Jab1/Csn5 in cleavage of Nedd8 from Cul1. Science 298:608–611

Decorsière A, Mueller H, Van Breugel PC, Abdul F, Gerossier L, Beran RK, Livingston CM, Niu C, Fletcher SP, Hantz O, Strubin M (2016) Hepatitis B virus X protein identifies the Smc5/6 complex as a host restriction factor. Nature 531:386–389

Dong C, Zhang H, Li L, Tempel W, Loppnau P, Min J (2018) Molecular basis of GID4-mediated recognition of degrons for the Pro/N-end rule pathway. Nat Chem Biol 14:466–473

Donovan KA, An J, Nowak RP, Yuan JC, Fink EC, Berry BC, Ebert BL, Fischer ES (2018) Thalidomide promotes degradation of SALL4, a transcription factor implicated in Duane Radial Ray syndrome. Elife 7

Duda DM, Borg LA, Scott DC, Hunt HW, Hammel M, Schulman BA (2008) Structural insights into NEDD8 activation of cullin-RING ligases: conformational control of conjugation. Cell 134:995–1006

Duda DM, Olszewski JL, Tron AE, Hammel M, Lambert LJ, Waddell MB, Mittag T, Decaprio JA, Schulman BA (2012) Structure of a glomulin-RBX1-CUL1 complex: inhibition of a RING E3 ligase through masking of its E2-binding surface. Mol Cell 47:371–382

Enchev RI, Scott DC, da Fonseca PC, Schreiber A, Monda JK, Schulman BA, Peter M, Morris EP (2012) Structural basis for a reciprocal regulation between SCF and CSN. Cell Rep 2:616–627

Fischer ES, Scrima A, Böhm K, Matsumoto S, Lingaraju GM, Faty M, Yasuda T, Cavadini S, Wakasugi M, Hanaoka F, Iwai S, Gut H, Sugasawa K, Thomä NH (2011) The molecular basis of CRL4DDB2/CSA ubiquitin ligase architecture, targeting, and activation. Cell 147:1024–1039

Fukutomi T, Takagi K, Mizushima T, Ohuchi N, Yamamoto M (2014) Kinetic, thermodynamic, and structural characterizations of the association between Nrf2-DLGex degron and Keap1. Mol Cell Biol 34:832–846
Geyer R, Wee S, Anderson S, Yates J, Wolf DA (2003) BTB/POZ domain proteins are putative substrate adaptors for cullin 3 ubiquitin ligases. Mol Cell 12:783–790
Godinho SI, Maywood ES, Shaw L, Tucci V, Barnard AR, Busino L, Pagano M, Kendall R, Quwailid MM, Romero MR, O'Neill J, Chesham JE, Brooker D, Lalanne Z, Hastings MH, Nolan PM (2007) The after-hours mutant reveals a role for Fbxl3 in determining mammalian circadian period. Science 316:897–900
Goldenberg SJ, Cascio TC, Shumway SD, Garbutt KC, Liu J, Xiong Y, Zheng N (2004) Structure of the Cand1-Cul1-Roc1 complex reveals regulatory mechanisms for the assembly of the multisubunit cullin-dependent ubiquitin ligases. Cell 119:517–528
Guo Y, Dong L, Qiu X, Wang Y, Zhang B, Liu H, Yu Y, Zang Y, Yang M, Huang Z (2014) Structural basis for hijacking CBF-β and CUL5 E3 ligase complex by HIV-1 Vif. Nature 505:229–233
Hao B, Zheng N, Schulman BA, Wu G, Miller JJ, Pagano M, Pavletich NP (2005) Structural basis of the Cks1-dependent recognition of p27(Kip1) by the SCF(Skp2) ubiquitin ligase. Mol Cell 20:9–19
Hao B, Oehlmann S, Sowa ME, Harper JW, Pavletich NP (2007) Structure of a Fbw7-Skp1-cyclin E complex: multisite-phosphorylated substrate recognition by SCF ubiquitin ligases. Mol Cell 26:131–143
Hori T, Osaka F, Chiba T, Miyamoto C, Okabayashi K, Shimbara N, Kato S, Tanaka K (1999) Covalent modification of all members of human cullin family proteins by NEDD8. Oncogene 18:6829–6834
Horvath CM (2004) Weapons of STAT destruction. Interferon evasion by paramyxovirus V protein. Eur J Biochem 271:4621–4628
Ivan M, Kondo K, Yang H, Kim W, Valiando J, Ohh M, Salic A, Asara JM, Lane WS, Kaelin WG (2001) HIFalpha targeted for VHL-mediated destruction by proline hydroxylation: implications for O2 sensing. Science 292:464–468
Jin J, Cardozo T, Lovering RC, Elledge SJ, Pagano M, Harper JW (2004) Systematic analysis and nomenclature of mammalian F-box proteins. Genes Dev 18:2573–2580
Jin J, Arias EE, Chen J, Harper JW, Walter JC (2006) A family of diverse Cul4-Ddb1-interacting proteins includes Cdt2, which is required for S phase destruction of the replication factor Cdt1. Mol Cell 23:709–721
Kaelin WG (2005) Proline hydroxylation and gene expression. Annu Rev Biochem 74:115–128
Kim DY, Kwon E, Hartley PD, Crosby DC, Mann S, Krogan NJ, Gross JD (2013) CBFβ stabilizes HIV Vif to counteract APOBEC3 at the expense of RUNX1 target gene expression. Mol Cell 49:632–644
Koren I, Timms RT, Kula T, Xu Q, Li MZ, Elledge SJ (2018) The eukaryotic proteome is shaped by E3 ubiquitin ligases targeting C-terminal degrons. Cell
Krönke J, Udeshi ND, Narla A, Grauman P, Hurst SN, McConkey M, Svinkina T, Heckl D, Comer E, Li X, Ciarlo C, Hartman E, Munshi N, Schenone M, Schreiber SL, Carr SA, Ebert BL (2014) Lenalidomide causes selective degradation of IKZF1 and IKZF3 in multiple myeloma cells. Science 343:301–305
Krönke J, Fink EC, Hollenbach PW, Macbeth KJ, Hurst SN, Udeshi ND, Chamberlain PP, Mani DR, Man HW, Gandhi AK, Svinkina T, Schneider RK, Mcconkey M, Järås M, Griffiths E, Wetzler M, Bullinger L, Cathers BE, Carr SA, Chopra R, Ebert BL (2015) Lenalidomide induces ubiquitination and degradation of CK1α in del(5q) MDS. Nature 523:183–188
Kurz T, Ozlü N, Rudolf F, O'Rourke SM, Luke B, Hofmann K, Hyman AA, Bowerman B, Peter M (2005) The conserved protein DCN-1/Dcn1p is required for cullin neddylation in C. elegans and S. cerevisiae. Nature 435:1257–1261
Li T, Chen X, Garbutt KC, Zhou P, Zheng N (2006) Structure of DDB1 in complex with a paramyxovirus V protein: viral hijack of a propeller cluster in ubiquitin ligase. Cell 124:105–117
Li T, Robert EI, Van Breugel PC, Strubin M, Zheng N (2010) A promiscuous alpha-helical motif anchors viral hijackers and substrate receptors to the CUL4-DDB1 ubiquitin ligase machinery. Nat Struct Mol Biol 17:105–111
Liakopoulos D, Doenges G, Matuschewski K, Jentsch S (1998) A novel protein modification pathway related to the ubiquitin system. EMBO J 17:2208–2214
Lin HC, Ho SC, Chen YY, Khoo KH, Hsu PH, Yen HC (2015) SELENOPROTEINS. CRL2 aids elimination of truncated selenoproteins produced by failed UGA/Sec decoding. Science 349:91–95
Lin HC, Yeh CW, Chen YF, Lee TT, Hsieh PY, Rusnac DV, Lin SY, Elledge SJ, Zheng N, Yen HS (2018) C-terminal end-directed protein elimination by CRL2 ubiquitin ligases. Mol Cell 70:602–613.e3
Lingaraju GM, Bunker RD, Cavadini S, Hess D, Hassiepen U, Renatus M, Fischer ES, Thomä NH (2014) Crystal structure of the human COP9 signalosome. Nature 512:161–165
Liu J, Furukawa M, Matsumoto T, Xiong Y (2002) NEDD8 modification of CUL1 dissociates p120 (CAND1), an inhibitor of CUL1-SKP1 binding and SCF ligases. Mol Cell 10:1511–1518
Liu X, Reitsma JM, Mamrosh JL, Zhang Y, Straube R, Deshaies RJ (2018) Cand1-mediated adaptive exchange mechanism enables variation in F-box protein expression. Mol Cell 69:773–786.e6
Lu G, Middleton RE, Sun H, Naniong M, Ott CJ, Mitsiades CS, Wong KK, Bradner JE, Kaelin WG (2014) The myeloma drug lenalidomide promotes the cereblon-dependent destruction of Ikaros proteins. Science 343:305–309

Mahrour N, Redwine WB, Florens L, Swanson SK, Martin-Brown S, Bradford WD, Staehling-Hampton K, Washburn MP, Conaway RC, Conaway JW (2008) Characterization of Cullin-box sequences that direct recruitment of Cul2-Rbx1 and Cul5-Rbx2 modules to Elongin BC-based ubiquitin ligases. J Biol Chem 283:8005–8013

Matta-Camacho E, Kozlov G, Li FF, Gehring K (2010) Structural basis of substrate recognition and specificity in the N-end rule pathway. Nat Struct Mol Biol 17:1182–1187

Matyskiela ME, Lu G, Ito T, Pagarigan B, Lu CC, Miller K, Fang W, Wang NY, Nguyen D, Houston J, Carmel G, Tran T, Riley M, Nosaka L, Lander GC, Gaidarova S, Xu S, Ruchelman AL, Handa H, Carmichael J, Daniel TO, Cathers BE, Lopez-Girona A, Chamberlain PP (2016) A novel cereblon modulator recruits GSPT1 to the CRL4(CRBN) ubiquitin ligase. Nature 535:252–257

Matyskiela ME, Couto S, Zheng X, Lu G, Hui J, Stamp K, Drew C, Ren Y, Wang M, Carpenter A, Lee CW, Clayton T, Fang W, Lu CC, Riley M, Abdubek P, Blease K, Hartke J, Kumar G, Vessey R, Rolfe M, Hamann LG, Chamberlain PP (2018) SALL4 mediates teratogenicity as a thalidomide-dependent cereblon substrate. Nat Chem Biol 14:981–987

Min JH, Yang H, Ivan M, Gertler F, Kaelin WG, Pavletich NP (2002) Structure of an HIF-1alpha -pVHL complex: hydroxyproline recognition in signaling. Science 296:1886–1889

Mosadeghi R, Reichermeier KM, Winkler M, Schreiber A, Reitsma JM, Zhang Y, Stengel F, Cao J, Kim M, Sweredoski MJ, Hess S, Leitner A, Aebersold R, Peter M, Deshaies RJ, Enchev RI (2016) Structural and kinetic analysis of the COP9-Signalosome activation and the cullin-RING ubiquitin ligase deneddylation cycle. Elife 5

Nangle S, Xing W, Zheng N (2013) Crystal structure of mammalian cryptochrome in complex with a small molecule competitor of its ubiquitin ligase. Cell Res 23:1417–1419

Nguyen HC, Yang H, Fribourgh JL, Wolfe LS, Xiong Y (2015) Insights into Cullin-RING E3 ubiquitin ligase recruitment: structure of the VHL-EloBC-Cul2 complex. Structure 23:441–449

Osaka F, Kawasaki H, Aida N, Saeki M, Chiba T, Kawashima S, Tanaka K, Kato S (1998) A new NEDD8-ligating system for cullin-4A. Genes Dev 12:2263–2268

Padmanabhan B, Tong KI, Ohta T, Nakamura Y, Scharlock M, Ohtsuji M, Kang MI, Kobayashi A, Yokoyama S, Yamamoto M (2006) Structural basis for defects of Keap1 activity provoked by its point mutations in lung cancer. Mol Cell 21:689–700

Petzold G, Fischer ES, Thomä NH (2016) Structural basis of lenalidomide-induced CK1α degradation by the CRL4(CRBN) ubiquitin ligase. Nature 532:127–130

Pierce NW, Lee JE, Liu X, Sweredoski MJ, Graham RL, Larimore EA, Rome M, Zheng N, Clurman BE, Hess S, Shan SO, Deshaies RJ (2013) Cand1 promotes assembly of new SCF complexes through dynamic exchange of F box proteins. Cell 153:206–215

Pintard L, Willis JH, Willems A, Johnson JL, Srayko M, Kurz T, Glaser S, Mains PE, Tyers M, Bowerman B, Peter M (2003) The BTB protein MEL-26 is a substrate-specific adaptor of the CUL-3 ubiquitin-ligase. Nature 425:311–316

Reitsma JM, Liu X, Reichermeier KM, Moradian A, Sweredoski MJ, Hess S, Deshaies RJ (2017) Composition and regulation of the cellular repertoire of SCF ubiquitin ligases. Cell 171:1326–1339.e14

Rusnac DV, Lin HC, Canzani D, Tien KX, Hinds TR, Tsue AF, Bush MF, Yen HS, Zheng N (2018) Recognition of the diglycine C-end degron by CRL2. Mol Cell 72:813–822.e4

Saha A, Deshaies RJ (2008) Multimodal activation of the ubiquitin ligase SCF by Nedd8 conjugation. Mol Cell 32:21–31

Schmalen I, Reischl S, Wallach T, Klemz R, Grudziecki A, Prabu JR, Benda C, Kramer A, Wolf E (2014) Interaction of circadian clock proteins CRY1 and PER2 is modulated by zinc binding and disulfide bond formation. Cell 157:1203–1215

Schulman BA, Carrano AC, Jeffrey PD, Bowen Z, Kinnucan ER, Finnin MS, Elledge SJ, Harper JW, Pagano M, Pavletich NP (2000) Insights into SCF ubiquitin ligases from the structure of the Skp1-Skp2 complex. Nature 408:381–386

Scott DC, Monda JK, Bennett EJ, Harper JW, Schulman BA (2011) N-terminal acetylation acts as an avidity enhancer within an interconnected multiprotein complex. Science 334:674–678

Scott DC, Sviderskiy VO, Monda JK, Lydeard JR, Cho SE, Harper JW, Schulman BA (2014) Structure of a RING E3 trapped in action reveals ligation mechanism for the ubiquitin-like protein NEDD8. Cell 157:1671–1684

Scott DC, Rhee DY, Duda DM, Kelsall IR, Olszewski JL, Paulo JA, de Jong A, Ovaa H, Alpi AF, Harper JW, Schulman BA (2016) Two distinct types of E3 ligases work in unison to regulate substrate ubiquitylation. Cell 166:1198–1214.e24

Shabek N, Zheng N (2014) Plant ubiquitin ligases as signaling hubs. Nat Struct Mol Biol 21:293–296

Sheard LB, Tan X, Mao H, Withers J, Ben-Nissan G, Hinds TR, Kobayashi Y, Hsu FF, Sharon M, Browse J, He SY, Rizo J, Howe GA, Zheng N (2010) Jasmonate perception by inositol-phosphate-potentiated COI1-JAZ co-receptor. Nature 468:400–405

Shirogane T, Jin J, Ang XL, Harper JW (2005) SCFbeta-TRCP controls clock-dependent transcription via casein kinase 1-dependent degradation of the mammalian period-1 (Per1) protein. J Biol Chem 280:26863–26872

Siepka SM, Yoo SH, Park J, Song W, Kumar V, Hu Y, Lee C, Takahashi JS (2007) Circadian mutant overtime reveals F-box protein FBXL3 regulation of

cryptochrome and period gene expression. Cell 129:1011–1023
Skowyra D, Craig KL, Tyers M, Elledge SJ, Harper JW (1997) F-box proteins are receptors that recruit phosphorylated substrates to the SCF ubiquitin-ligase complex. Cell 91:209–219
Stebbins CE, Kaelin WG, Pavletich NP (1999) Structure of the VHL-Elongin C-Elongin B complex: implications for VHL tumor suppressor function. Science 284:455–461
Takahashi JS (2017) Transcriptional architecture of the mammalian circadian clock. Nat Rev Genet 18:164–179
Tan X, Calderon-Villalobos LI, Sharon M, Zheng C, Robinson CV, Estelle M, Zheng N (2007) Mechanism of auxin perception by the TIR1 ubiquitin ligase. Nature 446:640–645
Tron AE, Arai T, Duda DM, Kuwabara H, Olszewski JL, Fujiwara Y, Bahamon BN, Signoretti S, Schulman BA, Decaprio JA (2012) The glomuvenous malformation protein Glomulin binds Rbx1 and regulates cullin RING ligase-mediated turnover of Fbw7. Mol Cell 46:67–78
Wei N, Chamovitz DA, Deng XW (1994) Arabidopsis COP9 is a component of a novel signaling complex mediating light control of development. Cell 78:117–124
Welcker M, Clurman BE (2008) FBW7 ubiquitin ligase: a tumour suppressor at the crossroads of cell division, growth and differentiation. Nat Rev Cancer 8:83–93
Wild R, Gerasimaite R, Jung JY, Truffault V, Pavlovic I, Schmidt A, Saiardi A, Jessen HJ, Poirier Y, Hothorn M, Mayer A (2016) Control of eukaryotic phosphate homeostasis by inositol polyphosphate sensor domains. Science 352:986–990
Wu G, Xu G, Schulman BA, Jeffrey PD, Harper JW, Pavletich NP (2003) Structure of a beta-TrCP1-Skp1-beta-catenin complex: destruction motif binding and lysine specificity of the SCF(beta-TrCP1) ubiquitin ligase. Mol Cell 11:1445–1456
Wu S, Zhu W, Nhan T, Toth JI, Petroski MD, Wolf DA (2013) CAND1 controls in vivo dynamics of the cullin 1-RING ubiquitin ligase repertoire. Nat Commun 4:1642
Xing W, Busino L, Hinds TR, Marionni ST, Saifee NH, Bush MF, Pagano M, Zheng N (2013) SCF(FBXL3) ubiquitin ligase targets cryptochromes at their cofactor pocket. Nature 496:64–68
Xu L, Wei Y, Reboul J, Vaglio P, Shin TH, Vidal M, Elledge SJ, Harper JW (2003) BTB proteins are substrate-specific adaptors in an SCF-like modular ubiquitin ligase containing CUL-3. Nature 425:316–321
Yamamoto M, Kensler TW, Motohashi H (2018) The KEAP1-NRF2 system: a thiol-based sensor-effector apparatus for maintaining redox homeostasis. Physiol Rev 98:1169–1203
Yamoah K, Oashi T, Sarikas A, Gazdoiu S, Osman R, Pan ZQ (2008) Autoinhibitory regulation of SCF-mediated ubiquitination by human cullin 1's C-terminal tail. Proc Natl Acad Sci U S A 105:12230–12235
Yu X, Yu Y, Liu B, Luo K, Kong W, Mao P, Yu XF (2003) Induction of APOBEC3G ubiquitination and degradation by an HIV-1 Vif-Cul5-SCF complex. Science 302:1056–1060
Zemla A, Thomas Y, Kedziora S, Knebel A, Wood NT, Rabut G, Kurz T (2013) CSN- and CAND1-dependent remodelling of the budding yeast SCF complex. Nat Commun 4:1641
Zheng N, Shabek N (2017) Ubiquitin ligases: structure, function, and regulation. Annu Rev Biochem 86:129–157
Zheng J, Yang X, Harrell JM, Ryzhikov S, Shim EH, Lykke-Andersen K, Wei N, Sun H, Kobayashi R, Zhang H (2002a) CAND1 binds to unneddylated CUL1 and regulates the formation of SCF ubiquitin E3 ligase complex. Mol Cell 10:1519–1526
Zheng N, Schulman BA, Song L, Miller JJ, Jeffrey PD, Wang P, Chu C, Koepp DM, Elledge SJ, Pagano M, Conaway RC, Conaway JW, Harper JW, Pavletich NP (2002b) Structure of the Cul1-Rbx1-Skp1-F boxSkp2 SCF ubiquitin ligase complex. Nature 416:703–709
Zhuang M, Calabrese MF, Liu J, Waddell MB, Nourse A, Hammel M, Miller DJ, Walden H, Duda DM, Seyedin SN, Hoggard T, Harper JW, White KP, Schulman BA (2009) Structures of SPOP-substrate complexes: insights into molecular architectures of BTB-Cul3 ubiquitin ligases. Mol Cell 36:39–50

Assembly and Regulation of CRL Ubiquitin Ligases

3

Kankan Wang, Raymond J. Deshaies, and Xing Liu

Abstract

Cullin-RING ubiquitin ligases (CRLs) determine the substrate specificity of ubiquitination reactions, and substrates are recruited to the cullin core through binding to their cognate substrate receptor modules. Because a family of substrate receptors compete for the same cullin core, the assembly and activity of CRLs are dynamically regulated to fulfill the needs of the cell to adapt to the changing pool of proteins demanding ubiquitination. Cullins are modified by NEDD8, a ubiquitin-like protein. This process, referred to as neddylation, promotes the E3 activity of CRLs by inducing conformational rearrangement in the Cullin-RING catalytic core. Cand1 is a cullin-associated protein whose binding is excluded by cullin neddylation. Although early biochemical studies suggested that Cand1 inhibits CRL activity, genetic studies revealed its positive role in ubiquitination. Emerging evidence from kinetic and quantitative proteomic studies demonstrated that Cand1 stimulates assembly of new Skp1-Cul1-F-box protein (SCF) complexes by exchanging the Skp1-F-box protein substrate receptor modules. Furthermore, aided by refined experimental design as well as computational simulation, an attractive model has been developed in which substrate, neddylation cycle and Cand1-mediated "adaptive exchange" collaborate to maintain the dynamics of the cellular SCF repertoire. Here, we review and discuss recent advances that have deepened our understanding of CRL regulation.

Keywords

NEDD8 · Neddylation · Cullin · CRLs · SCF · Cand1 · Ubiquitination · F-box proteins

Abbreviations

4HB	Four-helix bundle
APP-BP1	Amyloid-β precursor protein binding protein 1
CAND1	Cullin-associated NEDD8-dissociated protein 1
CAND2	Cullin-associated NEDD8-dissociated protein 2
CRLs	Cullin-RING ubiquitin ligases
CSN	COP9 signalosome
CTD	C-terminal domain

K. Wang
Department of Biochemistry, Purdue University, West Lafayette, IN, USA

R. J. Deshaies
Amgen, One Amgen Center Way, Thousand Oaks, CA, USA

X. Liu (✉)
Department of Biochemistry, Purdue University, West Lafayette, IN, USA

Center for Plant Biology, Purdue University, West Lafayette, IN, USA
e-mail: xingliu@purdue.edu

Y. Sun et al. (eds.), *Cullin-RING Ligases and Protein Neddylation*, Advances in Experimental Medicine and Biology 1217, https://doi.org/10.1007/978-981-15-1025-0_3

DCN1	Defective in cullin neddylation 1
DEN1	Deneddylase 1
DKO	Double knockout
FBPs	F-box proteins
FRET	Fluorescence resonance energy transfer
GEFs	Guanine nucleotide exchange factors
HEAT	Huntingtin-elongation-A subunit-TOR
NAE	NEDD8-activating enzyme
NEDD8	Neural precursor cell expressed, developmentally downregulated 8
NMR	Nuclear magnetic resonance
NTD	N-terminal domain
PONY	Potentiating neddylation
Rbx1	RING-box protein 1
Rbx2	RING-box protein 2
SCF	Skp1-Cul1-F-box
TBP	TATA-binding protein
TNFα	Tumor necrosis factor alpha
UBA3	Ubiquitin-activating enzyme 3
UCHL3	Ubiquitin C-terminal hydrolase L3
UFD	Ubiquitin-fold domain

3.1 Introduction to NEDD8

*N*eural precursor cell expressed, developmentally downregulated *8* (NEDD8) is a highly conserved protein across eukaryotic species (83% identity between humans and *Arabidopsis thaliana*), although different copy numbers for the *NEDD8* gene were found in genomes of several plant species (Rao-Naik et al. 1998; Mergner and Schwechheimer 2014). NEDD8 was first reported as a highly expressed mRNA in embryonic mouse brain (Kumar et al. 1992). Although NEDD8 mRNA exhibits a broad expression pattern in adult tissues, it is highly accumulated in the heart and skeletal muscle in humans (Kamitani et al. 1997). Its expression is increased during the early stage of mouse embryogenesis but decreased in the brain at late developmental stages (Kumar et al. 1992; Kamitani et al. 1997). In terms of subcellular localization, exogenous NEDD8 showed strong expression in the nucleus and much weaker expression in the cytoplasm (Kamitani et al. 1997).

NEDD8 is a ubiquitin-like protein. Among all of the ubiquitin-like proteins, NEDD8 possesses the greatest similarity with ubiquitin (~60% identical to human ubiquitin). It even contains the same C-terminal Leu-Arg-Gly-Gly residues, which are crucial for the conjugation of ubiquitin to its substrate proteins. Interestingly, both ubiquitin and Nedd8 are produced as precursors that need to be processed at the C-terminus to reveal the terminal glycine (Kamitani et al. 1997). Cdc53, the budding yeast Cul1, was the first protein identified to be covalently modified by NEDD8, suggesting a regulatory role of neddylation in the ubiquitin-proteasome system (Lammer et al. 1998; Liakopoulos et al. 1998). Genetic studies have demonstrated that neddylation is essential in multiple model organisms including mice (Tateishi et al. 2001), Arabidopsis (Dharmasiri et al. 2003), Drosophila (Ou et al. 2002), and fission yeast (Osaka et al. 2000), suggesting a conserved and important function of neddylation in development.

Given the high sequence similarity, it is not surprising that the overall structure of NEDD8 is similar with that of ubiquitin (Rao-Naik et al. 1998; Whitby et al. 1998). NEDD8 is comprised of a globular core and a C-terminal tail that is flexible in solution. Ubiquitin displays an asymmetric distribution of charged residues, thereby forming the acidic and basic patches (Wilkinson 1988). The globular domain of NEDD8 maintains a similar arrangement of acidic and basic patches, although charged residues making up the patches are only moderately conserved. Similarly, two surface-exposed hydrophobic patches, which are crucial for protein-protein interactions in ubiquitin (Shih et al. 2000; Sloper-Mould et al. 2001; Hu et al. 2002), are both conserved in NEDD8 (Enchev et al. 2015). However, NEDD8 and ubiquitin each have specific functions due to small differences in their structures. Seven unique residues that are conserved across NEDD8 orthologues but differ from the ones in ubiquitin are responsible for NEDD8-specific interactions. One such residue, Ala-72, was shown to play a critical role in preventing the interaction of NEDD8 with the ubiquitin E1 enzyme (Whitby et al. 1998). The other six charged residues present on the surface of NEDD8 are essential for regulating ubiquitin ligase activities by NEDD8 (Wu et al. 2002).

3.2 The NEDD8 Conjugation System

NEDD8 is initially translated as a precursor that requires proteolytic processing to expose the C-terminal Gly residue. *U*biquitin *C*-terminal *h*ydrolase *L3* (UCHL3), a member of the C12 family peptidases, displays a dual specificity for efficient processing of both NEDD8 and ubiquitin (Wada et al. 1998; Johnston et al. 1999; Linghu et al. 2002). In contrast, the deneddylase 1 (DEN1) from the C48 family peptidases acts exclusively on NEDD8. DEN1 can mediate the proteolytic processing of NEDD8 precursor as well as the deconjugation of NEDD8 from some protein substrates (Gan-Erdene et al. 2003; Mendoza et al. 2003; Wu et al. 2003). Of note, UCHL3 knockout in mice or DEN1 knockout in Drosophila does not result in neddylation defects (Kurihara et al. 2000; Chan et al. 2008), suggesting redundant function between both enzymes.

The processed NEDD8 is covalently conjugated to substrate proteins in a manner highly similar to ubiquitin conjugation, through a cascade of enzymes comprising E1 *N*EDD8-*a*ctivating *e*nzyme (NAE), E2 NEDD8-conjugating enzyme, and E3 NEDD8 ligase (Fig. 3.1). Initially, via the C-terminal Gly-76, NEDD8 forms a thioester bond with NAE. The human NAE complex is composed of two subunits, *a*myloid-β *p*recursor *p*rotein *b*inding *p*rotein *1* (APP-BP1) and *ub*iquitin-*a*ctivating enzyme *3* (UBA3) (Osaka et al. 1998; Gong and Yeh 1999). Interestingly, the APP-BP1 and UBA3 are highly homologous to the amino- and carboxy-terminal regions of ubiquitin E1, respectively. In Arabidopsis, the corresponding NAE has also been identified, with AXR1 as the counterpart of APP-BP1 and Ecr1 as the orthologue of UBA3 (Pozo 1998). The Arabidopsis *axr1* mutant exhibits abnormal response to the essential plant hormone auxin (Lincoln et al. 1990; Leyser et al. 1993), further demonstrating the importance of neddylation. Two active sites in the heterodimeric

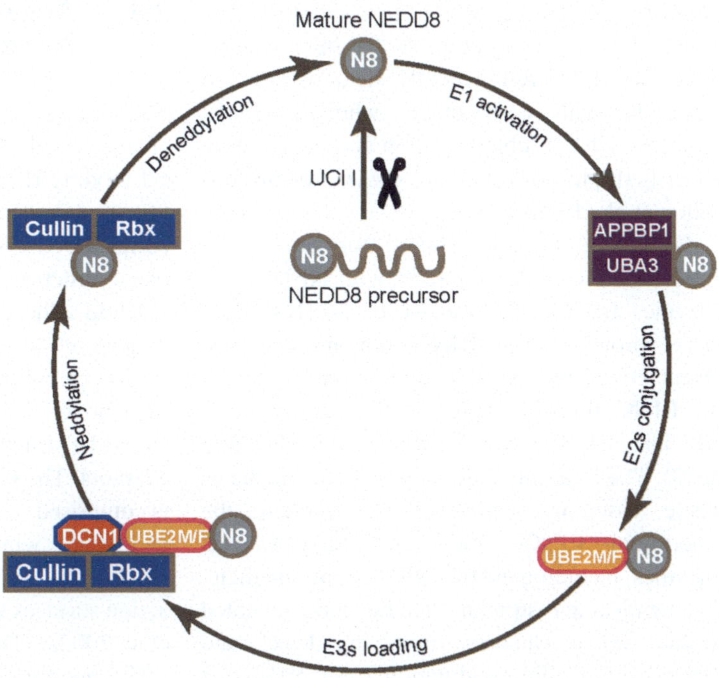

Fig. 3.1 Simplified scheme of cullin neddylation. Following precursor processing of NEDD8 by UCH, mature NEDD8 was attached to cullins in consecutive steps including E1 (UBA3/APPBP1)-mediated activation, E2 (UBE2M/F)-mediated conjugation, and E3 (Rbx/DCN1)-mediated ligation. NEDD8 isopeptidases such as the COP9 signalosome are able to catalyze the deneddylation of cullins

NAE complex, an adenylation domain in UBA3, and a catalytic Cys domain, are required for efficient activation and conjugation of NEDD8 (Walden et al. 2003a). The C-terminal *u*biquitin-*f*old *d*omain (UFD) in UBA3 plays an important role in mediating interactions with the E2 NEDD8-conjugating enzymes (Huang et al. 2005; Walden et al. 2003b).

Once activated by NAE, NEDD8 is transferred to a conserved cysteine on the E2 NEDD8-conjugating enzyme via a transthiolation reaction. To date, two NEDD8 E2-conjugating enzymes have been identified: the well-studied Ubc12 (also known as UBE2M) and less-characterized UBE2F (Liakopoulos et al. 1998; Osaka et al. 1998; Gong and Yeh 1999; Huang et al. 2009). Structural studies showed that the N-terminal extension and catalytic core domain of either Ubc12 or UBE2F bind to UBA3's hydrophobic groove and UFD of NAE, respectively (Huang et al. 2004, 2005, 2007, 2009). Although they interact with NAE in similar fashions, Ubc12 and UBE2F exhibit distinct substrate specificities for neddylation. Ubc12 pairs with *R*ING-*box* protein *1* (Rbx1) to regulate neddylation of Cullins 1--4, whereas UBE2F specifically interacts with *R*ING-*box* protein *2* (Rbx2) to mediate Cullin 5 neddylation (Huang et al. 2009). Interestingly, these two E2s were recently demonstrated to cross-talk with each other, wherein Ubc12 promotes UBE2F ubiquitylation and degradation under both physiological and stressed conditions (Zhou et al. 2018).

Following thioester bond formation, the E2~NEDD8 complex interacts with NEDD8 E3 enzymes for specific transfer of NEDD8 onto target proteins. The RING domain proteins, characterized by the zinc coordination sites in the RING domain, represent the first reported NEDD8 E3 ligases (Deshaies and Joazeiro 2009). The N-terminal domain of Rbx1 mediates stable assembly with cullins, whereas the C-terminal RING binds and activates E2 enzymes, including the E2~NEDD8 intermediate. Biochemical and structural studies have revealed a dual E3 mechanism—involving Rbx1 and *D*efective in *C*ullin *N*eddylation *1* (DCN1)—that mediates cullin neddylation (Kurz et al. 2008; Scott et al. 2010, 2011; Kim et al. 2008). Early studies demonstrated that Ubc12 stimulates cullin neddylation in an Rbx1-dependent manner (Kamura et al. 1999; Gray 2002), while DCN1 is required for optimal neddylation of cullin in *C. elegans* and *S. cerevisiae* (Kurz et al. 2005). Furthermore, in the presence of N-terminally acetylated E2, DCN proteins dramatically stimulate cullin neddylation in vitro (Monda et al. 2013; Scott et al. 2014). Thus, DCN1 is considered a co-E3, which is capable of binding both cullin and the acetylated N-terminus of Ubc12 via its potentiating neddylation (PONY) domain, restricting the flexible RBX1-bound Ubc12~NEDD8 to a catalytically competent orientation. Similarly, N-terminal acetylation in UBE2F also promotes DCN1-dependent neddylation of Cul5 that specifically employs Rbx2 as the RING protein (Monda et al. 2013). More recently, a sophisticated neddylation model was revealed, in which the NEDD8 and substrate protein synergistically regulate neddylation specificity by toggling E2/E3 conformations (Scott et al. 2014).

3.3 Activation of CRLs by Neddylation

The family of *c*ullin-*R*ING *l*igases (CRLs) is characterized by a cullin core linking an E2-binding RING and a substrate-binding receptor module. Early crystallographic studies of cullin-RING complexes revealed a highly elongated structure in which cullin serves as a rigid scaffold (Zheng et al. 2002a; Duda et al. 2008; Angers et al. 2006). The N-terminal domain (NTD) of cullin, which consists of three repeats of a novel five-helix structural motif, adopts a stalk-like structure that recruits substrate-binding adaptors. The C-terminal domain (CTD) of cullin is comprised of a four-helix bundle (4HB), an α/β and two winged-helix (WHA and WHB) subdomains. In terms of Rbx1 binding, two interaction surfaces were identified in the CTD (Zheng et al. 2002a). The Rbx1 RING domain docks into a V-shaped groove formed by the cullin α/β and WHB domains, whereas the N-terminal β-strand of Rbx1 is incorporated into the β-sheet of the cullin α/β domain. Importantly, in the structural model of a E2~ubiquitin thioester bound to Rbx1 in a CRL complex, the catalytic cysteine of E2 is positioned approximately 50 Å from the tip of the substrate receptor protein. Furthermore, the rigidity of the Cul1 scaffold that imposes this gap is required for E3 activity, as evidenced by the finding that introducing flexibility into the N-terminal repeat domain disrupts ubiquitination of a specific substrate (Zheng et al. 2002a). These observations thus beg the question of how ubiquitin is transferred to the substrate and to the continually elongating ubiquitin chain.

The cullin family members are the best-characterized neddylation substrates. Neddylation of cullin takes place on a highly conserved lysine residue, which is located in the C-terminal WHB subdomain (Wada et al. 1999). Mutagenesis studies indicate that cullin neddylation efficiently promotes the ubiquitination activity of *S*kp1-*C*ul1-*F*-box protein (SCF), the founding member of CRLs, resulting in rapid turnover of protein substrates (Wu et al. 2000; Read et al. 2000; Morimoto et al. 2000; Podust et al. 2000). In vitro biochemical evidence indicates that neddylation stimulates recruitment of ubiquitin charged E2 to the SCF complex (Kawakami et al. 2001). Interestingly, the proximity of the cullin neddylation site to the RING domain of Rbx1 suggests that neddylation might stimulate the activity of CRLs by regulating the interaction between NEDD8 and Rbx1-bound E2~ubiquitin (Wu et al. 2002; Zheng et al. 2002a; Read et al. 2000; Kawakami et al. 2001).

The crystal structure of neddylated Cul5^{ctd}-Rbx1 complex revealed that neddylation induces a striking reorientation of the cullin WHB subdomain (Duda et al. 2008), which frees the RING domain of Rbx1 from the compact WHB-RING architecture in the unneddylated cullin. As a result, Rbx1 adopts flexible conformations, providing potential catalytic geometries for ubiquitin transfer. Besides the C-terminus of NEDD8, the Leu8/Ile44/His68/Val70 face is also involved in the interaction between NEDD8 and Cul5 (Duda et al. 2008). However, a prior study reported that the hydrophobic patch around Ile44 of NEDD8 binds ubiquitin E2, but not the NEDD8 E2, in nuclear magnetic resonance (NMR) spectroscopy (Sakata et al. 2007). This disagreement between the two structural studies suggests that in the presence of E2~ubiquitin under conditions of ubiquitin transfer, structural rearrangements may occur to favor contacts between NEDD8 and E2~ubiquitin (see Chap. 2 for full discussions on the structure of CRLs with and without neddylation). Detailed insights into the multiple mechanisms by which NEDD8 activates SCF emerged from a kinetic study (Saha and Deshaies 2008). First, neddylation enhances E2 recruitment to SCF as well as the rate of ubiquitin transfer to substrates. Second, neddylation improves the rate of ubiquitin chain elongation. Third, neddylation enables crosslinking of substrate to E2 and enhances amide bond formation in the E2 active site. Finally, neddylation increases the fraction of ubiquitin-charged substrate molecules as well as the average number of ubiquitin molecules attached to a modified substrate. Overall, two distinct effects of neddylation on CRL activity were proposed: (1) enhancement of ubiquitin chain initiation by bridging the gap between the SCF substrate and the E2~ubiquitin and (2) improvement of chain elongation via stimulating E2 recruitment and enhancing E2 activity. Consistent with the structural and kinetic findings, unneddylated SCF complexes containing a mutant Cul1 with WHB deletion, which were engineered to mimic neddylation by constitutively releasing the RING, exhibit increased activity in vitro (Duda et al. 2008; Yamoah et al. 2008) and in vivo (Boh et al. 2011). Thus, the unneddylated CRLs can be considered as autoinhibited or off, and neddylation activates the complex by inducing profound conformational rearrangements.

In addition to mediating RING domain rearrangement, neddylation also disrupts the binding interface between Cul1-Rbx1 and CAND1 (cullin-associated NEDD8-dissociated protein 1), the substrate receptor exchange factor (see below). Moreover, both WHB and NEDD8 are reoriented to contact the cullin NTD in the neddylated CRL models, evoking the hypothesis that the CTD-NTD interface might be remodeled by WHB and/or NEDD8 to favor CRL activity (Duda et al. 2008) (see Fig. 2.4 in Chap. 2 for structural details). Further studies are required to provide new insight into the architecture of neddylated full-length CRLs.

The NEDD8 modification is reversible. Once cullins are neddylated, they can be rapidly deneddylated by the COP9 signalosome (CSN) in which CSN5 serves as the catalytic subunit (see Chap. 4 for full discussions on CSN and its role in regulating CRLs). CSN specifically recognizes neddylated cullins as its substrates. Recent studies have shown that neddylated Cul1 and Cul4 have a much higher affinity for CSN than their unneddylated forms (Cavadini et al. 2016; Mosadeghi et al. 2016). A key aspect of CSN activity is that it is inhibited by substrate binding to the CRL (Fischer et al. 2011; Emberley et al. 2012; Enchev et al. 2012). This is due to a steric clash between the bound substrate and CSN (Cavadini et al. 2016; Fischer et al. 2011; Enchev et al. 2012). As a consequence, a CRL in vivo should be most efficiently deneddylated when there is little or no substrate available. The mutually exclusive binding of substrate and CSN to a CRL lies at the heart of how differential assembly states of different CRL complexes is controlled.

Based on its biochemical activity, it is not surprising that CSN inhibits the activity of SCF ubiquitin ligase in vitro (Lyapina et al. 2001; Zhou et al. 2003). However, multiple lines of genetic evidence revealed that CSN promotes degradation of substrates mediated by SCF and other CRLs in vivo, suggesting a positive role of CSN in regulating CRL activity (Schwechheimer et al. 2001; Cope et al. 2002; Pintard et al. 2003). This apparent paradox about CSN function was partially resolved by the finding that the inhibitory effects of CSN exhibited in vitro prevent the autoubiquitination and degradation of CRL substrate receptors in vivo, thereby promoting CRL activity (Zhou et al. 2003; Wee et al. 2005; He et al. 2005). More importantly, deneddylation is also required for Cand1-mediated dynamic exchange of CRL substrate receptors, a process critical for sustaining continuous cycles of CRL assembly and disassembly (see later portion of this chapter).

3.4 CAND1: An Inhibitor or Activator of CRLs?

Cand1, formerly known as *T*ATA-*b*inding *p*rotein (TBP)-*i*nteracting *p*rotein *120A* (TIP120A), was originally identified as a TBP-interacting protein in rat liver nuclear extracts (Yogosawa et al. 1996). A subsequent study showed that Cand1 globally stimulates transcription driven by three classes of eukaryotic RNA polymerases (Makino et al. 1999). To date, the significance of these functions remains unknown. A milestone for understanding the role of Cand1 is the finding by four groups that Cand1 interacts with cullins (Zheng et al. 2002b; Liu et al. 2002; Hwang et al. 2003; Min et al. 2003; Oshikawa et al. 2003). Both the Skp1-binding N-terminus and the unneddylated C-terminal domain of cullins are required for Cand1-cullin association. Reciprocally, deletion of either N-terminal or C-terminal residues from Cand1 abolished its binding with Cul1, suggesting that Cand1-cullin interaction may involve both ends of Cand1. The crystal structure of Cand1-Cul1-Rbx1 brought a more comprehensive understanding of the Cand1-cullin complex architecture (Goldenberg et al. 2004). By virtue of 27 tandem *h*untingtin-*e*longation-*A* subunit-*T*OR (HEAT) repeats, Cand1 forms a highly sinuous superhelical structure that coils into a U-shaped belt that cradles Cul1-Rbx1. In the complex, Cand1 and Cul1 interact with each other in a head-to-tail pattern. Overall, like a clamp, Cand1 uses its N-terminal and C-terminal arches as two prongs to grip both ends of Cul1. Importantly, a β-hairpin motif protruding from the Cand1 main body interacts with the Skp1-binding helices located on the Cul1 N-terminus. Distally, two helical repeats on the N-terminus of Cand1 occupy a cleft on the Cul1 CTD where the neddylation acceptor lysine residue resides. These findings are consistent with the biochemical data that showed that either Skp1 or neddylation dissociated Cand1 from cullins (Zheng et al. 2002b; Liu et al. 2002; Hwang et al. 2003; Min et al. 2003; Oshikawa et al. 2003). In turn, the binding of Cand1 to Cul1 dissociated Skp1 from Cul1 and inhibited the ubiquitination of p27 in vitro (Zheng et al.

2002b). All these findings converge to the conclusion that Cand1 inhibits CRL activity in vitro.

Several lines of evidence, however, revealed that Cand1 is required for optimal CRL activity in vivo. Although RNAi-mediated silencing of Cand1 in human cells significantly increases Cul1-Skp1 association, the level of substrate protein also increases slightly (Zheng et al. 2002b). The Arabidopsis Cand1 loss-of-function mutants exhibit defects in CRL regulated signaling events and reduced turnover rates of CRL substrate proteins (Feng et al. 2004; Chuang et al. 2004; Cheng and Dai 2004). All these observations are reminiscent of the CSN paradox, and it was naturally hypothesized that Cand1 helps maintain the stability of CRL substrate receptors by sequestering cullin cores, thus promoting CRL activity in vivo (Zheng et al. 2002b; Cope and Deshaies 2003; Min et al. 2005). In support of this, a slight decrease in the level of Skp2 is observed in Cand1 knockdown Hela cells (Zheng et al. 2002b). However, several lines of evidence argued against the model of Cand1-mediated adaptor stabilization (Schmidt et al. 2009; Chua et al. 2011), and a few studies reported that Cand1 can promote CRL activity independently of adaptor stability (Lo and Hannink 2006; Bosu et al. 2010). Cand1 deletion does not affect the stability of at least a subset of substrate receptors in fission yeast (Schmidt et al. 2009), and Cand1 knockdown human cells exhibit a decrease in CRL3^{Keap1} activity despite an increase in the level of assembled CRL3^{Keap1} complexes (Lo and Hannink 2006). So, the question still remains: how does Cand1 promote CRL activity in vivo? An important hint came from the biochemical data demonstrating that human Skp2-Skp1 promotes the dissociation of Cand1 from Cul1. Furthermore, as mentioned in the prior section, substrate can inhibit binding and deneddylation of CRLs by CSN and thus maintaining neddylation (Cavadini et al. 2016; Fischer et al. 2011; Emberley et al. 2012; Enchev et al. 2012; Chew and Hagen 2007). Schmidt et al. suggested a refined model in which transient interaction of Cand1 with Cul1-Rbx1 scaffold drives dynamic cycles between association and displacement of different substrate receptors. This cycle would enable rapid remodeling of the CRL repertoire in the absence of substrates, and when substrates are available, they induce the neddylation of their cognate CRLs to promote their ubiquitination. Upon substrate consumption, CSN-mediated deneddylation would direct the CRLs again into the Cand1 cycle. Though this model can explain how Cand1 activates CRLs while competing against substrate receptors for binding to cullins, there was no direct biochemical evidence for the dynamic recycling of CRL complexes nor how Cand1 could promote such recycling.

3.5 CAND1: The Substrate Receptor Exchange Factor

A key breakthrough in the current understanding of Cand1 emerged from a kinetic analysis of SCF assembly and disassembly (Pierce et al. 2013). Using a *f*luorescence *r*esonance *e*nergy *t*ransfer (FRET) assay, the binding dynamics between an F-box protein and the Cul1-Rbx1 scaffold were monitored in a real-time manner. With a K_D in the picomolar range, Fbxw7-Skp1 forms an extremely tight complex with Cul1-Rbx1 regardless of the neddylation state. Importantly, addition of Cand1 to pre-assembled and unneddylated SCF^{FBXW7} complexes increases the dissociation rate of Fbxw7-Skp1 by more than one million fold, with a maximum rate of 1.3 s^{-1}. As expected, neddylation of Cul1 abolishes the effect of Cand1 due to its ability to block binding of Cand1 to Cul1. Reciprocally, the spontaneous dissociation of Cand1 from Cul1-Rbx1 is extremely slow, and this dissociation rate is accelerated greatly by Fbxw7-Skp1. Furthermore, a Skp1 mutant lacking a loop that is predicted to clash with a β-hairpin of Cand1 (Goldenberg et al. 2004) fails to displace Cand1 from Cul1-Rbx1, and the Cand1 mutant lacking the β-hairpin also fails to displace F-box–Skp1 from Cul1-Rbx1 (Liu et al. 2018). When either the Skp1 loop or the β-hairpin of Cand1 is deleted, Skp1 and Cand1 can form a stable ternary complex with Cul1 (Goldenberg et al. 2004; Liu et al. 2018).

The biochemical and kinetic results described above can be explained by hypothesizing the existence of a transient ternary complex consisting of Cand1, substrate receptor module, and Cul1 that upon dissociation of either Cand1 or the substrate receptor module would yield a stable SCF or Cul1-Cand1 complex. Given the ability of both Cand1 and the substrate receptor module to destabilize each other's association with Cul1-Rbx1, it was proposed that Cand1 mediates the exchange of substrate receptors on the Cul1-Rbx1 scaffold, which exactly mimics how *g*uanine nucleotide *e*xchange *f*actors (GEFs) act on Ras-like GTPases (Klebe et al. 1995; Goody and Hofmann-Goody 2002; Guo et al. 2005). In support of this hypothesis, addition of Cand1 to a mixture of pre-assembled $SCF^{\beta\text{-}TrCP}$ and free Fbxw7-Skp1 significantly increases the ubiquitylation of the SCF^{Fbxw7} substrate, CycE (Pierce et al. 2013). In contrast to the previous in vitro finding that Cand1 inhibits the ubiquitin ligase activity of SCF, these results showed for the first time that Cand1 promotes the ubiquitination of SCF substrates through accelerating the assembly of specific substrate receptors with Cul1 recycled from the pre-existing pool of SCFs. Based on its biochemical mechanism of action, Cand1 is defined as a "substrate receptor exchange factor" (Pierce et al. 2013).

Besides the few substrate receptors tested in vitro, quantitative mass spectrometry revealed that the cellular repertoire of SCF is dramatically altered by Cand1 depletion, in both human (Pierce et al. 2013) and yeast cells (Wu et al. 2013). Furthermore, in budding yeast lacking Cand1, Skp1-Grr1 failed to associate with Cdc53 (yeast Cul1) in response to glucose, likely due to inefficient release of Cdc53 from pre-existing SCF complexes (Zemla et al. 2013). To further reveal the power of Cand1-mediated exchange of substrate receptors, metabolic pulse-labelling assays were designed to examine SCF dynamics in cells (Pierce et al. 2013; Wu et al. 2013). As expected, several newly synthesized F-box proteins exhibit reduced assembly into Cul1 in the absence of Cand1. Importantly, the level of these F-box proteins in total cell lysate is unaffected, indicating that the rate of synthesis is not responsible for the reduced incorporation. Taken together, these data indicate that Cand1 modulates the SCF repertoire by acting as an exchange factor that equilibrates Cul1-Rbx1 with the entire pool of substrate receptors. This exchange activity of Cand1 ensures that the cellular repertoire of CRLs remains dynamic, and when specific substrates arise, they can stabilize their cognate CRLs to sustain their efficient degradation.

3.6 Regulation of CRL Assembly by Both Cand1 and Neddylation

Through blocking access of Cand1 to cullins, neddylation inhibits Cand1-mediated exchange, and therefore, neddylation activity should also regulate the assembly of CRLs. Surprisingly, a couple of early studies found that when neddylation was eliminated by MLN4924 (pevonedistat), a potent inhibitor of NAE (Soucy et al. 2009), the CRL network was marginally affected (Bennett et al. 2010; Lee et al. 2011). In light of the updated knowledge about the working mechanism of Cand1, methods used to probe cellular CRL complexes were re-evaluated. In a conventional immunoprecipitation assay, the CRL complexes may have been rearranged due to post-lysis exchange of substrate receptors by Cand1. This possibility was tested using an assay based on SILAC mass spectrometry (Reitsma et al. 2017). HEK293 cells with (heavy) or without (light) a 3xFLAG tag on the endogenous Cul1 are grown in media with and without heavy-labeled amino acids as indicated. Immediately before lysis, cells are mixed and the percentage exchange of F-box proteins is determined by immunoprecipitation of 3XFLAG-tagged Cul1 followed by quantitative mass spectrometry. Strikingly, robust exchange of F-box proteins is observed in the shortest immunoprecipitation performed (10 min). Importantly, this exchange is greatly suppressed for Skp1 and almost all F-box proteins in the lysate of Cand1/2 double-knockout (DKO) cells, suggesting that Cand1/2 accounts for the fast post-lysis re-equilibration of SCF complexes. Inhibition of neddylation by pevonedistat further enhances exchange for all FBPs, apparently through exposing Cand1 binding sites on Cul1. These findings point out that to precisely capture the landscape of cellular CRLs, it is necessary to inhibit the Cand1-mediated post-lysis exchange of substrate receptors.

Inspired by the finding that both SCF and Cul1-Cand1 complexes are extremely stable (Pierce et al. 2013; Reitsma et al. 2017), it was proposed that an excess of recombinant Cul1-Rbx1 in the lysis buffer, like a molecular sponge, could suppress exchange by sequestering any free Cand1/2 and F-box proteins. As predicted, exchange of most F-box proteins is drastically diminished in the immunoprecipitates when recombinant Cul1-Rbx1 is added to the lysis buffer (Reitsma et al. 2017). By virtue of this molecular sponge, it is now possible to investigate how the cellular SCF repertoire changes in response to various perturbations. Experiments using this new immunoprecipitation system revealed that F-box proteins show huge variations in their percentage association with Cul1, suggesting a non-equilibrium pool of SCF complexes, and most F-box proteins exhibit low percentages of Cul1 association. In contrast to previous reports that only a small fraction of Cul1 stably associates with Cand1 in 293T cells (Bennett et al. 2010), the new assay revealed that nearly half of the Cul1 assembles with Cand1 in $HEK293^{3xFLAG\text{-}Cul1}$ cells, underscoring the potential of Cand1 to rigorously regulate the SCF repertoire. Importantly, depletion of Cand1/2 leads to an increase in net assembly of F-box proteins, whereas pevonedistat treatment elicits the opposite effect in a Cand1/2-dependent manner. These results indicate that neddylation together with Cand1/2 activity maintain a non-equilibrium pool of SCF complexes.

Since the cellular concentrations of F-box proteins do not correlate with their percentage association with Cul1, what determines the landscape of cellular SCFs? As mentioned previously, the antagonistic binding of substrates and CSN to Cul1 (Cavadini et al. 2016; Fischer et al. 2011; Enchev et al. 2012) suggested that substrates play a key role in sculpting the dynamic SCF repertoire. To test this idea, the assembly state of individual SCF complexes was explored upon induction of substrate. As expected, assembly of a specific SCF complex is induced when its cognate substrate becomes available. Furthermore, this process depends on both Cand1/2 activity and neddylation (Liu et al. 2018; Reitsma et al. 2017).

A perplexing phenomenon found in both Arabidopsis and human Cand1-deficient cells is that increased assembly of a specific CRL complex occurs jointly with inefficient degradation of its substrate (Chuang et al. 2004; Lo and Hannink 2006; Pierce et al. 2013; Zhang et al. 2008). By exploring *T*umor *N*ecrosis *F*actor alpha (TNFα)-induced degradation of IκBα via $SCF^{\beta\text{-}TrCP}$-dependent ubiquitination, a study using quantitative approaches provided important insights into this paradox (Liu et al. 2018). Despite a higher level of $SCF^{\beta\text{-}TrCP}$ in unstimulated DKO cells, the degradation rate of IκBα is substantially reduced compared to WT cells. Immunoprecipitation assays revealed that phosphorylated IκBα (pIκBα, the substrate of $SCF^{\beta\text{-}TrCP}$) in DKO cells binds β-TrCP as normal, but the level of pIκBα-bound β-TrCP recruited to Cul1 decreases. These results suggest that substrates can bind both free and Cul1-bound F-box proteins, but substrates bound to free F-box proteins cannot gain access to Cul1 for efficient ubiquitination in cells lacking Cand1/2 activity. Consistent with this explanation, β-TrCP overexpression, which leads to an increased level of $SCF^{\beta\text{-}TrCP}$ but no change in the percentage association of β-TrCP with Cul1, fails to increase the degradation rate of IκBα in DKO cells, simply because the fraction of pIκBα bound to free β-TrCP still cannot gain access to Cul1. In contrast, Cul1 overexpression rescued the degradation defect of IκBα in DKO cells, because almost all β-TrCP molecules were driven to form SCF complexes. Taken together, these results demonstrate that dynamic exchange of F-box proteins associated with Cul1 is required for efficient substrate degradation, and restricted access of substrate to Cul1 represents the major deficiency in DKO cells.

Interestingly, although Cand1 inhibits neddylation of cullins, it promotes the binding of DCN1, the NEDD8 E3, to Cul1-Rbx1, and Cand1 decreased the K_D of DCN1-Cul1 by 36 fold (Kim et al. 2008; Monda et al. 2013; Liu et al. 2018; Keuss et al. 2016). Moreover, the DCN1-Cul1 complex could not be detected in DKO cells through co-immunoprecipitation. This finding suggests that DCN1 prefers Cand1-bound Cul1, which enables immediate neddylation of Cul1 upon the removal of Cand1 by F-box-Skp1. Indeed, in the presence of F-box-Skp1 modules, Cand1-bound Cul1 exhibited a higher neddylation rate than free Cul1 (Liu et al. 2018). In addition, neddylation biases the exchange reaction toward the formation of a stable SCF by preventing Cand1 from re-binding the Cul1 and thus acting as the energy input that drives Cul1 toward the formation of new SCF. Furthermore, this mechanism may also ensure that each new SCF is neddylated and activated coincident with its formation.

3.7 Assembly and Disassembly of SCF: The Cul1 Cycle

Based on all the available kinetic and quantitative studies of the components and regulators of SCFs, a mathematical model was developed to elucidate and investigate the dynamics of the SCF system (Liu et al. 2018). In this model, Cul1 constantly cycles through the exchange, neddylation and deneddylation stages, while substrate binding through an F-box protein locks Cul1 in the neddylated state, and thus the SCF is stabilized to allow substrate ubiquitination (Fig. 3.2). In addition to simulating and predicting

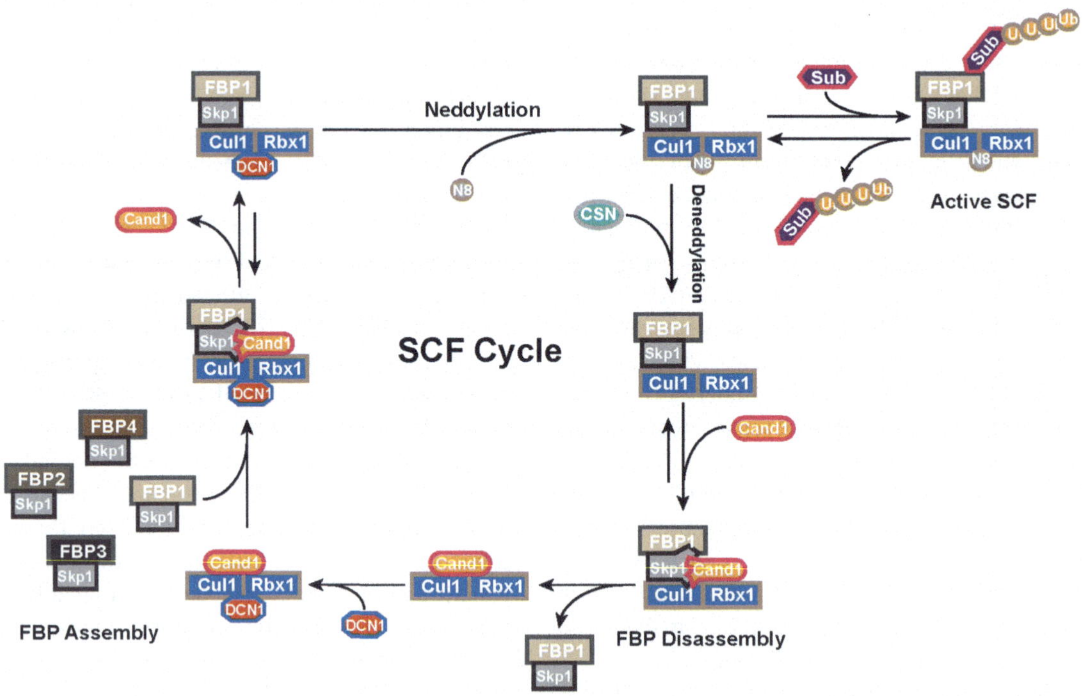

Fig. 3.2 Schematic illustration of adaptive SCF assembly. Cand1-mediated exchange of F-box proteins (FBP), together with the neddylation/deneddylation cycle, sustains the dynamicity of the SCF system. Binding of Cand1 to a pre-existing SCF complex forms an unstable ternary intermediate that expels either Cand1 or the Skp1-FBP module. After the Skp1-FBP has dissociated, the resulting Cand1-Cul1 promptly retrieves another Skp1-FBP module from the cellular pool. Disassociation of Cand1 from the newly formed ternary complexes generates a new SCF complex. Due to the preferential association of DCN1 with Cul1 that is bound by Cand1, neddylation occurs immediately after Cand1 dissociation. Binding of substrate stabilizes this neddylated SCF complex, resulting in efficient ubiquitylation of the substrate. In the absence of substrates, CSN catalyzes the deneddylation of Cul1, and the deneddylated SCF complex reenters the cycle

changes in substrate degradation upon genetic and chemical alterations, the model provides novel insights into the significance of the Cand1-mediated exchange mechanism. First, the model calculated that a Cul1 molecule with no bound substrate goes through the entire cycle with an average time of ~1.5 min. This rate is consistent with the rate of Cul1 neddylation and deneddylation measured in human cells (Liu et al. 2018). Considering the molar ratio of Skp1 to Cul1, an F-box protein could gain access to Cul1 every ~4 min. This rapid cycle apparently offers a more efficient way to remodel the SCF repertoire compared to synthesis-dependent regulation. Second, the total F-box protein concentration was predicted as one of the two most sensitive parameters affecting the substrate half-life in the DKO but not the WT cells. Indeed, overexpression of a single F-box protein, including a truncated F-box protein that binds only Cul1 but no substrate protein, dramatically increases the stability of multiple proteins that are substrates of SCFs only in the DKO cells. These cells also display dramatically lower cell proliferation rates and an increased level of apoptosis marker proteins. These observations led to the hypothesis that Cand1-mediated exchange permits the SCF system to tolerate large variations in the level of individual F-box proteins. Because such large changes can naturally occur during development, loss of Cand1 could have profound effects especially on multicellular organisms. In support of this, in the absence of Cand1, Arabidopsis exhibited severe defects throughout its life cycle, whereas cultured cells and yeasts were only modestly affected. More broadly, this Cand1-mediated "adaptive exchange" mechanism could confer organism tolerance to variations in the number of expressed F-box proteins, providing a foundation for large expansion or contraction of the F-box protein gene repertoire during the evolution of different species (Liu et al. 2018).

3.8 Closing Remarks

While the attractive picture of SCF dynamics has been increasingly revealed, important aspects of the assembly and regulation of CRLs remain to be explored. First, studies of dynamic exchange largely rely on the Cul1-based SCF system. However, Cand1 also associates with other cullins, and at steady state, Cand1 is not equally distributed across cullins, with more Cand1 associated with Cul1 and Cul5 than that with Cul2 and Cul4 (Bennett et al. 2010). Biochemical studies akin to those performed in SCF are required to answer whether Cand1 plays a similar role in regulating the dynamics of other CRLs and, if so, how important is the dynamic exchange in regulating the activity of these CRLs. Second, how is the distribution of Cand1 over different cullins determined and coordinated? Third, although the net assembly of SCF complexes was increased in the DKO cells and was decreased when neddylation was inhibited by pevonedistat, quite a few F-box proteins showed different trends in their assembly status (Reitsma et al. 2017). It is thus very interesting to study what roles Cand1 and neddylation play in the regulation of these F-box proteins. Furthermore, Cand1-mediated exchange has been hypothesized to confer tolerance to large variations in substrate receptor levels. Investigating specific effects of Cand1 at different developmental stages in multiple eukaryotic organisms will further elucidate the importance of this exchange mechanism.

References

Angers S, Li T, Yi X et al (2006) Molecular architecture and assembly of the DDB1-CUL4A ubiquitin ligase machinery. Nature 443(7111):590–593

Bennett EJ, Rush J, Gygi SP et al (2010) Dynamics of cullin-RING ubiquitin ligase network revealed by systematic quantitative proteomics. Cell 143(6):951–965

Boh BK, Smith PG, Hagen T (2011) Neddylation-induced conformational control regulates cullin RING ligase activity in vivo. J Mol Biol 409(2):136–145

Bosu DR, Feng H, Min K et al (2010) C. elegans CAND-1 regulates cullin neddylation, cell proliferation and morphogenesis in specific tissues. Dev Biol 346 (1):113–126

Cavadini S, Fischer ES, Bunker RD et al (2016) Cullin-RING ubiquitin E3 ligase regulation by the COP9 signalosome. Nature 531(7596):598–603

Chan Y, Yoon J, Wu JT et al (2008) DEN1 deneddylates non-cullin proteins in vivo. J Cell Sci 121 (Pt 19):3218–3223

Cheng Y, Dai X, Zhao Y (2004) AtCAND1, a HEAT-repeat protein that participates in auxin signaling in Arabidopsis. Plant Physiol 135(2):1020–1026
Chew EH, Hagen T (2007) Substrate-mediated regulation of cullin neddylation. J Biol Chem 282 (23):17032–17040
Chua YS, Boh BK, Ponyeam W et al (2011) Regulation of cullin RING E3 ubiquitin ligases by CAND1 in vivo. PLoS One 6(1):e16071
Chuang HW, Zhang W, Gray WM (2004) Arabidopsis ETA2, an apparent ortholog of the human cullin-interacting protein CAND1, is required for auxin responses mediated by the SCF(TIR1) ubiquitin ligase. Plant Cell 16(7):1883–1897
Cope GA, Deshaies RJ (2003) COP9 signalosome. Cell 114(6):663–671
Cope GA, Suh GS, Aravind L et al (2002) Role of predicted metalloprotease motif of Jab1/Csn5 in cleavage of Nedd8 from Cul1. Science 298(5593):608–611
Deshaies RJ, Joazeiro CA (2009) RING domain E3 ubiquitin ligases. Annu Rev Biochem 78:399–434
Dharmasiri S, Dharmasiri N, Hellmann H et al (2003) The RUB/Nedd8 conjugation pathway is required for early development in Arabidopsis. EMBO J 22 (8):1762–1770
Duda DM, Borg LA, Scott DC et al (2008) Structural insights into NEDD8 activation of cullin-RING ligases: conformational control of conjugation. Cell 134(6):995–1006
Emberley ED, Mosadeghi R, Deshaies RJ (2012) Deconjugation of Nedd8 from Cul1 is directly regulated by Skp1-F-box and substrate, and the COP9 signalosome inhibits deneddylated SCF by a noncatalytic mechanism. J Biol Chem 287 (35):29679–29689
Enchev RI, Scott DC, Da Fonseca PC et al (2012) Structural basis for a reciprocal regulation between SCF and CSN. Cell Rep 2(3):616–627
Enchev RI, Schulman BA, Peter M (2015) Protein neddylation: beyond cullin-RING ligases. Nat Rev Mol Cell Biol 16(1):30–44
Feng S, Shen Y, Sullivan JA et al (2004) Arabidopsis CAND1, an unmodified CUL1-interacting protein, is involved in multiple developmental pathways controlled by ubiquitin/proteasome-mediated protein degradation. Plant Cell 16(7):1870–1882
Fischer ES, Scrima A, Bohm K et al (2011) The molecular basis of CRL4DDB2/CSA ubiquitin ligase architecture, targeting, and activation. Cell 147 (5):1024–1039
Gan-Erdene T, Nagamalleswari K, Yin L et al (2003) Identification and characterization of DEN1, a deneddylase of the ULP family. J Biol Chem 278 (31):28892–28900
Goldenberg SJ, Cascio TC, Shumway SD et al (2004) Structure of the Cand1-Cul1-Roc1 complex reveals regulatory mechanisms for the assembly of the multisubunit cullin-dependent ubiquitin ligases. Cell 119(4):517–528
Gong L, Yeh ETH (1999) Identification of the activating and conjugating enzymes of the NEDD8 conjugation pathway. J Biol Chem 274(17):12036–12042
Goody RS, Hofmann-Goody W (2002) Exchange factors, effectors, GAPs and motor proteins: common thermodynamic and kinetic principles for different functions. Eur Biophys J: EBJ 31(4):268–274
Gray WM (2002) Role of the Arabidopsis RING-H2 protein RBX1 in RUB modification and SCF function. Plant Cell Online 14(9):2137–2144
Guo Z, Ahmadian MR, Goody RS (2005) Guanine nucleotide exchange factors operate by a simple allosteric competitive mechanism. Biochemistry 44 (47):15423–15429
He Q, Cheng P, He Q et al (2005) The COP9 signalosome regulates the Neurospora circadian clock by controlling the stability of the SCFFWD-1 complex. Genes Dev 19(13):1518–1531
Hu M, Li P, Li M et al (2002) Crystal structure of a UBP-family deubiquitinating enzyme in isolation and in complex with ubiquitin aldehyde. Cell 111 (7):1041–1054
Huang DT, Miller DW, Mathew R et al (2004) A unique E1-E2 interaction required for optimal conjugation of the ubiquitin-like protein NEDD8. Nat Struct Mol Biol 11(10):927–935
Huang DT, Paydar A, Zhuang M et al (2005) Structural basis for recruitment of Ubc12 by an E2 binding domain in NEDD8's E1. Mol Cell 17(3):341–350
Huang DT, Hunt HW, Zhuang M et al (2007) Basis for a ubiquitin-like protein thioester switch toggling E1-E2 affinity. Nature 445(7126):394–398
Huang DT, Ayrault O, Hunt HW et al (2009) E2-RING expansion of the NEDD8 cascade confers specificity to cullin modification. Mol Cell 33(4):483–495
Hwang J-W, Min K-W, Tamura T-A et al (2003) TIP120A associates with unneddylated cullin 1 and regulates its neddylation. FEBS Lett 541(1–3):102–108
Johnston SC, Riddle SM, Cohen RE et al (1999) Structural basis for the specificity of ubiquitin C-terminal hydrolases. EMBO J 18(14):3877–3887
Kamitani T, Kito K, Nguyen HP et al (1997) Characterization of NEDD8, a developmentally down-regulated ubiquitin-like protein. J Biol Chem 272 (45):28557–28562
Kamura T, Conrad MN, Yan Q et al (1999) The Rbx1 subunit of SCF and VHL E3 ubiquitin ligase activates Rub1 modification of cullins Cdc53 and Cul2. Genes Dev 13(22):2928–2933
Kawakami T, Chiba T, Suzuki T et al (2001) NEDD8 recruits E2-ubiquitin to SCF E3 ligase. EMBO J 20 (15):4003–4012
Keuss MJ, Thomas Y, McArthur R et al (2016) Characterization of the mammalian family of DCN-type NEDD8 E3 ligases. J Cell Sci 129(7):1441–1454
Kim AY, Bommelje CC, Lee BE et al (2008) SCCRO (DCUN1D1) is an essential component of the E3 complex for neddylation. J Biol Chem 283 (48):33211–33220

Klebe C, Prinz H, Wittinghofer A et al (1995) The kinetic mechanism of ran-nucleotide exchange catalyzed by RCC1. Biochemistry 34(39):12543–12552

Kumar S, Tomooka Y, Noda M (1992) Identification of a set of genes with developmentally down-regulated expression in the mouse brain. Biochem Biophys Res Commun 185(3):1155–1161

Kurihara LJ, Semenova E, Levorse JM et al (2000) Expression and functional analysis of Uch-L3 during mouse development. Mol Cell Biol 20(7):2498–2504

Kurz T, Ozlu N, Rudolf F et al (2005) The conserved protein DCN-1/Dcn1p is required for cullin neddylation in C. elegans and S. cerevisiae. Nature 435(7046):1257–1261

Kurz T, Chou YC, Willems AR et al (2008) Dcn1 functions as a scaffold-type E3 ligase for cullin neddylation. Mol Cell 29(1):23–35

Lammer D, Mathias N, Laplaza JM et al (1998) Modification of yeast Cdc53p by the ubiquitin-related protein rub1p affects function of the SCFCdc4 complex. Genes Dev 12(7):914–926

Lee JE, Sweredoski MJ, Graham RL et al (2011) The steady-state repertoire of human SCF ubiquitin ligase complexes does not require ongoing Nedd8 conjugation. Mol Cell Proteomics: MCP 10(5):M110.006460

Leyser HM, Lincoln CA, Timpte C et al (1993) Arabidopsis auxin-resistance gene AXR1 encodes a protein related to ubiquitin-activating enzyme E1. Nature 364(6433):161–164

Liakopoulos D, Doenges G, Matuschewski K et al (1998) A novel protein modification pathway related to the ubiquitin system. EMBO J 17(8):2208–2214

Lincoln C, Britton JH, Estelle M (1990) Growth and development of the axr1 mutants of Arabidopsis. Plant Cell 2(11):1071–1080

Linghu B, Callis J, Goebl MG (2002) Rub1p processing by Yuh1p is required for wild-type levels of Rub1p conjugation to Cdc53p. Eukaryot Cell 1(3):491–494

Liu J, Furukawa M, Matsumoto T et al (2002) NEDD8 modification of CUL1 dissociates p120CAND1, an inhibitor of CUL1-SKP1 binding and SCF ligases. Mol Cell 10(6):1511–1518

Liu X, Reitsma JM, Mamrosh JL et al (2018) Cand1-mediated adaptive exchange mechanism enables variation in F-box protein expression. Mol Cell 69 (5):773–86 e6

Lo SC, Hannink M (2006) CAND1-mediated substrate adaptor recycling is required for efficient repression of Nrf2 by Keap1. Mol Cell Biol 26(4):1235–1244

Lyapina S, Cope G, Shevchenko A et al (2001) Promotion of NEDD-CUL1 conjugate cleavage by COP9 signalosome. Science 292(5520):1382–1385

Makino Y, Yogosawa S, Kayukawa K et al (1999) TATA-binding protein-interacting protein 120, TIP120, stimulates three classes of eukaryotic transcription via a unique mechanism. Mol Cell Biol 19(12):7951–7960

Mendoza HM, Shen L-N, Botting C et al (2003) NEDP1, a highly conserved cysteine protease that deNEDDylates cullins. J Biol Chem 278(28):25637–25643

Mergner J, Schwechheimer C (2014) The NEDD8 modification pathway in plants. Front Plant Sci 5:103

Min KW, Hwang JW, Lee JS et al (2003) TIP120A associates with cullins and modulates ubiquitin ligase activity. J Biol Chem 278(18):15905–15910

Min KW, Kwon MJ, Park HS et al (2005) CAND1 enhances deneddylation of CUL1 by COP9 signalosome. Biochem Biophys Res Commun 334 (3):867–874

Monda JK, Scott DC, Miller DJ et al (2013) Structural conservation of distinctive N-terminal acetylation-dependent interactions across a family of mammalian NEDD8 ligation enzymes. Structure 21(1):42–53

Morimoto M, Nishida T, Honda R et al (2000) Modification of cullin-1 by ubiquitin-like protein Nedd8 enhances the activity of SCF(skp2) toward p27(kip1). Biochem Biophys Res Commun 270(3):1093–1096

Mosadeghi R, Reichermeier KM, Winkler M et al (2016) Structural and kinetic analysis of the COP9-signalosome activation and the cullin-RING ubiquitin ligase deneddylation cycle. Elife 5:e12102

Osaka F, Kawasaki H, Aida N et al (1998) A new NEDD8-ligating system for cullin-4A. Genes Dev 12 (15):2263–2268

Osaka F, Saeki M, Katayama S et al (2000) Covalent modifier NEDD8 is essential for SCF ubiquitin-ligase in fission yeast. EMBO J 19(13):3475–3484

Oshikawa K, Matsumoto M, Yada M et al (2003) Preferential interaction of TIP120A with Cul1 that is not modified by NEDD8 and not associated with Skp1. Biochem Biophys Res Commun 303(4):1209–1216

Ou CY, Lin YF, Chen YJ et al (2002) Distinct protein degradation mechanisms mediated by Cul1 and Cul3 controlling Ci stability in Drosophila eye development. Genes Dev 16(18):2403–2414

Pierce NW, Lee JE, Liu X et al (2013) Cand1 promotes assembly of new SCF complexes through dynamic exchange of F box proteins. Cell 153(1):206–215

Pintard L, Kurz T, Glaser S et al (2003) Neddylation and deneddylation of CUL-3 is required to target MEI-1/Katanin for degradation at the meiosis-to-mitosis transition in C. elegans. Curr Biol 13(11):911–921

Podust VN, Brownell JE, Gladysheva TB et al (2000) A Nedd8 conjugation pathway is essential for proteolytic targeting of p27Kip1 by ubiquitination. Proc Natl Acad Sci U S A 97(9):4579–4584

Pozo JC (1998) The ubiquitin-related protein RUB1 and auxin response in Arabidopsis. Science 280 (5370):1760–1763

Rao-Naik C, Delacruz W, Laplaza JM et al (1998) The rub family of ubiquitin-like proteins. J Biol Chem 273 (52):34976–34982

Read MA, Brownell JE, Gladysheva TB et al (2000) Nedd8 modification of Cul-1 activates SCFbeta TrCP-dependent ubiquitination of Ikappa Balpha. Mol Cell Biol 20(7):2326–2333

Reitsma JM, Liu X, Reichermeier KM et al (2017) Composition and regulation of the cellular repertoire of SCF ubiquitin ligases. Cell 171(6):1326–39 e14

Saha A, Deshaies RJ (2008) Multimodal activation of the ubiquitin ligase SCF by Nedd8 conjugation. Mol Cell 32(1):21–31
Sakata E, Yamaguchi Y, Miyauchi Y et al (2007) Direct interactions between NEDD8 and ubiquitin E2 conjugating enzymes upregulate cullin-based E3 ligase activity. Nat Struct Mol Biol 14(2):167–168
Schmidt MW, Mcquary PR, Wee S et al (2009) F-box-directed CRL complex assembly and regulation by the CSN and CAND1. Mol Cell 35(5):586–597
Schwechheimer C, Serino G, Callis J et al (2001) Interactions of the COP9 signalosome with the E3 ubiquitin ligase SCFTIRI in mediating auxin response. Science 292(5520):1379–1382
Scott DC, Monda JK, Grace CR et al (2010) A dual E3 mechanism for Rub1 ligation to Cdc53. Mol Cell 39 (5):784–796
Scott DC, Monda JK, Bennett EJ et al (2011) N-terminal acetylation acts as an avidity enhancer within an interconnected multiprotein complex. Science 334 (6056):674–678
Scott DC, Sviderskiy VO, Monda JK et al (2014) Structure of a RING E3 trapped in action reveals ligation mechanism for the ubiquitin-like protein NEDD8. Cell 157 (7):1671–1684
Shih SC, Sloper-Mould KE, Hicke L (2000) Monoubiquitin carries a novel internalization signal that is appended to activated receptors. EMBO J 19 (2):187–198
Sloper-Mould KE, Jemc JC, Pickart CM et al (2001) Distinct functional surface regions on ubiquitin. J Biol Chem 276(32):30483–30489
Soucy TA, Smith PG, Milhollen MA et al (2009) An inhibitor of NEDD8-activating enzyme as a new approach to treat cancer. Nature 458(7239):732–736
Tateishi K, Omata M, Tanaka K et al (2001) The NEDD8 system is essential for cell cycle progression and morphogenetic pathway in mice. J Cell Biol 155 (4):571–579
Wada H, Kito K, Caskey LS et al (1998) Cleavage of the C-terminus of NEDD8 by UCH-L3. Biochem Biophys Res Commun 251(3):688–692
Wada H, Yeh ET, Kamitani T (1999) Identification of NEDD8-conjugation site in human cullin-2. Biochem Biophys Res Commun 257(1):100–105
Walden H, Podgorski MS, Huang DT et al (2003a) The structure of the APPBP1-UBA3-NEDD8-ATP complex reveals the basis for selective ubiquitin-like protein activation by an E1. Mol Cell 12(6):1427–1437
Walden H, Podgorski MS, Schulman BA (2003b) Insights into the ubiquitin transfer cascade from the structure of the activating enzyme for NEDD8. Nature 422 (6929):330–334
Wee S, Geyer RK, Toda T et al (2005) CSN facilitates cullin-RING ubiquitin ligase function by counteracting autocatalytic adapter instability. Nat Cell Biol 7 (4):387–391
Whitby FG, Xia G, Pickart CM et al (1998) Crystal structure of the human ubiquitin-like protein NEDD8 and interactions with ubiquitin pathway enzymes. J Biol Chem 273(52):34983–34991
Wilkinson KD (1988) Purification and structural properties of ubiquitin. Ubiquitin. Springer, pp 5–38
Wu K, Chen A, Pan ZQ (2000) Conjugation of Nedd8 to CUL1 enhances the ability of the ROC1-CUL1 complex to promote ubiquitin polymerization. J Biol Chem 275(41):32317–32324
Wu K, Chen A, Tan P et al (2002) The Nedd8-conjugated ROC1-CUL1 core ubiquitin ligase utilizes Nedd8 charged surface residues for efficient polyubiquitin chain assembly catalyzed by Cdc34. J Biol Chem 277 (1):516–527
Wu K, Yamoah K, Dolios G et al (2003) DEN1 is a dual function protease capable of processing the C terminus of Nedd8 and deconjugating hyper-neddylated CUL1. J Biol Chem 278(31):28882–28891
Wu S, Zhu W, Nhan T et al (2013) CAND1 controls in vivo dynamics of the cullin 1-RING ubiquitin ligase repertoire. Nat Commun 4:1642
Yamoah K, Oashi T, Sarikas A et al (2008) Autoinhibitory regulation of SCF-mediated ubiquitination by human cullin 1's C-terminal tail. Proc Natl Acad Sci U S A 105(34):12230–12235
Yogosawa S, Makino Y, Yoshida T et al (1996) Molecular cloning of a novel 120-kDa TBP-interacting protein. Biochem Biophys Res Commun 229 (2):612–617
Zemla A, Thomas Y, Kedziora S et al (2013) CSN- and CAND1-dependent remodelling of the budding yeast SCF complex. Nat Commun 4:1641
Zhang W, Ito H, Quint M et al (2008) Genetic analysis of CAND1-CUL1 interactions in Arabidopsis supports a role for CAND1-mediated cycling of the SCFTIR1 complex. Proc Natl Acad Sci U S A 105 (24):8470–8475
Zheng N, Schulman BA, Song L et al (2002a) Structure of the Cul1-Rbx1-Skp1-F boxSkp2 SCF ubiquitin ligase complex. Nature 416(6882):703–709
Zheng J, Yang X, Harrell JM et al (2002b) CAND1 binds to unneddylated CUL1 and regulates the formation of SCF ubiquitin E3 ligase complex. Mol Cell 10 (6):1519–1526
Zhou C, Wee S, Rhee E et al (2003) Fission yeast COP9/ signalosome suppresses cullin activity through recruitment of the deubiquitylating enzyme Ubp12p. Mol Cell 11(4):927–938
Zhou W, Xu J, Tan M et al (2018) UBE2M is a stress-inducible dual E2 for neddylation and ubiquitylation that promotes targeted degradation of UBE2F. Mol Cell 70(6):1008–24 e6

Cullin-RING Ligase Regulation by the COP9 Signalosome: Structural Mechanisms and New Physiologic Players

4

Feng Rao, Hong Lin, and Yang Su

Abstract
The Cullin-RING E3 ligases (CRLs) are major ubiquitylation machineries regulated by reversible cycles of neddylation and deneddylation. The deneddylase COP9 Signalosome (CSN) terminates CRL catalytic cycle. CSN also provides a docking platform for several kinases and deubiquitinases that might play a role in regulating CRL. Recently, remarkable progress has been made in elucidating the biochemical principles and physiological implications of such exquisite regulation. The cryo-EM structures of CRL-CSN complexes provide the biochemical basis of their cognate interactions and reveal potential regulatory mechanisms during complex disassembly. Moreover, novel players beyond the canonical eight subunits of CSN were identified. This includes CSNAP, a potential 9th CSN subunit with regulatory functions, and the metabolite inositol hexakisphosphate (IP_6), which enhances CRL-CSN complex formation, with IP_6-metabolizing enzymes possibly instilling dynamics to the CRL-CSN system. Here, we review and summarize these new mechanistic insights along with progress in understanding CSN biology based on model organisms with genetically edited CSN subunits.

Keywords
Cullin-RING ubiquitin ligases · Neddylation/deneddylation · COP9 Signalosome · Inositol hexakisphosphate · IP6K · CSNAP

Abbreviations

CAND1	Cullin-associated NEDD8-dissociated protein 1
CK2	Casein kinase II
CRLs	Cullin-RING E3 Ligases
Cryo-EM	Electron cryomicroscopy
CSN	COP9 Signalosome
CSNAP	CSN acidic protein
DUB	Deubiquitinases
InsPs	Inositol triphosphate
IP3	Inositol triphosphate
IP3K	Inositol 1,4,5-trisphosphate 3-kinase
IP4	Inositol tetrakisphosphate
IP5	Inositol pentakisphosphate
IP6	Inositol hexakisphosphate
IP6K	Inositol hexakisphosphate kinase
IP7	5-Diphosphotidylinositol pentakisphosphate
IPK	Inositol phosphate kinases
IPK	Inositol phosphate kinases
IPMK	Inositol polyphosphate multikinase

F. Rao (✉) · H. Lin · Y. Su
Department of Biology, Guangdong Provincial Key Laboratory of Cell Microenvironment and Disease Research, Southern University of Science and Technology, Shenzhen, Guangdong, China
e-mail: raof@sustech.edu.cn

Y. Sun et al. (eds.), *Cullin-RING Ligases and Protein Neddylation*, Advances in Experimental Medicine and Biology 1217, https://doi.org/10.1007/978-981-15-1025-0_4

IPPK/ IP5K	Inositol 1,3,4,5,6-pentakisphosphate 2-kinase
ITPK1	Inositol 1,3,4-triphosphate 5/6-kinase
JAMM-motif	JAB1/MPN/Mov34-motif
kDa	Kilodalton
MPN domain	Mpr1/Pad1 N-terminal domain
PCI	Proteasome lid-CSN-initiation factor 3
PDB	Protein data bank
PKD	Protein kinase D
PPIP5K	Diphosphotidylinositol pentakisphosphate kinase
SCF	Skp1/cullin/F-box
Sgn	Signalosome
TNP	N2-(*m*-Trifluorobenzyl), N6-(*p*-nitrobenzyl)
UV	Ultraviolet

Protein homeostasis is controlled by the ubiquitin-proteasome system. In this system, a three-enzyme cascade (E1, E2, E3) for protein substrate ubiquitylation, which marks substrates for proteasomal degradation. The Cullin-RING ubiquitin ligases (CRLs) are the largest superfamily of E3 ligases, with over 200 known members in mammals (Petroski and Deshaies 2005; Lydeard et al. 2013). CRLs are generally composed of four core components: one of the seven canonical Cullin scaffold protein (Cul1, 2, 3, 4A/B, 5, or 7), an E2-interacting RING protein that forms cognate heterodimer with Cullins (Rbx1/2, also called Roc1/2), an adaptor protein specific for Cullin subfamily members, and a substrate receptor that recognizes the ubiquitylation targets. The substrate receptor module can be a separate protein, or, in the case of CRL3, a domain fused with the adapter as one protein. A significant portion of the proteome (estimated to be over 20%) is targeted for ubiquitylation by CRL (Soucy et al. 2009). These numerous ubiquitylation substrates implicate CRLs in a wide range of biological processes such as cell cycle progression, DNA repair, metabolism, autophagy, stemness, development, immunity, and carcinogenesis; some of these aspects are discussed in other chapters of this book. CRLs are being explored as therapeutic targets, especially in the field of cancer research (Skaar et al. 2014; Zhao and Sun 2013). In recently years, notable progress has been made by tailoring CRLs to degrade specific proteins of therapeutic relevance via the assistance of intermolecular "glue"-like small molecules (Bondeson and Crews 2017) (see Chap. 20 for full discussions on CRL-based drug development).

To avoid erroneous disruption of cellular activities, ubiquitin ligases need to be regulated (Vittal et al. 2015; Zheng and Shabek 2017). The E3 ligase activity of CRLs is tightly controlled by multiple mechanisms impinging on protein-protein interaction and posttranslational modifications on CRL or substrates (Lydeard et al. 2013). One major route of CRL regulation is through cycles of Cullin neddylation (modification by the ubiquitin-like protein Nedd8) and deneddylation. The unneddylated state permits the binding of CAND1 (cullin-associated NEDD8-dissociated protein 1) to exchange substrate receptor for surveying the substrate repertoire (Liu et al. 2018) (see Chap. 2 for detailed discussions on CRL regulation by CAND1). Neddylation at Cullin C-terminus enhances CRL activity by inducing productive E2 engagement with Rbx1/Roc1 (Wu et al. 2000; Duda et al. 2008). Conserved from plants to human, the COP9 Signalosome (CSN) is a deneddylase that targets all neddylated Cullins. In vitro, CSN inhibits CRL activity by catalyzing deneddylation and by sequestering deneddylated CRL (Lyapina et al. 2001; Enchev et al. 2012; Emberley et al. 2012). However, genetic disruption of CSN subunits diminishes, rather than enhances, CRL activity (Wolf et al. 2003). This conundrum, known to the field as the "CSN paradox," can be reconciled as CSN protects CRL substrate receptor components from auto-ubiquitylation and destruction due to prolonged E2 engagement (Wolf et al. 2003; Wee et al. 2005; Cope and Deshaies 2006). CSN therefore protects CRL from being aberrantly active under basal conditions while enabling proper CRL activation upon signal-dependent dissociation (Wolf et al. 2003). Thus, the assembly and disassembly of CRL-CSN complexes are critical steps of CRL regulation.

In this review, we summarize recent progress in understanding how CSN interacts with and regulates CRL, particularly the role of recently identified mediators like the inositol polyphosphate metabolites and CSNAP. Beginning with a brief introduction to the discovery of CSN, this review will focus on the mechanistic insights into CRL-CSN assembly interface and the multifaceted role of CSN as a central CRL regulator. Finally, the need for ongoing research on CRL-CSN disassembly mechanisms is discussed.

4.1 Identification of CSN

Deng and associates first purified CSN as the multi-subunit COP9 complex from *Arabidopsis thaliana*. The name COP9 derives from the *Co*nstitutive *P*hotomorphogenesis phenotype of its subunit mutants (Chamovitz et al. 1996; Wei et al. 1994). A separate attempt to identify 26S proteasome components purified human CSN as Signalosome (Sgn), which contains eight major subunits (Seeger et al. 1998). The name *COP9 Sign*alosome, abbreviated as CSN, was eventually adopted to incorporate both the plant and human name, and the eight CSN subunits are known as CSN1 to CSN8 with decreasing molecular weight. Soon after, CSN was found to interact with CRL1/SCF at biochemical and functional levels (Schwechheimer et al. 2001) and to catalyze Cullin deneddylation via its JAMM-motif (JAB1/MPN/Mov34 metalloenzyme) containing MPN (Mpr1/Pad1 N-terminal) domain located in subunit 5 (CSN5) (Lyapina et al. 2001; Cope et al. 2002). These early studies laid the foundation for CSN as a CRL deneddylase.

4.2 CSN as a Multifaceted CRL Regulator

When CSN was first identified as a CRL deneddylase, it immediately became apparent that genetic downregulation of CSN subunits in multiple model organisms dampens rather than enhances CRL activity, although Cullin neddylation is increased as expected. Thus, in the first report linking CSN to CRL regulation, the SCF^{TIR1} substrate PSIAA6 was in fact stabilized rather than degraded in CSN5 loss-of-function mutant Arabidopsis, leading to decreased response to the hormone auxin (Schwechheimer et al. 2001). This was puzzling, since CSN was validated as a biochemical CRL inhibitor at the same time (Lyapina et al. 2001; Cope et al. 2002). Adding to the puzzle, loss of CSN subunits delayed the degradation of other CRL substrates Sic1 in yeast and Katanin in C. elegans (Cope et al. 2002; Pintard et al. 2003), with consequent defects in temperature-sensitive growth and embryonic development, respectively. These puzzles were partially solved by the finding that active CRLs can auto-catalyze substrate receptor degradation (Zhou and Howley 1998; Galan and Peter 1999), which suggest that non-catalytic sequestration of CRL by CSN serves to prevent auto-destruction of CRL substrate receptors (Cope and Deshaies 2006).

Does CSN protect CRL also in mammals? CSN is essential in mouse, and global deletion of its subunits results in embryonic lethality (Lykke-Andersen et al. 2003; Tomoda et al. 2004; Menon et al. 2007). As such, hypomorphic, cell-specific, and inducible gene-targeting methods were applied to achieve spatiotemporal depletion of CSN subunits, which reveals important roles of CSN5 and CSN8 in cardiovascular homeostasis maintenance, immunological development, and hepatocyte proliferation (Menon et al. 2007; Su et al. 2011a, b; Lei et al. 2011; Asare et al. 2017). In these studies, Cullin neddylation was increased with CSN subunit deletion, but the stability of Cullin substrate receptors were not examined. Recently, Cornelius et al. reported that induced deletion of CSN5 specifically in the kidney leads to degradation of the CRL3 substrate receptor KLHL3. Consequently, WNK kinase, a $CRL3^{KLHL3}$ substrate, accumulates to cause familial hyperkalemic hypertension phenotype (Cornelius et al. 2018). Thus, the protective role of CSN on CRL substrate receptors also holds true in mammals.

Apart from substrate receptor stabilization, CSN could regulate CRL-based ubiquitylation machineries via several other mechanisms also independent of its deneddylase activity. First, the stability of Cullins is decreased in some CSN subunit deletion mutant models (He et al. 2005; Wu et al. 2005; Wang et al. 2010), extending the protective role of CSN from substrate receptor to Cullins themselves. Second, CSN protects the E2 UBC3/CDC34 from degradation catalyzed by $SCF^{\beta TrCP}$ (Fernandez-Sanchez et al. 2010), again suggesting that CSN is not merely a CRL inhibitor but also an integral component of the CRL activity cycle by keeping the E2 in place. Third, CSN associates with deubiquitinases (DUBs). The DUB USP15 (Ubp12 in yeast), for example, has been shown to stabilize the substrate receptor Pop1p of the SCF^{Pop1p} E3 ligase (Wee et al. 2005; Zhou et al. 2003), as well as the RING protein Rbx1 (Hetfeld et al. 2005), suggesting that USP15/Ubp12-mediated deubiquitylation can protect CRL from autocatalysis (Wu et al. 2006). On the other hand, CSN-associated DUBs can also prevent the degradation of certain CRL substrates, such as IkB (Schweitzer et al. 2007; Schweitzer and Naumann 2015), APC (Huang et al. 2009), and Mdm2 (Bai et al. 2019; Zou et al. 2014). These data suggest that CSN-associated DUBs might function as a CRL "rheostat," antagonizing excessive ubiquitylation of both CRL components and protein substrates. Last but not least, when CSN was first purified, it was found to contain kinase activities (Seeger et al. 1998), which was later attributed to other kinases that co-purifies with CSN, including casein kinase II (CK2), protein kinase D (PKD), and inositol 1,3,4-triphosphate 5/6-kinase (ITPK1) (Uhle et al. 2003; Sun et al. 2002; Harari-Steinberg and Chamovitz 2004). These kinases phosphorylate either CSN subunits and/or the CRL substrates such as c-Jun (Uhle et al. 2003; Sun et al. 2002). When CRL substrates are phosphorylated by CSN-associated kinases, their stability is altered (Berse et al. 2004; Bech-Otschir et al. 2001). In contrast, it is unclear how CSN subunit phosphorylation by its associated kinases contributes to CRL regulation.

The multifaceted role of CSN in CRL regulation underscores the potential of CSN as a docking platform integrating external stimuli to confer spatiotemporal CRL activation. In line with this notion, different types of DNA damage stimuli have been shown to trigger phosphorylation of various CSN subunits, with consequent alterations in CRL localization and activation (Meir et al. 2015; Fuzesi-Levi et al. 2014; Dubois et al. 2016), but detailed structure-function analysis of this regulation is lacking. In light of recent structural insights on the CRL-CSN complexes, as reviewed below, it might be worthwhile to reanalyze the biochemical data in current literature to come up with new hypothesis concerning mechanisms of dynamic CRL regulation by CSN, which could be specific to CRL subfamily.

4.3 Structure of the CSN Holo-Complexes

Crystal structures of various CRLs were solved since 2002 (Duda et al. 2008; Zheng et al. 2002; Angers et al. 2006; Zimmerman et al. 2010) and are reviewed in detail elsewhere in this book (see Chap. 3 for full discussions on the structural biology of CRL ubiquitin ligases). Briefly, Cullins serve as the scaffold, recruiting E2-conjugating enzyme at their C-terminus via the heterodimerizing RING domain proteins Rbx1/2 and recruiting ubiquitylation substrates via substrate receptor/adaptor bound to Cullin N-terminus (Fig. 4.1a). One hallmark of CRL structural features is their modularity, with similar overall subunit composition across Cullin families. Conceivably, this commonality allows generally conserved mechanisms of regulation by CSN.

Efforts made to solve CSN structures start with its subunits time (Lee et al. 2013; Echalier et al. 2013; Birol et al. 2014; Dessau et al. 2008) and culminate in the elucidation of holo-CSN crystal by Thoma and associates (Fig. 4.1b) (Lingaraju et al. 2014). The overall structural composition of the CSN complex is similar to the lid complex of the 19S regulatory particle of the 26S proteasome and to the eukaryotic initiation factor 3 complex (Hofmann and Bucher 1998). These three complexes, together known as the "Zomes"

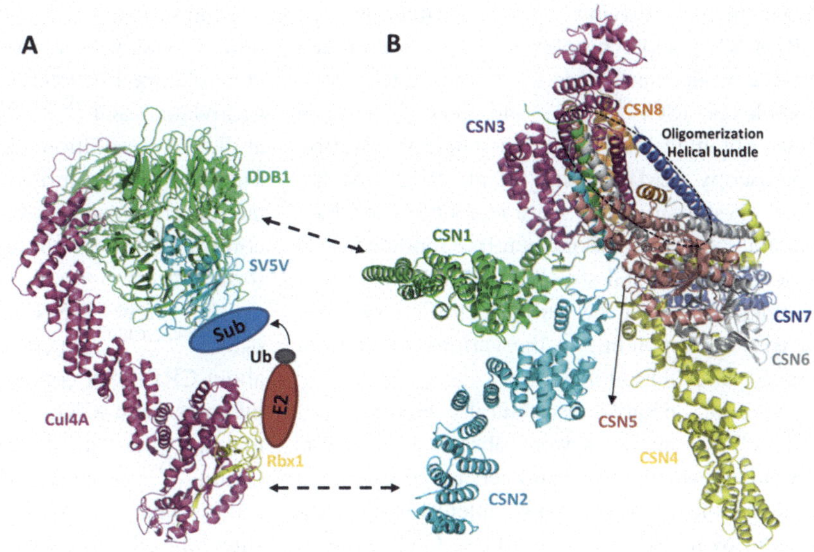

Fig. 4.1 Structural models of a CRL (CRL4^{SV5V}, PDB id: 2HYE) (Angers et al. 2006) (**a**) and the COP9 Signalosome (CSN, PDB id: 4D10) (Lingaraju et al. 2014) (**b**). Individual subunits are colored differently. The potential locations of E2 and substrate are shown in the scheme (**a**). The points of contact between CRL4 and CSN are represented as double-sided dashed arrows. The helical bundle formed by the C-terminal helix of all eight CSN subunits is highlighted in circle (**b**)

complexes (Alpi and Echalier 2017), are all involved in protein homeostasis regulation and are composed of subunits with MPN metalloprotease domain for isopeptide bond cleavage and proteasome lid-CSN-initiation factor 3 (PCI) domains for oligomerization. The structural similarity among "Zomes" complexes provokes the idea that they might be evolutionarily related.

The CSN holo-complex consists of two layers of organization. First, the PCI domains in CSN1, CSN2, CSN3, CSN4, CSN7, and CSN8 form edge-to-edge interactions, effectively generating a PCI hexamer (Sharon et al. 2009). Second, on top of the PCI platform, CSN5 and CSN6, the two non-PCI subunits, form a dimmer via interactions between their MPN domains. Third, the carboxy-terminal α-helices of all eight CSN subunits form a helical bundle, which drives the formation of the 8-subunit complex (Fig. 4.1b). The MPN domain of CSN5, but not CSN6, has a conserved metalloprotease active site for catalyzing deneddylation. However, unlike other JAMM family isopetidases, such as Rpn11 of the 19S lid complex of the 26S proteasome (Worden et al. 2014), CSN5 is catalytically inactive in the CSN complex. A CSN5 loop containing Glu104 inserts into its active site by interacting with the catalytic Zn^{2+} ion, thereby occluding the catalytic pocket. Thus, an activation mechanism is a prerequisite for CSN-catalyzed deneddylation.

To understand how auto-inhibited CSN is activated, computational modeling studies employing the EM map of CRL1/SCF-CSN complex (Enchev et al. 2012) were conducted. Base on the structural differences between CSN and the CSN-SCF complex, neddylated CRL binding likely induces movement in CSN4, which would then be expected to alter the CSN4-CSN6 contact interface, leading to a rearrangement of the CSN5-CSN6 dimer and eventually CSN5 activation. Although these predicted changes still require further structural validations, the stringent requirement of CRL binding for CSN activation suggests that CSN is dedicated to CRL deneddylation. Indeed, to our best knowledge, CSN has no other substrates but neddylated Cullins.

4.4 Structural Mechanisms of CRL-CSN Complex Assembly and Regulation

The recent progress in obtaining EM structures of CRL-CSN complexes has become a new cornerstone for our understanding on the mechanisms driving the formation of CRL-CSN supercomplexes. As mentioned above, molecular models of two CRL1/SCF complexes ($SCF^{Skp2/Cks1}$ and SCF^{Fbw7}) were first obtained by negative-staining electron microscopy (EM) (Enchev et al. 2012). At the time, crystal structures were available for CRLs but not CSN. The authors therefore modeled CSN based on the crystal structures of other "Zomes" complexes and successfully identified electron density corresponding to the various subunits in this 20 Å resolution map.

Several notable structural features of CRL-CSN complexes were observed from this EM map, which were validated by biochemical analysis and verified by subsequent higher resolution EM maps (see below). First, CSN binds to CRL primarily via embracement of the C-terminal portion of Cu1/Rbx1 ($Cul1^{CTD}$/Rbx1) by the N-terminal domain of CSN2. Consistently, deleting the N-terminal portion of CSN2 completely abolishes CRL binding. Second, CSN binding to $Cul1^{CTD}$/Rbx1 blocks the interaction between SCF and the E2 enzyme CDC34, consistent with decreased E2 activation in a substrate peptide ubiquitylation assay. Third, binding of catalytically inactive CSN still prevents CRL neddylation, suggesting that CSN also obstructs the neddylation E2, likely in a manner similar to ubiquitylation E2. Fourth, the contact between CSN1/CSN3 and substrate receptor would result in steric hindrance for incoming substrate. Consistently, CSN addition prevented phospho-p27 binding to $SCF^{Skp2/Cks1}$, and, reciprocally, substrate addition slows down $SCF^{Skp2/Cks1}$ deneddylation by CSN (Enchev et al. 2012; Emberley et al. 2012). Overall, these data lend further support for a multifactorial role of CSN in CRL regulation: besides catalyzing Cullin deneddylation, the tight binding of CSN to CRL also controls the access of ubiquitylation E2, substrates, as well as neddylation machineries.

More recently, cryo-EM maps of CRL4A-CSN and SCF-CSN complexes were obtained at resolutions possible to confidently identify protein secondary structures. Cavadini et al. reported structures of CSN in complex with neddylated $CRL4A^{DDB2}$ ($CSN\text{–}_{N8}CRL4A^{DDB2}$) at 8.3 Å resolution and $CSN\text{–}_{N8}CRL4A$ at 6.4 Å resolution (Cavadini et al. 2016), whereas Mosadeghi et al. obtained structures of CSN in complex with neddylated $SCF^{Skp2/Cks1}$ ($CSN\text{–}_{N8}SCF^{Skp2/Cks1}$) at 7.2 Å resolution (Mosadeghi et al. 2016). These structures further refined CRL-CSN contact interfaces and enabled better visualization of conformational changes during complex formation.

The new $CSN\text{–}_{N8}CRL4A$ and $CSN\text{–}_{N8}SCF^{Skp2/Cks1}$ structures revealed key insights about CRL-CSN contacts (Fig. 4.2a). First, CSN interacts with CRL at three sites, with CSN2 making the most extensive contact with $Cullin^{CTD}$. Second, Rbx1 RING domain is sandwiched by CSN2 and CSN4, thereby shielding the E2-binding site, which could explain the competition between CSN and E2 enzymes (Enchev et al. 2012). Third, CSN1 contacts the CRL4A adaptor DDB1, and CSN1 and CSN3 are in the vicinity of SCF adaptor/substrate receptor pair Skp1/Skp2/Cks1. Overall, these medium resolution structures agreed with each other and with the previous lower resolution models in outlining major CRL-CSN contact sites. Nonetheless, higher resolution structures are still required to visualize atomic details of the CRL-CSN interface. This need is further justified by the recent identification of small CSN cofactors such as CSNAP and inositol polyphosphates (see below), whose presence in CSN-CRL complexes may only be revealed by higher resolution structures.

Besides unveiling CRL-CSN contact interfaces, the $CSN\text{–}_{N8}CRL4A$ and $CSN\text{–}_{N8}SCF^{Skp2/Cks1}$ structures also showed that complex formation is associated with conformational changes in CRL and CSN. Several protein subunits can fit the respective map segments without alteration, whereas other subunits or their subdomains, specifically the N-terminal portions of CSN2 and CSN4 (Fig. 4.2a), the MPN-domains of CSN5 and CSN6, the RING domain of Rbx1, the WHB domain of Cullins, and Nedd8, undergo significant movements in order to fit their respective electron density. These conformational changes can be rationalized on two bases. First, CSN5 must undergo

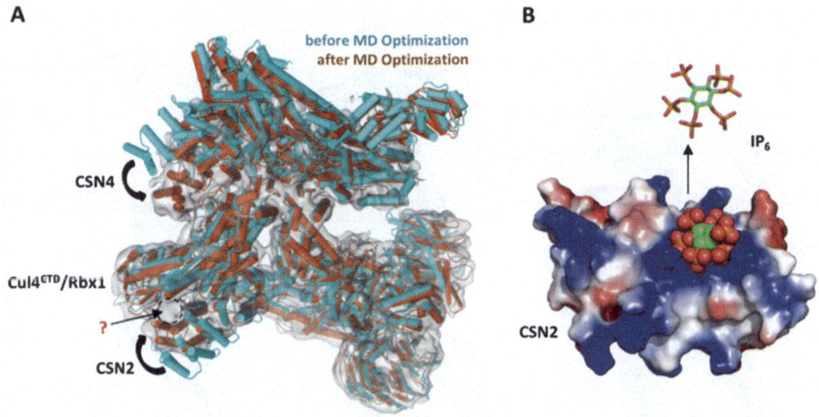

Fig. 4.2 Structural models of CRL4-CSN complex (**a**) and the IP_6-CSN2 complex (**b**). (**a**) Conformational changes in CSN induced by CRL4 binding. The crystal structures of CRL4A and CAN were fitted into the cryo-EM electron density map (EMD3314), followed by refinement with molecular dynamic flexible fitting (MDFF). Protein secondary structures are shown in pipes (helices) and sheets, with structures before and after MDFF shown in cyan and orange, respectively. The N-terminal helices of CSN2 and CSN4 move toward the C-terminal region of Cul4A and Roc1 during the fitting. Highlighted by a black dashed circle is the unmapped electron density located between CSN2 and $Cul4^{CTD}$/Rbx1, which possibly corresponds to IP_6. (**b**) Surface representation of the IP_6-CSN2 crystal structure. IP_6, shown as spheres with the inositol ring in cyan and the phosphate moieties in orange, bound to a basic pocket

conformational rearrangements in order to recognize the substrate Nedd8, as its catalytic active site is self-obstructed in the absence of substrates. Second, the rest of the segments with notable conformational rearrangements, i.e., the N-terminal portions of CSN2/CSN4 and the RING domain of Rbx1 and the WHB domain of Cullins, are in direct contact with each other. Conformational changes in these structural elements are consistent with an induced-fit mechanism during substrate-enzyme binding. Conceivably, CRL-CSN binding initiates conformational rearrangements changes at the N-terminal segments of CSN2 and CSN4, distant from the active site. These conformational are then transduced to the catalytic site in CSN5, rendering CSN an active deneddylase. While more structural dynamics evidences are needed to support this model, mutating the CSN4-CSN6 interface enhances the catalytic activity of CSN, consistent with the notion that substrate-induced CSN activation involves the CSN4-CSN6 interface. In conclusion, during CSN-CRL binding, significant conformational changes occur on both the substrate (CRL) and enzyme (CSN), inside and distant from the site of catalysis. Such binding-induced fit ensures the catalytic specificity of CSN.

Apart from revealing CRL-CSN assembly mechanisms, the cryo-EM structures of CRL-CSN complexes further provide clues on how CSN might be released for CRL activation. Specifically, the new structures support earlier biochemical observations that CSN and ubiquitylation substrates are sterically incompatible (Enchev et al. 2012; Emberley et al. 2012; Fischer et al. 2011). Consistently, the presence of CRL substrates (Bornstein et al. 2006), or a DNA fragment that binds substrate receptor (Cavadini et al. 2016), markedly inhibits Cullin deneddylation, conceivably by interfering with CSN-CRL complex formation. These findings suggest that CSN binds exclusively to substrate-free CRL. Accordingly, during the CRL activity cycle, CSN should come into play only after substrate has been ubiquitylated and delivered to the proteasome (Fig. 4.3). This opens up two more questions. First, neddylated CRLs are E2-bound during substrate degradation. Given that CSN competes with E2 (e.g., CDC34) for

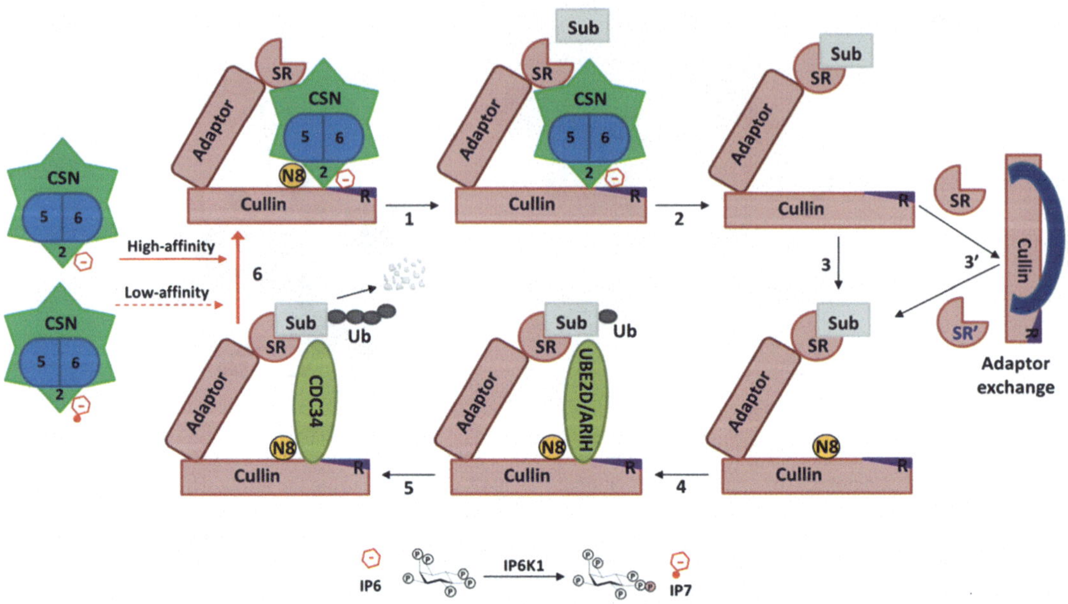

Fig. 4.3 Scheme depicting the catalytic cycle of CRL and the role of inositol polyphosphates in the assembly and disassembly of CRL-CSN complexes. CRL catalytic cycle is divided into six steps. CSN dissociation and re-association are drawn as the initiation and termination step of CRL catalytic cycle. **1**: CSN deneddylates bound CRL. **2**: Unneddylated, weaker CRL-CSN complex is amenable to substrate-induced CSN dissociation. **3/3′**: CSN-free, unneddylated CRL undergoes either direct re-neddylation (**3**) or after CAND1-mediated exchange of substrate receptor (**3′**) (Pierce et al. 2013). **4**: Neddylation induces conformational changes in Rbx1, activating CRL to recruit either UBE2D or ARIH and their homologs which initiate substrate ubiquitylation (Wu et al. 2010; Scott et al. 2016). **5**: CDC34, also called UBE2R, replaces UBE2D/ARIH to build a ubiquitin chain for signaling substrate degradation (Wu et al. 2010; Scott et al. 2016). **6**: After substrate ubiquitylation and removal, CSN competes off CDC34 from CRL to reassemble the inert CSN-CRL complex, thereby completing the CRL activity cycle. *R* Rbx1/Roc1, *SR* substrate receptor, *N* Nedd8, *Sub* substrate, *Ub* ubiquitin, *2* CSN2, *5* CSN5, *6* CSN6

CRL binding, what factor(s) then shifts CRL equilibrium from the active, E2-bound state to the inactive, CSN-bound state? Second, although neddylation strengthens CSN interaction, deneddylated CRL still binds CSN with nanomolar-level affinity (Mosadeghi et al. 2016); what factors then determines the disassembly of this inactive CRL-CSN complex? One possibility is that the incoming substrate sterically dislodges CSN. In the case of $CRL4^{DDB2}$, a damage-mimicking DNA duplex that binds avidly to DDB2 can indeed displace CSN (Lyapina et al. 2001). It remains unknown whether an ubiquitylation substrate protein can also displace prebound CSN. Another possibility is that CAND1, which wraps around deneddylated Cullins to release adaptors and substrate receptors (Goldenberg et al. 2004), could dissociate CRL-CSN by competing with CSN. However, so far, there is no evidence supporting that CAND1 significantly influence CRL-CSN complex stability (Bennett et al. 2010; Reitsma et al. 2017). For both of the above questions, the answer might partially involve the inositol polyphosphates and their metabolic enzymes, as discussed in detail below.

4.5 Inositol Polyphosphates and Their Synthases in the Assembly and Disassembly of CRL-CSN Complexes

The higher inositol polyphosphates (InsPs) originate from inositol triphosphate (IP_3), a 2nd messenger of the Gq-coupled GPCR pathway. A series of inositol phosphate kinases (IPKs), including inositol 1,4,5-trisphosphate 3-kinase (IP3K), inositol 1,3,4-triphosphate 5/6-kinase (ITPK1), inositol polyphosphate multikinase (IPMK), and inositol 1,3,4,5,6-pentakisphosphate 2-kinase (IPPK/IP5K), can phosphorylate IP_3 in an stepwise manner, leading to the synthesis of IP_4, IP_5, and IP_6 (Verbsky et al. 2005). IP_6, the most abundant inositol phosphate, can be further phosphorylated by IP6 kinases (IP6K) and PPIP5Ks to generate the inositol pyrophosphates, IP_7 and IP_8 (Fig. 4.3), which are potential signaling molecules containing energetic pyrophosphate bond(s) (Shears 2018). The higher inositol polyphosphates are found from yeast to human, but are not as well-studied as IP_3.

The link between CSN and the inositol polyphosphate system initiates from a serendipitous finding that CSN co-purifies with ITPK1, a versatile IPK that phosphorylates different IP_3 isomers (Wilson et al. 2001). This co-purification is grounded on direct ITPK1-CSN1 interaction (Sun et al. 2002). Because the purified ITPK1 can phosphorylate c-Jun, it was initially proposed that ITPK1 is the CSN-associated protein kinase. However, PKD and CK2 were subsequently identified as CSN-associated kinases (Uhle et al. 2003). Whether recombinant ITPK1 can act as a protein kinase remains unanswered, and the physiological relevance of its strong association with CSN also remains unknown.

Later, through an unbiased tandem affinity purification approach, inositol hexakisphosphate kinase-1 (IP6K1), another IPK that phosphorylates IP_6 to IP_7, was found to directly bind DDB1, the CRL4 adaptor (Rao et al. 2014a). DDB1 binding suppresses IP6K1 catalytic activity, but promotes its stability without affecting ubiquitylation, suggesting that IP6K1 is not a CRL4 substrate. Conversely, IP6K1 co-expression suppresses the expected increase in total cellular ubiquitylation levels caused by Cul4A-Rbx1-DDB1 overexpression, suggesting that IP6K1 and CRL4 form a mutually inhibited, inert complex. Mechanistically, IP6K1 links CRL4 to CSN. Thus, its knockdown dramatically diminishes CRL4-CSN complex formation and suggests an inert ternary complex consisting of IP6K1, CRL4, and CSN under basal conditions. Interestingly, UV radiation dissembles the IP6K1-CRL4-CSN complex, which is consistent with its known role to dissociate CSN-CRL4 for DNA repair (Scrima et al. 2011) and suggest a role for IP6K1 in stimuli-dependent CRL4 activation. This disassembly of IP6K1-CRL4-CSN complex releases IP6K1 from DDB1 sequestration evidenced by UV-induced rise in cellular IP_7 levels. Moreover, kinase-dead IP6K1 mutant, or the IP6K-specific inhibitor TNP, binds DDB1 with higher affinity under basal condition and inhibits UV-induced disassembly of CRL4-CSN complexes, suggesting that IP6K1-mediated conversion of IP_6 to IP_7 promotes CRL4-CSN dissociation.

In search for a molecular mechanism explaining the roles of IP_6 or IP_7 in CRL-CSN complex dynamics, purified proteins were employed to reconstitute key elements mediating CRL-CSN interactions. Two novel binding interfaces were identified between CRL-CSN (Scherer et al. 2016). First, there is electrostatic interaction between the N-terminal acidic tail of CSN2 and the conserved C-terminal basic canyon of Cullins. This is reminiscent of interactions between Cullin's basic canyon and the C-terminal acidic tail of the E2 CDC34 (Kleiger et al. 2009a, b), providing a plausible explanation for the earlier observation that CSN competes with CDC34 to inhibit CRL (Enchev et al. 2012). Second, IP5K, the IPK that synthesizes IP_6, also interacts with the CRL-CSN complex. Depleting IP5K alters the stability, neddylation, and activation of CRL1 and CRL4. IP_6, at nanomolar concentrations, enhances Cul4A/Roc1-CSN2 interaction in vitro.

How is IP_7 different from IP_6? In vitro, IP_7 can also promote CRL4A-CSN2 binding, but with threefold lower potency than IP_6. This data is unusual, since IP_7 often binds to an InsP-binding site with higher affinity than IP_6, due to its being more negatively charged (Rao et al. 2014b; Chakraborty et al. 2011). It is therefore tempting to hypothesize that IP6K1-catalyzed IP_6-IP_7 conversion could increase the percentage of lower-affinity CRL4-CSN complexes, rendering CRL-CSN complexes amenable to dissociation induced by incoming ubiquitylation substrates (Fig. 4.3). However, given that IP_7 can still promote CRL4-CSN interaction at concentrations comparable to endogenous levels, it remains possible that additional factors exist to shift CRL-CSN into a low affinity state, via either IP_7-depenendent or IP_7-independent mechanisms. The recently identified CSN subunit CSNAP is one such candidate (see below).

The above-discussed findings raise the tantalizing possibility that IP_6 and its metabolic enzymes comprise a signaling module in transmitting upstream stimuli to CRL-CSN complex dynamics and CRL activity control. However, some outstanding questions remain.

First, most of the above studies were conducted using Cul4A without substrate receptors. Therefore, it remains unclear whether CRL regulation by IP_6/IP_7 dynamics is generally applicable for all CRL4 ligases or only a subset with specific substrate receptors, and whether the same principle also applies to other Cullin members. In this regard, it was found that IP_6 can promote CSN binding to different Cullins to various extents, indicating that other CRL-CSN complexes may be similarly regulated by IP_6. By binding DDB1, the effect of IP6K1 is perhaps more restrict to CRL4. Note that there are other IP6K isoforms (IP6K2, IP6K3) and whether they interact and regulate other CRL family members and respond to other stimuli (e.g., cell cycle transition) is worth exploring.

Second, what is the structural basis and biochemical advantage to have CRL-CSN complex formation regulated by a small metabolite (IP_6)? Mechanistically, IP_6 could be an allosteric factor binding either CRL or CSN to induce conformational changes enhancing their interaction. Alternatively, IP_6 might act as intermolecular "glue" at the interface between CRL and CSN, thereby bridging the complex. We have recently obtained the crystal structure of IP_6 in complex with CSN2 (PDB id: 6A73, to be published) (Fig. 4.2b), the most critical CSN subunit in binding CRL. Moreover, in fitting the EM map of CSN–$_{N8}$CRL4A, we have also uncovered an extra electron density that fits IP_6 at the interface between CSN2 and Cul4A^{CTD} (Fig. 4.2a). These structural insights suggest that IP_6 is primarily a CSN cofactor recruiting CRL for deneddylation. Nonetheless, it is still puzzling why nature has evolved a highly conserved IP_6 binding pocket at the CRL-CSN interface instead of directly evolve a complementary binding interface between CRL and CSN. The integration of IP_6 into CSN could also be regulated as that of IP_7. Whether IP_6 is a switch determining the competition between CSN and E2 CDC34 also remains to be determined.

Finally, thus far, three IPKs, i.e., ITPK1, IP5K, and IP6K1, have also been found to interact directly with different components of the CRL-CSN complex. ITPK1 binds to CSN1 (Sun et al. 2002), whereas IP5K and IP6K1 bind to Cullin and the Cul4 adaptor DDB1, respectively (Rao et al. 2014a; Scherer et al. 2016). Given the known characteristic of classic metabolic enzymes in forming multi-subunit clusters (Castellana et al. 2014), the IPKs might also assemble an enzymatic chain to efficiently channel intermediates for IP_6/IP_7 production based on spatial and temporal demand. If this were the case, then any IPK in the synthetic pathway could be regulated in a stimuli-dependent manner to integrate CRL regulation into various cellular functions.

CRL neddylation is a valid anticancer target in clinical trials (Soucy et al. 2009; Li et al. 2014). The neddylation inhibitor Pevonedistat/MLN4924 is currently undergoing phase III clinical trial as an anticancer agent (NCT03268954) (Sekeres et al. 2018). Moreover, CRLs are also important tools in targeted protein degradation therapies (Bondeson and Crews 2017). It is therefore legitimate to propose that answering the above questions can help us understand the nature of CRL-CSN complex dynamics and also provide new venue to develop neddylation interference agents for therapeutic purposes.

4.6 CSNAP as a New CSN Subunit Regulating CRL Activity

Biochemical and structural analysis suggest that CSN is an active 8 subunit deneddylase, which is further supported by the fact that recombinant

CSN purified from *E. coli* is catalytically active with 8 subunits (Sharon et al. 2009). Interestingly, Sharon and associates identified a 6.2-kDa protein, named CSNAP (CSN acidic protein), as a stoichiometric CSN binding partner, making CSNAP the 9th CSN subunit. Mechanistically, the C-terminal F/D-rich region of CSNAP mediates its association with CSN, with Phe44 and Phe51 possibly playing direct roles in the interaction with a hydrophobic surface patch formed collectively by CSN3, CSN5, and CSN6. Notably, CSNAP shares structural homology to the 19S lid subunit DSS1 (Sem1p in yeast) of the 26S proteasome (Sone et al. 2004), thus completing a one-to-one correspondence between CSN and the 19S lid subcomplex. The identification of CSNAP therefore strengthens the evolutionary links between CSN and the other two Zomes complexes.

To further understand the biochemical and cellular role of CSNAP, the same group recently performed detailed proteomic and biochemical analysis (Fuzesi-Levi et al. 2019). Biochemically, CSNAP inhibits CRL-CSN binding both in vitro and in vivo. However, unlike IP_6, CSNAP regulation of CRL-CSN binding has minimal effects on Cullin neddylation. Rather, CSNAP deletion results in generally increased binding of CSN to CRL components (from Cullins, adaptors to substrate receptors), with only three substrate receptors showing an opposite pattern. Proteomic analysis suggests that the altered CRL binding landscape correlates with altered ubiquitination of hundreds of cellular proteins in CSNAP deleted cells. More proteins exhibit increased ubiquitylation, indicating augmented CRL activity, which would be consistent with CSNAP augmenting CSN's binding and inhibition of CRL. Ubiquitylation targets regulated by CSNAP are enriched for known CRL substrates and proteins involved in cell cycle and apoptosis pathways. Consequently, CSNAP depleted cells are stalled in S-phase of the cell cycle and display more severe DNA damage with compromised recovery, suggesting a role for CSNAP in DNA repair. Given that CSN as a direct CRL neddylation regulator is being targeted for anticancer therapy (Schlierf et al. 2016), CSNAP could be a novel therapeutic venue to inhibit CRL-based oncogenic adaptations.

4.7 Conclusion and Future Perspectives

In the past few years, emerging studies have begun to unveil the structural mechanisms and physiologic implications of CRL regulation by CSN. Furthermore, a new protein (CSNAP) and a small molecule factor (IP_6) linked to CRL-CSN interaction have been identified. These new factors are likely involved in CSN regulation. The spatiotemporal aspects of their mechanisms of action await further investigation. Such insights could broaden our understanding on how CRL-CSN complex affinity is dynamically fine-tuned to initiate and terminate the CRL activity cycle. Finally, the high neddylation status in cancer cells and the ongoing success of neddylation inhibitors in clinical trials support that the newly identified CSN regulators bare therapeutic potential.

4.8 Note Added in Proof

During the final stage of preparing this Chapter, cryo-EM structures of CSN-CRL2 were published (Faull et al. 2019). Overall, this study showed that the CSN-CRL2 binding interface and segments undergoing binding-induced structural changes are similar to that found for other CSN-CRL complexes, suggesting a conserved CSN activation mechanism. Further, this study reported tight binding of CSN to unneddylated CRL, which re-raises the question of how CSN-CRL complex is dissociated after Cullin deneddylation.

Acknowledgments This work was supported by grants from the National Science Foundation of China (31872798 and 91853129 to F.R.), the Shenzhen Municipal Government (Y01226126, JCYJ20170412153517422, and JCYJ20170817104311912 to F.R.), and the Department of Science and Technology of Guangdong Province (2018A030313207 to F.R. and 2017B030301018 to Guangdong Provincial Key Laboratory of Cell Microenviroment and Disease Research); the author wishes to thank Dr. Qiuyu Fu for the assistance in refitting the cryo-EM maps.

References

Alpi AF, Echalier A (2017) ZOMES: the intriguing interplay of PCI complexes and the ubiquitin in protein homeostasis. Cell Death Dis 8:e3021

Angers S et al (2006) Molecular architecture and assembly of the DDB1-CUL4A ubiquitin ligase machinery. Nature 443:590–593

Asare Y et al (2017) Inhibition of atherogenesis by the COP9 signalosome subunit 5 in vivo. Proc Natl Acad Sci U S A 114:E2766–E2775

Bai J et al (2019) SCF(FBXO22) targets HDM2 for degradation and modulates breast cancer cell invasion and metastasis. Proc Natl Acad Sci U S A 116: 11754–11763

Bech-Otschir D et al (2001) COP9 signalosome-specific phosphorylation targets p53 to degradation by the ubiquitin system. EMBO J 20:1630–1639

Bennett EJ, Rush J, Gygi SP, Harper JW (2010) Dynamics of cullin-RING ubiquitin ligase network revealed by systematic quantitative proteomics. Cell 143:951–965

Berse M et al (2004) Ubiquitin-dependent degradation of Id1 and Id3 is mediated by the COP9 signalosome. J Mol Biol 343:361–370

Birol M et al (2014) Structural and biochemical characterization of the Cop9 signalosome CSN5/CSN6 heterodimer. PLoS One 9:e105688

Bondeson DP, Crews CM (2017) Targeted protein degradation by small molecules. Annu Rev Pharmacol Toxicol 57:107–123

Bornstein G, Ganoth D, Hershko A (2006) Regulation of neddylation and deneddylation of cullin1 in SCFSkp2 ubiquitin ligase by F-box protein and substrate. Proc Natl Acad Sci U S A 103:11515–11520

Castellana M et al (2014) Enzyme clustering accelerates processing of intermediates through metabolic channeling. Nat Biotechnol 32:1011–1018

Cavadini S et al (2016) Cullin-RING ubiquitin E3 ligase regulation by the COP9 signalosome. Nature 531: 598–603

Chakraborty A, Kim S, Snyder SH (2011) Inositol pyrophosphates as mammalian cell signals. Sci Signal 4:re1

Chamovitz DA et al (1996) The COP9 complex, a novel multisubunit nuclear regulator involved in light control of a plant developmental switch. Cell 86: 115–121

Cope GA, Deshaies RJ (2006) Targeted silencing of Jab1/Csn5 in human cells downregulates SCF activity through reduction of F-box protein levels. BMC Biochem 7:1

Cope GA et al (2002) Role of predicted metalloprotease motif of Jab1/Csn5 in cleavage of Nedd8 from Cul1. Science 298:608–611

Cornelius RJ et al (2018) Renal COP9 signalosome deficiency alters CUL3-KLHL3-WNK signaling pathway. J Am Soc Nephrol 29:2627–2640

Dessau M et al (2008) The Arabidopsis COP9 signalosome subunit 7 is a model PCI domain protein with subdomains involved in COP9 signalosome assembly. Plant Cell 20:2815–2834

Dubois EL, Gerber S, Kisselev A, Harel-Bellan A, Groisman R (2016) UV-dependent phosphorylation of COP9/signalosome in UV-induced apoptosis. Oncol Rep 35:3101–3105

Duda DM et al (2008) Structural insights into NEDD8 activation of cullin-RING ligases: conformational control of conjugation. Cell 134:995–1006

Echalier A et al (2013) Insights into the regulation of the human COP9 signalosome catalytic subunit, CSN5/Jab1. Proc Natl Acad Sci U S A 110:1273–1278

Emberley ED, Mosadeghi R, Deshaies RJ (2012) Deconjugation of Nedd8 from Cul1 is directly regulated by Skp1-F-box and substrate, and the COP9 signalosome inhibits deneddylated SCF by a noncatalytic mechanism. J Biol Chem 287:29679–29689

Enchev RI et al (2012) Structural basis for a reciprocal regulation between SCF and CSN. Cell Rep 2:616–627

Faull SV et al (2019) Structural basis of Cullin 2 RING E3 ligase regulation by the COP9 signalosome. Nat Commun 10:3814

Fernandez-Sanchez ME, Sechet E, Margottin-Goguet F, Rogge L, Bianchi E (2010) The human COP9 signalosome protects ubiquitin-conjugating enzyme 3 (UBC3/Cdc34) from beta-transducin repeat-containing protein (betaTrCP)-mediated degradation. J Biol Chem 285:17390–17397

Fischer ES et al (2011) The molecular basis of CRL4DDB2/CSA ubiquitin ligase architecture, targeting, and activation. Cell 147:1024–1039

Fuzesi-Levi MG et al (2014) Dynamic regulation of the COP9 signalosome in response to DNA damage. Mol Cell Biol 34:1066–1076

Fuzesi-Levi MG et al (2019) CSNAP, the smallest CSN subunit, modulates proteostasis through cullin-RING ubiquitin ligases. Cell Death Differ

Galan JM, Peter M (1999) Ubiquitin-dependent degradation of multiple F-box proteins by an autocatalytic mechanism. Proc Natl Acad Sci U S A 96:9124–9129

Goldenberg SJ et al (2004) Structure of the Cand1-Cul1-Roc1 complex reveals regulatory mechanisms for the assembly of the multisubunit cullin-dependent ubiquitin ligases. Cell 119:517–528

Harari-Steinberg O, Chamovitz DA (2004) The COP9 signalosome: mediating between kinase signaling and protein degradation. Curr Protein Pept Sci 5:185–189

He Q, Cheng P, He Q, Liu Y (2005) The COP9 signalosome regulates the Neurospora circadian clock by controlling the stability of the SCFFWD-1 complex. Genes Dev 19:1518–1531

Hetfeld BK et al (2005) The zinc finger of the CSN-associated deubiquitinating enzyme USP15 is essential to rescue the E3 ligase Rbx1. Curr Biol 15:1217–1221

Hofmann K, Bucher P (1998) The PCI domain: a common theme in three multiprotein complexes. Trends Biochem Sci 23:204–205

Huang X, Langelotz C, Hetfeld-Pechoc BK, Schwenk W, Dubiel W (2009) The COP9 signalosome mediates beta-catenin degradation by deneddylation and blocks adenomatous polyposis coli destruction via USP15. J Mol Biol 391:691–702

Kleiger G, Saha A, Lewis S, Kuhlman B, Deshaies RJ (2009a) Rapid E2-E3 assembly and disassembly enable processive ubiquitylation of cullin-RING ubiquitin ligase substrates. Cell 139:957–968

Kleiger G, Hao B, Mohl DA, Deshaies RJ (2009b) The acidic tail of the Cdc34 ubiquitin-conjugating enzyme functions in both binding to and catalysis with ubiquitin ligase SCFCdc4. J Biol Chem 284:36012–36023

Lee JH et al (2013) Crystal structure and versatile functional roles of the COP9 signalosome subunit 1. Proc Natl Acad Sci U S A 110:11845–11850

Lei D et al (2011) COP9 signalosome subunit 8 is required for postnatal hepatocyte survival and effective proliferation. Cell Death Differ 18:259–270

Li L et al (2014) Overactivated neddylation pathway as a therapeutic target in lung cancer. J Natl Cancer Inst 106:dju083

Lingaraju GM et al (2014) Crystal structure of the human COP9 signalosome. Nature 512:161–165

Liu X et al (2018) Cand1-mediated adaptive exchange mechanism enables variation in F-box protein expression. Mol Cell 69:773–786 e6

Lyapina S et al (2001) Promotion of NEDD-CUL1 conjugate cleavage by COP9 signalosome. Science 292:1382–1385

Lydeard JR, Schulman BA, Harper JW (2013) Building and remodelling Cullin-RING E3 ubiquitin ligases. EMBO Rep 14:1050–1061

Lykke-Andersen K et al (2003) Disruption of the COP9 signalosome Csn2 subunit in mice causes deficient cell proliferation, accumulation of p53 and cyclin E, and early embryonic death. Mol Cell Biol 23:6790–6797

Meir M et al (2015) The COP9 signalosome is vital for timely repair of DNA double-strand breaks. Nucleic Acids Res 43:4517–4530

Menon S et al (2007) COP9 signalosome subunit 8 is essential for peripheral T cell homeostasis and antigen receptor-induced entry into the cell cycle from quiescence. Nat Immunol 8:1236–1245

Mosadeghi R et al (2016) Structural and kinetic analysis of the COP9-Signalosome activation and the cullin-RING ubiquitin ligase deneddylation cycle. Elife 5

Petroski MD, Deshaies RJ (2005) Function and regulation of cullin-RING ubiquitin ligases. Nat Rev Mol Cell Biol 6:9–20

Pierce NW et al (2013) Cand1 promotes assembly of new SCF complexes through dynamic exchange of F box proteins. Cell 153:206–215

Pintard L et al (2003) Neddylation and deneddylation of CUL-3 is required to target MEI-1/Katanin for degradation at the meiosis-to-mitosis transition in C. elegans. Curr Biol 13:911–921

Rao F et al (2014a) Inositol hexakisphosphate kinase-1 mediates assembly/disassembly of the CRL4-signalosome complex to regulate DNA repair and cell death. Proc Natl Acad Sci U S A 111:16005–16010

Rao F et al (2014b) Inositol pyrophosphates mediate the DNA-PK/ATM-p53 cell death pathway by regulating CK2 phosphorylation of Tti1/Tel2. Mol Cell 54: 119–132

Reitsma JM et al (2017) Composition and regulation of the cellular repertoire of SCF ubiquitin ligases. Cell 171:1326–1339 e14

Scherer PC et al (2016) Inositol hexakisphosphate (IP6) generated by IP5K mediates cullin-COP9 signalosome interactions and CRL function. Proc Natl Acad Sci U S A 113:3503–3508

Schlierf A et al (2016) Targeted inhibition of the COP9 signalosome for treatment of cancer. Nat Commun 7:13166

Schwechheimer C et al (2001) Interactions of the COP9 signalosome with the E3 ubiquitin ligase SCFTIRI in mediating auxin response. Science 292: 1379–1382

Schweitzer K, Naumann M (2015) CSN-associated USP48 confers stability to nuclear NF-kappaB/RelA by trimming K48-linked Ub-chains. Biochim Biophys Acta 1853:453–469

Schweitzer K, Bozko PM, Dubiel W, Naumann M (2007) CSN controls NF-kappaB by deubiquitinylation of IkappaBalpha. EMBO J 26:1532–1541

Scott DC et al (2016) Two distinct types of E3 ligases work in unison to regulate substrate ubiquitylation. Cell 166:1198–1214 e24

Scrima A et al (2011) Detecting UV-lesions in the genome: the modular CRL4 ubiquitin ligase does it best! FEBS Lett 585:2818–2825

Seeger M et al (1998) A novel protein complex involved in signal transduction possessing similarities to 26S proteasome subunits. FASEB J 12:469–478

Sekeres MA, Fram RJ, Hua Z, Ades L (2018) Phase 3 study of first line pevonedistat (PEV) + azacitidine (AZA) versus single-agent AZA in patients with higher-risk myelodysplastic syndromes (HR MDS), chronic myelomonocytic leukemia (CMML) or low-blast acute myelogenous leukemia (AML). J Clin Oncol 36

Sharon M et al (2009) Symmetrical modularity of the COP9 signalosome complex suggests its multifunctionality. Structure 17:31–40

Shears SB (2018) Intimate connections: Inositol pyrophosphates at the interface of metabolic regulation and cell signaling. J Cell Physiol 233:1897–1912

Skaar JR, Pagan JK, Pagano M (2014) SCF ubiquitin ligase-targeted therapies. Nat Rev Drug Discov 13:889–903

Sone T, Saeki Y, Toh-e A, Yokosawa H (2004) Sem1p is a novel subunit of the 26 S proteasome from Saccharomyces cerevisiae. J Biol Chem 279:28807–28816

Soucy TA et al (2009) An inhibitor of NEDD8-activating enzyme as a new approach to treat cancer. Nature 458:732–736

Su H, Li F, Ranek MJ, Wei N, Wang X (2011a) COP9 signalosome regulates autophagosome maturation. Circulation 124:2117–2128
Su H et al (2011b) Perturbation of cullin deneddylation via conditional Csn8 ablation impairs the ubiquitin-proteasome system and causes cardiomyocyte necrosis and dilated cardiomyopathy in mice. Circ Res 108:40–50
Sun Y, Wilson MP, Majerus PW (2002) Inositol 1,3,4-trisphosphate 5/6-kinase associates with the COP9 signalosome by binding to CSN1. J Biol Chem 277:45759–45764
Tomoda K, Yoneda-Kato N, Fukumoto A, Yamanaka S, Kato JY (2004) Multiple functions of Jab1 are required for early embryonic development and growth potential in mice. J Biol Chem 279:43013–43018
Uhle S et al (2003) Protein kinase CK2 and protein kinase D are associated with the COP9 signalosome. EMBO J 22:1302–1312
Verbsky JW, Chang SC, Wilson MP, Mochizuki Y, Majerus PW (2005) The pathway for the production of inositol hexakisphosphate in human cells. J Biol Chem 280:1911–1920
Vittal V, Stewart MD, Brzovic PS, Klevit RE (2015) Regulating the regulators: recent revelations in the control of E3 ubiquitin ligases. J Biol Chem 290: 21244–21251
Wang J et al (2010) Role of individual subunits of the Neurospora crassa CSN complex in regulation of deneddylation and stability of cullin proteins. PLoS Genet 6:e1001232
Wee S, Geyer RK, Toda T, Wolf DA (2005) CSN facilitates Cullin-RING ubiquitin ligase function by counteracting autocatalytic adapter instability. Nat Cell Biol 7:387–391
Wei N, Chamovitz DA, Deng XW (1994) Arabidopsis COP9 is a component of a novel signaling complex mediating light control of development. Cell 78:117–124
Wilson MP, Sun Y, Cao L, Majerus PW (2001) Inositol 1,3,4-trisphosphate 5/6-kinase is a protein kinase that phosphorylates the transcription factors c-Jun and ATF-2. J Biol Chem 276:40998–41004
Wolf DA, Zhou C, Wee S (2003) The COP9 signalosome: an assembly and maintenance platform for cullin ubiquitin ligases? Nat Cell Biol 5:1029–1033
Worden EJ, Padovani C, Martin A (2014) Structure of the Rpn11-Rpn8 dimer reveals mechanisms of substrate deubiquitination during proteasomal degradation. Nat Struct Mol Biol 21:220–227
Wu K, Chen A, Pan ZQ (2000) Conjugation of Nedd8 to CUL1 enhances the ability of the ROC1-CUL1 complex to promote ubiquitin polymerization. J Biol Chem 275:32317–32324
Wu JT, Lin HC, Hu YC, Chien CT (2005) Neddylation and deneddylation regulate Cul1 and Cul3 protein accumulation. Nat Cell Biol 7:1014–1020
Wu JT, Chan YR, Chien CT (2006) Protection of cullin-RING E3 ligases by CSN-UBP12. Trends Cell Biol 16:362–369
Wu K, Kovacev J, Pan ZQ (2010) Priming and extending: a UbcH5/Cdc34 E2 handoff mechanism for polyubiquitination on a SCF substrate. Mol Cell 37:784–796
Zhao Y, Sun Y (2013) Cullin-RING ligases as attractive anti-cancer targets. Curr Pharm Des 19: 3215–3225
Zheng N, Shabek N (2017) Ubiquitin ligases: structure, function, and regulation. Annu Rev Biochem 86:129–157
Zheng N et al (2002) Structure of the Cul1-Rbx1-Skp1-F boxSkp2 SCF ubiquitin ligase complex. Nature 416:703–709
Zhou P, Howley PM (1998) Ubiquitination and degradation of the substrate recognition subunits of SCF ubiquitin-protein ligases. Mol Cell 2:571–580
Zhou C et al (2003) Fission yeast COP9/signalosome suppresses cullin activity through recruitment of the deubiquitylating enzyme Ubp12p. Mol Cell 11: 927–938
Zimmerman ES, Schulman BA, Zheng N (2010) Structural assembly of cullin-RING ubiquitin ligase complexes. Curr Opin Struct Biol 20:714–721
Zou Q et al (2014) USP15 stabilizes MDM2 to mediate cancer-cell survival and inhibit antitumor T cell responses. Nat Immunol 15:562–570

Coordinated Actions Between p97 and Cullin-RING Ubiquitin Ligases for Protein Degradation

5

Wenbo Shi, Ran Ding, Pei Pei Zhou, Yuan Fang, Ruixi Wan, Yilin Chen, and Jianping Jin

Abstract

The cullin-RING ubiquitin ligases comprise the largest subfamily of ubiquitin ligases. They control ubiquitylation and degradation of a large number of protein substrates in eukaryotes. p97 is an ATPase domain-containing protein segregase. It plays essential roles in post-ubiquitylational events in the ubiquitin-proteasome pathway. Together with its cofactors, p97 collaborates with ubiquitin ligases to extract ubiquitylated substrates and deliver them to the proteasome for proteolysis. Here we review the structure, functions, and mechanisms of p97 in cellular protein degradation in coordination with its cofactors and the cullin-RING ubiquitin ligases.

Keywords

p97 · Cullin-RING ubiquitin ligase · Ubiquitylation · Protein degradation

Abbreviations

CMG	Cdc45-MCM-GINS helicase complex
CRL	Cullin-RING ubiquitin ligases
Cul1-9	Cullin 1-9
DUB	Deubiquitinating enzyme
ER	Endoplasmic reticulum
ERAD	Endoplasmic reticulum-associated degradation
HECT	Homologous to E6-AP carboxyl terminus
HIV	Human immunodeficiency virus
IL-1β	Interleukin-1β
MAD	Mitochondria-associated degradation
mitoTAD	Mitochondrial protein translocation-associated degradation
Pol II	RNA polymerase II
PTMs	Posttranslational modifications
PUB	PNGase/UBA or UBX-containing proteins
PUL	PLAP, Ufd3p, and Lub1p
RBR	RING-Between-RING
RCR	RING-Cys-relay
RING	Really Interesting New Gene
SCF	Skp1-Cullin1-F-box ubiquitin ligase
TNF-α	Tumor necrosis factor alpha
UBX	Ubiquitin regulatory X
UBXL	UBX-like
UPS	Ubiquitin-proteasome system
VBM	VCP-binding motif
VIM	VCP-interacting motif

Ubiquitylation is a posttranslational modification which can regulate almost every biological process in eukaryotes. Through an E1 (ubiquitin-

Wenbo Shi, Ran Ding, Pei Pei Zhou, Yuan Fang and Ruixi Wan contributed equally with all other contributors.

W. Shi · R. Ding · P. P. Zhou · Y. Fang · R. Wan · Y. Chen · J. Jin (✉)
Life Science Institute, Zhejiang University, HangZhou, China
e-mail: jianping_jin@zju.edu.cn

Y. Sun et al. (eds.), *Cullin-RING Ligases and Protein Neddylation*, Advances in Experimental Medicine and Biology 1217, https://doi.org/10.1007/978-981-15-1025-0_5

activating enzyme)-E2 (ubiquitin-conjugating enzyme)-E3 (ubiquitin ligase) enzymatic cascade, ubiquitin can be conjugated to protein substrates via a peptide bond between the last glycine residue of ubiquitin and an amino group of residues (either lysines or the N-terminal methionine) in protein substrates (Swatek and Komander 2016). The protein ubiquitylation machinery can not only couple single ubiquitin but also synthesize a chain of ubiquitin molecules to its substrates. Different conjugation formats of ubiquitin molecules to substrates result in distinctive destinies of ubiquitylated proteins. The main consequence of protein ubiquitylation is proteolysis by the 26S proteasome.

Protein ubiquitylation is a very specific process. The specificity is mainly determined by ubiquitin ligases which recognize substrates specifically. Ubiquitin ligases can be classified into four families: HECT (Homologous to E6-AP carboxyl terminus), RING finger (Really Interesting New Gene), RBR (RING-Between-RING), and RCR (RING-Cys-relay) domain ubiquitin ligases (Fig. 5.1) (Zheng and Shabek 2017; Huibregtse et al. 1995; Wenzel et al. 2011; Dove and Klevit 2017; Pao et al. 2018). RING finger ubiquitin ligases lack a catalytic cysteine and thus rely on activated E2 enzymes (E2~Ub) to conjugate ubiquitin directly to substrates (Zheng and Shabek 2017). However, HECT, RBR, and RCR ubiquitin ligases all have at least one catalytic cysteine that forms a covalent E3~Ub thioester intermediate with the C-terminal glycine residue of ubiquitin, prior to ubiquitin transfer to substrates (Zheng and Shabek 2017; Huibregtse et al. 1995; Wenzel et al. 2011; Dove and Klevit 2017; Pao et al. 2018).

5.1 Cullin-RING Ubiquitin Ligases

The RING finger family is the largest class of ubiquitin ligases. Among them, the cullin-RING ubiquitin ligases (CRLs) comprise the largest E3 subfamily which controls many biological events in eukaryotes. Hundreds of cellular proteins have been identified as ubiquitylation substrates of CRLs. CRLs contain multiple subunits, among which cullins function as scaffold proteins. Human genome encodes eight cullins (Cul1, 2, 3, 4A, 4B, 5, 7, 9) (Petroski and Deshaies 2005; Nguyen et al. 2017a; Genschik et al. 2013; Lee and Diehl 2014; Lee and Zhou 2007; Okumura et al. 2016; Sarikas et al. 2011; Cai and Yang 2016; Kamura et al. 2004; Sun and Li 2013). Each cullin forms their own modular ubiquitin ligases with different families of substrate adaptors which govern substrate specificity (Fig. 5.2). However, there is significant redundancy between Cul4A and Cul4B in terms of substrate adaptors, substrates, and biological activities. Moreover, Cul9 belongs to the RBR ubiquitin ligase and might form an atypical CRL ubiquitin ligase (Dove and Klevit 2017). Two E2 recruiting RING finger proteins are involved in complex composition of CRLs, including RBX1 (also called ROC1, HRT1), which is shared among CRL1, 2, 3, 4A, 4B, and 7 (Petroski and Deshaies 2005; Nguyen et al. 2017a), and RBX2 (also called SAG, RNF7) which interacts with Cul5 specifically (Kamura et al. 2004; Sun and Li 2013). The archetypal CRL is the SCF ubiquitin ligases (also called CRL1) which employ F-box proteins as substrate adaptors. Human cells express at least 69 F-box proteins

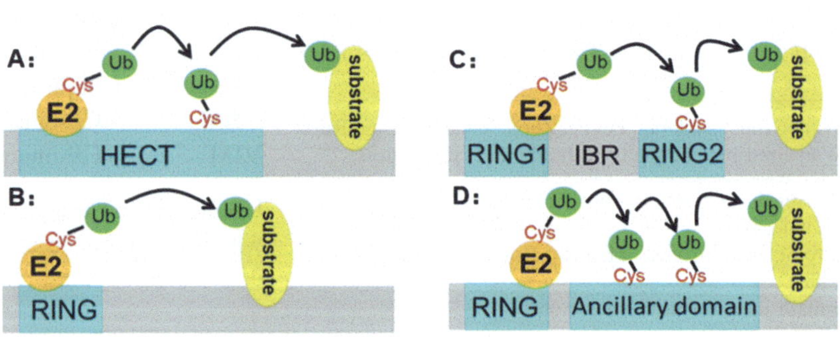

Fig. 5.1 Classification of ubiquitin ligases. Ubiquitin ligases can be classified into four families, including (**a**) the HECT domain ubiquitin ligase, (**b**) the RING finger ubiquitin ligase, (**c**) the RBR ubiquitin ligase, and (**d**) the RCR ubiquitin ligase

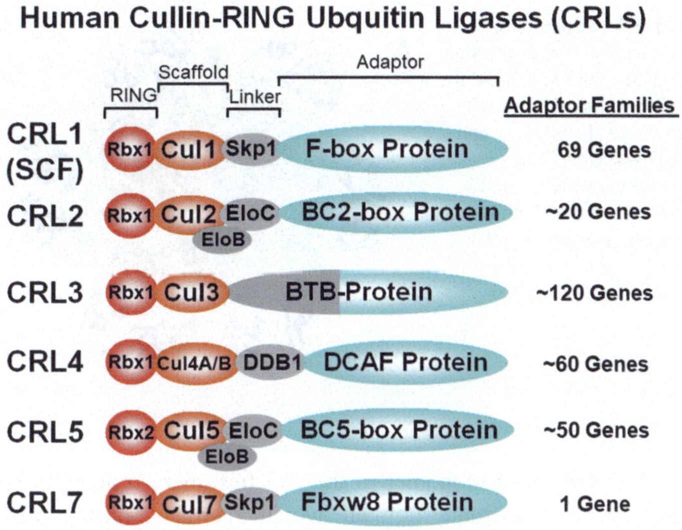

Fig. 5.2 The cullin-RING ubiquitin ligases (CRLs). Each CRL has its own unique composition of modular ubiquitin ligases

which regulate various biological activities (Jin et al. 2004).

The activity of CRLs is dynamically and tightly regulated. First, components of CRLs do not stay together all the time. Cullins and RBX usually bind with each other more stably. However, substrate adaptors do not always associate with cullin-RBX heterodimers. Moreover, recent studies indicated CAND1 and CAND2 are key to CRL assembly and disassembly (Reitsma et al. 2017; Liu et al. 2018; Pierce et al. 2013). In the case of F-box proteins, they interact with Skp1 more tightly. However, assembly rates of F-box proteins into SCF holoenzymes vary from 0% to 70% (Reitsma et al. 2017). Some F-box proteins even form a distinctive ubiquitin ligase rather than the SCF complex (Saiga et al. 2009). Second, the activity of CRLs is regulated by Nedd8, a ubiquitin-like protein. Nedd8 is conjugated to cullins via its own E1-E2-E3 enzymatic cascade (Deshaies et al. 2010; Duda et al. 2011; Hori et al. 1999). Nedd8 can be removed by the Cop9 signalosome, a deneddylation enzyme (Wei and Deng 2003; Lydeard et al. 2013; Cope and Deshaies 2003). The balance between neddylation and deneddylation of cullins is important for optimal activities of CRLs when they are needed to promote ubiquitylation and turnover of their corresponding substrates. Untimely neddylation of cullins increases the risk of autoubiquitylation and degradation of substrate adaptors (He et al. 2005; Cope and Deshaies 2006).

5.2 p97 Is a Master Regulator of the Ubiquitin-Proteasome System (UPS)

Ubiquitylated substrates of CRLs are often delivered to the 26S proteasome for degradation. However, there is no indication that CRLs could directly contact the proteasome to present ubiquitylated substrates. Recent studies further indicated that p97 (also called VCP, Cdc48, CDC-48, or TER94 in different organisms) and its associated cofactors play critical roles in recognizing and delivering ubiquitylated substrates to the proteasome (van den Boom and Meyer 2018; Meyer et al. 2012; Stach and Freemont 2017; Yeung et al. 2008; Buchberger et al. 2015; Hänzelmann and Schindelin 2017; Barthelme and Sauer 2016).

p97 is an evolutionarily conserved ATPase of the AAA+ protein superfamily (van den Boom and Meyer 2018; Meyer et al. 2012; Stach and Freemont 2017; Yeung et al. 2008; Buchberger et al. 2015; Hänzelmann and Schindelin 2017; Barthelme and Sauer 2016). Structurally, it forms a ring-like hexamer (Fig. 5.3) (Banerjee

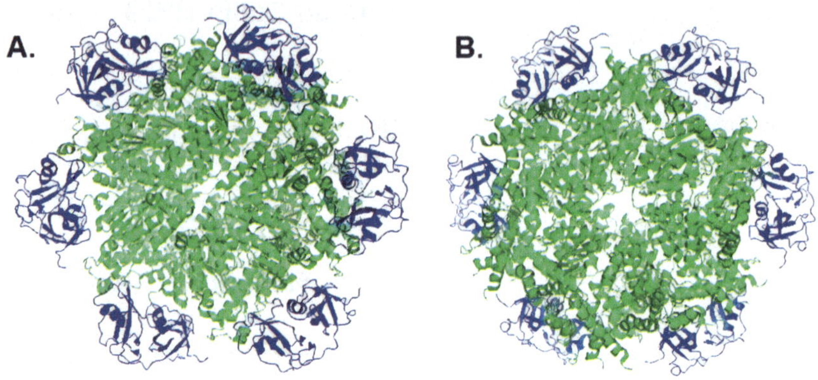

Fig. 5.3 Structure of human p97. The N-terminal domains are in blue and the ATPase domains are in green. (**a**) Front view and (**b**) back view. The structures were created using Pymol and amino acid coordinates of p97 published in Banerjee et al. (Banerjee et al. 2016)

et al. 2016). Each monomer of the p97 hexamer contains a N-terminal domain which recruits dozens of regulatory cofactors, two AAA+ ATPase domains (called D1 and D2) that ensure p97 to provide enough energy to carry out various activities, and a C-terminal tail ending with a HbYX motif (hydrophobic residue-tyrosine-X, X is a variable residue) which also interacts with a subset of p97 cofactors (Hänzelmann and Schindelin 2017). p97 is often considered as a protein segregase or unfoldase, referring to its biochemical property (van den Boom and Meyer 2018; Meyer et al. 2012; Stach and Freemont 2017; Yeung et al. 2008; Buchberger et al. 2015; Hänzelmann and Schindelin 2017; Barthelme and Sauer 2016).

p97 is widely expressed and very abundant in eukaryotes, where it plays multiple roles in various cellular activities. Cdc48, the yeast ortholog of human p97, was first identified to be involved in cell cycle progression among 18 cold-sensitive yeast mutants in a genetic screen (Moir et al. 1982). Mammalian p97 was then found to interact with clathrin and to modulate protein-protein interactions in membrane transport processes (Pleasure et al. 1993).

The function of p97 in the UPS was initially linked to the ERAD (endoplasmic reticulum-associated degradation) protein degradation pathway (Ye et al. 2001; Schuberth et al. 2004). Its role was then quickly expanded to the maintenance of intracellular proteostasis. Using its ATPase activity, p97, with the assistance of its cofactors, can extract ubiquitylated proteins from plasma membranes, organelles, or chromatin and then deliver them to the 26S proteasome for degradation (van den Boom and Meyer 2018; Meyer et al. 2012; Stach and Freemont 2017; Yeung et al. 2008; Buchberger et al. 2015; Hänzelmann and Schindelin 2017; Barthelme and Sauer 2016; Cooney et al. 2019; Twomey et al. 2019).

p97 consists of two ATPase domains and promotes post-ubiquitylation processing by acting as a protein segregase or unfoldase (van den Boom and Meyer 2018; Meyer et al. 2012; Stach and Freemont 2017; Yeung et al. 2008; Buchberger et al. 2015; Hänzelmann and Schindelin 2017; Barthelme and Sauer 2016; Cooney et al. 2019; Twomey et al. 2019). Bodnar et al. proposed a model to elucidate how p97 works together with its cofactors to process ubiquitylated substrates, based on their studies on Cdc48 and the Ufd1-Npl4 heterodimer in budding yeast (Bodnar and Rapoport 2017). In their model, the D1 domain of Cdc48 is in the ADP-bound state under the static condition. When ATP interacts with the D1 domain, the N-terminal domain of Cdc48 moves upward for binding to the Ufd1-Npl4 heterodimer. Ubiquitylated substrates then interact with the Ufd1-Npl4 heterodimer. This interaction triggers ATP hydrolysis in the D2 domain which uses the energy to generate a pulling force to move ubiquitylated protein substrates completely through the hexameric ring and to promote substrates unfolding. Subsequent ATP hydrolysis by the D1 ATPase domain provides the D1 domain with the energy to release substrates from the Cdc48-Ufd1-Npl4 core complex. Intriguingly, this release requires cooperation between Cdc48 and the deubiquitinating enzyme (DUB), Otu1, which trims polyubiquitin chain to an oligoubiquitin. Together, this proposal provided a general mechanism by which Cdc48 or its mammalian ortholog p97 works together with their cofactors to extract ubiquitylated substrates from membranes. However, it is unclear whether this model fits with actions of p97 in processing non-membranous ubiquitylated substrates.

Research efforts on p97 in the past two decades have built a complex p97 signaling network. Under different intracellular states and with external stimuli, p97 can accurately and timely mobilize different effective cofactors to mediate different signaling pathways. However, there are still many questions remained to be answered and functional mechanisms needed to be explored in the future.

5.3 Cofactors of p97

p97 typically requires different types of cofactors to execute its diverse roles in many critical signaling pathways (Stach and Freemont 2017; Yeung et al. 2008; Buchberger et al. 2015; Hänzelmann and Schindelin 2017; Barthelme and Sauer 2016). Thus far, dozens of cofactors have been reported in human (Table 5.1). Most of the cofactors bind to the N-terminus of p97, via their ubiquitin-related UBX (ubiquitin regulatory X), UBXL (UBX-like), VIM (VCP-interacting motif), VBM (VCP-binding motif), and SHP-binding (SHP box/BS1, also called binding site 1) domains (Stach and Freemont 2017; Yeung et al. 2008; Buchberger et al. 2015; Hänzelmann and Schindelin 2017; Barthelme and Sauer 2016). However, proteins with either PUB (PNGase/UBA or UBX-containing proteins, also known as PUG) or PUL (PLAP, Ufd3p, and Lub1p) domains can interact with the C-terminal HbYX motif of p97 (Stach and Freemont 2017; Yeung et al. 2008; Buchberger et al. 2015; Hänzelmann and Schindelin 2017; Barthelme and Sauer 2016). These proteins provide additional groups of p97 cofactors.

In order to reveal their interactions with either N-terminus or C-terminus of p97, major efforts have been made to solve structures of p97-interactive domains over the last decade (Hänzelmann et al. 2011; Kloppsteck et al. 2012; Schuberth and Buchberger 2008; Bodnar et al. 2018; Rao et al. 2017; Kim et al. 2014; Kim and Kim 2014; Lim et al. 2016; Liu et al. 2013; Hänzelmann and Schindelin 2011; Stapf et al. 2011; Huang et al. 2019; Greenblatt et al. 2011; Suzuki et al. 2001; Schaeffer et al. 2014; Elliott et al. 2014; Allen et al. 2006; Madsen et al. 2009; Qiu et al. 2010). These structural and biochemical studies have contributed significantly to our understanding of p97's biological functions and its regulations, which are summarized below.

5.3.1 The UBX Domain-Containing Cofactors

The UBX domain comprises ~80 amino acids with a ubiquitin-like fold (Hänzelmann et al. 2011; Kloppsteck et al. 2012; Schuberth and Buchberger 2008; Yi and Kaler 2018). It is the most common p97-interactive domain among p97 cofactors. Thus far, 14 UBX-containing cofactors have been identified in human genome (Hänzelmann et al. 2011; Kloppsteck et al. 2012; Schuberth and Buchberger 2008; Yi and Kaler 2018). Some of these cofactors also encode a UBA (ubiquitin-associating) domain which binds to polyubiquitin chains. The UBX domain interacts with the N-terminus of p97 via a conserved FPR motif. This signature motif is interpolated in the hydrophobic pocket formed between two subdomains of the N-terminal part of p97. A good example of the UBX-containing cofactors is UBXD7 which has been extensively studied in terms of its biological roles and underlying mechanisms (Alexandru et al. 2008; den Besten et al. 2012; Bandau et al. 2012) and will be discussed in detail in Sects. 5.6 and 5.8.3.

Table 5.1 Cofactors of p97 in human

Domain	Name	Gene ID
UBX	ASPSCR1 (TUG)	79058
	UBXD1	80700
	UBXD2	23190
	UBXD3	127733
	UBXD4	165324
	UBXD5	91544
	UBXD6	7993
	UBXD7	26043
	UBXD8	23197
	SAKS1	51035
	p47	55968
	p37	137886
	FAF1	11124
UBXL	ASPSCR1 (TUG)	79058
	VCIP135	80124
	YOD1	55432
	NPL4	55666
VIM	SVIP	258010
	gp78 (AMFR)	267
	UBXD1	80700
	ANKZF1 (Vms1)	55139
	AIRAPL (ZFAND2B)	130617
	VIMP (SELENOS, SELS)	55829
VBM	UBE4B (Ufd2)	10277
	HRD1 (SYVN1)	84447
	NUB1L (NUB1)	51667
	RHBDL4 (RHBDD1)	84236
	ATXN3	4287
SHP	Derlin 1	79139
	Derlin 2	51009
	DVC1 (Spartan)	83932
	Ufd1	7353
	p47	55968
	p37	137886
	UBXD4	165324
	UBXD5	91544
	ASPSCR1 (TUG)	79058
PUB	PNGase	55768
	RNF31 (HOIP)	55072
	UBXD1	80700
PUL	PLAA	9373
Others	UBE4A	9354
	ZFAND1	79752
	HDAC6	10013

UBXD1, although having a UBX domain, binds to p97 via its two other p97-interactive domains (Kern et al. 2009).

5.3.2 The UBXL Domain Cofactors

Human genome encodes only four UBXL domain-containing proteins. Interestingly, three of them have a protease domain, although the MPN domain on Npl4 is an inactive one which might contribute to ubiquitin binding (Hänzelmann and Schindelin 2017; Hänzelmann et al. 2011; Bodnar et al. 2018; Kim et al. 2014; Kim and Kim 2014). The UBXL domain adopts a similar structure as the UBX domain, although these two domains share low sequence identity. Structurally, both UBXL and UBX domains bind to the N-terminal part of p97 in a similar fashion (Hänzelmann and Schindelin 2017; Hänzelmann et al. 2011). In analogy to the FPR motif in the UBX proteins, the UBXL domain of yeast OTU1 (homolog of human YOD1) also features a YPP motif that inserts into the hydrophobic pocket between two subdomains of the N-terminal part of p97 (Kim et al. 2014; Kim and Kim 2014); however, the UBXL domain of Npl4 does not have such a motif, and it binds to p97 in a different manner (Bodnar et al. 2018). Npl4 forms a heterodimer with Ufd1. Together they regulate the majority of p97's functions in the UPS pathway and are therefore considered as the core cofactors of p97.

5.3.3 The VIM Domain Cofactors

The VIM domain has a signature motif with a conserved $Rx_2h_3AAx_2Rh$ sequence where h represents a hydrophobic residue and x is any amino acids (Liu et al. 2013; Hänzelmann and Schindelin 2011; Stapf et al. 2011). Structurally, the VIM domain adopts an α-helical motif which also interacts with the hydrophobic cleft between the two subdomains of the N-terminal part of p97. Two conserved arginine residues in the VIM domain are important for the interaction between the VIM domain and p97 via hydrogen bonds and electrostatic interactions. Thus far, the VIM domain has been found in six human proteins (Hänzelmann and Schindelin 2017), including gp78, one of the main ubiquitin ligases involved in the ERAD pathway (Joshi et al. 2017; Zhang et al. 2015a; Fang et al. 2001).

5.3.4 The VBM Domain Cofactors

The VBM domain contains a conserved motif of EhR_4Lxh2 sequences where h and x represent a hydrophobic residue and any amino acids, respectively (Buchberger et al. 2015; Hänzelmann et al. 2011). Similar to the VIM, the VBM domain also adopts an α-helical conformation. The complex structure between the VBM domain of human RHBDL4, an important protein in the ERAD pathway, and the N-terminal domain of p97 demonstrated similar interactions of VBM and VIM with p97 (Lim et al. 2016). However, the α-helices of VBM and VIM run in opposite directions in the complex with the N-terminal domain of p97. Thus far, five VBM-carrying proteins have been identified in human cells. Of notes, these cofactors are either ubiquitin ligases, DUB, or ubiquitin-binding proteins, including HRD1, another important ubiquitin ligase in the ERAD pathway (Gardner et al. 2001; Hampton 2002).

5.3.5 The SHP-Binding Motif Cofactors

The SHP-binding motif has been found as an extra p97-binding motif in nine human proteins, including five UBX-containing cofactors (Hänzelmann and Schindelin 2017; Hänzelmann et al. 2011; Greenblatt et al. 2011). The SHP-binding motif has a conserved sequence of $hx_{1-2}F/Wx_{0-1}GxGx_2L$ with two invariable glycine residues and a well-conserved aromatic residue (x is any amino acid and h represents a hydrophobic residue). Structurally, the SHP-binding motif contacts a different site from other p97-binding motifs on the N-terminal domain of p97. Ufd1, a component of the Ufd1-Npl4 heterodimer, contains a SHP-binding motif (Bodnar et al. 2018). Therefore, Ufd1 and Npl4 interact with the N-terminal domain of p97 at two separate sites (Bodnar et al. 2018), indicating that the heterodimer has much higher binding affinity to p97 than each subunit alone.

5.3.6 The PUB Domain Cofactors

The PUB domain is one of the only two domains found to interact with the C-terminus of p97 (Suzuki et al. 2001; Schaeffer et al. 2014; Elliott et al. 2014; Allen et al. 2006; Madsen et al. 2009). Thus far, only four human proteins have been identified to bear the PUB domain, including HOIP (also called RNF31) which synthesizes linear polyubiquitin chains and plays important roles in the NF-κB activation (Hänzelmann and Schindelin 2017; Schaeffer et al. 2014). The interaction between the PUB domain and p97 is regulated by electrostatic and hydrophobic contacts, especially with the hydrophobic Leu-804 and the aromatic side chain of Tyr-805 of p97 (Schaeffer et al. 2014). The side chains of these two residues point into a hydrophobic patch of the PUB domain. UBXD1, one of the UBX domain-containing p97 cofactors, has a PUB domain not far away from its UBX domain. Notably, the PUB and UBX domains could associate with p97 simultaneously, since these two domains bind to two separate parts of p97 (Kern et al. 2009).

5.3.7 The PUL Domain Cofactor

The PUL domain (PLAP, Ufd3p, and Lub1p) is the other domain that interacts with the extreme C-terminus of p97 (Qiu et al. 2010). Thus far, PLAA is the only human p97 cofactor with the PUL domain. Yeast Ufd3 (also called Doa1), the ortholog of human PLAA, employs its PUL domain and PFU, a ubiquitin-binding domain, to foster the formation of a Ufd3-Cdc48-ubiquitin ternary complex (Nishimasu et al. 2010; Mullally et al. 2006). Phosphorylation at Tyr-805 of p97 blocks the interaction between Ufd3 and p97, further supporting that Ufd3 binds to the extreme C-terminus of p97 (Li et al. 2008; Lavoie et al. 2000; Zhao et al. 2007).

5.3.8 Other p97 Cofactors

Thus far, dozens of p97-interacting proteins have been identified; however, we still don't comprehend the interactions of many cofactors with p97 or its counterpart, Cdc48 in yeast. Cdc48 interacts with Ufd2 via the last ten residues at its C-terminus. It is still not fully understood how Ufd2 interacts with Cdc48, although the G274D and C385Y mutations of Ufd2 could not bind to Cdc48 (Baek et al. 2011). UBE4B, a Ufd2 homolog in human, contains a VBM domain at its N-terminus (Morreale et al. 2009); however, no obvious p97-binding domain or motif has been identified in UBE4A, another Ufd2 homolog in human. Yeast Cuz1 protein and its human ortholog ZFAND1 contain a ubiquitin-like domain which interacts with p97 and modulates p97's function in the UPS pathway (Sá-Moura et al. 2013). This unique ubiquitin-like domain is required for Cuz1 binding to the N-terminal domain of Cdc48. Interestingly, ZFAND1 could recruits p97 and the proteasome to clear arsenite-induced stress granules (Turakhiya et al. 2018), indicating p97 and its cofactors play important roles in phase separation.

5.4 Regulation of the Interactions Between p97 and Its Cofactors by Posttranslational Modifications

The interactions between p97 and its cofactors could be regulated by posttranslational modifications (PTMs). Several comprehensive proteomic studies revealed that p97 is extensively modified by phosphorylation at 68 sites, ubiquitylation at 38 lysine sites, acetylation at 24 lysine sites, sumoylation at 7 lysine sites, and methylation at 7 lysine and arginine sites (PhosphoSitePlus, http://www.phosphosite.org) (Hornbeck et al. 2015). However, the biological functions of these PTMs and relevant enzymes in charge of these PTMs are largely unknown. Phosphorylation of Tyr805 of p97 by the c-Src kinase inhibits the interaction of p97 with either PUB or PUL domain (Li et al. 2008). Moreover, several phosphorylation sites have been identified in the N-terminal domain of p97 as well. Their phosphorylation potentially blocks the association of p97 with its cofactors. For instance, Tyr110 and Tyr143 are key residues for interactions of p97 with either UBX or UBXL domain of cofactors. Their phosphorylation potentially shuts down the interaction of p97 with either UBX or UBXL domain proteins. Conversely, phosphorylation in the p97-interacting domains of cofactors could also affect their interactions with p97. For example, both serine residues at Ser229 and Ser231 in the SHP domain of Ufd1 are phosphorylated and potentially regulate the interaction between Ufd1 and p97 (PhosphoSitePlus, http://www.phosphosite.org) (Hornbeck et al. 2015).

Hitherto, it is unclear how ubiquitylation, acetylation, sumoylation, and methylation of either p97 or cofactors could influence the interactions between p97 and its cofactors.

5.5 Synergistic Effect of p97 and Cofactors for Substrate Recognition

Two models have been proposed to explain the mechanism by which p97 recognizes its substrates (Hänzelmann et al. 2011). One model is that cofactors bind to ubiquitylated substrates via the UBA domain or other substrate-associating domain. The interactions between cofactors and substrates increase cofactors' affinity to p97. Typical examples are UBXD7, FAF1, and SAKS1 (Hänzelmann et al. 2011; McNeill et al. 2004; LaLonde and Bretscher 2011). Before binding to ubiquitylated substrates, their UBA domains interact with the intramolecular UBX domains; therefore, their UBX domains are unavailable for p97 interaction. Binding of ubiquitylated substrates to cofactors could induce a conformational change in cofactors that enhances their affinity to p97. In addition, binding of p97 to the Ufd1-Npl4 heterodimer could promote the interaction of p97 with FAF1 and UBXD7 (Hänzelmann et al. 2011), indicating a hierarchy in cofactor binding to p97 as well. The other model is that cofactors interact with p97 and then enhance their affinities to ubiquitylated substrates. One example is p47. Indeed, in a GST pull-down experiment, the p97-p47 complex has much higher affinity to GST-Ub than p97 or p47 alone (Meyer et al. 2002).

5.6 The Interactions Between p97 and CRL Ubiquitin Ligases

A breakthrough of p97's cellular functions in the ubiquitin-proteasome pathway came from a systematic study in Ray Deshaies' laboratory (Alexandru et al. 2008). Using a mass spectrometry approach, they found some p97 cofactors associate with large numbers of ubiquitin ligases, especially CRLs. These cofactors typically contain both UBX and UBA domains. Of them, UBXD7 is particularly interesting, because it appears to bind many components of CRL1-4, including Cullin1-4, RBX1, linker proteins, and many substrate adaptors (Alexandru et al. 2008). Moreover, they showed that UBXD7 regulates HIF1α proteolysis in coordination with the $CRL2^{VHL}$ ubiquitin ligase. UBXD7 contains four domains, UBA, UAS (ubiquitin-associated), UIM (ubiquitin-interacting motif), and UBX. Importantly, UBXD7 only interacts with the active (neddylated) form of cullins via its conserved UIM domain. Structurally, the UIM domain contacts the hydrophobic patch centering Leu8, Ile44, and Val70 residues of NEDD8 (den Besten et al. 2012; Bandau et al. 2012). Additionally, UBXD7 interacts with ubiquitylated protein substrates via its UBA domain and with the N-terminal domain of p97 using its UBX domain. Therefore, there is a coordinated action among the three domains of UBXD7.

Similarly, Ubx5, the ortholog of UBXD7 in yeast, also contains four domains, UBA, UAS, UIM, and UBX. Together with the Cdc48-Ufd1-Npl4 protein complex, Ubx5 mediates the degradation of Rpb1, the largest subunit of RNA Pol II, under UV irradiation (Verma et al. 2011). Ubx5 also directly interacts with Rub1 (an ortholog of human Nedd8 in yeast)-modified Cul3 and Cdc48 through its UIM domain and UBX domain, respectively (den Besten et al. 2012).

Taken together, the interaction between p97 and CRLs is mediated by UBXD7/Ubx5 docking on the active form (NEDD8- or Rub1-modified) of cullins through its UIM motif.

Thus far, there is no evidence to show either UBXD7 or Ubx5 regulates degradation of all CRL substrates. In some cases, the p97-Ufd1-Npl4 core complex is required, but no other p97 cofactors are necessary. Besides UBXD7, other UBA- and UBX-containing proteins also interact with certain CRL components (Alexandru et al. 2008; Hu et al. 2017; Zhang et al. 2012; Bennett et al. 2010). How these cofactors regulate substrate turnover of CRLs remains largely unknown. In one case, FAF1 was found to potentiate $SCF^{\beta\text{-TrCP}}$-controlled β-catenin ubiquitylation and turnover, because it partially promotes the interaction between β-catenin and β-TrCP, the F-box substrate adaptor in the $SCF^{\beta\text{-TrCP}}$ E3 complex (Zhang et al. 2012). In another case, UBXN-3, the ortholog of human FAF1 in both *Caenorhabditis elegans and Xenopus laevis*, is involved in replisome disassembly via ubiquitylation and Cdc48-dependent removal of the Cdc45-MCM-GINS helicase complex (CMG) from chromatin (Sonneville et al. 2017). In this case, the $CRL2^{LRR\text{-}1}$ ubiquitin ligase ubiquitylates CMG, specifically the MCM7 subunit, to trigger disassembly of CMG in early embryos of *C. elegans* and *Xenopus* egg extracts. However, the detailed mechanism by which FAF1 coordinates with p97 in substrate turnover remains obscure.

Other non-UBA domain-containing p97 cofactors interact with cullins as well (Hu et al. 2017; Bennett et al. 2010). For instance, both ASPL (also called ASPSCR1) and UBXN11 (also called UBXD5) could associate with Cullin1. However, the biological significance of these interactions remains to be investigated. It is also unclear how these proteins interact with Cullin1 and how they determine the fate of ubiquitylated substrates of the SCF ubiquitin ligases.

5.7 Regulation of CRLs by p97

As we mentioned previously, the activities of CRLs are tightly regulated via cullin neddylation. Besides, CRLs are dynamically assembled and disassembled, providing additional layers of regulation. Intriguingly, yeast Cdc48 appears to play an important role in cadmium stress-induced disassembly of the SCF ubiquitin ligase complex (Yen et al. 2012). In this case, Cdc48 promotes dissociation of the F-box subunit, Met30, from the SCF^{Met30} ubiquitin ligase to inhibit substrate ubiquitylation. The unanswered question is whether any cofactors are required to assist Cdc48 to segregate the SCF^{Met30} ubiquitin ligase.

5.8 Promoting Degradation of CRLs' Substrates by p97 and Its Cofactors

As an ATPase-containing protein segregase or unfoldase, p97 governs turnover of an ever-growing number of ubiquitylated substrates in a lot of intracellular signaling pathways. Overall, it coordinates with diverse cofactors and a large number of cullin-RING ubiquitin ligases to help process ubiquitylated proteins for recycling or degradation by the proteasome in many cellular contexts from yeast to mammals.

5.8.1 p97 and CRLs Coordinate the ERAD Pathway

In the endoplasmic reticulum (ER), misfolded proteins are recognized and ubiquitylated by the protein quality control machinery and then degraded via the ERAD pathway. Thus far, most ERAD substrates are found to be ubiquitylated by non-cullin ubiquitin ligases. However, the SCF ubiquitin ligases could ubiquitylate misfolded glycoproteins of the ER for degradation. In these cases, two F-box proteins, FBXO2 (also called Fbx2 or Fbs1) and FBXO6 (also called Fbx6 or Fbs2), recognize the oligosaccharide sugar chains of misfolded glycoproteins as degron using their FBA domains and then ubiquitylate misfolded glycoproteins for proteolysis via the p97-dependent proteasome pathway (Yoshida et al. 2002, 2003, 2005; Liu et al. 2012). Moreover, the ER-bound SCF^{FBXO2} complex also ubiquitylates and promotes degradation of BACE1 (Gong et al. 2010), the major β-secretase which participates in the generation of β-amyloid peptides in Alzheimer's disease. However, it is unclear whether p97 or any of its cofactors is involved in degradation of BACE1. Equally unclear, no information is available whether any p97 cofactors are involved in glycoprotein degradation.

5.8.2 p97 and CRLs Mediate Mitochondria-Associated Degradation (MAD)

In a process named mitochondria-associated degradation, p97 also extracts ubiquitylated proteins from mitochondrial outer membrane. Cohen et al. demonstrated that the budding yeast SCF^{Mdm30} ubiquitin ligase ubiquitylates Fzo1, a mitochondrial transmembrane GTPase (Fritz et al. 2003; Cohen et al. 2008), and then coordinates with the Cdc48-Vms1-Npl4 complex to trigger Fzo1 protein degradation via the 26S proteasome (Heo et al. 2010). Intriguingly, in the MAD, instead of Ufd4, Vms1 forms a p97-Vms1-Npl4 complex to extract ubiquitylated substrates from mitochondrial outer membrane for proteasomal degradation, indicating the uniqueness in MAD regulation. Moreover, the SCF^{Fbw7} ubiquitin ligase was found to mediate apoptosis via ubiquitylation and destruction of Mcl1, a mitochondrial membrane protein (Inuzuka et al. 2011). p97 was also shown to be required for the proteasomal turnover of Mcl1 (Xu et al. 2011), although no connection between p97 and SCF^{Fbw7} has been established.

More recently, Ubx2, a yeast p97 cofactor in the ERAD pathway, was shown to recruit budding yeast Cdc48 for removal of arrested precursor proteins from the TOM complex, which is the channel on mitochondria membrane for import of precursor proteins into mitochondria, in a process named the mitochondrial protein translocation-associated degradation (mitoTAD) (Mårtensson et al. 2019). This quality control process is important to surveil the TOM complex to prevent congesting of the TOM channel with mitochondria precursor proteins under non-stress conditions. Protein ubiquitylation plays an important role in the mitoTAD, as evidenced by that ubiquitylated proteins were observed to accumulate at the TOM complex when both Ubx2 and Vms1 are deleted. However, whether CRLs or non-cullin ubiquitin ligases are involved in this process is unknown. Considering that UBXD8, the ortholog of Ubx2 in human, interacts with several components of CRL ubiquitin ligases (Alexandru et al. 2008), it is conceivable that the CRL ubiquitin ligases could participate in the ubiquitylation step during the mitoTAD process.

5.8.3 p97 and CRLs Regulate Chromatin-Associated Degradation

p97-mediated chromatin-associated degradation plays essential roles in cell cycle, DNA replication, and DNA damage responses (Meyer et al. 2012). Cockayne syndrome group A (CSA) and B (CSB) proteins act in transcription-coupled DNA repair, a subpathway of nucleotide excision repair. p97 has been demonstrated to function in ultraviolet radiation-induced CSB degradation (He et al. 2016, 2017). In this case, CSB and the CRL4A^{CSA} ubiquitin ligase interact with the transcription pre-initiation complex and travel with RNA polymerase II (Pol II) as a dynamically functional unit during transcriptional elongation. Once the whole transcription machinery encounters UV-induced DNA lesions, the Pol II holoenzyme stalls at lesion sites and serves as a damage sensor. If the Pol II stalling is irreversible, Rbp1, the largest subunit of Pol II, is ubiquitylated in a CSB-dependent manner (He et al. 2016, 2017). In budding yeast, it appears that Cul3 either alone (Verma et al. 2011) or together with Rsp5 mediates UV-induced Rbp1 degradation (Harreman et al. 2009). Whereas in human cells, CRL5 is the ubiquitin ligase (Weems et al. 2017; Yasukawa et al. 2008). Simultaneously with Rbp1 ubiquitylation (He et al. 2016), CSB is also ubiquitylated by the CRL4A^{CSA} ubiquitin ligase. The p97 complex is then recruited to lesion sites and interacts with CRL4A^{CSA} and extracts ubiquitin-conjugated CSB from the chromatin for degradation under the assistance of UBXD7 (He et al. 2016). Moreover, p97 and its cofactors, Ufd1 and UBXD7, are required for UV-induced Rbp1 degradation (He et al. 2017).

Efficient proteolysis of Cdc25A via the UPS in response to DNA damage signals is essential for robust activation of the G2/M checkpoint (Jin et al. 2003, 2008; Busino et al. 2003; Melixetian et al. 2009), which prevents cells with damaged DNA from entering into mitosis to maintain genome stability. Cdc25A ubiquitylation is triggered via phosphorylation of Cdc25A by the checkpoint kinase Chk1 and Chk1-activated NEK11 kinase (Jin et al. 2003, 2008; Busino et al. 2003; Melixetian et al. 2009). Phosphorylated Cdc25A is then recognized by the β-TrCP subunit of the SCF$^{\beta\text{-TrCP}}$ ubiquitin ligase which conjugates Cdc25A with polyubiquitin chains for the p97-Ufd1-Npl4 complex to bind and to deliver to the proteasome for rapid degradation (Jin et al. 2003, 2008; Busino et al. 2003; Melixetian et al. 2009; Riemer et al. 2014).

To maintain genome integrity, chromosomal DNA has to be replicated once and only once per cell cycle (Havens and Walter 2011). This is controlled mainly by the replication licensing factor Cdt1. Therefore, after DNA replication is accomplished, the licensing activity of Cdt1 has to be inactivated. One way to inhibit Cdt1 activity is Cdt1 ubiquitylation by the CRL4^{Cdt2} ubiquitin ligase and degradation through the proteasome (Jin et al. 2006; Sansam et al. 2006; Higa et al. 2006). Using a genome-wide siRNA screen, Raman et al. identified multiple factors required for Cdt1 turnover upon UV irradiation in HeLa cells, including the CRL4^{Cdt2} ubiquitin ligase and the p97-Ufd1-Npl4 complex (Raman et al. 2011).

These studies highlighted the importance of the orchestrated roles between CRLs and the p97 protein complex in cell cycle, DNA replication, and DNA damage responses.

5.8.4 p97 and CRLs' Control Degradation of Soluble Cytoplasmic and Nuclear Protein

p97 and its cofactors also collaborate with the CRL ubiquitin ligases in cellular protein degradation.

Such a coordinated proteolysis is key for intracellular and extracellular signaling pathways.

With regard to extracellular signaling pathways, several studies have shown that the p97-Ufd1-Npl4 complex plays crucial roles in both canonical and noncanonical NF-κB signaling pathways (Li et al. 2014; Schweitzer et al. 2016; Zhang et al. 2015b; Yilmaz et al. 2014). In the canonical NF-κB pathway, the p97-Ufd1-Npl4 complex, in concert with the $SCF^{β\text{-}TrCP}$ ubiquitin ligase, promotes rapid proteolysis of IκBα, the major inhibitor of NF-κB transcription factors, by recognizing and delivering ubiquitylated IκBα to the 26S proteasome for turnover under cytokine stimulation (Li et al. 2014; Schweitzer et al. 2016). In this process, Ufd1 employs its polyubiquitin-binding domain to recognize polyubiquitin chains of IκBα. Meanwhile, p97 interacts with the $SCF^{β\text{-}TrCP}$ ubiquitin ligase. These protein-protein interactions are pivotal for the p97-Ufd1-Npl4 complex to specifically recognize ubiquitylated IκBα and to drive IκBα turnover under either tumor necrosis factor alpha (TNF-α) or interleukin-1β (IL-1β) treatment (Li et al. 2014). These events are essential for the cytokine-induced canonical NF-κB activation.

In the noncanonical NF-κB signaling pathway, the p97-Ufd1-Npl4 complex also works together with the $SCF^{β\text{-}TrCP}$ ubiquitin ligase to induce partial degradation of the p100 and p105 to generate the p52 and p50 subunits of NF-κB, respectively (Zhang et al. 2015b; Yilmaz et al. 2014).

In addition to the NF-κB signaling pathway, p97 cooperates with the CRL ubiquitin ligases to mediate diverse cellular events as well. For instance, UBXD7 links the p97-Ufd1-Npl4 complex with ubiquitylated HIF1α and stimulates its rapid turnover via the proteasome (Alexandru et al. 2008), thereby regulating the HIF1α target gene expression in response to change in oxygen concentration. In this process, UBXD7 uses its UIM domain to bind the active $CUL2^{VHL}$ ubiquitin ligase (den Besten et al. 2012; Bandau et al. 2012), as previously discussed.

In a more recent study, p97 was observed to negatively regulate NRF2, an oxidative stress-responsive transcription factor, by extracting ubiquitylated NRF2 from the $CRL3^{KEAP1}$ ubiquitin ligase, with the aid of the Ufd1-Npl4 heterodimer and UBXD7, for proteasomal degradation to ensure a low basal level of NRF2 in mammalian cells (Tao et al. 2017).

p97-mediated substrate degradation also associates with metabolism. Nguyen et al. revealed that one subunit of glutamine synthetase (GS) homodecamer can be ubiquitylated by $CRL4^{CRBN}$, then dissociated by the p97-Ufd1-Npl4 complex, and subsequently degraded by the proteasome in response to glutamine treatment (Nguyen et al. 2017b). Furthermore, they showed that p97 is also required for the small molecule-induced turnover of all four known $CRL4^{CRBN}$ substrates, including Ikaros family zinc finger proteins 1 (IKZF1) and 3 (IKZF3), casein kinase 1α (CK1α), and the translation termination factor GSPT1 in the presence of the small molecule inducers, thalidomide, lenalidomide, or pomalidomide, in mammalian cells (Nguyen et al. 2017b).

Moreover, p97 has been reported to play a role in germline-specific sex determination in *C. elegans* (Sasagawa et al. 2009, 2010). Sasagawa et al. revealed that p97 is involved in the switch from sperm to oocyte in the germline of the worm hermaphrodite. They showcased that the p97-Ufd1-Npl4 complex interacts with the CUL2 ubiquitin ligase via cofactor Npl4 binding to the Elongin C subunit of the ubiquitin ligase. More specifically, TRA-1A, the terminal effector of the sex determination pathway, is regulated by p97- and CRL2-mediated proteolysis (Sasagawa et al. 2009, 2010).

Intriguingly, Yen et al. elaborated an alternative mechanism of SCF regulation in which cadmium stress induces selective recruitment of Cdc48 to catalyze dissociation of the autoubiquitylated F-box subunit Met30 from SCF^{Met30} ubiquitin ligase complex to inhibit substrate ubiquitylation and trigger downstream events in yeast (Yen et al. 2012).

5.8.5 p97 in Viral Protein-Induced Cellular Protein Degradation

It is quite common for human immunodeficiency virus (HIV) to hijack the cellular ubiquitin machinery to degrade host proteins for their needs (Zheng and Shabek 2017). For instance, HIV Vpu protein associates with the SCF$^{\beta\text{-TrCP}}$ ubiquitin ligase to promote ubiquitylation and turnover of human CD4 protein (Margottin et al. 1998). Together with the Ufd1-Npl4 heterodimer, p97 extracts ubiquitylated CD4 protein and escorts it to the proteasome for degradation (Binette et al. 2007; Magadán et al. 2010).

5.9 Perspective

Although p97 has been known to coordinate with CRLs for protein turnover in many cases, several important questions remain unanswered. In addition to the p97-Ufd1-Npl4 core complex, some substrates require additional p97 cofactors for the action (Alexandru et al. 2008; Tao et al. 2017), but other substrates do not. For instance, UBXD7 interacts with most, if not all, cullins (Alexandru et al. 2008); however, several substrates of CRLs only need the p97-Ufd1-Npl4 complex for substrate degradation (Raman et al. 2011; Li et al. 2014; Schweitzer et al. 2016; Zhang et al. 2015b; Yilmaz et al. 2014). There is no evidence that any other p97 cofactors are directly involved in these processes. It appears that this selection is more related to ubiquitin ligases than to substrates. Therefore, the important question is how CRLs recruit the p97 complex. UBXD7 has been shown to interact with neddylated Cul2 and Cul4 using its UIM domain (den Besten et al. 2012; Bandau et al. 2012). If this is a general phenomenon, why isn't UBXD7 a general factor to assist all CRL ubiquitin ligases to process their ubiquitylated substrates? If UBXD7 is not a general factor, how does UBXD7 determine its specificity?

Besides UBXD7 and the Ufd1-Npl4 heterodimer, several other p97 cofactors interact with CRLs (Hu et al. 2017; Bennett et al. 2010). How these p97 cofactors are involved in CRL-mediated protein turnover is largely unknown.

Yeast Cdc48 has been found to promote dissociation of the F-box subunit, Met30, from the SCF$^{\text{Met30}}$ ubiquitin ligase to suppress substrate ubiquitylation (Yen et al. 2012). If p97 and its yeast counterpart, Cdc48, coordinate with ubiquitin ligases for substrate turnover, it is hard to imagine that disassembling an active CRL ubiquitin ligase is a common function of p97. The question is whether p97 could unload ubiquitin ligases from their substrates after ubiquitylation is done, in order to recycle ubiquitin ligases for further usage. This is extremely important, if the cellular concentration of a given ubiquitin ligase is much lower than concentration of its substrate and if the substrate proteolysis is a very quick action. Assembly of more active ubiquitin ligases could be an answer (Reitsma et al. 2017; Liu et al. 2018; Pierce et al. 2013) but might not be the only answer for the UPS to deal with this kind of emergent situation. In the future, it would be interesting to test whether p97 and its cofactors recycle CRLs or other ubiquitin ligases to promote protein ubiquitylation and then degradation.

References

Alexandru G, Graumann J, Smith GT, Kolawa NJ, Fang R, Deshaies RJ (2008) UBXD7 binds multiple ubiquitin ligases and implicates p97 in HIF1alpha turnover. Cell 134(5):804–816

Allen MD, Buchberger A, Bycroft M (2006) The PUB domain functions as a p97 binding module in human peptide N-glycanase. J Biol Chem 281(35): 25502–25508

Baek GH, Kim I, Rao H (2011) The Cdc48 ATPase modulates the interaction between two proteolytic factors Ufd2 and Rad23. Proc Natl Acad Sci U S A 108(33):13558–13563

Bandau S, Knebel A, Gage ZO, Wood NT, Alexandru G (2012) UBXN7 docks on neddylated cullin complexes using its UIM motif and causes HIF1α accumulation. BMC Biol 10:36

Banerjee S, Bartesaghi A, Merk A, Rao P, Bulfer SL, Yan Y, Green N, Mroczkowski B, Neitz RJ, Wipf P, Falconieri V, Deshaies RJ, Milne JL, Huryn D, Arkin M, Subramaniam S (2016) 2.3 Å resolution cryo-EM structure of human p97 and mechanism of allosteric inhibition. Science 351(6275):871–875

Barthelme D, Sauer RT (2016) Origin and functional evolution of the Cdc48/p97/VCP AAA+ protein unfolding and remodeling machine. J Mol Biol 428 (9 Pt B):1861–1869

Bennett EJ, Rush J, Gygi SP, Harper JW (2010) Dynamics of cullin-RING ubiquitin ligase network revealed by systematic quantitative proteomics. Cell 143 (6):951–965

Binette J, Dubé M, Mercier J, Halawani D, Latterich M, Cohen EA (2007) Requirements for the selective degradation of CD4 receptor molecules by the human immunodeficiency virus type 1 Vpu protein in the endoplasmic reticulum. Retrovirology 4:75

Bodnar NO, Rapoport TA (2017) Molecular mechanism of substrate processing by the Cdc48 ATPase complex. Cell 169(4):722–735

Bodnar NO, Kim KH, Ji Z, Wales TE, Svetlov V, Nudler E, Engen JR, Walz T, Rapoport TA (2018) Structure of the Cdc48 ATPase with its ubiquitin-binding cofactor Ufd1-Npl4. Nat Struct Mol Biol 25 (7):616–622

Buchberger A, Schindelin H, Hänzelmann P (2015) Control of p97 function by cofactor binding. FEBS Lett 589(19 Pt A):2578–2589

Busino L, Donzelli M, Chiesa M, Guardavaccaro D, Ganoth D, Dorrello NV, Hershko A, Pagano M, Draetta GF (2003) Degradation of Cdc25A by beta-TrCP during S phase and in response to DNA damage. Nature 426(6962):87–91

Cai W, Yang H (2016) The structure and regulation of Cullin 2 based E3 ubiquitin ligases and their biological functions. Cell Div 11(7)

Cohen MM, Leboucher GP, Livnat-Levanon N, Glickman MH, Weissman AM (2008) Ubiquitin-proteasome-dependent degradation of a mitofusin, a critical regulator of mitochondrial fusion. Mol Biol Cell 19 (6):2457–2464

Cooney I, Han H, Stewart MG, Carson RH, Hansen DT, Iwasa JH, Price JC, Hill CP, Shen PS (2019) Structure of the Cdc48 segregase in the act of unfolding an authentic substrate. Science 365(6452):502–505

Cope GA, Deshaies RJ (2003) COP9 signalosome: a multifunctional regulator of SCF and other cullin-based ubiquitin ligases. Cell 114(6):663–671

Cope GA, Deshaies RJ (2006) Targeted silencing of Jab1/Csn5 in human cells downregulates SCF activity through reduction of F-box protein levels. BMC Biochem 7:1

den Besten W, Verma R, Kleiger G, Oania RS, Deshaies RJ (2012) NEDD8 links cullin-RING ubiquitin ligase function to the p97 pathway. Nat Struct Mol Biol 19 (5):511–516

Deshaies RJ, Emberley ED, Saha A (2010) Control of cullin-ring ubiquitin ligase activity by nedd8. Subcell Biochem 54:41–56

Dove KK, Klevit RE (2017) RING-between-RING E3 ligases: emerging themes amid the variations. J Mol Biol 429(22):3363–3375

Duda DM, Scott DC, Calabrese MF, Zimmerman ES, Zheng N, Schulman BA (2011) Structural regulation of cullin-RING ubiquitin ligase complexes. Curr Opin Struct Biol 21(2):257–264

Elliott PR, Nielsen SV, Marco-Casanova P, Fiil BK, Keusekotten K, Mailand N, Freund SM, Gyrd-Hansen-M, Komander D (2014) Molecular basis and regulation of OTULIN-LUBAC interaction. Mol Cell 54 (3):335–348

Fang S, Ferrone M, Yang C, Jensen JP, Tiwari S, Weissman AM (2001) The tumor autocrine motility factor receptor, gp78, is a ubiquitin protein ligase implicated in degradation from the endoplasmic reticulum. Proc Natl Acad Sci U S A 98(25):14422–14427

Fritz S, Weinbach N, Westermann B (2003) Mdm30 is an F-box protein required for maintenance of fusion-competent mitochondria in yeast. Mol Biol Cell 14 (6):2303–2313

Gardner RG, Shearer AG, Hampton RY (2001) In vivo action of the HRD ubiquitin ligase complex: mechanisms of endoplasmic reticulum quality control and sterol regulation. Mol Cell Biol 21(13):4276–4291

Genschik P, Sumara I, Lechner E (2013) The emerging family of CULLIN3-RING ubiquitin ligases (CRL3s): cellular functions and disease implications. EMBO J 32(17):2307–2320

Gong B, Chen F, Pan Y, Arrieta-Cruz I, Yoshida Y, Haroutunian V, Pasinetti GM (2010) SCFFbx2-E3-ligase-mediated degradation of BACE1 attenuates Alzheimer's disease amyloidosis and improves synaptic function. Aging Cell 9(6):1018–1031

Greenblatt EJ, Olzmann JA, Kopito RR (2011) Derlin-1 is a rhomboid pseudoprotease required for the dislocation of mutant α-1 antitrypsin from the endoplasmic reticulum. Nat Struct Mol Biol 18(10):1147–1152

Hampton RY (2002) ER-associated degradation in protein quality control and cellular regulation. Curr Opin Cell Biol 14(4):476–482

Hänzelmann P, Schindelin H (2011) The structural and functional basis of the p97/valosin-containing protein (VCP)-interacting motif (VIM): mutually exclusive binding of cofactors to the N-terminal domain of p97. J Biol Chem 286(44):38679–38690

Hänzelmann P, Schindelin H (2017) The interplay of cofactor interactions and post-translational modifications in the regulation of the AAA+ ATPase p97. Front Mol Biosci 4:21

Hänzelmann P, Buchberger A, Schindelin H (2011) Hierarchical binding of cofactors to the AAA ATPase p97. Structure 19(6):833–843

Harreman M, Taschner M, Sigurdsson S, Anindya R, Reid J, Somesh B, Kong SE, Banks CA, Conaway RC, Conaway JW, Svejstrup JQ (2009) Distinct ubiquitin ligases act sequentially for RNA polymerase II polyubiquitylation. Proc Natl Acad Sci U S A 106 (49):20705–20710

Havens CG, Walter JC (2011) Mechanism of CRL4 (Cdt2), a PCNA-dependent E3 ubiquitin ligase. Genes Dev 25(15):1568–1582

He Q, Cheng P, He Q, Liu Y (2005) The COP9 signalosome regulates the Neurospora circadian clock by controlling the stability of the SCFFWD-1 complex. Genes Dev 19(13):1518–1531

He J, Zhu Q, Wani G, Sharma N, Wani AA (2016) Valosin-containing protein (VCP)/p97 segregase mediates proteolytic processing of Cockayne Syndrome Group B (CSB) in damaged chromatin. J Biol Chem 291(14):7396–7408

He J, Zhu Q, Wani G, Wani AA (2017) UV-induced proteolysis of RNA polymerase II is mediated by VCP/p97 segregase and timely orchestration by

Cockayne syndrome B protein. Oncotarget 8 (7):11004–11019
Heo JM, Livnat-Levanon N, Taylor EB, Jones KT, Dephoure N, Ring J, Xie J, Brodsky JL, Madeo F, Gygi SP, Ashrafi K, Glickman MH, Rutter J (2010) A stress-responsive system for mitochondrial protein degradation. Mol Cell 40(3):465–480
Higa LA, Banks D, Wu M, Kobayashi R, Sun H, Zhang H (2006) L2DTL/CDT2 interacts with the CUL4/DDB1 complex and PCNA and regulates CDT1 proteolysis in response to DNA damage. Cell Cycle 5 (15):1675–1680
Hori T, Osaka F, Chiba T, Miyamoto C, Okabayashi K, Shimbara N, Kato S, Tanaka K (1999) Covalent modification of all members of human cullin family proteins by NEDD8. Oncogene 18(48): 6829–6834
Hornbeck PV, Zhang B, Murray B, Kornhauser JM, Latham V, Skrzypek E (2015) PhosphoSitePlus, 2014: mutations, PTMs and recalibrations. Nucleic Acids Res 43(Database issue):D512–D520
Hu Y, O'Boyle K, Auer J, Raju S, You F, Wang P, Fikrig E, Sutton RE (2017) Multiple UBXN family members inhibit retrovirus and lentivirus production and canonical NFκB signaling by stabilizing IκBα. PLoS Pathog 13(2):e1006187
Huang R, Ripstein ZA, Rubinstein JL, Kay LE (2019) Cooperative subunit dynamics modulate p97 function. Proc Natl Acad Sci U S A 116(1):158–167
Huibregtse JM, Scheffner M, Beaudenon S, Howley PM (1995) A family of proteins structurally and functionally related to the E6-AP ubiquitin-protein ligase. Proc Natl Acad Sci U S A 92(7):2563–2567
Inuzuka H, Shaik S, Onoyama I, Gao D, Tseng A, Maser RS, Zhai B, Wan L, Gutierrez A, Lau AW, Xiao Y, Christie AL, Aster J, Settleman J, Gygi SP, Kung AL, Look T, Nakayama KI, DePinho RA, Wei W (2011) SCF(FBW7) regulates cellular apoptosis by targeting MCL1 for ubiquitylation and destruction. Nature 471 (7336):104–109
Jin J, Shirogane T, Xu L, Nalepa G, Qin J, Elledge SJ, Harper JW (2003) SCFbeta-TRCP links Chk1 signaling to degradation of the Cdc25A protein phosphatase. Genes Dev 17(24):3062–3074
Jin J, Cardozo T, Lovering RC, Elledge SJ, Pagano M, Harper JW (2004) Systematic analysis and nomenclature of mammalian F-box proteins. Genes Dev 18 (21):2573–2580
Jin J, Arias EE, Chen J, Harper JW, Walter JC (2006) A family of diverse Cul4-Ddb1-interacting proteins includes Cdt2, which is required for S phase destruction of the replication factor Cdt1. Mol Cell 23 (5):709–721
Jin J, Ang XL, Ye X, Livingstone M, Harper JW (2008) Differential roles for checkpoint kinases in DNA damage-dependent degradation of the Cdc25Aprotein phosphatase. J Biol Chem 283(28):19322–19328
Joshi V, Upadhyay A, Kumar A, Mishra A (2017) Gp78 E3 ubiquitin ligase: essential functions and contributions in proteostasis. Front Cell Neurosci 11:259
Kamura T, Maenaka K, Kotoshiba S, Matsumoto M, Kohda D, Conaway RC, Conaway JW, Nakayama KI (2004) VHL-box and SOCS-box domains determine binding specificity for Cul2-Rbx1 and Cul5-Rbx2 modules of ubiquitin ligases. Genes Dev 18 (24):3055–3065
Kern M, Fernandez-Sáiz V, Schäfer Z, Buchberger A (2009) UBXD1 binds p97 through two independent binding sites. Biochem Biophys Res Commun 380 (2):303–307
Kim SJ, Kim EE (2014) Crystallization and preliminary X-ray crystallographic analysis of the complex between the N-D1 domain of VCP from Homo sapiens and the N domain of OTU1 from Saccharomyces cerevisiae. Acta Crystallogr F Struct Biol Commun 70(Pt 8):1087–1089
Kim SJ, Cho J, Song EJ, Kim SJ, Kim HM, Lee KE, Suh SW, Kim EE (2014) Structural basis for ovarian tumor domain-containing protein 1 (OTU1) binding to p97/ valosin-containing protein (VCP). J Biol Chem 289 (18):12264–12274
Kloppsteck P, Ewens CA, Förster A, Zhang X, Freemont PS (2012) Regulation of p97 in the ubiquitin-proteasome system by the UBX protein-family. Biochim Biophys Acta 1823(1):125–129
LaLonde DP, Bretscher A (2011) The UBX protein SAKS1 negatively regulates endoplasmic reticulum-associated degradation and p97-dependent degradation. J Biol Chem 286(6):4892–4901
Lavoie C, Chevet E, Roy L, Tonks NK, Fazel A, Posner BI, Paiement J, Bergeron JJ (2000) Tyrosine phosphorylation of p97 regulates transitional endoplasmic reticulum assembly in vitro. Proc Natl Acad Sci U S A 97 (25):13637–13642
Lee EK, Diehl JA (2014) SCFs in the new millennium. Oncogene 33(16):2011–2018
Lee J, Zhou P (2007) DCAFs, the missing link of the CUL4-DDB1 ubiquitin ligase. Mol Cell 26 (6):775–780
Li G, Zhao G, Schindelin H, Lennarz WJ (2008) Tyrosine phosphorylation of ATPase p97 regulates its activity during ERAD. Biochem Biophys Res Commun 375 (2):247–251
Li JM, Wu H, Zhang W, Blackburn MR, Jin J (2014) The p97-UFD1L-NPL4 protein complex mediates cytokine-induced IκBα proteolysis. Mol Cell Biol 34 (3):335–347
Lim JJ, Lee Y, Ly TT, Kang JY, Lee JG, An JY, Youn HS, Park KR, Kim TG, Yang JK, Jun Y, Eom SH (2016) Structural insights into the interaction of p97 N-terminus domain and VBM in rhomboid protease, RHBDL4. Biochem J 473(18):2863–2880
Liu B, Zheng Y, Wang TD, Xu HZ, Xia L, Zhang J, Wu YL, Chen GQ, Wang LS (2012) Proteomic identification of common SCF ubiquitin ligase FBXO6-interacting glycoproteins in three kinds of cells. J Proteome Res 11(3):1773–1781

Liu S, Fu QS, Zhao J, Hu HY (2013) Structural and mechanistic insights into the arginine/lysine-rich peptide motifs that interact with P97/VCP. Biochim Biophys Acta 1834(12):2672–2678
Liu X, Reitsma JM, Mamrosh JL, Zhang Y, Straube R, Deshaies RJ (2018) Cand1-mediated adaptive exchange mechanism enables variation in F-box protein expression. Mol Cell 69(5):773–786
Lydeard JR, Schulman BA, Harper JW (2013) Building and remodelling Cullin-RING E3 ubiquitin ligases. EMBO Rep 14(12):1050–1061
Madsen L, Seeger M, Semple CA, Hartmann-Petersen R (2009) New ATPase regulators--p97 goes to the PUB. Int J Biochem Cell Biol 41(12):2380–2388
Magadán JG, Pérez-Victoria FJ, Sougrat R, Ye Y, Strebel K, Bonifacino JS (2010) Multilayered mechanism of CD4 downregulation by HIV-1 Vpu involving distinct ER retention and ERAD targeting steps. PLoS Pathog 6(4):e1000869
Margottin F, Bour SP, Durand H, Selig L, Benichou S, Richard V, Thomas D, Strebel K, Benarous R (1998) A novel human WD protein, h-beta TrCp, that interacts with HIV-1 Vpu connects CD4 to the ER degradation pathway through an F-box motif. Mol Cell 1 (4):565–574
Mårtensson CU, Priesnitz C, Song J, Ellenrieder L, Doan KN, Boos F, Floerchinger A, Zufall N, Oeljeklaus S, Warscheid B, Becker T (2019) Mitochondrial protein translocation-associated degradation. Nature 569 (7758):679–683
McNeill H, Knebel A, Arthur JS, Cuenda A, Cohen P (2004) A novel UBA and UBX domain protein that binds polyubiquitin and VCP and is a substrate for SAPKs. Biochem J 384(Pt 2):391–400
Melixetian M, Klein DK, Sørensen CS, Helin K (2009) NEK11 regulates CDC25A degradation and the IR-induced G2/M checkpoint. Nat Cell Biol 11 (10):1247–1253
Meyer HH, Wang Y, Warren G (2002) Direct binding of ubiquitin conjugates by the mammalian p97 adaptor complexes, p47 and Ufd1-Npl4. EMBO J 21 (21):5645–5652
Meyer H, Bug M, Bremer S (2012) Emerging functions of the VCP/p97 AAA-ATPase in the ubiquitin system. Nat Cell Biol 14(2):117–123
Moir D, Stewart SE, Osmond BC, Botstein D (1982) Cold-sensitive cell-division-cycle mutants of yeast: isolation, properties, and pseudoreversion studies. Genetics 100:547–563
Morreale G, Conforti L, Coadwell J, Wilbrey AL, Coleman MP (2009) Evolutionary divergence of valosin-containing protein/cell division cycle protein 48 binding interactions among endoplasmic reticulum-associated degradation proteins. FEBS J 276(5):1208–1220
Mullally JE, Chernova T, Wilkinson KD (2006) Doa1 is a Cdc48 adapter that possesses a novel ubiquitin binding domain. Mol Cell Biol 26(3):822–830
Nguyen HC, Wang W, Xiong Y (2017a) Cullin-RING E3 ubiquitin ligases: bridges to destruction. Subcell Biochem 83:323–347
Nguyen TV, Li J, Lu CJ, Mamrosh JL, Lu G, Cathers BE, Deshaies RJ (2017b) p97/VCP promotes degradation of CRBN substrate glutamine synthetase and neosubstrates. Proc Natl Acad Sci U S A 114(14): 3565–3571
Nishimasu R, Komori H, Higuchi Y, Nishimasu H, Hiroaki H (2010) Crystal structure of a PFU-PUL domain pair of Saccharomyces cerevisiae Doa1/Ufd3. Kobe J Med Sci 56(3):E125–E139
Okumura F, Joo-Okumura A, Nakatsukasa K, Kamura T (2016) The role of cullin 5-containing ubiquitin ligases. Cell Div 11(1)
Pao KC, Wood NT, Knebel A, Rafie K, Stanley M, Mabbitt PD, Sundaramoorthy R, Hofmann K, van Aalten DMF, Virdee S (2018) Activity-based E3 ligase profiling uncovers an E3 ligase with esterification activity. Nature 556(7701):381–385
Petroski MD, Deshaies RJ (2005) Function and regulation of cullin-RING ubiquitin ligases. Nat Rev Mol Cell Biol 6(1):9–20
Pierce NW, Lee JE, Liu X, Sweredoski MJ, Graham RL, Larimore EA, Rome M, Zheng N, Clurman BE, Hess S, Shan SO, Deshaies RJ (2013) Cand1 promotes assembly of new SCF complexes through dynamic exchange of F box proteins. Cell 153(1):206–215
Pleasure IT, Black MM, Keen JH (1993) Valosin-containing protein, VCP, is a ubiquitous clathrin-binding protein. Nature 365:459–462
Qiu L, Pashkova N, Walker JR, Winistorfer S, Allali-Hassani A, Akutsu M, Piper R, Dhe-Paganon S (2010) Structure and function of the PLAA/Ufd3-p97/Cdc48 complex. J Biol Chem 285(1):365–372
Raman M, Havens CG, Walter JC, Harper JW (2011) A genome-wide screen identifies p97 as an essential regulator of DNA damage-dependent CDT1 destruction. Mol Cell 44(1):72–84
Rao MV, Williams DR, Cocklin S, Loll PJ (2017) Interaction between the AAA$^+$ ATPase p97 and its cofactor ataxin3 in health and disease: Nucleotide-induced conformational changes regulate cofactor binding. J Biol Chem 292(45):18392–18407
Reitsma JM, Liu X, Reichermeier KM, Moradian A, Sweredoski MJ, Hess S, Deshaies RJ (2017) Composition and regulation of the cellular repertoire of SCF ubiquitin ligases. Cell 171(6):1326–1339
Riemer A, Dobrynin G, Dressler A, Bremer S, Soni A, Iliakis G, Meyer H (2014) The p97-Ufd1-Npl4 ATPase complex ensures robustness of the G2/M checkpoint by facilitatingCDC25A degradation. Cell Cycle 13 (6):919–927
Saiga T, Fukuda T, Matsumoto M, Tada H, Okano HJ, Okano H, Nakayama KI (2009) Fbxo45 forms a novel ubiquitin ligase complex and is required for neuronal development. Mol Cell Biol 29(13):3529–3543
Sá-Moura B, Funakoshi M, Tomko RJ Jr, Dohmen RJ, Wu Z, Peng J, Hochstrasser M (2013) A conserved protein with AN1 zinc finger and ubiquitin-like domains modulates Cdc48 (p97) function in the ubiquitin-proteasome pathway. J Biol Chem 288(47): 33682–33696

Sansam CL, Shepard JL, Lai K, Ianari A, Danielian PS, Amsterdam A, Hopkins N, Lees JA (2006) DTL/CDT2 is essential for both CDT1 regulation and the early G2/M checkpoint. Genes Dev 20(22):3117–3129

Sarikas A, Hartmann T, Pan ZQ (2011) The cullin protein family. Genome Biol 12(4):220

Sasagawa Y, Otani M, Higashitani N, Higashitani A, Sato K, Ogura T, Yamanaka K (2009) Caenorhabditis elegans p97 controls germline-specific sex determination by controlling the TRA-1 level in a CUL-2-dependent manner. J Cell Sci 122(Pt 20):3663–3672

Sasagawa Y, Yamanaka K, Saito-Sasagawa Y, Ogura T (2010) Caenorhabditis elegans UBX cofactors for CDC-48/p97 control spermatogenesis. Genes Cells 15 (12):1201–1215

Schaeffer V, Akutsu M, Olma MH, Gomes LC, Kawasaki M, Dikic I (2014) Binding of OTULIN to the PUB domain of HOIP controls NF-κB signaling. Mol Cell 54(3):349–361

Schuberth C, Buchberger A (2008) UBX domain proteins: major regulators of the AAA ATPase Cdc48/p97. Cell Mol Life Sci 65(15):2360–2371

Schuberth C, Richly H, Rumpf S, Buchberger A (2004) Shp1 and Ubx2 are adaptors of Cdc48 involved in ubiquitin-dependent protein degradation. EMBO Rep 5:818–824

Schweitzer K, Pralow A, Naumann M (2016) p97/VCP promotes Cullin-RING-ubiquitin-ligase/proteasome-dependent degradation of IκBα and the preceding liberation of RelA from ubiquitinated IκBα. J Cell Mol Med 20(1):58–70

Sonneville R, Moreno SP, Knebel A, Johnson C, Hastie CJ, Gartner A, Gambus A, Labib K (2017) CUL-2^{LRR-1} and UBXN-3 drive replisome disassembly during DNA replication termination and mitosis. Nat Cell Biol 19(5):468–479

Stach L, Freemont PS (2017) The AAA+ ATPase p97, a cellular multitool. Biochem J 474(17):2953–2976

Stapf C, Cartwright E, Bycroft M, Hofmann K, Buchberger A (2011) The general definition of the p97/valosin-containing protein (VCP)-interacting motif (VIM) delineates a new family of p97 cofactors. J Biol Chem 286(44):38670–38678

Sun Y, Li H (2013) Functional characterization of SAG/RBX2/ROC2/RNF7, an antioxidant protein and an E3 ubiquitin ligase. Protein Cell 4(2):103–116

Suzuki T, Park H, Till EA, Lennarz WJ (2001) The PUB domain: a putative protein-protein interaction domain implicated in the ubiquitin-proteasome pathway. Biochem Biophys Res Commun 287(5):1083–1087

Swatek KN, Komander D (2016) Ubiquitin modifications. Cell Res 26(4):399–422

Tao S, Liu P, Luo G, Rojo de la Vega M, Chen H, Wu T, Tillotson J, Chapman E, Zhang DD (2017) p97 negatively regulates NRF2 by extracting ubiquitylated NRF2 from the KEAP1-CUL3 E3 complex. Mol Cell Biol 37(8). pii: e00660-16

Turakhiya A, Meyer SR, Marincola G, Böhm S, Vanselow JT, Schlosser A, Hofmann K, Buchberger A (2018) ZFAND1 recruits p97 and the 26S proteasome to promote the clearance of arsenite-induced stress granules. Mol Cell 70(5):906–919

Twomey EC, Ji Z, Wales TE, Bodnar NO, Ficarro SB, Marto JA, Engen JR, Rapoport TA (2019) Substrate processing by the Cdc48 ATPase complex is initiated by ubiquitin unfolding. Science 365(6452)

van den Boom J, Meyer H (2018) VCP/p97-mediated unfolding as a principle in protein homeostasis and signaling. Mol Cell 69(2):182–194

Verma R, Oania R, Fang R, Smith GT, Deshaies RJ (2011) Cdc48/p97 mediates UV-dependent turnover of RNA Pol II. Mol Cell 41(1):82–92

Weems JC, Slaughter BD, Unruh JR, Boeing S, Hall SM, McLaird MB, Yasukawa T, Aso T, Svejstrup JQ, Conaway JW, Conaway RC (2017) Cockayne syndrome B protein regulates recruitment of the Elongin A ubiquitin ligase to sites of DNA damage. J Biol Chem 292(16):6431–6437

Wei N, Deng XW (2003) The COP9 signalosome. Annu Rev Cell Dev Biol 19:261–286

Wenzel DM, Lissounov A, Brzovic PS, Klevit RE (2011) UBCH7 reactivity profile reveals parkin and HHARI to be RING/HECT hybrids. Nature 474(7349):105–108

Xu S, Peng G, Wang Y, Fang S, Karbowski M (2011) The AAA-ATPase p97 is essential for outer mitochondrial membrane protein turnover. Mol Biol Cell 22(3): 291–300

Yasukawa T, Kamura T, Kitajima S, Conaway RC, Conaway JW, Aso T (2008) Mammalian Elongin A complex mediates DNA-damage-induced ubiquitylation and degradation of Rpb1. EMBO J 27 (24):3256–3266

Ye Y, Meyer HH, Rapoport TA (2001) The AAA ATPase Cdc48/p97 and its partners transport proteins from the ER into the cytosol. Nature 414:652–656

Yen JL, Flick K, Papagiannis CV, Mathur R, Tyrrell A, Ouni I, Kaake RM, Huang L, Kaiser P (2012) Signal-induced disassembly of the SCF ubiquitin ligase complex by Cdc48/p97. Mol Cell 48(2):288–297

Yeung HO, Kloppsteck P, Niwa H, Isaacson RL, Matthews S, Zhang X, Freemont PS (2008) Insights into adaptor binding to the AAA protein p97. Biochem Soc Trans 36(Pt 1):62–67

Yi L, Kaler SG (2018) Interaction between the AAA ATPase p97/VCP and a concealed UBX domain in the copper transporter ATP7A is associated with motor neuron degeneration. J Biol Chem 293(20): 7606–7617

Yilmaz ZB, Kofahl B, Beaudette P, Baum K, Ipenberg I, Weih F, Wolf J, Dittmar G, Scheidereit C (2014) Quantitative dissection and modeling of the NF-κB p100-p105 module reveals interdependent precursor proteolysis. Cell Rep 9(5):1756–1769

Yoshida Y, Chiba T, Tokunaga F, Kawasaki H, Iwai K, Suzuki T, Ito Y, Matsuoka K, Yoshida M, Tanaka K, Tai T (2002) E3 ubiquitin ligase that recognizes sugar chains. Nature 418(6896):438–442

Yoshida Y, Tokunaga F, Chiba T, Iwai K, Tanaka K, Tai T (2003) Fbs2 is a new member of the E3 ubiquitin ligase family that recognizes sugar chains. J Biol Chem 278 (44):43877–43884

Yoshida Y, Adachi E, Fukiya K, Iwai K, Tanaka K (2005) Glycoprotein-specific ubiquitin ligases recognize N-glycans in unfolded substrates. EMBO Rep 6 (3):239–244

Zhang L, Zhou F, Li Y, Drabsch Y, Zhang J, van Dam H, ten Dijke P (2012) Fas-associated factor 1 is a scaffold protein that promotes β-transducin repeat-containing protein(β-TrCP)-mediated β-catenin ubiquitination and degradation. J Biol Chem 287 (36):30701–30710

Zhang T, Xu Y, Liu Y, Ye Y (2015a) gp78 functions downstream of Hrd1 to promote degradation of misfolded proteins of the endoplasmic reticulum. Mol Biol Cell 26(24):4438–4450

Zhang Z, Wang Y, Li C, Shi Z, Hao Q, Wang W, Song X, Zhao Y, Jiao S, Zhou Z (2015b) The transitional endoplasmic reticulum ATPase p97 regulates the alternative nuclear factor NF-κB signaling via partial degradation of the NF-κB subunit p100. J Biol Chem 290(32): 19558–19568

Zhao G, Zhou X, Wang L, Li G, Schindelin H, Lennarz WJ (2007) Studies on peptide:N-glycanase-p97 interaction suggest that p97 phosphorylation modulates endoplasmic reticulum-associated degradation. Proc Natl Acad Sci U S A 104(21):8785–8790

Zheng N, Shabek N (2017) Ubiquitin ligases: structure, function, and regulation. Annu Rev Biochem 86: 129–157

Regulation of Stem Cells by Cullin-RING Ligase

6

Lin Fu, Chun-Ping Cui, and Lingqiang Zhang

Abstract

Stem cells can remain quiescent, self-renewal, and differentiate into many types of cells and even cancer stem cells. The coordination of these complex processes maintains the homeostasis of the organism. Ubiquitination is an important posttranslational modification process that regulates protein stability and activity. The ubiquitination levels of stem cell-associated proteins are closely related with stem cell characteristics. Cullin-RING Ligases (CRLs) are the largest family of E3 ubiquitin ligases, accounting for approximately 20% of proteins degraded by proteasome. In this review, we discuss the role of CRLs in stem cell homeostasis, self-renewal, and differentiation and expound their ubiquitination substrates. In addition, we also discuss the effect of CRLs on the formation of cancer stem cells that may provide promising therapy strategies for cancer.

Keywords

Stem cells · Self-renewal · Differentiation · Ubiquitination · CRLs

L. Fu
Institute of Chronic Disease, Qingdao Municipal Hospital, Qingdao University, Qingdao, China

C.-P. Cui · L. Zhang (✉)
State Key Laboratory of Proteomics, National Center for Protein Sciences (Beijing), Beijing Institute of Lifeomics, Beijing, China
e-mail: zhanglq@nic.bmi.ac.cn

Abbreviations

ASB2	Ankyrin-repeat SOCS box-containing protein 2
BTB	Broad-Complex, tramtrack, and bric à brac
BTBD6A	BTB/POZ domain-containing protein 6
CBs	Cystoblasts
CCA	Cholangiocarcinoma
CML	Chronic myelogenous leukemia
COP1	Constitutive photomorphogenic 1
CRBN	Cereblon
CRLs	Cullin-RING Ligases
CSCs	Cancer stem cells
D4R	Dopamine D4 receptor
DCAFs	DDB1- and CUL4-associated factors
DCX	DDB1-CUL4-X-box
DDB1	DNA damage-binding protein 1
DLL1	Delta-like ligand 1
Dpp	Decapentaplegic
DSH	Disheveled
ESCs	Embryonic stem cells
FBW7	F-box and WD repeat domain-containing protein 7
FLNA and FLNB	Proteins filamin A and B
FWD1	F-box and WD repeat-containing protein 1A
GSCs	Germline stem cells
HECT	Homologous to E6AP C-terminus

Y. Sun et al. (eds.), *Cullin-RING Ligases and Protein Neddylation*, Advances in Experimental Medicine and Biology 1217, https://doi.org/10.1007/978-981-15-1025-0_6

HIF1α	Hypoxia-inducible factor α
HSCs	Hematopoietic stem cells
iPSC	Induced pluripotent stem cell
ISCs	Intestinal stem cells
KBTBD8	Kelch repeat and BTB domain-containing 8
KCTD2/10	Potassium channel tetramerization domain-containing protein 2/10
KEAP1	Kelch-like ECH-associated protein 1
KLHL12	Kelch-like family member 12
LICs	Leukemia-initiating cells
LRR	Leucine-rich repeats
MaSCs	Mammary stem cells
MB	Medulloblastoma
MSCs	Muscle stem cells
NF1	Neurofibromatosis type 1
NPC	Nasopharyngeal carcinoma
OSCs	Osteoblast stem cells
OTUD7B	OTU domain-containing protein 7B
PLZF	Promyelocytic leukemia zinc finger
POZ	Poxvirus and zinc finger
PrxIII	Peroxiredoxin III
Ptc1	Patched1
RA	Retinoic acid
RBR	Ring-in-between-Ring
RBX1	Ring-box 1
RhoA	Ras homolog family member A
RHOGEF	RHO guanine nucleotide exchange factor
RING	Really interesting new gene
ROS	Reactive oxygen species
Shh	Sonic hedgehog
SIMs	Substrate-interacting motifs
Skp1	S-phase kinase-associated protein 1
SOCS	Suppressors of cytokine signaling
Sox2	Sex-determining region Y-box 2
T-ALL	T-cell acute lymphocytic leukemia
VHL	Von Hippel-Lindau

6.1 Introduction

Stem cells are a class of pluripotent cells with self-renewing ability. Under certain conditions, stem cells can differentiate into multiple functional cells (Fuchs and Chen 2013). Besides, stem cells can differentiate into downstream cell types upon appropriate environmental stimuli. Stem cells can divide symmetrically or asymmetrically (Neumuller and Knoblich 2009). Symmetric cell division ensures that all elements between two identical stem cells are evenly distributed, and usually only one daughter cell differentiation occurs later. On the other hand, asymmetric cell division leads to unequal division of stem cell components, a process that involves the correct localization of the mitotic spindle (Gonczy 2008). Thus, one cell is still a stem cell and the other uses a different cell fate. Upon asymmetric division, a daughter cell may move from a relative position to a niche, causing it to differentiate.

According to the developmental stage in which stem cells are located, stem cells are divided into embryonic stem cells (ES cells) and adult stem cells (somatic stem cells) (Becker et al. 1963; Thomson et al. 1998). Stem cells help maintain or repair tissues that are susceptible to damage in adult organisms (Leblond and Walker 1956). It is important to maintain the number and function of stem cells because its abnormalities lead to many diseases ranging from birth defects to cancer.

Although progenitor cells also have the ability to self-renew, they are usually short-lived (He et al. 2009). The ability of stem cells to undergo long-term self-renewal is critical to the entire organism. Somatic stem cells are found in a special microenvironment called niche which usually exhibits varying degrees of quiescence due to specific tissue characteristics. For example, HSCs are dormant, whereas MaSCs appear to be cycling (Passegue and Wagers 2006) and ISCs proliferate rapidly (Barker et al. 2007).

Cancer stem cells (CSCs) are a minority population of cells in tumors and usually have a low rate of proliferation (Reiman et al. 2010). They may be derived from stem cells, which acquire cancer characteristics through genetic and epigenetic changes. Alternatively, they may be derived from transformed progenitor cells with self-renewal capabilities (Takebe and Ivy 2010). Current cancer treatments often fail to eliminate advanced tumors, probably because they are not effective against the CSCs population. There is also a crosstalk between CSCs and niche. For example, CSCs guide the formation of niche, while niche controls the proliferation, differentiation, and metastasis of CSCs (Nguyen et al. 2012). Stem cell induction can be achieved by activating transcription factors involved in the induction of reprogramming of iPSCs such as Oct4, Sox2, Nanog, and KLF4 (Heddleston et al. 2010).

Quiescence and self-renewal prevent stem cell exhaustion, whereas differentiation can induce different cell lineages. Improper control can lead to stem cell disorder and even cancers (Warr et al. 2011). This regulation is achieved at two different levels. First, in the stem cell niche, chemokines, cytokines, growth factors, and other secretory molecules constitute an extracellular network that controls SC maintenance (Trumpp et al. 2010). Second, proteins and signaling pathways involved in cell cycle, proliferation, growth, and survival are essential in SC homeostasis (Warr et al. 2011).

The ubiquitin system is crucial for stem cell functions (Yau and Rape 2016). Ubiquitin is covalently linked to the target protein by an isopeptide bond between its C-terminal glycine and mainly the lysine residue of the acceptor substrates. Ubiquitin modification of proteins is carried out by an enzyme cascade including the ubiquitin-activating (E1), ubiquitin-conjugating (E2), and ubiquitin-ligating (E3) enzymes (Weissman 2001).

There are two main types of E3s in eukaryotes, defined as the presence of either a HECT (homologous to E6AP C-terminus) or a RING (really interesting new gene) domain. Most E3 ligases share a characteristic RING domain that depends on a ubiquitin-charged E2 for catalysis (Plechanovova et al. 2012; Saha et al. 2011). Other sets of E3 ligases possess either a HECT domain or HECT-like RBR (Ring-in-between-Ring) domain, which carries an active cysteine with ubiquitin before being transferred to the substrates (Wenzel et al. 2011). The HECT family is composed of large proteins, each of which interacts directly with E2 enzymes and its specific substrates.

Cullin-RING Ligases (CRLs) occupied the majority in RING E3 ligases. All CRLs share the similar core architecture with a curved cullin protein acting as a molecular scaffold (He et al. 2006; Jin et al. 2004; Linossi and Nicholson 2012; Petroski and Deshaies 2005; Stogios et al. 2005). CRLs consist of eight cullins designated CUL1, CUL2, CUL3, CUL4A, CUL4B, CUL5, CUL7, and CUL9 (also known as PARC), which serve as scaffolds (Zhao and Sun 2013). The C-terminal part of cullins binds tightly to the RING-H2 domain proteins RBX1 or RBX2 (also known as ROC1 and ROC2/SAG, respectively) which transfer ubiquitin from the ubiquitin-conjugated E2 to the substrates. Cullins interact with their specific substrate adapter or receptor proteins via N-terminal domains (Wang et al. 2014b).

In this review, we analyze the role of CRLs in embryonic stem cells and somatic stem cell and discuss the underlying mechanisms that control the balance between quiescence, self-renewal, and differentiation. We also discuss how dysregulation of CRL-mediated protein ubiquitination leads to the development of cancer stem cells.

6.2 CRL1 and Stem Cells

CRL1, also known as SCF complex (Skp1-CUL1-F-box), is the most representative member of CRLs (Cardozo and Pagano 2004). It consists of three invariable components: Cullin1 (CUL1, scaffold protein), RBX1, and Skp1 (S-phase kinase-associated protein 1, adaptor protein). CUL1 consists of three repeats of a five-helix motif at the N-terminus and a globular domain at the C-terminus. Skp1 and RBX1 bind to the N-terminus and the C-terminus of CUL1, respectively. Then, F-box receptor binds to Skp1 via its F-box motif and specifically recognizes substrates through different substrate-interacting motifs (SIMs) (Nakayama and Nakayama 2005). F-box proteins are divided into three subfamilies: FBXW proteins which contain WD40 repeats, FBXL proteins which contain leucine-rich repeats (LRR), and FBXO proteins which do not contain the above two types of domains (Cardozo and Pagano 2004). They are involved in cell cycle regulation and stem cell self-renewal and differentiation (Fig. 6.1).

6.2.1 FBW7

F-box and WD repeat domain-containing protein 7 (FBW7), also known as FBXW7, SEL-10, hCdc4, or hAgo, is one of the best characterized F-box proteins. Germline deletion of FBW7 causes embryonic lethality at around E10.5 which indicates its broad and essential functions during development (Tetzlaff et al. 2004). As an important adaptor of SCF complex, FBW7 mediates the ubiquitination of many key regulators (Fig. 6.1).

6.2.1.1 FBW7 in Embryonic Stem Cells and Somatic Stem Cells

Unlike other defined transcription factors, c-Myc is an unstable protein exhibiting a half-life of about 20–30 min (Hann and Eisenman 1984). Ubiquitin-proteasome system is responsible for the degradation of many short-lived regulatory proteins in vivo. Loss of c-Myc induces an irreversible transition toward differentiation (Reed 2003). FBW7 is also elevated in hESCs and is required for their proliferation as well as for iPSC reprogramming (Buckley et al. 2012). Thus, FBW7 plays important roles in the maintenance of embryonic stem cells (Takeishi and Nakayama 2014).

SAG/RBX/ROC protein is an essential RING component of SCF E3 ubiquitin ligase (Duan et al. 1999). SAG inactivation causes neurofibromatosis type 1 (NF1) accumulation and Ras inhibition, which blocks embryonic stem cells from undergoing endothelial differentiation and inhibits angiogenesis and proliferation in teratomas. NF1 is targeted for ubiquitination and degradation by SAG-CUL1-FBW7 E3 ligase and establishes an ubiquitin-dependent regulatory mechanism for the NF1-Ras pathway during embryogenesis (Tan et al. 2011).

FBW7 also acts as a key regulator of the maintenance and differentiation of neural stem cells in the brain. FBW7 plays a central role in the degradation of Notch family members (Matsumoto et al. 2011). It controls neurogenesis by antagonizing Notch and c-Jun N-terminal kinase (JNK)/c-Jun signaling (Hoeck et al. 2010).

Notch is also important for directing lymphoid lineage cell fate determination and has been implicated in HSC self-renewal (Nie et al. 2008). Notch is expressed by HSCs, while its ligand, Jagged, is expressed by the HSC niche, and increased Jagged/Notch activation results in increased HSC number and niche expansion (Duncan et al. 2005). Therefore, FBW7 could regulate HSC self-renewal through targeting Notch for degradation.

In long-term HSCs, SCF^{FBW7} can also ubiquitinate c-Myc (Yada et al. 2004). Deletion of FBW7 strongly causes an increase in the frequency of actively cycling LSK (Lin^-, $Sca\text{-}1^+$, $c\text{-}Kit^+$) population, which ultimately leads to exhaustion of HSCs (Matsuoka et al. 2008) that can be rescued by simultaneous losing of a single allele of the c-Myc (Reavie et al. 2010). As a result, a considerable proportion of FBW7-knockout mice developed anemia (Matsuoka et al. 2008; Onoyama et al. 2007). Furthermore, $FBW7^{-/-}$ LSK cells downregulate genes involved in HSC quiescence (Thompson et al. 2008).

SOX9 is an early development transcription factor necessary for the maintenance of stem and progenitor cell populations (Kadaja et al. 2014). It can also be ubiquitinated and degraded by FBW7 E3 ubiquitin ligase (Hong et al. 2016).

In skin epithelia, the p63 isoform $\Delta Np63\alpha$ promotes proliferation of basal cells. SCF^{FBW7} was proposed to target p63 for degradation during keratinocyte differentiation (Galli et al. 2010).

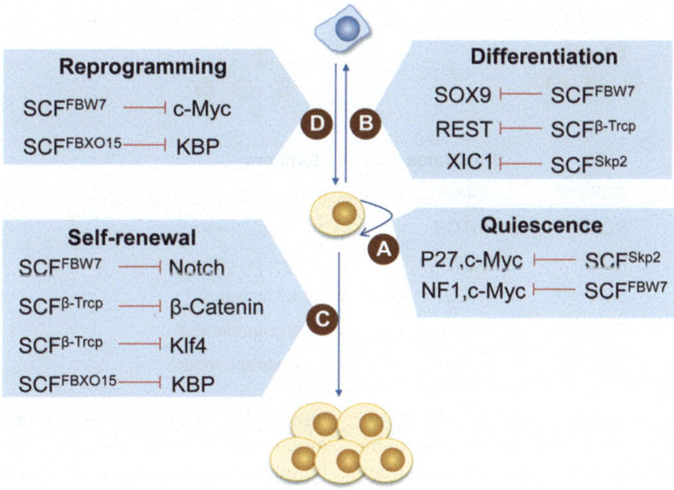

Fig. 6.1 CRL1 complexes maintain stem cell quiescence, self-renewal, differentiation, and reprogramming. E3 ligase-CRL1, also called SCF complex, can medicate the ubiquitination and degradations of multiple proteins which are important in stem cell quiescence(A), differentiation (B), self-renewal(C), and reprogramming(D). Examples of CRL1 and their substrates discussed in this review are shown in the figure

Thus, SCF^{FBW7} could regulate the proliferation of epidermal stem cells. Further, downregulated SCF^{FBW7} promotes the differentiation of keratinocyte (Onoyama et al. 2011).

In addition, FBW7 plays key roles in regulating lipogenesis, cell proliferation, and differentiation in the liver. Moreover, FBW7 deficiency skewed the differentiation of liver stem cells toward the cholangiocyte lineage rather than the hepatocyte lineage (Candi et al. 2006).

6.2.1.2 FBW7 in Cancer Stem Cell

Because we have previously introduced FBW7 can target several oncoproteins for degradation (Fig. 6.1), FBW7 is not surprising as a tumor suppressor (Fig. 6.2). In fact, FBW7 is located on chromosomal region 4q32, which is frequently lost in tumors. Moreover, highest mutation frequencies of FBW7 were found in tumors of the bile duct (cholangiocarcinoma, 35%), blood (T-cell acute lymphocytic leukemia, T-ALL 31%), endometrium (9%), colon (9%), and stomach (6%) (Akhoondi et al. 2007; Lee et al. 2006; Malyukova et al. 2007).

Ablation of FBW7 in the hematopoietic compartment results in T-ALL. In the hematopoietic compartment, FBW7 mutant mouse can promote the formation of cancer cells (King et al. 2013). Besides, due to the stabilization of c-Myc, some aged mice develop T-ALL (Onoyama et al. 2007). Interestingly, treating those mice with c-Myc inhibitors, like a cell-permeable small molecule (JQ1) that binds competitively to acetyl-lysine recognition motifs, or bromodomains, will lead to T-ALL remission (Filippakopoulos et al. 2010).

FBW7 regulates cell cycle via degrading cyclin E and Notch. Cyclin E is frequently dysregulated in cancer and can drive cells from the G0 or G1 phase to the S phase (Spruck et al. 1999). Besides, the aberrant activation of Notch promotes leukemia stem cell self-renewal (Rathinam et al. 2008). It is possible that loss of FBW7 promotes the development of T-ALL by stabilizing c-Myc, cyclin E, and Notch (O'Neil et al. 2007).

Loss of FBW7 accelerates intestinal tumorigenesis through promoting the accumulation of β-catenin in adenomas. Intestinal alterations and susceptibility to adenoma formation suggest that lack of FBW7 leads to stem cell crypt expansion (Babaei-Jadidi et al. 2011).

The mTOR is another downstream target of FBW7. Treatment of colon cancer cells with the mTOR inhibitor rapamycin suppresses their migration and tumor sphere formation.

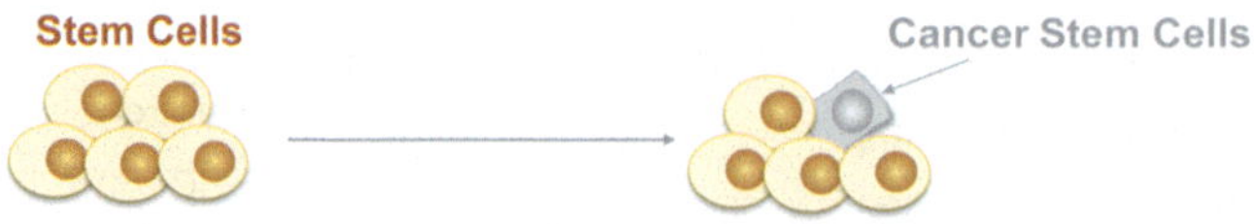

E3	Substrates	Cancers
SCF^{FBW7}	c-Myc	T-ALL CML
SCF^{FBW7}	mTOR	CCA, colon cancer
$SCF^{\beta\text{-}Trcp}$	β-Catenin	gastric cancer, prostate cancer
$SCF^{\beta\text{-}Trcp}$	REST	medulloblastoma
$SCF^{\beta\text{-}Trcp}$	IκBα	pancreatic carcinoma
SCF^{Skp2}	Akt	NPC, prostate cancer

Fig. 6.2 CRL1 complexes exhibit paradoxical functions in cancer stem cells. Cancer stem cells are thought to be derived from normal stem cells. CRL1 plays different roles in different kinds of cancers through ubiquitinate various substrates. In this figure, green indicates CRL complexes play roles in tumor suppression in the corresponding tumors. Red represents CRL complexes play roles in promoting cancer in the relevant tumors. *T-ALL* t-cell acute lymphocytic leukemia, *CML* chronic myelogenous leukemia, *CCA* cholangiocarcinoma, *NPC* nasopharyngeal carcinoma

Downregulated FBW7 expression promotes EMT through increasing expression of mesenchymal stem cell markers, SOX2, OCT4, and NANOG, leading to invasive phenotype, greater tumor-initiating potential, and non-adherent growth ability (Wang et al. 2013).

Loss of FBW7 is also associated with stem-like competence in cholangiocarcinoma (CCA), and rapamycin treatment of CCA cells suppresses invasion, metastatic potential, and tumor sphere formation (Yang et al. 2015).

Compared with T-ALL, FBW7 plays an important role in the development and the progression of CML (Fig. 6.2). FBW7 is critical in the maintenance of LICs in CML. Although c-Myc can promote tumor progression, overexpressed oncogenic c-Myc inhibits tumor growth by inducing p53-dependent apoptosis. Deletion of FBW7 increases c-Myc expression in LICs; decreases colony formation potential; eliminates leukemic cell infiltration in peripheral blood, the spleen, the liver, and the lungs; and leads to induction of apoptosis in a p53-dependent manner. Thus, the ablation of FBW7 leads to the apoptosis in LICs and eventually inhibits the progression of CML (Reavie et al. 2013). Moreover, knockdown of FBW7 causes LICs to differentiate and enter the cell cycle becoming sensitized to imatinib and cytosine arabinoside drug treatments (Takeishi et al. 2013).

6.2.2 β-TrCP

β-TrCP, also known as FWD1, is a versatile F-box protein in the SCF E3 ligase complex, with multiple substrates as degradation targets, such as Wnt, REST, and Klf4 (Fig. 6.1).

6.2.2.1 Wnt

Intestinal stem cells can replenish every cell type in this rapidly cycling tissue. Alterations in these stem cell functions might lead to the expansion of the stem-like cells, adenomas, and even cancer. Besides, Wnt signaling can regulate the self-renewal of crypt base columnar cells (Clevers et al. 2014).

In the absence of Wnt, β-catenin is phosphorylated, which mediates the recognition of it by the $SCF^{\beta\text{-}TrCP}$ for ubiquitination and degradation (Aberle et al. 1997; Hart et al. 1999). Wnt signals could inhibit the phosphorylation of β-catenin and finally drive a transcriptional program that promotes self-renewal (Hernandez et al. 2012). Mutations in β-TrCP lead to β-catenin stabilization and nuclear translocation, which subsequently leads to colon cancer (Ilyas et al. 1997).

6.2.2.2 REST

REST is expressed in non-neuronal cells and stem/progenitor neural cells, in which it inhibits neuronal differentiation by blocking the expression of neuron-specific genes (Ballas et al. 2005; Ooi and Wood 2007). The transition from embryonic stem cell to NSCs requires the proteasome-mediated degradation of REST. $SCF^{\beta\text{-}Trcp}$ can target REST for proteasomal degradation (Westbrook et al. 2008). Knockdown β-TrCP inhibits the differentiation of ESCs into neuronal. Consistently, the level of β-TrCP is upregulated during neuronal differentiation.

6.2.2.3 Klf4

Ubiquitinated Klf4 conjugates are observed at a high level in proliferating cells as compared with serum-starved cells (Chen et al. 2005). Klf4 expression is downregulated in response to TGF-β signaling (Hu and Wan 2011). Klf4 phosphorylation by ERK1 recruits β-TrCP to its N-terminal region for protein degradation (Kim et al. 2012). Thus, treatment with inhibitors of E3 ligases of β-TrCP might enhance self-renewal capacity and enable Klf4 to reprogram embryonic fibroblasts more efficiently.

6.2.2.4 Paradoxical Functions of β-TrCP in CSC

On one hand, β-TrCP can degrade β-catenin and thus inhibits Wnt signaling pathway, which is critical for cancer stem cell proliferation (Jiang and Struhl 1998). The genetic alterations of β-TrCP and the accumulation of β-catenin have also been confirmed in studies of gastric cancer and prostate cancer (Gerstein et al. 2002). REST is found to be overexpressed in human MB which arises from undifferentiated NSCs present in the cerebellum (Majumder 2006). GLI1 promotes the generation of many kinds of cancer stem cells. The protein stability of GLI1 is also regulated by E3 ubiquitin ligases $SCF^{\beta\text{-}TrCP}$ (Jiang 2006). Therefore, upregulation of β-TrCP may be a promising approach to the treatment of these diseases (Fig. 6.2).

On the other hand, β-TrCP can also accelerate the turnover of IκBα, activate the NF-κB pathway in pancreatic carcinoma cells, and enhance the chemoresistance of cancer cells (Fig. 6.2) (Muerkoster et al. 2005).

6.2.3 Skp2

Skp2 (S-phase kinase-associated protein 2, also known as FBXL1) is another well-studied F-box protein. It can recognize substrates via the LRR motif. Skp2 targets various proteins such as c-Myc, MEF, Xic1, p27, and Akt (Fig. 6.1).

6.2.3.1 c-Myc

Skp2 is the first identified E3 ligase for c-Myc in yeast cells. Skp2-mediated ubiquitination of c-Myc has been shown to regulate c-Myc transcriptional activity. In turn, Skp2-induced activation of several promoters requires c-Myc. The interaction between Skp2 and c-Myc leads to diminished c-Myc protein levels. In particular, Skp2-mediated c-Myc turnover is observed at the G1 to S-phase transition during the activation of resting lymphocytes. Apart from the Skp2 promotion of c-Myc degradation, Skp2 is also involved in regulating c-Myc's cellular function by enhancing c-Myc-induced S-phase entry (Kim et al. 2003).

6.2.3.2 MEF

Myeloid Elf-1-Like Factor (MEF) (also known as ELF4) is an ETS-related transcription factor with strong transcriptional activating activity that affects hematopoietic stem cell behavior and is required for normal NK cell and NK T-cell development. MEF is a short-lived protein whose expression level also peaks during late G1 phase. Overexpression of MEF drives cells through the G1/S transition, thereby promoting cell proliferation. Skp2 targets MEF for ubiquitination and proteolysis. In this way, Skp2 regulates HSC proliferation (Liu et al. 2006).

6.2.3.3 Xic1

Xenopus cyclin-dependent kinase inhibitor (Xic1) is essential for primary neurogenesis at an early stage and before these cells exit the cell cycle (Vernon et al. 2003). Skp2 degrades Xic1 in embryos, and this contributes to the ability of Skp2 to regulate neurogenesis. As a result, Skp2 plays important roles in differentiation of these neuronal cells (Boix-Perales et al. 2007).

6.2.3.4 p27

Skp2 knockout mice are viable and show a reduced growth rate and increased apoptosis (Nakayama et al. 2000). G1/S cyclin-dependent kinase inhibitor p27 is a tumor suppressor. Skp2 regulates apoptosis, cell cycle progression, and proliferation through promoting targeting p27 for ubiquitination and degradation (Nakayama et al. 2004). Depletion of Skp2 in LT-HSCs promotes proliferation (Rodriguez et al. 2011). In the Skp2-null cells, the expression levels of p27 are normal, but cyclin D1 is upregulated. Intriguingly, cyclin D1 isn't a target of Skp2.

6.2.3.5 Akt

The dysregulation of cell-cycle and aerobic glycolysis are important mechanisms for the self-renewal and proliferation of CSCs. Skp2 mediates the nonproteolytic K63-linked ubiquitination of Akt and promotes Akt-mediated glycolysis and tumorigenesis (Chan et al. 2012).

6.2.3.6 Skp2 in Cancer Stem Cells

It is not surprising that overexpression of Skp2 correlates with poor prognosis of nasopharyngeal carcinoma (NPC) and prostate cancer (Chan et al. 2013). Knockdown or pharmacological inhibition of Skp2 dramatically reduces the $ALDH^+$ CSC population and downregulates the sphere formation capability of NPC CSCs (Wang et al. 2014a). In prostate cancer, a small molecule inhibitor of Skp2 reduces prostate CSC population through p53-independent cellular senescence and inhibition of aerobic glycolysis. Moreover, Skp2 inhibitors sensitize prostate cancer cells to doxorubicin and cyclophosphamide drug therapy (Fig. 6.2) (Chan et al. 2013).

In addition to these three well-studied F-box proteins, there are other proteins that play roles in stem cell regulation (Fig. 6.1). Fbxo15 may reduce the exposure of ESCs to ROS (Donato et al. 2017). The TDH-GCN5L1-Fbxo15-KBP axis limits mitochondrial biogenesis in mouse ESCs. Therefore, SCF^{FBXO15} was also used as a marker for iPSCs (Donato et al. 2017). KDM2B (also known as FBXL10) is reported to control hematopoietic stem cell self-renewal, somatic cell reprogramming and senescence, and tumorigenesis (Ueda et al. 2015).

6.3 CRL2/5 and Stem Cell

ECS complex is another major type of RING finger E3 ligases, including three invariable subunits: Elongin B/C (adaptor protein), Cullin2/5 (scaffold protein), and Rbx1/2 (RING finger protein, interacting with E2). It has a variable substrate recognition element that binds to Elongin B/C via BC-box domain. There are two main groups in the BC-box proteins, von Hippel-Lindau (VHL-box) proteins and suppressors of cytokine signaling-box (SOCS box) proteins (Kamura et al. 2004).

6.3.1 VHL-box and CRL2

The VHL-box proteins include VHL, LRR-1, and FEM1B (Pozzebon et al. 2013). VHL was first reported as a tumor suppressor that is inactivated in the familial kidney cancer syndrome VHL disease (Latif et al. 1993). About 57% of sporadic clear cell cancers of the kidney contain VHL-inactivating mutations, and 98% lose heterozygosity at the VHL locus (Gnarra et al. 1994).

Just like progenitor cells in the hypoxic environment of bone marrow, stem cells are relatively deficient in oxygen, thereby protecting long-term HSCs from damage by ROS. Under hypoxic conditions, HIF1α transcription factor stabilizes stem cell quiescence and promotes cell survival under hypoxic conditions (Fig. 6.3).

Depletion of VHL stabilizes HIF1α protein and leads to HSC quiescence, which is determined by an increase in the number of LSK cells in G0 phase and a decrease in peripheral blood differentiation status (Takubo et al. 2010). When HSCs differentiate, they leave the HSCs niche, and then HIF1α is degraded by VHL. As a

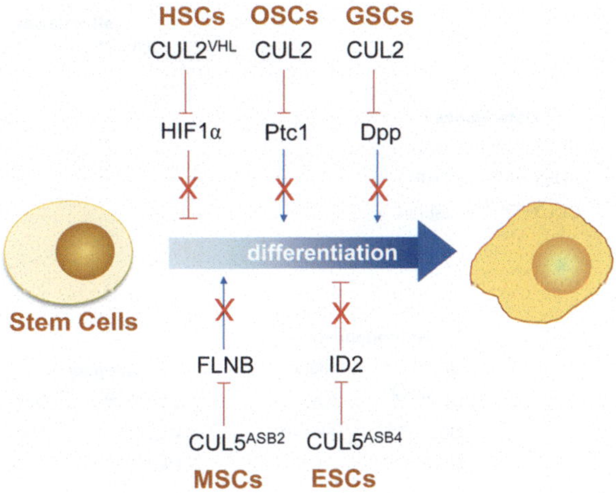

Fig. 6.3 CRL2/5 complexes control stem cell differentiation in multiple ways. CRL2 and CRL5 have similar structures. They adopt different BC-box proteins to facilitate the ubiquitination of substrates. As is shown in figure, CRL2 and CRL5 are involved in the differentiations of both embryonic stem cells and adult stem cells. *HSCs* hematopoietic stem cells, *OSCs* osteoblast stem cells, *GSCs* germline stem cells, *MSCs* muscle stem cells, *ESCs* embryonic stem cells

result, differentiation of cells in the peripheral blood is attenuated.

Furthermore, a hypoxic environment and stable HIF1α expression expand the subpopulation of cancer cells positive for CSC markers. Besides, hypoxia indirectly leads to chemotherapy resistance (Maugeri-Sacca et al. 2011). Therefore, introducing VHL into cancer cells may be an effective approach to reduce therapeutic resistance.

The sonic hedgehog (Shh) signaling pathway plays a crucial role in cell proliferation and differentiation via Patched1 (Ptc1), a 12-transmembrane receptor protein. The C-terminal cytoplasmic tail of Ptc1 can be cleaved to release the seventh intracellular domain (ICD7). Ptc1 ICD7 interacted with most components of the CUL2-based E3 ligase complex, including Elongin B/C, ZYG11B, and CUL2 itself. CUL2 knockdown inhibits the Shh-induced osteoblast stem cell differentiation (Yamaki et al. 2016).

Decapentaplegic (Dpp) is essential for the maintenance of germline stem cells (GSCs) in the *Drosophila* ovary (Xie and Spradling 1998). Depletion of CUL2 in somatic cells results in upregulation of Dpp signal and production of additional accumulation of extra GSC-like cells in the ovarian primordial germ chamber. Ectopic expression of CUL2 also regulates Dpp and thus controls the differentiation of GSCs to cystoblasts (CBs) (Ayyub et al. 2015).

6.3.2 SOCS Box and CRL5

SOCS box protein can be subdivided into four classes, SOCS, WSB, SSB, and ASB, which, respectively, contain SH2 domain, WD-40 repeats, SPRY domain, and Ankyrin repeats N terminal to the SOCS box (Hilton et al. 1998). They degrade many components of cytokine signaling and negatively regulate cell proliferation (Fig. 6.3) (Kamura et al. 2004).

Ankyrin-repeat SOCS box-containing protein 2 (ASB2) targets the actin-binding proteins filamin A and B (FLNA and FLNB) for proteasomal degradation. Reducing ASB2 in leukemia cells delays retinoic acid-induced differentiation and filamin degradation. Conversely, ASB2 expression induces filamin degradation. ASB2 may regulate hematopoietic cell differentiation by modulating cell proliferation and actin remodeling through targeting filamins for degradation (Heuze et al. 2008). During induced differentiation of C2C12 cells, knockdown of ASB2 delays FLNB degradation as well as myoblast

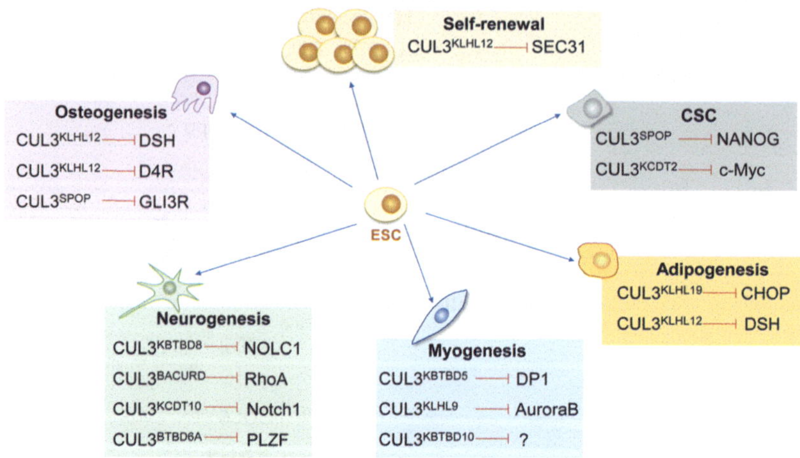

Fig. 6.4 CRL3 complexes regulate stem cell self-renewal and differentiation. CUL3 recruits different substrates through multiple BTB proteins. Embryonic stem cells undergo self-renewal and differentiation via different proteins that can serve as substrates of CRL3. Here, the functions of CRL3 in self-renewal, osteogenesis, neurogenesis, myogenesis, adipogenesis, and cancer stem cell development are exemplified

fusion and expression of muscle contractile proteins (Bello et al. 2009).

The ID proteins (ID1 to ID4) are known to mediate differentiation and cell cycle control, thereby affecting cellular functions such as metastasis, angiogenesis, apoptosis, and maintenance of trophoblast stem cells (TBSCs) (Zebedee and Hara 2001).The ubiquitin ligase Ankyrin-repeat, SOCS box-containing 4 (ASB4) is abundantly expressed in the developing placenta and is highly upregulated during the differentiation of embryonic stem cells into endothelial cell lineages (Ferguson et al. 2007). ASB4 promotes embryonic stem cell differentiation into vascular lineages by degrading ID2 (Townley-Tilson et al. 2014).

6.4 CRL3 and Stem Cell

CUL3 interacts with BTB (broad-complex, tramtrack, and bric à brac) proteins, which act as both substrate adaptors and receptors. The BTB domain is also known as the POZ domain (poxvirus and zinc finger) (Staller et al. 2003). The BTB domain mediates CUL3 binding, and an adjacent protein-interaction domain recruits substrate for ubiquitination (Zhuang et al. 2009). The BTB protein family are mainly composed of five subfamilies, including proteins with the BTB domain alone, the BTB domain with one or more zinc fingers (BTB-ZF), the BTB domain with one or more kelch repeats (BTB-Kelch), the BTB domain with the pipsqueak domain (BTB-PSQ), and the BTB domain with another functional domain (Chaharbakhshi and Jemc 2016). CUL3 knockout mice are lethality at E7.5 (Singer et al. 1999). CRL3 exhibit multiple functions in self-renewal, osteogenesis, neurogenesis, myogenesis, adipogenesis, and cancer stem cell development (Fig. 6.4).

6.4.1 Multiple Roles of KLHL12

A genetic screen first linked kelch-like family member 12 (KLHL12) complex with secretion of collagen, an essential component for stem cell niche (Jin et al. 2012). $CUL3^{KLHL12}$ monoubiquitinates the COPII vesicle protein SEC31, which forms a sufficiently large COPII vesicle and envelops the secreted collagen molecules into the extracellular space (Jin et al. 2012).

DSH can block the β-catenin destruction complex. In addition, the canonical Wnt pathway can stimulate osteogenesis and inhibit adipogenesis (James 2013). $CUL3^{KLHL12}$ targets DSH for degradation and inhibits Wnt signaling, which indicates a potential role in stem cell regulation (Angers et al. 2006).

D4R belongs to the G protein-coupled receptors and positively regulates osteogenesis (Lee et al. 2015). $CUL3^{KLHL12}$ promotes ubiquitination of the D4R and negatively regulates osteogenesis (Skieterska et al. 2016).

6.4.2 Multiple Roles of SPOP

Speckle type BTB/POZ protein (SPOP), also known as BTBD32, is one of the genes with the highest mutation frequency in prostate cancer. The mutation rate is 10–15%. The SPOP mutant prostate cancer has a higher degree of malignancy and a worse prognosis (Wang et al. 2019).

NANOG, an ESC transcription factor, maintains the self-renewal and pluripotency of ESCs (Chambers et al. 2003). NANOG is highly expressed in various cancers and plays a pleiotropic role in the tumorigenesis cascade, such as CSC population modulation (Chen et al. 2016). NANOG can be ubiquitinated by the Rbx1-CUL3-SPOP E3 ubiquitin ligase complex, which inhibits the self-renewal and tumorigenic ability of prostate cancer stem cells (Wang et al. 2019).

IHH signaling is required for chondrocyte and osteoblast differentiation. $CRL3^{SPOP}$ degrades the downstream transcription repressor GLI3R and thus promotes IHH signaling (Cai and Liu 2016).

RA controls differentiation of stem cells through RARs and transcriptional coactivators like the p160 family member SRC-3 (Ferry et al. 2011). SRC-3 is recruited to target promoters together with RARs but is subsequently phosphorylated and degraded by $CUL3^{SPOP}$ (Li et al. 2011a).

6.4.3 BTB Proteins in Neurogenesis

Neural crest cells are capable of differentiating to produce a variety of specialized cell types such as chondrocytes, melanocytes, and glial cells (Betancur et al. 2010; Simoes-Costa and Bronner 2015). Kelch repeat and BTB domain-containing 8 (KBTBD8) is an essential regulator of human and Xenopus tropicalis neural crest specification. $CUL3^{KBTBD8}$ monoubiquitinates NOLC1 and its paralogue TCOF1, which drives formation of a TCOF1-NOLC1 platform that connects RNA polymerase I with ribosome modification enzymes and remodels the translational program of differentiating cells in favor of neural crest (Dixon et al. 2006).

During development, neurons need to migrate to specific locations in the embryonic cerebral cortex and then establish dendritic spine connections. Rho GTPases can provide driving forces for cell migration (Ridley 2006). $CUL3^{BACURD}$ complexes were originally discovered in *Drosophila* which mediate the degradation of RhoA and controlled actin cytoskeleton structure and cell movement (Chen et al. 2009).

RND2 and RND3 also belong to Rho GTPase families. BACURD1 and BACURD2 reorganize the actin cytoskeleton via RND2 and RND3, to coordinate early steps in cortical neurogenesis which is critical for cortical neuron development (Gladwyn-Ng et al. 2016). Further, BACURD2 is identified as an interaction partner of RND2 (Gladwyn-Ng et al. 2015).

Notch signaling pathway consists of four receptors: Notch1–Notch4 (Takebe et al. 2015). Potassium channel tetramerization domain-containing 10 (KCDT10, also known as HBARUCD3) interacts with Notch1 and ubiquitinates and degrades Notch1 (Ren et al. 2014). KCTD10 may play roles in the development of neuroepithelium of neural tube and dorsal root ganglion (Sun et al. 2007).

Mutations of KLHL16 cause massive disorganization of the intermediate filaments (Cleveland et al. 2009). $CUL3^{KLHL20}$ can target RHOGEF for ubiquitination and degradation and then facilitate neurite outgrowth (Lin et al. 2011).

Promyelocytic leukemia zinc finger (PLZF, also known as ZBTB16, ZNF145) is a negative regulator of differentiation. BTBD6A targets PLZF for ubiquitination and degradation and then re-localizes PLZF from the nucleus to the cytoplasm in zebra fish (Sobieszczuk et al. 2010). Interestingly, the highly conserved PLZF is involved in many differentiation processes (Liu et al. 2016). It still needs further investigation.

6.4.4 BTB Proteins in Myogenesis

Mutations in KBTBD5, KBTBD10, as well as KBTBD13 are associated with nemaline myopathy (Sambuughin et al. 2012). Mutation in KLHL9 causes distal myopathy (Cirak et al. 2010).

KBTBD5 promotes the ubiquitination and degradation of DP1 (Gong et al. 2015). The transcription factor E2F1 forms a heterodimer complex with DP1 and regulates the expression of essential cell division genes involved in G1-S transition (Rowland and Bernards 2006). $CRL3^{KBTBD5}$ targets DP1 for degradation, thereby inhibiting E2F1-DP1-driven transcription, which is essential for skeletal muscle myogenesis (Gong et al. 2015). Besides, in C2C12 myoblasts, knockdown KBTBD10 inhibits myotube formation (du Puy et al. 2012). Moreover, KLHL9 can target Aurora B for ubiquitination and degradation (Sumara et al. 2007). Because Aurora B can regulate the assembly and disassembly of type III intermediate filaments, including vimentin and desmin, deficiency of KLHL9 destroys skeletal muscle functions (Gupta and Beggs 2014).

6.4.5 BTB Proteins in Other Stem Cell Regulation

Potassium channel tetramerization domain-containing protein 2 (KCTD2) interacts with c-Myc and promotes c-Myc protein ubiquitination and degradation. Besides, KCTD2 expression is markedly decreased in patient-derived glioma stem cells. As a result, KCTD2 depletion acquires glioma stem cell features and affects gliomagenesis by destabilizing c-Myc (Kim et al. 2017).

C/EBPα is considered as a primary transcription factor that mediates adipogenesis. Its activity is delayed by CHOP (also called growth arrest-DNA damage-induced 153, GADD153), a dominant negative form of C/EBP family members (Li et al. 2006). KLHL19 targets CHOP for degradation, thereby initiating adipogenesis (Huang et al. 2012).

6.5 CRL4 and Stem Cell

CUL4 proteins interact with the E2 enzyme via the RING finger protein Rbx1 at their C-terminus. Similar to CUL1, at their N-terminus, they employ DDB1 (DNA damage-binding protein 1), as an adaptor protein, which in turn recruits different substrate receptor proteins known as DDB1- and CUL4-associated factors (DCAFs) (Jackson and Xiong 2009). These DCAFs have been suggested to determine the substrate specificity in different CUL4-DDB1-based E3 ligase complexes (Lee and Zhou 2007). Most DCAFs have a WD40-repeat domain and mediate the binding of DCAF to DDB1 through a conserved "WDXR" or "DXR" structural region.

6.5.1 DDB1

Germline CUL4A-deleted mice are viable and display no gross abnormality (Liu et al. 2009), possibly due to redundancy with CUL4B, whereas DDB1 deletion is embryonic lethal, and embryos are not seen past E12.5 (Cang et al. 2006).

DDB1 plays a critical role in ESC self-renewal, and silencing of DDB1 leads ESCs to differentiate (Buckley et al. 2012). Silencing of DDB1 results in a significant reduction in the expression level of the pluripotency marker gene Oct3/4, coupled with morphological abnormalities in ESCs (Buckley et al. 2012). In a similar manner, the E3 ligase $CUL4A^{DDB1}$ supports the self-renewal of hematopoietic precursors. Hematopoietic stem cells highly express DDB1 which monitors stem cell divisions by ubiquitination and proteasomal degradation of p53 (Fig. 6.5) (Gao et al. 2015). Specific deletion of DDB1 in the brain results in elimination of neuronal progenitor cells, hemorrhages in the brain, and neonatal lethality (Cang et al. 2006).

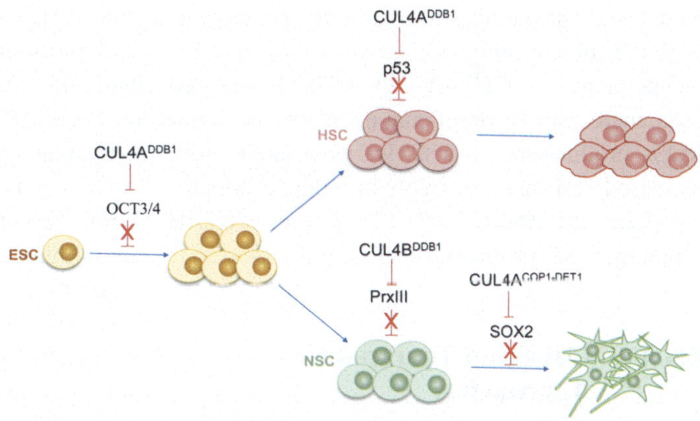

Fig. 6.5 CRL4 complexes modulate stem cell self-renewal and differentiation. CRL4 regulates self-renewal stem cells through target OCT3/4, p53, and PrxIII for degradations. Besides, CUL4A$^{COP1-DET1}$ destroyed the stemness of neural stem cells through interacts and degrade SOX2. *HSCs* hematopoietic stem cells, *NSCs* neural stem cells, *ESCs* embryonic stem cells

6.5.2 CRL4 in Spermatogonial Stem Cells

Mammalian spermatogenesis is a complex and highly regulated process. The pluripotent spermatogonial stem cells proliferate through mitotic cell divisions, differentiate into primary spermatocytes, and then undergo two meiotic divisions to produce haploid round spermatids, which subsequently differentiate into highly specialized spermatozoa through spermiogenesis (de Kretser et al. 1998). CUL4A and CUL4B have been shown to play distinct and essential roles during spermatogenesis (Kopanja et al. 2011; Lin et al. 2016; Yin et al. 2011, 2016). These two proteins display complementary expression patterns in adult mouse testis, where CUL4A is mainly present in meiosis-stage spermatocytes, while CUL4B is predominantly expressed in Sertoli cells, spermatogonia, and spermatids (Kopanja et al. 2011; Yin et al. 2011). Disruption of CUL4A caused male infertility because of the defective meiotic progression (Kopanja et al. 2011; Yin et al. 2011), whereas the male infertility phenotype resulted from CUL4B deletion due to abnormal postmeiotic sperm development (Lin et al. 2016; Yin et al. 2016).

In addition, DDB1- and CUL4-associated factor 17 (DCAF17) show high expression in testis. There exists a gradual increase in DCAF17 mRNA levels with the age. Deletion of DCAF17 does not have any effect on female fertility but causes spermatogenesis defects and male infertility in mice (Ali et al. 2018). Disruption of DCAF17 causes asymmetric acrosome capping, impaired nuclear compaction, and abnormal round spermatid to elongated spermatid transition. Therefore, DCAF17 is essential for spermiogenesis.

6.5.3 CRL4 in Neural Stem Cells

CUL4A expression is upregulated during neural stem cell differentiation, and CUL4A overexpression can increase Sox2 ubiquitination level and promote its degradation (Cui et al. 2018). The sex-determining region Y-box 2 (Sox2) is a key factor for maintaining NPC and ESC pluripotency (Graham et al. 2003). The DCAF family member constitutive photomorphogenic 1 (COP1) can directly interact with Sox2 and form a ligase complex CUL4A$^{COP1-DET1}$ to promote the differentiation of neural stem cells by increasing the ubiquitination of Sox2 (Cui et al. 2018). The deubiquitinase OTUD7B (OTU domain-containing protein 7B) antagonizes this effect and maintains the stemness of NPCs through deubiquitinating Sox2 (Cui et al. 2018).

Peroxiredoxin III (PrxIII) is a reactive oxygen species (ROS) scavenger. High endogenous ROS levels are required for NSC proliferation. CUL4B targets PrxIII for degradation (Li et al. 2011b). DDB1 and ROC1 in the DDB1-CUL4B-ROC1 complex are also indispensable for the proteolysis of PrxIII. In addition, the degradation of PrxIII is independent of CUL4A. In CUL4B-silenced cells, there exists a significant decrease in cellular ROS production, which is associated with increased resistance to hypoxia-induced apoptosis (Lee and Zhou 2007). Therefore, CUL4B promotes NSC proliferation (Fig. 6.5).

6.5.4 CUL4 and Therapeutic Approaches

Thalidomide interferes with limb development. Thalidomide and its derivatives lenalidomide and pomalidomide (together known as Immunomodulatory Drugs: IMiDs) emerged as effective treatments for multiple myeloma and 5q-dysplasia. CRBN (cereblon) is a substrate recognition component of a DCX (DDB1-CUL4-X-box) E3 protein ligase complex that mediates the ubiquitination and subsequent proteasomal degradation of target protein MEIS2. IMiDs block endogenous substrates MEIS2 from binding to $CRL4^{CRBN}$ E3 ubiquitin ligase and promote the ubiquitination of Ikaros/Aiolos (Ito et al. 2010). By interacting with both the CUL4-cereblon complexes, IMiDs can induce degradations that are beneficial for leukemia treatment (Fischer et al. 2014; Kronke et al. 2014; Lu et al. 2014). Similar to thalidomide, antitumor sulfonamides induce ubiquitination and proteasomal degradation of the splice regulator RBM39 through recruit CUL4-DCAF15 (Han et al. 2017). By elucidating the role of ubiquitination in stem cells and early human development, new and safe treatments will be developed to correct many of the genetic diseases that result from germline mutations.

6.6 Conclusions and Perspectives

Cullin-RING E3 ligases, like other E3s, can regulate different biological processes proteolytically or non-proteolytically. It is not surprising that CRLs have emerged as an important regulator of proteins which are critical for stem cell niche or transcriptional controls. Here, we summarize the roles of various CRLs in stem cell pluripotency, reprogramming, self-renewal, and differentiations. In addition, the deregulations of these CRLs are highly associated with developmental diseases and cancers. It provides us many candidate biomarkers and promising therapeutic targets. Moreover, stem cells are essential and limited, and many diseases can be treated by generating more effective iPSCs. Through analyzing the regulations of CRLs in certain known key transcriptional factors, we may extend the clinical applications of iPSCs. On the other hand, CRLs that can improve the efficiency of cellular reprogramming may also be potential targets. However, since only a very small portion of CRLs are well characterized in stem cell regulatory system, more work is needed. Identifying the roles and substrates of other functional CRLs will help us better understand the characteristics of stem cell and further gain more therapeutic approaches.

References

Aberle H, Bauer A, Stappert J, Kispert A, Kemler R (1997) beta-catenin is a target for the ubiquitin-proteasome pathway. EMBO J 16:3797–3804

Akhoondi S, Sun D, von der Lehr N, Apostolidou S, Klotz K, Maljukova A, Cepeda D, Fiegl H, Dafou D, Marth C et al (2007) FBXW7/hCDC4 is a general tumor suppressor in human cancer. Cancer Res 67: 9006–9012

Ali A, Mistry BV, Ahmed HA, Abdulla R, Amer HA, Prince A, Alazami AM, Alkuraya FS, Assiri A (2018) Deletion of DDB1- and CUL4- associated factor-17 (Dcaf17) gene causes spermatogenesis defects and male infertility in mice. Sci Rep 8:9202

Angers S, Thorpe CJ, Biechele TL, Goldenberg SJ, Zheng N, MacCoss MJ, Moon RT (2006) The KLHL12-Cullin-3 ubiquitin ligase negatively regulates the Wnt-beta-catenin pathway by targeting Dishevelled for degradation. Nat Cell Biol 8:348–357

Ayyub C, Banerjee KK, Joti P (2015) Reduction of Cullin-2 in somatic cells disrupts differentiation of germline stem cells in the Drosophila ovary. Dev Biol 405: 269–279

Babaei-Jadidi R, Li N, Saadeddin A, Spencer-Dene B, Jandke A, Muhammad B, Ibrahim EE, Muraleedharan R, Abuzinadah M, Davis H et al (2011) FBXW7 influences murine intestinal homeostasis and cancer, targeting Notch, Jun, and DEK for degradation. J Exp Med 208:295–312

Ballas N, Grunseich C, Lu DD, Speh JC, Mandel G (2005) REST and its corepressors mediate plasticity of

neuronal gene chromatin throughout neurogenesis. Cell 121:645–657
Barker N, van Es JH, Kuipers J, Kujala P, van den Born M, Cozijnsen M, Haegebarth A, Korving J, Begthel H, Peters PJ et al (2007) Identification of stem cells in small intestine and colon by marker gene Lgr5. Nature 449:1003–1007
Becker AJ, Mc CE, Till JE (1963) Cytological demonstration of the clonal nature of spleen colonies derived from transplanted mouse marrow cells. Nature 197: 452–454
Bello NF, Lamsoul I, Heuze ML, Metais A, Moreaux G, Calderwood DA, Duprez D, Moog-Lutz C, Lutz PG (2009) The E3 ubiquitin ligase specificity subunit ASB2beta is a novel regulator of muscle differentiation that targets filamin B to proteasomal degradation. Cell Death Differ 16:921–932
Betancur P, Bronner-Fraser M, Sauka-Spengler T (2010) Assembling neural crest regulatory circuits into a gene regulatory network. Annu Rev Cell Dev Biol 26: 581–603
Boix-Perales H, Horan I, Wise H, Lin HR, Chuang LC, Yew PR, Philpott A (2007) The E3 ubiquitin ligase skp2 regulates neural differentiation independent from the cell cycle. Neural Dev 2:27
Buckley SM, Aranda-Orgilles B, Strikoudis A, Apostolou E, Loizou E, Moran-Crusio K, Farnsworth CL, Koller AA, Dasgupta R, Silva JC et al (2012) Regulation of pluripotency and cellular reprogramming by the ubiquitin-proteasome system. Cell Stem Cell 11:783–798
Cai H, Liu A (2016) Spop promotes skeletal development and homeostasis by positively regulating Ihh signaling. Proc Natl Acad Sci U S A 113:14751–14756
Candi E, Rufini A, Terrinoni A, Dinsdale D, Ranalli M, Paradisi A, De Laurenzi V, Spagnoli LG, Catani MV, Ramadan S et al (2006) Differential roles of p63 isoforms in epidermal development: selective genetic complementation in p63 null mice. Cell Death Differ 13:1037–1047
Cang Y, Zhang J, Nicholas SA, Bastien J, Li B, Zhou P, Goff SP (2006) Deletion of DDB1 in mouse brain and lens leads to p53-dependent elimination of proliferating cells. Cell 127:929–940
Cardozo T, Pagano M (2004) The SCF ubiquitin ligase: insights into a molecular machine. Nat Rev Mol Cell Biol 5:739–751
Chaharbakhshi E, Jemc JC (2016) Broad-complex, tramtrack, and bric-a-brac (BTB) proteins: Critical regulators of development. Genesis 54:505–518
Chambers I, Colby D, Robertson M, Nichols J, Lee S, Tweedie S, Smith A (2003) Functional expression cloning of Nanog, a pluripotency sustaining factor in embryonic stem cells. Cell 113:643–655
Chan CH, Li CF, Yang WL, Gao Y, Lee SW, Feng Z, Huang HY, Tsai KK, Flores LG, Shao Y et al (2012) The Skp2-SCF E3 ligase regulates Akt ubiquitination, glycolysis, herceptin sensitivity, and tumorigenesis. Cell 149:1098–1111
Chan CH, Morrow JK, Li CF, Gao Y, Jin G, Moten A, Stagg LJ, Ladbury JE, Cai Z, Xu D et al (2013) Pharmacological inactivation of Skp2 SCF ubiquitin ligase restricts cancer stem cell traits and cancer progression. Cell 154:556–568
Chen CL, Uthaya Kumar DB, Punj V, Xu J, Sher L, Tahara SM, Hess S, Machida K (2016) NANOG metabolically reprograms tumor-initiating stem-like cells through tumorigenic changes in oxidative phosphorylation and fatty acid metabolism. Cell Metab 23:206–219
Chen Y, Yang Z, Meng M, Zhao Y, Dong N, Yan H, Liu L, Ding M, Peng HB, Shao F (2009) Cullin mediates degradation of RhoA through evolutionarily conserved BTB adaptors to control actin cytoskeleton structure and cell movement. Mol Cell 35:841–855
Chen ZY, Wang X, Zhou Y, Offner G, Tseng CC (2005) Destabilization of Kruppel-like factor 4 protein in response to serum stimulation involves the ubiquitin-proteasome pathway. Cancer Res 65:10394–10400
Cirak S, von Deimling F, Sachdev S, Errington WJ, Herrmann R, Bonnemann C, Brockmann K, Hinderlich S, Lindner TH, Steinbrecher A et al (2010) Kelch-like homologue 9 mutation is associated with an early onset autosomal dominant distal myopathy. Brain 133:2123–2135
Cleveland DW, Yamanaka K, Bomont P (2009) Gigaxonin controls vimentin organization through a tubulin chaperone-independent pathway. Hum Mol Genet 18:1384–1394
Clevers H, Loh KM, Nusse R (2014) Stem cell signaling. An integral program for tissue renewal and regeneration: Wnt signaling and stem cell control. Science 346:1248012
Cui CP, Zhang Y, Wang C, Yuan F, Li H, Yao Y, Chen Y, Li C, Wei W, Liu CH et al (2018) Dynamic ubiquitylation of Sox2 regulates proteostasis and governs neural progenitor cell differentiation. Nat Commun 9:4648
de Kretser DM, Loveland KL, Meinhardt A, Simorangkir D, Wreford N (1998) Spermatogenesis. Hum Reprod 13(Suppl 1):1–8
Dixon J, Jones NC, Sandell LL, Jayasinghe SM, Crane J, Rey JP, Dixon MJ, Trainor PA (2006) Tcof1/Treacle is required for neural crest cell formation and proliferation deficiencies that cause craniofacial abnormalities. Proc Natl Acad Sci USA 103:13403–13408
Donato V, Bonora M, Simoneschi D, Sartini D, Kudo Y, Saraf A, Florens L, Washburn MP, Stadtfeld M, Pinton P et al (2017) The TDH-GCN5L1-Fbxo15-KBP axis limits mitochondrial biogenesis in mouse embryonic stem cells. Nat Cell Biol 19:341–351
du Puy L, Beqqali A, van Tol HT, Monshouwer-Kloots J, Passier R, Haagsman HP, Roelen BA (2012) Sarcosin (Krp1) in skeletal muscle differentiation: gene expression profiling and knockdown experiments. Int J Dev Biol 56:301–309
Duan H, Wang Y, Aviram M, Swaroop M, Loo JA, Bian J, Tian Y, Mueller T, Bisgaier CL, Sun Y (1999) SAG, a

novel zinc RING finger protein that protects cells from apoptosis induced by redox agents. Mol Cell Biol 19:3145–3155

Duncan AW, Rattis FM, DiMascio LN, Congdon KL, Pazianos G, Zhao C, Yoon K, Cook JM, Willert K, Gaiano N et al (2005) Integration of Notch and Wnt signaling in hematopoietic stem cell maintenance. Nat Immunol 6:314–322

Ferguson JE 3rd, Wu Y, Smith K, Charles P, Powers K, Wang H, Patterson C (2007) ASB4 is a hydroxylation substrate of FIH and promotes vascular differentiation via an oxygen-dependent mechanism. Mol Cell Biol 27:6407–6419

Ferry C, Gaouar S, Fischer B, Boeglin M, Paul N, Samarut E, Piskunov A, Pankotai-Bodo G, Brino L, Rochette-Egly C (2011) Cullin 3 mediates SRC-3 ubiquitination and degradation to control the retinoic acid response. Proc Natl Acad Sci USA 108: 20603–20608

Filippakopoulos P, Qi J, Picaud S, Shen Y, Smith WB, Fedorov O, Morse EM, Keates T, Hickman TT, Felletar I et al (2010) Selective inhibition of BET bromodomains. Nature 468:1067–1073

Fischer ES, Bohm K, Lydeard JR, Yang H, Stadler MB, Cavadini S, Nagel J, Serluca F, Acker V, Lingaraju GM et al (2014) Structure of the DDB1-CRBN E3 ubiquitin ligase in complex with thalidomide. Nature 512:49–53

Fuchs E, Chen T (2013) A matter of life and death: self-renewal in stem cells. EMBO Rep 14:39–48

Galli F, Rossi M, D'Alessandra Y, De Simone M, Lopardo T, Haupt Y, Alsheich-Bartok O, Anzi S, Shaulian E, Calabro V et al (2010) MDM2 and Fbw7 cooperate to induce p63 protein degradation following DNA damage and cell differentiation. J Cell Sci 123:2423–2433

Gao J, Buckley SM, Cimmino L, Guillamot M, Strikoudis A, Cang Y, Goff SP, Aifantis I (2015) The CUL4-DDB1 ubiquitin ligase complex controls adult and embryonic stem cell differentiation and homeostasis. Elife 4:e07539

Gerstein AV, Almeida TA, Zhao G, Chess E, Shih Ie M, Buhler K, Pienta K, Rubin MA, Vessella R, Papadopoulos N (2002) APC/CTNNB1 (beta-catenin) pathway alterations in human prostate cancers. Genes Chromosomes Cancer 34:9–16

Gladwyn-Ng I, Huang L, Ngo L, Li SS, Qu Z, Vanyai HK, Cullen HD, Davis JM, Heng JI (2016) Bacurd1/Kctd13 and Bacurd2/Tnfaip1 are interacting partners to Rnd proteins which influence the long-term positioning and dendritic maturation of cerebral cortical neurons. Neural Dev 11:7

Gladwyn-Ng IE, Li SS, Qu Z, Davis JM, Ngo L, Haas M, Singer J, Heng JI (2015) Bacurd2 is a novel interacting partner to Rnd2 which controls radial migration within the developing mammalian cerebral cortex. Neural Dev 10:9

Gnarra JR, Tory K, Weng Y, Schmidt L, Wei MH, Li H, Latif F, Liu S, Chen F, Duh FM et al (1994) Mutations of the VHL tumour suppressor gene in renal carcinoma. Nat Genet 7:85–90

Gonczy P (2008) Mechanisms of asymmetric cell division: flies and worms pave the way. Nat Rev Mol Cell Biol 9:355–366

Gong W, Gohla RM, Bowlin KM, Koyano-Nakagawa N, Garry DJ, Shi X (2015) Kelch Repeat and BTB Domain Containing Protein 5 (Kbtbd5) Regulates Skeletal Muscle Myogenesis through the E2F1-DP1 Complex. J Biol Chem 290:15350–15361

Graham V, Khudyakov J, Ellis P, Pevny L (2003) SOX2 functions to maintain neural progenitor identity. Neuron 39:749–765

Gupta VA, Beggs AH (2014) Kelch proteins: emerging roles in skeletal muscle development and diseases. Skelet Muscle 4:11

Han T, Goralski M, Gaskill N, Capota E, Kim J, Ting TC, Xie Y, Williams NS, Nijhawan D (2017) Anticancer sulfonamides target splicing by inducing RBM39 degradation via recruitment to DCAF15. Science 356

Hann SR, Eisenman RN (1984) Proteins encoded by the human c-myc oncogene: differential expression in neoplastic cells. Mol Cell Biol 4:2486–2497

Hart M, Concordet JP, Lassot I, Albert I, del los Santos R, Durand H, Perret C, Rubinfeld B, Margottin F, Benarous R et al (1999) The F-box protein beta-TrCP associates with phosphorylated beta-catenin and regulates its activity in the cell. Curr Biol 9:207–210

He S, Nakada D, Morrison SJ (2009) Mechanisms of stem cell self-renewal. Annu Rev Cell Dev Biol 25:377–406

He YJ, McCall CM, Hu J, Zeng Y, Xiong Y (2006) DDB1 functions as a linker to recruit receptor WD40 proteins to CUL4-ROC1 ubiquitin ligases. Genes Dev 20:2949–2954

Heddleston JM, Li Z, Lathia JD, Bao S, Hjelmeland AB, Rich JN (2010) Hypoxia inducible factors in cancer stem cells. Br J Cancer 102:789–795

Hernandez AR, Klein AM, Kirschner MW (2012) Kinetic responses of beta-catenin specify the sites of Wnt control. Science 338:1337–1340

Heuze ML, Lamsoul I, Baldassarre M, Lad Y, Leveque S, Razinia Z, Moog-Lutz C, Calderwood DA, Lutz PG (2008) ASB2 targets filamins A and B to proteasomal degradation. Blood 112:5130–5140

Hilton DJ, Richardson RT, Alexander WS, Viney EM, Willson TA, Sprigg NS, Starr R, Nicholson SE, Metcalf D, Nicola NA (1998) Twenty proteins containing a C-terminal SOCS box form five structural classes. Proc Natl Acad Sci U S A 95:114–119

Hoeck JD, Jandke A, Blake SM, Nye E, Spencer-Dene B, Brandner S, Behrens A (2010) Fbw7 controls neural stem cell differentiation and progenitor apoptosis via Notch and c-Jun. Nat Neurosci 13:1365–1372

Hong X, Liu W, Song R, Shah JJ, Feng X, Tsang CK, Morgan KM, Bunting SF, Inuzuka H, Zheng XF et al (2016) SOX9 is targeted for proteasomal degradation by the E3 ligase FBW7 in response to DNA damage. Nucl Acids Res 44:8855–8869

Hu D, Wan Y (2011) Regulation of Kruppel-like factor 4 by the anaphase promoting complex pathway is involved in TGF-beta signaling. J Biol Chem 286: 6890–6901
Huang X, Ordemann J, Muller JM, Dubiel W (2012) The COP9 signalosome, cullin 3 and Keap1 supercomplex regulates CHOP stability and adipogenesis. Biol Open 1:705–710
Ilyas M, Tomlinson IP, Rowan A, Pignatelli M, Bodmer WF (1997) Beta-catenin mutations in cell lines established from human colorectal cancers. Proc Natl Acad Sci U S A 94:10330–10334
Ito T, Ando H, Suzuki T, Ogura T, Hotta K, Imamura Y, Yamaguchi Y, Handa H (2010) Identification of a primary target of thalidomide teratogenicity. Science 327:1345–1350
Jackson S, Xiong Y (2009) CRL4s: the CUL4-RING E3 ubiquitin ligases. Trends Biochem Sci 34:562–570
James AW (2013) Review of signaling pathways governing MSC osteogenic and adipogenic differentiation. Scientifica (Cairo) 2013:684736
Jiang J (2006) Regulation of Hh/Gli signaling by dual ubiquitin pathways. Cell Cycle 5:2457–2463
Jiang J, Struhl G (1998) Regulation of the Hedgehog and Wingless signalling pathways by the F-box/WD40-repeat protein Slimb. Nature 391:493–496
Jin J, Cardozo T, Lovering RC, Elledge SJ, Pagano M, Harper JW (2004) Systematic analysis and nomenclature of mammalian F-box proteins. Genes Dev 18:2573–2580
Jin L, Pahuja KB, Wickliffe KE, Gorur A, Baumgartel C, Schekman R, Rape M (2012) Ubiquitin-dependent regulation of COPII coat size and function. Nature 482:495–500
Kadaja M, Keyes BE, Lin M, Pasolli HA, Genander M, Polak L, Stokes N, Zheng D, Fuchs E (2014) SOX9: a stem cell transcriptional regulator of secreted niche signaling factors. Genes Dev 28:328–341
Kamura T, Maenaka K, Kotoshiba S, Matsumoto M, Kohda D, Conaway RC, Conaway JW, Nakayama KI (2004) VHL-box and SOCS-box domains determine binding specificity for Cul2-Rbx1 and Cul5-Rbx2 modules of ubiquitin ligases. Genes Dev 18:3055–3065
Kim EJ, Kim SH, Jin X, Jin X, Kim H (2017) KCTD2, an adaptor of Cullin3 E3 ubiquitin ligase, suppresses gliomagenesis by destabilizing c-Myc. Cell Death Differ 24:649–659
Kim MO, Kim SH, Cho YY, Nadas J, Jeong CH, Yao K, Kim DJ, Yu DH, Keum YS, Lee KY et al (2012) ERK1 and ERK2 regulate embryonic stem cell self-renewal through phosphorylation of Klf4. Nat Struct Mol Biol 19:283–290
Kim SY, Herbst A, Tworkowski KA, Salghetti SE, Tansey WP (2003) Skp2 regulates Myc protein stability and activity. Mol Cell 11:1177–1188
King B, Trimarchi T, Reavie L, Xu L, Mullenders J, Ntziachristos P, Aranda-Orgilles B, Perez-Garcia A, Shi J, Vakoc C et al (2013) The ubiquitin ligase FBXW7 modulates leukemia-initiating cell activity by regulating MYC stability. Cell 153:1552–1566
Kopanja D, Roy N, Stoyanova T, Hess RA, Bagchi S, Raychaudhuri P (2011) Cul4A is essential for spermatogenesis and male fertility. Dev Biol 352:278–287
Kronke J, Udeshi ND, Narla A, Grauman P, Hurst SN, McConkey M, Svinkina T, Heckl D, Comer E, Li X et al (2014) Lenalidomide causes selective degradation of IKZF1 and IKZF3 in multiple myeloma cells. Science 343:301–305
Latif F, Tory K, Gnarra J, Yao M, Duh FM, Orcutt ML, Stackhouse T, Kuzmin I, Modi W, Geil L et al (1993) Identification of the von Hippel-Lindau disease tumor suppressor gene. Science 260:1317–1320
Leblond CP, Walker BE (1956) Renewal of cell populations. Physiol Rev 36:255–276
Lee DJ, Tseng HC, Wong SW, Wang Z, Deng M, Ko CC (2015) Dopaminergic effects on in vitro osteogenesis. Bone Res 3:15020
Lee J, Zhou P (2007) DCAFs, the missing link of the CUL4-DDB1 ubiquitin ligase. Mol Cell 26:775–780
Lee JW, Soung YH, Kim HJ, Park WS, Nam SW, Kim SH, Lee JY, Yoo NJ, Lee SH (2006) Mutational analysis of the hCDC4 gene in gastric carcinomas. Eur J Cancer 42:2369–2373
Li C, Ao J, Fu J, Lee DF, Xu J, Lonard D, O'Malley BW (2011a) Tumor-suppressor role for the SPOP ubiquitin ligase in signal-dependent proteolysis of the oncogenic co-activator SRC-3/AIB1. Oncogene 30:4350–4364
Li X, Huang HY, Chen JG, Jiang L, Liu HL, Liu DG, Song TJ, He Q, Ma CG, Ma D et al (2006) Lactacystin inhibits 3T3-L1 adipocyte differentiation through induction of CHOP-10 expression. Biochem Biophys Res Commun 350:1–6
Li X, Lu D, He F, Zhou H, Liu Q, Wang Y, Shao C, Gong Y (2011b) Cullin 4B protein ubiquitin ligase targets peroxiredoxin III for degradation. J Biol Chem 286:32344–32354
Lin CY, Chen CY, Yu CH, Yu IS, Lin SR, Wu JT, Lin YH, Kuo PL, Wu JC, Lin SW (2016) Human X-linked intellectual disability factor CUL4B Is required for post-meiotic sperm development and male fertility. Sci Rep 6:20227
Lin MY, Lin YM, Kao TC, Chuang HH, Chen RH (2011) PDZ-RhoGEF ubiquitination by Cullin3-KLHL20 controls neurotrophin-induced neurite outgrowth. J Cell Biol 193:985–994
Linossi EM, Nicholson SE (2012) The SOCS box-adapting proteins for ubiquitination and proteasomal degradation. IUBMB Life 64:316–323
Liu L, Lee S, Zhang J, Peters SB, Hannah J, Zhang Y, Yin Y, Koff A, Ma L, Zhou P (2009) CUL4A abrogation augments DNA damage response and protection against skin carcinogenesis. Mol Cell 34:451–460
Liu TM, Lee EH, Lim B, Shyh-Chang N (2016) Concise review: balancing stem cell self-renewal and differentiation with PLZF. Stem Cells 34:277–287
Liu Y, Hedvat CV, Mao S, Zhu XH, Yao J, Nguyen H, Koff A, Nimer SD (2006) The ETS protein MEF is

regulated by phosphorylation-dependent proteolysis via the protein-ubiquitin ligase SCFSkp2. Mol Cell Biol 26:3114–3123
Lu G, Middleton RE, Sun H, Naniong M, Ott CJ, Mitsiades CS, Wong KK, Bradner JE, Kaelin WG Jr (2014) The myeloma drug lenalidomide promotes the cereblon-dependent destruction of Ikaros proteins. Science 343:305–309
Majumder S (2006) REST in good times and bad: roles in tumor suppressor and oncogenic activities. Cell Cycle 5:1929–1935
Malyukova A, Dohda T, von der Lehr N, Akhoondi S, Corcoran M, Heyman M, Spruck C, Grander D, Lendahl U, Sangfelt O (2007) The tumor suppressor gene hCDC4 is frequently mutated in human T-cell acute lymphoblastic leukemia with functional consequences for Notch signaling. Cancer Res 67: 5611–5616
Matsumoto A, Onoyama I, Sunabori T, Kageyama R, Okano H, Nakayama KI (2011) Fbxw7-dependent degradation of Notch is required for control of "stemness" and neuronal-glial differentiation in neural stem cells. J Biol Chem 286:13754–13764
Matsuoka S, Oike Y, Onoyama I, Iwama A, Arai F, Takubo K, Mashimo Y, Oguro H, Nitta E, Ito K et al (2008) Fbxw7 acts as a critical fail-safe against premature loss of hematopoietic stem cells and development of T-ALL. Genes Dev 22:986–991
Maugeri-Sacca M, Vigneri P, De Maria R (2011) Cancer stem cells and chemosensitivity. Clin Cancer Res 17:4942–4947
Muerkoster S, Arlt A, Sipos B, Witt M, Grossmann M, Kloppel G, Kalthoff H, Folsch UR, Schafer H (2005) Increased expression of the E3-ubiquitin ligase receptor subunit betaTRCP1 relates to constitutive nuclear factor-kappaB activation and chemoresistance in pancreatic carcinoma cells. Cancer Res 65:1316–1324
Nakayama K, Nagahama H, Minamishima YA, Matsumoto M, Nakamichi I, Kitagawa K, Shirane M, Tsunematsu R, Tsukiyama T, Ishida N et al (2000) Targeted disruption of Skp2 results in accumulation of cyclin E and p27(Kip1), polyploidy and centrosome overduplication. EMBO J 19:2069–2081
Nakayama K, Nagahama H, Minamishima YA, Miyake S, Ishida N, Hatakeyama S, Kitagawa M, Iemura S, Natsume T, Nakayama KI (2004) Skp2-mediated degradation of p27 regulates progression into mitosis. Dev Cell 6:661–672
Nakayama KI, Nakayama K (2005) Regulation of the cell cycle by SCF-type ubiquitin ligases. Semin Cell Dev Biol 16:323–333
Neumuller RA, Knoblich JA (2009) Dividing cellular asymmetry: asymmetric cell division and its implications for stem cells and cancer. Genes Dev 23: 2675–2699
Nguyen LV, Vanner R, Dirks P, Eaves CJ (2012) Cancer stem cells: an evolving concept. Nat Rev Cancer 12:133–143
Nie L, Perry SS, Zhao Y, Huang J, Kincade PW, Farrar MA, Sun XH (2008) Regulation of lymphocyte development by cell-type-specific interpretation of Notch signals. Mol Cell Biol 28:2078–2090
O'Neil J, Grim J, Strack P, Rao S, Tibbitts D, Winter C, Hardwick J, Welcker M, Meijerink JP, Pieters R et al (2007) FBW7 mutations in leukemic cells mediate NOTCH pathway activation and resistance to gamma-secretase inhibitors. J Exp Med 204:1813–1824
Onoyama I, Tsunematsu R, Matsumoto A, Kimura T, de Alboran IM, Nakayama K, Nakayama KI (2007) Conditional inactivation of Fbxw7 impairs cell-cycle exit during T cell differentiation and results in lymphomatogenesis. J Exp Med 204:2875–2888
Onoyama I, Suzuki A, Matsumoto A, Tomita K, Katagiri H, Oike Y, Nakayama K, Nakayama KI (2011) Fbxw7 regulates lipid metabolism and cell fate decisions in the mouse liver. J Clin Invest 121:342–354
Ooi L, Wood IC (2007) Chromatin crosstalk in development and disease: lessons from REST. Nat Rev Genet 8:544–554
Passegue E, Wagers AJ (2006) Regulating quiescence: new insights into hematopoietic stem cell biology. Dev Cell 10:415–417
Petroski MD, Deshaies RJ (2005) Function and regulation of cullin-RING ubiquitin ligases. Nat Rev Mol Cell Biol 6:9–20
Plechanovova A, Jaffray EG, Tatham MH, Naismith JH, Hay RT (2012) Structure of a RING E3 ligase and ubiquitin-loaded E2 primed for catalysis. Nature 489:115–120
Pozzebon ME, Varadaraj A, Mattoscio D, Jaffray EG, Miccolo C, Galimberti V, Tommasino M, Hay RT, Chiocca S (2013) BC-box protein domain-related mechanism for VHL protein degradation. Proc Natl Acad Sci U S A 110:18168–18173
Rathinam C, Thien CB, Langdon WY, Gu H, Flavell RA (2008) The E3 ubiquitin ligase c-Cbl restricts development and functions of hematopoietic stem cells. Genes Dev 22:992–997
Reavie L, Della Gatta G, Crusio K, Aranda-Orgilles B, Buckley SM, Thompson B, Lee E, Gao J, Bredemeyer AL, Helmink BA et al (2010) Regulation of hematopoietic stem cell differentiation by a single ubiquitin ligase-substrate complex. Nat Immunol 11:207–215
Reavie L, Buckley SM, Loizou E, Takeishi S, Aranda-Orgilles B, Ndiaye-Lobry D, Abdel-Wahab O, Ibrahim S, Nakayama KI, Aifantis I (2013) Regulation of c-Myc ubiquitination controls chronic myelogenous leukemia initiation and progression. Cancer Cell 23:362–375
Reed SI (2003) Ratchets and clocks: the cell cycle, ubiquitylation and protein turnover. Nat Rev Mol Cell Biol 4:855–864
Reiman JM, Knutson KL, Radisky DC (2010) Immune promotion of epithelial-mesenchymal transition and generation of breast cancer stem cells. Cancer Res 70:3005–3008

Ren K, Yuan J, Yang M, Gao X, Ding X, Zhou J, Hu X, Cao J, Deng X, Xiang S et al (2014) KCTD10 is involved in the cardiovascular system and Notch signaling during early embryonic development. PLoS One 9:e112275

Ridley AJ (2006) Rho GTPases and actin dynamics in membrane protrusions and vesicle trafficking. Trends Cell Biol 16:522–529

Rodriguez S, Wang L, Mumaw C, Srour EF, Lo Celso C, Nakayama K, Carlesso N (2011) The SKP2 E3 ligase regulates basal homeostasis and stress-induced regeneration of HSCs. Blood 117:6509–6519

Rowland BD, Bernards R (2006) Re-evaluating cell-cycle regulation by E2Fs. Cell 127:871–874

Saha A, Lewis S, Kleiger G, Kuhlman B, Deshaies RJ (2011) Essential role for ubiquitin-ubiquitin-conjugating enzyme interaction in ubiquitin discharge from Cdc34 to substrate. Mol Cell 42:75–83

Sambuughin N, Swietnicki W, Techtmann S, Matrosova V, Wallace T, Goldfarb L, Maynard E (2012) KBTBD13 interacts with Cullin 3 to form a functional ubiquitin ligase. Biochem Biophys Res Commun 421:743–749

Simoes-Costa M, Bronner ME (2015) Establishing neural crest identity: a gene regulatory recipe. Development 142:242–257

Singer JD, Gurian-West M, Clurman B, Roberts JM (1999) Cullin-3 targets cyclin E for ubiquitination and controls S phase in mammalian cells. Genes Dev 13:2375–2387

Skieterska K, Shen A, Clarisse D, Rondou P, Borroto-Escuela DO, Lintermans B, Fuxe K, Xiang YK, Van Craenenbroeck K (2016) Characterization of the interaction between the dopamine D4 receptor, KLHL12 and beta-arrestins. Cell Signal 28:1001–1014

Sobieszczuk DF, Poliakov A, Xu Q, Wilkinson DG (2010) A feedback loop mediated by degradation of an inhibitor is required to initiate neuronal differentiation. Genes Dev 24:206–218

Spruck CH, Won KA, Reed SI (1999) Deregulated cyclin E induces chromosome instability. Nature 401: 297–300

Staller P, Sulitkova J, Lisztwan J, Moch H, Oakeley EJ, Krek W (2003) Chemokine receptor CXCR4 downregulated by von Hippel-Lindau tumour suppressor pVHL. Nature 425:307–311

Stogios PJ, Downs GS, Jauhal JJ, Nandra SK, Prive GG (2005) Sequence and structural analysis of BTB domain proteins. Genome Biol 6:R82

Sumara I, Quadroni M, Frei C, Olma MH, Sumara G, Ricci R, Peter M (2007) A Cul3-based E3 ligase removes Aurora B from mitotic chromosomes, regulating mitotic progression and completion of cytokinesis in human cells. Dev Cell 12:887–900

Sun JK, Zhang B, Zhang J, Zhou JL (2007) Preparation of mouse KCTD10 antibody and expression analysis of KCTD10 in neuroepithelium of neural tube and dorsal root ganglion. Sheng Wu Gong Cheng Xue Bao 23:1011–1016

Takebe N, Ivy SP (2010) Controversies in cancer stem cells: targeting embryonic signaling pathways. Clin Cancer Res 16:3106–3112

Takebe N, Miele L, Harris PJ, Jeong W, Bando H, Kahn M, Yang SX, Ivy SP (2015) Targeting Notch, Hedgehog, and Wnt pathways in cancer stem cells: clinical update. Nat Rev Clin Oncol 12:445–464

Takeishi S, Nakayama KI (2014) Role of Fbxw7 in the maintenance of normal stem cells and cancer-initiating cells. Br J Cancer 111:1054–1059

Takeishi S, Matsumoto A, Onoyama I, Naka K, Hirao A, Nakayama KI (2013) Ablation of Fbxw7 eliminates leukemia-initiating cells by preventing quiescence. Cancer Cell 23:347–361

Takubo K, Goda N, Yamada W, Iriuchishima H, Ikeda E, Kubota Y, Shima H, Johnson RS, Hirao A, Suematsu M et al (2010) Regulation of the HIF-1alpha level is essential for hematopoietic stem cells. Cell Stem Cell 7:391–402

Tan M, Zhao Y, Kim SJ, Liu M, Jia L, Saunders TL, Zhu Y, Sun Y (2011) SAG/RBX2/ROC2 E3 ubiquitin ligase is essential for vascular and neural development by targeting NF1 for degradation. Dev Cell 21: 1062–1076

Tetzlaff MT, Yu W, Li M, Zhang P, Finegold M, Mahon K, Harper JW, Schwartz RJ, Elledge SJ (2004) Defective cardiovascular development and elevated cyclin E and Notch proteins in mice lacking the Fbw7 F-box protein. Proc Natl Acad Sci U S A 101: 3338–3345

Thompson BJ, Jankovic V, Gao J, Buonamici S, Vest A, Lee JM, Zavadil J, Nimer SD, Aifantis I (2008) Control of hematopoietic stem cell quiescence by the E3 ubiquitin ligase Fbw7. J Exp Med 205:1395–1408

Thomson JA, Itskovitz-Eldor J, Shapiro SS, Waknitz MA, Swiergiel JJ, Marshall VS, Jones JM (1998) Embryonic stem cell lines derived from human blastocysts. Science 282:1145–1147

Townley-Tilson WH, Wu Y, Ferguson JE 3rd, Patterson C (2014) The ubiquitin ligase ASB4 promotes trophoblast differentiation through the degradation of ID2. PLoS One 9:e89451

Trumpp A, Essers M, Wilson A (2010) Awakening dormant haematopoietic stem cells. Nat Rev Immunol 10:201–209

Ueda T, Nagamachi A, Takubo K, Yamasaki N, Matsui H, Kanai A, Nakata Y, Ikeda K, Konuma T, Oda H et al (2015) Fbxl10 overexpression in murine hematopoietic stem cells induces leukemia involving metabolic activation and upregulation of Nsg2. Blood 125:3437–3446

Vernon AE, Devine C, Philpott A (2003) The cdk inhibitor p27Xic1 is required for differentiation of primary neurones in Xenopus. Development 130:85–92

Wang J, Huang Y, Guan Z, Zhang JL, Su HK, Zhang W, Yue CF, Yan M, Guan S, Liu QQ (2014a) E3-ligase Skp2 predicts poor prognosis and maintains cancer stem cell pool in nasopharyngeal carcinoma. Oncotarget 5:5591–5601

Wang X, Jin J, Wan F, Zhao L, Chu H, Chen C, Liao G, Liu J, Yu Y, Teng H et al (2019) AMPK promotes SPOP-mediated NANOG degradation to regulate prostate cancer cell stemness. Dev Cell 48:345–360 e347

Wang Y, Liu Y, Lu J, Zhang P, Wang Y, Xu Y, Wang Z, Mao JH, Wei G (2013) Rapamycin inhibits FBXW7 loss-induced epithelial-mesenchymal transition and cancer stem cell-like characteristics in colorectal cancer cells. Biochem Biophys Res Commun 434:352–356

Wang Z, Liu P, Inuzuka H, Wei W (2014b) Roles of F-box proteins in cancer. Nat Rev Cancer 14:233–247

Warr MR, Pietras EM, Passegue E (2011) Mechanisms controlling hematopoietic stem cell functions during normal hematopoiesis and hematological malignancies. Wiley Interdiscip Rev Syst Biol Med 3:681–701

Weissman AM (2001) Themes and variations on ubiquitylation. Nat Rev Mol Cell Biol 2:169–178

Wenzel DM, Lissounov A, Brzovic PS, Klevit RE (2011) UBCH7 reactivity profile reveals parkin and HHARI to be RING/HECT hybrids. Nature 474:105–108

Westbrook TF, Hu G, Ang XL, Mulligan P, Pavlova NN, Liang A, Leng Y, Maehr R, Shi Y, Harper JW et al (2008) SCFbeta-TRCP controls oncogenic transformation and neural differentiation through REST degradation. Nature 452:370–374

Xie T, Spradling AC (1998) decapentaplegic is essential for the maintenance and division of germline stem cells in the Drosophila ovary. Cell 94:251–260

Yada M, Hatakeyama S, Kamura T, Nishiyama M, Tsunematsu R, Imaki H, Ishida N, Okumura F, Nakayama K, Nakayama KI (2004) Phosphorylation-dependent degradation of c-Myc is mediated by the F-box protein Fbw7. The EMBO J 23:2116–2125

Yamaki Y, Kagawa H, Hatta T, Natsume T, Kawahara H (2016) The C-terminal cytoplasmic tail of hedgehog receptor Patched1 is a platform for E3 ubiquitin ligase complexes. Mol Cell Biochem 414:1–12

Yang H, Lu X, Liu Z, Chen L, Xu Y, Wang Y, Wei G, Chen Y (2015) FBXW7 suppresses epithelial-mesenchymal transition, stemness and metastatic potential of cholangiocarcinoma cells. Oncotarget 6: 6310–6325

Yau R, Rape M (2016) The increasing complexity of the ubiquitin code. Nat Cell Biol 18:579–586

Yin Y, Lin C, Kim ST, Roig I, Chen H, Liu L, Veith GM, Jin RU, Keeney S, Jasin M et al (2011) The E3 ubiquitin ligase Cullin 4A regulates meiotic progression in mouse spermatogenesis. Dev Biol 356:51–62

Yin Y, Liu L, Yang C, Lin C, Veith GM, Wang C, Sutovsky P, Zhou P, Ma L (2016) Cell autonomous and nonautonomous function of CUL4B in mouse spermatogenesis. J Biol Chem 291:6923–6935

Zebedee Z, Hara E (2001) Id proteins in cell cycle control and cellular senescence. Oncogene 20:8317–8325

Zhao Y, Sun Y (2013) Cullin-RING Ligases as attractive anti-cancer targets. Curr Pharm Des 19: 3215–3225

Zhuang M, Calabrese MF, Liu J, Waddell MB, Nourse A, Hammel M, Miller DJ, Walden H, Duda DM, Seyedin SN et al (2009) Structures of SPOP-substrate complexes: insights into molecular architectures of BTB-Cul3 ubiquitin ligases. Mol Cell 36:39–50

Viral Manipulations of the Cullin-RING Ubiquitin Ligases

7

Ying Liu and Xu Tan

Abstract

Cullin-RING ubiquitin ligases (CRLs) are efficient and diverse toolsets of the cells to regulate almost every biological process. However, these characteristics have also been usurped by many viruses to optimize for their replication. CRLs are often at the forefront of the arms races in the coevolution of viruses and hosts. Here we review the modes of actions and functional consequences of viral manipulations of host cell CRLs. We also discuss the therapeutic applications to target these viral manipulations for treating viral infections.

Keywords

Cullins · Virus infection · CRL · Degradation · Ubiquitination

Abbreviations

CRL	Cullin-RING ubiquitin ligases
Cul1-5	Cullin 1-5
DCAF	DDB1-Cul4-associated factors
HBV	Hepatitis B virus
HIV	Human immunodeficiency virus
KSHV	Kaposi's sarcoma-associated herpesvirus
RSV	Respiratory syncytial virus
SCF	SKP1-Cullin1-Fbox E3 ligase
SR	Substrate receptor

Viruses rely on the host cell machinery to infect and replicate. It is remarkable that viruses can use their small repertoire of proteins to control the host cells and redirect almost every cellular apparatus toward massive production of viral particles. The protein degradation machinery is not spared by the virus. In fact, viruses heavily manipulate the ubiquitin proteasome system to optimize for viral replication (Hershko and Ciechanover 1998; Schwartz and Ciechanover 1999). Cullin-RING ubiquitin ligases, the largest family of eukaryote ubiquitin ligases, are hijacked by viruses that are very different taxonomically to evade different human defense mechanisms (Mahon et al. 2014; Barry and Fruh 2006). For example, paramyxovirus, which belongs to the *Paramyxoviridae*, and Hepatitis B virus (HBV), which belongs to the *Hepadnaviridae*, both use a small protein (V protein for paramyxovirus and HBx for HBV) to hijack the Cul4-DDB1 (CRL4) E3 ligase to target multiple host proteins for ubiquitination and degradation (Li et al. 2006; Decorsiere et al. 2016). As another example of the widespread viral hijacking of Cullin-RING, HIV use multiple accessory proteins, Vpu, Vpr, Vif, and Vpx (Vpx

Y. Liu · X. Tan (✉)
MOE Key Laboratory of Bioorganic Phosphorus Chemistry & Chemical Biology, Beijing Advanced Innovation Center for Structural Biology, Beijing Frontier Research Center for Biological Structures, School of Pharmaceutical Sciences, Tsinghua University, Beijing, China
e-mail: xutan@tsinghua.edu.cn

Y. Sun et al. (eds.), *Cullin-RING Ligases and Protein Neddylation*, Advances in Experimental Medicine and Biology 1217, https://doi.org/10.1007/978-981-15-1025-0_7

is HIV-2 specific), to hijack multiple Cul1-, Cul4-, and Cul5-based CRLs (Sauter and Kirchhoff 2018; Malim and Bieniasz 2012). These viral proteins usually function as adaptors between the CRLs and the protein targets, which are often involved in host antiviral responses. The accessory proteins possess multiple protein binding sites that can interact with both the CRL substrate receptor and the target proteins, thereby recruiting the target proteins to the CRLs for ubiquitination and subsequent degradation.

Evidence is accumulating to support that a large number of proteins are affected by the viral hijacking. For example, a recent study showed that hundreds of proteins might be downregulated by HIV Vpr, including proteins from a variety of function categories such as RNA-binding proteins and DNA-binding proteins (Greenwood et al. 2019). Our own study quantitated the proteome of HIV-1-infected primary CD4+ T cells and uncovered protein level changes of hundreds of human proteins. Importantly, these protein changes are not associated with mRNA level changes determined by RNA-seq in the same study, supporting widespread alterations of protein stability by the viral proteins (Liu et al. 2019). It is expected that proteins targeted by the virus-hijacked ubiquitin ligases would be ubiquitinated and degraded, explaining the downregulated protein levels. But what is the mechanism of the upregulation of protein levels? An analysis of a published list of physiological substrates of CRL5 E3s shows that most of them are upregulated by HIV infection (Liu et al. 2019). This phenomenon is likely explained by HIV Vif's hijacking of CRL5, which competes away the physiological substrates from the E3s. The breadth and complexity of the cellular proteins destabilized/stabilized by viral hijacking of the host ubiquitin ligases highlight the importance of this hijacking to the viral replication and pathogenesis. The importance of the viral hijacking is also reflected in the positive selection of the amino acid sequences involved in the binding sites between the viral protein and the host target proteins (Daugherty and Malik 2012). Positive selection in amino acid sequences is a hallmark of the evolutionary arms race between a host protein and its viral antagonist protein, providing strong evidence that the interaction is critical to the survival of both the virus and the host during the coevolution of the two parties (Daugherty and Malik 2012).

The widespread viral hijacking of CRLs is not accidental. Ubiquitination-mediated protein degradation is involved in almost every biological process, regulating important protein turnover that often controls the progression and magnitude of the process in a rapid fashion (Deshaies and Joazeiro 2009; Zheng and Shabek 2017). The target proteins can be massively degraded within a few hours, which are critical for viruses to evade the cellular defense system to expedite their replication (Weekes et al. 2014; Matheson et al. 2015). In addition to speed, the catalytic nature of ubiquitination and proteasomal degradation ensures that regulation of the ubiquitin ligases can have a magnifying effect due to the low reaction stoichiometry, i.e., one ligase molecule can ubiquitinate many substrate molecules (Harper and Tan 2012). When hijacked by a virus, this stoichiometry ratio can maximize the impact of the hijacked ligase. A third attribute of CRLs is that they are highly modular, thereby making it straightforward for the viral protein to exploit or interfere (Deshaies and Joazeiro 2009; Zheng and Shabek 2017). Usually, substrate specificity of CRLs are solely determined by individual substrate receptor (SR) proteins. In addition, the catalytic mechanism of the ubiquitination reaction is through induced proximity (Deshaies and Joazeiro 2009). These properties facilitate viral hijacking by simply introducing new protein-protein interaction between SR and a host defense protein. The viral protein can bind both the SR and the host defense protein and functions as a linker between the two (Mahon et al. 2014). The formation of this complex is sufficient to enable the ubiquitination of the host defense protein. The high modularity and proximity-based catalysis of CRLs provide great versatility to evolve new functionality in host evolution, which likely account for the great diversity of the current CRL family members. However, this property has also been taken advantaged by a variety of viruses to work against the hosts. On the other hand, the widespread viral hijacking of CRLs and the importance of the hijacking to viral replication indicate that targeting these hijacking events might provide opportunities to discover effective antiviral drugs (Huang and Dixit 2016; Votteler and Schubert 2008).

In addition to viral hijacking of the CRL ligases, viral proteins can affect ligase activities in many other ways. For example, they can affect phosphorylation of the substrate proteins to promote their ubiquitination, as in the case of KSHV V-cyclin's phosphorylation of p27, which leads to p27 ubiquitination by CRL1-SKP2 ligase and subsequent degradation (Ashizawa et al. 2012; Ellis et al. 1999; Liu et al. 2007). As another sample, Epstein-Barr virus (EBV) encodes a protease named BPLF1, which can function as a deneddylase to remove Nedd8 protein from Cullins and inhibit CRL activity (Gastaldello et al. 2010). Below, we will discuss in detail a number of examples of viral manipulations of the different family of CRLs and the therapeutic implications of these molecular events. We will emphasize on HIV-1 given its many well-studied examples of hijacking multiple CRLs.

7.1 Cullin 1-RING Ubiquitin Ligases (CRL1)

CRL1 E3 ligases are the prototype of the whole CRL family. The invariable components of CRL1 E3s are Cullin 1, Rbx1 (also named Roc1), and the adaptor protein SKP1. SKP1 recognizes different SRs through binding to the highly conserved F-box domain of the SRs (Zheng and Shabek 2017; Bai et al. 1996; Skowyra et al. 1997; Feldman et al. 1997). Therefore, CRL1s have traditionally been termed as SCF E3 ligases (Skp1-Cullin1-F-box). There are about 70 F-box proteins in the human genome, targeting a large number of proteins in the cells given that each F-box can target multiple proteins for ubiquitination (Jin et al. 2004). The functions of different CRL1s are widely involved in cell cycle regulation; cell growth; cell death, development, and differentiation etc (Skaar et al. 2013). The diversity of the CRL1 family also makes them facile targets to be hijacked by many viruses, including herpesvirus, rotavirus, HIV-1, Hepatitis E virus, etc. (Ashizawa et al. 2012; Liu et al. 2007; Rodrigues et al. 2009; Graff et al. 2009; Surjit et al. 2012; Collins and Collins 2014). We will describe several well-studied examples below (Fig. 7.1).

HIV-1 Vpu had been well known to promote virus release from infected cells. However, it took a 20-year long research to unveil the underlying mechanism. A human membrane protein named tetherin was identified to be an HIV restriction factor that is counteracted by Vpu (Neil et al. 2008; Van Damme et al. 2008). Tetherin, as its name indicates, can function as a tether between nascent virions and infected cells, as well as between nascent virions, the net effect of which is to block the release of the nascent virions. Tetherin accomplishes this interesting function through its two membrane anchoring domains, one at its N-terminus and one at the C-terminus (Perez-Caballero 2009). The expression level of tetherin is induced by interferon (Neil et al. 2008; Van Damme et al. 2008). Therefore, it is an effector of the interferon-mediated antiviral response, and it was later shown to be a general mechanism for defense against many different viruses (Evans et al. 2010). Vpu is also a transmembrane protein that interacts with tetherin through its transmembrane domain (Vigan and Neil 2010). In addition, the intracellular domain of Vpu interacts with the F-box protein β-TrCP, which is the SR for the E3 $SCF^{\beta\text{-}TrCP}$ (Margottin

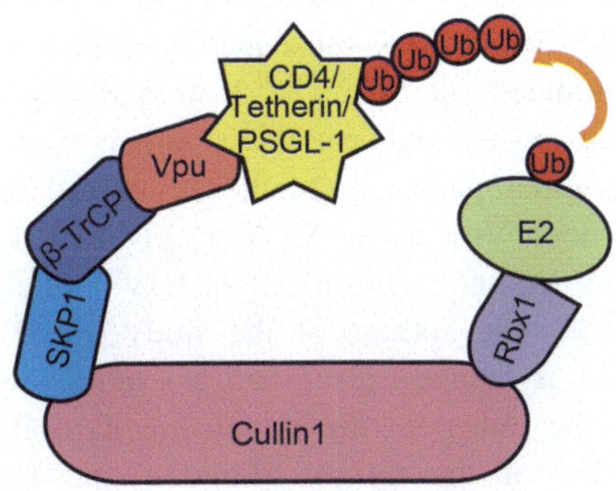

Fig. 7.1 Examples of viral hijacking of CRL1. HIV-1 Vpu protein recruits CD4 or tetherin or PSGL-1 to SCF-β-TrCP E3 ligases for ubiquitination

et al. 1998). This interaction links tetherin to an E3 ligase, which results in its ubiquitination and degradation from the cell surface due to ubiquitination-induced endosomal degradation (Roy et al. 2014). Vpu is not present in the incoming viruses and is only produced late in the HIV-1 life cycle, which suits the timing to antagonize tetherin and promote virus release (Neil et al. 2008).

Another target of $SCF^{\beta\text{-TrCP}}$-Vpu E3 ligase is the CD4 molecule on the surface of T cells. Distinct from antagonism of tetherin, Vpu targets CD4 molecule using its intracellular domain, and the degradation of CD4 after ubiquitination is through the proteasomal pathway (Roy et al. 2014; Dube et al. 2010). The physiological significance of Vpu-induced CD4 degradation is thought to avoid the re-adsorption of the nascent virions on the surface due to the contact of envelop protein and CD4. But the exact significance of this degradation remains to be demonstrated since CD4 is also the receptor for HIV, which complicates genetic studies to confirm the function of the degradation (Dube et al. 2010; Chaudhuri et al. 2007).

A comprehensive study to quantitatively measure membrane protein abundance during HIV infection in CD4+ T cells has revealed more potential targets of Vpu (Matheson et al. 2015). By comparing the differences in proteomic changes in infections with wild-type HIV or Vpu-deficient HIV, this study identified over 100 potential protein targets of Vpu-mediated degradation. From this list, Matheson et al. found an amino acid transporter protein named SNAT1, which was subsequently shown to be ubiquitinated by $SCF^{\beta\text{-TrCP}}$-Vpu and degraded by the endosomal pathway, rather than the proteasomal pathway (Matheson et al. 2015). Why does Vpu target SNAT1, given that this protein does not seem to affect HIV-1 infection *in vitro*? The authors of the study found that SNAT1 is required for T cell activation, which might be linked to the establishment of latent reservoir of the viruses (Sugden and Cohen 2015). But the exact mechanistic explanation of the Vpu-SNAT1 antagonism requires further study.

Our group recently identified another HIV restriction factor and target of $SCF^{\beta\text{-TrCP}}$-Vpu E3 ligase: PSGL-1, from a genome-wide proteomic profiling in human primary CD4+ T cells during HIV-1 infection (Liu et al. 2019). PSGL-1 is significantly downregulated during HIV-1 infection in Vpu-dependent fashion. PSGL-1 is a transmembrane protein and exerts anti-HIV-1 function mainly by associating with the nascent virions and potently inhibiting their infectivity. Vpu can partially relieve the restriction by inducing the ubiquitination and proteasomal degradation of PSGL-1 via $SCF^{\beta\text{-TrCP2}}$. Interestingly, only β-TrCP2, but not β-TrCP1, is responsible for the ubiquitination, which is different from the cases of CD4 and tetherin where both β-TrCP1 and β-TrCP2 can mediate Vpu-dependent CD4 and tetherin ubiquitination (Liu et al. 2019). Also different from tetherin, PSGL-1 is specifically expressed in T cells and monocytes and is induced by interferon γ, but not interferon α. Importantly, PSGL-1 is a key mediator of interferon γ's anti-HIV effects in human CD4+ T cells (Liu et al. 2019). Therefore, Vpu's hijacking of $SCF^{\beta\text{-TrCP2}}$ is key to resist interferon γ's anti-HIV effects.

Remarkably, Vpu, a small transmembrane protein with a size smaller than 10 kilodalton, can perform such a diversity of functions to recruit very different membrane proteins for ubiquitination (Sauter and Kirchhoff 2018). This is a great example demonstrating the versatility and modularity CRLs and the diverse functions of viral hijacking of CRLs in viral replication.

7.2 Cullin 2-RING Ubiquitin Ligases (CRL2)

Cullin 2- and Cullin 5-based CRLs share the same adaptor module: a protein complex consisting of a pair of small proteins—elongin B and elongin C (EloB/EloC). However, the substrate receptor of CRL2s and CRL5s has so-called VHL box and SOCS box sequence motifs, respectively, that differentiate between Cullin 2 and Cullin 5, conferring specificity to these two CRL families (Sarikas et al. 2011). CRL2 prototype is the Cul2-Rbx1-EloB/EloC-VHL E3 ligase that targets the hypoxia-induced factor 1α (HIF-1α), which is a key transcription factor in regulation of cellular response to hypoxia (Ke and Costa 2006). VHL binds to EloB/EloC using its so-called BC box motif, which can also recognize Cul2. VHL can also bind to HIF-1α and promote the ubiquitination and degradation of HIF-1α under normal oxygen conditions. HIF-1α binding to VHL is dependent on the hydroxylation of two proline residues of HIF-1α, but the hydroxyl groups are removed under hypoxia conditions; therefore HIF-1α degradation is abolished, and HIF-1α is released into the nucleus to promote the transcription of hypoxia response genes (Kaelin 2002). Mutations in VHL can lead to von Hipple-Lindau disease, which is characterized by visceral cysts and benign tumors with potential for malignant transformation (Kaelin 2002).

Respiratory syncytial virus (RSV) is a common respiratory virus that causes cold-like symptoms especially in children below age of 2 years old. It has been reported that the nonstructural protein NS1 of RSV can hijack CRL2 to target human STAT2 for ubiquitination and degradation (Elliott et al. 2007). STAT2, together with STAT1 and IRF-9, form a key transcription factor mediating the signaling transduction of type I interferon pathway, which is a master regulator of the antiviral innate immune responses (Schneider et al. 2014). By degrading STAT2, RSV shuts down the whole type I interferon responses to evade from the human innate defense system. The NS1 protein binds to elongin C and Cul2 to form the CRL2-NS1 E3 ligase to recruit STAT2 for ubiquitination (Elliott et al. 2007) (Fig. 7.2). Inhibition of the functions of STAT1 or STAT2 proteins has been a well-used strategy for viral evasion of the interferon-mediated antiviral responses. For example, HCV nonstructural protein 5A (NS5A) can inhibit STAT1 phosphorylation, which is required for the interaction between STAT1 and STAT2 and subsequent transcription activation (Lan et al. 2007). Influenza virus can also use its NS1 protein to inhibit the phosphorylation of STAT1 as well as STAT2 to block both the type I and type II interferon pathway (Jia et al. 2010). In contrast,

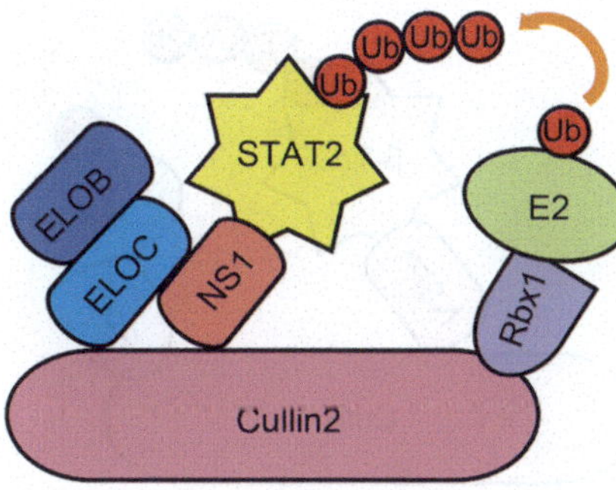

Fig. 7.2 An example of viral hijacking of CRL2. RSV NS1 protein recruits host cell STAT2 protein to CRL2 ligase for ubiquitination

RSV's degradation of STAT2 would be a thorough strategy to block the interferon pathway since it is permanent removal of the protein. Very interestingly, it has also been reported that RSV can stabilize the protein level of HIF-1α, which was attributed to the release of nitric oxide (Kilani et al. 2004). RSV's hijack of Cul2-EloB/EloC complex might be another contributing factor to the stabilization, since the hijack might exclude HIF-1α from ubiquitination and subsequent proteasomal degradation.

7.3 Cullin 3-RING Ubiquitin Ligases (CRL3)

CRL3s usually use one single protein, a member of the so-called BTB family proteins, to perform the functions of adaptor and substrate receptor, which is unique among CRLs (Zhuang et al. 2009). Human BTB family has close to 200 members, and they usually share a BTB domain that binds to Cul3 and a protein-protein interaction domain such as Kelch-like domain or MATH domain that binds to ubiquitination substrates (Stogios et al. 2005). There are relatively fewer known examples of CRL3s involved in viral infection.

One of the examples is that rotavirus uses its nonstructural protein NSP1 to recruit Cul3-Rbx1 to target β-TrCP for ubiquitination and degradation (Fig. 7.3) (Lutz et al. 2016; Davis and Patton 2017). As mentioned above, β-TrCP is the SR for

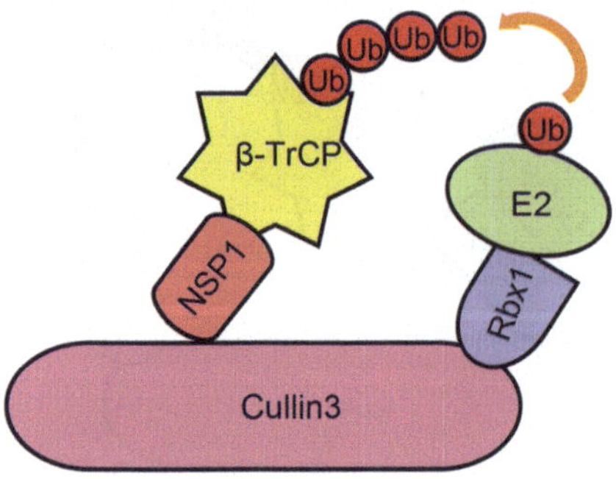

Fig. 7.3 An example of viral hijacking of CRL3: Rotavirus NS1 protein recruits host cell β-TrCP protein to CRL3 ligase for ubiquitination

CRL1, and one of its ubiquitination targets is IκB (Frescas and Pagano 2008). Ubiquitination of IκB is required for the activation of the NF-κB pathway, a key pathway for stimulating cytokine and interferon production to elicit host defenses (Chen 2005). Therefore, rotavirus NSP1's ubiquitination and consequent degradation of β-TrCP lead to inactivation of NF-κB to suppress host antiviral responses. Interestingly, in this case, there is no requirement for a BTB protein since Cul3 mutations that abolish BTB binding still can mediate this interaction. NSP1 binds Cul3 directly to recruit β-TrCP for ubiquitination (Davis and Patton 2017). Another interesting aspect of this targeting is that the ubiquitination is not only dependent on Cul3 but also dependent on Cul1, demonstrating a cross talk between different CRL families (Davis et al. 2017). A third unique aspect of NSP1 as an E3 ligase is that the viral protein itself possesses a RING domain so it can function as an E3 ligase on its own (Davis and Patton 2017). NSP1 from simian and murine rotavirus strains can directly bind IRF3, a key transcription factor for innate immune responses, and promote IRF3 ubiquitination and degradation independent of Cul3 (Sen et al. 2009) . The functional diversity of NSP1 as an E3 ligase highlights the versatility of viral hijacking of the ubiquitin-proteasome pathway.

Another example of viral hijacking of CRL3 is illustrated by human papilloma virus (HPV), one of the most common sexually transmitted viruses that can cause warts and precancerous lesions in persistently infected patients. Among the 150 types of HPV, about a dozen types are known to be associated with different human cancers (Nowinska et al. 2017). HPV encodes E6 and E7 proteins that are potent agents capable of transforming cells into immortalized cells. Their transformation abilities are associated with their hijacking of the human E3 ligases to degrade two key tumor suppressors: p53 and RB, respectively (Ajay et al. 2012). E6 associates with the HECT-type E3 ligase E6AP to recruit p53 for ubiquitination, whereas E7 of the HPV type 16 has been known to associate with Cul2-EloB/EloC E3 ligase to target RB for ubiquitination and proteasomal degradation (Yim and Park 2005). More recent protein-protein interaction profiling studies have revealed that E7 proteins from many different HPV types are associated with Cul3 instead of Cul2 (White et al. 2012). It is hypothesized that CRL3s might be involved in RB degradation in those HPV types. E7 proteins from different HPV types hijack different CRLs, again underscoring the versatility of the hijacking mechanisms.

7.4 Cullin 4-RING Ubiquitin Ligases (CRL4)

CRL4s employ a large, 125-kilodalton protein-DDB1 as the adaptor, which recruits a family of so called DCAF (DDB1-Cul4 -associated factors) proteins as substrate receptors (Zheng and Shabek 2017). CRL4s are widely hijacked by different viral proteins to promote viral infections. Viral proteins can either bind to DDB1 and directly recruit their target proteins or interact with DCAF proteins to change their substrate specificity. Simian virus 5, belonging to the paramyxovirus family, encodes a V protein to bind to DDB1 and recruit human STAT1 protein for ubiquitination and proteasomal degradation (Li et al. 2006). This hijacking leads to the dampening of the interferon responses to viral infection. Similarly, hepatitis B virus encodes a small protein called HBx to bind to DDB1 and recruit a DNA-binding protein complex Smc5-Smc6 for ubiquitination and degradation (Fig. 7.4) (Decorsiere et al. 2016; Murphy et al.

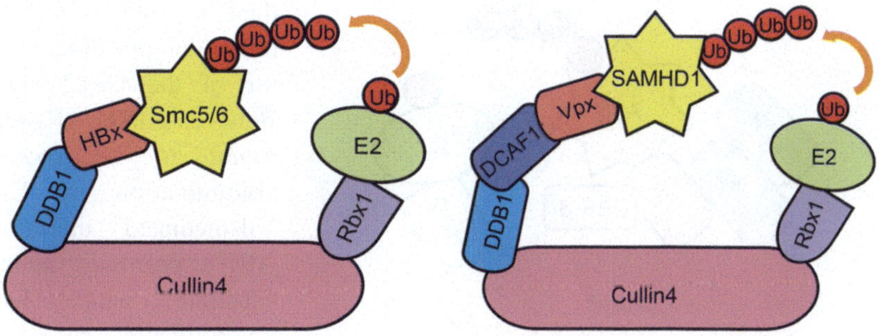

Fig. 7.4 Examples of viral hijacking of CRL4. HIV-2 or SIV Vpx protein recruits host cell SAMHD1 protein to CRL4 ligase for ubiquitination. HBV HBx protein recruits host cell Smc5-Smc6 proteins to CRLS4 ligase for ubiquitination

2016). HBx has long been known to be important for the transcription of HBV genome, but the mechanism was elusive. It had also been known that HBx associates with CRL4 complex, but the substrate protein was unknown until 2016. In that year, two papers reported the identification of Smc5-Smc6 complex as a target of HBx using immunoprecipitation-coupled mass spectrometry and showed that HBx overcomes Smc5-Smc6 inhibition of HBV transcription (Decorsiere et al. 2016; Murphy et al. 2016).

Another way for the viral protein to hijack CRL4 is to bind to one of the DCAFs and change its substrate specificity. A well-known example is the association between HIV-1 accessory protein Vpr and DCAF1 (also called VprBP). The function of Vpr in HIV-1 replication has been enigmatic, but its association with CRL4-DCAF1 provided a clue. Another well-known phenomenon of Vpr is that it can induce cell cycle arrest at the G2/M phase (He et al. 1995). A recent proteomic profiling has revealed that hundreds of proteins are downregulated by Vpr, directly or indirectly (Greenwood et al. 2019). A number of these proteins have been validated to contribute to the cell cycle arrest induced by Vpr, supporting the promiscuous substrate recruitment of Vpr. However, the functional significance of the cell cycle arrest induced by Vpr is still not clear. In comparison, a Vpr homolog protein called Vpx, which is encoded by Simian immunodeficiency virus (SIV) and HIV-2 but not HIV-1, has yielded its secret. Vpx, like Vpr, hijacks CRLs but they target a very different set of targets. One of the targets is SAMHD1, a dNTP hydrolase that is highly expressed in myeloid cells and resting CD4+ T cells (Laguette et al. 2011; Hrecka et al. 2011). SAMHD1 has been shown to potently inhibit lentivirus reverse transcription by reducing the concentration of dNTPs, the building blocks of lentiviral DNA (Lahouassa et al. 2012). Vpx recruits CRL4-VprBP to ubiquitinate SAMHD1 to alleviate this inhibition (Fig. 7.4) (Laguette et al. 2011; Hrecka et al. 2011; Ahn et al. 2012). It is interesting that HIV-1 does not have such a counteracting mechanism. It is speculated that HIV-1 thus avoids infection of myeloid cells such as dendritic cells to evade detection by these cells (Lim and Emerman 2011). The high homology of Vpr and Vpx showcases how evolution can reprogram a viral protein to target different substrate proteins for ubiquitination.

7.5 Cullin 5-RING Ubiquitin Ligases (CRL5)

As described above, CRL5 and CRL2 share the adaptor protein complex, EloB/EloC. Different from CRL2, CRL5 employs the so-called SOCS box protein as SR to recruit substrate proteins. In addition, CRL5 is unique among all CRLs by preferentially using Rbx2 (Roc2/Sag) rather than Rbx1(Roc1) as the catalytic module.

HIV-1 Vif protein hijacks CRL5 to target an HIV restriction factor APOBEC3G for ubiquitination and degradation (Fig. 7.5), which is the first example of HIV-1 accessory protein targeting a restriction factor to an E3 (Yu et al. 2003). APOBEC3G was identified from a comparison of the cDNAs of two closely related cell

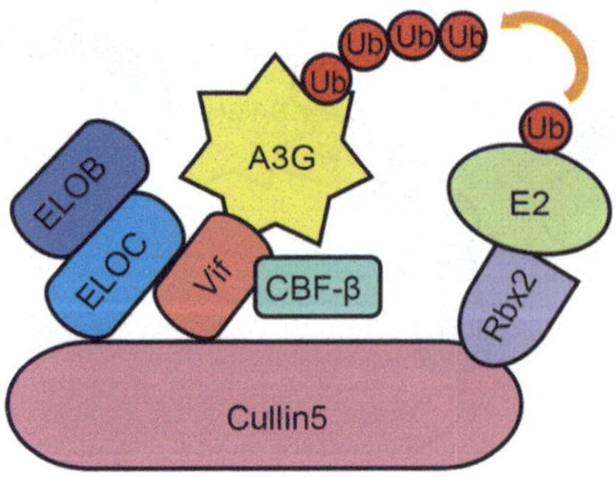

Fig. 7.5 An example of viral hijacking of CRL5. HIV Vif protein, with the host cell cofactor CBFβ, recruits host cell APOBEC3G (A3G) protein to CRL5 ligase for ubiquitination

lines, one susceptible to both Vif-deficient and wild-type HIV-1 and the other only susceptible to the wild-type virus. APOBEC3G, originally named CEM15, is specifically expressed in the cell line that is resistant to Vif-deficient virus (Sheehy et al. 2002). APOBEC3G is a DNA-editing enzyme that associates with the HIV-1 reverse transcriptase to be packaged into nascent virions (Malim and Bieniasz 2012). During the next round of infection, APOBEC3G will extensively mutate the newly synthesized viral DNA to block viral infection (Zhang et al. 2003; Lecossier et al. 2003; Mangeat et al. 2003; Harris et al. 2003). Biochemical work has identified that Vif hijacks CRL5 to recruit APOBEC3G for ubiquitination and proteasomal degradation (Yu et al. 2003). Interestingly, later work uncovered another cofactor CBFβ that is required for Vif's hijacking of CRL5 (Jager et al. 2012; Zhang et al. 2011). Structural studies have revealed that Vif impressively organizes the formation of this E3 ligase complex by simultaneously interacting with four proteins: Cul5, EloC, CBFβ, and APOBEC3G (Guo et al. 2014). Interestingly, CBFβ's physiological function is to bind to the transcription factor RUNX to facilitate the transcription of target genes of RUNX (Kim et al. 2013). Vif's hijacking of CBFβ excludes RUNX from binding to CBFβ and inhibits RUNX-mediated gene transcription, which further benefits the viral replication since many of these target genes are involved in immune responses (Kim et al. 2013). Therefore, this hijacking has dual advantages for the virus. More recently, proteomic profiling has identified another regulator of the CRL5 ligases. ARIH2, a RING-Between-RING (RBR) E3 ligase, has been found to be required for the efficient ubiquitination by CRL5 ligases, including Vif-mediated ubiquitination of APOBEC3. ARIH2 promotes the initial ubiquitination of the substrate, which facilitates the following ubiquitination chain elongation mediated by CRL5s (Huttenhain et al. 2019). ARIH2 is yet another Vif cofactor in promoting APOBEC3 ubiquitination. This intricate structural organization by Vif underlines the amazing capability of viral proteins to evolve complex functions within a compact genome.

7.6 Therapeutic Targeting of the Viral Hijacking of CRLs

Many lines of functional evidence suggest that we might be able to develop new antiviral therapies by blocking the viral hijacking of CRLs. Evolution history has also proven that these hijacking events are vital to the replication of the viruses. Targeting the ubiquitin ligases for drug development has not been as straightforward as targeting other enzymes, since E3 ligases do not have natural binding sites for small molecules. Instead, E3 ligases function mostly through protein-protein interaction, which is traditionally thought to be challenging to target. Nevertheless, there are more and more examples to show that protein-protein interaction might not be as "undruggable" as previously thought. Vif-APOBEC3G, the first pair of HIV accessory protein and its cognizant restriction factor, has been intensively targeted for drug screening, and several studies have identified specific inhibitors. RN-18 was the first reported Vif inhibitor that was identified from a cell-based screen to search for small molecules that stabilizes APOBEC3G in the presence of Vif (Nathans et al. 2008). RN-18 downregulated Vif protein level and increased APOBEC3G protein level. Several RN-18 analogs were developed to optimize its potency and water solubility (Zhou et al. 2017; Bennett et al. 2018). Recently, the same group have identified a single mutation on Vif that confers resistance to these inhibitors, further proving that the inhibitors indeed target Vif in a specific manner (Sharkey et al. 2019). A different class of Vif inhibitor were identified using a similar method, but these compounds more likely target APOBEC3G based on binding assay (Cen et al. 2010). VEC-5, a compound discovered based on structural model of the Vif-EloB/EloC complex, has been shown to stabilize APOBEC3G (Zuo et al. 2012). More recently, a compound named N.41 was identified from a screen using a FRET-based biochemical assay to search for inhibitor of Vif-APOBEC3G interaction (Pery et al. 2015). N.41 has been shown to possess strong anti-HIV activity in PBMCs in an APOBEC3G-dependent fashion. In general, the multiple protein-protein interfaces in the complex of CRL5-Vif-CBFβ-APOBEC3G provide many potential small-molecule target sites, some of which could be very unpredictable. Additional potential allosteric binding sites in the complex could also be targeted by small molecules. These published Vif inhibitors provided proofs of concept for therapeutic targeting of viral hijacking of CRL E3 ligases. Future development of therapeutics targeting the viral hijacking of E3 ligases should present many new opportunities to treat a variety of pathogenic viruses.

7.7 Perspectives on Future Research

For the past two decades, research on virus-host interaction has revealed remarkable roles of the CRLs E3 ligases in viral infection and pathogenesis. These results were built on the advancement of our understanding of the functions and mechanisms of the ubiquitin-proteasome pathway in general but also significantly enrich and inform about the pathway. Currently knowledge on this topic is only the tip of the iceberg as recent systematic approaches such as mass spectrometry profiling and genome-wide functional genomic screening are suggesting a much larger number of genes involved in the interaction between viruses and hosts (McDougall et al. 2018; Shah et al. 2015). These systematic approaches would continue to reveal new genes, protein complexes, and signaling networks regulating protein degradation in the contexts of viral infections and immune responses. From the application point of view, following the great stride in targeting the CRLs for cancer drug development such as the PROTAC approach, targeting the CRLs for antiviral drug development is expected to reap the benefit of a better mechanistic understanding of the functions and mechanisms of CRLs in viral infections.

Acknowledgments We apologize to colleagues whose works are not included in this review due to space limitation. This work was supported by the China National Funds for Excellent Young Scientists (31722030) and a grant (Grant No. 31670777) from the National Natural Science Foundation of China and from the Beijing Advanced Innovation Center for Structural Biology.

References

Ahn J et al (2012) HIV/simian immunodeficiency virus (SIV) accessory virulence factor Vpx loads the host cell restriction factor SAMHD1 onto the E3 ubiquitin ligase complex CRL4DCAF1. J Biol Chem 287:12550–12558

Ajay AK, Meena AS, Bhat MK (2012) Human papillomavirus 18 E6 inhibits phosphorylation of p53 expressed in HeLa cells. Cell Biosci 2:2–2

Ashizawa A et al (2012) The ubiquitin system and Kaposi's sarcoma-associated herpesvirus. Front Microbiol 3:66–66

Bai C et al (1996) SKP1 connects cell cycle regulators to the ubiquitin proteolysis machinery through a novel motif, the F-box. Cell 86:263–274. https://doi.org/10.1016/s0092-8674(00)80098-7

Barry M, Fruh K (2006) Viral modulators of cullin RING ubiquitin ligases: culling the host defense. Sci STKE 2006:pe21. https://doi.org/10.1126/stke.3352006pe21

Bennett RP, Salter JD, Smith HCA (2018) New class of antiretroviral enabling innate immunity by protecting APOBEC3 from HIV Vif-dependent degradation. Trends Mol Med 24:507–520

Cen et al (2010) Small molecular compounds inhibit HIV-1 replication through specifically stabilizing APOBEC3G. J Biol Chem 285:16546–16562. https://doi.org/10.1074/jbc.M109.085308

Chaudhuri R, Lindwasser OW, Smith W, Hurley JH, Bonifacino JS (2007) Downregulation of CD4 by human immunodeficiency virus type 1 Nef is dependent on Clathrin and involves direct interaction of Nef with the AP2 Clathrin adaptor. J Virol 81:3877–3890

Chen ZJ (2005) Ubiquitin signaling in the NF-κB pathway. Nat Cell Biol 7:758
Collins DR, Collins KL (2014) HIV-1 accessory proteins adapt cellular adaptors to facilitate immune evasion. PLoS Pathog 10
Daugherty MD, Malik HS (2012) Rules of engagement: molecular insights from host-virus arms races. Annu Rev Genet 46:677–700. https://doi.org/10.1146/annurev-genet-110711-155522
Davis K, Morelli M, Patton J (2017) Rotavirus NSP1 requires casein kinase II-mediated phosphorylation for hijacking of Cullin-RING ligases. 8. https://doi.org/10.1128/mBio.01213-17
Davis KA, Patton JT (2017) Shutdown of interferon signaling by a viral-hijacked E3 ubiquitin ligase. Microbial Cell 4:387–389
Decorsiere A et al (2016) Hepatitis B virus X protein identifies the Smc5/6 complex as a host restriction factor. Nature 531:386–389
Deshaies RJ, Joazeiro CA (2009) RING domain E3 ubiquitin ligases. Annu Rev Biochem 78:399–434. https://doi.org/10.1146/annurev.biochem.78.101807.093809
Dube M, Bego MG, Paquay C, Cohen EA (2010) Modulation of HIV-1-host interaction: role of the Vpu accessory protein. Retrovirology 7:114–114
Elliott J et al (2007) Respiratory syncytial virus NS1 protein degrades STAT2 by using the Elongin-Cullin E3 ligase. J Virol 81:3428–3436. https://doi.org/10.1128/JVI.02303-06
Ellis M et al (1999) Degradation of p27 Kip cdk inhibitor triggered by Kaposi's sarcoma virus cyclin-cdk6 complex. EMBO J 18:644–653
Evans DT, Serra-Moreno R, Singh RK, Guatelli JC (2010) BST-2/tetherin: a new component of the innate immune response to enveloped viruses. Trends Microbiol 18:388–396. https://doi.org/10.1016/j.tim.2010.06.010
Feldman RM, Correll CC, Kaplan KB, Deshaies RJ (1997) A complex of Cdc4p, Skp1p, and Cdc53p/cullin catalyzes ubiquitination of the phosphorylated CDK inhibitor Sic1p. Cell 91:221–230. https://doi.org/10.1016/s0092-8674(00)80404-3
Frescas D, Pagano M (2008) Deregulated proteolysis by the F-box proteins SKP2 and beta-TrCP: tipping the scales of cancer. Nat Rev Cancer 8:438–449. https://doi.org/10.1038/nrc2396
Gastaldello S et al (2010) A deneddylase encoded by Epstein-Barr virus promotes viral DNA replication by regulating the activity of cullin-RING ligases. Nat Cell Biol 12:351–361. https://doi.org/10.1038/ncb2035
Graff JW, Ettayebi K, Hardy ME (2009) Rotavirus NSP1 inhibits NFκB activation by inducing proteasome-dependent degradation of β-TrCP: a novel mechanism of IFN antagonism. PLoS Pathog 5:e1000280
Greenwood EJD et al (2019) Promiscuous targeting of cellular proteins by Vpr drives systems-level proteomic remodeling in HIV-1 infection. Cell Rep 27:1579–1596 e1577. https://doi.org/10.1016/j.celrep.2019.04.025
Guo Y et al (2014) Structural basis for hijacking CBF-beta and CUL5 E3 ligase complex by HIV-1 Vif. Nature 505:229–233. https://doi.org/10.1038/nature12884
Harper JW, Tan MK (2012) Understanding cullin-RING E3 biology through proteomics-based substrate identification. Mol Cell Proteomics MCP 11:1541–1550. https://doi.org/10.1074/mcp.R112.021154
Harris RS et al (2003) DNA deamination mediates innate immunity to retroviral infection. Cell 113:803–809. https://doi.org/10.1016/s0092-8674(03)00423-9
He J et al (1995) Human immunodeficiency virus type 1 viral protein R (Vpr) arrests cells in the G2 phase of the cell cycle by inhibiting p34cdc2 activity. J Virol 69:6705–6711
Hershko A, Ciechanover A (1998) The ubiquitin system. Annu Rev Biochem 67:425–479
Hrecka K et al (2011) Vpx relieves inhibition of HIV-1 infection of macrophages mediated by the SAMHD1 protein. Nature 474:658–661
Huang X, Dixit VM (2016) Drugging the undruggables: exploring the ubiquitin system for drug development. Cell Res 26:484–498
Huttenhain R et al (2019) ARIH2 Is a Vif-Dependent regulator of CUL5-Mediated APOBEC3G degradation in HIV infection. Cell Host Microbe 26:86–99 e87. https://doi.org/10.1016/j.chom.2019.05.008
Jager S et al (2012) Vif hijacks CBF-β to degrade APOBEC3G and promote HIV-1 infection. Nature 481:371–375
Jia D et al (2010) Influenza virus non-structural protein 1 (NS1) disrupts interferon signaling. PLoS One 5:e13927. https://doi.org/10.1371/journal.pone.0013927
Jin J et al (2004) Systematic analysis and nomenclature of mammalian F-box proteins. Genes Dev 18:2573–2580. https://doi.org/10.1101/gad.1255304
Kaelin WG Jr (2002) Molecular basis of the VHL hereditary cancer syndrome. Nat Rev Cancer 2:673–682. https://doi.org/10.1038/nrc885
Ke Q, Costa M (2006) Hypoxia-Inducible Factor-1 (HIF-1). Mol Pharmacol 70:1469–1480
Kilani MM, Mohammed KA, Nasreen N, Tepper RS, Antony VB (2004) RSV causes HIF-1α stabilization via NO release in primary bronchial epithelial cells. Inflammation 28:245–251
Kim DY et al (2013) CBFbeta stabilizes HIV Vif to counteract APOBEC3 at the expense of RUNX1 target gene expression. Mol Cell 49:632–644. https://doi.org/10.1016/j.molcel.2012.12.012
Laguette N et al (2011) SAMHD1 is the dendritic- and myeloid-cell-specific HIV-1 restriction factor counteracted by Vpx. Nature 474:654–657. https://doi.org/10.1038/nature10117
Lahouassa H et al (2012) SAMHD1 restricts the replication of human immunodeficiency virus type 1 by depleting the intracellular pool of deoxynucleoside triphosphates. Nat Immunol 13:223–228. https://doi.org/10.1038/ni.2236
Lan KH et al (2007) HCV NS5A inhibits interferon-alpha signaling through suppression of STAT1

phosphorylation in hepatocyte-derived cell lines. J Hepatol 46:759–767. https://doi.org/10.1016/j.jhep.2006.11.013

Lecossier D, Bouchonnet F, Clavel F, Hance AJ (2003) Hypermutation of HIV-1 DNA in the absence of the Vif protein. Science 300:1112. https://doi.org/10.1126/science.1083338

Li T, Chen X, Garbutt KC, Zhou P, Zheng N (2006) Structure of DDB1 in complex with a paramyxovirus V protein: viral hijack of a propeller cluster in ubiquitin ligase. Cell 124:105–117

Lim ES, Emerman M (2011) HIV: going for the watchman. Nature 474:587–588. https://doi.org/10.1038/474587a

Liu J et al (2007) Kaposi's sarcoma-associated herpesvirus LANA protein downregulates nuclear glycogen synthase kinase 3 activity and consequently blocks differentiation. J Virol 81:4722–4731. https://doi.org/10.1128/JVI.02548-06

Liu Y et al (2019) Proteomic profiling of HIV-1 infection of human CD4(+) T cells identifies PSGL-1 as an HIV restriction factor. Nat Microbiol 4:813–825. https://doi.org/10.1038/s41564-019-0372-2

Lutz LM, Pace CR, Arnold MM (2016) Rotavirus NSP1 associates with components of the cullin RING ligase family of E3 ubiquitin ligases. J Virol 90:6036–6048

Mahon C, Krogan NJ, Craik CS, Pick E (2014) Cullin E3 ligases and their rewiring by viral factors. Biomol Ther 4:897–930. https://doi.org/10.3390/biom4040897

Malim MH, Bieniasz PD (2012) HIV restriction factors and mechanisms of evasion. Cold Spring Harb Perspect Med 2:a006940. https://doi.org/10.1101/cshperspect.a006940

Mangeat B et al (2003) Broad antiretroviral defence by human APOBEC3G through lethal editing of nascent reverse transcripts. Nature 424:99–103. https://doi.org/10.1038/nature01709

Margottin F et al (1998) A novel human WD protein, h-beta TrCp, that interacts with HIV-1 Vpu connects CD4 to the ER degradation pathway through an F-box motif. Mol Cell 1:565–574

Matheson NJ et al (2015) Cell surface proteomic map of HIV infection reveals antagonism of amino acid metabolism by Vpu and Nef. Cell Host Microbe 18:409–423

McDougall WM, Perreira JM, Reynolds EC, Brass AL (2018) CRISPR genetic screens to discover host-virus interactions. Curr Opin Virol 29:87–100. https://doi.org/10.1016/j.coviro.2018.03.007

Murphy CM et al (2016) Hepatitis B virus X protein promotes degradation of SMC5/6 to enhance HBV replication. Cell Rep 16:2846–2854

Nathans RS et al (2008) Small-molecule inhibition of HIV-1 Vif. Nat Biotechnol 26:1187–1192

Neil SJ, Zang T, Bieniasz PD (2008) Tetherin inhibits retrovirus release and is antagonized by HIV-1 Vpu. Nature 451:425–430. https://doi.org/10.1038/nature06553

Nowinska K, Ciesielska U, Podhorska-Okolow M, Dziegiel P (2017) The role of human papillomavirus in oncogenic transformation and its contribution to the etiology of precancerous lesions and cancer of the larynx: a review. Adv Clin Exp Med 26:539–547. https://doi.org/10.17219/acem/67461

Perez-Caballero D et al (2009) Tetherin inhibits HIV-1 release by directly tethering virions to cells. Cell 139:499–511. https://doi.org/10.1016/j.cell.2009.08.039

Pery E et al (2015) Identification of a novel HIV-1 inhibitor targeting Vif-dependent degradation of human APOBEC3G protein. J Biol Chem 290:10504–10517

Rodrigues L et al (2009) Termination of NF-kappaB activity through a gammaherpesvirus protein that assembles an EC5S ubiquitin-ligase. EMBO J 28:1283–1295. https://doi.org/10.1038/emboj.2009.74

Roy N, Pacini G, Berlioz-Torrent C, Janvier K (2014) Mechanisms underlying HIV-1 Vpu-mediated viral egress. Front Microbiol 5:177. https://doi.org/10.3389/fmicb.2014.00177

Sarikas A, Hartmann T, Pan ZQ (2011) The cullin protein family. Genome Biol 12:220. https://doi.org/10.1186/gb-2011-12-4-220

Sauter D, Kirchhoff F (2018) Multilayered and versatile inhibition of cellular antiviral factors by HIV and SIV accessory proteins. Cytokine Growth Factor Rev 40:3–12. https://doi.org/10.1016/j.cytogfr.2018.02.005

Schneider WM, Chevillotte MD, Rice CM (2014) Interferon-stimulated genes: a complex web of host defenses. Annu Rev Immunol 32:513–545. https://doi.org/10.1146/annurev-immunol-032713-120231

Schwartz AL, Ciechanover A (1999) The ubiquitin-proteasome pathway and pathogenesis of human diseases. Annu Rev Med 50:57–74. https://doi.org/10.1146/annurev.med.50.1.57

Sen A, Feng N, Ettayebi K, Hardy ME, Greenberg HB (2009) IRF3 inhibition by rotavirus NSP1 is host cell and virus strain dependent but independent of NSP1 proteasomal degradation. J Virol 83:10322–10335

Shah PS, Wojcechowskyj JA, Eckhardt M, Krogan NJ (2015) Comparative mapping of host-pathogen protein-protein interactions. Curr Opin Microbiol 27:62–68. https://doi.org/10.1016/j.mib.2015.07.008

Sharkey et al (2019) HIV-1 escape from small-molecule antagonism of Vif. MBio 26:e00144–19. https://doi.org/10.1128/mBio.00144-19

Sheehy AM, Gaddis NC, Choi JD, Malim MH (2002) Isolation of a human gene that inhibits HIV-1 infection and is suppressed by the viral Vif protein. Nature 418:646–650

Skaar JR, Pagan JK, Pagano M (2013) Mechanisms and function of substrate recruitment by F-box proteins. Nat Rev Mol Cell Biol 14:369–381. https://doi.org/10.1038/nrm3582

Skowyra D, Craig KL, Tyers M, Elledge SJ, Harper JW (1997) F-box proteins are receptors that recruit phosphorylated substrates to the SCF ubiquitin-ligase complex. Cell 91:209–219. https://doi.org/10.1016/s0092-8674(00)80403-1

Stogios PJ, Downs GS, Jauhal JJ, Nandra SK, Prive GG (2005) Sequence and structural analysis of BTB

domain proteins. Genome Biol 6:R82. https://doi.org/10.1186/gb-2005-6-10-r82
Sugden SM, Cohen EA (2015) Attacking the supply lines: HIV-1 restricts alanine uptake to prevent T cell activation. Cell Host Microbe 18:514–517
Surjit M, Varshney B, Lal SK (2012) The ORF2 glycoprotein of hepatitis E virus inhibits cellular NF-κB activity by blocking ubiquitination mediated proteasomal degradation of IκBα in human hepatoma cells. BMC Biochem 13:7–7
Van Damme N et al (2008) The interferon-induced protein BST-2 restricts HIV-1 release and is downregulated from the cell surface by the viral Vpu protein. Cell Host Microbe 3:245–252. https://doi.org/10.1016/j.chom.2008.03.001
Vigan R, Neil SJ (2010) Determinants of tetherin antagonism in the transmembrane domain of the human immunodeficiency virus type 1 Vpu protein. J Virol 84:12958–12970. https://doi.org/10.1128/JVI.01699-10
Votteler J, Schubert US (2008) Ubiquitin ligases as therapeutic targets in HIV-1 infection. Expert Opin Ther Targets 12:131–143
Weekes MP et al (2014) Quantitative temporal viromics: an approach to investigate host-pathogen interaction. Cell 157:1460–1472. https://doi.org/10.1016/j.cell.2014.04.028
White EA et al (2012) Systematic identification of interactions between host cell proteins and E7 oncoproteins from diverse human papillomaviruses. Proc Natl Acad Sci U S A 109:E260–E267
Yim E, Park J (2005) The role of HPV E6 and E7 Oncoproteins in HPV-associated cervical carcinogenesis. Cancer Res Treat 37:319–324
Yu X et al (2003) Induction of APOBEC3G ubiquitination and degradation by an HIV-1 Vif-Cul5-SCF complex. Science 302:1056–1060. https://doi.org/10.1126/science.1089591
Zhang H et al (2003) The cytidine deaminase CEM15 induces hypermutation in newly synthesized HIV-1 DNA. Nature 424:94–98
Zhang W, Du J, Evans SL, Yu Y, Yu XF (2011) T-cell differentiation factor CBF-beta regulates HIV-1 Vif-mediated evasion of host restriction. Nature 481:376–379. https://doi.org/10.1038/nature10718
Zheng N, Shabek N (2017) Ubiquitin ligases: structure, function, and regulation. Annu Rev Biochem 86:129–157. https://doi.org/10.1146/annurev-biochem-060815-014922
Zhou M et al (2017) Synthesis, biological evaluation and molecular docking study of N-(2-methoxyphenyl)-6-((4-nitrophenyl)sulfonyl)benzamide derivatives as potent HIV-1 Vif antagonists. Eur J Med Chem 129:310–324. https://doi.org/10.1016/j.ejmech.2017.01.010
Zhuang M et al (2009) Structures of SPOP-substrate complexes: insights into molecular architectures of BTB-Cul3 ubiquitin ligases. Mol Cell 36:39–50
Zuo T et al (2012) Small-molecule inhibition of human immunodeficiency virus type 1 replication by targeting the interaction between Vif and ElonginC. J Virol 86:5497–5507. https://doi.org/10.1128/JVI.06957-11

The Biology of F-box Proteins: The SCF Family of E3 Ubiquitin Ligases

8

Khai M. Nguyen and Luca Busino

Abstract

F-box proteins function as substrate adaptors for the S-phase kinase-associated protein 1 (SKP1)-cullin 1 (CUL1)-F-box protein (SCF) ubiquitin ligase complexes, which mediate the proteasomal degradation of a diverse range of regulatory proteins. 20 years since the F-box protein family has been discovered, our understanding of substrate-recognition regulation and the roles F-box proteins play in cellular processes has continued to expand. Here, we provide an introduction to the discovery and classification of F-box proteins, the overall structural assembly of SCF complexes, the varied mechanisms by which F-box proteins recognize their substrates, and the role F-box proteins play in diseases and their potentials in targeted therapies.

Keywords

SCF ubiquitin ligase · F-box · Phosphorylation · Posttranslational modification · Cell cycle

K. M. Nguyen
Department of Cancer Biology, University of Pennsylvania, Philadelphia, PA, USA

Perelman School of Medicine, University of Pennsylvania, Philadelphia, PA, USA

Abramson Family Cancer Research Institute, University of Pennsylvania, Philadelphia, PA, USA

Biochemistry Program, Earlham College, Richmond, IN, USA

L. Busino (✉)
Department of Cancer Biology, University of Pennsylvania, Philadelphia, PA, USA

Perelman School of Medicine, University of Pennsylvania, Philadelphia, PA, USA

Abramson Family Cancer Research Institute, University of Pennsylvania, Philadelphia, PA, USA
e-mail: businol@upenn.edu

Abbreviations

CASH	Carbohydrate-binding/sugar hydrolysis domain
Cdc4	Cell division control protein 4
CDK	Cyclin-dependent kinase
CH	Calponin homology
CTD	C-terminal domain
FBP	F-box protein
GSK3	Glycogen synthase kinase 3
LRR	Leucine-rich repeats
NTD	N-terminal domain
PCKD	Polycystic kidney disease
PTM	Posttranslational modification
SBD	Substrate-binding domain
SCF	SKP1-CULLIN1-F-box protein ubiquitin ligase
SKP1	S-phase kinase-associated protein 1
TDL	Traf domain-like

Y. Sun et al. (eds.), *Cullin-RING Ligases and Protein Neddylation*, Advances in Experimental Medicine and Biology 1217, https://doi.org/10.1007/978-981-15-1025-0_8

8.1 Discovery and Classification of F-box Proteins

The region of homology that would eventually became known as the F-box domain was first observed in cyclin F. Encoded by the CCNF gene, cyclin F was discovered by serendipity during a study on the polycystic kidney disease (PCKD) gene locus (Kraus et al. 1994). Cyclin F was later found to suppress the yeast Cdc4 mutant (Cdc4^{mut}) that exhibited arrest at the G1/S transition and degradation deficiency of Sic1, the inhibitor of Cdk1-Clb (B-type cyclins) complexes (Bai et al. 1994). The effort to find other suppressors of Cdc4^{mut} resulted in the discovery of SKP1, whose inactivation promoted cell cycle arrest both in G1 and G2 phases. These cell cycle arrests also appeared to correspond to stabilization of CDK regulators, and thus, SKP1 was shown to be necessary for ubiquitin-mediated degradation of these proteins (Bai et al. 1996). Binding between SKP1 and cyclin F, as well as other proteins, occurred via a conserved 40-amino acid domain. Since this domain was originally seen in cyclin F, it was named the F-box domain, marking the birth of the F-box protein family (Bai et al. 1994, 1996). A year later, it was found that SKP1, CULLIN1, and Cdc4 assembled to form an E3 ligase protein complex required to ubiquitylate Sic1, which became known as the SKP1-CULLIN1-F-box (SCF) ubiquitin ligase (Feldman et al. 1997).

In mammals, naming of F-box proteins (FBPs) has been based on the structural class of their substrate-binding domains (SBDs) (Winston et al. 1999; Cenciarelli et al. 1999). Mammalian FBPs are, in general, classified into three subgroups: FBWs, FBLs, and FBXs, comprised of 17, 22, and 39 members, respectively. FBWs are named as such due to the presence of WD40 repeat domains ("FB" for F-box and "W" for WD40 repeat domain) in their structures. A β-propeller arrangement found in many protein-protein interactions characterizes FBWs' SBDs (Smith et al. 1999). FBLs are distinguished by their leucine-rich repeats (LRRs), hence the "L"; here, the SBD is an arc-shaped α-β-repeat structure, also found in many other protein-binding schemes (Kobe and Kajava 2001; Enkhbayar et al. 2003). The third class consists of FBPs that contain neither WD40 repeats nor LRRs, but other protein-binding moieties such as CH (calponin homology), cyclin box, TDL (Traf domain-like), CASH (carbohydrate-interacting), Sec7, proline-rich domains, or zinc-finger domains (Yoshida et al. 2002). This variation in protein-binding moieties indicates that, apart from phosphorylation and glycosylation, FBPs regulation might occur via a spectrum of post-translational modifications, including methylation, ribosylation, farnesylation, and acetylation, implicating involvement in a variety of organ-specific functions.

8.2 Structure and Assembly

F-box proteins assemble with SKP1, CULLIN1, and RBX1 to form the Cullin-RING ligase 1 (CRL1) complex, also known as the SCF (SKP1-CULLIN1-F-box) E3 ubiquitin ligase (Fig. 8.1). SKP1, CULLIN1, and RBX1 weigh 18.7, 89.7, and 12.3 kilodaltons, respectively, while the F-box protein can have a mass ranging from 47 to over 110 kilodaltons (Zheng et al. 2002a). RBX1 contains the RING domain, via which it interacts with the C-terminus of

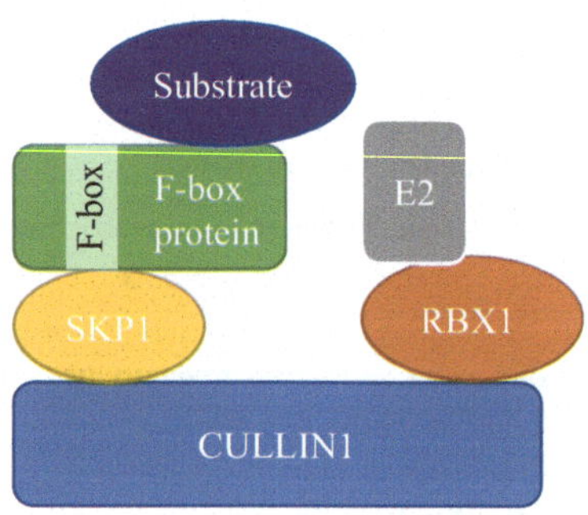

Fig. 8.1 Schematic view of the SCF complex

CULLIN1 to collectively form a catalytic core complex that recruits an E2 enzyme. SKP1 acts as an adapter that bridges the N-terminus of CULLIN1 and the FBP, which is the substrate-recognition subunit (Bai et al. 1996).

The four subunits (RBX1, CULLIN1, SKP1, and FBP) are held together via CULLIN1's scaffolding function. CULLIN1 consists of an N-terminal helical region (NTD) that is 415 amino acids long and a C-terminal globular α/β domain (CTD). Three Cullin repeats—that is, a five-helix structural motif arranged in a regularly repeated manner, culminating in an arc-shaped architecture stretching ~100 angstroms—make up the NTD. Interaction with the SKP1-F-box module occurs at the N-terminus of the first repeat. The CULLIN1 CTD, on the other hand, assembles in a globular manner surrounding the RBX1 protein. This is achieved through the formation of five intermolecular β-strands, of which the second strand is the N-terminus of RBX1. The Rbx RING domain is embedded into a ~30 angstrom-wide groove, formed from the remainder of the CULLIN1 CTD. Remarkably, the structural rigidity of the CULLIN1 scaffold is necessary for the SCF function, as shown by loss of catalytic activity when the NTD and CTD of CULLIN1 were linked by a flexible linker (Zheng et al. 2002a).

The N-terminus of the first Cullin repeat of CULLIN1 interacts with both SKP1 and the F-box motif. The interface of this binding consists of highly conserved amino acid residues, present in the orthologues of both CULLIN1 (Met 43, Tyr 46, Thr 47, Tyr 50, Tyr 139, Arg 142) and SKP1 (Asn 49 and 108, Tyr 109). RBX1 interacts with CULLIN1 through a conserved 16-residue β-strand and its RING domain, which are both inserted into a V-shaped groove on CULLIN1.

As is also the case for other Cullin-RING ligases, SCF complexes are regulated by neddylation, whereby the ubiquitin-like protein NEDD8 is conjugated to the Cullin protein through a cascade of E1, E2, and E3 enzymes similar to that of ubiquitination (Huang et al. 2004). Notably, Cullins recruit a NEDD8 E2 enzyme (Ubc12) that functions as its own NEDD8 E3-ligase. Cullin neddylation is regulated by the multi-subunit complex COP9 signalosome which hydrolyzes the NEDD8 moiety from cullins (Lyapina et al. 2001). Collectively, this system creates a balance between neddylation/deneddylation, thereby regulating CRL activity and assembly. NEDD8 conjugation to cullins induces flexibility in the RING finger subunit, which brings the E2 enzyme closer to the substrate and triggers ubiquitination (Duda et al. 2008). Cullin neddylation also regulates binding of the Cullin-associated and neddylation-dissociated 1 (CAND1) protein, which binds only unneddylated Cullins and potentiates exchange of F-box proteins in SCFs or substrate adaptors in other CRLs (Zheng et al. 2002b; Pierce et al. 2013).

8.3 Mechanism of Phospho-Dependent Substrate Recognition

F-box proteins recruit their substrates in response to a range of stimuli (Fig. 8.2). As such, the activity of specific SCF complexes is stringently regulated. The best-characterized model of substrate binding by FBPs is the *canonical phosphodegron* model, in which the FBP recognizes short, defined amino acid sequences. For instance, β-TrCP1 and β-TrCP2 (also known as FBXW1 and FBXW11, respectively) bind a Asp-pSer-Gly-X-X-pSer motif (X and pSer representing any amino acid and phosphorylated serine, respectively), while FBXW7 interacts with the consensus sequence pThr-Pro-Pro-X-pSer (pThr denoting phosphorylated threonine) (Lau and Fukushima 2012; Frescas and Pagano 2008) (Fig. 8.2).

In the FBW-subfamily proteins, substrate binding occurs at the WD40 domain (Hao et al. 2007). FBXW7 binding to CYCLIN E1 at its WD40 β-propeller assembly is partially stabilized by interactions involving residues outside the degron as well as those between phosphorylated residues. A left-handed polyproline helix from the three residues preceding the CYCLIN E1 degron guides it on top of the WD40 propeller blades, while the three residues between pThr380 and

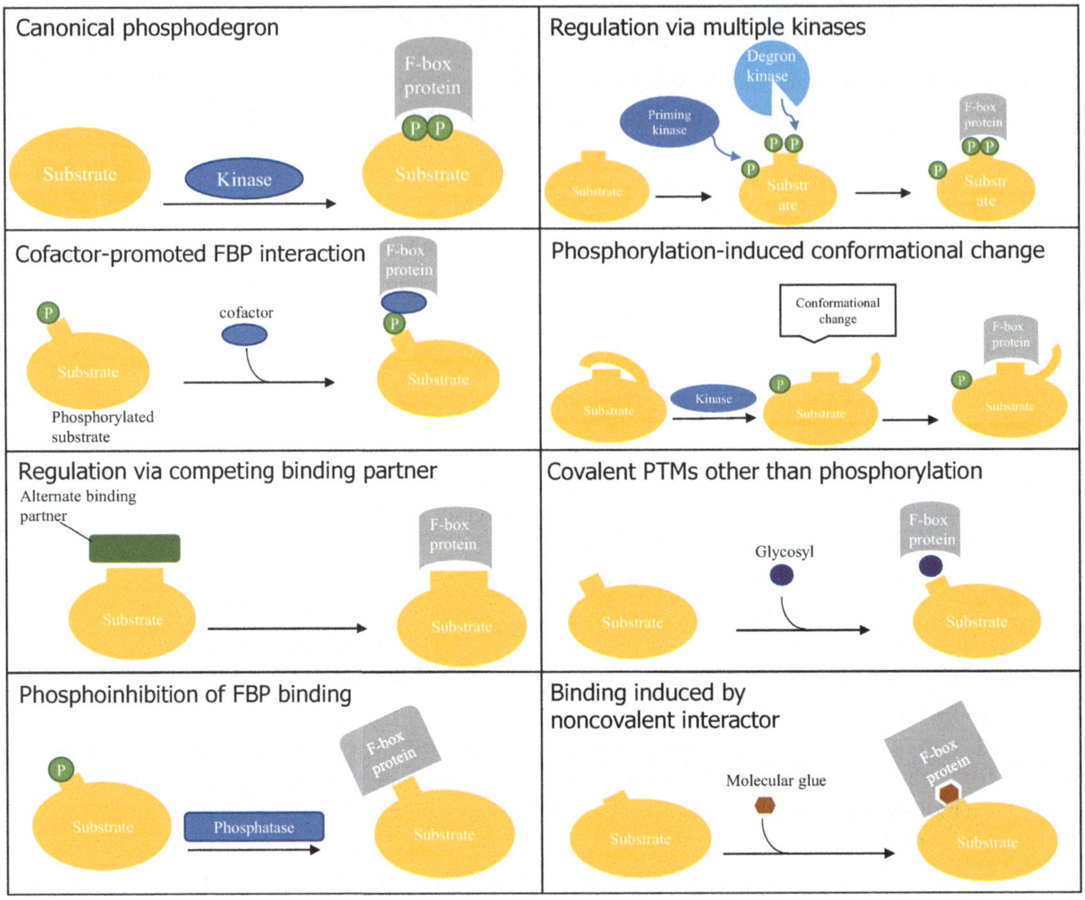

Fig. 8.2 Mechanisms of substrate recognition by F-box proteins

pSer384 form contacts to FBXW7 via backbone hydrogen bonding and side chain interactions through another left-handed polyproline helix. Phosphodegron recognition is conserved among different substrates: CYCLIN E1, NOTCH1, c-MYC, and c-JUN share similar phosphodegron as well as binding characteristics—the motif containing the 375th to 379th residues on CYCLIN E1, which is buried deeply in a hydrophobic cage on FBXW7, is maintained in c-MYC, NOTCH1, and c-JUN. It is thought that these conserved residues form van der Waals interactions with FBXW7. In a similar manner, β-TrCP utilizes the top of its WD40 domain to recognize β-catenin (Wu et al. 2003).

Phosphorylation of simple phosphodegrons can be carried out by a single kinase; in a similar manner, degrons containing multiple phosphorylation sites are controlled by multiple kinases. For instance, glycogen synthase kinase 3 (GSK3) and cyclin-dependent kinase 2 (CDK2) are responsible for the phosphorylation of different residues on the CYCLIN E1 degron, while in the case of the c-JUN degron, a prior phosphorylation by another kinase is required for GSK3-mediated phosphorylation of the degron (Welcker et al. 2003; Wei et al. 2005). Priming phosphorylation, a phosphorylation event near but not at the degron, is epitomized by β-TrCP-mediated degradation of β-catenin. In the case of β-catenin, CK1α first phosphorylates Ser45, before GSK3 phosphorylates Thr41, Ser37, and Ser33. Ser45 is not recognized by β-TrCP; however, its phosphorylation by CK1α is required for subsequent phosphorylation of residues within the degron (Liu et al. 2002).

A second mode of substrate recognition, of which SKP2 is an archetype, presents a deviation from canonical phosphorylation-dependent binding of substrate. In an example of *cofactor-dependent substrate recognition,* SKP2 binds p27 following Thr187 phosphorylation by a CDK, potentiating its ubiquitination and degradation. This ubiquitination event requires CSK1 (CDK regulatory subunit 1), which binds both p27 and SKP2, as it was found initially that in vitro reconstitution of the reaction without CKS1 displayed poor activity. Crystal structure of the SKP2-CKS1-p27 complex further revealed that CKS1, but not SKP2, binds the phospho-threonine 187 on $p27^{Kip1}$, establishing CKS1's integral role in SKP2-mediated binding and degradation of p27. The noncovalent interaction between CKS1 and SKP2 is evolutionarily conserved, which was explained by the high sequence homology between CKS1 isoforms from humans and other species (Hao et al. 2005; Ganoth et al. 2001).

A benefit to phosphorylation being a prerequisite for SCF-mediated degradation is that substrate recognition can be fine-tuned via adjustment of kinase activity in relation to a regulatory pathway. Degradation of the yeast CDK inhibitor and Cdc4 substrate Sic1, for instance, is required for G1-S transition; Cdc4 requires CDK-dependent phosphorylation of at least six out of nine amino acids on the Sic1 degron for it be recognized. As CDK activity increases during G1 phase of the cell cycle, Sic1 recognition by SCF^{Cdc4} and subsequent proteasomal degradation is increased, allowing S-phase entry (Feldman et al. 1997).

8.4 Mechanism of Substrate Recognition Independent of Phosphorylation

8.4.1 Recognition of Unmodified Degrons

Although degron phosphorylation potentiates rapid and selective regulation of substrates by SCF complexes, studies have also shown FBPs' ability to recognize degrons that have not been phosphorylated. For instance, the cyclin-homology domain on cyclin F recognizes an Arg-X-(Ile/Leu) motif in its substrates, in a comparable manner to how other cyclins recognize substrates for phosphorylation by CDKs. This mechanism of recognition is identical between different substrates of CYCLIN F, including CP110 and RRM2 (D'Angiolella et al. 2010, 2012). Since the Arg-X-Ile/Leu degron is not modifiable, substrates with this motif are degraded continuously if other regulatory mechanisms are not in place. As a result, oscillations in CYCLIN F levels throughout the cell cycle would account for timely degradation of its targets.

8.4.2 Control of F-box Protein-Substrate Recognition via Physical Obstruction

Physical access to the degron can be controlled by phosphorylation of the substrate at sites other than the degron, where the phosphorylation initiates a conformational change that opens up the degron for FBP recognition. Regulation of CYCLIN F-dependent RRM2 ubiquitination epitomizes this regulatory mechanism. Access to the Arg-X-Ile degron of RRM2 is blocked in the absence of CDK-mediated phosphorylation at Thr33, a residue outside of the degron. Hence, increased CDK activity in G2 phase of the cell cycle triggers RRM2 degradation as it is no longer needed to produce deoxyribonucleotides needed for the synthesis of DNA in S-phase (D'Angiolella et al. 2012).

Substrate recruitment by F-box proteins is not confined to recognition of short, defined amino acid sequences. A number of reports have shown FBP-substrate interactions on the basis of conserved domain structures. For instance, FBXL3 recognizes the substrates cryptochromes 1 and 2 (CRY1 and CRY2) through extended contacts between its LRRs and the CRY protein surface. The FBXL3 C-terminus is buried into a conserved pocket in CRY1. In the absence of FBXL3 binding, this pocket is filled with the endogenous metabolite flavin adenine dinucleotide (FAD). Hence, FAD functions as a competitive antagonist to regulate FBXL3-CRY1 interaction, thereby presenting a novel mode of regulating substrate binding (Busino et al. 2007; Godinho et al. 2007; Siepka et al. 2007; Xing et al. 2013).

8.4.3 F-box Protein Control via Posttranslational Modifications Other than Phosphorylation

Covalent modifications other than phosphorylation also trigger recognition by FBPs. Some substrates of FBXO2 and FBXO6, such as pre-integrin B1 and α-chain of T-cell receptor, respectively, are glycosylated (Yoshida et al. 2002). This interaction occurs via F-box-associated (FBA) domains. FBXO2- and FBXO6-mediated recognition of glycosylated substrates is consistent with their role in endoplasmic reticulum-associated degradation of incorrectly folded proteins, since proteins are frequently glycosylated when folding in the ER. Binding to glycosylated degrons appears not to be mutually exclusive to other modes of substrate recognition. FBXO6 also targets checkpoint kinase 1 (CHK1) for degradation, which has not been reported to be glycosylated (Zhang et al. 2009).

FBP-dependent protein degradation may also be regulated by means of acetylation and deacetylation. PRMT1, for instance, is polyubiquitinated by FBXL17 via recognition of an IKxxxIK motif that spans its 199th to 205th amino acid residues. Interestingly, recognition of this motif is contingent upon deacetylation of K200 and acetylation of K205. Endogenously, degradation of PRMT1 is thus promoted via a concerted mechanism involving the deacetylase SIRT1 and acetyltransferase p300. SIRT1 first deacetylates both K200 and K205, following which p300 selectively acetylates K205, resulting in an acetyldegron recognized for polyubiquitination by FBXL17 (Lai et al. 2017).

Methylation is another posttranslational means by which FBP-substrate recognition can be regulated. The transcription factor FOXO1 is ubiquitinated by the FBP Skp2, which results in its degradation by the proteasome (Huang et al. 2005). In this process, methylation of FOXO1 by the histone methyltransferase G9a promotes binding with Skp2 and consequently Skp2-mediated ubiquitination and degradation (Chae et al. 2019).

PTMs usually, but not always, promote substrate recognition by FBPs. An unmodified degron on CDT2 is recognized by FBXO11; consequently, CDK-dependent phosphorylation of this degron prevents binding of FBXO11 (Rossi et al. 2013; Abbas et al. 2013). In a similar manner, phosphorylation of p85B also prevents its binding by FBXL2. In these cases, phosphatases, as opposed to kinases, potentiate substrate recognition, as shown in the case of p85B with the tyrosine phosphatase PTPL1 (Kuchay et al. 2013).

PTM control of proteolysis is not exclusive to the substrate. The E3 ligase β-TrCP has been found to be modulated by ADP-ribosylation. Specifically, mono-ADP-ribosylation of β-TrCP by the ADP-ribosyltransferase PARP11 results in β-TrCP-dependent ubiquitination and degradation of IFNα/β receptor subunit 1 (IFNAR1), highlighting a potential target to increase the efficacy of interferon antiviral therapy (Guo et al. 2019).

PTMs also mediate FBP localization, which consequently mediate another layer of FBP regulation. FBXL20, for instance, undergoes isoprenylation at its Cys-Ala-Ala-X (CAAX) motif, which directs it to the membrane; CAAX-mediated membrane insertion of FBXL20 is required for RIM1 degradation (Yao et al. 2007). Likewise, geranylgeranylated (and thus membrane-bound) FBXL2 is required for HCV replication, via NS5A-dependent degradation of IP3R3 (Kuchay et al. 2018). In *Legionella pneumophila*, the F-box containing protein AnkB anchors to the *Legionella*-containing vacuole membrane, via host farnesyltransferase-mediated CAAX farnesylation. This localized FBP subsequently exploits the ubiquitin proteasome pathway of the protozoan and mammalian hosts, allowing for propagation and proliferation of *L. pneumophila* cells (Price et al. 2010).

8.4.4 Localization-Dependent Substrate Recognition

As discussed in the previous section, FBP-mediate substrate recognition may also be regulated via intracellular localization of both the substrate and the E3 ubiquitin ligase. In the case of RRM2, CYCLIN F is a nuclear protein, whereas RRM2 is cytoplasmic. Thus, RRM2 nuclear import is required for its CYCLIN F-dependent degradation (D'Angiolella et al. 2012). The same applies to other substrates of CYCLIN F. CP110 regulates centrosome duplication and as such requires centrosome-localized CYCLIN F for its degradation (D'Angiolella et al. 2010).

FBP localization regulation is not an absolute posttranslational affair. In the case of FBXW7, localization control begins with alternative splicing. Three isoforms of FBXW7, named α, β, and γ, are produced from mRNAs with distinguishable 5′ exons. These isoforms are translated into proteins that localize into different subcellular compartments; the α, β, and γ isoforms localize to the nucleus, cytoplasm, and nucleolus, respectively (Welcker et al. 2004). Moreover, expression of each specific isoform differs across tissue types and is regulated in accordance to the cell cycle as well as p53 activity (Matsumoto et al. 2006). FBXW7α is expressed in most tissue samples, whereas FBXW7β is found only in the brain and testes and FBXW7γ only in cardiac and striated muscle. FBXW7α mRNA was found to be high during G0 phase but is rapidly depleted as cells enter the cell cycle (Matsumoto et al. 2006). FBXW7β upregulation was detected upon X-irradiation; this event was however not observed in *p53*-deficient mice. These reports indicate yet another mode of spatial and temporal regulation of F-box protein activity.

8.4.5 Modulation of SCF Activity via Proteostasis of F-box Proteins

SCF activity is also modulated by expression levels of the FBPs. Cyclin F, for instance, is controlled during the cell cycle transcriptionally and via β-TrCP- and casein kinase II-dependent degradation (Mavrommati et al. 2018). FBXL7 levels are regulated via FBXL18-mediated polyubiquitination, and FBXO5 (also known as EMI1—early mitotic inhibitor 1) is regulated via degradation by $SCF^{\beta\text{-TrCP}}$ throughout the cell cycle (Liu et al. 2015; Margottin-Goguet et al. 2003; Guardavaccaro et al. 2003). Likewise, the $SCF^{(SKP2\text{-}Cks1)}$ complex is regulated by APC/C (Cdh1)-mediated degradation of SKP2 (Bashir et al. 2004; Wei et al. 2004).

FBP proteostasis may also be maintained by SCF auto-ubiquitination of FBPs. Evidence has emerged demonstrating SCF-mediated autocatalytic ubiquitination of SKP1-unbound F-box proteins and subsequent proteasomal degradation, as shown in a model based on yeast Met30 where FBP auto-ubiquitination by the CULLIN1-RBX1 complex functions independently of SKP1 (Mathur et al. 2015).

8.4.6 Ligand-Dependent Substrate Recognition

FBP degradation is not always tied to the cell cycle. FBXL5-mediated degradation of iron regulatory protein 2 (IRP2) occurs under high iron and oxygen levels; in turn, IRP2 acts upon iron-responsive mRNAs and mediates their translation and stability (Chollangi et al. 2012). This regulatory mechanism operates through the hemerythrin domain of FBXL5, which binds two iron ions (Shu et al. 2012). As such, this domain loses stability in iron-depleted conditions, resulting in unfolding and subsequent degradation by HERC2 (Moroishi et al. 2014). Taken together, regulation of SCF ubiquitin ligases via FBP levels occurs in spatial, temporal, and stimuli-dependent manners.

Iron-dependent regulation of FBXL5 also highlights a mode of regulation in response to noncovalent signals. This mechanism was first elucidated via studies of the plant hormones auxin and jasmonate. The auxin and jasmonate receptors (TIR and COI1, respectively) are FBPs that degrade the AUX-IAA and JAZ families of transcription repressors (Tan et al. 2007; Sheard et al. 2010). These hormones function as "molecular glue"—in that, they stabilize binding between the FBP and its substrate by filling the gap between the substrate and FBP. Furthermore, both TIR1 and COL1 employ a small-molecule cofactor in their structures: inositol hexakisphosphate and inositol pentakisphosphate, respectively. In the case of COL1, particularly, inositol pentakisphosphate plays a role in promoting interaction with an essential carboxyl group on jasmonate, suggesting similar interplay involving inositol hexakisphosphate (Sheard et al. 2010).

Taken together, mechanisms of substrate recognition by F-box proteins are becoming more and more diversified. Although SCF E3 ligases are the best characterized ubiquitin ligase complex, there remain modes of substrate recruitment yet to be elucidated, especially those that deviate from the phosphodegron model. Nevertheless, our understanding of substrate-ligase pairing continues to expand, bringing novel possibilities for SCF-targeted therapies.

8.5 Overall Function and Significance of SCF Ligases in Disease

The diversity of protein-binding motifs present in the F-box protein family members implicates interactions with a variety of different substrates and involvement in a wide range of tissue-specific functions. Since the start of the FBP field, it has been well-documented that FBP-dependent ubiquitination events play a role in a plethora of pathologies, which include sleep and mood disorders, diabetes, Parkinson's disease, and bacterial and viral infections. Nevertheless, due to the SCF complex's lasting association with the cell cycle and cell proliferation, studies of FBPs in the context of disease have mainly focused on cancer. In particular, extensive characterization of SKP2, FBXW7, and β-TrCP have allowed for insights into how disturbances of FBP function mediate neoplastic malignancies (Table 8.1).

SKP2 promotes S-phase entry by targeting the CDK inhibitor p27 for proteasomal-dependent degradation. This function renders SKP2 an epitome of an oncogenic FBP. In the context of cancer, overexpression of SKP2 is associated with poor prognosis in a variety of cancers; this role in tumorigenesis has been confirmed via studies in mouse models (Lin et al. 2010; Zhao et al. 2013; Wang et al. 2009). Inactivation of SKP2 induces p53-independent cellular senescence and suppresses tumorigenesis, and SKP2 has been shown to be required for survival of cells with aberrant Rb expression. Lastly, SKP2-mediated degradation of growth suppressors such as p27, p21 and p57 provides a rationale for its role as an oncoprotein (Frescas and Pagano 2008).

On the other hand, FBXW7 loss-of-function, but not overexpression, has been found in cancer, thereby implicating its antitumor function (Welcker and Clurman 2008). Frequently

Table 8.1 Mechanisms of F-box protein function disruption in human diseases

Mechanisms of F-box protein function disruption in disease	
Mode of interference	Consequence
Overexpression of F-box protein or amplification of FBP function	Increased degradation of substrate
Deleted F-box protein	Stabilization of substrate
Mutation in F-box protein-substrate-recognition domain	Stabilization of substrate
Mutation in substrate degron	Stabilization of substrate
Microbial F-box protein	Degradation of substrate via exogenous FBP-mediated ubiquitylation
Hijack of endogenous F-box protein by microbial protein	F-box protein retargeted for degradation of an alternative host substrate
Inhibition of endogenous F-box protein function by microbial protein	Stabilization of substrate

mutated and deleted in tumors, FBXW7 is estimated to be deregulated in 17 types of human cancers. Although mutations in FBXW7 are most often found in T-cell acute lymphoblastic leukemia (T-ALL; 31%), they are also found in solid tumors of the breast, intestine, and bone (Akhoondi et al. 2007). FBXW7's function as a tumor suppressor has been shown in mouse models, where conditional deletion of *Fbxw7* results in hematological malignancies, namely, thymic lymphoma and T-ALL (Maser et al. 2007). Since FBXW7 normally functions to degrade growth-promoting factors (MYC, JUN, CYCLIN E, NOTCH), its deletion triggers cell cycle entry and p53-dependent apoptosis, resulting in premature loss of hematopoietic stem cells (Onoyama et al. 2007; Thompson et al. 2007). As such, FBXW7 and p53 form a synergism to prevent tumor formation. Furthermore, loss of FBXW7 function, in some cases, potentiates chemotherapeutic resistance likely due to upregulation of MCL1 (myeloid cell leukemia sequence 1), an antiapoptotic factor and FBXW7 substrate (Inuzuka et al. 2011; Wertz et al. 2011).

Mutations in F-box proteins that hamper function of their respective SCF complex tend to affect the FBP-substrate interaction. The majority of such mutations in FBXW7 are missense mutations in residues essential for substrate binding, which interfere with its normal functioning; in fact, mutations of the two Arg residues of FBXW7 that come into direct contact with the substrate phosphodegron make up 43% of oncogenic FBXW7 mutations (Akhoondi et al. 2007). Mutations of substrates also affect the protein-substrate interface. The MYC phosphodegron, for instance, is often mutated in Burkitt's lymphoma, while mutations in NOTCH that disrupt its interaction with FBXW7 are frequently found in T-ALL (Gregory and Hann 2000; Weng et al. 2004).

Nevertheless, the effects of these mutations can be context-dependent, due to the variety of substrates of these FBPs as well as the contextual roles of each respective substrate. For instance, despite functioning as a tumor suppressor in many cancers, FBXW7 also facilitates a pro-survival mechanism in multiple myeloma via the degradation of NF-kB2 (p100); but this role appears to be B-cell specific (Busino et al. 2012).

In another example, the β-TrCP substrate DEPTOR (DEP-domain-containing mammalian target of rapamycin (mTOR)-interacting protein) has both tumor suppressing and promoting roles, the former via inhibiting mTOR, protein synthesis, and cell proliferation and the latter largely through maintaining PI3K-AKT activation (Duan et al. 2011; Gao et al. 2011; Zhao et al. 2011).

Apart from their roles in cancer, the functions of F-box proteins in other disease contexts are an emerging area of research. Sleep and mood disorders have been associated with β-TrCP- and FBXL3-mediated regulation of circadian rhythm, which operate by way of degrading the BMAL-CLOCK regulators PER1, PER2, CRY1, and CRY2 (Busino et al. 2007; Takahashi et al. 2008; Keers et al. 2012). FBPs are also hijacked in viral infections. For instance, the HIV protein U (Vpu) retargets β-TrCP for ubiquitination of CD4 and BST2 (bone marrow stromal antigen 2), which potentiates viral release and dispersion from the cell (Akari et al. 2001; Bour et al. 2001). In a similar manner, FBXL2 interacts with the HCV protein NS5A and is required for replication of HCV virions (Kuchay et al. 2018). A number of viruses and bacteria also encode their own FBPs that function in conjunction with the remaining SCF components from the host. The Agrobacterium FBP VirF, for instance, utilizes the host plant's Skp1-Cul1-Rbx1 complex to mediate degradation of host proteins VIP1 and VirE2 (Tzfira et al. 2004). Similarly, the Polerovirus FBP P0 degrades the *Arabidopsis* RNA-slicer AGO1, which impairs host cells' posttranscriptional gene silencing (Bortolamiol et al. 2007).

Lastly, FBXO7 has been found to mediate ubiquitination and degradation of MFN1 in a PINK1-dependent manner and participate in mitochondrial recruitment of PARKIN and mitophagy, providing a rationale for the mutations in FBXO7 found in Parkinson's disease (Burchell et al. 2013). This indicates potential extensions of our understanding of FBP implications in other diseases in the coming future, as the field migrates away from cancer biology.

8.6 Brief Conclusion

As the F-box protein field continues to expand 20 years after its initial emergence, so too does our appreciation of SCF ligase functional variety. Although the phosphodegron model remains the most ubiquitous mode of FBP-substrate pairing, others have emerged and diversified our understanding of FBP function. Characterization of novel substrates enriches the known biochemical and biological roles of FBPs and overall highlights the broad activity scope of E3 ligases. Last but not least, extensive characterization of SCF E3 ligases has provided insights applicable toward developing clinical therapies that target this system.

References

Abbas T et al (2013) CRL1-FBXO11 promotes Cdt2 ubiquitylation and degradation and regulates Pr-Set7/Set8-mediated cellular migration. Mol Cell 49:1147–1158

Akari H, Bour S, Kao S, Adachi A, Strebel K (2001) The human immunodeficiency virus Type 1 accessory protein vpu induces apoptosis by suppressing the nuclear factor kappaB-dependent expression of antiapoptotic factors. J Exp Med 194:1299–1311

Akhoondi S et al (2007) FBXW7/hCDC4 is a general tumor suppressor in human cancer. Cancer Res 67:9006–9012

Bai C, Richman R, Elledge SJ (1994) Human cyclin F. EMBO J 13:6087–6098

Bai C et al (1996) SKP1 connects cell cycle regulators to the ubiquitin proteolysis machinery through a novel motif, the F-box. Cell 86:263–274

Bashir T, Dorrello NV, Amador V, Guardavaccaro D, Pagano M (2004) Control of the SCFSkp2–Cks1 ubiquitin ligase by the APC/CCdh1 ubiquitin ligase. Nature 428:190–193

Bortolamiol D, Pazhouhandeh M, Marrocco K, Genschik P, Ziegler-Graff V (2007) The polerovirus F box protein P0 targets ARGONAUTE1 to suppress RNA silencing. Curr Biol 17:1615–1621

Bour S, Akari H, Strebel K (2001) The human immunodeficiency virus Type 1 Vpu protein inhibits NF- kB activation by interfering with B TrCP-mediated degradation of IkB. J Biol Chem 276:15920–15928

Burchell VS et al (2013) The Parkinson's disease – linked proteins Fbxo7 and Parkin interact to mediate mitophagy. Nat Neurosci:1–11. https://doi.org/10.1038/nn.3489

Busino L et al (2007) SCFFbxl3 controls the oscillation of the circadian clock by directing the degradation of cryptochrome proteins. Science 316:900–904

Busino L et al (2012) Fbxw7 α - and GSK3-mediated degradation of p100 is a pro-survival mechanism in multiple myeloma. Nat Cell Biol 14:375–385

Cenciarelli C et al (1999) Identification of a family of human F-box proteins F-box proteins are an expanding family of eukaryotic proteins characterized by an approximately 40 amino- acid motif, the F box (so named because cyclin F was SCFs). The other three subunits are the Skp. Curr Biol 9:1177–1179

Chae YC et al (2019) FOXO1 degradation via G9a-mediated methylation promotes cell proliferation in colon cancer. Nucleic Acids Res 47:1692–1705

Chollangi S, Thompson JW, Ruiz JC, Gardner KH, Bruick RK (2012) Hemerythrin-like domain within F-box and leucine-rich repeat protein 5 (FBXL5) communicates cellular iron and oxygen availability by distinct mechanisms. J Biol Chem 287:23710–23717

D'Angiolella V et al (2010) SCF Cyclin F controls centrosome homeostasis and mitotic fidelity through CP110 degradation. Nature 466:138–142

D'Angiolella V et al (2012) Cyclin F-mediated degradation of ribonucleotide reductase M2 controls genome integrity and DNA repair. Cell 149:1023–1034

Duan S et al (2011) mTOR generates an auto-amplification loop by triggering the b TrCP- and CK1 a -dependent degradation of DEPTOR. Mol Cell 44:317–324

Duda DM et al (2008) Structural insights into NEDD8 activation of Cullin-RING ligases: conformational control of conjugation. Cell 134:995–1006

Enkhbayar P, Kamiya M, Osaki M, Matsumoto T, Matsushima N (2003) Structural principles of Leucine-Rich Repeat (LRR) proteins. Proteins 403:394–403

Feldman RMR, Correll CC, Kaplan KB, Deshaies RJA (1997) Complex of Cdc4p , Skp1p, and Cdc53p/Cullin catalyzes ubiquitination of the phosphorylated CDK inhibitor Sic1p. Cell 91:221–230

Frescas D, Pagano M (2008) Deregulated proteolysis by the F-box proteins SKP2 and β-TrCP: tipping the scales of cancer. Nat Rev Cancer 8:438–449

Ganoth D et al (2001) The cell-cycle regulatory protein Cks1 is required for SCF Skp2 -mediated ubiquitinylation of p27. Nat Cell Biol 3:321–324

Gao D et al (2011) mTOR drives its own activation via SCF b TrCP -dependent degradation of the mTOR inhibitor DEPTOR. Mol Cell 44:290–303

Godinho SIH et al (2007) The after-hours mutant reveals a role for Fbx13 in determining mammalian circadian period. Science 316:897–900

Gregory MA, Hann SR (2000) c-Myc proteolysis by the ubiquitin-proteasome pathway: stabilization of c-Myc

in Burkitt's lymphoma cells. Mol Cell Biol 20:2423–2435
Guardavaccaro D et al (2003) Control of meiotic and mitotic progression by the F box protein β-Trcp1 in vivo. Dev Cell 4:799–812
Guo T et al (2019) ADP-ribosyltransferase PARP11 modulates the interferon antiviral response by mono-ADP-ribosylating the ubiquitin E3 ligase β-TrCP. Nat Microbiol. https://doi.org/10.1038/s41564-019-0428-3
Hao B et al (2005) Structural basis of the Cks1-dependent recognition of p27 Kip1 by the SCF Skp2 ubiquitin ligase. Mol Cell 20:9–19
Hao B, Oehlmann S, Sowa ME, Harper JW, Pavletich NP (2007) Structure of a Fbw7-Skp1-Cyclin E complex: multisite-phosphorylated substrate recognition by SCF ubiquitin ligases. Mol Cell 26:131–143
Huang DT et al (2004) A unique E1-E2 interaction required for optimal conjugation of the ubiquitin-like protein NEDD8. Nat Struct Mol Biol 11:927–935
Huang H et al (2005) Skp2 inhibits FOXO1 in tumor suppression through ubiquitin-mediated degradation. Proc Natl Acad Sci 102:1649–1654
Inuzuka H et al (2011) SCFFBW7 regulates cellular apoptosis by targeting MCL1 for ubiquitylation and destruction. Nature 471:104–109
Keers R et al (2012) Reduced anxiety and depression-like behaviours in the circadian period mutant mouse afterhours. PLoS One 7:1–10
Kobe B, Kajava AV (2001) The leucine-rich repeat as a protein recognition motif. Curr Opin Struct Biol 11: 725–732
Kraus B et al (1994) A novel cyclin gene (CCNF) in the region of the polycystic kidney disease gene (PKD1). Genomics 24:27–33
Kuchay S et al (2013) FBXL2- and PTPL1-mediated degradation of p110-free p85β regulatory subunit controls the PI(3)K signalling cascade. Nat Cell Biol 15: 472–480
Kuchay S et al (2018) NS5A promotes constitutive degradation of IP3R3 to counteract apoptosis induced by hepatitis C virus report NS5A promotes constitutive degradation of IP3R3 to counteract apoptosis induced by hepatitis C virus. Cell Rep 25:833–840.e3
Lai Y, Li J, Li X, Zou C (2017) Lipopolysaccharide modulates p300 and Sirt1 to promote PRMT1 stability via an SCFFbxl17-recognized acetyldegron. J Cell Sci 130:3578–3587
Lau AW, Fukushima H (2012) The Fbw7 and Beta-TRCP E3 ubiquitin ligases and their roles in tumorigenesis. Front Biosci 17:2197–2212
Lin H et al (2010) Skp2 targeting suppresses tumorigenesis by Arf-p53- independent cellular senescence. Nature 464:374–379
Liu C et al (2002) Control of Beta-catenin phosphorylation/degradation by a dual-kinase mechanism. Cell 108:837–847
Liu Y et al (2015) F-box protein Fbxl18 mediates polyubiquitylation and proteasomal degradation of the pro-apoptotic SCF subunit Fbxl7. Cell Death Dis 6:e1630–e1639
Lyapina S et al (2001) Promotion of NEDD8-CUL1 conjugate cleavage by COP9 signalosome. Science 292: 1382–1385
Margottin-Goguet F et al (2003) Prophase destruction of Emi1 by the SCFβTrCP/Slimb ubiquitin ligase activates the anaphase promoting complex to allow progression beyond prometaphase. Dev Cell 4: 813–826
Maser RS et al (2007) Chromosomally unstable mouse tumours have genomic alterations similar to diverse human cancers. Nature 447:966–971
Mathur R, Yen JL, Kaiser P (2015) Skp1 independent function of Cdc53/Cul1 in F-box protein homeostasis. PLoS Genet 11:1–21
Matsumoto A, Onoyama I, Nakayama KI (2006) Expression of mouse Fbxw7 isoforms is regulated in a cell cycle- or p53-dependent manner. Biochem Biophys Res Commun 350:114–119
Mavrommati I et al (2018) β-TrCP- and Casein kinase II-mediated degradation of cyclin F controls timely mitotic progression. Cell Rep 24:3404–3412
Moroishi T, Yamauchi T, Nishiyama M, Nakayama KI (2014) HERC2 targets the iron regulator FBXL5 for degradation and modulates iron metabolism. J Biol Chem 289:16430–16441
Onoyama I et al (2007) Conditional inactivation of Fbxw7 impairs cell-cycle exit during T cell differentiation and results in lymphomatogenesis. J Exp Med 204: 2875–2888
Pierce NW et al (2013) Cand1 promotes assembly of new SCF complexes through dynamic exchange of F box proteins. Cell 153:206–215
Price CTD, Al-Quadan T, Santic M, Jones SC, Abu Kwaik Y (2010) Exploitation of conserved eukaryotic host cell farnesylation machinery by an F-box effector of legionella pneumophila. J Exp Med 207:1713–1726
Rossi M et al (2013) Regulation of the CRL4Cdt2 ubiquitin ligase and cell-cycle exit by the SCFFbxo11 ubiquitin ligase. Mol Cell 49:1159–1166
Sheard LB et al (2010) Jasmonate perception by inositol-phosphate-potentiated COI1-JAZ co-receptor. Nature 468:400–407
Shu C et al (2012) The structural basis of iron sensing by the human F-box protein FBXL5. Chembiochem 13:788–791
Siepka SM et al (2007) Circadian mutant overtime reveals F-box protein FBXL3 regulation of cryptochrome and period gene expression. Cell 129:1011–1023
Smith TF, Gaitatzes C, Saxena K, Neer EJ (1999) The WD repeat: a common architecture for diverse functions. Trends Biochem Sci 0004:181–185
Takahashi JS, Hong HK, Ko CH, Mcdearmon EL (2008) The genetics of mammalian circadian order and disorder: implications for physiology and disease. Nat Rev Genet 9
Tan X et al (2007) Mechanism of auxin perception by the TIR1 ubiquitin ligase. Nature 446:640–645
Thompson BJ et al (2007) The SCF FBW7 ubiquitin ligase complex as a tumor suppressor in T cell leukemia. J Exp Med 204:1825–1835

Tzfira T, Vaidya M, Citovsky V (2004) Involvement of targeted proteolysis in plant genetic transformation by agrobacterium. Nature 431:87–92

Wang H et al (2009) Skp2 is required for survival of aberrantly proliferating Rb1 -deficient cells and for tumorigenesis in Rb1 + / − mice. Nat Genet 42:83–88

Wei W et al (2004) Degradation of the SCF component Skp2 in cell-cycle phase G1 by the anaphase-promoting complex. Nature 428:194–198

Wei W, Jin J, Schlisio S, Harper JW, Kaelin WG (2005) The v-Jun point mutation allows c-Jun to escape GSK3-dependent recognition and destruction by the Fbw7 ubiquitin ligase. Cancer Cell 8:25–33

Welcker M, Clurman BE (2008) FBW7 ubiquitin ligase: a tumour suppressor at the crossroads of cell division, growth and differentiation. Nat Rev Cancer 8: 83–93

Welcker M et al (2003) Multisite phosphorylation by Cdk2 and GSK3 controls cyclin E degradation. Mol Cell 12:381–392

Welcker M, Orian A, Grim JA, Eisenman RN, Clurman BE (2004) A nucleolar isoform of the Fbw7 ubiquitin ligase regulates c-Myc and cell size. Curr Biol 14:1852–1857

Weng AP et al (2004) Activating mutations of NOTCH1 in human T cell acute lymphoblastic Leukemia. Science 306:269–271

Wertz IE et al (2011) Sensitivity to antibtubulin chemotherapeutics is regulated by MCL1 and FBW7. Nature 471:110–114

Winston JT et al (1999) A family of mammalian F-box proteins. Curr Biol 9:1180–1182

Wu G et al (2003) Structure of a β-TrCP1-Skp1-β-catenin complex: destruction motif binding and lysine specificity of the SCFβ-TrCP1 ubiquitin ligase. Mol Cell 11:1445–1456

Xing W et al (2013) SCF FBXL3 ubiquitin ligase targets cryptochromes at their cofactor pocket. Nature 496: 64–68

Yao I et al (2007) SCRAPPER-dependent ubiquitination of active zone protein RIM1 regulates synaptic vesicle release. Cell 130:943–957

Yoshida Y, Chiba T, Tokunaga F (2002) E3 ubiquitin ligase that recognizes sugar chains. Nature 418:438–442

Zhang YW et al (2009) The F box protein Fbx6 regulates Chk1 stability and cellular sensitivity to replication stress. Mol Cell 35:442–453

Zhao H et al (2013) Skp2 deletion unmasks a p27 safeguard that blocks tumorigenesis in the absence of pRb and p53 tumor suppressors. Cancer Cell 24:645–659

Zhao Y, Xiong X, Sun Y (2011) DEPTOR, an mTOR inhibitor, is a physiological substrate of SCF b TrCP E3 ubiquitin ligase and regulates survival and autophagy. Mol Cell 44:304–316

Zheng J et al (2002b) CAND1 binds to unneddylated CUL1 and regulates the formation of SCF ubiquitin E3 ligase complex. Mol Cell 10:1519–1526

Zheng N et al (2002a) Structure of the Cul1–Rbx1–Skp1–F box Skp2 SCF ubiquitin ligase complex. Nature 416:703–709

Targeting SCF E3 Ligases for Cancer Therapies

9

Jing Liu, Yunhua Peng, Jinfang Zhang, Jiangang Long, Jiankang Liu, and Wenyi Wei

Abstract

SKP1-cullin-1-F-box-protein (SCF) E3 ubiquitin ligase complex is responsible for the degradation of proteins in a strictly regulated manner, through which it exerts pivotal roles in regulating various key cellular processes including cell cycle and division, apoptosis, and differentiation. The substrate specificity of the SCF complex largely depends on the distinct F-box proteins, which function in either tumor promotion or suppression or in a context-dependent manner. Among the 69 F-box proteins identified in human genome, FBW7, SKP2, and β-TRCP have been extensively investigated among various types of cancer in respective of their roles in cancer development, progression, and metastasis. Moreover, several specific inhibitors have been developed to target those E3 ligases, and their efficiency in tumors has been determined. In this review, we provide a summary of the roles of SCF E3 ligases in cancer development, as well as the potential application of miRNA or specific inhibitors for cancer therapy.

Keywords

SCF E3 ligase · SKP2 · FBW7 · β-TRCP · Cancer

J. Liu · J. Long · J. Liu (✉)
Center for Mitochondrial Biology and Medicine, The Key Laboratory of Biomedical Information Engineering of Ministry of Education, School of Life Science and Technology and Frontier Institute of Science and Technology, Xi'an Jiaotong University, Xi'an, China
e-mail: j.liu@mail.xjtu.edu.cn

Y. Peng
Center for Mitochondrial Biology and Medicine, The Key Laboratory of Biomedical Information Engineering of Ministry of Education, School of Life Science and Technology and Frontier Institute of Science and Technology, Xi'an Jiaotong University, Xi'an, China

Department of Pathology, Beth Israel Deaconess Medical Center, Harvard Medical School, Boston, MA, USA

J. Zhang · W. Wei (✉)
Department of Pathology, Beth Israel Deaconess Medical Center, Harvard Medical School, Boston, MA, USA
e-mail: wwei2@bidmc.harvard.edu

Abbreviation

5gg	1,2,3,4,6-Penta-O-galloyl-beta-D-glucose pentagalloylglucose
AIB1	Amplified in breast cancer 1
APC	Adenomatous polyposis coli
APC/C^{Cdh1}	Anaphase-promoting complex/cyclosome Cdh1 complex
ATRA	All-trans retinoic acid
BLM	Bloom
C/EBPδ	CCAAT enhancer binding protein δ
CAND1	Cullin-associated Nedd8-dissociated protein 1
CDC25A	Cell division cycle 25 homologue A
CDH1	E-cadherin

Y. Sun et al. (eds.), *Cullin-RING Ligases and Protein Neddylation*, Advances in Experimental Medicine and Biology 1217, https://doi.org/10.1007/978-981-15-1025-0_9

CLL	Chronic lymphocytic leukemia
CML	Chronic myeloid leukemia
CPD	Cdc4 phosphodegron
CRL	Cullin-RING ubiquitin ligase
CSN	COP9 signalosome complex
CTD	Carboxy-terminal domain
D-box	Destruction box
DD	Dimerization domain
DEPTOR	DEP domain-containing mTOR-interacting protein
EBP2	EBNA1-binding protein 2
EGCG	Epigallocatechin-3-gallate
ER	Estrogen receptor
FBW7	F-box and WD repeat domain-containing 7
FBXO1	F-box only 1, also known as cyclin F
FDA	Food and Drug Administration
FOXO1	Forkhead box O1
FOXP3	Forkhead box P3
HCC	Hepatocellular carcinoma
HECT	Homologous to E6-associated protein C-terminus
HES5	Hes family bHLH transcription factor 5
HSC	Hematopoietic stem cells
IκB	Inhibitor of nuclear factor-κB
KLF5	Krüppel-like factor 5
LRR	Leucine-rich repeat
miRNA	MicroRNA
MMTV	Mouse mammary tumor virus
NLS	Nuclear localization signal
NPM	Nucleophosmin
NSCLC	Non-small cell lung cancer
ORC1	Origin recognition complex 1
p21/CIP	Cyclin-dependent kinase inhibitor 1A, CDKN1A
p27/KIP	Cyclin-dependent kinase inhibitor 1B, CDKN1B
p57/KIP2	Cyclin-dependent kinase inhibitor 1C, CDKN1C
PDCD4	Programmed cell death protein
PGC1α	Peroxisome proliferator-activated receptor gamma coactivator 1α
PHB2	Prohibitin 2, also known as REA
PIN 1	Peptidylprolyl cis/trans isomerases
PS1	Presenilin 1
PSA	Prostate-specific antigen
RB1	Retinoblastoma 1
RBL2	Retinoblastoma-like protein
RBX1	RING box 1, also known as regulator of cullins-1 or ROC1
Rictor	Rapamycin-insensitive companion of mTOR
RING	Really interesting new gene
SCC	Squamous cell cancer
SCF	SKP1-cullin-1-F-box-protein
SKP1	S-phase kinase-associated protein 1
SKP2	S-phase kinase-associated protein 2, also known as FBXL1
SLP-1	Stomatin-like protein 1
SREBP1	Sterol regulatory element-binding transcription factor 1
T-ALL	T-cell acute lymphoblastic leukemia
TGIF1	5′-TG-3′-interacting factor 1
Tob1	Transducer of ERBB2
Topo IIα	DNA topoisomerase 2-alpha
UBA	E1 ubiquitin-activating enzyme
UBC	E2 ubiquitin-conjugating enzyme
UPS	Ubiquitin-proteasome system
VEGFR2	Vascular endothelial growth factor receptor 2
WHB	Winged-helix B
YAP1	Yes-associated protein 1
β-TRCP	Beta-transducin repeat-containing protein

9.1 Introduction

The ubiquitin-proteasome system (UPS) is responsible for the targeted degradation of particular proteins in a tightly regulated way, which requires the E1/E2/E3 enzymatic cascade to label the specific substrates with ubiquitin chain and subsequently destructed by the 26S proteasome (Bedford et al. 2011). The E1 ubiquitin-activating enzyme (UBA) activates ubiquitin, a 76-amino acid protein which is broadly expressed among all eukaryotic cells, and then the activated ubiquitin is transferred onto the E2 ubiquitin-conjugating enzyme (UBC). The specific transferring of ubiquitin onto the substrates requires the E3 ubiquitin ligase, which either functions as receptors of ubiquitin and then directly transfers ubiquitin onto substrates (such as homologous to E6-associated protein C-terminus, HECT E3 ligases) (Bernassola et al. 2008) or as scaffolding proteins that bridge E2-Ub and substrates (such as U-box domain (Cyr et al. 2002) and really interesting new gene, RING (Deshaies and Joazeiro 2009)). The SKP1-cullin-1-F-box-protein (SCF) type of E3 ubiquitin ligase complexes belong to the largest RING finger E3 ligase family, cullin-RING ubiquitin ligase (CRL), and dictates the degradation of proteins in respect to distinct intracellular functions, including cell cycle progression, signal transduction, and transcription (Wang et al. 2014a).

CRL is inhibited by the assembly inhibitor CAND1 (cullin-associated Nedd8-dissociated protein 1) (Liu et al. 2002; Zheng et al. 2002), and this inhibitory effect can be released by the conjunction of NEDD8 (Duda et al. 2008), a ubiquitin-like protein, on the conserved lysine residue(s) in the WHB (winged-helix B) subdomain of cullins, resulting in the activation of CRL for subsequent ubiquitination cascade (Morimoto et al. 2003; Hori et al. 1999). The NEDD8 molecule can be dissociated from cullins by the deneddylase, CSN (COP9 signalosome complex) (Cavadini et al. 2016).

In the SCF multi-subunit ubiquitin ligase complexes, cullin-1 functions as a rigid scaffold protein to interact with SKP1 (S-phase kinase-associated protein 1) and the RBX1 (RING box 1, also known as regulator of cullins-1 or ROC1) subunits through its N- and C-terminus, respectively (Cardozo and Pagano 2004). SKP1 is an adaptor protein for bridging the F-box protein and cullin-1, while RBX1 is essential for recruiting ubquitin-E2 enzyme and subsequently transferring ubiquitin onto the substrates which is specifically recognized by F-box protein (Cardozo and Pagano 2004) (Fig. 9.1).

In the human genome, there are a total of 69 putative F-box proteins. Each of them contains at least two main functional domains, an F-box domain and a variable carboxy-terminal domain (CTD). The homologous F-box domain, first identified in FBXO1 (F-box only 1, also known as cyclin F), is essential for the direct docking onto SKP1 (Bai et al. 1996). According to the different substrate recognizing domains in their carboxy-terminal, F-box proteins can be further subclassified into three different subfamilies, namely, FBXW subfamily, FBXL subfamily, and FBXO subfamily. The FBXW subfamily comprises ten F-box members that contain a WD40 repeat domain, the 22 FBXL subfamily members contain leucine-rich repeat (LRR) domain, and the other FBXO proteins contain various domains (such as kelch repeats or proline-rich motifs) that are not fully characterized.

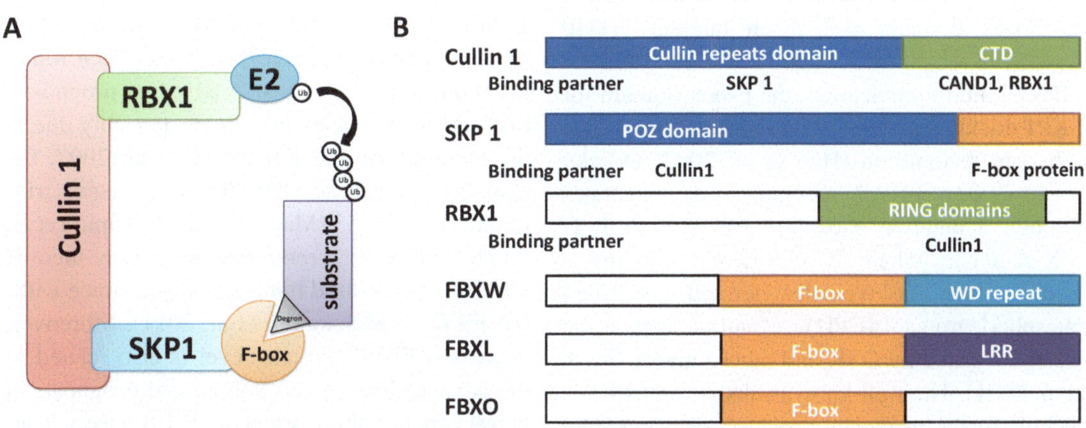

Fig. 9.1 A schematic diagram of SCF E3 ligase complex and its subunits. *SKP1* S-phase kinase-associated protein 1, *RBX1* RING box 1, *Ub* ubiquitin, *CTD* carboxy-terminal domain, *POZ* poxvirus and zinc finger, *RING* really interesting new gene, *WD* WD40 domain, *LRR* leucine-rich repeat

The best characterized F-box proteins in human cancers are SKP2 (S-phase kinase-associated protein 2), FBW7 (F-box and WD repeat domain-containing 7), and β-TRCP (beta-transducin repeat-containing protein), among which SKP2 functions as an oncogene, FBW7 as a tumor suppressor, and β-TRCP as an oncogene or tumor suppressor in a context-dependent manner.

9.2 FBW7 as a Target for Cancer Therapy

9.2.1 FBW7 Is a Haplo-insufficient Tumor Suppressor

FBW7 has been extensively investigated during the past decades and regarded as a tumor suppressor in various human cancers (Welcker and Clurman 2008). FB*W7* locates on chromosome 4q32, a region that is highly mutated in cancers, and encodes three different FBW7 isoforms (FBW7α, β, and γ) via alternative splicing with differences only in their first exons. The unique N-terminus enables different subcellular distribution of these FBW7 isoforms, among which FBW7α localizes to the nucleoplasm in human tissues and cells, FBW7β mainly in the cytoplasm, and FBW7γ in the nucleolus (Matsumoto et al. 2006).

FBW7 contains a dimerization domain (DD), an F-box domain, and seven tandem WD40 repeats, in which the DD domain is essential for FBW7 homo-dimerization, the F-box domain for SKP1 docking, and the WD40 repeats for specific substrate recognition (Hao et al. 2007; Orlicky et al. 2003). FBW7 recognizes its substrates through a putative phospho-motif (I/L-I/L/P-T/S-X-X-S/T/E, where X represents any amino acid), which is typically termed as Cdc4 phosphodegron (CPD) and can be phosphorylated by GSK and other kinases (Nash et al. 2001). The well-known substrates of FBW7 include many oncogenic proteins, such as cyclin E (Koepp et al. 2001; Tetzlaff et al. 2004; Strohmaier et al. 2001), c-Myc (Yada et al. 2004), c-Jun (Nateri et al. 2004; Wei et al. 2005), NOTCH 1 (Tetzlaff et al. 2004), MCL1 (Inuzuka et al. 2011; Wertz et al. 2011), mTOR (Mao et al. 2008), Aurora A (Finkin et al. 2008), Aurora B (Teng et al. 2012), and others (Fig. 9.2).

The first notion to show that FBW7 might be a tumor suppressor came from the fact that its oncogenic substrate cyclin E is activated in various cancers (Koepp et al. 2001). After identification of the loss-of-function *FBW7* gene mutations in breast and endometrial cancers (Strohmaier et al. 2001; Moberg et al. 2001), extensive studies had focused on pinpointing the mutations in *FBW7* gene in a wide spectrum of human cancers. From a comprehensive screen study, approximately 6% of all primary human cancers harbor *FBW7* mutations, and nearly half of these are missense mutations at three arginine residues within the WD40 repeats (R465, R479, and R505), whereas other mutations usually result in premature termination of FBW7 translation (Akhoondi et al. 2007). *FBW7* missense mutations have been found predominantly in T-ALL (T-cell acute lymphoblastic leukemia, approximately 31%) and cholangiocarcinoma (35%), followed by primary endometrial cancer (9%), colorectal cancer (9%), and stomach cancer (6%) (Akhoondi et al. 2007; Maser et al. 2007).

Apart from those mutations of *FBW7* in the abovementioned cancer, the deregulation of FBW7 protein expression has also been reported in various cancers, such as breast cancer and T-ALL (Koepp et al. 2001; Maser et al. 2007). Downregulation of FBW7β has also been found in gliomas and is considered as a prognostic marker for glioblastoma patients, possibly due to the mutation in p53 (Hagedorn et al. 2007; Gu et al. 2007), because FBW7β is a direct transcriptional target of p53 (Mao et al. 2004; Kimura et al. 2003). However, it remains elusive why there is absence of increased tumorigenesis in mice without FBW7β (Matsumoto et al. 2011). Moreover, decreased FBW7 protein level is also related to the advanced stage, metastasis, and prognosis in breast cancer (Akhoondi et al. 2010), gastric cancer (Yokobori et al. 2009), and prostate cancer (Koh et al. 2006).

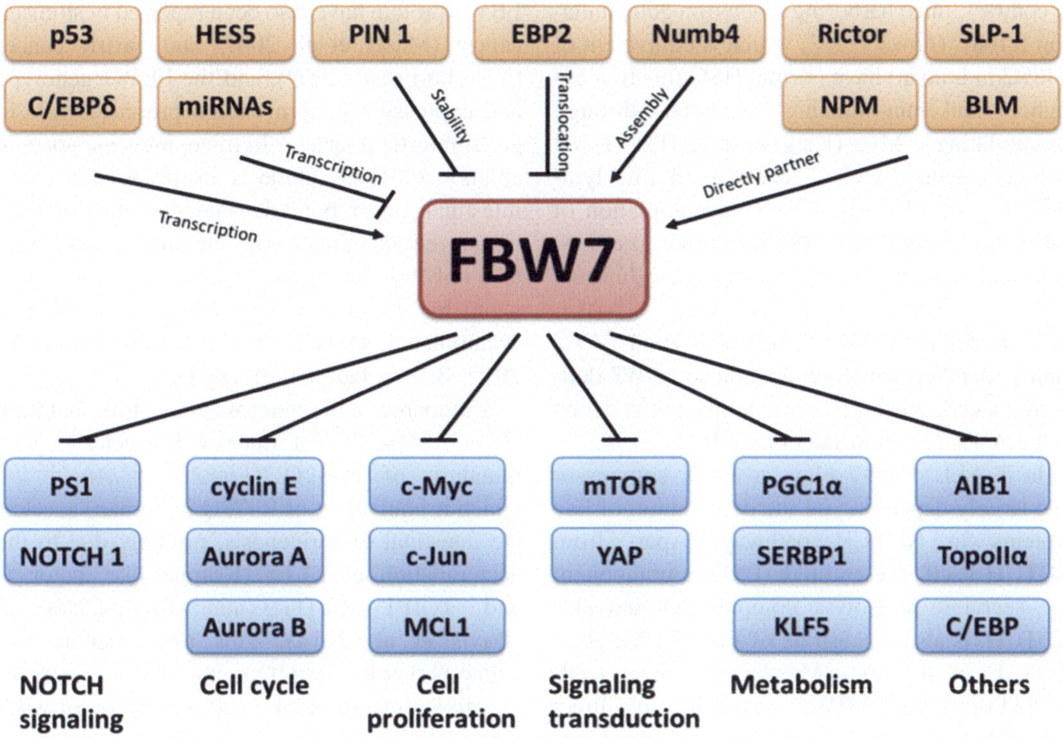

Fig. 9.2 A schematic diagram for SCF^{FBW7} E3 ligase complex and its upstream regulators and substrates. *SKP1* S-phase kinase-associated protein 1, *RBX1* RING box 1, *Ub* ubiquitin, *CPD* Cdc4 phosphodegron, *C/EBP* CCAAT enhancer binding protein, *HES5* hes family bHLH transcription factor 5, *PIN 1* peptidylprolyl cis/trans isomerases, *EBP2* EBNA1-binding protein 2, *Rictor* rapamycin-insensitive companion of mTOR, *SLP-1* stomatin-like protein 1, *NPM* nucleophosmin, *BLM* bloom, *PS1* presenilin 1, *MCL1* induced myeloid leukemia cell differentiation protein, *PGC1* peroxisome proliferator-activated receptor gamma coactivator 1α, *SERBP1* SERPINE1 mRNA-binding protein 1, *KLF5* Krüppel-like factor 5, *TGIF1* 5′-TG-3′-interacting factor 1, *AIB1* amplified in breast cancer 1 protein

Mice with ablation of double *Fbw7* alleles die at embryonic day 10.5 due to defects in hematopoietic and vascular development, which can be attributed to the inefficient degradation of its substrates NOTCH1 and NOTCH4 (Nateri et al. 2004; Tsunematsu et al. 2004). *Fbw7*$^{+/-}$ mice are viable and fertile, but they are susceptible to radiation-induced tumorigenesis, approximately 70% of irradiated *Fbw7*$^{+/-}$/*Tp53*$^{+/-}$ mice develop lymphomas, and some mice also display multiple tumors in epithelial tissues such as the lung, liver, and ovary (Mao et al. 2004). In order to decipher the role of FBW7 in tumorigenesis, mice with tissue-specific deletion of *FBW7* have been developed in various tissues, including the bone marrow, T-cell lineage, intestine, liver, breast, and brain. These studies suggest that the FBW7 tumor suppressor might function in a haplo-insufficient manner (Onoyama et al. 2007; Thompson et al. 2008; Matsuoka et al. 2008).

9.2.2 FBW7 in T-Cell Acute Lymphoblastic Leukemia

FBW7 mutations and decreased expression have been found in T-ALL (Maser et al. 2007), suggesting a predominant tumor suppressor role in leukemia. In mice, ablation of double alleles of *Fbw7* in the bone marrow causes premature depletion of hematopoietic stem cells (HSC) due to accumulation of c-Myc and NOTCH1, as well as activation of p53-dependent apoptosis (Thompson et al. 2008; Matsuoka et al. 2008).

In contrast, mice harboring a heterozygous mutation R468C (*Fbw7*$^{R468C/+}$, which is equivalent of R465C in human) have normal HSC function but higher leukemia-initiating potential through accumulating c-Myc (King et al. 2013). T-cell lineage-specific *Fbw7*$^{-/-}$ leads to thymic lymphoma, partly due to excessive accumulation of c-Myc (Onoyama et al. 2007). In order to escape from cell death in the setting of high c-Jun and c-Myc levels, *FBW7*-deficient T-ALL accumulates more MCL1, a pro-survival BCL2 family member, which renders those *FBW7*-deficient cancers sensitive to sorafenib, but resistant to the ABT-737 (Inuzuka et al. 2011).

In T-ALL, FBW7 plays a tumor suppressor role largely depending on the degradation of key proteins in NOTCH pathway. Apart from NOTCH itself, presenilin 1 (PS1), a component of γ-secretase proteolytic complex that activates NOTCH, is also a substrate of FBW7 (Wu et al. 1998; Li et al. 2002). Moreover, another well-studied substrate of FBW7, c-Myc, is also a direct target gene of NOTCH pathway in several types of human cancers, including T-ALL (Haque et al. 2017; Herranz et al. 2014; Weng et al. 2006), mammary cancer (Klinakis et al. 2006), and others. Another target gene of the NOTCH pathway, HES5 (hes family bHLH transcription factor 5), directly represses transcription of FBW7β, forming a positive feedback regulatory loop (Sancho et al. 2013). Moreover, Numb4, a membrane-bound protein that associates with NOTCH1, promotes FBW7 activation and subsequent degradation of NOTCH (Jiang et al. 2012a). All those studies suggest that targeting FBW7 should be a promising strategy for treating T-ALL patients.

9.2.3 FBW7 in Gastrointestinal Cancer

FBW7 has been identified as the fourth most frequently mutated gene of human colorectal carcinomas and has been described as a poor prognosis marker in human colorectal carcinoma (Rajagopalan et al. 2004; Wood et al. 2007). *FBW7* mutations and deregulated expression of FBW7 protein have also been reported in stomach cancer (Maser et al. 2007) and gastric cancer (Yokobori et al. 2009), and the FBW7 genotype and expression pattern have a prognosis role in gastrointestinal cancer. In mice, intestine-specific ablation of *Fbw7* alone is insufficient to cause intestinal cancer, but it does increase the intestinal tumorigenesis in mice with mutation in adenomatous polyposis coli (*Apc*$^{Min/+}$ mice), which is triggered by aberrant Wnt signaling in part due to accumulated NOTCH1 and c-Jun (Sancho et al. 2010; Babaei-Jadidi et al. 2011).

Compared with heterozygous null mutation (*Fbw7*$^{+/-}$/*Apc*$^{1322T/+}$), mice with a heterozygous mutation of R482Q (*Fbw7*$^{R482Q/+}$/*Apc*$^{1322T/+}$, which is equivalent of R479Q in human) accelerate intestinal tumorigenesis, possibly due to the accumulation of KLF5 (Krüppel-like factor 5) and TGIF1 (5′-TG-3′-interacting factor 1) (Davis et al. 2014). This phenomenon also suggests that the R482Q mutant has a dominant-negative effect to block the dimerization of wild-type FBW7 homo-dimer, and FBW7 functions as a haploid-insufficient tumor suppressor. As a canonic substrate of FBW7, c-Myc is frequently misregulated in colorectal cancer (Muzny et al. 2012), attributed to either *FBW7* deficiency or higher expression of USP28, the deubiquitinase that antagonizes FBW7-mediated c-Myc ubiquitination. On the other hand, USP28 is also a c-Myc target gene, which forms a positive feedback to facilitate high c-Myc level in those tumors (Diefenbacher et al. 2014). Moreover, in gastric cancer, miRNA-25 is highly expressed, binds to the 3′ UTR of FBW7 mRNA, and represses FBW7 expression (Gong et al. 2015).

9.2.4 FBW7 in Hepatocellular Carcinoma (HCC)

Apart from the abovementioned oncogenes, other substrates of FBW7 include SREBP1 (sterol regulatory element-binding transcription factor 1) (Sundqvist et al. 2005) and PGC1α (peroxisome proliferator-activated receptor gamma coactivator 1α) (Olson et al. 2008), both of which are key factors in the regulation of metabolism of glucose and lipid in the liver. It has been reported that *FBW7* expression is decreased in hepatocellular carcinoma (HCC) (Tu et al. 2012; Imura et al. 2014), partly through upregulating c-Myc, cyclin E, and YAP1 (yes-associated protein 1) to facilitate liver tumorigenesis (Tu et al. 2012, 2014).

The mice with one allele deletion of *Fbw7* and *p53* ($p53^{+/-}Fbw7^{+/-}$) develop HCC (Mao et al. 2004), and liver-specific ablation of *Fbw7* results in hamartomas (Onoyama et al. 2011), indicating a key role of FBW7 in the development of liver cancer in vivo. For the detail mechanism, Topo IIα (DNA topoisomerase 2-alpha) (Chen et al. 2011a) and AIB1 (amplified in breast cancer 1) (Liu et al. 2012) have been reported as FBW7 substrates that are involved in HCC. Besides, in HCC, miRNA-27b, miRNA-115-3P, and miRNA-770 are highly expressed, which repress FBW7 expression (Gong et al. 2015; Sun et al. 2016; Tang et al. 2016; Wu et al. 2016).

9.2.4.1 Upstream Regulators of FBW7

FBW7α is transcriptionally regulated by C/EBPδ (CCAAT enhancer binding protein δ). Conversely, FBW7α also targets C/EBPδ for degradation in a GSK-3β-dependent manner, thus forming a negative feedback loop (Balamurugan et al. 2013). Moreover, FBW7β is a bona fide transcriptional target of p53 (Mao et al. 2004; Kimura et al. 2003), and the first exon 1b, but not exon 1a or 1c, of *FBW7* contains a putative p53 binding site with p53-dependent transcriptional activity (Kimura et al. 2003). Besides, the transcription of FBW7β can be directly repressed by HES5, a downstream factor in NOTCH signaling (Sancho et al. 2013). This regulation, together with a FBW7β-dependent NOTCH degradation, functions in a feedback control of NOTCH signaling. On the posttranslational level, FBW7 can be negatively regulated by the PIN1 (peptidylprolyl cis/trans isomerases) in a phosphorylation-dependent manner (Min et al. 2012). Specifically, Pin1 recognizes phosphorylated FBW7 and promotes its self-ubiquitination and destruction by disrupting FBW7 dimerization (Min et al. 2012).

Moreover, FBW7 expression can be repressed by miRNA (microRNA), including miR-27a (Spruck 2011; Lerner et al. 2011), miR-223 (Xu et al. 2010a; Kurashige et al. 2012), miR-25 (Gong et al. 2015; Lu et al. 2012), miR-27b (Sun et al. 2016), miR-24 (Zhao et al. 2016), miR-32 (Xia et al. 2017; Hua et al. 2016), miR-92a (Jiang et al. 2017), miR-129-5p (Hasler et al. 2012), miR-155-3p (Tang et al. 2016), miR-367 (Xu et al. 2017), and miR-770 (Wu et al. 2016). Their role in gastric cancer (Gong et al. 2015), tongue squamous cell carcinoma (Zhao et al. 2016), esophageal squamous cell carcinoma (Kurashige et al. 2012), breast cancer (Xia et al. 2017), non-small cell lung cancer (NSCLC) (Xu et al. 2017), HCC (Sun et al. 2016; Tang et al. 2016; Wu et al. 2016), osteosarcoma (Jiang et al. 2017), and multiple myeloma (Hua et al. 2016) has been evidenced.

Furthermore, FBW7 can be trapped by a decoy pseudo-substrate EBP2 (EBNA1-binding protein 2) that leads to nuclear translocation of FBW7 so as to block its E3 ligase function (Welcker et al. 2011). Apart from those factors that regulate FBW7 abundance and activation, other protein partners, such as Rictor (rapamycin-insensitive companion of mTOR) (Guo et al. 2012), SLP-1 (stomatin-like protein 1) (Zhang et al. 2012), NPM (nucleophosmin) (Bonetti et al. 2008), and BLM (bloom) (Chandra et al. 2013), directly bind to FBW7 to facilitate substrate ubiquitination. Besides these mechanisms, Numb4 was reported to promote FBW7 assembly and activation (Jiang et al. 2012a) (Fig. 9.2).

9.2.4.2 Targeting FBW7 for Cancer Therapy

As mentioned above, Pin1 is a key regulator for FBW7 protein stability, and deletion of Pin1 leads to elevated FBW7, which facilitates MCL1 degradation, sensitizing cancer cells to chemotherapy Taxol (Min et al. 2012). Thus, inhibitor of Pin1 should have potential in targeting FBW7 for cancer treatment. Similarly, other reagents or drugs that antagonize miRNA (such as miR-27a, miR-223, and miR-25) should also be useful in cancers with FBW7 deficiency. For example, genistein, a natural dietary agent that can inhibit miR-223 expression and subsequently induce FBW7, promotes apoptosis of pancreatic cancer cells (Ma et al. 2013).

9.2.5 SKP2 as a Target for Cancer Therapy

9.2.5.1 SKP2 Is an Oncogene

SKP2, also known as FBXL1 (F-box and leucine-rich repeat protein 1), is a member of the FBXL subfamily of F-box proteins and plays a pivotal role in cell cycle progression and proliferation. Structurally, SKP2 consists of an F-box domain, ten tandem LRRs, a D-box (destruction BOX, $RxxLx_{(2-5)}N/E/D$, where X represents any amino acid), and an NLS (nuclear localization signal).

The oncogenic role of SKP2 has been reported in various types of cancers, including lymphomas (Adamson et al. 2001), skin cancer (Qu et al. 2014), lung cancer (Yokoi et al. 2002; Zhu et al. 2004), HCC (Lu et al. 2009), oral squamous cell cancer (SCC) (Kudo et al. 2001; Tosco et al. 2011), laryngeal cancer (Dong et al. 2003), salivary gland mucoepidermoid carcinoma (Handra-Luca et al. 2006), pancreatic cancer (Einama et al. 2006), breast cancer (Voduc et al. 2008; Radke et al. 2005; Sonoda et al. 2006; Chan et al. 2013), prostate cancer (Lin et al. 2010; Nguyen et al. 2011; Wang et al. 2012), gastric cancer (Masuda et al. 2002), and nasopharyngeal carcinoma (Wang et al. 2014b).

Mice with double allele ablation of *SKP2* have smaller littermates (Nakayama et al. 2000), are resistant to one allele deletion of *Pten* ($Pten^{+/-}$), induced development of adrenal or prostate tumor, as well as $Arf^{-/-}$ induced development of sarcomas and lymphomas, through triggering immature cell senescence (Lin et al. 2010). Ablation of *Skp2* not only abolishes spontaneous pituitary tumorigenesis in retinoblastoma 1 (RB1) deficient ($Rb1^{+/-}$) mice (Wang et al. 2010) but also blocks tumorigenesis induced by pituitary-specific double depletion of tumor suppressor *RB1* and *p53* ($Skp2^{-/-}Pomc^{Cre}Rb1^{lox/lox}Trp53^{lox/lox}$). However, *Skp2* deficiency is insufficient to block pituitary tumor in mice without *Rb* and *p27* ($Skp2^{-/-}Pomc^{Cre}Cdkn1b^{-/-}$) (Zhao et al. 2013). *Skp2* deficiency delays breast cancer formation in *MMTV-Neu* mice (Chan et al. 2012), and SKP2 serves as a marker for poor prognosis in Her2-positive breast cancer though regulating AKT1 ubiquitination and activation (Chan et al. 2012). Moreover, ablation of *SKP2* leads to resistance to DMBA-TPA-induced skin tumorigenesis, which cannot be overturned by deleting *p27* (Sistrunk et al. 2013), and the epidermis has elevated apoptosis possibly due to an accumulation of p300 (Sistrunk et al. 2013). *SKP2* deficiency also increases HSC populations and rescues one allele *PTEN* loss ($PTEN^{+/-}$)-induced defect in long-term reconstitution ability of HSCs, through inducing cyclin D1 expression (Wang et al. 2011). Taken together, these lines of evidence support the notion that SKP2 is a bona fide proto-oncoprotein.

SKP2 has a wide spectrum of substrates in respect to various intracellular functions, including p21/CIP (cyclin-dependent kinase inhibitor 1A, CDKN1A) (Yu et al. 1998), p27/KIP (cyclin-dependent kinase inhibitor 1B, CDKN1B) (Tsvetkov et al. 1999; Kossatz et al. 2004), p57/KIP2 (cyclin-dependent kinase inhibitor 1C, CDKN1C) (Kamura et al. 2003), p130 (retinoblastoma-like protein, RBL2) (Tedesco et al. 2002), CDH1 (E-cadherin) (Inuzuka et al. 2012), FOXO1(forkhead box O1) (Huang et al. 2005), MYC (von der Lehr et al. 2003; Kim et al. 2003), E2F (Marti et al. 1999), Tob1 (transducer of ERBB2) (Hiramatsu et al. 2006), TAL1 (Nie et al. 2008), and PHB2 (prohibitin 2, also known as REA) (Umanskaya et al. 2007), ORC1 (origin recognition complex 1) (Mendez et al. 2002), and CDT1 (Nishitani et al. 2006) (Fig. 9.3).

9.2.5.2 SKP2 in Leukemia

SKP2 expression is an independent prognostic factor for AML (Min et al. 2004) and lymphoma (Seki et al. 2010; Uddin et al. 2008; Latres et al. 2001), whereas CKS1 binds SKP2 to facilitate the degradation of p27 in multiple myelomas (Shaughnessy 2005; Zhan et al. 2007). Furthermore, transgenic mice that expressed SKP2 together with activated N-Ras in the T-lymphoid lineage accelerate the development of T-cell lymphomas (Latres et al. 2001). In addition, BCR-ABL fusion oncogene, the most common

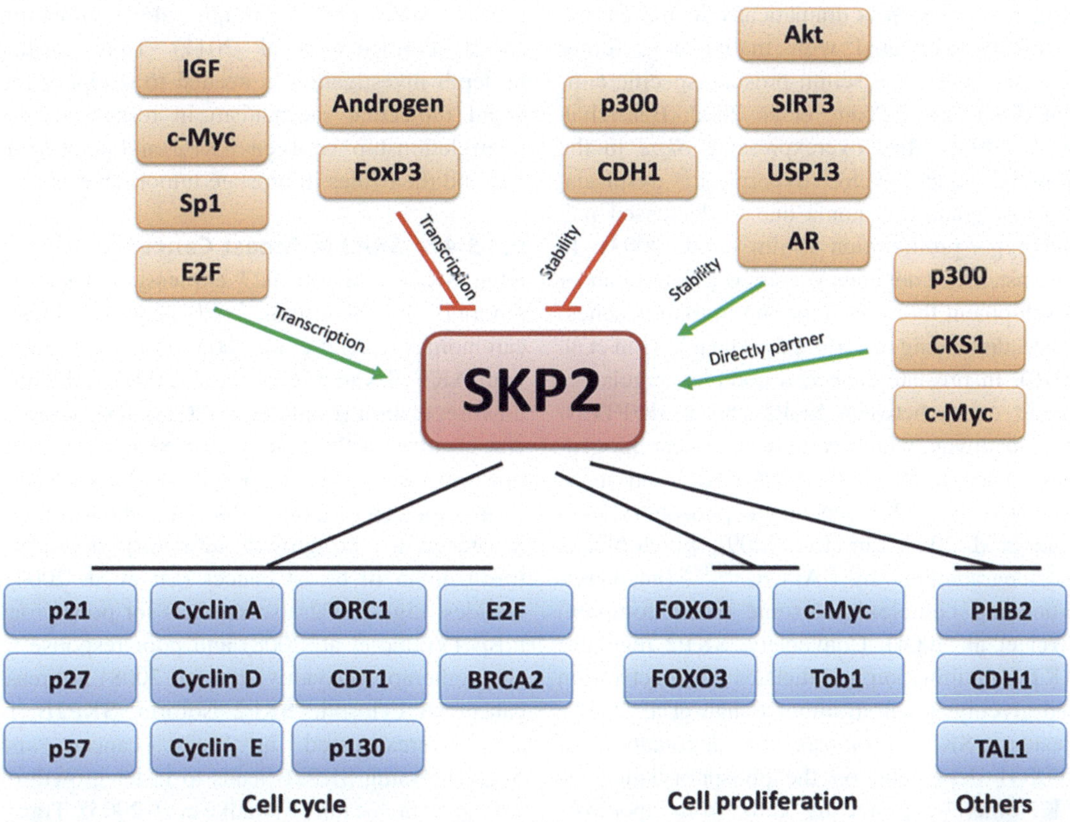

Fig. 9.3 A schematic diagram for the upstream regulators and substrates of the SCF^{SKP2} E3 ligase complex. *IGF* insulin-like growth factor, *CDH1* E-cadherin, *USP13* ubiquitin carboxyl-terminal hydrolase 13, *FOXO1* forkhead box O1, *FOXO3* forkhead box O3, *FOXP3* forkhead box P3, *SIRT3* regulatory protein SIR2 homologue 3, *CKS1* cyclin-dependent kinase regulatory subunit 1, *p21* cyclin-dependent kinase inhibitor 1A, CDKN1A, *p27* cyclin-dependent kinase inhibitor 1B, CDKN1B, *57* cyclin-dependent kinase inhibitor 1C, CDKN1C, *p130* retinoblastoma-like protein, RBL2, *Tob1* transducer of ERBB2, *TAL1* T-cell acute lymphocytic leukemia protein 1, *PHB2* prohibitin 2, also known as REA, *ORC1* origin recognition complex 1, *BRCA2* breast cancer type 2 susceptibility protein

genetic abnormality found in chronic myeloid leukemia (CML), upregulates SKP2 expression (Andreu et al. 2005) via the PI3K/AKT/Sp1 pathway (Chen et al. 2009), so as to promote p27 degradation and cell proliferation. Besides, MYC and SKP2 mRNA levels are correlated in human CML samples. Mechanistically, MYC binds to the E-box in *SKP2* promoter to induce SKP2 transcription in human myeloid leukemia cells (Bretones et al. 2011).

Bortezomib, also known as Velcade, is the first proteasome inhibitor approved by the Food and Drug Administration (FDA) for treating multiple myeloma (Cavo 2006), but the development of resistance limits its long-term use, which is partly due to SKP2 (Driscoll and Malek 2015; Iskandarani et al. 2016). Through disrupting the incorporation of Skp2 into the SCF^{SKP2} complex, a specific inhibitor DT204 enhances bortezomib-induced apoptosis and inhibits myeloma tumorigenesis in vivo (Malek et al. 2017). In contrast to other cancer types, p27 is highly expressed in chronic lymphocytic leukemia (CLL), which is partly due to low c-Myc and its transcription target SKP2 (Caraballo et al. 2014).

9.2.5.3 SKP2 in Prostate Cancer

In human prostate cancer, SKP2 level is a predictor of tumor aggressiveness and survival, and its protein expression is dramatically increased and positively correlated with malignancy, tumor aggressiveness, and serum prostate-specific antigen (PSA) level (Yang et al. 2002; Ben-Izhak et al. 2003). Mice overexpressing *Skp2* in the prostate gland develop hyperplasia, dysplasia, and low-grade carcinoma due to decreased p27 and hyper-proliferation (Shim et al. 2003). In contrast, *SKP2* deficiency restricts prostate cancer development through triggering immature senescence, depending on Atf4, p27, and p21 (Lin et al. 2010). In prostate cancer, a complex regulatory circuit exists between SKP2 and PI3K/PTEN/AKT pathway, which participates in carcinogenesis (Wang et al. 2006). First, SKP2 activity is regulated by AKT-dependent phosphorylation (Gao et al. 2009; Lin et al. 2009), which blocks its degradation by APC/C^{Cdh1} (anaphase-promoting complex/cyclosome Cdh1 complex) (Wei et al. 2004). Conversely, SKP2 regulates AKT ubiquitination and dictates AKT activation and glycolysis within tumor (Chan et al. 2012). Besides, SKP2 promotes the degradation of FOXO1 depending on the phosphorylation by AKT, thereby protecting cells from apoptosis (Huang et al. 2005). Intriguingly, the regulation of p27 by the PI3K/PTEN/AKT signaling depends on SKP2 in PC3 and DU145 prostate cancer cell lines, but not in LNCaP and PC346 (van Duijn and Trapman 2006), suggesting that additional regulating mechanism exists in those prostate cancer cell lines.

Moreover, the level of SKP2 is inversely correlated to BRCA2 in human prostate cancer (Arbini et al. 2011), and SKP2 controls the degradation of BRCA2, facilitating cancer metastasis (Moro et al. 2006). Other substrates, such as c-Myc, also participate in the development of prostate cancer (Rebello et al. 2017). It is possible that SKP2 cooperates with c-Myc to induce RhoA transcription independently of its E3 ligase activity (Chan et al. 2010). In addition, the expression level of SKP2 can be repressed by androgen, leading to accumulation of p27 (Lu et al. 2002; Jiang et al. 2012b; Chuu et al. 2011), while androgen receptor (AR) participates in controlling the stability of SKP2 (Wang et al. 2008). However, it has been also reported that androgen depletion causes senescence through downregulating SKP2 (Pernicova et al. 2011). Thus, further in-depth investigation is needed to decipher the detail molecular mechanism in respect to the interrelationship between SKP2 and androgen/AR and their roles in prostate tumorigenesis.

9.2.5.4 SKP2 in Breast Cancer

High SKP2 and low p27 expression has been reported in high-grade, ER-negative breast carcinomas (Zheng et al. 2005) and is correlated with AKT elevation (Gao et al. 2009) and Cdh1 downregulation (Fujita et al. 2008). The latter is consistent with the finding that SKP2 is a bona fide substrate of Cdh1 (Wei et al. 2004). SKP2 expression can be induced by IGF, which confers resistance to the growth inhibitory action of trastuzumab in breast cancer (Lu et al. 2004). Besides, high SKP2 also predicts for poor prognosis (Voduc et al. 2008) and poor response to chemotherapy (Davidovich et al. 2008) in breast cancer. A cytosolic SKP2 isoform, SKP2B, is also overexpressed in breast cancer, and overexpressing SKP2B leads to faster growth of xenograft breast tumor (Radke et al. 2005). Transgenic mice overexpressing SKP2B develop mammary tumors, partly due to reduced prohibitin (PHB2), which is an inhibitor of the estrogen receptor (ER) (Umanskaya et al. 2007). Through degrading PHB2, SKP2B also leads to p53 inactivation to impact tumorigenesis (Chander et al. 2010).

SKP2 also promotes proliferation of normal breast cell, while ablation of *SKP2* restricts the growth of breast cancer cells (Fujita et al. 2008). Furthermore, SKP2 inhibits apoptosis through forming a complex with p300, which blocks p300-mediated acetylation and activation of p53 (Kitagawa et al. 2008). Consistently, depletion of *SKP2* leads to higher level of apoptosis and sensitizes breast tumor cells to epirubicin (Sun et al. 2007). Furthermore, SKP2 facilitates cell migration, invasion, and breast cancer metastasis by inducing RhoA GTPase transcription (Chan et al. 2010). Mechanistically, SKP2 interacts with c-Myc to recruit Miz1 and p300 onto the RhoA promoter, which is independent of SKP2 E3 ligase activity. Ablation of *SKP2* profoundly restricted metastasis of breast cancer to the lung (Chan et al. 2010). In human breast cancer, SKP2 transcription can be repressed by FoxP3 (forkhead box P3) (Zuo et al. 2007a), and deletion, mutations, and downregulation of FoxP3 have been commonly found in human breast cancer (Zuo et al. 2007b).

By targeting SKP2, several natural compounds have been reported to be efficient in inhibiting breast cancer cell proliferation. ATRA (all-trans retinoic acid) promotes SKP2 degradation, which leads to cell cycle arrest of breast cancer cell lines (Dow et al. 2001). EGCG (epigallocatechin-3-gallate), a main component in green tea, synergistically enhanced the growth inhibitory effect of tamoxifen and paclitaxel, through decreasing SKP2 expression and accumulating p27 (Huang et al. 2008). Gallic acid decreases SKP2 expression and SKP/p27 association, leading to p27 accumulation and cell cycle arrest in G2/M phase (Hsu et al. 2011). Pentagalloylglucose, quercetin, curcumin, and lycopene decrease SKP2 expression, leading to cell cycle arrest partly due to FOXO1 accumulation (Roy et al. 2007).

9.2.5.5 Upstream Regulators of SKP2

c-Myc has been reported as a bona fide transcription factor for SKP2 (Bretones et al. 2011), and SKP2 also targets c-Myc for ubiquitination and degradation (von der Lehr et al. 2003; Kim et al. 2003), thus forming a feedback loop to restrict the functions of both SKP2 and c-Myc. Similarly, by binding to a functional E2F response element within the promoter of *SKP2*, E2F is a transcription factor for SKP2 (Zhang and Wang 2006) and also a substrate of SKP2 (Marti et al. 1999), forming another layer of feedback loop for SKP2 regulation. Moreover, SKP2 transcription can be promoted by IGF1 (Lu et al. 2004) and Sp1 (Chen et al. 2009) and repressed by androgen (Jiang et al. 2012b) and transcriptional repressor FoxP3 (Zuo et al. 2007a).

At posttranslational level, SKP2 can be phosphorylated by oncogenic AKT1 (Gao et al. 2009; Lin et al. 2009), which facilitates its cytosolic translocation and escapes from APC/Cdh1-mediated degradation (Gao et al. 2009; Wei et al. 2004). SIRT3 and p300 function oppositely on regulating the acetylation, translocation, and stability of SKP2, where SIRT3 enhances its stability and p300 promotes its degradation (Inuzuka et al. 2012). Besides these regulatory mechanisms, D-box-dependent degradation of SKP2 by APC/C^{Cdh1} can be blocked by AR (Wang et al. 2008) and antagonized by deubiquitinase USP13 (Chen et al. 2011b). Through inhibiting mTOR, rapamycin decreases SKP2 expression and promotes SKP2 degradation, suggesting a key role of mTOR in SKP2 regulation (Shapira et al. 2006).

9.2.5.6 Targeting SKP2 for Cancer Therapy

Given the pivotal role of SKP2 in tumorigenesis, targeting SKP2 could be promising for the treatment of various cancers with deregulation of SKP2. To date, several selective small molecule inhibitors for SKP2 have been developed by targeting the binding interface between SKP2 and its substrate p27 or between SKP2 and SKP1. These inhibitors include compound A/CpdA (Chen et al. 2008) and its analogue (Shouksmith et al. 2015), SMIP0001 and SMIP0004 (small molecule inhibitors of p27 depletion) (Rico-Bautista et al. 2010), compound 1/SKPin C1 (Wu et al. 2012), compound 25/SZL-P1-41 and its derivatives (Chan et al. 2013), NSC689857 and NSC681152 (Ungermannova et al. 2013), and DT204 (Malek et al. 2017) (Fig. 9.4).

Mechanistically, CpdA has been identified by a high-throughput screening, and it interferes the incorporation of SKP2 into SCF^{SKP2} ligase, leading to substrate accumulation and G1/S cell cycle arrest (Chen et al. 2008). Based on the structure of CpdA, several derivatives with potential in blocking SKP2-mediated p27 degradation have been synthesized (Shouksmith et al. 2015). However, their efficiency in vivo needs to be investigated in the future. SMIP0001 and SMIP0004 were identified in a cell-based high-throughput screening by measuring endogenous p27 protein level (Rico-Bautista et al. 2010). Both compounds lead to G1 delay and inhibition of colony growth, through reducing SKP2 and

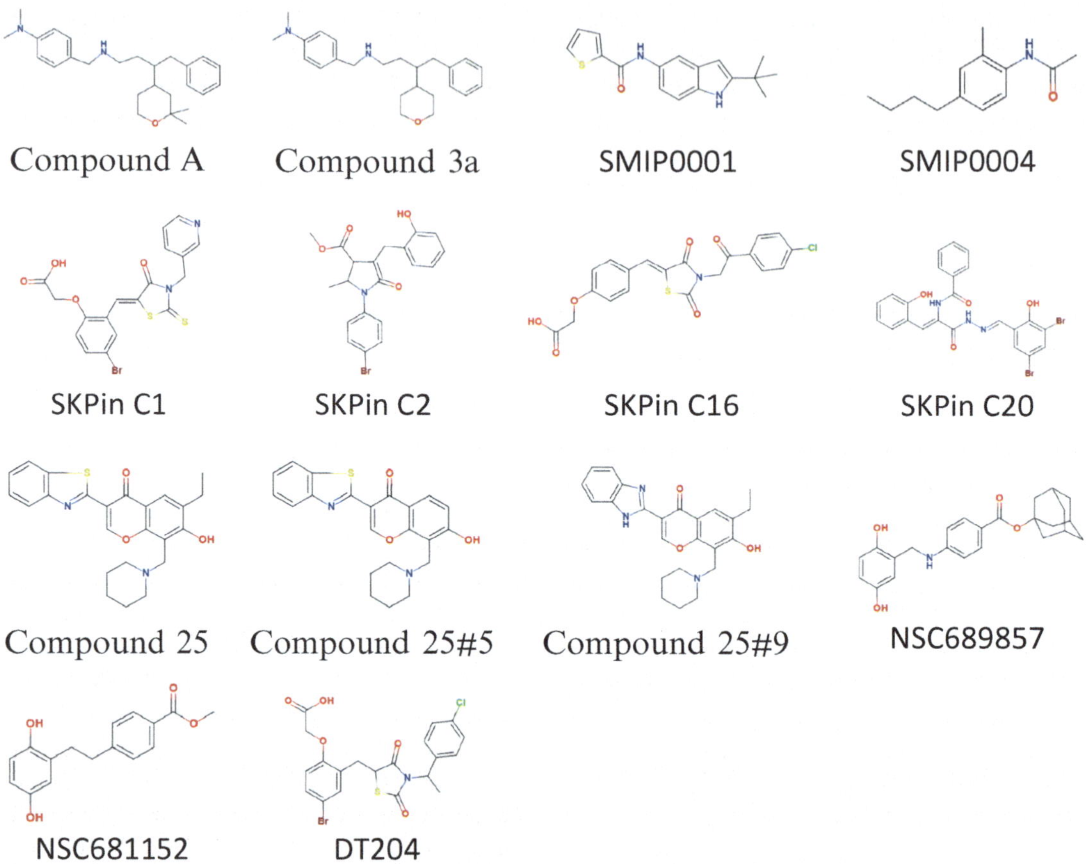

Fig. 9.4 Specific inhibitors for SKP2. Compound A, Compound 3a, SMIP0001, SMIP0004, and DT-204 were screened from a cell-based assay for accumulating p27. SKPin C1, C2, C16, C20 were identified from an in silico high-throughput screening targeted to the binding interface between SKP2 and p27. Compound 25, 25#5, and 25#9 were identified through an in silico high-throughput screening for interrupting Skp2-Skp1 interface. NSC689857 and NSC681152 were identified through a high-throughput AlphaScreen assay based on the interface between SKP2 and Cks1

stabilizing p27 and p21 (Rico-Bautista et al. 2010). However, their role may not be strictly dependent on p27 and requires further studies.

SKPin C1, together with C2, C16, and C20, have been identified through an in silico high-throughput screening targeted to the binding interface between SKP2 and p27 (Wu et al. 2012). They selectively inhibit SKP2-mediated p27 degradation and lead to cell cycle arrest in the G1 or G2/M phases (Wu et al. 2012). Compound 25 and its analogues (25#5 and 25#9) have been identified through an in silico high-throughput screening based on the Skp2-Skp1 interface in SCF^{SKP2} complex (Chan et al. 2013). Those compounds selectively inhibit SKP2 E3 ligase activity and suppress tumor growth in vitro and in vivo (Chan et al. 2013). NSC689857 and NSC681152 have been identified in a high-throughput AlphaScreen assay by targeting SKP2-Cks1 interface, and those compounds are efficient in interrupting protein-protein interaction between SKP2 and Cks1, leading to p27 accumulation (Ungermannova et al. 2013), and their roles in vivo need more investigations.

Apart from those selective inhibitors, several natural compounds, such as 1,2,3,4,6-penta-O-galloyl-beta-D-glucose (pentagalloylglucose, 5gg), curcumin, quercetin, lycopene, silibinin (Roy et al. 2007), EGCG (Huang et al. 2008), and Vitamin D (Yang and Burnstein 2003), are also effective in reducing SKP2 expression to prevent tumorigenesis. Besides, rapamycin, an inhibitor of mTOR, not only significantly decreases SKP2 expression but also promotes SKP2 degradation, leading to cell cycle arrest in G1 phase (Shapira et al. 2006). Taken together, these lines of evidence suggest that targeting SKP2 could be a promising strategy for treating cancer with deregulation of SKP2.

9.2.6 β-TRCP as a Target for Cancer Therapy

9.2.6.1 β-TRCP Functions as an Oncogene or Tumor Suppressor in a Context-Dependent Manner

Different from SKP2 and FBW7, the role of β-TRCP in tumorigenesis is context- and tissue-dependent, though it has been regarded mainly as an oncoprotein. In human, two β-TRCP homologues (β-TRCP1, also known as FBW1, and β-TRCP2, also known as FBW11) exist with only noticeable difference in the N-terminal sequences and are encoded by two distinct genes, *β-TRCP1* and *β-TRCP2*. The biochemical properties of β-TRCP1 and β-TRCP2 are not indistinguishable, and they might compensate each other as mice lacking β-TRCP1 develop normally (Nakayama et al. 2003). Similar to FBW7, β-TRCP also consists of a dimerization domain, an F-box motif at its N-terminus, and seven WD-40 repeats at the C-terminus (Margottin et al. 1998). β-TRCP recognizes its substrates through a phospho-motif (DSGXXS, where X represents any amino acid), which is termed as phosphodegron and can be phosphorylated by GSK3 (Zhou et al. 2004), CHK (Busino et al. 2003), and other kinases.

The substrates of β-TRCP include Emi1 (Margottin-Goguet et al. 2003), CDC25A (cell division cycle 25 homologue A) (Busino et al. 2003; Jin et al. 2003), VEGFR2 (vascular endothelial growth factor receptor 2), DEPTOR (DEP domain-containing mTOR-interacting protein) (Zhao et al. 2011; Gao et al. 2011; Duan et al. 2011), IκB (inhibitor of nuclear factor-κB) (Yaron et al. 1998; Suzuki et al. 2000; Strack et al. 2000), β-catenin (Hart et al. 1999), Wee1A (Watanabe et al. 2004), cyclin D1 (Wei et al. 2008), BTG (Sasajima et al. 2012), PDCD4 (programmed cell death protein 4) (Dorrello et al. 2006), pro-caspase-3 (Tan et al. 2006), MCL1 (Ding et al. 2007), BimEL (Dehan et al. 2009), p53 (Xia et al. 2009), Snail (Zhou et al. 2004; Xu et al. 2010b), and fibronectin (Ray et al. 2006), whose functions are involved in cell cycle, apoptosis, and cell migration. Apart from its role in promoting the degradation of its substrates, β-TRCP also adds polyubiquitin chain onto its non-degradation substrate c-Myc to antagonize the degradation of c-Myc medicated by FBW7 (Popov et al. 2010).

9.2.6.2 β-TRCP Deregulation in Cancers

Mutations and overexpression of β-TRCP1 have been reported in several human cancers, including prostate cancer (Gerstein et al. 2002), colon cancer (Spiegelman et al. 2000), breast cancer (Wood et al. 2007), gastric cancer (Saitoh and Katoh 2001; Kim et al. 2007), and pancreatic cancer (Muerkoster et al. 2005). The mammary glands of *Btrc1*$^{-/-}$ mice display a hypoplastic phenotype, while mice with epithelial cell-specific overexpression β-TRCP1 under a mouse mammary tumor virus (MMTV) promoter show increased mammary gland hyperplasia, and nearly one third of mice develop mammary, ovarian, and/or uterine cancers, partly through NF-kB transactivation (Kudo et al. 2004). Transgenic mice that specifically express β-TRCP1 in the intestine, liver, and kidney under a rat calbindin-D9K promoter develop intestinal carcinoma and hepatic or urothelial tumors, partly through β-catenin activation (Belaidouni et al. 2005). *Btrc1*$^{-/-}$ mouse fibroblasts have increased genetic instability (Margottin-Goguet et al. 2003; Watanabe et al. 2004; Guardavaccaro et al. 2003).

Among the substrates of β-TRCP, β-catenin is an oncogene, while IκB, p53, and FOXO3 are tumor suppressors. Thus, it is still elusive whether β-TRCP is a bona fide oncogene or tumor suppressor, and additional genetic models are required to validate the context-dependent role of β-TRCP in tumorigenesis.

9.2.6.3 Target β-TRCP for Cancer Therapy

In line with the predominant oncogenic role of β-TRCP in certain contexts, targeting β-TRCP may be a therapeutic strategy in a defined subset of human tumors. Erioflorin, a small molecule isolated from *Eriophyllum lanatum*, has recently been demonstrated to disrupt the interaction between β-TRCP and its substrate PDCD4, thereby preventing PDCD4 from degradation (Blees et al. 2012). However, β-TRCP may not serve as a broad anticancer drug target largely because it functions as an oncogene or tumor suppressor in a context-dependent manner.

9.3 Discussion and Perspective

In viewing the clinic application of bortezomib in treating multiple myeloma, drugs targeting the ubiquitin-proteasome pathways have become a promising strategy for anticancer therapy (Richardson et al. 2003; Chanan-Khan et al. 2016). Due to the universal role of such inhibitors to degrade non-specific proteins, compounds that target unique F-box E3 ligase should have overwhelming strength and specificity, and thus those F-box E3 ligases are perfect protein baits for high-throughput screening of small molecular inhibitors.

Given the predominant oncogenic role of SKP2 in various human cancers, more specific inhibitor should be screened based on different strategies, either blocking SKP2 activation by disassociating it from SCF complex or disrupting the binding of SKP2 to one specific substrate in a given cancer context without affecting the abundance of other substrates. The latter should have more potential as it overcomes the side effect because the general SKP2 inhibitor (not specific for a given substrate) could lead to the accumulation of other substrates. On the other hand, for tumors deficient in tumor suppressive F-box proteins such as FBW7, a "synthetic lethality" strategy should be applied to explore inhibiting which downstream oncogenic substrate of FBW7 could achieve targeted killing of *FBW7*-deficient cells. To this end, our previous study showed that in the T-ALL disease setting, MCL1 inhibitor could specifically kill *FBW7*-deficient but not WT T-ALL cells (Inuzuka et al. 2011). Similar approach can be applied to other tumor suppressive F-box proteins to identify novel targeted therapies. Lastly, for F-box protein with context-dependent role in tumorigenesis such as β-TRCP, caution should be taken toward simple application of β-TRCP inhibitor as different tissues might respond differently, given its role in promoting versus suppressing tumorigenesis in a tissue or cellular context-dependent manner.

References

Adamson A, Lin Z, Perez-Ordonez B, Jordan R, Tripp S, Perkins S et al (2001) Expression of Skp2, a p27 ubiquitin ligase in malignant lymphoma: correlation with p27 and proliferation index. Mod Pathol 14 (1):154a. https://doi.org/10.1182/blood.V100.8.2950

Akhoondi S, Sun D, von der Lehr N, Apostolidou S, Klotz K, Maljukova A et al (2007) FBXW7/hCDC4 is a general tumor suppressor in human cancer. Cancer Res 67(19):9006–9012. https://doi.org/10.1158/0008-5472.CAN-07-1320

Akhoondi S, Lindstrom L, Widschwendter M, Corcoran M, Bergh J, Spruck C et al (2010) Inactivation of FBXW7/hCDC4-beta expression by promoter hypermethylation is associated with favorable prognosis in primary breast cancer. Breast Cancer Res 12(6). https://doi.org/10.1186/Bcr2788

Andreu EJ, Lledo E, Poch E, Ivorra C, Albero MP, Martinez-Climent JA et al (2005) BCR-ABL induces the expression of Skp2 through the PI3K pathway to promote p27 (Kip1) degradation and proliferation of chronic myelogenous leukemia cells. Cancer Res 65 (8):3264–3272. https://doi.org/10.1158/0008-5472

Arbini AA, Greco M, Yao JL, Bourne P, Marra E, Hsieh JT et al (2011) Skp2 overexpression is associated with loss of BRCA2 protein in human prostate cancer. Am J Pathol 178(5):2367–2376. https://doi.org/10.1016/j.ajpath.2011.01.050

Babaei-Jadidi R, Li NN, Saadeddin A, Spencer-Dene B, Jandke A, Muhammad B et al (2011) FBXW7 influences murine intestinal homeostasis and cancer, targeting notch, Jun, and DEK for degradation. J Exp Med 208(2):295–312. https://doi.org/10.1084/jem.20100830

Bai C, Sen P, Hofmann K, Ma L, Goebl M, Harper JW et al (1996) SKP1 connects cell cycle regulators to the ubiquitin proteolysis machinery through a novel motif, the F-box. Cell 86(2):263–274. https://doi.org/10.1016/S0092-8674(00)80098-7

Balamurugan K, Sharan S, Klarmann KD, Zhang Y, Coppola V, Summers GH et al (2013) FBXW7 alpha attenuates inflammatory signalling by downregulating C/EBP delta and its target gene Tlr4. Nat Commun 4. https://doi.org/10.1038/Ncomms2677

Bedford L, Lowe J, Dick LR, Mayer RJ, Brownell JE (2011) Ubiquitin-like protein conjugation and the ubiquitin-proteasome system as drug targets. Nat Rev Drug Discov 10(1):29–46. https://doi.org/10.1038/nrd3321

Belaidouni N, Peuchmaur M, Perret C, Florentin A, Benarous R, Besnard-Guerin C (2005) Overexpression of human beta TrCP1 deleted of its F box induces tumorigenesis in transgenic mice. Oncogene 24(13):2271–2276. https://doi.org/10.1038/sj.onc.1208418

Ben-Izhak O, Lahav-Baratz S, Meretyk S, Ben-Eliezer S, Sabo E, Dirnfeld M et al (2003) Inverse relationship between Skp2 ubiquitin ligase and the cyclin dependent kinase inhibitor p27(Kip1) in prostate cancer. J Urol 170(1):241–245. https://doi.org/10.1097/01.ju.0000072113.34524.a7

Bernassola F, Karin M, Ciechanover A, Melino G (2008) The HECT family of E3 ubiquitin ligases: multiple players in cancer development. Cancer Cell 14 (1):10–21. https://doi.org/10.1016/j.ccr.2008.06.001

Blees JS, Bokesch HR, Rubsamen D, Schulz K, Milke L, Bajer MM et al (2012) Erioflorin stabilizes the tumor suppressor Pdcd4 by inhibiting its interaction with the E3-ligase beta-TrCP1. PLoS One 7(10). https://doi.org/10.1371/journal.pone.0046567

Bonetti P, Davoli T, Sironi C, Amati B, Pelicci PG, Colombo E (2008) Nucleophosmin and its AML-associated mutant regulate c-Myc turnover through Fbw7 gamma. J Cell Biol 182(1):19–26. https://doi.org/10.1083/jcb.200711040

Bretones G, Acosta JC, Caraballo JM, Ferrandiz N, Gomez-Casares MT, Albajar M et al (2011) SKP2 oncogene is a direct MYC target gene and MYC down-regulates p27(KIP1) through SKP2 in human leukemia cells. J Biol Chem 286(11):9815–9825. https://doi.org/10.1074/jbc.M110.165977

Busino L, Donzelli M, Chiesa M, Guardavaccaro D, Ganoth D, Dorrello NV et al (2003) Degradation of Cdc25A by beta-TrCP during S phase and in response to DNA damage. Nature 426(6962):87–91. https://doi.org/10.1038/nature02082

Caraballo JM, Acosta JC, Cortes MA, Albajar M, Gomez-Casares MT, Batlle-Lopez A et al (2014) High p27 protein levels in chronic lymphocytic leukemia are associated to low Myc and Skp2 expression, confer resistance to apoptosis and antagonize Myc effects on cell cycle. Oncotarget 5(13):4694–4708. https://doi.org/10.18632/oncotarget.2100

Cardozo T, Pagano M (2004) The SCF ubiquitin ligase: insights into a molecular machine. Nat Rev Mol Cell Biol 5(9):739–751. https://doi.org/10.1038/nrm1471

Cavadini S, Fischer ES, Bunker RD, Potenza A, Lingaraju GM, Goldie KN et al (2016) Cullin-RING ubiquitin E3 ligase regulation by the COP9 signalosome. Nature 531(7596):598–603. https://doi.org/10.1038/nature17416

Cavo M (2006) Proteasome inhibitor bortezomib for the treatment of multiple myeloma. Leukemia 20 (8):1341–1352. https://doi.org/10.1038/sj.leu.2404278

Chan CH, Lee SW, Li CF, Wang J, Yang WL, Wu CY et al (2010) Deciphering the transcriptional complex critical for RhoA gene expression and cancer metastasis. Nat Cell Biol 12(5):457–U103. https://doi.org/10.1038/ncb2047

Chan CH, Li CF, Yang WL, Gao Y, Lee SW, Feng ZZ et al (2012) The Skp2-SCF E3 ligase regulates Akt Ubiquitination, glycolysis, Herceptin sensitivity, and tumorigenesis. Cell 149(5):1098–1111. https://doi.org/10.1016/j.cell.2012.02.065

Chan CH, Morrow JK, Li CF, Gao Y, Jin GX, Moten A et al (2013) Pharmacological inactivation of Skp2 SCF ubiquitin ligase restricts cancer stem cell traits and cancer progression. Cell 154(3):556–568. https://doi.org/10.1016/j.cell.2013.06.048

Chanan-Khan AA, Lentzsch S, Quach H, Horvath N, Capra M, Ovilla R et al (2016) Daratumumab, bortezomib and dexamethasone versus bortezomib and dexamethasone alone for relapsed or refractory multiple myeloma based on prior treatment exposure: updated efficacy analysis of castor. Blood 128(22):3313

Chander H, Halpern M, Resnick-Silverman L, Manfredi JJ, Germain D (2010) Skp2B attenuates p53 function by inhibiting prohibitin. EMBO Rep 11(3):220–225. https://doi.org/10.1038/embor.2010.2

Chandra S, Priyadarshini R, Madhavan V, Tikoo S, Hussain M, Mudgal R et al (2013) Enhancement of c-Myc degradation by BLM helicase leads to delayed tumor initiation. J Cell Sci 126(16):3782–3795. https://doi.org/10.1242/jcs.124719

Chen JY, Wang MC, Hung WC (2009) Transcriptional activation of Skp2 by BCR-ABL in K562 chronic myeloid leukemia cells. Leuk Res 33(11):1520–1524. https://doi.org/10.1016/j.leukres.2009.03.007

Chen MC, Chen CH, Chuang HC, Kulp SK, Teng CM, Chen CS (2011a) Novel mechanism by which histone deacetylase inhibitors facilitate topoisomerase II alpha degradation in hepatocellular carcinoma cells. Hepatology 53(1):148–159. https://doi.org/10.1002/hep.23964

Chen MF, Gutierrez GJ, Ronai ZA (2011b) Ubiquitin-recognition protein Ufd1 couples the endoplasmic reticulum (ER) stress response to cell cycle control. Proc Natl Acad Sci U S A 108(22):9119–9124. https://doi.org/10.1073/pnas.1100028108

Chen Q, Xie WL, Kuhn DJ, Voorhees PM, Lopez-Girona-A, Mendy D et al (2008) Targeting the p27 E3 ligase SCFSkp2 results in p27-and Skp2-mediated cell-cycle arrest and activation of autophagy. Blood 111 (9):4690–4699. https://doi.org/10.1182/blood-2007-09-112904

Chuu CP, Kokontis JM, Hiipakka RA, Fukuchi J, Lin HP, Lin CY et al (2011) Androgen suppresses proliferation of castration-resistant LNCaP 104-R2 prostate cancer cells through androgen receptor, Skp2, and c-Myc. Cancer Sci 102(11):2022–2028. https://doi.org/10.1111/j.1349-7006.2011.02043.x

Cyr DM, Hohfeld J, Patterson C (2002) Protein quality control: U-box-containing E3 ubiquitin ligases join the fold. Trends Biochem Sci 27(7):368–375. https://doi.org/10.1016/S0968-0004(02)02125-4

Davidovich S, Ben-Izhak O, Shapira M, Futerman B, Hershko DD (2008) Over-expression of Skp2 is associated with resistance to preoperative doxorubicin-based chemotherapy in primary breast cancer. Breast Cancer Res 10(4). https://doi.org/10.1186/Bcr2122

Davis H, Lewis A, Behrens A, Tomlinson I (2014) Investigation of the atypical FBXW7 mutation spectrum in human tumours by conditional expression of a heterozygous propeller tip missense allele in the mouse intestines. Gut 63(5):792–799. https://doi.org/10.1136/gutjnl-2013-304719

Dehan E, Bassermann F, Guardavaccaro D, Vasiliver-Shamis G, Cohen M, Lowes KN et al (2009) Beta TrCP- and Rsk1/2-mediated degradation of BimEL inhibits apoptosis. Mol Cell 33(1):109–116. https://doi.org/10.1016/j.molcel.2008.12.020

Deshaies RJ, Joazeiro CAP (2009) RING Domain E3 Ubiquitin Ligases. Annu Rev Biochem 78:399–434. https://doi.org/10.1146/annurev.biochem.78.101807.093809

Diefenbacher ME, Popov N, Blake SM, Schulein-Volk C, Nye E, Spencer-Dene B et al (2014) The deubiquitinase U5P28 controls intestinal homeostasis and promotes colorectal cancer. J Clin Investig 124 (8):3407–3418. https://doi.org/10.1172/JC173733

Ding QQ, He XG, Hsu JM, Xia WY, Chen CT, Li LY et al (2007) Degradation of Mcl-1 by beta-TrCP mediates glycogen synthase kinase 3-induced tumor suppression and chemosensitization. Mol Cell Biol 27 (11):4006–4017. https://doi.org/10.1128/Mcb.00620-06

Dong YY, Sui L, Watanabe Y, Sugimoto K, Tokuda M (2003) S-phase kinase-associated protein 2 expression in laryngeal squamous cell carcinomas and its prognostic implications. Oncol Rep 10(2):321–325. https://doi.org/10.3892/or.10.2.321

Dorrello NV, Peschiaroli A, Guardavaccaro D, Colburn NH, Sherman NE, Pagano M (2006) S6K1- and beta TRCP-mediated degradation of PDCD4 promotes protein translation and cell growth. Science 314 (5798):467–471. https://doi.org/10.1126/science.1130276

Dow R, Hendley J, Pirkmaier A, Musgrove EA, Germain D (2001) Retinoic acid-mediated growth arrest requires ubiquitylation and degradation of the F-box protein Skp2. J Biol Chem 276(49):45945–45951. https://doi.org/10.1074/jbc.M103593200

Driscoll JJ, Malek E (2015) Pharmacogenomics of Bortezomib in multiple myeloma patients reveals that the ubiquitin ligase SCF-Skp2 promotes drug resistance. Blood 126(23):3021

Duan S, Skaar JR, Kuchay S, Toschi A, Kanarek N, Ben-Neriah Y et al (2011) mTOR generates an auto-amplification loop by triggering the βTrCP-and CK1-α-dependent degradation of DEPTOR. Mol Cell 44 (2):317–324. https://doi.org/10.1016/j.molcel.2011.09.005

Duda DM, Borg LA, Scott DC, Hunt HW, Hammel M, Schulman BA (2008) Structural insights into NEDD8 activation of Cullin-RING ligases: conformational control of conjugation. Cell 134(6):995–1006. https://doi.org/10.1016/j.cell.2008.07.022

Einama T, Kagata Y, Tsuda H, Morita D, Ogata S, Ueda S et al (2006) High-level Skp2 expression in pancreatic ductal adenocarcinoma: correlation with the extent of lymph node metastasis, higher histological grade, and poorer patient outcome. Pancreas 32(4):376–381. https://doi.org/10.1097/01.mpa.0000220862.78248.c4

Finkin S, Aylon Y, Anzi S, Oren M, Shaulian E (2008) Fbw7 regulates the activity of endoreduplication mediators and the p53 pathway to prevent drug-induced polyploidy. Oncogene 27(32):4411–4421. https://doi.org/10.1038/onc.2008.77

Fujita T, Liu WJ, Doihara H, Date H, Wan Y (2008) Dissection of the ApC(Cdh1)-Skp2 cascade in breast cancer. Clin Cancer Res 14(7):1966–1975. https://doi.org/10.1158/1078-0432.CCR-07-1585

Gao D, Inuzuka H, Tan M-KM, Fukushima H, Locasale JW, Liu P et al (2011) mTOR drives its own activation via SCF βTrCP-dependent degradation of the mTOR inhibitor DEPTOR. Mol Cell 44(2):290–303. https://doi.org/10.1016/j.molcel.2011.08.030

Gao DM, Inuzuka H, Tseng A, Chin RY, Toker A, Wei WY (2009) Phosphorylation by Akt1 promotes cytoplasmic localization of Skp2 and impairs APC-Cdh1-mediated Skp2 destruction. Nat Cell Biol 11(4):397–U392. https://doi.org/10.1038/ncb1847

Gerstein AV, Almeida TA, Zhao GJ, Chess E, Shih IM, Buhler K et al (2002) APC/CTNNB1 (beta-catenin) pathway alterations in human prostate cancers. Genes Chromosomes Cancer 34(1):9–16. https://doi.org/10.1002/gcc.10037

Gong JH, Cui Z, Li L, Ma Q, Wang QF, Gao YH et al (2015) MicroRNA-25 promotes gastric cancer proliferation, invasion, and migration by directly targeting F-box and WD-40 domain protein 7, FBXW7. Tumor Biol 36(10):7831–7840. https://doi.org/10.1007/s13277-015-3510-3

Gu ZD, Inomata K, Ishizawa K, Horii A (2007) The FBXW7 beta-form is suppressed in human glioma cells. Biochem Biophys Res Commun 354 (4):992–998. https://doi.org/10.1016/j.bbrc.2007.07.080

Guardavaccaro D, Kudo Y, Boulaire J, Barchi M, Busino L, Donzelli M et al (2003) Control of meiotic and mitotic progression by the F box protein beta-Trcp1 in vivo. Dev Cell 4(6):799–812. https://doi.org/10.1016/S1534-5807(03)00154-0

Guo Z, Zhou YN, Evers BM, Wang QD (2012) Rictor regulates FBXW7-dependent c-Myc and cyclin E degradation in colorectal cancer cells. Biochem Biophys Res Commun 418(2):426–432. https://doi.org/10.1016/j.bbrc.2012.01.054

Hagedorn M, Delugin M, Abraldes I, Allain N, Belaud-Rotureau MA, Turmo M et al (2007) FBXW7/hCDC4 controls glioma cell proliferation in vitro and is a prognostic marker for survival in glioblastoma patients. Cell Div 2. https://doi.org/10.1186/1747-1028-2-9

Handra-Luca A, Ruhin B, Lesty C, Fouret P (2006) P27, SKP2, and extra-cellular signal-related kinase signalling in human salivary gland mucoepidermoid carcinoma. Oral Oncol 42(10):1005–1010. https://doi.org/10.1016/j.oraloncology.2005.12.022

Hao B, Oehlmann S, Sowa ME, Harper JW, Pavletich NP (2007) Structure of a Fbw7-Skp1-cyclin E complex: multisite-phosphorylated substrate recognition by SCF ubiquitin ligases. Mol Cell 26(1):131–143. https://doi.org/10.1016/j.molcel.2007.02.022

Haque R, Song JY, Haque M, Lei FY, Sandhu P, Ni B et al (2017) C-Myc-induced survivin is essential for promoting the notch-dependent T cell differentiation from hematopoietic stem cells. Genes 8(3). https://doi.org/10.3390/Genes8030097

Hart M, Concordet JP, Lassot I, Albert I, del los Santos R, Durand H et al (1999) The F-box protein beta-TrCP associates with phosphorylated beta-catenin and regulates its activity in the cell. Curr Biol 9 (4):207–210. https://doi.org/10.1016/S0960-9822(99)80091-8

Hasler R, Jacobs G, Till A, Grabe N, Cordes C, Nikolaus S et al (2012) Microbial pattern recognition causes distinct functional micro-RNA signatures in primary human monocytes. PLoS One 7(2). https://doi.org/10.1371/journal.pone.0031151

Herranz D, Ambesi-Impiombato A, Palomero T, Schnell SA, Belver L, Wendorff AA et al (2014) A NOTCH 1-driven MYC enhancer promotes T cell development, transformation and acute lymphoblastic leukemia. Nat Med 20(10):1130–1137. https://doi.org/10.1038/nm.3665

Hiramatsu Y, Kitagawa K, Suzuki T, Uchida C, Hattori T, Kikuchi H et al (2006) Degradation of Tob1 mediated by SCF (Skp2)-dependent ubiquitination. Cancer Res 66(17):8477–8483. https://doi.org/10.1158/0008-5472.CAN-06-1603

Hori T, Osaka F, Chiba T, Miyamoto C, Okabayashi K, Shimbara N et al (1999) Covalent modification of all members of human cullin family proteins by NEDD8. Oncogene 18(48):6829–6834. https://doi.org/10.1038/sj.onc.1203093

Hsu JD, Kao SH, Ou TT, Chen YJ, Li YJ, Wang CJ (2011) Gallic acid induces G2/M phase arrest of breast cancer cell MCF-7 through stabilization of p27(Kip1) attributed to disruption of p27(Kip1)/Skp2 complex. J Agric Food Chem 59(5):1996–2003. https://doi.org/10.1021/jf103656v

Hua JS, Ding TL, Yang LJ (2016) Dysfunction of microRNA-32 regulates ubiquitin ligase FBXW7 in multiple myeloma disease. Onco Targets Ther 9:6573–6579. https://doi.org/10.2147/Ott.S105945

Huang H, Regan KM, Wang F, Wang DP, Smith DI, van Deursen JMA et al (2005) Skp2 inhibits FOXO1 in tumor suppression through ubiquitin-mediated degradation. Proc Natl Acad Sci USA 102(5):1649–1654. https://doi.org/10.1073/pnas.0406789102

Huang HC, Way TD, Lin CL, Lin JK (2008) EGCG stabilizes p27(kip1) in E2-stimulated MCF-7 cells through down-regulation of the Skp2 protein. Endocrinology 149(12):5972–5983. https://doi.org/10.1210/en.2008-0408

Imura S, Tovuu LO, Utsunomiya T, Morine Y, Ikemoto T, Arakawa Y et al (2014) The role of Fbxw7 expression in hepatocellular carcinoma and adjacent non-tumor liver tissue. J Gastroenterol Hepatol 29 (10):1822–1829. https://doi.org/10.1111/jgh.12623

Inuzuka H, Shaik S, Onoyama I, Gao DM, Tseng A, Maser RS et al (2011) SCFFBW7 regulates cellular apoptosis by targeting MCL1 for ubiquitylation and destruction. Nature 471(7336):104–U128. https://doi.org/10.1038/nature09732

Inuzuka H, Gao D, Finley LW, Yang W, Wan L, Fukushima H et al (2012) Acetylation-dependent regulation of Skp2 function. Cell 150(1):179–193. https://doi.org/10.1016/j.cell.2012.05.038

Iskandarani A, Bhat AA, Siveen KS, Prabhu KS, Kuttikrishnan S, Khan MA et al (2016) Bortezomib-mediated downregulation of S-phase kinase protein-2 (SKP2) causes apoptotic cell death in chronic myelogenous leukemia cells. J Transl Med 14. https://doi.org/10.1186/s12967-016-0823-y

Jiang JT, Pan YQ, Regan KM, Wu CP, Zhang XG, Tindall DJ et al (2012b) Androgens repress expression of the f-box protein Skp2 via p107 dependent and independent mechanisms in LNCaP prostate cancer cells. Prostate 72(2):225–232. https://doi.org/10.1002/pros.21430

Jiang XL, Xing HY, Kim TM, Jung YC, Huang W, Yang HW et al (2012a) Numb regulates glioma stem cell fate and growth by altering epidermal growth factor receptor and Skp1-Cullin-F-box ubiquitin ligase activity. Stem Cells 30(7):1313–1326. https://doi.org/10.1002/stem.1120

Jiang XS, Li XF, Wu FF, Gao HL, Wang GR, Zheng H et al (2017) Overexpression of miR-92a promotes the tumor growth of osteosarcoma by suppressing F-box and WD repeat-containing protein 7. Gene 606:10–16. https://doi.org/10.1016/j.gene.2017.01.002

Jin JP, Shirogane T, Xu L, Nalepa G, Qin J, Elledge SJ et al (2003) SCF beta-TRCP links Chk1 signaling to degradation of the Cdc25A protein phosphatase. Genes Dev 17(24):3062–3074. https://doi.org/10.1101/gad.1157503

Kamura T, Hara T, Kotoshiba S, Yada M, Ishida N, Imaki H et al (2003) Degradation of p57(Kip2) mediated by SCFSkp2 - dependent ubiquitylation. Proc Natl Acad Sci USA 100(18):10231–10236. https://doi.org/10.1073/pnas.1831009100

Kim CJ, Song JH, Cho YG, Kim YS, Kim SY, Nam SW et al (2007) Somatic mutations of the beta-TrCP gene in gastric cancer. APMIS 115(2):127–133. https://doi.org/10.1111/j.1600-0463.2007.apm_562.x

Kim SY, Herbst A, Tworkowski KA, Salghetti SE, Tansey WP (2003) Skp2 regulates Myc protein stability and activity. Mol Cell 11(5):1177–1188. https://doi.org/10.1016/S1097-2765(03)00173-4

Kimura T, Gotoh M, Nakamura Y, Arakawa H (2003) hCDC4b, a regulator of cyclin E, as a direct transcriptional target of p53. Cancer Sci 94(5):431–436. https://doi.org/10.1111/j.1349-7006.2003.tb01460.x

King B, Trimarchi T, Reavie L, Xu LY, Mullenders J, Ntziachristos P et al (2013) The ubiquitin ligase FBXW7 modulates leukemia-initiating cell activity by regulating MYC stability. Cell 153(7):1552–1566. https://doi.org/10.1016/j.cell.2013.05.041

Kitagawa M, Lee SH, McCormick F (2008) Skp2 suppresses p53-dependent apoptosis by inhibiting p300. Mol Cell 29(2):217–231. https://doi.org/10.1016/j.molcel.2007.11.036

Klinakis A, Szaboics M, Politi K, Kiaris H, Artavanis-Tsakonas S, Efstratiadis A (2006) Myc is a Notch1 transcriptional target and a requisite for Notch1-induced mammary tumorigenesis in mice. Proc Natl Acad Sci U S A 103(24):9262–9267. https://doi.org/10.1073/pnas.0603371103

Koepp DM, Schaefer LK, Ye X, Keyomarsi K, Chu C, Harper JW et al (2001) Phosphorylation-dependent ubiquitination of cyclin E by the SCFFbw7 ubiquitin ligase. Science 294(5540):173–177. https://doi.org/10.1126/science.1065203

Koh MS, Ittmann M, Kadmon D, Thompson TC, Leach FS (2006) CDC4 gene expression as potential biomarker for targeted therapy in prostate cancer. Cancer Biol Ther 5(1):78–83. https://doi.org/10.4161/cbt.5.1.2290

Kossatz U, Dietrich N, Zender L, Buer J, Manns MP, Malek NP (2004) Skp2-dependent degradation of p27kip1 is essential for cell cycle progression. Genes Dev 18(21):2602–2607. https://doi.org/10.1101/gad.321004

Kudo Y, Kitajima S, Sato S, Miyauchi M, Ogawa I, Takata T (2001) High expression of S-phase kinase-interacting protein 2, human F-box protein, correlates with poor prognosis in oral squamous cell carcinomas. Cancer Res 61(19):7044–7047

Kudo Y, Guardavaccaro D, Santamaria PG, Koyama-Nasu R, Latres E, Bronson R et al (2004) Role of F-box protein beta Trcp1 in mammary gland development and tumorigenesis. Mol Cell Biol 24 (18):8184–8194. https://doi.org/10.1128/Mcb.24.18.8184-8194.2004

Kurashige J, Watanabe M, Iwatsuki M, Kinoshita K, Saito S, Hiyoshi Y et al (2012) Overexpression of microRNA-223 regulates the ubiquitin ligase FBXW7 in oesophageal squamous cell carcinoma. Br J Cancer 106(1):182–188. https://doi.org/10.1038/bjc.2011.509

Latres E, Chiarle R, Schulman BA, Pavletich NP, Pellicer A, Inghirami C et al (2001) Role of the F-box protein Skp2 in lymphomagenesis. Proc Natl Acad Sci USA 98(5):2515–2520. https://doi.org/10.1073/pnas.041475098

Lerner M, Lundgren J, Akhoondi S, Jahn A, Ng HF, Moqadam FA et al (2011) MiRNA-27a controls FBW7/hCDC4-dependent cyclin E degradation and cell cycle progression. Cell Cycle 10(13):2172–2183. https://doi.org/10.4161/cc.10.13.16248

Li JH, Pauley AM, Myers RL, Shuang RQ, Brashler JR, Yan RQ et al (2002) SEL-10 interacts with presenilin 1, facilitates its ubiquitination, and alters A-beta peptide production. J Neurochem 82(6):1540–1548. https://doi.org/10.1046/j.1471-4159.2002.01105.x

Lin HK, Wang GC, Chen ZB, Teruya-Feldstein J, Liu Y, Chan CH et al (2009) Phosphorylation-dependent regulation of cytosolic localization and oncogenic function of Skp2 by Akt/PKB. Nat Cell Biol 11(4):420–U144. https://doi.org/10.1038/ncb1849.

Lin HK, Chen ZB, Wang GC, Nardella C, Lee SW, Chan CH et al (2010) Skp2 targeting suppresses tumorigenesis by Arf-p53-independent cellular senescence. Nature 464(7287):374–U366. https://doi.org/10.1038/nature08815

Liu JD, Furukawa M, Matsumoto T, Xiong Y (2002) NEDD8 modification of CUL1 short article dissociates p120(CAND1), an inhibitor of CULl-SKP1 binding and SCIF ligases. Mol Cell 10(6):1511–1518. https://doi.org/10.1016/S1097-2765(02)00783-9

Liu YH, Tong ZW, Li T, Chen Q, Zhuo LT, Li WG et al (2012) Hepatitis B virus X protein stabilizes amplified in breast cancer 1 protein and cooperates with it to promote human hepatocellular carcinoma cell invasiveness. Hepatology 56(3):1015–1024. https://doi.org/10.1002/hep.25751

Lu D, Davis MPA, Abreu-Goodger C, Wang W, Campos LS, Siede J et al (2012) MiR-25 regulates Wwp2 and Fbxw7 and promotes reprogramming of mouse fibroblast cells to iPSCs. PLoS One 7(8). https://doi.org/10.1371/journal.pone.0040938

Lu LF, Schulz H, Wolf DA (2002) The F-box protein SKP2 mediates androgen control of p27 stability in LNCaP human prostate cancer cells. BMC Cell Biol 3. https://doi.org/10.1186/1471-2121-3-22

Lu MD, Ma JB, Xue WQ, Cheng C, Wang Y, Zhao YM et al (2009) The expression and prognosis of FOXO3a and Skp2 in human hepatocellular carcinoma. Pathol Oncol Res 15(4):679–687. https://doi.org/10.1007/s12253-009-9171-z

Lu YH, Zi XL, Pollak M (2004) Molecular mechanisms underlying IGF-I-induced attenuation of the growth-inhibitory activity of trastuzumab (Herceptin) on SKBR3 breast cancer cells. Int J Cancer 108 (3):334–341. https://doi.org/10.1002/ijc.11445

Ma J, Cheng L, Liu H, Zhang J, Shi Y, Zeng FP et al (2013) Genistein down-regulates miR-223 expression in pancreatic cancer cells. Curr Drug Targets 14

(10):1150–1156. https://doi.org/10.2174/13894501113149990187

Malek E, Abdel-Malek MAY, Jagannathan S, Vad N, Karns R, Jegga AG et al (2017) Pharmacogenomics and chemical library screens reveal a novel SCFSKP2 inhibitor that overcomes Bortezomib resistance in multiple myeloma. Leukemia 31(3):645–653. https://doi.org/10.1038/leu.2016.258

Mao JH, Perez-Iosada J, Wu D, DelRosario R, Tsunematsu R, Nakayama KI et al (2004) Fbxw7/Cdc4 is a p53-dependent, haploinsufficient tumour suppressor gene. Nature 432(7018):775–779. https://doi.org/10.1038/nature03155

Mao JH, Kim IJ, Wu D, Climent J, Kang HC, DelRosario R et al (2008) FBXW7 targets mTOR for degradation and cooperates with PTEN in tumor suppression. Science 321(5895):1499–1502. https://doi.org/10.1126/science.1162981

Margottin F, Bour SP, Durand H, Selig L, Benichou S, Richard V et al (1998) A novel human WD protein, h-beta TrCP, that interacts with HIV-1 Vpu connects CD4 to the ER degradation pathway through an F-box motif. Mol Cell 1(4):565–574. https://doi.org/10.1016/S1097-2765(00)80056-8

Margottin-Goguet F, Hsu JY, Loktev A, Hsieh HM, Reimann JDR, Jackson PK (2003) Prophase destruction of Emi1 by the SCF beta TrCP/Slimb ubiquitin ligase activates the anaphase promoting complex to allow progression beyond prometaphase. Dev Cell 4 (6):813–826. https://doi.org/10.1016/S1534-5807(03)00153-9

Marti A, Wirbelauer C, Scheffner M, Krek W (1999) Interaction between ubiquitin-protein ligase SCFSKP2 and E2F-1 underlies the regulation of E2F-1 degradation. Nat Cell Biol 1(1):14–19. https://doi.org/10.1038/8984

Maser RS, Choudhury B, Campbell PJ, Feng B, Wong KK, Protopopov A et al (2007) Chromosomally unstable mouse tumours have genomic alterations similar to diverse human cancers. Nature 447(7147):966–U963. https://doi.org/10.1038/nature05886

Masuda T, Inoue H, Sonoda H, Mine S, Yoshikawa Y, Nakayama K et al (2002) Clinical and biological significance of S-phase kinase-associated protein 2 (Skp2) gene expression in gastric carcinoma: modulation of malignant phenotype by Skp2 overexpression, possibly via p27 proteolysis. Cancer Res 62(13):3819–3825

Matsumoto A, Onoyama I, Nakayama KI (2006) Expression of mouse Fbxw7 isoforms is regulated in a cell cycle- or p53-dependent manner. Biochem Biophys Res Commun 350(1):114–119. https://doi.org/10.1016/j.bbrc.2006.09.003

Matsumoto A, Tateishi Y, Onoyama I, Okita Y, Nakayama K, Nakayama KI (2011) Fbxw7 beta resides in the endoplasmic reticulum membrane and protects cells from oxidative stress. Cancer Sci 102 (4):749–755. https://doi.org/10.1111/j.1349-7006.2011.01851.x

Matsuoka S, Oike Y, Onoyama I, Iwama A, Arai F, Takubo K et al (2008) Fbxw7 acts as a critical fail-safe against premature loss of hematopoietic stem cells and development of T-ALL. Genes Dev 22 (8):986–991. https://doi.org/10.1101/gad.1621808

Mendez J, Zou-Yang XH, Kim SY, Hidaka M, Tansey WP, Stillman B (2002) Human origin recognition complex large subunit is degraded by ubiquitin-mediated proteolysis after initiation of DNA replication. Mol Cell 9(3):481–491. https://doi.org/10.1016/S1097-2765(02)00467-7

Min SH, Lau AW, Lee TH, Inuzuka H, Wei S, Huang PY et al (2012) Negative regulation of the stability and tumor suppressor function of Fbw7 by the Pin1 prolyl isomerase. Mol Cell 46(6):771–783. https://doi.org/10.1016/j.molcel.2012.04.012

Min YH, Cheong JW, Lee MH, Kim JY, Lee ST, Hahn JS et al (2004) Elevated S-phase kinase-associated protein 2 protein expression in acute myelogenous leukemia: its association with constitutive phosphorylation of phosphatase and tensin homologue protein and poor prognosis. Clin Cancer Res 10(15):5123–5130. https://doi.org/10.1158/1078-0432.Ccr-04-0136

Moberg KH, Bell DW, Wahrer DCR, Haber DA, Hariharan IK (2001) Archipelago regulates Cyclin E levels in drosophila and is mutated in human cancer cell lines. Nature 413(6853):311–316. https://doi.org/10.1038/35095068

Morimoto M, Nishida T, Nagayama Y, Yasuda H (2003) Nedd8-modification of Cul1 is promoted by Roc1 as a Nedd8-E3 ligase and regulates its stability. Biochem Biophys Res Commun 301(2):392–398. https://doi.org/10.1016/S0006-291x(02)03051-6

Moro L, Arbini AA, Marra E, Greco M (2006) Up-regulation of Skp2 after prostate cancer cell adhesion to basement membranes results in BRCA2 degradation and cell proliferation. J Biol Chem 281 (31):22100–22107. https://doi.org/10.1074/jbc.M604636200

Muerkoster S, Arlt A, Sipos B, Witt M, Grossmann M, Kloppel G et al (2005) Increased expression of the E3-ubiquitin ligase receptor subunit beta TRCP1 relates to constitutive nuclear factor-kappa B activation and chemoresistance in pancreatic carcinoma cells. Cancer Res 65(4):1316–1324. https://doi.org/10.1158/0008-5472.Can-04-1626

Muzny DM, Bainbridge MN, Chang K, Dinh HH, Drummond JA, Fowler G et al (2012) Comprehensive molecular characterization of human colon and rectal cancer. Nature 487(7407):330–337. https://doi.org/10.1038/nature11252

Nakayama K, Nagahama H, Minamishima YA, Matsumoto M, Nakamichi I, Kitagawa K et al (2000) Targeted disruption of Skp2 results in accumulation of cyclin E and p27(Kip1), polyploidy and centrosome overduplication. EMBO J 19(9):2069–2081. https://doi.org/10.1093/emboj/19.9.2069

Nakayama K, Hatakeyama S, Maruyama S, Kikuchi A, Onoe K, Good RA et al (2003) Impaired degradation of inhibitory subunit of NF-kappa B (I kappa B) and beta-catenin as a result of targeted disruption of the beta-TrCP1 gene. Proc Natl Acad Sci USA 100 (15):8752–8757. https://doi.org/10.1073/pnas.1133216100

Nash P, Tang XJ, Orlicky S, Chen QH, Gertler FB, Mendenhall MD et al (2001) Multisite phosphorylation of a CDK inhibitor sets a threshold for the onset of DNA replication. Nature 414(6863):514–521. https://doi.org/10.1038/35107009

Nateri AS, Riera-Sans L, Da Costa C, Behrens A (2004) The ubiquitin ligase SCFFbw7 antagonizes apoptotic JNK signaling. Science 303(5662):1374–1378. https://doi.org/10.1126/science.1092880

Nguyen PL, Lin DI, Lei JY, Fiorentino M, Mueller E, Weinstein MH et al (2011) The impact of Skp2 overexpression on recurrence-free survival following radical prostatectomy. Urol Oncol 29(3):302–308. https://doi.org/10.1016/j.urolonc.2009.03.022

Nie L, Wu H, Sun XH (2008) Ubiquitination and degradation of Tal1/SCL are induced by notch signaling and depend on Skp2 and CHIP. J Biol Chem 283 (2):684–692. https://doi.org/10.1074/jbc.M704981200

Nishitani H, Sugimoto N, Roukos V, Nakanishi Y, Saijo M, Obuse C et al (2006) Two E3 ubiquitin ligases, SCF-Skp2 and DDB1-Cul4, target human Cdt1 for proteolysis. EMBO J 25(5):1126–1136. https://doi.org/10.1038/sj,emboj.7601002

Olson BL, Hock MB, Ekholm-Reed S, Wohlschlegel JA, Dev KK, Kralli A et al (2008) SCFCdc4 acts antagonistically to the PGC-1 alpha transcriptional coactivator by targeting it for ubiquitin-mediated proteolysis. Genes Dev 22(2):252–264. https://doi.org/10.1101/gad.1624208

Onoyama I, Tsunematsu R, Matsumoto A, Kimura T, de Alboran IM, Nakayama K et al (2007) Conditional inactivation of Fbxw7 impairs cell-cycle exit during T cell differentiation and results in lymphomatogenesis. J Exp Med 204(12):2875–2888. https://doi.org/10.1084/Jem.20062299

Onoyama I, Suzuki A, Matsumoto A, Tomita K, Katagiri H, Oike Y et al (2011) Fbxw7 regulates lipid metabolism and cell fate decisions in the mouse liver. J Clin Investig 121(1):342–354. https://doi.org/10.1172/JCI40725

Orlicky S, Tang XJ, Willems A, Tyers M, Sicheri F (2003) Structural basis for phosphodependent substrate selection and orientation by the SCFCdc4 ubiquitin ligase. Cell 112(2):243–256. https://doi.org/10.1016/S0092-8674(03)00034-5

Pernicova Z, Slabakova E, Kharaishvili G, Bouchal J, Kral M, Kunicka Z et al (2011) Androgen depletion induces senescence in prostate cancer cells through down-regulation of Skp2. Neoplasia 13(6):526–536. https://doi.org/10.1593/neo.11182

Popov N, Schulein C, Jaenicke LA, Eilers M (2010) Ubiquitylation of the amino terminus of Myc by SCF beta-TrCP antagonizes SCFFbw7-mediated turnover. Nat Cell Biol 12(10):973–981. https://doi.org/10.1038/ncb2104

Qu X, Shen LL, Zheng Y, Cui Y, Feng ZH, Liu F et al (2014) A signal transduction pathway from TGF-beta 1 to SKP2 via Akt1 and c-Myc and its correlation with progression in human melanoma. J Investig Dermatol 134(1):159–167. https://doi.org/10.1038/jid.2013.281

Radke S, Pirkmaier A, Germain D (2005) Differential expression of the F-box proteins Skp2 and Skp2B in breast cancer. Oncogene 24(21):3448–3458. https://doi.org/10.1038/sj.onc.1208328

Rajagopalan H, Jallepalli PV, Rago C, Velculescu VE, Kinzler KW, Vogelstein B et al (2004) Inactivation of hCDC4 can cause chromosomal instability. Nature 428 (6978):77–81. https://doi.org/10.1038/nature02313

Ray D, Osmundson EC, Kiyokawa H (2006) Constitutive and UV-induced fibronectin degradation is a ubiquitination-dependent process controlled by beta-TrCP. J Biol Chem 281(32):23060–23065. https://doi.org/10.1074/jbc.M604311200

Rebello RJ, Pearson RB, Hannan RD, Furic L (2017) Therapeutic approaches targeting MYC-driven prostate cancer. Genes 8(2). https://doi.org/10.3390/Genes8020071

Richardson PG, Barlogie B, Berenson J, Singhal S, Jagannath S, Irwin D et al (2003) A phase 2 study of bortezomib in relapsed, refractory myeloma. N Engl J Med 348(26):2609–2617. https://doi.org/10.1056/Nejmoa030288

Rico-Bautista E, Yang CC, Lu LF, Roth GP, Wolf DA (2010) Chemical genetics approach to restoring p27Kip1 reveals novel compounds with antiproliferative activity in prostate cancer cells. BMC Biol 8. https://doi.org/10.1186/1741-7007-8-153

Roy S, Kaur M, Agarwal C, Tecklenburg M, Sclafani RA, Agarwal R (2007) P21 and p27 induction by silibinin is essential for its cell cycle arrest effect in prostate carcinoma cells. Mol Cancer Ther 6(10):2696–2707. https://doi.org/10.1158/1535-7163.MCT-07-0104

Saitoh T, Katoh M (2001) Expression profiles of beta TRCP1 and beta TRCP2, and mutation analysis of beta TRCP2 in gastric cancer. Int J Oncol 18(5):959–964

Sancho R, Jandke A, Davis H, Diefenbacher ME, Tomlinson I, Behrens A (2010) F-box and WD repeat domain-containing 7 regulates intestinal cell lineage commitment and is a haploinsufficient tumor suppressor. Gastroenterology 139(3):929–941. https://doi.org/10.1053/j.gastro.2010.05.078

Sancho R, Blake SM, Tendeng C, Clurman BE, Lewis J, Behrens A (2013) Fbw7 repression by Hes5 creates a feedback loop that modulates notch-mediated intestinal and neural stem cell fate decisions. PLoS Biol 11(6). https://doi.org/10.1371/journal.pbio.1001586

Sasajima H, Nakagawa K, Kashiwayanagi M, Yokosawa H (2012) Polyubiquitination of the B-cell translocation gene 1 and 2 proteins is promoted by the SCF ubiquitin ligase complex containing beta TrCP. Biol Pharm Bull 35(9):1539–1545. https://doi.org/10.1248/bpb.b12-00330

Seki R, Ohshima K, Fujisaki T, Uike N, Kawano F, Gondo H et al (2010) Prognostic significance of S-phase kinase-associated protein 2 and p27(kip1) in patients with diffuse large B-cell lymphoma: effects of rituximab. Ann Oncol 21(4):833–841. https://doi.org/10.1093/annonc/mdp481

Shapira M, Kakiashvili E, Rosenberg T, Hershko DD (2006) The mTOR inhibitor rapamycin down-regulates the expression of the ubiquitin ligase subunit Skp2 in breast cancer cells. Breast Cancer Res 8(4):R46. https://doi.org/10.1186/Bcr1533

Shaughnessy J (2005) Amplification and overexpression of CKS1B at chromosome band 1q21 is associated with reduced levels of p27(Kip1) and an aggressive clinical course in multiple myeloma. Hematology 10:117–126. https://doi.org/10.1080/10245330512331390140

Shim EH, Johnson L, Noh HL, Kim YJ, Sun H, Zeiss C et al (2003) Expression of the F-box protein SKP2 induces hyperplasia, dysplasia, and low-grade carcinoma in the mouse prostate. Cancer Res 63 (7):1583–1588

Shouksmith AE, Evans LE, Tweddle DA, Miller DC, Willmore E, Newell DR et al (2015) Synthesis and activity of putative small-molecule inhibitors of the F-box protein SKP2. Aust J Chem 68(4):660–679. https://doi.org/10.1071/CH14586

Sistrunk C, Kim SH, Wang X, Lee SH, Kim Y, Macias E et al (2013) Skp2 deficiency inhibits chemical skin tumorigenesis independent of p27(Kip1) accumulation. Am J Pathol 182(5):1854–1864. https://doi.org/10.1016/j.ajpath.2013.01.016

Sonoda H, Inoue H, Ogawa K, Utsunomiya T, Masuda TA, Mori M (2006) Significance of Skp2 expression in primary breast cancer. Clin Cancer Res 12 (4):1215–1220. https://doi.org/10.1158/1078-0432.CCR-05-1709

Spiegelman VS, Slaga TJ, Pagano M, Minamoto T, Ronai Z, Fuchs SY (2000) Wnt/beta-catenin signaling induces the expression and activity of beta TrCP ubiquitin ligase receptor. Mol Cell 5(5):877–882. https://doi.org/10.1016/S1097-2765(00)80327-5

Spruck C (2011) miR-27a regulation of SCFFbw7 in cell division control and cancer. Cell Cycle 10 (19):3232–3232. https://doi.org/10.4161/cc.10.19.17125

Strack P, Caligiuri M, Pelletier M, Boisclair M, Theodoras A, Beer-Romero P et al (2000) SCF beta-TRCP and phosphorylation dependent ubiquitination of I kappa B alpha catalyzed by Ubc3 and Ubc4. Oncogene 19(31):3529–3536. https://doi.org/10.1038/sj.onc.1203647

Strohmaier H, Spruck CH, Kaiser P, Won KA, Sangfelt O, Reed SI (2001) Human F-box protein hCdc4 targets cyclin E for proteolysis and is mutated in a breast cancer cell line. Nature 413(6853):316–322. https://doi.org/10.1038/35095076

Sun LC, Cai L, Yu Y, Meng QW, Cheng XS, Zhao YB et al (2007) Knockdown of s-phase kinase-associated protein-2 expression in MCF-7 inhibits cell growth and enhances the cytotoxic effects of epirubicin. Acta Biochim Biophys Sin 39(12):999–1007. https://doi.org/10.1111/j.1745-7270.2007.00361.x

Sun XF, Sun JP, Hou HT, Li K, Liu X, Ge QX (2016) MicroRNA-27b exerts an oncogenic function by targeting Fbxw7 in human hepatocellular carcinoma. Tumor Biol 37(11):15325–15332. https://doi.org/10.1007/s13277-016-5444-9

Sundqvist A, Bengoechea-Alonso MT, Ye X, Lukiyanchuk V, Jin JP, Harper JW et al (2005) Control of lipid metabolism by phosphorylation-dependent degradation of the SREBP family of transcription factors by SCFFbw7. Cell Metab 1(6):379–391. https://doi.org/10.1016/j.cmet.2005.04.010

Suzuki H, Chiba T, Suzuki T, Fujita T, Ikenoue T, Omata M et al (2000) Homodimer of two F-box proteins beta TrCP1 or beta TrCP2 binds to I kappa B alpha for signal-dependent ubiquitination. J Biol Chem 275 (4):2877–2884. https://doi.org/10.1074/jbc.275.4.2877

Tan MJ, Gallegos JR, Gu QY, Huang YH, Li J, Jin YT et al (2006) SAG/ROC-SCF beta-TrCP E3 ubiquitin ligase promotes pro-caspase-3 degradation as a mechanism of apoptosis protection. Neoplasia 8 (12):1042–1054. https://doi.org/10.1593/neo.06568

Tang B, Lei BA, Qi GY, Liang XS, Tang F, Yuan SG et al (2016) MicroRNA-155-3p promotes hepatocellular carcinoma formation by suppressing FBXW7 expression. J Exp Clin Cancer Res 35. https://doi.org/10.1186/s13046-016-0371-6

Tedesco D, Lukas J, Reed SI (2002) The pRb-related protein p130 is regulated by phosphorylation-dependent proteolysis via the protein-ubiquitin ligase SCFSkp2. Genes Dev 16(22):2946–2957. https://doi.org/10.1101/gad.1011202

Teng CL, Hsieh YC, Phan L, Shin J, Gully C, Velazquez-Torres G et al (2012) FBXW7 is involved in Aurora B degradation. Cell Cycle 11(21):4059–4068. https://doi.org/10.4161/cc.22381

Tetzlaff MT, Yu W, Li MM, Zhang PM, Finegold M, Mahon K et al (2004) Defective cardiovascular development and elevated cyclin E and notch proteins in mice lacking the Fbw7 F-box protein. Proc Natl Acad Sci USA 101(10):3338–3345. https://doi.org/10.1073/pnas.0307875101

Thompson BJ, Jankovic V, Gao J, Buonamici S, Vest A, Lee JM et al (2008) Control of hematopoietic stem cell quiescence by the E3 ubiquitin ligase Fbw7. J Exp Med 205(6):1395–1408. https://doi.org/10.1084/jem.20080277

Tosco P, Maggiore GML, Forni P, Berrone S, Chiusa L, Garzino-Demo P (2011) Correlation between Skp2 expression and nodal metastasis in stage I and II oral squamous cell carcinomas. Oral Dis 17(1):102–108. https://doi.org/10.1111/j.1601-0825.2010.01713.x

Tsunematsu R, Nakayama K, Oike Y, Nishiyama M, Ishida N, Hatakeyama S et al (2004) Mouse Fbw7/Sel-10/Cdc4 is required for notch degradation during vascular development. J Biol Chem 279 (10):9417–9423. https://doi.org/10.1074/jbc.M312337200

Tsvetkov LM, Yeh KH, Lee SJ, Sun H, Zhang H (1999) P27(Kip1) ubiquitination and degradation is regulated by the SCFSkp2 complex through phosphorylated Thr187 in p27. Curr Biol 9(12):661–664. https://doi.org/10.1016/S0960-9822(99)80290-5

Tu KS, Zheng X, Zan XF, Han SS, Yao YM, Liu QG (2012) Evaluation of Fbxw7 expression and its correlation with the expression of c-Myc, cyclin E and p53 in human hepatocellular carcinoma. Hepatol Res 42 (9):904–910. https://doi.org/10.1111/j.1872-034X.2012.01005.x

Tu KS, Yang W, Li C, Zheng X, Lu ZT, Guo C et al (2014) Fbxw7 is an independent prognostic marker and induces apoptosis and growth arrest by regulating YAP abundance in hepatocellular carcinoma. Mol Cancer 13. https://doi.org/10.1186/1476-4598-13-110

Uddin S, Hussain A, Ahmed M, Belgaumi A, Al-Dayel F, Ajarim D et al (2008) S-phase kinase protein 2 is an attractive therapeutic target in a subset of diffuse large B-cell lymphoma. J Pathol 216(4):483–494. https://doi.org/10.1002/path.2433

Umanskaya K, Radke S, Chander H, Monardo R, Xu X, Pan ZQ et al (2007) Skp2B stimulates mammary gland development by inhibiting REA, the repressor of the estrogen receptor. Mol Cell Biol 27(21):7615–7622. https://doi.org/10.1128/Mcb.01239-07

Ungermannova D, Lee J, Zhang G, Dallmann HG, McHenry CS, Liu XD (2013) High-throughput screening AlphaScreen assay for identification of small-molecule inhibitors of ubiquitin E3 ligase SCFSkp2-Cks1. J Biomol Screen 18(8):910–920. https://doi.org/10.1177/1087057113485789

van Duijn PW, Trapman J (2006) PI3K/Akt signaling regulates p27(kip) expression via Skp2 in PC3 and DU145 prostate cancer cells, but is not a major factor in p27(kip) regulation in LNCaP and PC346 cells. Prostate 66(7):749–760. https://doi.org/10.1002/pros.20398

Voduc D, Nielsen TO, Cheang MC, Foulkes WD (2008) The combination of high cyclin E and Skp2 expression in breast cancer is associated with a poor prognosis and the basal phenotype. Hum Pathol 39(10):1431–1437. https://doi.org/10.1016/j.humpath.2008.03.004

von der Lehr N, Johansson S, Wu S, Bahram F, Castell A, Cetinkaya C et al (2003) The F-box protein Skp2 participates in c-Myc proteosomal degradation and acts as a cofactor for c-Myc-regulated transcription. Mol Cell 11(5):1189–1200. https://doi.org/10.1016/S1097-2765(03)00193-X

Wang HB, Sun DQ, Ji P, Mohler J, Zhu L (2008) An AR-Skp2 pathway for proliferation of androgen-dependent prostate-cancer cells. J Cell Sci 121 (15):2578–2587. https://doi.org/10.1242/jcs.030742

Wang HB, Bauzon F, Ji P, Xu XL, Sun DQ, Locker J et al (2010) Skp2 is required for survival of aberrantly proliferating Rb1-deficient cells and for tumorigenesis in Rb1(+/−) mice. Nat Genet 42(1):83–U105. https://doi.org/10.1038/ng.498

Wang J, Han F, Wu J, Lee SW, Chan CH, Wu CY et al (2011) The role of Skp2 in hematopoietic stem cell quiescence, pool size, and self-renewal. Blood 118 (20):5429–5438. https://doi.org/10.1182/blood-2010-10-312785

Wang J, Huang Y, Guan Z, Zhang JL, Su HK, Zhang W et al (2014b) E3-ligase Skp2 predicts poor prognosis and maintains cancer stem cell pool in nasopharyngeal carcinoma. Oncotarget 5(14):5591–5601. https://doi.org/10.18632/oncotarget.2149

Wang SY, Garcia AJ, Wu M, Lawson DA, Witte ON, Wu H (2006) Pten deletion leads to the expansion of a prostatic stem/progenitor cell subpopulation and tumor initiation. Proc Natl Acad Sci U S A 103 (5):1480–1485. https://doi.org/10.1073/pnas0510652103

Wang ZW, Gao DM, Fukushima H, Inuzuka H, Liu PD, Wan LX et al (2012) Skp2: a novel potential therapeutic target for prostate cancer. Biochim Biophys Acta 1825(1):11–17. https://doi.org/10.1016/j.bbcan.2011.09.002

Wang ZW, Liu PD, Inuzuka H, Wei WY (2014a) Roles of F-box proteins in cancer. Nat Rev Cancer 14 (4):233–247. https://doi.org/10.1038/nrc3700

Watanabe N, Arai H, Nishihara Y, Taniguchi M, Watanabe N, Hunter T et al (2004) M-phase kinases induce phospho-dependent ubiquitination of somatic Wee1 by SCF beta-TrCP. Proc Natl Acad Sci USA 101(13):4419–4424. https://doi.org/10.1073/pnas.0307700101

Wei S, Yang HC, Chuang HC, Yang J, Kulp SK, Lu PJ et al (2008) A novel mechanism by which thiazolidinediones facilitate the proteasomal degradation of cyclin D1 in cancer cells. J Biol Chem 283 (39):26759–26770. https://doi.org/10.1074/jbc.M802160200

Wei W, Ayad NG, Wan Y, Zhang G-J, Kirschner MW, Kaelin WG (2004) Degradation of the SCF component Skp2 in cell-cycle phase G1 by the anaphase-promoting complex. Nature 428(6979):194–198. https://doi.org/10.1038/nature02381

Wei WY, Jin JP, Schlisio S, Harper JW, Kaelin WG (2005) The v-Jun point mutation allows c-Jun to escape GSK3-dependent recognition and destruction by the Fbw7 ubiquitin ligase. Cancer Cell 8 (1):25–33. https://doi.org/10.1016/j.ccr.2005.06.005

Welcker M, Clurman BE (2008) FBW7 ubiquitin ligase: a tumour suppressor at the crossroads of cell division, growth and differentiation. Nat Rev Cancer 8 (2):83–93. https://doi.org/10.1038/nrc2290

Welcker M, Larimore EA, Frappier L, Clurman BE (2011) Nucleolar targeting of the Fbw7 ubiquitin ligase by a pseudosubstrate and glycogen synthase kinase 3. Mol Cell Biol 31(6):1214–1224. https://doi.org/10.1128/Mcb.01347-10

Weng AP, Millholland JM, Yashiro-Ohtani Y, Arcangeli ML, Lau A, Wai C et al (2006) C-Myc is an important direct target of Notch1 in T-cell acute lymphoblastic leukemia/lymphoma. Genes Dev 20(15):2096–2109. https://doi.org/10.1101/gad.1450406

Wertz IE, Kusam S, Lam C, Okamoto T, Sandoval W, Anderson DJ et al (2011) Sensitivity to antitubulin chemotherapeutics is regulated by MCL1 and FBW7. Nature 471(7336):110–114. https://doi.org/10.1038/nature09779

Wood LD, Parsons DW, Jones S, Lin J, Sjoblom T, Leary RJ et al (2007) The genomic landscapes of human breast and colorectal cancers. Science 318 (5853):1108–1113. https://doi.org/10.1126/science.1145720

Wu GY, Hubbard EJA, Kitajewski JK, Greenwald I (1998) Evidence for functional and physical association between Caenorhabditis elegans SEL-10, a Cdc4p-related protein, and SEL-12 presenilin. Proc Natl Acad Sci USA 95(26):15787–15791. https://doi.org/10.1073/pnas.95.26.15787

Wu L, Grigoryan AV, Li YF, Hao B, Pagano M, Cardozo TJ (2012) Specific small molecule inhibitors of Skp2-mediated p27 degradation. Chem Biol 19 (12):1515–1524. https://doi.org/10.1016/j.chembiol.2012.09.015

Wu WJ, Shi J, Hu G, Yu X, Lu H, Yang ML et al (2016) Wnt/beta-catenin signaling inhibits FBXW7 expression by upregulation of microRNA-770 in hepatocellular carcinoma. Tumor Biol 37(5):6045–6051. https://doi.org/10.1007/s13277-015-4452-5

Xia W, Zhou JY, Luo HB, Liu YZ, Peng CC, Zheng WL et al (2017) MicroRNA-32 promotes cell proliferation, migration and suppresses apoptosis in breast cancer cells by targeting FBXW7. Cancer Cell Int 17. https://doi.org/10.1186/s12935-017-0383-0

Xia YF, Padre RC, De Mendoza TH, Bottero V, Tergaonkar VB, Verma IM (2009) Phosphorylation of p53 by I kappa B kinase 2 promotes its degradation by beta-TrCP. Proc Natl Acad Sci U S A 106 (8):2629–2634. https://doi.org/10.1073/pnas.0812256106

Xu J, Wu WB, Wang J, Huang CJ, Wen W, Zhao F et al (2017) miR-367 promotes the proliferation and invasion of non-small cell lung cancer via targeting FBXW7. Oncol Rep 37(2):1052–1058. https://doi.org/10.3892/or.2016.5314

Xu Y, Lee SH, Kim HS, Kim NH, Piao S, Park SH et al (2010b) Role of CK1 in GSK3 beta-mediated phosphorylation and degradation of snail. Oncogene 29 (21):3124–3133. https://doi.org/10.1038/onc.2010.77

Xu YF, Sengupta T, Kukreja L, Minella AC (2010a) MicroRNA-223 regulates cyclin E activity by modulating expression of F-box and WD-40 domain protein 7. J Biol Chem 285(45):34439–34446. https://doi.org/10.1074/jbc.M110.152306

Yada M, Hatakeyama S, Kamura T, Nishiyama M, Tsunematsu R, Imaki H et al (2004) Phosphorylation-dependent degradation of c-Myc is mediated by the F-box protein Fbw7. EMBO J 23(10):2116–2125. https://doi.org/10.1038/sj.emboj.7600217

Yang ES, Burnstein KL (2003) Vitamin D inhibits G(1) to S progression in LNCaP prostate cancer cells through p27(Kip1) stabilization and Cdk2 mislocalization to the cytoplasm. J Biol Chem 278(47):46862–46868. https://doi.org/10.1074/jbc.M306340200

Yang G, Ayala G, De Marzo A, Tian WH, Frolov A, Wheeler TM et al (2002) Elevated Skp2 protein expression in human prostate cancer: association with loss of the cyclin-dependent kinase inhibitor p27 and PTEN and with reduced recurrence-free survival. Clin Cancer Res 8(11):3419–3426

Yaron A, Hatzubai A, Davis M, Lavon I, Amit S, Manning AM et al (1998) Identification of the receptor component of the I kappa B alpha-ubiquitin ligase. Nature 396(6711):590–594. https://doi.org/10.1038/25159

Yokobori T, Mimori K, Iwatsuki M, Ishii H, Onoyama I, Fukagawa T et al (2009) p53-altered FBXW7 expression determines poor prognosis in gastric cancer cases. Cancer Res 69(9):3788–3794. https://doi.org/10.1158/0008-5472.CAN-08-2846

Yokoi S, Yasui K, Saito-Ohara F, Koshikawa K, Iizasa T, Fujisawa T et al (2002) A novel target gene, SKP2, within the 5p13 amplicon that is frequently detected in small cell lung cancers. Am J Pathol 161(1):207–216. https://doi.org/10.1016/S0002-9440(10)64172-7

Yu ZK, Gervais JLM, Zhang H (1998) Human CUL-1 associates with the SKP1/SKP2 complex and regulates p21(CIP1/WAF1) and cyclin D proteins. Proc Natl Acad Sci U S A 95(19):11324–11329. https://doi.org/10.1073/pnas.95.19.11324

Zhan FH, Colla S, Wu XS, Chen BZ, Stewart JP, Kuehl WM et al (2007) CKS1B, overexpressed in aggressive disease, regulates multiple myeloma growth and survival through SKP2- and p27(Kip1)-dependent and -independent mechanisms. Blood 109 (11):4995–5001. https://doi.org/10.1182/blood-2006-07-038703

Zhang L, Wang C (2006) F-box protein Skp2: a novel transcriptional target of E2F. Oncogene 25 (18):2615–2627. https://doi.org/10.1038/sj.onc.1209286

Zhang W, MacDonald EM, Koepp DM (2012) The stomatin-like protein SLP-1 and Cdk2 interact with the F-box protein Fbw7-gamma. PLoS One 7(10). https://doi.org/10.1371/journal.pone.0047736

Zhao HL, Bauzon F, Fu H, Lu ZL, Cui JH, Nakayama K et al (2013) Skp2 deletion unmasks a p27 safeguard that blocks tumorigenesis in the absence of pRb and p53 tumor suppressors. Cancer Cell 24(5):645–659. https://doi.org/10.1016/j.ccr.2013.09.021

Zhao JZ, Hu CX, Chi JD, Li JS, Peng C, Yun XW et al (2016) miR-24 promotes the proliferation, migration and invasion in human tongue squamous cell carcinoma by targeting FBXW7. Oncol Rep 36 (2):1143–1149. https://doi.org/10.3892/or.2016.4891

Zhao Y, Xiong X, Sun Y (2011) DEPTOR, an mTOR inhibitor, is a physiological substrate of SCF βTrCP E3 ubiquitin ligase and regulates survival and autophagy. Mol Cell 44(2):304–316. https://doi.org/10.1016/j.molcel.2011.08.029

Zheng JY, Yang XM, Harrell JM, Ryzhikov S, Shim EH, Lykke-Andersen K et al (2002) CAND1 binds to unneddylated CUL1 and regulates the formation of SCF ubiquitin E3 ligase complex. Mol Cell 10

(6):1519–1526. https://doi.org/10.1016/S1097-2765(02)00784-0

Zheng WQ, Zheng JM, Ma R, Meng FF, Ni CR (2005) Relationship between levels of Skp2 and P27 in breast carcinomas and possible role of Skp2 as targeted therapy. Steroids 70(11):770–774. https://doi.org/10.1016/j.steroids.2005.04.012

Zhou BHP, Deng J, Xia WY, Xu JH, Li YM, Gunduz M et al (2004) Dual regulation of snail by GSK-3 beta-mediated phosphorylation in control of epithelial-mesenchymal transition. Nat Cell Biol 6 (10):931–940. https://doi.org/10.1038/ncb1173

Zhu CQ, Blackhall FH, Pintilie M, Iyengar P, Liu N, Ho J et al (2004) Skp2 gene copy number aberrations are common in non-small cell lung carcinoma, and its overexpression in tumors with ras mutation is a poor prognostic marker. Clin Cancer Res 10(6):1984–1991. https://doi.org/10.1158/1078-0432.Ccr-03-0470

Zuo T, Liu RH, Zhang HM, Chang X, Liu Y, Wang LZ et al (2007a) FOXP3 is a novel transcriptional repressor for the breast cancer oncogene SKP2. J Clin Investig 117(12):3765–3773. https://doi.org/10.1172/JCI32538

Zuo T, Wang LZ, Morrison C, Chang X, Zhang HM, Li WQ et al (2007b) FOXP3 is an X-linked breast cancer suppressor gene and an important repressor of the HER-2/ErbB2 oncogene. Cell 129(7):1275–1286. https://doi.org/10.1016/j.cell.2007.04.034

Knockout Mouse Models Provide Insight into the Biological Functions of CRL1 Components

10

Tadashi Nakagawa, Keiko Nakayama, and Keiichi I. Nakayama

Abstract

The CRL1 complex, also known as the SCF complex, is a ubiquitin ligase that in mammals consists of an adaptor protein (SKP1), a scaffold protein (CUL1), a RING finger protein (RBX1, also known as ROC1), and one of about 70 F-box proteins. Given that the F-box proteins determine the substrate specificity of the CRL1 complex, the variety of these proteins allows the generation of a large number of ubiquitin ligases that promote the degradation or regulate the function of many substrate proteins and thereby control numerous key cellular processes. The physiological and pathological functions of these many CRL1 ubiquitin ligases have been studied by the generation and characterization of knockout mouse models that lack specific CRL1 components. In this chapter, we provide a comprehensive overview of these mouse models and discuss the role of each CRL1 component in mouse physiology and pathology.

Keywords

Knockout mouse models · CUL1 · SKP1 · F-box proteins

Abbreviations

APC/C	Anaphase-promoting complex/cyclosome
ASK1	Apoptotic signal-regulating kinase 1
BACE1	β-site amyloid precursor protein-cleaving enzyme 1
BMP	Bone morphogenetic protein
CDK1	Cyclin-dependent kinase 1
CML	Chronic myeloid leukemia
CP110	Centrosomal protein 110
CRL	Cullin-RING ubiquitin ligase
CRY	Cryptochrome
CUL1	Cullin-1
DiPIUS	Differential proteomics-based identification of ubiquitylation substrates
DKO	Double knockout
DP	Double positive
EMI1	Early mitotic inhibitor 1
EMI2	Early mitotic inhibitor 2
F-box	Cyclin F-box
FBXL	LRR-containing F-box proteins
FBXO	Other domain-containing F-box proteins
FBXW	WD40 domain-containing F-box proteins
GC	Germinal center

T. Nakagawa · K. Nakayama (✉)
Division of Cell Proliferation, ART, Graduate School of Medicine, Tohoku University, Sendai, Japan
e-mail: nakayak2@med.tohoku.ac.jp

K. I. Nakayama (✉)
Department of Molecular and Cellular Biology, Medical Institute of Bioregulation, Kyushu University, Fukuoka, Japan
e-mail: nakayak1@bioreg.kyushu-u.ac.jp

Y. Sun et al. (eds.), *Cullin-RING Ligases and Protein Neddylation*, Advances in Experimental Medicine and Biology 1217, https://doi.org/10.1007/978-981-15-1025-0_10

GSTP1	Glutathione S-transferase π1
HECT	Homologous to E6-associated protein C-terminus
HSC	Hematopoietic stem cell
IRE	Iron-responsive element
IRP	Iron-regulatory protein
JNK	c-Jun NH2-terminal kinase
KO	Knockout
LAP	Leukemia-associated protein
LIC	Leukemia-initiating cell
LRR	Leucine-rich repeat
MAFbx	Muscle atrophy F-box protein
MAPK	Mitogen-activated protein kinase
mTOR	Mammalian target of rapamycin
NASH	Nonalcoholic steatohepatitis
NSPC	Neural stem/progenitor cell
PD-1	Programmed cell death-1
PD-L1	Programmed cell death-ligand 1
PER	Period
PHD	Plant homeodomain
PI3K	Phosphoinositide 3-kinase
PRC	Polycomb repressive complex
PTEN	Phosphatase and tensin homolog
RBR	RING between RING
RBX1	RING box 1, also known as regulator of Cullins 1 or ROC1
RIM1	Rab3-interacting molecule 1
RING	Really interesting new gene
ROS	Reactive oxygen species
RRM2	Ribonucleotide reductase 2
SASP	Senescence-associated secretory phenotype
SCF	SKP1–CUL1–F-box protein
Scg2	Secretogranin 2
SKP1	S-phase kinase-associated protein 1
SKP2	S-phase kinase-associated protein 2
SREBP	Sterol regulatory element-binding protein
SSC	Spermatogonial stem cell
T-ALL	T-cell acute lymphoblastic leukemia
U-box	UFD2-box
β-TrCP	β-transducin repeat-containing protein

10.1 Introduction

Ubiquitylation is a posttranslational protein modification that has many regulatory roles in eukaryotic cells, with one prominent role being the marking of target proteins for degradation by the 26S proteasome and the consequent activation or inhibition of a particular cellular process (Nakagawa and Nakayama 2015; Swatek and Komander 2016). Protein ubiquitylation is mediated by three enzymatic reactions: ubiquitin activation by an E1 ubiquitin-activating enzyme, ubiquitin conjugation to an E2 ubiquitin-conjugating enzyme, and transfer of ubiquitin from E2 to the substrate protein by an E3 ubiquitin ligase (Kleiger and Mayor 2014). The specificity of ubiquitylation is conferred by the E3 ubiquitin ligases, which physically interact with their cognate substrates.

E3 enzymes are categorized into three major classes on the basis of their domain structure and catalytic mechanism (Metzger et al. 2012): (1) RING (really interesting new gene) finger or RING- related PHD (plant homeodomain), LAP (leukemia-associated protein), or U (UFD2)-box domain families (Deshaies and Joazeiro 2009), (2) the HECT (homologous to E6-associated protein C-terminus) domain family (Rotin and Kumar 2009), and (3) the RBR (RING between RING) finger domain family (Smit and Sixma 2014). Whereas some E3 ligases function as a monomer, others exist as a complex that includes both an E2 binding component and a substrate binding component.

The CRL1 (Cullin-RING ubiquitin ligase 1) complex, also known as the SCF (SKP1–CUL1–F-box protein) complex, is a multisubunit RING finger-type ubiquitin ligase and a founding member of the CRL family, which also includes CRL2, CRL3, CRL4A, CRL4B, CRL5, CRL7, CRL9, and APC/C (anaphase-promoting complex/cyclosome) (Petroski and Deshaies 2005). The core of the CRL1 complex consists of three proteins: the scaffold protein CUL1 (Cullin-1), the RING finger protein RBX1 (RING box protein 1; also known as regulator of Cullins 1, or ROC1), and the adaptor protein SKP1 (S-phase kinase-associated protein 1). SKP1 links CUL1 to one of the F-box proteins that direct the CRL1 complex to its targets for ubiquitylation. Each F-box protein contains an F-box (cyclin F-box) domain that is necessary for the association with SKP1, as well as a substrate binding domain such as a WD40 domain or LRR (leucine-rich repeat) domain. The human and mouse genomes encode ~70 F-box proteins, which have been classified into three categories: (1) FBXWs (WD40 domain-containing F-box proteins), (2) FBXLs (LRR-containing F-box proteins), and (3) FBXOs (other domain-containing F-box proteins) (Jin et al. 2004). Several knockout (KO) mouse models for components of the CRL1 complex have been established in order to facilitate investigation of their in vivo roles and suitability as drug targets. We here provide a comprehensive overview of such mice model and discuss how they have shed light on the physiological and pathological functions of the CRL1 complex.

10.2 CUL1

CUL1 serves as a scaffold protein of CRL1 by connecting the RBX1–E2 module and the SKP1–F-box protein module for substrate binding, and it is thus an essential component. Studies of *Cul1*-deleted mice were first reported in 1999 by two research groups (Dealy et al. 1999; Wang et al. 1999). Heterozygous mice manifested no visible pathology, but no $Cul1^{-/-}$ mice were born, indicating that CUL1 is required for embryonic development. Analysis of pregnant females revealed that $Cul1^{-/-}$ embryos underwent implantation in the uterine wall but that they failed to develop further and died between embryonic day (E) 6.5 and E7.5 before gastrulation. Accumulation of cyclin E protein, without a concomitant increase in the amount of cyclin E mRNA, was apparent in the $Cul1^{-/-}$ embryos and blastocysts, indicating that cyclin E is a physiological substrate of CRL1 in early embryos. Whether or how the accumulated cyclin E is responsible for the embryonic lethality by CUL1 KO remains to be tested.

10.3 SKP1

SKP1 is an adaptor protein that mediates the association of CUL1 with an F-box protein. As far as we are aware, a SKP1 KO mouse has not yet been described, but our unpublished results indicated that SKP1 KO mice died in utero early in the embryonic development. A transgenic mouse that expresses an NH_2-terminal SKP1-binding fragment of CUL1 (CUL1-N252) under the control of the *Cd4* gene enhancer and promoter in T lymphocytes has been analyzed to provide insight into the in vivo function of SKP1 (Piva et al. 2002). CUL1-N252 binds to SKP1 but lacks the RBX1 binding region, and it thus inhibits the interaction of SKP1 with endogenous CUL1-RBX1. The transgenic animals at 6–8 weeks of age did not show a defect in T-cell differentiation but manifested a decrease in the number as well as hypoplasia of T cells in lymphoid organs as a result of their impaired proliferation. The accumulation of cyclin E, β-catenin, and the cyclin-dependent kinase inhibitor p27 was detected in the thymus of these mice, again implicating these proteins as physiological targets of CRL1. Curiously, despite the low index of T-cell proliferation in young adults, >80% of CUL1-N252 transgenic mice developed T-cell lymphoma and died between 4 and 16 months of age. These lymphomas showed marked karyotype heterogeneity associated with amplification and overexpression of the *Myc* gene, suggesting that SKP1 plays a key role in the maintenance of genome stability and prevention of neoplastic transformation. The mechanistic link between the accumulation of CRL1 substrates and T-cell depletion or *Myc* amplification remains uncharacterized.

10.4 RBX1/ROC1

RBX1 (also known as ROC1) serves as an E2 module for multiple CRL complexes. Targeted inactivation of RBX1 in mice resulted in embryonic death at ~E7.5 due to failure of cell proliferation in association with the accumulation of p27, a known substrate of CUL1 and CUL4 (Tan et al. 2009; Zhou et al. 2013). Simultaneous ablation of p27 extended the embryonic life span to E9.5, indicating that p27-mediated proliferation arrest contributes to the early death of RBX1 KO embryos. The fact that these embryos did not survive to term, however, also indicates that the role of p27 is no longer critical beyond E9.5 and that the accumulation of other substrates of CRLs that rely on RBX1 plays a role in late embryonic mortality.

10.5 FBXW1 (β-TrCP1) and FBXW11 (β-TrCP2)

β-TrCP (β-transducin repeat-containing protein) is encoded by a single gene in invertebrates, but mammalian genomes harbor two different *β-TrCP* genes that encode proteins—β-TrCP1 (FBXW1) and β-TrCP2 (FBXW11)—that share 88% amino acid sequence identity. Molecular characterization of β-TrCP was accelerated by the identification in substrate proteins of a short peptide motif—the β-TrCP degron (DSGXXS)—that is recognized by the WD40 domain of β-TrCP when its two serine residues are phosphorylated. To date, >50 proteins have been identified as targets of β-TrCP-mediated ubiquitylation, with such ubiquitylation playing a role in key signaling pathways including the Wnt (through β-catenin degradation), Sonic hedgehog (through Gli processing and degradation), and NF-κB (through IκB degradation and NF-κB/Rel processing) signaling pathways (Frescas and Pagano 2008; Nakayama and Nakayama 2006), suggesting that the loss of β-TrCP would be detrimental to normal physiology in mice. However, β-TrCP1 KO mice showed no gross tissue abnormalities up to 16 months of age, with only moderate disruption of spermatogenesis and fertility being apparent in male mice (Guardavaccaro et al. 2003; Nakayama et al. 2003) as well as loss of a specific cell type in the retina in both sexes (Baguma-Nibasheka and Kablar 2009). Given that the biochemical functions of β-TrCP1 and β-TrCP2 were thought to be indistinguishable, one interpretation of these findings could be that β-TrCP2 continued to target β-TrCP substrates for ubiquitylation in the absence of β-TrCP1. To test this hypothesis, we then generated β-TrCP2 KO mice and surprisingly found that these mice were not recovered at birth but were absorbed before E9.5 (Nakagawa et al. 2015), indicating that β-TrCP1 could not compensate for the loss of β-TrCP2 during embryonic development. As is often observed for embryos that die around this time, we detected an abnormal vasculature in the yolk sac and massive apoptosis throughout the embryo itself, probably due to malnutrition. However, the molecular targets of β-TrCP2 that are responsible for this phenotype remain to be identified.

Simultaneous β-TrCP1 KO and doxycycline-inducible β-TrCP2 knockdown throughout the body of adult mice was found to result in a pronounced testicular phenotype characterized by impairment of spermatogenesis without any notable abnormality in other tissues, with this phenotype being attributed to accumulation of the β-TrCP substrate SNAIL in the testis (Kanarek et al. 2010). However, the widespread expression of β-TrCP1/2 in the testis, including in male germ cells and Sertoli cells, made it difficult to elucidate the molecular mechanism underlying the contribution of β-TrCP to spermatogenesis. We therefore generated male germ cell-specific and Sertoli cell-specific β-TrCP1/2 DKO (double knockout) mice by combining whole-body β-TrCP1 KO and Cre-mediated, cell type-specific β-TrCP2 KO with the use of a *Stra8–Cre* transgene for germ cells or an *Amh–Cre* transgene for Sertoli cells. Loss of β-TrCP1/2 in male germ cells did not result in the accumulation of the transcription factor SNAIL in these cells, even though the mice were completely sterile (Nakagawa et al. 2017). To identify the β-TrCP substrate (or substrates) responsible for this failure of spermatogenesis, we searched for proteins annotated as spermatogenic regulators that contain the β-TrCP degron. This search identified DMRT1 (doublesex- and Mab3-related transcription factor 1) as a novel substrate of β-TrCP at the mitosis-to-meiosis transition, and we showed that the impaired degradation of DMRT1 inhibits entry into meiosis and was thus responsible, at least in part, for the spermatogenic failure in the β-TrCP1/2 DKO mice. The lack of a change in SNAIL protein level in our germ cell-specific β-TrCP1/2 KO mice led us to hypothesize that SNAIL might be regulated by β-TrCP1/2 in

Sertoli cells and that β-TrCP1/2 in such cells also might play a key role in male germ cell development. Indeed, Sertoli cell-specific β-TrCP1/2 DKO male mice were sterile and showed accumulation of SNAIL in Sertoli cells as well as downregulation of the expression of E-cadherin—which is encoded by a SNAIL target gene—at the periphery of seminiferous tubules (Morohoshi et al. 2019). These results indicate that β-TrCP1/2 regulate the interaction between Sertoli cells and germ cells through degradation of SNAIL in Sertoli cells, with such regulation being critical for male germ cell development.

Intestinal epithelium-specific and neuron-specific β-TrCP1/2 DKO mice have also been investigated. Given that the accumulation of β-catenin as a result of mutation of its β-TrCP degron has been shown to drive colorectal carcinogenesis (Kuipers et al. 2015; Sparks et al. 1998), loss of β-TrCP in the intestinal epithelium was also expected to give rise to colorectal neoplasia. However, tamoxifen-induced loss of β-TrCP1/2 in the intestinal epithelium of adult mice (achieved with the use of a *villin–Cre–ER* transgene) resulted in death within 1 week associated with disruption of the colonic epithelial barrier and severe inflammation dependent on the induction of interleukin-1β (Kanarek et al. 2014). The mutant mice showed an abnormal mitotic index as well as DNA damage in the intestine, indicating that chromosomal instability was responsible for the excessive induction of interleukin-1β. How the loss of β-TrCP1/2 results in chromosomal instability remains to be elucidated, however.

Analysis of neuron-specific DKO mice has also revealed that β-TrCP1/2 play a critical role in the generation of circadian rhythm in the brain. The circadian clock in mammals is dependent on a cell-autonomous, negative-feedback loop in which the clock proteins CLOCK and BMAL1 activate the expression of PER1 (Period 1) and PER2 as well as CRY1 (cryptochrome 1) and CRY2. PER1/2 and CRY1/2 then repress the activity of CLOCK/BMAL1, closing the autoregulatory loop (Takahashi 2017). In addition to transcriptional regulation, PER1/2 and CRY1/2 are subjected to proteasomal degradation directed by F-box proteins so as to ensure the accuracy of the 24-hour rhythm. PER1/2 are targeted for degradation by β-TrCP, whereas CRY1/2 are targeted by FBXL3. Consistent with this scenario, mice with only one (*β-TrCP1*$^{+/-}$*β-TrCP2*$^{-/-}$) or no (*β-TrCP1*$^{-/-}$-*β-TrCP2*$^{-/-}$) *β-TrCP* allele as a result of tamoxifen-induced systemic deletion of the *β-TrCP2* gene on the *β-TrCP1*$^{+/-}$ or *β-TrCP1*$^{-/-}$ background, respectively, showed longer circadian periods and arrhythmicity in association with the stabilization and accumulation of PER proteins (D'Alessandro et al. 2017). The same phenotypes were also observed in neuron-specific β-TrCP1/β-TrCP2 DKO mice generated with the use of an *Scg2 (secretogranin 2)–TA* and *TetO–Cre* system driven by the neuron-specific promoter of the *Scg2* gene, indicating that β-TrCP regulation of PER in neurons is key to the robustness of the circadian rhythm in mice. Whether β-TrCP substrates other than PER proteins also contribute to circadian rhythmicity remains to be determined.

10.6 FBXW5

NASH (nonalcoholic steatohepatitis) is the most prevalent liver disease and a major cause of cirrhosis and hepatocellular carcinoma (Younossi et al. 2018). ASK1 (apoptotic signal-regulating kinase 1, also known as MAP 3K5) has been identified as a therapeutic target for NASH, with an ASK1 inhibitor currently undergoing phase III trials (Sumida and Yoneda 2018). FBXW5 was recently found to function as an activator of ASK1 by mediating its Lys^{63}-linked ubiquitylation. Its contribution to the pathogenesis of NASH was also demonstrated by the observation that hepatocyte-specific FBXW5 KO mice (generated with an *albumin–Cre* transgene) are resistant to the development of hepatic steatosis, fibrosis, and inflammation induced by a high-fat, high-cholesterol diet (Bai et al. 2019). Molecular analysis revealed that the activation of ASK1 and its downstream kinases JNK (c-Jun NH_2-terminal kinase) and p38 MAPK (mitogen-activated protein kinase) induced by this diet in the liver of wild-type mice was attenuated in the KO mice, indicating that FBXW5 is required for aberrant activation of ASK1 in response to metabolic stress in the liver. Ablation of ASK1 in a hepatocyte cell line markedly inhibited the enhancing effects of FBXW5 overexpression on the activation of JNK and p38 MAPK as well as on lipid accumulation induced by exposure to palmitic acid and oleic acid. These observations support the notion that suppression of FBXW5 is also a promising strategy for the treatment of NASH. A detailed phenotypic analysis of whole-body FBXW5 KO mice is awaited for the assessment of the side effects of FBXW5 inhibition.

10.7 FBXW7 (SEL-10, hCDC4, or hAGO)

FBXW7 (also known as FBW7, SEL-10, hCDC4, or hAGO) is best characterized as a tumor suppressor protein. Consistently, it has been demonstrated that FBXW7 ubiquitylates and degrades several oncogenic proteins, and loss-of-function mutations in *Fbxw7* gene are frequently observed in a variety of human cancers (Davis et al. 2014; Shimizu et al. 2018). KO mouse studies not only confirmed anti-oncogenic properties and its mechanisms of FBXW7 but also revealed its contribution to cellular differentiation in relation to tissue and cancer stem cell maintenance.

Fbxw7$^{-/-}$ embryos died in utero at E10.5 with impaired vascular development in the brain and the yolk sac (Tetzlaff et al. 2004b; Tsunematsu et al. 2004). This impairment in vascular development was attributed to a defect in endothelial cells, given that Fbxw7 mRNA was detected predominantly in endothelial lineages at E9.5 and also mesoderm-derived para-aortic splanchnopleural explants isolated from *Fbxw7*$^{-/-}$ embryos did not form vascular network on stromal cells in vitro (Tsunematsu et al. 2004). Since FBXW7 substrate NOTCH4 was accumulated in extracts prepared from *Fbxw7*$^{-/-}$ embryos and its downstream target Hey1 mRNA was markedly increased in the vessels of these embryos, endothelial abnormalities are likely resulted from the dysregulation of Notch signaling. However, genetic rescue experiments were not performed to confirm that NOTCH4 accumulation is the primary cause of embryonic lethality. *Fbxw7*$^{+/-}$ mice appeared healthy and fertile but showed increased susceptibility to radiation-induced tumorigenesis, especially in the background of *p53*$^{+/-}$ or *Pten*$^{+/-}$ (Kwon et al. 2012; Mao et al. 2004), supporting the notion that FBXW7 is a tumor suppressor.

A high frequency of *Fbxw7* gene mutations is observed in T-ALL (T-cell acute lymphoblastic leukemia) (Akhoondi et al. 2007; Malyukova et al. 2007; Maser et al. 2007; O'Neil et al. 2007; Thompson et al. 2007). In agreement with this observation, mice in which *Fbxw7* gene was conditionally deleted in T lymphocytes with the use of a *Lck–Cre* or a *Cd4–Cre* transgene exhibited a thymic hyperplasia and subsequent development of T-cell lymphoma (Onoyama et al. 2007; Thompson et al. 2008). NOTCH1, NOTCH3, and MYC, positive regulators of cell cycle progression, were accumulated in FBXW7-deficient immature T cells, and these cells could not stop proliferation at CD4 and CD8 DP (double positive) stage when wild-type cells normally exit cell cycle. Importantly, this defect of cell cycle arrest was rescued by deletion of *Myc* gene, demonstrating that the accumulation of MYC in FBXW7-deficient DP thymocytes is responsible for the failure of cell cycle exit and the consequent hyperproliferation phenotype. In contrast to immature T cells, mature FBXW7-deficient T cells underwent apoptosis in response to mitogenic stimulation with the accumulation of MYC and p53. This abnormal apoptosis could be corrected by the additional deletion of *p53* gene. Since p53 did not seem to be a substrate of FBXW7, it is likely that accumulation of p53 was caused by increased MYC protein level (Zindy et al. 1998). These data demonstrated that FBXW7 regulates proliferation and apoptosis in a manner dependent on the differentiation stage in T-cell lineage.

Conditional deletion of *Fbxw7* in murine HSCs (hematopoietic stem cells) with the use of *Mx1–Cre* transgene and poly(I:C) injection led to severe leukopenia resulting from loss of self-renewal activity and exhaustion of quiescent HSCs with accumulation of NOTCH1 and MYC (Matsuoka et al. 2008; Thompson et al. 2008). Since enforced MYC expression in HSCs also led to loss of self-renewal activity (Wilson et al. 2004), MYC appears to be the main substrate of FBXW7 in HSC maintenance. Interestingly, most of the HSC-specific FBXW7-deficient mice that did not exhibit leukopenia showed extrathymic development of DP T cells and T-ALL in which p53 protein level is reduced (Matsuoka et al. 2008). In addition, loss of p53 significantly promoted T-cell leukemogenesis in these mice. These findings suggest that deletion of FBXW7 provides a selective advantage to hematopoietic cells that harbor suppressed p53 function. How FBXW7-deficient cells evaded cell death in the setting of elevated MYC protein level has remained elusive, but anti-apoptotic MCL-1 might be involved because it was shown to be a substrate of FBXW7 and accumulated in thymic lymphoma and acute lymphoblastic leukemia cells from mice with the T-cell lineage-specific deletion of FBXW7 (Inuzuka et al. 2011).

The role of FBXW7 in tumor metastasis through regulating microenvironment was also revealed by the analysis of mice in which FBXW7 is depleted in HSCs with the use of a *Mx1–Cre* transgene and poly(I:C) injection (Yumimoto et al. 2015). These mice were susceptible to metastasis of injected melanoma and lung cancer cells and breast cancer cells to the lung. Mechanistically, it was demonstrated that FBXW7 KO in bone marrow-derived stromal cells resulted in the accumulation of NOTCH1, leading to the upregulation of CCL2 chemokine which likely promoted the formation of metastatic niches through recruitment of monocytic myeloid-derived suppressor cells and macrophages. These results suggest that FBXW7 antagonizes cancer development in both cell-autonomous and non-cell-autonomous manners.

Mice with HSC population expressing BCR–ABL, a fusion protein found in the CML (chronic myeloid leukemia) patients, develop CML, and thus these mice serve as a mouse model of CML pathogenesis (Pear et al. 1998). Intriguingly, loss of FBXW7 in BCR–ABL-expressing cells (generated by retroviral introduction of BCR–ABL to isolated HSCs from *Mx1–Cre; Fbxw7*$^{flox/flox}$ mice combined with tamoxifen injection) was shown to suppress progression of CML through enhanced apoptosis resulting from accumulation of MYC and p53 (Reavie et al. 2013; Takeishi et al. 2013). Analysis of FBXW7-ablated CML leukemic cells further elucidated the reduction of quiescent LICs (leukemia-initiating cells), indicating that FBXW7 is essential for the maintenance of LIC dormancy. Consistent with the hypothesis that LIC dormancy is the mechanism by which CML shows resistance to anticancer drugs, combination of FBXW7 KO and an anticancer drug (either imatinib or cytosine arabinoside) was found to markedly enhance the anticancer effect, supporting the notion that the "wake-up" therapeutic strategy combining anticancer agents with FBXW7 inhibition which sensitizes LICs to these drugs by putting LICs out of dormant state is potentially effective for eradicating LICs (Takeishi and Nakayama 2016).

In contrast to T-ALL, HSC-specific FBXW7-deficient mice did not develop B-cell lymphoma. Consistently, the mutation rate of *Fbxw7* gene is extremely low in human B-cell lymphoma (Akhoondi et al. 2007; Song et al. 2008). To examine the difference in sensitivity to *Fbxw7* mutation in T lymphocytes and B lymphocyte, we recently generated B-lymphocyte-specific FBXW7 KO mice with the use of a *Cd19–Cre* transgene and found that FBXW7 is essential for B-lymphocyte survival (unpublished data), at least in part by destabilizing nuclear NF-κB2/p100 which was previously reported to be a substrate of FBXW7 (Arabi et al. 2012; Busino et al. 2012; Fukushima et al. 2012).

Fbxw7 mutations are also found in other human cancers, including early stage human colon adenomas (Rajagopalan et al. 2004). Deletion of FBXW7 in mouse intestinal epithelium with the use of a *villin–Cre* transgene was reported to cause an increase in transit-amplifying progenitor cells accompanied by a reduction of goblet and Paneth cells (Babaei-Jadidi et al. 2011; Sancho et al. 2010). These mice developed polyp-like structures and adenomas, but not carcinomas, indicating that loss of FBXW7 alone is not sufficient to cause intestinal carcinomas. When *Apc*^{min} mono-allele bearing a nonsense mutation at codon 850 of the *Apc* gene was introduced (Fodde 2002), adenoma was developed with much shortened latency. Even in this condition, neoplasms did not progress beyond the adenoma stage and were neither invasive nor metastatic. Intestinal extracts prepared from FBXW7-deficient mice exhibited increased abundance of NOTCH1, NOTCH4, and c-JUN. Importantly, *c-Jun* gene deletion in *Apc*$^{min/+}$, *villin–Cre*, and *Fbxw7*$^{flox/flox}$ mice reduced tumor area to the size comparable with *Apc*$^{min/+}$ tumors, indicating that accumulated c-JUN is responsible for the increase in tumor size resulting from loss of FBXW7 (Sancho et al. 2010). In contrast to heterozygous deletion of *Apc*, *p53* codeletion with *Fbxw7* in the intestinal epithelium caused penetrant, aggressive, and metastatic adenocarcinomas (Grim et al. 2012). These tumors exhibited a sign of chromosomal instability, a commonly observed phenotype in human colorectal cancers. Although the molecular mechanisms were not evident, it was also noted that loss of p53 reverted the number of Paneth cells, but not goblet cells (Grim et al. 2012). These results suggested that FBXW7 contributes to not only proliferation arrest but also differentiation of intestinal cells, some of which depend on p53 activity.

The significant role of FBXW7 in cellular differentiation was also illustrated by the analysis of hepatocyte-, neural cell-, or spermatogonium-specific FBXW7 KO mice. Hepatocyte-specific FBXW7 KO mice with the use of an *albumin–Cre* or a *Mx1–Cre* transgene combined with poly (I:C) injection exhibited hepatomegaly and steatohepatitis probably as a result of stabilization of SREBP (sterol regulatory element-binding protein) and alteration of its downstream target gene expression (Onoyama et al. 2011). Loss of FBXW7 also caused skewed hepatic differentiation toward the cholangiocyte lineage rather than the hepatocyte lineage with accumulation of NOTCH1. Concomitant deletion of NOTCH cofactor RBPJ could rescue this defect, indicating that the skewed developmental orientation of hepatic stem cells to the cholangiocyte lineage is dependent on NOTCH1 accumulation induced by the loss of FBXW7.

Mice with NSPC (neural stem/progenitor cell)-specific deletion of FBXW7 with the use of a *nestin–Cre* transgene exhibited impaired stem cell differentiation to the neurons and increased progenitor cell death. These mice died perinatally due to the absence of suckling behavior (Hoeck et al. 2010; Matsumoto et al. 2011). Genetic and pharmacologic rescue experiments demonstrated that accumulated NOTCH1 was responsible for differentiation defects, while increase in c-JUN contributed to neural cell viability.

Deletion of FBXW7 in SSCs (spermatogonial stem cells) with the use of a *Stra8–Cre* transgene led to accumulation of undifferentiated spermatogonia with concomitant reduction of differentiating and mature germ cells (Kanatsu-Shinohara et al. 2014). Among the FBXW7 substrates tested, accumulation of MYC and cyclin E1 was noted in germ cells from FBXW7-deficient pup testes, and overexpression of MYC, but not cyclin E1, phenocopied effects of FBXW7 loss to enhance colonization of germ line stem cells. These data demonstrated that FBXW7 counteracts with positive regulators of self-renewal by degrading MYC to induce differentiation of SSCs.

Taken together, these various observations suggest that FBXW7 plays a critical role in proliferation and tumorigenesis as well as stem cell differentiation in different tissues by targeting tissue-specific substrates for degradation.

10.8 FBXW8

FBXW8 is a placenta- and embryo-specific F-box protein that binds to CUL1 through SKP1 but also interacts directly with CUL7. Although the significance of this association with two cullins is unknown, the fact that *FBXW8* mutations have not been identified in individuals with 3M syndrome, a rare growth disorder that is thought to be caused by the loss of CUL7 function, has suggested that FBXW8 is not absolutely required for CUL7 function (Li et al. 2014; Yan et al.

2014). This conclusion has also been supported by studies of KO mice.

About two-thirds of FBXW8 KO mouse embryos were found to die in utero beginning at E12.5 as a result of abnormal placental development and growth retardation, whereas the remaining one-third of embryos grew to adulthood but were smaller than their littermates throughout life (Tsunematsu et al. 2006, Tsutsumi et al. 2008). Histological analysis revealed aberrant development of spongiotrophoblasts in the placenta of the KO mice. Although CUL7 KO mice also exhibited an abnormally thin spongiotrophoblast layer, they did not show marked embryonic mortality. However, CUL7 KO neonates died soon after birth as a result of respiratory failure (Arai et al. 2003), which was not apparent in FBXW8 KO mice. Together with the observation that the stability of the FBXW8 protein was decreased in the absence of CUL7 (Arai et al. 2003; Tsunematsu et al. 2006), these results suggest that the role of CUL7 in placental development relies, at least in part, on FBXW8, whereas that in neonatal, lung function is independent of FBXW8. The FBXW8 substrates responsible for defective placental development in FBXW8 KO mice remain to be identified.

Phenotypes of FBXW protein KO mice described herein with identified primary substrates are summarized in Table 10.1.

10.9 FBXL1 (SKP2)

SKP2 (S-phase kinase-associated protein 2), also known as FBXL1, is the prototypical and best-characterized mammalian FBXL protein. SKP2 promotes cell proliferation by targeting negative regulators of the cell cycle for degradation, and its role as an oncoprotein has been supported by the fact that it is overexpressed in many cancers (Frescas and Pagano 2008; Nakayama and Nakayama 2006). SKP2 KO mice were found to be born approximately in the expected Mendelian ratio, but they were smaller than their wild-type littermates (Nakayama et al. 2000). The cellular phenotypes of these mice included nuclear enlargement and polyploidy in the liver, lung, kidney, and testis as well as an increased number of centrosomes in mouse embryonic fibroblasts. Providing support for a critical role of SKP2 in tumor progression, SKP2 KO mice were shown to be resistant to tumor development induced by loss of either the tumor suppressor proteins $p19^{ARF}$ or PTEN (phosphatase and tensin homolog) (Lin et al. 2010).

Although many substrates of SKP2 have been identified in cultured cells (Frescas and Pagano 2008; Nakayama and Nakayama 2006), p27 seems to be a key substrate, given that SKP2 KO mice showed marked accumulation of p27 (Nakayama et al. 2000) and that prominent cellular phenotypes of these mice were no longer apparent in SKP2 and p27 DKO mice (Kossatz et al. 2004; Nakayama et al. 2004).

10.10 FBXL3 and FBXL21

Screening of randomly mutagenized mice to detect genes related to circadian rhythm led to the identification of the *after-hours* and *overtime* mutants, both of which manifested a prolonged circadian cycle. Genetic analysis of both these mice revealed mutations in *Fbxl3*, the loss of function of which was found to result in the accumulation of CRY1/2 proteins (Busino et al. 2007; Godinho et al. 2007; Siepka et al. 2007).

Interestingly, FBXL21, a paralog of FBXL3, interacts more strongly with CRY1/2 than does FBXL3 and protects CRY1/2 from $CRL1^{FBXL3}$ activity, resulting in CRY1/2 stabilization (Hirano et al. 2013; Yoo et al. 2013). Importantly, the lengthening of the circadian cycle apparent in FBXL3 KO mice was partially reversed by simultaneous deletion of *Fbxl21*, providing support for the notion that FBXL21 antagonizes the function of FBXL3 in regulation of CRY1/2 stability and circadian rhythm in mice (Hirano et al. 2013). In contrast to FBXL3 KO mice, FBXL21 single KO mice did not exhibit abnormal rhythmic behaviors with a period indistinguishable from that of wild-type mice. However, these mice exhibited a decrease in wheel-running activities near the

Table 10.1 Phenotypes of FBXW protein KO mice including identified primary substrates

F-box protein	Systemic or conditional KO	Phenotype of KO mice	Primary substrate	References
β-TrCP1 (FBXW1)	Whole body	Male subfertility, abnormal retinal development	ND	Baguma-Nibasheka and Kablar (2009), Guardavaccaro et al. (2003) and Nakayama et al. (2003)
β-TrCP2 (FBXW11)	Whole body	Embryonic lethality	ND	Nakagawa et al. (2015)
β-TrCP1+2	Spermatogonia	Male sterility	DMRT1	Nakagawa et al. (2017)
β-TrCP1+2	Sertoli cells	Male sterility	SNAIL	Morohoshi et al. (2019)
β-TrCP1+2	Intestinal epithelial cells	Disruption of colonic epithelial barrier	ND	Kanarek et al. (2014)
β-TrCP1+2	Neuronal cells	Abnormal circadian rhythm	PER1/2	D'Alessandro et al. (2017)
FBXW5	Hepatocytes	Resistance to nonalcoholic steatohepatitis	ASK1	Bai et al. (2019)
FBXW7	Whole body	Embryonic lethality	ND	Tetzlaff et al. (2004b) and Tsunematsu et al. (2006)
FBXW7	T cells	Development of T-cell lymphoma	MYC	Onoyama et al. (2007) and Thompson et al. (2008)
FBXW7	Hematopoietic stem cells	Loss of self-renewal activity	ND	Matsuoka et al. (2008) and Thompson et al. (2008)
FBXW7	Bone marrow stromal cells	Increased susceptibility to cancer metastasis	NOTCH1	Yumimoto et al. (2015)
FBXW7	Chronic myeloid leukemia cells	Reduction of quiescent leukemia-initiating cells	MYC	Reavie et al. (2013) and Takeishi et al. (2013)
FBXW7	Intestinal epithelial cells	Development of adenoma	c-JUN	Babaei-Jadidi et al. (2011) and Sancho et al. (2010)
FBXW7	Hepatocytes	Steatohepatitis	SREBP	Onoyama et al. (2011)
FBXW7	Hepatocytes	Skewed differentiation to the cholangiocytes	NOTCH1	Onoyama et al. (2011)
FBXW7	Neural stem/ progenitor cells	Impaired differentiation to the neurons	NOTCH1	Hoeck et al. (2010) and Matsumoto et al. (2011)
FBXW7	Neural stem/ progenitor cells	Reduced viability	c-JUN	Hoeck et al. (2010) and Matsumoto et al. (2011)
FBXW7	Spermatogonial stem cells	Impaired differentiation	MYC	Kanatsu-Shinohara et al. (2014)
FBXW8	Whole body	Partial embryonic lethality, abnormal placental development, small size	ND	Tsunematsu et al. (2006) and Tsutsumi et al. (2008)

ND not determined, *NA* not applicable

subjective dawn, though the underlying mechanism has yet to be elucidated (Hirano et al. 2013).

10.11 FBXL5

FBXL5 plays a central role in cellular iron metabolism. Cellular and organismal iron levels are strictly controlled by hormonal as well as cell-intrinsic mechanisms because iron is required for various biological processes but in excess contributes to the generation of toxic ROS (reactive oxygen species) (Hentze et al. 2010; Muckenthaler et al. 2017). In particular, the IRP (iron-regulatory protein) and IRE (iron-responsive element) system functions prominently to maintain the cellular iron level in balance (Wilkinson and Pantopoulos 2014). In response to iron deprivation, IRP1 loses its [4Fe–4S] cluster, and the apo-IRP1 protein then binds to IREs present in the mRNAs for various proteins related to iron trafficking, storage, or utilization. This binding modulates production of the encoded proteins and thereby increases the availability of intracellular iron. In contrast to IRP1, IRP2 does not contain an Fe–S cluster that can function in iron sensing, but its abundance is regulated by FBXL5 in an iron-dependent manner. Under iron-replete conditions, FBXL5 constitutively ubiquitylates IRP2 and thereby targets it for proteasomal degradation. A decline in iron availability results in destabilization of FBXL5 and the consequent stabilization of IRP2. This scenario suggests that loss of FBXL5 would result in the accumulation of IRP2 even in the presence of sufficient iron, which in turn would lead to a buildup of iron, the excessive generation of ROS, and the induction of cellular damage. Disruption of *Fbxl5* in mice indeed resulted in constitutive accumulation of IRP2 and misexpression of its target genes (Moroishi et al. 2011). The FBXL5 KO mice died during embryogenesis at ~E8.5 as a result of overwhelming oxidative stress, most noticeably in extraembryonic tissues in which iron is supplied from the mother. Importantly, concomitant deletion of *Irp2* completely rescued the embryonic lethality induced by FBXL5 loss, showing that IRP2 is a primary substrate of FBXL5.

Specific deletion of FBXL5 in the liver (with the use of an *albumin–Cre* transgene), which controls systemic iron levels, also resulted in IRP2 accumulation in this organ, leading to iron buildup, liver damage, and the development of steatohepatitis (Moroishi et al. 2011). The serum iron concentration was also increased in these mice, whereas the secretion of hepcidin, which negatively regulates iron availability, was unexpectedly decreased, with this effect being attributable to a decrease in BMP (bone morphogenetic protein) signaling of unknown cause and contributing to systemic iron overload and increased sensitivity to a high-iron diet.

Given that iron overload is linked to several neurodegenerative diseases (Rouault 2013), NSPC-specific FBXL5 KO mice were generated with the use of a *nestin–Cre* transgene in order to investigate this relation (Yamauchi et al. 2017). These mice died within a day of delivery as a result of respiratory failure, indicative of disruption of the control of respiration by the nervous system. The brain of these FBXL5 KO mouse embryos revealed increased proliferation and attenuated neuronal differentiation of NSPCs. It also manifested increased levels of IRP2 and iron, resulting in the generation of ROS and activation of the PI3K (phosphoinositide 3-kinase)–AKT–mTOR (mammalian target of rapamycin) signaling pathway. Importantly, treatment with either a PI3K inhibitor or an mTOR inhibitor rescued the proliferation defect of FBXL5-deficient NSPCs in vitro, indicating that activation of ROS–PI3K–AKT–mTOR signaling by IRP2-mediated iron overload is responsible for the dysfunction of these cells.

Hematopoiesis is particularly sensitive to iron levels (Camaschella 2015; Fleming and Ponka 2012). Ablation of FBXL5 specifically in HSCs of mice with the use of an *Mx1–Cre* transgene and poly(I:C) injection resulted in impairment of the repopulation ability of these cells (Muto et al. 2017). This impairment was due to the accumulation of IRP2, iron overload, and oxidative stress that promoted exit of the cells from quiescence and their premature exhaustion. Importantly, ablation of IRP2 restored the repopulation ability of the FBXL5-deficient HSCs, indicating that FBXL5 supports HSC function primarily through suppression of IRP2 activity.

10.12 FBXL10 (KDM2B) and FBXL11 (KDM2A)

FBXL10 (also known as KDM2B) binds to SKP1 through its F-box domain, but it does not appear to associate with CUL1 (Koyama-Nasu et al. 2007). Instead, FBXL10 seems to function primarily in epigenetic chromatin regulation. FBXL10 binds to CpG islands in the genome through a CXXC domain and recruits PRC1 (polycomb repressive complex 1), which catalyzes ubiquitylation of histone H2A. In addition, FBXL10 removes methyl groups attached to Lys^{36} of histone H3 (H3K36) in a reaction mediated by its JmjC domain. FBXL10 exists as two isoforms as a result of alternative use of exon 1, with the short isoform (isoform 2) lacking the JmjC domain. This situation affords the opportunity to investigate the function of the H3K36 demethylase activity of FBXL10 by targeted exon disruption. Mice lacking both isoforms of FBXL10 died in utero at ~E11.5 to E13.5 manifesting severe developmental abnormalities including small size, failure of neural tube closure, and limb and craniofacial malformation (Andricovich et al. 2016). The embryonic death was attributed primarily to defective hematopoiesis, given that the vasculature of the yolk sac was absent and that loss of FBXL10 only in the hematopoietic lineage (achieved with the use of a *Tie2–Cre* transgene) also induced death at the same embryonic stage as did systemic ablation. FBXL10 KO in adult hematopoietic stem cells with the use of a *Vav1–Cre* transgene or the *Mx1–Cre* transgene and poly(I:C) injection also resulted in a reduction in the number of long-term HSCs and defective lymphopoiesis, suggesting that FBXL10 is required for normal development of the hematopoietic lineage in both embryonic and adult mice. Transcriptomic analysis of hematopoietic progenitors of these HSC-specific FBXL10 KO mice revealed upregulation of the expression of PRC target genes as well as downregulation of that of transcription factor genes related to hematopoietic development. How many of these genes are direct targets of FBXL10 remains unknown, but chromatin immunoprecipitation sequencing analysis of human leukemic cell lines detected FBXL10 at the promoters of some genes whose expression was altered in the FBXL10 KO cells.

About one-half of mice with targeted disruption of the long isoform of FBXL10 (isoform 1) manifested failure of neural tube closure and exencephaly and died shortly after birth (Fukuda et al. 2011), indicating that isoform 2 is able to compensate for the loss of isoform 1 in embryonic hematopoiesis, which therefore must be independent of JmjC domain-catalyzed H3K36 demethylation. The molecular mechanism underlying these brain phenotypes is currently unknown. Disruption of only isoform 2, with isoform 1 remaining intact, also resulted in partial perinatal lethality associated with craniofacial abnormalities, cleft palate, and the eyes-open-at-birth phenotype (Boulard et al. 2016). Transcriptome-wide analysis of female embryos, which showed more severe phenotypes than did male embryos, revealed dysregulation of X-linked genes, possibly as a result of abnormal overexpression of *Xist*, which plays a key role in inactivation of one of the two X chromosomes in females. These observations implicated FBXL10 in X chromosome regulation, but again the molecular mechanism by which FBXL10 regulates *Xist* expression is not clear.

FBXL11 (also designated as KDM2A) has a structure similar to that of its paralog FBXL10, but it has not been shown to associate with PRC1 (Wu et al. 2013). FBXL11 deficiency in mice led to severe growth retardation with embryonic death at E10.5 to E12.5. The brain of the mutant mice manifested reduced cell proliferation, increased apoptosis, and arrested neuronal differentiation (Kawakami et al. 2015). Although expression of FBXL11 was found to be ubiquitous in the embryo proper, it was not examined in extraembryonic tissues. The mechanism by which loss of FBXL11 leads to embryonic death therefore remains to be investigated.

10.13 FBXL12

Four-fifths of FBXL12 KO mice were found to die soon after birth as a result of intrauterine growth retardation and a consequent inability to compete with their wild-type or heterozygous littermates for suckling (Nishiyama et al. 2015). The remaining one-fifth of the homozygous mutant embryos grew to adulthood but were smaller compared with $Fbxl12^{+/-}$ or $Fbxl12^{+/+}$ mice. FBXL12 is expressed specifically in the junctional zone of the mouse placenta, and loss of FBXL12 led to a reduction in the numbers of both spongiotrophoblasts and glycogen cells in this zone. A search for substrates of FBXL12 in a human trophoblast cell line with the use of the DiPIUS (differential proteomics-based identification of ubiquitylation substrates) system (Yumimoto et al. 2012) identified ALDH3A1 and ALDH3A2, enzymes that oxidize aldehydes to generate carboxylic acids. Indeed, these proteins were found to accumulate in the FBXL12 KO placenta, and inhibition of ALDH3 activity rescued the defective differentiation of trophoblast stem cells, indicating that accumulation of ALDH3 family proteins and a consequent increase in the concentration of toxic aldehydes were at least partially responsible for the placental abnormality of FBXL12 KO mice.

In adult mice, FBXL12 is highly expressed in the thymus, and examination of FBXL12 KO mice that survived the neonatal stage revealed impaired differentiation of CD4/CD8 DP T cells into CD4 or CD8 single-positive cells in association with the accumulation of ALDH3 (Nita et al. 2016), indicating that FBXL12 contributes to T-cell differentiation in adult mice by regulating the abundance of ALDH3.

10.14 FBXL20 (SCRAPPER)

FBXL20, also known as SCRAPPER, is highly expressed in the brain, where its localization at synapses is mediated by a fatty acid attached to its CAAX domain (Yao et al. 2007). FBXL20 directly binds to and ubiquitylates the presynaptic protein RIM1 (Rab3-interacting molecule 1) both in vitro and in vivo, and FBXL20 KO mice were found to show increased neurotransmitter release, probably as a result of the accumulation of RIM1. The studied mice had a hybrid 129Sv × C57BL/6 background and showed stochastic death shortly after birth, a reduced life span for those that survived the neonatal period, and a smaller body size compared with wild-type littermates. A subsequent study found that the lethality of *Fbxl20* deletion was more pronounced for mice backcrossed to the C57BL/6J background (Yao et al. 2011), indicating that genetic background markedly affects this phenotype of FBXL20 KO mice, although the factors contributing to this effect remain unknown.

Behavioral analysis revealed a lower level of freezing induced by foot shock or by subsequent placement in the apparatus where the foot shock had been delivered for $Fbxl20^{+/-}$ mice compared with wild-type littermates. Combined with the observed high level of expression of FBXL20 in the hippocampus (Yao et al. 2007), these results suggested that FBXL20 plays a role in the formation of contextual fear memories in this region of the brain. Whether or how RIM1 accumulation contributes to this phenotype awaits clarification.

Phenotypes of FBXL protein KO mice with identified primary substrates are summarized in Table 10.2.

10.15 FBXO1 (Cyclin F)

The F-box domain was first identified in cyclin F, also known as FBXO1. Similar to cyclin A, the expression of cyclin F begins in S phase of the cell cycle, peaks in G_2, and declines as cells enter mitosis (Bai et al. 1994). These changes in expression level during the cell cycle and the identification of cell cycle-related substrates for cyclin F such as RRM2 (ribonucleotide reductase 2) and CP110 (centrosomal protein 110) indicated that cyclin F contributes to regulation of cell proliferation (D'Angiolella et al. 2013). Loss of cyclin F in mice induced embryonic death between E9.5 and E10.5 as a result of extraembryonic defects (Tetzlaff et al. 2004a). Cyclin F was found to be expressed at a high level in the chorionic trophoblast layer, and the absence of cyclin F led to a reduced level of cell proliferation in this region and impairment of chorioallantoic fusion, which connects the mother to the embryo to allow the exchange of nutrients and metabolic waste products. Whether the attenuated trophoblast proliferation was causal for or merely correlative with the placental failure was not determined, and the cyclin F substrates responsible for this phenotype were not identified. Although cyclin F is expressed in all dividing tissues in both embryos and adult mice, conditional ablation of cyclin F in the eye (with the use of a *Pax6–Cre* transgene), bone (with an *α1-collagen–Cre* transgene), or gut and bladder (with an *FABP–Cre* transgene) did not give rise to notable defects, indicating that cyclin F is not essential for the development and physiology of at least these tissues (Tetzlaff et al. 2004a).

Table 10.2 Phenotypes of FBXL protein KO mice including identified primary substrates

SKP2 (FBXL1)	Whole body	Small size, polyploidy	p27	Kossatz et al. (2004) and Nakayama et al. (2000, 2004)
FBXL3	Whole body	Abnormal circadian rhythm	CRY1/2	Godinho et al. (2007) and Siepka et al. (2007)
FBXL5	Whole body	Embryonic lethality, abnormal placental development	IRP2	Moroishi et al. (2011)
FBXL5	Hepatocytes	Steatohepatitis	IRP2	Moroishi et al. (2011)
FBXL5	Neural stem/progenitor cells	Neonatal lethality, abnormal neuronal development	IRP2	Yamauchi et al. (2017)
FBXL5	Hematopoietic stem cells	Reduced number of hematopoietic stem cells	IRP2	Muto et al. (2017)
FBXL10 (KDM2B)	Whole body	Embryonic lethality	ND	Andricovich et al. (2016), Boulard et al. (2016) and Fukuda et al. (2011)
FBXL10	Hematopoietic stem cells	Reduced number of hematopoietic stem cells	ND	Andricovich et al. (2016)
FBXL11 (KDM2A)	Whole body	Embryonic lethality	ND	Kawakami et al. (2015)
FBXL12	Whole body	Partial neonatal lethality, abnormal placental development, small size	ALDH3	Nishiyama et al. (2015)
FBXL12	T cells	Abnormal T-cell development	ALDH3	Nita et al. (2016)
FBXL20	Whole body	Partial neonatal lethality, abnormal memory formation	ND	Yao et al. (2007, 2011)
FBXL21	Whole body	Decreased activities near the subjective dawn	CRY1/2	Hirano et al. (2013)

ND not determined, *NA* not applicable

Interest in cyclin F was recently ignited by the identification of mutations in its gene associated with familial and sporadic amyotrophic lateral sclerosis and frontotemporal dementia (Williams et al. 2016). Mouse models expressing these mutant forms of cyclin F have not yet been described, but their establishment is likely to provide important insight into pathological mechanisms.

10.16 FBXO2

FBXO2 is expressed in the brain, ears, and testis (Erhardt et al. 1998; Thalmann et al. 1997; Yoshida et al. 2003), specifically recognizes N-linked high-mannose oligosaccharides attached to proteins, and targets for ubiquitylation and degradation N-glycosylated proteins such as β1-integrin (Yoshida et al. 2002), the NR1 subunit of the NMDA subtype of glutamate receptors (Kato et al. 2005), and BACE1 (β-site amyloid precursor protein-cleaving enzyme 1), the latter of which contributes to amyloidosis in Alzheimer's disease (Gong et al. 2010). Mice with targeted deletion of FBXO2 developed accelerated age-related hearing loss starting at ~2 months of age without noticeable pathology in the brain (Nelson et al. 2007). Cellular degeneration was found to begin in the epithelial support cells of the organ of Corti and progressed to hair cells and the spiral ganglion. These observations thus indicated that FBXO2 is essential for homeostasis of the inner ear and that it acts to prevent age-related hearing loss. Of note, the abundance of SKP1 in the cochlea was reduced in parallel with the loss of FBXO2, suggesting that FBXO2 somehow stabilizes SKP1 in the ear. The relation between hearing loss and the potential accumulation of N-glycosylated proteins in the FBXO2 KO mice, however, remains unclear.

10.17 FBXO4

FBXO4 binds to αB-crystallin (also known as HSPB5), and the two proteins assemble with the other CRL1 components to form an active ubiquitin ligase that has been thought to target cyclin D1 for degradation and thereby to suppress oncogenesis (Barbash et al. 2008; Lin et al. 2006). *Fbxo4*$^{+/-}$ and *Fbxo4*$^{-/-}$ mice were thus found to accumulate cyclin D1 and to develop multiple types of tumor, including lymphoma, histiocytic sarcoma, and, less frequently, mammary and hepatocellular carcinoma (Vaites et al. 2011). In contrast to these findings, FBXO4 KO mice generated by our group did not show accumulation of cyclin D1 or development of tumors for up to 1 year of age (Kanie et al. 2012). We also obtained evidence that CRL1 does not play a role in cyclin D1 degradation, despite several previous studies having indicated that FBXO4 (Barbash et al. 2008; Lin et al. 2006), FBXW8 (Okabe et al. 2006), SKP2 (Yu et al. 1998), and FBXO31 (Santra et al. 2009) ubiquitylate cyclin D1 for proteasomal degradation. The reason of these discrepancies remains unknown.

10.18 FBXO5 (EMI1) and FBXO43 (EMI2)

EMI1 (early mitotic inhibitor 1), also known as FBXO5, plays a key role in cell cycle progression. Although EMI1 forms a CRL1 ubiquitin ligase (Marzio et al. 2019; Reitsma et al. 2017), the primary function of EMI1 is thought to be inhibition of APC/C ubiquitin ligase activity in G_1 phase in order to promote DNA replication, in S phase to prevent DNA rereplication, and in G_2-M phase to ensure proper mitotic progression (Lara-Gonzalez et al. 2017). The loss of EMI1 would therefore be expected to impair cell cycle progression, an effect not compatible with embryonic development. Indeed, EMI1 KO mouse embryos were found to die at the preimplantation stage, with the EMI1-deficient cells manifesting abnormal mitosis (Lee et al. 2006). However, EMI1-deficient embryos showed no obvious defects in entry into or progression through S phase, indicating that EMI1 plays a nonredundant role in M-phase progression during early embryonic development but that its role in S-phase regulation is redundant and can be undertaken by other proteins.

The EMI1 paralog EMI2 (early mitotic inhibitor 2, also known as FBXO43) contributes specifically to germ cell development. Although the ability of EMI2 to inhibit APC/C activity appears to be redundant with that of EMI1, the expression of EMI2 is restricted to reproductive organs where germ cells are produced (Shoji et al. 2006). In frog and mouse oocytes, the role of EMI2 as a component of cytostatic factor, which is required for the establishment and maintenance of meiotic metaphase II arrest, has been well characterized (Madgwick and Jones 2007). Consistent with this role, female EMI2 KO mice were found to be viable but infertile, with their oocytes showing defects in meiosis II (Gopinathan et al. 2017). Male EMI2 KO mice were also sterile, indicating that EMI2 is essential for meiosis in both males and females. In spermatogenesis, EMI2 was found to be essential for meiosis I progression at the early diplotene stage. Increased APC/C activity in the absence of EMI2 would be expected to result in excessive degradation of its substrate cyclin B and consequent inactivation of CDK1 (cyclin-dependent kinase 1). Importantly, crossing of EMI2 KO mice with a knock-in mouse line with elevated CDK1 activity rescued the meiosis I arrest apparent in EMI2-deficient male germ cells, supporting the notion that APC/C inhibition is the primary function of EMI2 in early spermatogenesis.

10.19 FBXO7

A genome-wide association study identified autosomal recessive mutations in *FBXO7* (also known as *PARK15*) in individuals with a juvenile form of familial Parkinson's disease (Shojaee et al. 2008). Parkinson's disease is typically characterized by motor deficits due to loss of midbrain dopaminergic neurons with intracellular inclusions containing aggregates of α-synuclein (Kalia and Lang 2015; Poewe et al. 2017). Support for a causative role of FBXO7 loss in this disease was provided by analysis of FBXO7 KO mice (Vingill et al. 2016). Systemic deletion of FBXO7 in mice thus gave rise to defects in locomotor activity and death within 1 month after birth. Conditional

FBXO7 KO in the forebrain with the use of a *Nex–Cre* transgene also resulted in the development of motor defects, whereas loss of FBXO7 in catecholaminergic (including dopaminergic) neurons induced with a *tyrosine hydroxylase–Cre* transgene gave rise to much milder motor coordination problems that were apparent only in older mice challenged with forced movement. These results demonstrated a key role for FBXO7 expressed in forebrain non-dopaminergic neurons in coordination with accurate movement in mice. The same study identified the PSMA2 core subunit of the proteasome as a substrate of FBXO7. $CRL1^{FBXO7}$ was thus found to mediate the polyubiquitylation of PSMA2 via a ubiquitin-Lys^{63} linkage, but this modification did not result in the degradation of PSMA2. Loss of FBXO7 resulted in the attenuation of the assembly and activity of the proteasome, indicating that FBXO7 promotes proteasome function through ubiquitylation of PSMA2. Although proteasome dysfunction has been linked to Parkinson's disease, it remains unclear whether dysregulation of the proteasome is the primary cause of the motor deficits in FBXO7 KO mice.

10.20 FBXO8

Downregulation of FBXO8 expression is frequently observed in glioma as well as in hepatocellular, gastric, and colorectal cancer (Wang et al. 2013, 2017; Wu et al. 2015; Yu et al. 2014). FBXO8 KO mice were found to be more susceptible than their wild-type littermates to the development of colorectal cancer induced by azoxymethane and dextran sodium sulfate (FeiFei et al. 2019). Proteomics analysis of proteins associated with FBXO8 identified GSTP1 (glutathione S-transferase π1), which was also shown to be ubiquitylated by FBXO8 and thereby targeted for degradation. The abundance of GSTP1 was shown to be negatively correlated with that of FBXO8 in both human and mouse colorectal cancer. In addition, FBXO8 knockdown increased cell proliferation and attenuated apoptosis in human colorectal cancer cells, and both of these effects were prevented by simultaneous knockdown of GSTP1. These results thus implicated GSTP1 as a primary target of FBXO8 for suppression of colorectal carcinogenesis.

10.21 FBXO11 (UBR6)

A screen for deafness in a large-scale mouse mutagenesis program identified the *Jeff* line with a mutation in *Fbxo11* (also known as *Ubr6*) that resulted in a glutamine-to-leucine substitution at amino acid position 491 (Hardisty-Hughes et al. 2006; Nolan et al. 2000). Deafness in *Jeff* heterozygotes was attributed to chronic inflammation of the middle ear that developed at weaning and raised the threshold for a cochlear nerve response (Hardisty et al. 2003). All *Jeff* homozygotes died shortly after birth as a result of respiratory problems (Hardisty-Hughes et al. 2006).

Biochemical analysis to identify FBXO11 substrates pinpointed BCL6, which plays an important role in the initiation and maintenance of the GC (germinal center) reaction in the spleen (Duan et al. 2012). Targeted deletion of *Fbxo11* in GC-derived B cells with the use of a *Cγ1–Cre* transgene resulted in the accumulation of BCL6 and an increase in the number of GC B cells, leading to the development of lymphoproliferative disease (Schneider et al. 2016). Given that these phenotypes are also observed in mice that overexpress BCL6 as a result of a chromosomal translocation that juxtaposes the BCL6 and immunoglobulin heavy chain genes and which is analogous to that detected in humans with diffuse large B-cell lymphoma (Cattoretti et al. 2005), BCL6 can be regarded as a primary substrate of FBXO11 for maintenance of GC homeostasis. Whether BCL6 also contributes to hearing loss in *Jeff* mice remains to be determined.

De novo heterozygous loss-of-function mutations in *FBXO11* were recently identified in individuals with syndromic intellectual disability (Fritzen et al. 2018; Gregor et al. 2018; Jansen et al. 2019). Further studies of FBXO11 KO mice focusing on the nervous system may shed light on the pathogenesis of this disease.

10.22 FBXO15

The *FBXO15* gene was identified as a target of the transcription factors Oct3/4 and Sox2, which are essential for maintenance of undifferentiated embryonic stem cells (Tokuzawa et al. 2003). Although FBXO15 is specifically expressed in early embryos as well as in adult reproductive organs, FBXO15 KO mice showed no gross abnormalities, with embryonic development and fertility appearing to be unaffected (Tokuzawa et al. 2003), indicating that FBXO15 is dispensable for normal development and physiology in mice.

10.23 FBXO22

Cellular senescence has emerged as a mechanism of proliferative arrest that limits tumorigenesis but also drives aging-related pathologies as a result, at in least in part, of cytokine secretion related to the SASP (senescence-associated secretory phenotype) (Hinds and Pietruska 2017). Transcriptome-wide analysis of senescent cells identified FBXO22 mRNA as a transcript that was upregulated at later time points of senescence induction (Johmura et al. 2016). The *FBXO22* gene was shown to be a transcriptional target of p53, a critical player in the induction of cellular senescence, and the FBXO22 protein was found to ubiquitylate p53 and thereby target it for degradation. This negative feedback loop serves to maintain an adequate level of p53 and to result in the upregulation of the cell cycle inhibitor $p16^{INK4A}$ and promotion of SASP. FBXO22 KO mice were found to be born but to be smaller than their wild-type littermates and to manifest accumulation of p53 and a concomitant reduction in the amount of p16 in all tissues tested. The SASP and aging-related phenotypes of these mice as well as their susceptibility to tumorigenesis remain to be investigated.

10.24 FBXO30

FBXO30 KO mice showed no gross abnormalities with the exception of a nursing problem evident in female mice (Liu et al. 2016). Histological analysis of pregnant FBXO30 KO mice revealed mammary gland atrophy with few glands containing milk droplets as a result of a defect in the production of mature luminal epithelial cells. Mass spectrometric analysis of FBXO30-associated proteins identified the motor protein Eg5 (also known as KIF11) as a substrate for FBXO30-mediated ubiquitylation and degradation, which occur specifically in S phase of the cell cycle, when FBXO30 is highly expressed. The accumulation of Eg5 apparent in FBXO30-deficient mammary gland epithelial cells in vitro was shown to result in centrosome amplification, the formation of multipolar or multiple bipolar spindles, abnormal separation of chromosomes, and cell cycle arrest. These phenotypes were rescued by depletion of Eg5 or treatment of these cells with an Eg5 inhibitor. Furthermore, Eg5 accumulated in the mammary glands of FBXO30 KO mice, and treatment of these animals with an Eg5 inhibitor rescued the defect in mammopoiesis. These observations thus supported the notion that Eg5 is a primary substrate of FBXO30 in mammary gland development.

10.25 FBXO32

FBXO32, also known as MAFbx (muscle atrophy F box protein) or atrogin-1, is a striated muscle-specific F-box protein whose expression in mice is induced by stimuli that trigger muscle atrophy such as food deprivation, immobilization, denervation, hind limb suspension, glucocorticoid treatment, and cachexia (Bodine and Baehr 2014; Bodine et al. 2001; Gomes et al. 2001). FBXO32 KO mice were found to grow to adulthood without obvious abnormalities, but they died at ~16–18 months of age as a result of congestive heart failure (Zaglia et al. 2014). Cutting of the sciatic nerve as a model of denervation results in a loss of gastrocnemius and tibialis anterior muscle mass, and this effect was attenuated in FBXO32 KO mice, further implicating FBXO32 in muscle atrophy (Bodine et al. 2001). Identified substrates of FBXO32 include the transcription factor MyoD (Tintignac et al. 2005) and eukaryotic translation initiation factor 3 subunit F (Lagirand-Cantaloube et al. 2008), but both of these proteins were identified in a cultured myoblast cell line, with further in vivo analyses being required to determine whether either plays a role in muscle sparing in FBXO32 KO mice.

10.26 FBXO38

Cancer cells acquire resistance to attack by cytotoxic T cells through expression of ligands that bind to receptors on the T cells and thereby suppress their function. One of the best characterized receptor–ligand pairs in this regard is PD-1 (programmed cell death-1) and PD-L1 (programmed cell death-ligand 1), and inhibitory antibodies to these proteins have recently been introduced for cancer immunotherapy (Sharma and Allison 2015). Proteome-wide analysis of PD-1 binding proteins identified FBXO38, which was shown to ubiquitylate PD-1 for degradation (Meng et al. 2018). Generation of T-cell-specific FBXO38 KO mice with the use of a *Cd4–Cre* transgene revealed that FBXO38 deficiency resulted in increased expression of PD-1 in tumor-infiltrating cytotoxic T cells and consequent acceleration of tumor growth. PD-1 blockade normalized the rate of tumor growth in these mice, supporting the notion that PD-1 is a primary substrate of FBXO38 in suppression of tumor progression.

10.27 FBXO45

FBXO45 contains an F-box domain that binds to SKP1, but, like FBXL10, it does not appear to associate with CUL1. Instead, it relies on another RING finger domain-containing protein—PAM in human, PHR in mouse—for the E2 recruitment required for ubiquitin ligase activity (Saiga et al. 2009). FBXO45 is specifically expressed in the nervous system, and FBXO45 KO mice die during embryogenesis as a result of abnormal innervation of the diaphragm, impaired synapse formation, and aberrant development of axon tracts in the brain, implicating FBXO45 in neural development (Saiga et al. 2009). FBXO45 has been shown to regulate neurotransmission by inducing the degradation of the synaptic protein Munc13-1 (Tada et al. 2010), suggesting that Munc13-1 accumulation may be responsible for the abnormal development of the nervous system in FBXO45 KO mice.

Phenotypes of FBXO protein KO mice with identified primary substrates are summarized in Table 10.3.

10.28 Summary

Although the in vivo functions of more than half of all F-box proteins remain to be investigated with KO mouse models, the available data for the other F-box proteins reviewed in this chapter indicate that the CRL1 ubiquitin ligase plays a key role in a wide variety of physiological and pathological processes in mice.

Consistent with the estimate that 25%–30% of all KO mouse models show embryonic or neonatal mortality (Adams et al. 2013), deletion of 11 of the 27 F-box protein genes examined to date has been found to result in death in utero as a result of extraembryonic (β-TrCP2, FBXW8, FBXL5, FBXL10, FBXL12, cyclin F), embryonic (EMI1, FBXO45), or both extraembryonic and embryonic (FBXW7) abnormalities. FBXL11 and FBXL20 KO mice remain to be further analyzed to determine whether extraembryonic or embryonic defects are responsible for the embryonic lethal phenotype.

Characterization of several KO mouse models (β-TrCP1+2, FBXL3, FBXL5, FBXL21, FBXO2, FBXO7, FBXO11, and FBXO45) has revealed a key role for CRL1 in nervous system development or function. The association of mutations in several human F-box protein genes with neurological diseases—such as mutations of the cyclin F gene with amyotrophic lateral sclerosis and frontotemporal dementia, of *FBXO7* with Parkinson's disease, and of *FBXO11* with syndromic intellectual disability—emphasizes the importance of CRL1 in the physiology and pathology of the brain. The roles of F-box proteins in cancer have also been well studied, with KO mice having been invaluable tools for

Table 10.3 Phenotypes of FBXO protein KO mice including identified primary substrates

Cyclin F (FBXO1)	Whole body	Embryonic lethality	ND	Tetzlaff et al. (2004a)
Cyclin F	Retinal cells	Not detected	NA	Tetzlaff et al. (2004a)
Cyclin F	Osteoblasts	Not detected	NA	Tetzlaff et al. (2004a)
Cyclin F	Intestinal and bladder epithelial cells	Not detected	NA	Tetzlaff et al. (2004a)
FBXO2	Whole body	Deafness	ND	Nelson et al. (2007)
FBXO4	Whole body	Not detected, or tumor development	Cyclin D1	Vaites et al. (2011) and Kanie et al. (2012)
FBXO5 (EMI1)	Whole body	Embryonic lethality	ND	Lee et al. (2006)
FBXO7	Whole body	Parkinsonism	ND	Vingill et al. (2016)
FBXO8	Whole body	Promotion of colorectal cancer	GSTP1	FeiFei et al. (2019)
FBXO11 (UBR6)	Whole body	Deafness (heterozygous), perinatal lethality (homozygous)	ND	Hardisty-Hughes et al. (2006) and Hardisty et al. (2003)
FBXO11	Germinal center B cells	Lymphoproliferative disease	BCL6	Schneider et al. (2016)
FBXO15	Whole body	Not detected	NA	Tokuzawa et al. (2003)
FBXO22	Whole body	Small size	p53	Johmura et al. (2016)
FBXO30	Whole body	Female nursing problem, abnormal mammary gland development	Eg5	Liu et al. (2016)
FBXO32	Whole body	Resistance to muscle atrophy, congestive heart failure	ND	Bodine et al. (2001) and Zaglia et al. (2014)
FBXO38	T cells	Enhanced tumor growth	PD-1	Meng et al. (2018)
FBXO43 (EMI2)	Whole body	Male and female sterility	ND	Gopinathan et al. (2017)
FBXO45	Whole body	Neonatal lethality, abnormal neural development	ND	Saiga et al. (2009)

ND not determined, *NA* not applicable

the evaluation of in which tissues and how these proteins contribute to tumor suppression or promotion (Wang et al. 2014).

Given that most F-box proteins remain functional orphans, the generation and characterization of KO mice for these proteins combined with biochemical analysis to identify relevant substrates will expand the range of physiological functions of CRL1. Further, such studies for all F-box proteins have the potential to provide a basis for the development of drugs that target F-box proteins themselves or their regulators or substrates. Such drugs may prove effective for the treatment of various diseases including neurodegenerative disorders and cancer.

Acknowledgment We thank H. Inuzuka and W. Wei for the discussion. This study was funded in part by KAKENHI grants (18H05215) from the Ministry of Education, Culture, Sports, Science and Technology (MEXT) of Japan.

References

Adams D, Baldock R, Bhattacharya S, Copp AJ, Dickinson M, Greene ND, Henkelman M, Justice M, Mohun T, Murray SA et al (2013) Bloomsbury report on mouse embryo phenotyping: recommendations from the IMPC workshop on embryonic lethal screening. Dis Model Mech 6:571–579

Akhoondi S, Sun D, von der Lehr N, Apostolidou S, Klotz K, Maljukova A, Cepeda D, Fiegl H, Dafou D, Marth C et al (2007) FBXW7/hCDC4 is a general tumor suppressor in human cancer. Cancer Res 67:9006–9012

Andricovich J, Kai Y, Peng W, Foudi A, Tzatsos A (2016) Histone demethylase KDM2B regulates lineage commitment in normal and malignant hematopoiesis. J Clin Invest 126:905–920

Arabi A, Ullah K, Branca RM, Johansson J, Bandarra D, Haneklaus M, Fu J, Aries I, Nilsson P, Den Boer ML et al (2012) Proteomic screen reveals Fbw7 as a modulator of the NF-kappaB pathway. Nat Commun 3:976

Arai T, Kasper JS, Skaar JR, Ali SH, Takahashi C, DeCaprio JA (2003) Targeted disruption of p185/

Cul7 gene results in abnormal vascular morphogenesis. Proc Natl Acad Sci U S A 100:9855–9860
Babaei-Jadidi R, Li N, Saadeddin A, Spencer-Dene B, Jandke A, Muhammad B, Ibrahim EE, Muraleedharan R, Abuzinadah M, Davis H et al (2011) FBXW7 influences murine intestinal homeostasis and cancer, targeting Notch, Jun, and DEK for degradation. J Exp Med 208:295–312
Baguma-Nibasheka M, Kablar B (2009) Abnormal retinal development in the Btrc null mouse. Dev Dyn 238:2680–2687
Bai C, Richman R, Elledge SJ (1994) Human cyclin F. EMBO J 13:6087–6098
Bai L, Chen MM, Chen ZD, Zhang P, Tian S, Zhang Y, Zhu XY, Liu Y, She ZG, Ji YX et al (2019) F-box/WD repeat-containing protein 5 mediates the ubiquitination of apoptosis signal-regulating kinase 1 and exacerbates nonalcoholic steatohepatitis in mice. Hepatology
Barbash O, Zamfirova P, Lin DI, Chen X, Yang K, Nakagawa H, Lu F, Rustgi AK, Diehl JA (2008) Mutations in Fbx4 inhibit dimerization of the SCF (Fbx4) ligase and contribute to cyclin D1 overexpression in human cancer. Cancer Cell 14:68–78
Bodine SC, Baehr LM (2014) Skeletal muscle atrophy and the E3 ubiquitin ligases MuRF1 and MAFbx/atrogin-1. Am J Physiol Endocrinol Metab 307:E469–E484
Bodine SC, Latres E, Baumhueter S, Lai VK, Nunez L, Clarke BA, Poueymirou WT, Panaro FJ, Na E, Dharmarajan K et al (2001) Identification of ubiquitin ligases required for skeletal muscle atrophy. Science 294:1704–1708
Boulard M, Edwards JR, Bestor TH (2016) Abnormal X chromosome inactivation and sex-specific gene dysregulation after ablation of FBXL10. Epigenetics Chromatin 9:22
Busino L, Bassermann F, Maiolica A, Lee C, Nolan PM, Godinho SI, Draetta GF, Pagano M (2007) SCFFbxl3 controls the oscillation of the circadian clock by directing the degradation of cryptochrome proteins. Science 316:900–904
Busino L, Millman SE, Scotto L, Kyratsous CA, Basrur V, O'Connor O, Hoffmann A, Elenitoba-Johnson KS, Pagano M (2012) Fbxw7alpha- and GSK3-mediated degradation of p100 is a pro-survival mechanism in multiple myeloma. Nat Cell Biol 14:375–385
Camaschella C (2015) Iron-deficiency anemia. N Engl J Med 372:1832–1843
Cattoretti G, Pasqualucci L, Ballon G, Tam W, Nandula SV, Shen Q, Mo T, Murty VV, Dalla-Favera R (2005) Deregulated BCL6 expression recapitulates the pathogenesis of human diffuse large B cell lymphomas in mice. Cancer Cell 7:445–455
D'Alessandro M, Beesley S, Kim JK, Jones Z, Chen R, Wi J, Kyle K, Vera D, Pagano M, Nowakowski R et al (2017) Stability of wake-sleep cycles requires robust degradation of the PERIOD protein. Curr Biol 27:3454–3467 e3458
D'Angiolella V, Esencay M, Pagano M (2013) A cyclin without cyclin-dependent kinases: cyclin F controls genome stability through ubiquitin-mediated proteolysis. Trends Cell Biol 23:135–140
Davis RJ, Welcker M, Clurman BE (2014) Tumor suppression by the Fbw7 ubiquitin ligase: mechanisms and opportunities. Cancer Cell 26:455–464
Dealy MJ, Nguyen KV, Lo J, Gstaiger M, Krek W, Elson D, Arbeit J, Kipreos ET, Johnson RS (1999) Loss of Cul1 results in early embryonic lethality and dysregulation of cyclin E. Nat Genet 23:245–248
Deshaies RJ, Joazeiro CA (2009) RING domain E3 ubiquitin ligases. Annu Rev Biochem 78:399–434
Duan S, Cermak L, Pagan JK, Rossi M, Martinengo C, di Celle PF, Chapuy B, Shipp M, Chiarle R, Pagano M (2012) FBXO11 targets BCL6 for degradation and is inactivated in diffuse large B-cell lymphomas. Nature 481:90–93
Erhardt JA, Hynicka W, DiBenedetto A, Shen N, Stone N, Paulson H, Pittman RN (1998) A novel F box protein, NFB42, is highly enriched in neurons and induces growth arrest. J Biol Chem 273:35222–35227
FeiFei W, HongHai X, YongRong Y, PingXiang W, JianHua W, XiaoHui Z, JiaoYing L, JingBo S, Kun Z, XiaoLi R et al (2019) FBX8 degrades GSTP1 through ubiquitination to suppress colorectal cancer progression. Cell Death Dis 10:351
Fleming RE, Ponka P (2012) Iron overload in human disease. N Engl J Med 366:348–359
Fodde R (2002) The APC gene in colorectal cancer. Eur J Cancer 38:867–871
Frescas D, Pagano M (2008) Deregulated proteolysis by the F-box proteins SKP2 and beta-TrCP: tipping the scales of cancer. Nat Rev Cancer 8:438–449
Fritzen D, Kuechler A, Grimmel M, Becker J, Peters S, Sturm M, Hundertmark H, Schmidt A, Kreiss M, Strom TM et al (2018) De novo FBXO11 mutations are associated with intellectual disability and behavioural anomalies. Hum Genet 137:401–411
Fukuda T, Tokunaga A, Sakamoto R, Yoshida N (2011) Fbxl10/Kdm2b deficiency accelerates neural progenitor cell death and leads to exencephaly. Mol Cell Neurosci 46:614–624
Fukushima H, Matsumoto A, Inuzuka H, Zhai B, Lau AW, Wan L, Gao D, Shaik S, Yuan M, Gygi SP et al (2012) SCF(Fbw7) modulates the NFkB signaling pathway by targeting NFkB2 for ubiquitination and destruction. Cell Rep 1:434–443
Godinho SI, Maywood ES, Shaw L, Tucci V, Barnard AR, Busino L, Pagano M, Kendall R, Quwailid MM, Romero MR et al (2007) The after-hours mutant reveals a role for Fbxl3 in determining mammalian circadian period. Science 316:897–900
Gomes MD, Lecker SH, Jagoe RT, Navon A, Goldberg AL (2001) Atrogin-1, a muscle-specific F-box protein highly expressed during muscle atrophy. Proc Natl Acad Sci U S A 98:14440–14445
Gong B, Chen F, Pan Y, Arrieta-Cruz I, Yoshida Y, Haroutunian V, Pasinetti GM (2010) SCFFbx2-E3-ligase-mediated degradation of BACE1 attenuates Alzheimer's disease amyloidosis and improves synaptic function. Aging Cell 9:1018–1031

Gopinathan L, Szmyd R, Low D, Diril MK, Chang HY, Coppola V, Liu K, Tessarollo L, Guccione E, van Pelt AMM et al (2017) Emi2 is essential for mouse spermatogenesis. Cell Rep 20:697–708

Gregor A, Sadleir LG, Asadollahi R, Azzarello-Burri S, Battaglia A, Ousager LB, Boonsawat P, Bruel AL, Buchert R, Calpena E et al (2018) De Novo variants in the F-box protein FBXO11 in 20 individuals with a variable neurodevelopmental disorder. Am J Hum Genet 103:305–316

Grim JE, Knoblaugh SE, Guthrie KA, Hagar A, Swanger J, Hespelt J, Delrow JJ, Small T, Grady WM, Nakayama KI et al (2012) Fbw7 and p53 cooperatively suppress advanced and chromosomally unstable intestinal cancer. Mol Cell Biol 32:2160–2167

Guardavaccaro D, Kudo Y, Boulaire J, Barchi M, Busino L, Donzelli M, Margottin-Goguet F, Jackson PK, Yamasaki L, Pagano M (2003) Control of meiotic and mitotic progression by the F box protein beta-Trcp1 in vivo. Dev Cell 4:799–812

Hardisty RE, Erven A, Logan K, Morse S, Guionaud S, Sancho-Oliver S, Hunter AJ, Brown SD, Steel KP (2003) The deaf mouse mutant Jeff (Jf) is a single gene model of otitis media. J Assoc Res Otolaryngol 4:130–138

Hardisty-Hughes RE, Tateossian H, Morse SA, Romero MR, Middleton A, Tymowska-Lalanne Z, Hunter AJ, Cheeseman M, Brown SD (2006) A mutation in the F-box gene, Fbxo11, causes otitis media in the Jeff mouse. Hum Mol Genet 15:3273–3279

Hentze MW, Muckenthaler MU, Galy B, Camaschella C (2010) Two to tango: regulation of Mammalian iron metabolism. Cell 142:24–38

Hinds P, Pietruska J (2017) Senescence and tumor suppression. F1000Res 6:2121

Hirano A, Yumimoto K, Tsunematsu R, Matsumoto M, Oyama M, Kozuka-Hata H, Nakagawa T, Lanjakornsiripan D, Nakayama KI, Fukada Y (2013) FBXL21 regulates oscillation of the circadian clock through ubiquitination and stabilization of cryptochromes. Cell 152:1106–1118

Hoeck JD, Jandke A, Blake SM, Nye E, Spencer-Dene B, Brandner S, Behrens A (2010) Fbw7 controls neural stem cell differentiation and progenitor apoptosis via Notch and c-Jun. Nat Neurosci 13:1365–1372

Inuzuka H, Shaik S, Onoyama I, Gao D, Tseng A, Maser RS, Zhai B, Wan L, Gutierrez A, Lau AW et al (2011) SCF(FBW7) regulates cellular apoptosis by targeting MCL1 for ubiquitylation and destruction. Nature 471:104–109

Jansen S, van der Werf IM, Innes AM, Afenjar A, Agrawal PB, Anderson IJ, Atwal PS, van Binsbergen E, van den Boogaard MJ, Castiglia L et al (2019) De novo variants in FBXO11 cause a syndromic form of intellectual disability with behavioral problems and dysmorphisms. Eur J Hum Genet

Jin J, Cardozo T, Lovering RC, Elledge SJ, Pagano M, Harper JW (2004) Systematic analysis and nomenclature of mammalian F-box proteins. Genes Dev 18:2573–2580

Johmura Y, Sun J, Kitagawa K, Nakanishi K, Kuno T, Naiki-Ito A, Sawada Y, Miyamoto T, Okabe A, Aburatani H et al (2016) SCF(Fbxo22)-KDM4A targets methylated p53 for degradation and regulates senescence. Nat Commun 7:10574

Kalia LV, Lang AE (2015) Parkinson's disease. Lancet 386:896–912

Kanarek N, Horwitz E, Mayan I, Leshets M, Cojocaru G, Davis M, Tsuberi BZ, Pikarsky E, Pagano M, Ben-Neriah Y (2010) Spermatogenesis rescue in a mouse deficient for the ubiquitin ligase SCF{beta}-TrCP by single substrate depletion. Genes Dev 24:470–477

Kanarek N, Grivennikov SI, Leshets M, Lasry A, Alkalay I, Horwitz E, Shaul YD, Stachler M, Voronov E, Apte RN et al (2014) Critical role for IL-1beta in DNA damage-induced mucositis. Proc Natl Acad Sci U S A 111:E702–E711

Kanatsu-Shinohara M, Onoyama I, Nakayama KI, Shinohara T (2014) Skp1-Cullin-F-box (SCF)-type ubiquitin ligase FBXW7 negatively regulates spermatogonial stem cell self-renewal. Proc Natl Acad Sci U S A 111:8826–8831

Kanie T, Onoyama I, Matsumoto A, Yamada M, Nakatsumi H, Tateishi Y, Yamamura S, Tsunematsu R, Matsumoto M, Nakayama KI (2012) Genetic reevaluation of the role of F-box proteins in cyclin D1 degradation. Mol Cell Biol 32:590–605

Kato A, Rouach N, Nicoll RA, Bredt DS (2005) Activity-dependent NMDA receptor degradation mediated by retrotranslocation and ubiquitination. Proc Natl Acad Sci U S A 102:5600–5605

Kawakami E, Tokunaga A, Ozawa M, Sakamoto R, Yoshida N (2015) The histone demethylase Fbxl11/Kdm2a plays an essential role in embryonic development by repressing cell-cycle regulators. Mech Dev 135:31–42

Kleiger G, Mayor T (2014) Perilous journey: a tour of the ubiquitin-proteasome system. Trends Cell Biol 24:352–359

Kossatz U, Dietrich N, Zender L, Buer J, Manns MP, Malek NP (2004) Skp2-dependent degradation of p27kip1 is essential for cell cycle progression. Genes Dev 18:2602–2607

Koyama-Nasu R, David G, Tanese N (2007) The F-box protein Fbl10 is a novel transcriptional repressor of c-Jun. Nat Cell Biol 9:1074–1080

Kuipers EJ, Grady WM, Lieberman D, Seufferlein T, Sung JJ, Boelens PG, van de Velde CJ, Watanabe T (2015) Colorectal cancer. Nat Rev Dis Primers 1:15065

Kwon YW, Kim IJ, Wu D, Lu J, Stock WA Jr, Liu Y, Huang Y, Kang HC, DelRosario R, Jen KY et al (2012) Pten regulates Aurora-A and cooperates with Fbxw7 in modulating radiation-induced tumor development. Mol Cancer Res 10:834–844

Lagirand-Cantaloube J, Offner N, Csibi A, Leibovitch MP, Batonnet-Pichon S, Tintignac LA, Segura CT, Leibovitch SA (2008) The initiation factor eIF3-f is a major target for atrogin1/MAFbx function in skeletal muscle atrophy. EMBO J 27:1266–1276

Lara-Gonzalez P, Kim T, Desai A (2017) Taming the beast: control of APC/C(Cdc20)-dependent destruction. Cold Spring Harb Symp Quant Biol 82:111–121
Lee H, Lee DJ, Oh SP, Park HD, Nam HH, Kim JM, Lim DS (2006) Mouse emi1 has an essential function in mitotic progression during early embryogenesis. Mol Cell Biol 26:5373–5381
Li Z, Pei XH, Yan J, Yan F, Cappell KM, Whitehurst AW, Xiong Y (2014) CUL9 mediates the functions of the 3M complex and ubiquitylates survivin to maintain genome integrity. Mol Cell 54:805–819
Lin DI, Barbash O, Kumar KG, Weber JD, Harper JW, Klein-Szanto AJ, Rustgi A, Fuchs SY, Diehl JA (2006) Phosphorylation-dependent ubiquitination of cyclin D1 by the SCF(FBX4-alphaB crystallin) complex. Mol Cell 24:355–366
Lin HK, Chen Z, Wang G, Nardella C, Lee SW, Chan CH, Yang WL, Wang J, Egia A, Nakayama KI et al (2010) Skp2 targeting suppresses tumorigenesis by Arf-p53-independent cellular senescence. Nature 464:374–379
Liu Y, Wang Y, Du Z, Yan X, Zheng P, Liu Y (2016) Fbxo30 regulates mammopoiesis by targeting the bipolar mitotic kinesin Eg5. Cell Rep 15:1111–1122
Madgwick S, Jones KT (2007) How eggs arrest at metaphase II: MPF stabilisation plus APC/C inhibition equals cytostatic factor. Cell Div 2:4
Malyukova A, Dohda T, von der Lehr N, Akhoondi S, Corcoran M, Heyman M, Spruck C, Grander D, Lendahl U, Sangfelt O (2007) The tumor suppressor gene hCDC4 is frequently mutated in human T-cell acute lymphoblastic leukemia with functional consequences for Notch signaling. Cancer Res 67: 5611–5616
Mao JH, Perez-Losada J, Wu D, Delrosario R, Tsunematsu R, Nakayama KI, Brown K, Bryson S, Balmain A (2004) Fbxw7/Cdc4 is a p53-dependent, haploinsufficient tumour suppressor gene. Nature 432:775–779
Marzio A, Puccini J, Kwon Y, Maverakis NK, Arbini A, Sung P, Bar-Sagi D, Pagano M (2019) The F-box domain-dependent activity of EMI1 regulates PARPi sensitivity in triple-negative breast cancers. Mol Cell 73:224–237 e226
Maser RS, Choudhury B, Campbell PJ, Feng B, Wong KK, Protopopov A, O'Neil J, Gutierrez A, Ivanova E, Perna I et al (2007) Chromosomally unstable mouse tumours have genomic alterations similar to diverse human cancers. Nature 447:966–971
Matsumoto A, Onoyama I, Sunabori T, Kageyama R, Okano H, Nakayama KI (2011) Fbxw7-dependent degradation of Notch is required for control of "stemness" and neuronal-glial differentiation in neural stem cells. J Biol Chem 286:13754–13764
Matsuoka S, Oike Y, Onoyama I, Iwama A, Arai F, Takubo K, Mashimo Y, Oguro H, Nitta E, Ito K et al (2008) Fbxw7 acts as a critical fail-safe against premature loss of hematopoietic stem cells and development of T-ALL. Genes Dev 22:986–991
Meng X, Liu X, Guo X, Jiang S, Chen T, Hu Z, Liu H, Bai Y, Xue M, Hu R et al (2018) FBXO38 mediates PD-1 ubiquitination and regulates anti-tumour immunity of T cells. Nature 564:130–135
Metzger MB, Hristova VA, Weissman AM (2012) HECT and RING finger families of E3 ubiquitin ligases at a glance. J Cell Sci 125:531–537
Morohoshi A, Nakagawa T, Nakano S, Nagasawa Y, Nakayama K (2019) The ubiquitin ligase subunit beta-TrCP in Sertoli cells is essential for spermatogenesis in mice. Dev Biol 445:178–188
Moroishi T, Nishiyama M, Takeda Y, Iwai K, Nakayama KI (2011) The FBXL5-IRP2 axis is integral to control of iron metabolism in vivo. Cell Metab 14:339–351
Muckenthaler MU, Rivella S, Hentze MW, Galy B (2017) A red carpet for iron metabolism. Cell 168:344–361
Muto Y, Nishiyama M, Nita A, Moroishi T, Nakayama KI (2017) Essential role of FBXL5-mediated cellular iron homeostasis in maintenance of hematopoietic stem cells. Nat Commun 8:16114
Nakagawa T, Nakayama K (2015) Protein monoubiquitylation: targets and diverse functions. Genes Cells 20:543–562
Nakagawa T, Araki T, Nakagawa M, Hirao A, Unno M, Nakayama K (2015) S6 kinase- and beta-TrCP2-dependent degradation of p19Arf is required for cell proliferation. Mol Cell Biol 35:3517–3527
Nakagawa T, Zhang T, Kushi R, Nakano S, Endo T, Nakagawa M, Yanagihara N, Zarkower D, Nakayama K (2017) Regulation of mitosis-meiosis transition by the ubiquitin ligase beta-TrCP in male germ cells. Development 144:4137–4147
Nakayama KI, Nakayama K (2006) Ubiquitin ligases: cell-cycle control and cancer. Nat Rev Cancer 6:369–381
Nakayama K, Nagahama H, Minamishima YA, Matsumoto M, Nakamichi I, Kitagawa K, Shirane M, Tsunematsu R, Tsukiyama T, Ishida N et al (2000) Targeted disruption of Skp2 results in accumulation of cyclin E and p27(Kip1), polyploidy and centrosome overduplication. EMBO J 19:2069–2081
Nakayama K, Hatakeyama S, Maruyama S, Kikuchi A, Onoe K, Good RA, Nakayama KI (2003) Impaired degradation of inhibitory subunit of NF-kappa B (I kappa B) and beta-catenin as a result of targeted disruption of the beta-TrCP1 gene. Proc Natl Acad Sci U S A 100:8752–8757
Nakayama K, Nagahama H, Minamishima YA, Miyake S, Ishida N, Hatakeyama S, Kitagawa M, Iemura S, Natsume T, Nakayama KI (2004) Skp2-mediated degradation of p27 regulates progression into mitosis. Dev Cell 6:661–672
Nelson RF, Glenn KA, Zhang Y, Wen H, Knutson T, Gouvion CM, Robinson BK, Zhou Z, Yang B, Smith RJ et al (2007) Selective cochlear degeneration in mice lacking the F-box protein, Fbx2, a glycoprotein-specific ubiquitin ligase subunit. J Neurosci 27: 5163–5171
Nishiyama M, Nita A, Yumimoto K, Nakayama KI (2015) FBXL12-mediated degradation of ALDH3 is essential

for trophoblast differentiation during placental development. Stem Cells 33:3327–3340

Nita A, Nishiyama M, Muto Y, Nakayama KI (2016) FBXL12 regulates T-cell differentiation in a cell-autonomous manner. Genes Cells 21:517–524

Nolan PM, Peters J, Strivens M, Rogers D, Hagan J, Spurr N, Gray IC, Vizor L, Brooker D, Whitehill E et al (2000) A systematic, genome-wide, phenotype-driven mutagenesis programme for gene function studies in the mouse. Nat Genet 25:440–443

O'Neil J, Grim J, Strack P, Rao S, Tibbitts D, Winter C, Hardwick J, Welcker M, Meijerink JP, Pieters R et al (2007) FBW7 mutations in leukemic cells mediate NOTCH pathway activation and resistance to gamma-secretase inhibitors. J Exp Med 204: 1813–1824

Okabe H, Lee SH, Phuchareon J, Albertson DG, McCormick F, Tetsu O (2006) A critical role for FBXW8 and MAPK in cyclin D1 degradation and cancer cell proliferation. PLoS One 1:e128

Onoyama I, Tsunematsu R, Matsumoto A, Kimura T, de Alboran IM, Nakayama K, Nakayama KI (2007) Conditional inactivation of Fbxw7 impairs cell-cycle exit during T cell differentiation and results in lymphomatogenesis. J Exp Med 204:2875–2888

Onoyama I, Suzuki A, Matsumoto A, Tomita K, Katagiri H, Oike Y, Nakayama K, Nakayama KI (2011) Fbxw7 regulates lipid metabolism and cell fate decisions in the mouse liver. J Clin Invest 121: 342–354

Pear WS, Miller JP, Xu L, Pui JC, Soffer B, Quackenbush RC, Pendergast AM, Bronson R, Aster JC, Scott ML et al (1998) Efficient and rapid induction of a chronic myelogenous leukemia-like myeloproliferative disease in mice receiving P210 bcr/abl-transduced bone marrow. Blood 92:3780–3792

Petroski MD, Deshaies RJ (2005) Function and regulation of cullin-RING ubiquitin ligases. Nat Rev Mol Cell Biol 6:9–20

Piva R, Liu J, Chiarle R, Podda A, Pagano M, Inghirami G (2002) In vivo interference with Skp1 function leads to genetic instability and neoplastic transformation. Mol Cell Biol 22:8375–8387

Poewe W, Seppi K, Tanner CM, Halliday GM, Brundin P, Volkmann J, Schrag AE, Lang AE (2017) Parkinson disease. Nat Rev Dis Primers 3:17013

Rajagopalan H, Jallepalli PV, Rago C, Velculescu VE, Kinzler KW, Vogelstein B, Lengauer C (2004) Inactivation of hCDC4 can cause chromosomal instability. Nature 428:77–81

Reavie L, Buckley SM, Loizou E, Takeishi S, Aranda-Orgilles B, Ndiaye-Lobry D, Abdel-Wahab O, Ibrahim S, Nakayama KI, Aifantis I (2013) Regulation of c-Myc ubiquitination controls chronic myelogenous leukemia initiation and progression. Cancer Cell 23:362–375

Reitsma JM, Liu X, Reichermeier KM, Moradian A, Sweredoski MJ, Hess S, Deshaies RJ (2017) Composition and regulation of the cellular repertoire of SCF ubiquitin ligases. Cell 171:1326–1339 e1314

Rotin D, Kumar S (2009) Physiological functions of the HECT family of ubiquitin ligases. Nat Rev Mol Cell Biol 10:398–409

Rouault TA (2013) Iron metabolism in the CNS: implications for neurodegenerative diseases. Nat Rev Neurosci 14:551–564

Saiga T, Fukuda T, Matsumoto M, Tada H, Okano HJ, Okano H, Nakayama KI (2009) Fbxo45 forms a novel ubiquitin ligase complex and is required for neuronal development. Mol Cell Biol 29:3529–3543

Sancho R, Jandke A, Davis H, Diefenbacher ME, Tomlinson I, Behrens A (2010) F-box and WD repeat domain-containing 7 regulates intestinal cell lineage commitment and is a haploinsufficient tumor suppressor. Gastroenterology 139:929–941

Santra MK, Wajapeyee N, Green MR (2009) F-box protein FBXO31 mediates cyclin D1 degradation to induce G1 arrest after DNA damage. Nature 459: 722–725

Schneider C, Kon N, Amadori L, Shen Q, Schwartz FH, Tischler B, Bossennec M, Dominguez-Sola D, Bhagat G, Gu W et al (2016) FBXO11 inactivation leads to abnormal germinal-center formation and lymphoproliferative disease. Blood 128:660–666

Sharma P, Allison JP (2015) The future of immune checkpoint therapy. Science 348:56–61

Shimizu K, Nihira NT, Inuzuka H, Wei W (2018) Physiological functions of FBW7 in cancer and metabolism. Cell Signal 46:15–22

Shojaee S, Sina F, Banihosseini SS, Kazemi MH, Kalhor R, Shahidi GA, Fakhrai-Rad H, Ronaghi M, Elahi E (2008) Genome-wide linkage analysis of a Parkinsonian-pyramidal syndrome pedigree by 500 K SNP arrays. Am J Hum Genet 82:1375–1384

Shoji S, Yoshida N, Amanai M, Ohgishi M, Fukui T, Fujimoto S, Nakano Y, Kajikawa E, Perry AC (2006) Mammalian Emi2 mediates cytostatic arrest and transduces the signal for meiotic exit via Cdc20. EMBO J 25:834–845

Siepka SM, Yoo SH, Park J, Song W, Kumar V, Hu Y, Lee C, Takahashi JS (2007) Circadian mutant overtime reveals F-box protein FBXL3 regulation of cryptochrome and period gene expression. Cell 129: 1011–1023

Smit JJ, Sixma TK (2014) RBR E3-ligases at work. EMBO Rep 15:142–154

Song JH, Schnittke N, Zaat A, Walsh CS, Miller CW (2008) FBXW7 mutation in adult T-cell and B-cell acute lymphocytic leukemias. Leuk Res 32:1751–1755

Sparks AB, Morin PJ, Vogelstein B, Kinzler KW (1998) Mutational analysis of the APC/beta-catenin/Tcf pathway in colorectal cancer. Cancer Res 58:1130–1134

Sumida Y, Yoneda M (2018) Current and future pharmacological therapies for NAFLD/NASH. J Gastroenterol 53:362–376

Swatek KN, Komander D (2016) Ubiquitin modifications. Cell Res 26:399–422

Tada H, Okano HJ, Takagi H, Shibata S, Yao I, Matsumoto M, Saiga T, Nakayama KI, Kashima H, Takahashi T et al (2010) Fbxo45, a novel ubiquitin

ligase, regulates synaptic activity. J Biol Chem 285:3840–3849
Takahashi JS (2017) Transcriptional architecture of the mammalian circadian clock. Nat Rev Genet 18:164–179
Takeishi S, Nakayama KI (2016) To wake up cancer stem cells, or to let them sleep, that is the question. Cancer Sci 107:875–881
Takeishi S, Matsumoto A, Onoyama I, Naka K, Hirao A, Nakayama KI (2013) Ablation of Fbxw7 eliminates leukemia-initiating cells by preventing quiescence. Cancer Cell 23:347–361
Tan M, Davis SW, Saunders TL, Zhu Y, Sun Y (2009) RBX1/ROC1 disruption results in early embryonic lethality due to proliferation failure, partially rescued by simultaneous loss of p27. Proc Natl Acad Sci U S A 106:6203–6208
Tetzlaff MT, Bai C, Finegold M, Wilson J, Harper JW, Mahon KA, Elledge SJ (2004a) Cyclin F disruption compromises placental development and affects normal cell cycle execution. Mol Cell Biol 24:2487–2498
Tetzlaff MT, Yu W, Li M, Zhang P, Finegold M, Mahon K, Harper JW, Schwartz RJ, Elledge SJ (2004b) Defective cardiovascular development and elevated cyclin E and Notch proteins in mice lacking the Fbw7 F-box protein. Proc Natl Acad Sci U S A 101:3338–3345
Thalmann R, Henzl MT, Thalmann I (1997) Specific proteins of the organ of Corti. Acta Otolaryngol 117:265–268
Thompson BJ, Buonamici S, Sulis ML, Palomero T, Vilimas T, Basso G, Ferrando A, Aifantis I (2007) The SCFFBW7 ubiquitin ligase complex as a tumor suppressor in T cell leukemia. J Exp Med 204: 1825–1835
Thompson BJ, Jankovic V, Gao J, Buonamici S, Vest A, Lee JM, Zavadil J, Nimer SD, Aifantis I (2008) Control of hematopoietic stem cell quiescence by the E3 ubiquitin ligase Fbw7. J Exp Med 205:1395–1408
Tintignac LA, Lagirand J, Batonnet S, Sirri V, Leibovitch MP, Leibovitch SA (2005) Degradation of MyoD mediated by the SCF (MAFbx) ubiquitin ligase. J Biol Chem 280:2847–2856
Tokuzawa Y, Kaiho E, Maruyama M, Takahashi K, Mitsui K, Maeda M, Niwa H, Yamanaka S (2003) Fbx15 is a novel target of Oct3/4 but is dispensable for embryonic stem cell self-renewal and mouse development. Mol Cell Biol 23:2699–2708
Tsunematsu R, Nakayama K, Oike Y, Nishiyama M, Ishida N, Hatakeyama S, Bessho Y, Kageyama R, Suda T, Nakayama KI (2004) Mouse Fbw7/Sel-10/Cdc4 is required for notch degradation during vascular development. J Biol Chem 279:9417–9423
Tsunematsu R, Nishiyama M, Kotoshiba S, Saiga T, Kamura T, Nakayama KI (2006) Fbxw8 is essential for Cul1-Cul7 complex formation and for placental development. Mol Cell Biol 26:6157–6169
Tsutsumi T, Kuwabara H, Arai T, Xiao Y, Decaprio JA (2008) Disruption of the Fbxw8 gene results in pre- and postnatal growth retardation in mice. Mol Cell Biol 28:743–751
Vaites LP, Lee EK, Lian Z, Barbash O, Roy D, Wasik M, Klein-Szanto AJ, Rustgi AK, Diehl JA (2011) The Fbx4 tumor suppressor regulates cyclin D1 accumulation and prevents neoplastic transformation. Mol Cell Biol 31:4513–4523
Vingill S, Brockelt D, Lancelin C, Tatenhorst L, Dontcheva G, Preisinger C, Schwedhelm-Domeyer N, Joseph S, Mitkovski M, Goebbels S et al (2016) Loss of FBXO7 (PARK15) results in reduced proteasome activity and models a parkinsonism-like phenotype in mice. EMBO J 35:2008–2025
Wang Y, Penfold S, Tang X, Hattori N, Riley P, Harper JW, Cross JC, Tyers M (1999) Deletion of the Cul1 gene in mice causes arrest in early embryogenesis and accumulation of cyclin E. Curr Biol 9:1191–1194
Wang F, Qiao Y, Yu J, Ren X, Wang J, Ding Y, Zhang X, Ma W, Ding Y, Liang L (2013) FBX8 acts as an invasion and metastasis suppressor and correlates with poor survival in hepatocellular carcinoma. PLoS One 8:e65495
Wang Z, Liu P, Inuzuka H, Wei W (2014) Roles of F-box proteins in cancer. Nat Rev Cancer 14:233–247
Wang FF, Zhang XJ, Yan YR, Zhu XH, Yu J, Ding Y, Hu JL, Zhou WJ, Zeng ZC, Liao WT et al (2017) FBX8 is a metastasis suppressor downstream of miR-223 and targeting mTOR for degradation in colorectal carcinoma. Cancer Lett 388:85–95
Wilkinson N, Pantopoulos K (2014) The IRP/IRE system in vivo: insights from mouse models. Front Pharmacol 5:176
Williams KL, Topp S, Yang S, Smith B, Fifita JA, Warraich ST, Zhang KY, Farrawell N, Vance C, Hu X et al (2016) CCNF mutations in amyotrophic lateral sclerosis and frontotemporal dementia. Nat Commun 7:11253
Wilson A, Murphy MJ, Oskarsson T, Kaloulis K, Bettess MD, Oser GM, Pasche AC, Knabenhans C, Macdonald HR, Trumpp A (2004) c-Myc controls the balance between hematopoietic stem cell self-renewal and differentiation. Genes Dev 18:2747–2763
Wu X, Johansen JV, Helin K (2013) Fbxl10/Kdm2b recruits polycomb repressive complex 1 to CpG islands and regulates H2A ubiquitylation. Mol Cell 49:1134–1146
Wu P, Wang F, Wang Y, Men H, Zhu X, He G, Ma W, Xin S, Wu J, Liao W et al (2015) Significance of FBX8 in progression of gastric cancer. Exp Mol Pathol 98:360–366
Yamauchi T, Nishiyama M, Moroishi T, Kawamura A, Nakayama KI (2017) FBXL5 inactivation in mouse brain induces aberrant proliferation of neural stem progenitor cells. Mol Cell Biol 37
Yan J, Yan F, Li Z, Sinnott B, Cappell KM, Yu Y, Mo J, Duncan JA, Chen X, Cormier-Daire V et al (2014) The 3M complex maintains microtubule and genome integrity. Mol Cell 54:791–804
Yao I, Takagi H, Ageta H, Kahyo T, Sato S, Hatanaka K, Fukuda Y, Chiba T, Morone N, Yuasa S et al (2007) SCRAPPER-dependent ubiquitination of active zone protein RIM1 regulates synaptic vesicle release. Cell 130:943–957

Yao I, Takao K, Miyakawa T, Ito S, Setou M (2011) Synaptic E3 ligase SCRAPPER in contextual fear conditioning: extensive behavioral phenotyping of Scrapper heterozygote and overexpressing mutant mice. PLoS One 6:e17317

Yoo SH, Mohawk JA, Siepka SM, Shan Y, Huh SK, Hong HK, Kornblum I, Kumar V, Koike N, Xu M et al (2013) Competing E3 ubiquitin ligases govern circadian periodicity by degradation of CRY in nucleus and cytoplasm. Cell 152:1091–1105

Yoshida Y, Chiba T, Tokunaga F, Kawasaki H, Iwai K, Suzuki T, Ito Y, Matsuoka K, Yoshida M, Tanaka K et al (2002) E3 ubiquitin ligase that recognizes sugar chains. Nature 418:438–442

Yoshida Y, Tokunaga F, Chiba T, Iwai K, Tanaka K, Tai T (2003) Fbs2 is a new member of the E3 ubiquitin ligase family that recognizes sugar chains. J Biol Chem 278:43877–43884

Younossi Z, Anstee QM, Marietti M, Hardy T, Henry L, Eslam M, George J, Bugianesi E (2018) Global burden of NAFLD and NASH: trends, predictions, risk factors and prevention. Nat Rev Gastroenterol Hepatol 15:11–20

Yu ZK, Gervais JL, Zhang H (1998) Human CUL-1 associates with the SKP1/SKP2 complex and regulates p21(CIP1/WAF1) and cyclin D proteins. Proc Natl Acad Sci U S A 95:11324–11329

Yu Y, Sun L, Ren N, Li Y, Rong L, Zhao G (2014) Downexpression of F box only protein 8 correlates with tumor grade and poor prognosis in human glioma. Int J Clin Exp Pathol 7:8071–8076

Yumimoto K, Matsumoto M, Oyamada K, Moroishi T, Nakayama KI (2012) Comprehensive identification of substrates for F-box proteins by differential proteomics analysis. J Proteome Res 11:3175–3185

Yumimoto K, Akiyoshi S, Ueo H, Sagara Y, Onoyama I, Ueo H, Ohno S, Mori M, Mimori K, Nakayama KI (2015) F-box protein FBXW7 inhibits cancer metastasis in a non-cell-autonomous manner. J Clin Invest 125:621–635

Zaglia T, Milan G, Ruhs A, Franzoso M, Bertaggia E, Pianca N, Carpi A, Carullo P, Pesce P, Sacerdoti D et al (2014) Atrogin-1 deficiency promotes cardiomyopathy and premature death via impaired autophagy. J Clin Invest 124:2410–2424

Zhou W, Wei W, Sun Y (2013) Genetically engineered mouse models for functional studies of SKP1-CUL1-F-box-protein (SCF) E3 ubiquitin ligases. Cell Res 23:599–619

Zindy F, Eischen CM, Randle DH, Kamijo T, Cleveland JL, Sherr CJ, Roussel MF (1998) Myc signaling via the ARF tumor suppressor regulates p53-dependent apoptosis and immortalization. Genes Dev 12: 2424–2433

The Roles of Cullin-2 E3 Ubiquitin Ligase Complex in Cancer

11

Xijuan Liu, Giada Zurlo, and Qing Zhang

Abstract

Posttranslational protein modifications play an important role in regulating protein stability and cellular function. There are at least eight Cullin family members. Among them, Cullin-2 forms a functional E3 ligase complex with elongin B, elongin C, RING-box protein 1 (RBX1, also called ROC1), as well as the substrate recognition subunit (SRS) to promote the substrate ubiquitination and degradàtion. In this book chapter, we will review Cullin-2 E3 ligase complexes that include various SRS proteins, including von Hippel Lindau (pVHL), leucine-rich repeat protein-1 (LRR-1), preferentially expressed antigen of melanoma (PRAME), sex-determining protein FEM-1 and early embryogenesis protein ZYG-11. We will focus on the VHL signaling pathway in clear cell renal cell carcinoma (ccRCC), which may reveal various therapeutic avenues in treating this lethal cancer.

Keywords

VHL · HIF · Cancer · Hypoxia · Cullin-2 · E3 ubiquitin ligase · ccRCC · pVHL

X. Liu
Lineberger Comprehensive Cancer Center, University of North Carolina School of Medicine, Chapel Hill, NC, USA

G. Zurlo
Lineberger Comprehensive Cancer Center, University of North Carolina School of Medicine, Chapel Hill, NC, USA

Department of Pathology, UT Southwestern Medical Center, Dallas, TX, USA

Q. Zhang (✉)
Lineberger Comprehensive Cancer Center, University of North Carolina School of Medicine, Chapel Hill, NC, USA

Department of Pharmacology, University of North Carolina, Chapel Hill, NC, USA

Department of Pathology and Laboratory Medicine, University of North Carolina, Chapel Hill, NC, USA

Department of Pathology, UT Southwestern Medical Center, Dallas, TX, USA
e-mail: Qing.Zhang@UTSouthwestern.edu

Abbreviations

Ankrd37	Ankyrin repeat domain 37
APC/C	Anaphase-promoting complex/cyclosome
ARNT	Aryl hydrocarbon receptor nuclear translocator
ccRCC	Clear cell renal cell carcinoma
ChIP-Seq	ChIP assays followed by deep sequencing
CKI	Cip/Kip CDK-inhibitor
COL4α2	Collagen IV alpha 2
CRL	Cullin-based ringlike
ECM	Extracellular matrix
EglNs	Egl nine homologs
GLUT1	Glucose uptake and metabolism
HREs	Hypoxia-responsive elements

Y. Sun et al. (eds.), *Cullin-RING Ligases and Protein Neddylation*, Advances in Experimental Medicine and Biology 1217, https://doi.org/10.1007/978-981-15-1025-0_11

LRR-1	Leucine-rich repeat protein-1
MDM2	Double minute 2 homolog
PD1	Programmed cell death protein 1
PDL1	Programmed cell death protein 1 ligand 1
PI3K	Phosphoinositide 3-kinase
PKB	Protein kinase B
PRAME	Preferentially expressed antigen of melanoma
pVHL	Von Hippel–Lindau protein
RBX1	RING-box protein 1
SRS	Substrate recognition subunit
ZHX2	Zinc fingers and homeoboxes 2

Posttranslational protein modifications are crucial for the regulation of protein stability and function (Wang et al. 2014). Protein stability is controlled mainly by Cullin-based ringlike (CRL) E3 ligase complexes. Cullin(s) serve as important scaffold components to form the functional E3 ubiquitin ligase, and there are at least eight Cullin family members reported so far, including Cullin-1, Cullin-2, Cullin-3, Cullin-4A, Cullin-4B, Cullin-5, Cullin-7, and Cullin-9 (Deshaies and Joazeiro 2009). Among them, Cullin-1-based E3 ligase is the most well studied, showed to be involved in various human diseases. However, until recently, the role of Cullin-2 in diseases has been understudied.

Cullin-2 forms a functional E3 ligase complex with elongin B, elongin C, RING-box protein 1 (RBX1, also called ROC1), as well as the substrate recognition subunit (SRS) (Cai and Yang 2016). Elongin B and C were originally reported as the key components for the transcriptional elongation, until they were found to form the complex with RBX1, Cullin-2 and one SRS called von Hippel-Lindau (pVHL) (Bradsher et al. 1993). In this case, Cullin-2 N-terminus binds with elongin B, elongin C, and pVHL, while the C-terminus binds with RBX1 (Kamura et al. 2004; Nguyen et al. 2015). RBX1 then promotes the neddylation of Cullin-2, which is essential for activating the CRL-mediated ubiquitin transfer from the ubiquitin-conjugating enzyme E2, and eventual ubiquitin chain formation on the substrate executed by E3 (Xu et al. 2015).

Besides pVHL, there are a few other SRS proteins that have been discovered to form a functional E3 ligase complex with Cullin-2. Interestingly, most of these proteins contain the domain called "VHL box," which is composed of an N-terminal BC box and a C-terminal Cullin-2 box. The BC box is mainly responsible for binding with elongin B and C. This box has a consensus sequence of (S,T,P)LXXX(C,S,A)XXXΦ, where X means any amino acid, while Φ refers to hydrophobic amino acids (Mahrour et al. 2008; Conaway et al. 1998; Kamura et al. 1998). The most important amino acid is the Leucine at the 2nd position. The Cullin-2 box has a consensus sequence of ΦPXXΦXXXΦ, which is mainly responsible for binding with Cullin-2 (Mahrour et al. 2008; Conaway et al. 1998; Kamura et al. 1998; Wang et al. 2016). In this book chapter, the function of these SRS proteins and their disease relevance will be discussed.

11.1 LRR-1

Leucine-rich repeat protein-1 (LRR-1) was originally found to be a nuclear protein in *C. elegans* that contains the canonical elongin BC and Cullin-2 boxes (Burger et al. 2013; Starostina et al. 2010). Therefore, it was considered that LRR-1 may form a functional E3 ligase. Further research showed that LRR-1 could promote the degradation of Cip/Kip CDK-inhibitor (CKI) p21 in *C. elegans* (Starostina et al. 2010). Mutation of Cullin-2 in germ cells led to p21 accumulation and G1-phase cell cycle arrest. Interestingly, human LRR-1 also degrades p21, but does not affect cell cycle. It was later determined that human LRR-1 only promotes the degradation of cytoplasmic but not nuclear p21. Cytoplasmic p21 leads to decreased Rho/ROCK/LIMK pathway, which is responsible for cofilin phosphorylation (Starostina et al. 2010). Therefore, LRR-1 depletion in human cells leads to increased cytoplasmic p21 accumulation and decreased cofilin phosphorylation, therefore promoting cell motility. In addition, LRR-1 was previously reported to degrade the HORMA domain (named after the Hop1p, Rev7p, and MAD2 proteins) containing protein HTP-3 (Burger et al. 2013). HTP-3 is important to regulate progression through meiotic prophase in *C. elegans* (Lui and Colaiacovo 2013). Therefore, Cullin-2 may play an important role in the development of the germline in nematodes, which remains to be determined.

11.2 PRAME

Similar to pVHL and LRR-1, preferentially expressed antigen of melanoma (PRAME) also contains a VHL box. PRAME was found to be overexpressed in many different cancers and exert its oncogenic function partially by repressing retinoic acid receptor signaling activity in these cells (Kilpinen et al. 2008; Epping et al. 2005). High PRAME expression correlates with worse prognosis in several different cancers, including breast cancer and neuroblastoma (Doolan et al. 2008; Epping et al. 2008; Oberthuer et al. 2004). By performing mass spectrometry for identifying the potential interactors of PRAME, Cullin-2, elongin B, C, and RING finger protein RBX1 were found, suggesting that PRAME is a functional component of the Cullin-2-based E3 ligase complex (Costessi et al. 2011). On the other hand, it was found that PRAME preferentially binds with chromatin. ChIP assays followed by deep sequencing (ChIP-Seq) revealed that PRAME frequently occupies a transcriptionally active promoter region that is often bound by NFY complexes, including NFYA and NFYB, which is frequently expressed in cycling cells (Maity and de Crombrugghe 1998). In addition, PRAME was found to be enriched at enhancer regions nearby, suggesting that the role of PRAME in transcriptional regulation could be multifaceted. Interestingly, the potential substrates of PRAME remain to be determined. Their identification would likely uncover the molecular mechanism by which PRAME serves as a potential oncogene in many types of human cancers.

11.3 FEM-1

By performing mass spectrometry to identify Cullin-2-associated proteins in *C. elegans*, it was found that FEM-1 binds with Cullin-2 (Starostina et al. 2007). In addition, FEM-1 also contains a VHL-box motif. Interestingly, Cullin-2 null mutant males exhibit a feminization phenotype, mainly due to the accumulation of TRA-1 as a result of lack of degradation by Cullin-2-based E3 ligase complex, which includes FEM-1 as the SRS, and FEM-2/3 as cofactors (Starostina et al. 2007). Therefore, TRA-1 is a ubiquitin substrate for FEM-1 in *C. elegans*. FEM1b and zinc finger protein GLI1 are mammalian orthologs for FEM-1 and TRA-1, respectively. Consistently, FEM1b was found to be the critical component of Cullin-2-based E3 ligase, promoting GLI1 ubiquitination and degradation (Gilder et al. 2013). Additionally, mouse FEM1b interacts with ankyrin repeat domain 37 (Ankrd37) and promotes its degradation (Shi et al. 2011).

11.4 ZYG-11

There are two ZYG-11 family members in *C. elegans*, including ZYG-11 and ZER-1. Both family members contain a VHL-box on the N-terminus that binds to elongin C and Cullin-2- (Vasudevan et al. 2007). There are three human family members, including ZYG11a, ZYG11b, and ZYG11bl. Among them, ZYG11b and ZYG11bl have a VHL-box motif and bind to elongin C and Cullin-2 (Wang et al. 2016). It is likely that ZYG-11 family members can form the functional CRL E3 ligase complex. Recently, ZYG-11 in *C. elegans* was reported to promote cyclin B1 ubiquitination and degradation (Balachandran et al. 2016). This regulation is also conserved in human cells. It is important to point out that ZYG11a and ZYG11b promote cyclin B1 ubiquitination and degradation during mitosis, which is independent of canonical cyclin B1 degradation controlled by the anaphase-promoting complex/cyclosome (APC/C) E3 ubiquitin ligase complex (Balachandran et al. 2016). These results show that ZYG-11 family members may play important roles in regulating mitosis of human cells.

11.5 pVHL

Von Hippel–Lindau (*VHL*) gene, located on chromosome 3p25–26 (Chittiboina and Lonser 2015), was first identified and cloned in 1993 (Latif et al. 1993). *VHL* consists of three exons and is widely expressed in both fetal and adult tissues. Von Hippel–Lindau protein (pVHL), product of *VHL*, is composed of 213 amino acid residues and has a molecular weight of 24–30 kDa. A smaller pVHL isoform of approximately 19 kDa is produced as a result of internal translational initiation from codon 54 (Iliopoulos et al. 1998). pVHL is a critical tumor suppressor in clear cell renal cell carcinoma (ccRCC), which is the predominant form of kidney cancer accounting for more than 85% of all cases (Zhang et al. 2018). Located primarily in the cytoplasm, pVHL can also be found in the nucleus, the mitochondria, and in association with the endoplasmic reticulum (Kaelin 2002). pVHL shuttles between the nucleus and the cytoplasm and cannot suppress tumor growth when the nuclear-cytoplasmic trafficking is artificially inhibited (Lee et al. 1999). pVHL displays no enzymatic activity, but forms a complex, called $CRL2^{VHL}$, with elongin B, elongin C, Cullin-2, and RBX1 which has E3 ubiquitin ligase activity. pVHL consists of an α- and a β-domain; the former binds to elongin C, which in turn binds to elongin B and Cullin-2, whereas the latter recognizes and binds to substrates for ubiquitination and degradation (Cardote et al. 2017).

11.5.1 VHL Disease

VHL disease, named after Eugen von Hippel and Arvid Lindau, was first described in the early 1990s (Robinson and Ohh 2014). Affecting approximately 1 in 35,000 people, it is a hereditary cancer syndrome characterized by one or more of the following types of tumor: retinal or central nervous system hemangioblastoma, pheochromocytoma, clear cell renal cell carcinoma (ccRCC), pancreatic cystadenoma, and endolymphatic sac tumor of the inner ear (Lee et al. 1998). VHL disease is caused by mutations in the *VHL* tumor-suppressor gene: more than 700 *VHL* mutations have been identified, located throughout the three exons (Nordstrom-O'Brien et al. 2010). Approximately 30% to 38% of VHL patients have missense mutations, 20% to 37% large or partial germline deletions, and 23% to 27% nonsense or frameshift mutations (Stolle et al. 1998). The tumors associated with VHL disease are typically benign with the exception of ccRCC, which remains the major cause of both morbidity and mortality for VHL patients. In addition, *VHL* mutations are also implicated in sporadic ccRCC and hemangioblastoma development (Tarade and Ohh 2018).

VHL patients are typically *VHL* heterozygotes, carrying both the wild-type and the inactivated *VHL* allele. Somatic inactivation or loss of the remaining wild-type *VHL* allele could lead to tumor or cyst (Kim and Kaelin 2004). Clinically, multiple subtypes of VHL disease can be subdivided according to the risk level to develop hemangioblastoma, pheochromocytoma, or ccRCC. Patients are first subdivided in Type 1 and Type 2 based on their propensity, low and high, respectively, to develop pheochromocytoma. Type 1 patients frequently harbor *VHL* deletions, nonsense, or frameshift mutations and develop hemangioblastoma and ccRCC. Type 2 VHL disease is associated with missense mutations and can be further subclassified based on low (Type 2A) and high (Type 2B) risk of renal cell carcinoma. Some type 2 VHL disease develop pheochromocytoma without the other stigmata of VHL disease (type 2C) (Tarade and Ohh 2018; Kim and Kaelin 2004). However, practitioners do not rely on this classification because families can move between clinical subtypes (Kim and Zschiedrich 2018). Besides, *VHL* mutations can also drive autosomal-recessive polycythemia: for instance, R200W *VHL* mutation causes Chuvash polycythemia endemic to the Chuvash Autonomous Republic in Russia (Ang et al. 2002), and H191D *VHL* causes an additional polycythemia disorder endemic to Croatia. These *VHL* mutation-mediated polycythemia disorders are sometimes referred to as Type 3 VHL disease (Pastore et al. 2003).

11.5.2 pVHL E3 Ubiquitin Ligase-Dependent Function

11.5.2.1 pVHL and the α Subunits of Hypoxia-Inducible Factor HIF-α

The canonical targets of pVHL are the α subunits of hypoxia-inducible factor (HIF), HIF-1α, HIF-2α, and HIF-3α, which contribute substantially to the transforming phenotype of ccRCC (Robinson and Ohh 2014). HIF-αs are unstable and form the HIF transcriptional factor with aryl hydrocarbon receptor nuclear translocator (ARNT, or HIF-1β) (Cai and Yang 2016). The role of pVHL in the regulation of HIF has been well documented. Under normoxic condition, HIF-αs are hydroxylated, recognized by the pVHL E3 ubiquitin ligase complex, polyubiquitinated, and degraded by the proteasome (Vriend and Reiter 2016). HIF hydroxylation on proline residues 402 and 564, which is vital for pVHL binding, depends on the α-ketoglutarate-dependent dioxygenases Egl nine homologs (EglNs) (Choudhry and Harris 2018) (Fig. 11.1). Of the three EglN family members EglN1, EglN2, and EglN3 (alternatively called PHD2, PHD1 and PHD3, respectively), EglN1/PHD2 is the main enzyme responsible of HIF-1α hydroxylation (Zurlo et al. 2016). Under hypoxia, EglN activity is inhibited and HIF-1α are not hydroxylated. Non-hydroxylated HIF-1α accumulates and translocates to the nucleus. Here, it heterodimerizes with ARNT (HIF-1β)

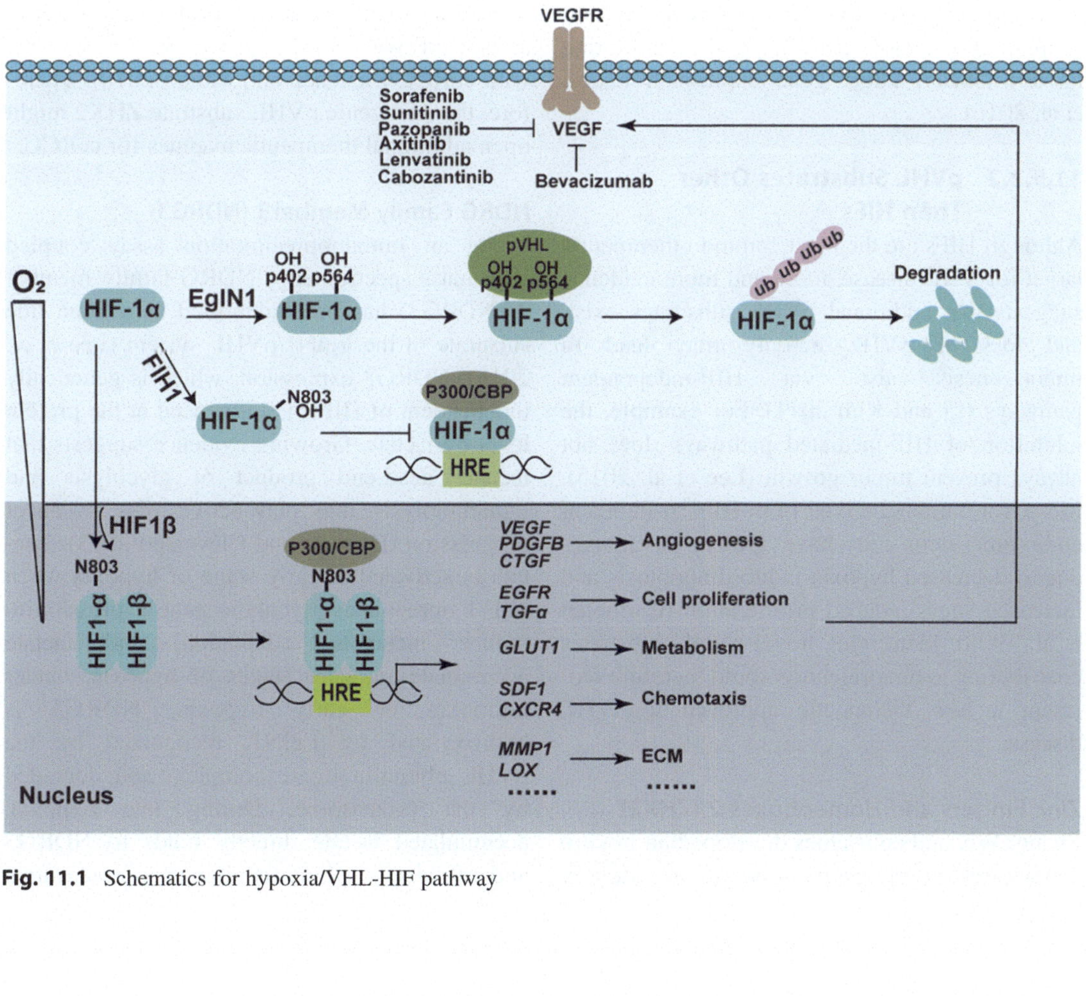

Fig. 11.1 Schematics for hypoxia/VHL-HIF pathway

and associates with coactivator(s) p300/CBP to form an active complex that binds to hypoxia-responsive elements (HREs). In this case, HIF-α can lead to tumorigenesis by activating its downstream cascade, including proteins responsible for cell proliferation (e.g., transforming growth factor TGFα and epidermal growth factor receptor EGFR), angiogenesis (e.g., vascular endothelial growth factor VEGF, platelet-derived growth factor B PDGFB, and connective tissue growth factor CTGF), glucose uptake and metabolism (GLUT1), and chemotaxis (stromal cell-derived factor SDF1 and its receptor C-X-C chemokine receptor 4 CXCR4) (Semenza 2003) (Fig. 11.1). Additionally, HIF can activate genes that affect the tumor microenvironment, such as the extracellular matrix (ECM) formation and turnover (e.g., matrix metallopeptidase 1 MMP1 and lysyl oxidase LOX) (Lee et al. 1998). Importantly, HIF-1α can also be hydroxylated on asparagine-803 by factor-inhibiting HIF (FIH), which prevents HIF-1α association with the transcriptional activators p300/CBP, thus inhibiting its downstream target gene expression (Zurlo et al. 2016).

11.5.2.2 pVHL Substrates Other Than HIFs

Although HIFs are the most common therapeutic target for VHL disease, more and more evidence suggests that additional pVHL substrates exist, and loss of pVHL activity may lead to tumorigenesis also via HIF-independent pathways (Li and Kim 2011). For example, the inhibition of HIF-mediated pathways does not always prevent tumor growth (Lee et al. 2015). Moreover, tumors derived from HIF-1α-deficient embryonic stem cells have growth advantages due to decreased hypoxia-induced apoptosis and increased stress-induced proliferation (Carmeliet et al. 1998). Identifying novel pVHL substrates contributing to tumorigenesis could therefore represent a new therapeutic approach for VHL diseases.

Zinc Fingers and Homeoboxes 2 (ZHX2)

Zhang, Wu, and colleagues developed an in vitro genome-wide screening strategy to identify proteins binding with pVHL complexes which competed with hydroxylated but not non-hydroxylated HIF-1α peptide. A genome-wide human cDNA library, which was divided into approximately 700 pools with 24 cDNAs/pool, allowed the discovery of proteins of which the binding to pVHL can be displaced by the p-OH HIF1α peptide. Based on this strategy, the zinc fingers and homeoboxes 2 (ZHX2) transcription factor was identified as a pVHL substrate (Zhang and Zhang 2018). ZHX2 accumulates in the nucleus of pVHL-deficient ccRCCs, but not in pVHL–wild-type ccRCCs or non-tumor tissues, and decreases pVHL-deficient ccRCC cell growth in vitro and in vivo. Mechanistically, ZHX2 binds to the NF-κB RELA/p65 subunit and controls RELA/p65 nuclear localization, thus promoting the expression of canonical NF-κB target genes (Fig. 11.2). Of note, ChIP-sequencing studies demonstrated the co-occupancy of ZHX2 and RELA/p65 at sites that were enriched for NF-κB consensus motifs in active promoters of genes associated with worse prognosis for patients with ccRCC (Sanchez and Simon 2018). Therefore, the oncogenic pVHL substrate ZHX2 might open additional therapeutic avenues for ccRCC.

NDRG Family Member3 (NDRG3)

Using an immunoprecipitation assay coupled with mass spectrometry, NDRG family member 3 (NDRG3) has been identified as a bona fide substrate of the EglN1/pVHL system (Lee et al. 2015). NDRG3 expression, which is genetically independent of HIFs, is influenced at the protein level by lactate. Growing evidence suggests that lactate, dead-end product of glycolysis and glutaminolysis, may play active roles in cancer progression (Doherty and Cleveland 2013). Lactate is activated at early stage of hypoxia when HIF-1 upregulates glycolytic gene expression to induce metabolic adaptation, and lactate accumulates at later stages of hypoxia. Under normoxia or early hypoxia, NDRG3 is hydroxylated by EglN1, recognized by the pVHL ubiquitin ligase complex, and degraded by the proteasome. During late hypoxia, accumulated lactate directly binds to NDRG3 and inhibits its ubiquitination and proteasomal

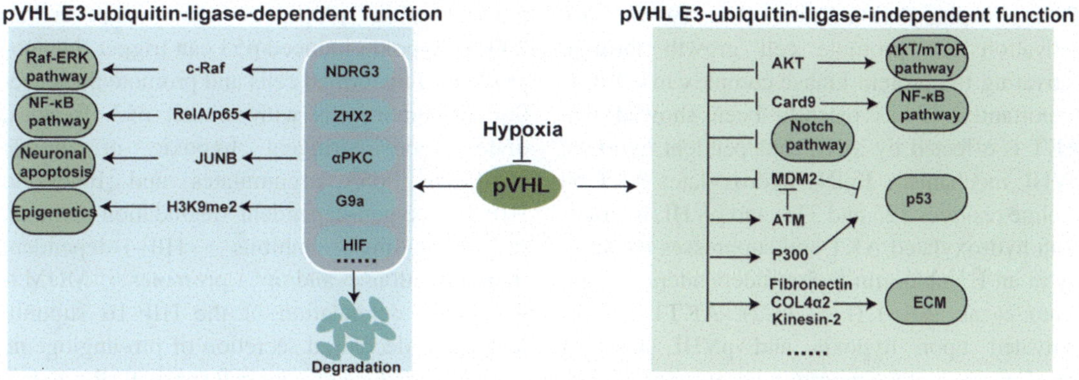

Fig. 11.2 Schematics for pVHL E3 ubiquitin-dependent/ pVHL E3 ubiquitin-independent functions

degradation by disrupting the NDRG3-pVHL interaction. Stabilized NDRG3 interacts with the proto-oncogene c-Raf and induces its phosphorylation, therefore promoting the activation of downstream extracellular signal-regulated kinase ERK1/2 signaling contributing to hypoxic cell growth and angiogenesis. Thus, NDRG3 acts as a lactate sensor that triggers downstream kinase signaling in a hypoxia-dependent manner, and the NDRG3-Raf-ERK axis provides the genetic basis for the lactate-induced hypoxia responses (Fig. 11.2). This study provides an additional model where metabolism wires to signal transduction for downstream gene regulation and corresponding cell responses. The lactate-NDRG3-Raf-ERK signaling pathway provides an extended mechanistic clue of disorders caused by *VHL* mutations (Park et al. 2015). A combinatorial targeting of HIF and NDRG3 might prove effective in cancer therapy.

Other substrates of pVHL have also been identified, including the atypical protein kinase C aPKC (Okuda et al. 2001), which is important for JUNB regulation and is involved in pheochromocytoma (Lee et al. 2005), erythropoietin receptor EPOR (Heir et al. 2016), B-Myb (Okumura et al. 2016), filamin A FLNA (Segura et al. 2016), centrosomal protein 68 Cep68 (Yin et al. 2017), retinal binding protein 1 RBP1 (Kuznetsova et al. 2003), retinal binding protein 7 RPB7 (Na et al. 2003), sprouty2 (Anderson et al. 2011), the β2-adrenergic receptor (Xie et al. 2009), ceramide kinase-like (CERKL) (Chen et al. 2015), and G9a methyltransferase (Casciello et al. 2017) (Fig. 11.2). To understand the physiological functions of pVHL, new substrates and the detailed molecular mechanism need to be further investigated.

11.5.3 pVHL E3 Ubiquitin Ligase-Independent Function

Not all proteins recognized by pVHL are labeled for proteasomal degradation (Fig. 11.2). In fact, certain *VHL* mutations associated with cancer pathogenesis encode for a pVHL displaying normal ubiquitination function. Moreover, type 2C pVHL mutants, associated with pheochromocytoma, maintain the ability to poly-ubiquitinate HIF (Hoffman et al. 2001), indicating that pVHL has other targets that are critical for tumor suppression.

11.5.3.1 Akt

The Ser/Thr kinase AKT, also known as protein kinase B (PKB), is present in three isoforms conserved across mammalian genomes: AKT1 (PKBα), AKT2 (PKBβ), and AKT3 (PKBγ). AKT1 is phosphorylated by phosphoinositide 3-kinase (PI3K) on two key residues, T308 and S473. Phosphorylation also occurs on corresponding residues in AKT2 (T309 and S474) and AKT3 (T305 and S472) (Manning and Toker 2017). AKT1 maximal activation requires S473 phosphorylation, which stabilizes

T308 phosphorylation (Alessi et al. 1996). AKT activation can promote cell growth through activating the protein kinase complex mTORC1. Importantly, it has recently been showed that AKT is affected by a HIF-independent hypoxia/pVHL mechanism. EglN1 hydroxylates AKT on proline residues 125 and 313, and pVHL interacts with hydroxylated AKT and suppresses its activity in an E3 ubiquitin ligase-independent fashion (Guo et al. 2016) (Fig. 11.2). AKT1 is thus activated upon hypoxia and pVHL loss. A Pan-Cancer proteogenomic atlas in over 10,000 human cancers across 32 types also revealed that *VHL* mutations are associated with highly active PI3K-AKT-mTOR pathway (Zhang et al. 2017). These studies demonstrate that hypoxia and *VHL* defects lead to AKT activation, suggesting that targeting the AKT oncogenic pathway may be effective in patients with pVHL-deficient or hypoxic tumors.

11.5.3.2 p53

TP53, the most commonly mutated gene in human cancers, encodes a transcription factor, named p53, which mediates cell cycle arrest or apoptosis in response to DNA damage and other stimuli (Levine et al. 2006). Oncoprotein double minute 2 homolog (MDM2) shuttles p53 from the nucleus to the cytosol and polyubiquitinates p53 for degradation (Mendoza et al. 2014). In response to DNA damage, p53 phosphorylation by the ATM serine/threonine kinase prevents MDM2 binding, thus activating p53 activity. In addition, acetylation of p53 on several C-terminal lysine residues by the histone acetyltransferase p300 promotes p53 stabilization and transcriptional activation (Kobet et al. 2000). Roe and colleagues demonstrated that pVHL α domain interacts with p53, therefore blocking the MDM2-mediated p53 ubiquitination and nuclear export in a HIF-independent manner. pVHL also independently interacts with ATM and p300 to promote their enzymatic modification of p53 (Fig. 11.2). Consequently, absence of wild-type pVHL leads to reduced p53 protein levels and p53-mediated transcription (Roe et al. 2006). Another study also revealed that hypoxic stress mediates p53 accumulation, but not its transactivation properties (Koumenis et al. 2001). Hypoxia-induced p53 can trigger the apoptosis of transformed cells and promote the selection of tumor cells with mutant p53. Besides, under more stringent hypoxic or anoxic conditions, p53 accumulates and binds to HIF-1α oxygen-dependent degradation domain. p53 binding inhibits HIF-1-dependent transactivation and/or promotes MDM2-dependent degradation of the HIF-1α subunit, leading to decreased secretion of pro-angiogenic actors and reduced tumor cell survival (Ravi et al. 2000). These studies, showing loss of p53 activity in the absence of pVHL, may partially explain why RCCs are relatively resistant to chemotherapy and radiation despite the fact that many lack *TP53* mutations. These findings highlight a new mechanism through which p53 can affect cancer progression and response to treatment.

11.5.3.3 Tumor Microenvironment

HIFs activate the expression of genes that affect the tumor microenvironment, such as the extracellular matrix (ECM). However, accumulating evidence shows that pVHL directly regulates the tumor microenvironment in a HIF-independent manner. pVHL interacts with fibronectin, a glycoprotein that interacts with integrin proteins to regulate the ECM, and loss of the pVHL-fibronectin interaction is associated with defective ECM formation. Moreover, the reintroduction of pVHL in pVHL-negative renal cancer cells, where the lack of ECM is thought to promote and maintain tumor angiogenesis, partially restores their ability to form a fibronectin matrix (Ohh et al. 1998). pVHL also binds to collagen IV alpha 2 (COL4α2), and pVHL loss is associated with a loss of COL4α2 from the ECM (Kurban et al. 2008; Grosfeld et al. 2007). Besides, pVHL associates and binds to microtubules and inhibits their depolymerization (Hergovich et al. 2003). Lolkema et al. have suggested that pVHL interacts with microtubules indirectly via through kinesin-2 (Lolkema et al. 2007) (Fig. 11.2). Loss of pVHL has also been implicated in the loss of primary cilia, which is associated with renal cyst formation and a common occurrence in VHL disease (Lubensky et al. 1996). The implications of these regulatory functions of pVHL for VHL disease remain to be determined.

11.5.3.4 Other Functions

pVHL has also been associated to the regulation of nuclear factor-κB (NF-κB). In fact, pVHL interacts with and negatively regulates NF-κB agonists, IκB kinase β, and Card9. pVHL acts as an adaptor protein and prevents activating phosphorylation or promotes inhibitory phosphorylation. pVHL loss results in increased NF-κB activity, which subsequently promotes oncogenesis in part via the activation of anti-apoptotic and proliferative pathways (Cummins et al. 2006; Yang et al. 2007). Further studies demonstrated that pVHL is also implicated in Notch signal pathway, with its loss resulting in constitutive activation of Notch in ccRCC (Sjolund et al. 2008) (Fig. 11.2). The physiological and pathophysiological mechanisms of these targets need to be investigated further for a deeper understanding of pVHL functions.

11.5.4 pVHL and Epigenetics

In recent years, the role of pVHL in epigenetics has gained more attention. McRonald and colleagues investigated promoter-specific methylation in RCC patient tissues and found significant changes in their DNA methylation levels compared to normal tissue (McRonald et al. 2009). Histone methylation is dynamically regulated by histone methyltransferases and demethylases. These modifications are important epigenetic phenomena, resulting either in active or inactive chromatin states. G9a can methylate histone H3 on lysine 9 (H3K9) from an unmodified state to a dimethylated state (H3K9me2). In breast cancer, G9a hydroxylation is inhibited and VHL cannot interact with G9a under hypoxia, which leads to increased H3K9 dimethylation (Casciello et al. 2017). In RCC, the mechanism between G9a and pVHL needs further research. pVHL loss causes the transcriptional activation of HIF target genes, including many genes that encode histone demethylases, such as Jumonji proteins. HIF-1α directly binds to the promoters of Jumonji family proteins under a hypoxic environment, and pVHL inactivation has been shown to modulate protein levels of several histone modifiers, including JARID1C, HDAC1, JMJD1A, JMJD2B, and JMJD2C (Xia et al. 2009; Yao et al. 2017). Chakraborty and colleagues found that pVHL-defective ccRCC tumors display an increased acetylation, and a shift toward mono- or un-methylation, of histone H3 on lysine 27 (H3K27). Moreover, pVHL-defective ccRCC cells are dependent largely on the H3K27 methyltransferase EZH1 for survival: pharmacological EZH inhibition preferentially kills $VHL^{-/-}$ ccRCC (Chakraborty et al. 2017). Therefore, EZH1 could be a therapeutic target in ccRCC.

11.5.5 Therapeutic Treatment for Renal Cell Carcinoma

VHL patients typically develop the first manifestation of their disease in their 20s, and the penetrance of VHL has been reported to be nearly 100% at age 60 years (Binderup 2018). VHL patients may develop ccRCC as well as multiple kidney cysts, which may have malignant potential. A threshold size of 3–4 cm RCC can be surgically removed before progression to metastatic disease, but repeated surgery for multifocal bilateral disease could lead to high risk of end-stage renal disease, requiring renal transplantation or dialysis. For metastatic ccRCC, drugs targeting HIF target genes have been employed. Cytokines such as interferon-α and high-dose IL-2 have been used since the 1990s, but both drugs have severe toxicity, low response rates, and typically benefit only a small subset of patients. Some drugs have significantly improved for metastatic ccRCC, such as sorafenib, sunitinib, pazopanib, axitinib, lenvatinib, and cabozantinib, which target the VEGF pathway (Motzer et al. 2007, 2013; Rini et al. 2011). In addition, the first recombinant human anti-VEGF monoclonal antibody bevacizumab shows clinical efficacy and is approved for use with interferon-α (Motzer et al. 2007) (Fig. 11.1). Since HIF-2α is constitutively activated in ccRCC, the novel HIF-2α inhibitor PT2399, which prevents binding of HIF-2α to ARNT/HIF-1β, was developed. Clinical trials with another HIF-2α inhibitor, PT2385, are currently in progress (NCT02293980 and NCT03108066) (Kim and Zschiedrich 2018). Many studies are now investigating immunotherapy agents for metastatic RCC, such as avelumab and atezolizumab, which target programmed cell death protein 1 ligand 1 (PDL1) and nivolumab and pembrolizumab, which target programmed cell death protein 1 (PD1) (Hsieh et al. 2017). Single-agent treatment shows limited efficacy, such as nivolumab, which showed an overall survival benefit compared with everolimus, but the response rate to nivolumab was only 25% (5% for everolimus). In addition, most patients treated with nivolumab did not experience significant tumor shrinkage (Motzer et al. 2015). Despite the availability of multiple systemic therapies, metastatic ccRCC is still difficult to cure (Jilg et al. 2012). Some tumors pose a diagnostic problem because they show a combination of features that are characteristic of different subtypes (Reuter and Tickoo 2010). Due to the above reasons, it is necessary to employ multiagent therapy. Ongoing clinical trials are investigating combinations of anti-VEGF treatments with immunotherapy, as well as combinations of the immune modulators nivolumab and ipilimumab, an inhibitor of the cytotoxic T-lymphocyte-associated antigen CTLA-4 (Kim and Zschiedrich 2018).

Importantly, *VHL* loss alone is insufficient to induce ccRCC: for instance, *VHL* loss in mice is unable to induce ccRCC. Several inactivating mutations in genes including *PBRM1*, *SETD2*, *BAP1*, *KDM5C*, *TP53*, and *MTOR* have been revealed in ccRCC (Cancer Genome Atlas Research Network 2013). New treatments targeting the downstream events of these mutations may help cure RCC: for instance, complex mTORC1 inhibitors everolimus and temsirolimus are effective for metastatic ccRCC (Hudes et al. 2007; Motzer et al. 2010).

11.6 Conclusions

Over the past two decades, researchers from different groups have made significant contributions to the molecular understanding of pVHL function. However, the lack of a truly effective treatment for VHL disease indicates that more research is needed to uncover more facets of this fascinating protein. Discovering the full repertoire of structural and functional regulations between pVHL and its substrates/targets is the main goal to deciphering *VHL*-related cancers at the molecular level. Recently, pVHL function has been broadened to include several HIF-independent mechanisms of action, and it is likely that many more still need to be revealed. The deep understanding of pVHL machinery may provide novel effective therapeutic modalities for *VHL*-related cancers.

References

Alessi DR, Andjelkovic M, Caudwell B, Cron P, Morrice N, Cohen P, Hemmings BA (1996) Mechanism of activation of protein kinase b by insulin and igf-1. EMBO J 15:6541–6551

Anderson K, Nordquist KA, Gao XL, Hicks KC, Zhai B, Gygi SP, Patel TB (2011) Regulation of cellular levels of sprouty2 protein by prolyl hydroxylase domain and von Hippel-Lindau proteins. J Biol Chem 286:42027–42036

Ang SO, Chen H, Hirota K, Gordeuk VR, Jelinek J, Guan YL, Liu EL, Sergueeva AI, Miasnikova GY, Mole D, Maxwell PH, Stockton DW, Semenza GL, Prchal JT (2002) Disruption of oxygen homeostasis underlies congenital chuvash polycythemia. Nat Genet 32:614–621

Balachandran RS, Heighington CS, Starostina NG, Anderson JW, Owen DL, Vasudevan S, Kipreos ET (2016) The ubiquitin ligase crl2zyg11 targets cyclin b1 for degradation in a conserved pathway that facilitates mitotic slippage. J Cell Biol 215:151–166

Binderup MLM (2018) Von Hippel-Lindau disease: diagnosis and factors influencing disease outcome. Dan Med J 65

Bradsher JN, Jackson KW, Conaway RC, Conaway JW (1993) Rna polymerase ii transcription factor siii. I Identification, purification, and properties. J Biol Chem 268:25587–25593

Burger J, Merlet J, Tavernier N, Richaudeau B, Arnold A, Ciosk R, Bowerman B, Pintard L (2013) Crl2(lrr-1) e3-ligase regulates proliferation and progression through meiosis in the caenorhabditis elegans germline. PLoS Genet 9:e1003375

Cai W, Yang H (2016) The structure and regulation of cullin 2 based e3 ubiquitin ligases and their biological functions. Cell Div 11(7)

Cancer Genome Atlas Research Network (2013) Comprehensive molecular characterization of clear cell renal cell carcinoma. Nature 499:43–49

Cardote TAF, Gadd MS, Ciulli A (2017) Crystal structure of the cul2-rbx1-elobc-vhl ubiquitin ligase complex. Structure 25:901–911 e903

Carmeliet P, Dor Y, Herbert JM, Fukumura D, Brusselmans K, Dewerchin M, Neeman M, Bono F, Abramovitch R, Maxwell P, Koch CJ, Ratcliffe P, Moons L, Jain RK, Collen D, Keshert E (1998) Role of hif-1alpha in hypoxia-mediated apoptosis, cell proliferation and tumour angiogenesis. Nature 394:485–490

Casciello F, Al-Ejeh F, Kelly G, Brennan DJ, Ngiow SF, Young A, Stoll T, Windloch K, Hill MM, Smyth MJ, Gannon F, Lee JS (2017) G9a drives hypoxia-mediated gene repression for breast cancer cell survival and tumorigenesis. Proc Natl Acad Sci U S A 114:7077–7082

Chakraborty AA, Nakamura E, Qi J, Creech A, Jaffe JD, Paulk J, Novak JS, Nagulapalli K, McBrayer SK, Cowley GS, Pineda J, Song JX, Wang YYE, Carr SA, Root DE, Signoretti S, Bradner JE, Kaelin WG (2017) Hif activation causes synthetic lethality between the vhl tumor suppressor and the ezh1 histone methyltransferase. Sci Transl Med 9

Chen JX, Liu F, Li H, Archacld S, Gao M, Liu Y, Liao SJ, Huang M, Wang JX, Yu SS, Li C, Tang ZH, Liu MG (2015) Pvhl interacts with ceramide kinase like (cerkl) protein and ubiquitinates it for oxygen dependent proteasomal degradation. Cell Signal 27:2314–2323

Chittiboina P, Lonser RR (2015) Von Hippel-Lindau disease. Handb Clin Neurol 132:139–156

Choudhry H, Harris AL (2018) Advances in hypoxia-inducible factor biology. Cell Metab 27:281–298

Conaway JW, Kamura T, Conaway RC (1998) The elongin bc complex and the von Hippel-Lindau tumor suppressor protein. Biochim Biophys Acta 1377: M49–M54

Costessi A, Mahrour N, Tijchon E, Stunnenberg R, Stoel MA, Jansen PW, Sela D, Martin-Brown S, Washburn MP, Florens L, Conaway JW, Conaway RC, Stunnenberg HG (2011) The tumour antigen prame is a subunit of a cul2 ubiquitin ligase and associates with active nfy promoters. EMBO J 30:3786–3798

Cummins EP, Berra E, Comerford KM, Ginouves A, Fitzgerald KT, Seeballuck F, Godson C, Nielsen JE, Moynagh P, Pouyssegur J, Taylor CT (2006) Prolyl hydroxylase-1 negatively regulates ikappab kinase-beta, giving insight into hypoxia-induced nfkappab activity. Proc Natl Acad Sci U S A 103:18154–18159

Deshaies RJ, Joazeiro CA (2009) Ring domain e3 ubiquitin ligases. Annu Rev Biochem 78:399–434

Doherty JR, Cleveland JL (2013) Targeting lactate metabolism for cancer therapeutics. J Clin Invest 123:3685–3692

Doolan P, Clynes M, Kennedy S, Mehta JP, Crown J, O'Driscoll L (2008) Prevalence and prognostic and predictive relevance of prame in breast cancer. Breast Cancer Res Treat 109:359–365

Epping MT, Wang L, Edel MJ, Carlee L, Hernandez M, Bernards R (2005) The human tumor antigen prame is a dominant repressor of retinoic acid receptor signaling. Cell 122:835–847

Epping MT, Hart AA, Glas AM, Krijgsman O, Bernards R (2008) Prame expression and clinical outcome of breast cancer. Br J Cancer 99:398–403

Gilder AS, Chen YB, Jackson RJ 3rd, Jiang J, Maher JF (2013) Fem1b promotes ubiquitylation and suppresses transcriptional activity of gli1. Biochem Biophys Res Commun 440:431–436

Grosfeld A, Stolze IP, Cockman ME, Pugh CW, Edelmann M, Kessler B, Bullock AN, Ratcliffe PJ, Masson N (2007) Interaction of hydroxylated collagen iv with the von Hippel-Lindau tumor suppressor. J Biol Chem 282:13264–13269

Guo JP, Chakraborty AA, Liu PD, Gan WJ, Zheng XN, Inuzuka H, Wang B, Zhang JF, Zhang LL, Yuan M, Novak J, Cheng JQ, Toker A, Signoretti S, Zhang Q, Asara JM, Kaelin WG, Wei WY (2016) Pvhl suppresses kinase activity of akt in a proline-hydroxylation-dependent manner. Science 353:929–932

Heir P, Srikumar T, Bikopoulos G, Bunda S, Poon BP, Lee JE, Raught B, Ohh M (2016) Oxygen-dependent regulation of erythropoietin receptor turnover and signaling. J Biol Chem 291:7357–7372

Hergovich A, Lisztwan J, Barry R, Ballschmieter P, Krek W (2003) Regulation of microtubule stability by the von Hippel-Lindau tumour suppressor protein pvhl. Nat Cell Biol 5:64–70

Hoffman MA, Ohh M, Yang HF, Klco JM, Ivan M, Kaelin WG (2001) Von Hippel-Lindau protein mutants linked to type 2c vhl disease preserve the ability to downregulate hif. Hum Mol Genet 10:1019–1027

Hsieh JJ, Purdue MP, Signoretti S, Swanton C, Albiges L, Schmidinger M, Heng DY, Larkin J, Ficarra V (2017) Renal cell carcinoma. Nat Rev Dis Primers 3

Hudes G, Carducci M, Tomczak P, Dutcher J, Figlin R, Kapoor A, Staroslawska E, Sosman J, McDermott D, Bodrogi I, Kovacevic Z, Lesovoy V, Schmidt-Wolf IGH, Barbarash O, Gokmen E, O'Toole T, Lustgarten S, Moore L, Motzer RJ (2007) Temsirolimus, interferon alfa, or both for advanced renal-cell carcinoma. New Engl J Med 356:2271–2281

Iliopoulos O, Ohh M, Kaelin WG Jr (1998) Pvhl19 is a biologically active product of the von Hippel-Lindau gene arising from internal translation initiation. Proc Natl Acad Sci U S A 95:11661–11666

Jilg CA, Neumann HPH, Glasker S, Schafer O, Leiber C, Ardelt PU, Schwardt M, Schultze-Seemann W (2012) Nephron sparing surgery in von Hippel-Lindau associated renal cell carcinoma; clinicopathological long-term follow-up. Fam Cancer 11:387–394

Kaelin WG (2002) Molecular basis of the vhl hereditary cancer syndrome. Nat Rev Cancer 2:673–682

Kamura T, Sato S, Haque D, Liu L, Kaelin WG Jr, Conaway RC, Conaway JW (1998) The elongin bc complex interacts with the conserved socs-box motif present in members of the socs, ras, wd-40 repeat, and ankyrin repeat families. Genes Dev 12:3872–3881

Kamura T, Maenaka K, Kotoshiba S, Matsumoto M, Kohda D, Conaway RC, Conaway JW, Nakayama KI (2004) Vhl-box and socs-box domains determine binding specificity for cul2-rbx1 and cul5-rbx2 modules of ubiquitin ligases. Genes Dev 18:3055–3065

Kilpinen S, Autio R, Ojala K, Iljin K, Bucher E, Sara H, Pisto T, Saarela M, Skotheim RI, Bjorkman M, Mpindi

JP, Haapa-Paananen S, Vainio P, Edgren H, Wolf M, Astola J, Nees M, Hautaniemi S, Kallioniemi O (2008) Systematic bioinformatic analysis of expression levels of 17,330 human genes across 9,783 samples from 175 types of healthy and pathological tissues. Genome Biol 9:R139

Kim WY, Kaelin WG (2004) Role of vhl gene mutation in human cancer. J Clin Oncol 22:4991–5004

Kim E, Zschiedrich S (2018) Renal cell carcinoma in von Hippel-Lindau disease-from tumor genetics to novel therapeutic strategies. Front Pediatr 6

Kobet E, Zeng X, Zhu Y, Keller D, Lu H (2000) Mdm2 inhibits p300-mediated p53 acetylation and activation by forming a ternary complex with the two proteins. Proc Natl Acad Sci U S A 97:12547–12552

Koumenis C, Alarcon R, Hammond E, Sutphin P, Hoffman W, Murphy M, Derr J, Taya Y, Lowe SW, Kastan M, Giaccia A (2001) Regulation of p53 by hypoxia: dissociation of transcriptional repression and apoptosis from p53-dependent transactivation. Mol Cell Biol 21:1297–1310

Kurban G, Duplan E, Ramlal N, Hudon V, Sado Y, Ninomiya Y, Pause A (2008) Collagen matrix assembly is driven by the interaction of von Hippel-Lindau tumor suppressor protein with hydroxylated collagen iv alpha 2. Oncogene 27:1004–1012

Kuznetsova AV, Meller J, Schnell PO, Nash JA, Ignacak ML, Sanchez Y, Conaway JW, Conaway RC, Czyzyk-Krzeska MF (2003) Von Hippel-Lindau protein binds hyperphosphorylated large subunit of rna polymerase ii through a proline hydroxylation motif and targets it for ubiquitination. Proc Natl Acad Sci U S A 100:2706–2711

Latif F, Tory K, Gnarra J, Yao M, Duh FM, Orcutt ML, Stackhouse T, Kuzmin I, Modi W, Geil L et al (1993) Identification of the von Hippel-Lindau disease tumor suppressor gene. Science 260:1317–1320

Lee JY, Dong SM, Park WS, Yoo NJ, Kim CS, Jang JJ, Chi JG, Zbar B, Lubensky IA, Linehan WM, Vortmeyer AO, Zhuang Z (1998) Loss of heterozygosity and somatic mutations of the vhl tumor suppressor gene in sporadic cerebellar hemangioblastomas. Cancer Res 58:504–508

Lee S, Neumann M, Stearman R, Stauber R, Pause A, Pavlakis GN, Klausner RD (1999) Transcription-dependent nuclear-cytoplasmic trafficking is required for the function of the von Hippel-Lindau tumor suppressor protein. Mol Cell Biol 19:1486–1497

Lee S, Nakamura E, Yang H, Wei W, Linggi MS, Sajan MP, Farese RV, Freeman RS, Carter BD, Kaelin WG Jr, Schlisio S (2005) Neuronal apoptosis linked to egln3 prolyl hydroxylase and familial pheochromocytoma genes: developmental culling and cancer. Cancer Cell 8:155–167

Lee DC, Sohn HA, Park ZY, Oh S, Kang YK, Lee KM, Kang M, Jang YJ, Yang SJ, Hong YK, Noh H, Kim JA, Kim DJ, Bae KH, Kim DM, Chung SJ, Yoo HS, Yu DY, Park KC, Yeom YI (2015) A lactate-induced response to hypoxia. Cell 161:595–609

Levine AJ, Hu W, Feng Z (2006) The p53 pathway: what questions remain to be explored? Cell Death Differ 13:1027–1036

Li M, Kim WY (2011) Two sides to every story: the hif-dependent and hif-independent functions of pvhl. J Cell Mol Med 15:187–195

Lolkema MP, Mans DA, Snijckers CM, van Noort M, van Beest M, Voest EE, Giles RH (2007) The von Hippel-Lindau tumour suppressor interacts with microtubules through kinesin-2. FEBS Lett 581:4571–4576

Lubensky IA, Gnarra JR, Bertheau P, Walther MM, Linehan WM, Zhuang ZP (1996) Allelic deletions of the vhl gene detected in multiple microscopic clear cell renal lesions in von Hippel-Lindau disease patients. Am J Pathol 149:2089–2094

Lui DY, Colaiacovo MP (2013) Meiotic development in caenorhabditis elegans. Adv Exp Med Biol 757:133–170

Mahrour N, Redwine WB, Florens L, Swanson SK, Martin-Brown S, Bradford WD, Staehling-Hampton K, Washburn MP, Conaway RC, Conaway JW (2008) Characterization of cullin-box sequences that direct recruitment of cul2-rbx1 and cul5-rbx2 modules to elongin bc-based ubiquitin ligases. J Biol Chem 283:8005–8013

Maity SN, de Crombrugghe B (1998) Role of the ccaat-binding protein cbf/nf-y in transcription. Trends Biochem Sci 23:174–178

Manning BD, Toker A (2017) Akt/pkb signaling: navigating the network. Cell 169:381–405

McRonald FE, Morris MR, Gentle D, Winchester L, Baban D, Ragoussis J, Clarke NW, Brown MD, Kishida T, Yao M, Latif F, Maher ER (2009) Cpg methylation profiling in vhl related and vhl unrelated renal cell carcinoma. Mol Cancer 8

Mendoza M, Mandani G, Momand J (2014) The mdm2 gene family. Biomol Concepts 5:9–19

Motzer RJ, Hutson TE, Tomczak P, Michaelson MD, Bukowski RM, Rixe O, Oudard S, Negrier S, Szczylik C, Kim ST, Chen I, Bycott PW, Baum CM, Figlin RA (2007) Sunitinib versus interferon alfa in metastatic renal-cell carcinoma. New Engl J Med 356:115–124

Motzer RJ, Escudier B, Oudard S, Hutson TE, Porta C, Bracarda S, Grunwald V, Thompson JA, Figlin RA, Hollaender N, Kay A, Ravaud A, Grp R-S (2010) Phase 3 trial of everolimus for metastatic renal cell carcinoma final results and analysis of prognostic factors. Cancer-Am Cancer Soc 116:4256–4265

Motzer RJ, McCann L, Deen K (2013) Pazopanib versus sunitinib in renal cancer reply. New Engl J Med 369

Motzer RJ, Escudier B, McDermott DF, George S, Hammers HJ, Srinivas S, Tykodi SS, Sosman JA, Procopio G, Plimack ER, Castellano D, Choueiri TK, Gurney H, Donskov F, Bono P, Wagstaff J, Gauler TC, Ueda T, Tomita Y, Schutz FA, Kollmannsberger C, Larkin J, Ravaud A, Simon JS, Xu LA, Waxman IM, Sharma P, Investigators C (2015) Nivolumab versus everolimus in advanced renal-cell carcinoma. New Engl J Med 373:1803–1813

Na X, Duan HO, Messing EM, Schoen SR, Ryan CK, di Sant'Agnese PA, Golemis EA, Wu G (2003) Identification of the rna polymerase ii subunit hsrpb7 as a novel target of the von Hippel-Lindau protein. EMBO J 22:4249–4259

Nguyen HC, Yang H, Fribourgh JL, Wolfe LS, Xiong Y (2015) Insights into cullin-ring e3 ubiquitin ligase recruitment: structure of the vhl-elobc-cul2 complex. Structure 23:441–449

Nordstrom-O'Brien M, van der Luijt RB, van Rooijen E, van den Ouweland AM, Majoor-Krakauer DF, Lolkema MP, van Brussel A, Voest EE, Giles RH (2010) Genetic analysis of von Hippel-Lindau disease. Hum Mutat 31:521–537

Oberthuer A, Hero B, Spitz R, Berthold F, Fischer M (2004) The tumor-associated antigen prame is universally expressed in high-stage neuroblastoma and associated with poor outcome. Clin Cancer Res 10:4307–4313

Ohh M, Yauch RL, Lonergan KM, Whaley JM, Stemmer-Rachamimov AO, Louis DN, Gavin BJ, Kley N, Kaelin WG, Iliopoulos O (1998) The von Hippel-Lindau tumor suppressor protein is required for proper assembly of an extracellular fibronectin matrix. Mol Cell 1:959–968

Okuda H, Saitoh K, Hirai S, Iwai K, Takaki Y, Baba M, Minato N, Ohno S, Shuin T (2001) The von hippel-lindau tumor suppressor protein mediates ubiquitination of activated atypical protein kinase c. J Biol Chem 276:43611–43617

Okumura F, Uematsu K, Byrne SD, Hirano M, Joo-Okumura A, Nishikimi A, Shuin T, Fukui Y, Nakatsukasa K, Kamura T (2016) Parallel regulation of von Hippel-Lindau disease by pvhl-mediated degradation of b-myb and hypoxia-inducible factor alpha. Mol Cell Biol 36:1803–1817

Park KC, Lee DC, Yeom YI (2015) Ndrg3-mediated lactate signaling in hypoxia. BMB Rep 48:301–302

Pastore Y, Jedlickova K, Guan YL, Liu EL, Fahner J, Hasle H, Prchal JF, Prchal JT (2003) Mutations of von Hippel-Lindau tumor-suppressor gene and congenital polycythemia. Am J Hum Genet 73:412–419

Ravi R, Mookerjee B, Bhujwalla ZM, Sutter CH, Artemov D, Zeng QW, Dillehay LE, Madan A, Semenza GL, Bedi A (2000) Regulation of tumor angiogenesis by p53-induced degradation of hypoxia-inducible factor 1 alpha. Genes Dev 14:34–44

Reuter VE, Tickoo SK (2010) Differential diagnosis of renal tumours with clear cell histology. Pathology 42:374–383

Rini BI, Escudier B, Tomczak P, Kaprin A, Szczylik C, Hutson TE, Michaelson MD, Gorbunova VA, Gore ME, Rusakov IG, Negrier S, Ou YC, Castellano D, Lim HY, Uemura H, Tarazi J, Cella D, Chen C, Rosbrook B, Kim S, Motzer RJ (2011) Comparative effectiveness of axitinib versus sorafenib in advanced renal cell carcinoma (axis): a randomised phase 3 trial. Lancet 378:1931–1939

Robinson CM, Ohh M (2014) The multifaceted von Hippel-Lindau tumour suppressor protein. FEBS Lett 588:2704–2711

Roe JS, Kim H, Lee SM, Kim ST, Cho EJ, Youn HD (2006) P53 stabilization and transactivation by a von Hippel-Lindau protein. Mol Cell 22:395–405

Sanchez DJ, Simon MC (2018) Transcriptional control of kidney cancer. Science 361:226–227

Segura I, Lange C, Knevels E, Moskalyuk A, Pulizzi R, Eelen G, Chaze T, Tudor C, Boulegue C, Holt M, Daelemans D, Matondo M, Ghesquiere B, Giugliano M, Ruiz de Almodovar C, Dewerchin M, Carmeliet P (2016) The oxygen sensor phd2 controls dendritic spines and synapses via modification of filamin a. Cell Rep 14:2653–2667

Semenza GL (2003) Targeting hif-1 for cancer therapy. Nat Rev Cancer 3:721–732

Shi YQ, Liao SY, Zhuang XJ, Han CS (2011) Mouse fem1b interacts with and induces ubiquitin-mediated degradation of ankrd37. Gene 485:153–159

Sjolund J, Johansson M, Manna S, Norin C, Pietras A, Beckman S, Nilsson E, Ljungberg B, Axelson H (2008) Suppression of renal cell carcinoma growth by inhibition of notch signaling in vitro and in vivo. J Clin Investig 118:217–228

Starostina NG, Lim JM, Schvarzstein M, Wells L, Spence AM, Kipreos ET (2007) A cul-2 ubiquitin ligase containing three fem proteins degrades tra-1 to regulate c. Elegans sex determination. Dev Cell 13:127–139

Starostina NG, Simpliciano JM, McGuirk MA, Kipreos ET (2010) Crl2(lrr-1) targets a cdk inhibitor for cell cycle control in c. Elegans and actin-based motility regulation in human cells. Dev Cell 19:753–764

Stolle C, Glenn G, Zbar B, Humphrey JS, Choyke P, Walther M, Pack S, Hurley K, Andrey C, Klausner R, Linehan WM (1998) Improved detection of germline mutations in the von Hippel-Lindau disease tumor suppressor gene. Hum Mutat 12:417–423

Tarade D, Ohh M (2018) The hif and other quandaries in vhl disease. Oncogene 37:139–147

Vasudevan S, Starostina NG, Kipreos ET (2007) The caenorhabditis elegans cell-cycle regulator zyg-11 defines a conserved family of cul-2 complex components. EMBO Rep 8:279–286

Vriend J, Reiter RJ (2016) Melatonin and the von Hippel-Lindau/hif-1 oxygen sensing mechanism: a review. BBA-Rev Cancer 1865:176–183

Wang Z, Liu P, Inuzuka H, Wei W (2014) Roles of f-box proteins in cancer. Nat Rev Cancer 14:233–247

Wang S, Xia W, Qiu M, Wang X, Jiang F, Yin R, Xu L (2016) Atlas on substrate recognition subunits of crl2 e3 ligases. Oncotarget 7:46707–46716

Xia XB, Lemieux ME, Li W, Carroll JS, Brown M, Liu XS, Kung AL (2009) Integrative analysis of hif binding and transactivation reveals its role in maintaining histone methylation homeostasis. Proc Natl Acad Sci U S A 106:4260–4265

Xie L, Xiao KH, Whalen EJ, Forrester MT, Freeman RS, Fong GH, Gygi SP, Lefkowitz RJ, Stamler JS (2009) Oxygen-regulated beta(2)-adrenergic receptor hydroxylation by egln3 and ubiquitylation by pvhl. Sci Signal 2

Xu J, Li L, Yu G, Ying W, Gao Q, Zhang W, Li X, Ding C, Jiang Y, Wei D, Duan S, Lei Q, Li P, Shi T, Qian X, Qin J, Jia L (2015) The neddylation-cullin 2-rbx1 e3 ligase axis targets tumor suppressor rhob for degradation in liver cancer. Mol Cell Proteomics 14:499–509

Yang H, Minamishima YA, Yan Q, Schlisio S, Ebert BL, Zhang X, Zhang L, Kim WY, Olumi AF, Kaelin WG (2007) Pvhl acts as an adaptor to promote the inhibitory phosphorylation of the nf-kappa b agonist card9 by ck2. Mol Cell 28:15–27

Yao XS, Tan J, Lim KJ, Koh J, Ooi WF, Li ZM, Huang DC, Xing MJ, Chan YS, Qu JZ, Tay ST, Wijaya G, Lam YN, Hong JH, Lee-Lim AP, Guan PY, Ng MSW, He CZ, Lin JS, Nandi T, Qamra A, Xu C, Myint SS, Davies JOJ, Goh JY, Loh G, Tan BC, Rozen SG, Yu Q, Tan IBH, Cheng CWS, Li S, Chang KTE, Tan PH, Silver DL, Lezhava A, Steger G, Hughes JR, Teh BT, Tan P (2017) Vhl deficiency drives enhancer activation of oncogenes in clear cell renal cell carcinoma. Cancer Discov 7:1284–1305

Yin H, Zheng L, Liu W, Zhang D, Li W, Yuan L (2017) Rootletin prevents cep68 from vhl-mediated proteasomal degradation to maintain centrosome cohesion. Biochim Biophys Acta Mol Cell Res 1864:645–654

Zhang J, Zhang Q (2018) Vhl and hypoxia signaling: beyond hif in cancer. Biomedicines 6

Zhang Y, Kwok-Shing Ng P, Kucherlapati M, Chen F, Liu Y, Tsang YH, de Velasco G, Jeong KJ, Akbani R, Hadjipanayis A, Pantazi A, Bristow CA, Lee E, Mahadeshwar HS, Tang J, Zhang J, Yang L, Seth S, Lee S, Ren X, Song X, Sun H, Seidman J, Luquette LJ, Xi R, Chin L, Protopopov A, Westbrook TF, Shelley CS, Choueiri TK, Ittmann M, Van Waes C, Weinstein JN, Liang H, Henske EP, Godwin AK, Park PJ, Kucherlapati R, Scott KL, Mills GB, Kwiatkowski DJ, Creighton CJ (2017) A pan-cancer proteogenomic atlas of pi3k/akt/mtor pathway alterations. Cancer Cell 31:820–832 e823

Zhang J, Wu T, Simon J, Takada M, Saito R, Fan C, Liu XD, Jonasch E, Xie L, Chen X, Yao X, Teh BT, Tan P, Zheng X, Li M, Lawrence C, Fan J, Geng J, Liu X, Hu L, Wang J, Liao C, Hong K, Zurlo G, Parker JS, Auman JT, Perou CM, Rathmell WK, Kim WY, Kirschner MW, Kaelin WG Jr, Baldwin AS, Zhang Q (2018) Vhl substrate transcription factor zhx2 as an oncogenic driver in clear cell renal cell carcinoma. Science 361:290–295

Zurlo G, Guo JP, Takada M, Wei WY, Zhang Q (2016) New insights into protein hydroxylation and its important role in human diseases. BBA-Rev Cancer 1866:208–220

Cullin 3 and Its Role in Tumorigenesis 12

Ruey-Hwa Chen

Abstract

Cullin 3 (Cul3) family of ubiquitin ligases comprises three components, the RING finger protein RBX1, the Cul3 scaffold, and a Bric-a-brac/Tramtrack/Broad complex (BTB) protein. The BTB protein serves as a bridge to connect Cul3 to substrate and is functionally equivalent to the combination of substrate adaptor and linker in other Cullin complexes. Human genome encodes for ~180 BTB proteins, implying a broad spectrum of ubiquitination signals and substrate repertoire. Accordingly, Cul3 ubiquitin ligases are involved in diverse cellular processes, including cell division, differentiation, cytoskeleton remodeling, stress responses, and nerve cell functions. Emerging evidence has pointed to the prominent role of Cul3 ubiquitin ligases in cancer. This chapter will describe recent advances on the roles of Cul3 E3 ligase complexes in regulating various cancer hallmarks and therapeutic responses and the mutation/dysregulation of Cul3 substrate adaptors in cancer. In particular, we will focus on several extensively studied substrate adaptors, such as Keap1, SPOP, KLHL20, and LZTR1, and will also discuss other recently identified Cul3 adaptors with oncogenic or tumor-suppressive functions. We conclude that Cul3 ubiquitin ligases represent master regulators of human malignancies and highlight the importance of developing modulating agents for oncogenic/tumor-suppressive Cul3 E3 ligase complexes to prevent or intervene tumorigenesis.

Keywords

Cullin3 ubiquitin ligase · Cancer · Keap1 · SPOP · KLHL20 · LZTR1

Abbreviations

AR	Androgen receptor
BTB	Bric-a-brac/Tramtrack/Broad
ccRCC	Clear cell renal cell carcinoma
Cul3	Cullin 3
DAPK	Death-associated protein kinase
DLBCL	Diffuse large B-cell lymphoma
DMF	Dimethyl fumarate
GLS	Glutaminase
HIF-1	Hypoxia-inducible factor-1
HRE	Hypoxia-responsive element
IFN	Interferon
LLPS	Liquid-liquid phase separation
ME1	Malic enzyme 1
Nrf2	NF-E2-related factor 2
NS	Noonan syndrome
PHR	PAM, Highwire, and RPM-1
PIN	Prostatic intraepithelial neoplasia

R.-H. Chen (✉)
Institute of Biological Chemistry, Academia Sinica, Taipei, Taiwan
e-mail: rhchen@gate.sinica.edu.tw

Y. Sun et al. (eds.), *Cullin-RING Ligases and Protein Neddylation*, Advances in Experimental Medicine and Biology 1217, https://doi.org/10.1007/978-981-15-1025-0_12

PPP	Pentose phosphate pathways
ROS	Reactive oxygen species
SBC	SPOP binding consensus
SPOP	Speckle-type BTB/POZ protein
TME	Tumor microenvironment
ZF	Zn finger

12.1 Introduction

The Cullin 3 (Cul3) ubiquitin ligase family adopts a similar architecture to other Cullin containing ubiquitin E3 ligase complexes (Genschik et al. 2013; Pintard et al. 2004). The C-terminal part of Cul3 tightly associates with the RING finger protein RBX1, which serves as the catalytic subunit. The N-terminus of Cul3 binds to Bric-a-brac/Tramtrack/Broad complex (BTB) domain-containing proteins as substrate-specific adaptors. The BTB domain is structurally similar to other linker proteins in the Cullin complex, such as Skp1 and EloC, and therefore binds Cul3 via motifs analogous to those in the Skp1-Cul1 complex (Stogios et al. 2005). However, Cul3 complex differs from other Cullin complexes by two aspects. First, due to the dimerization ability of BTB domain, Cul3 complex can adopt a dimeric architecture containing two units of RBX1-Cul3-BTB protein complex (Canning et al. 2013; Zhuang et al. 2009). Second, many BTB proteins also contain additional protein interaction domains for substrate recruitment. Therefore, a single BTB protein in the Cul3 complex confers the functional properties of both linker and substrate adaptor, unlike other Cullin complex (Lydeard et al. 2013; Zimmerman et al. 2010).

The human genome encodes ~180 BTB proteins, which can be subdivided into subfamilies based on the existence of additional protein interaction domains, such as MATH, kelch, Zn finger (ZF), as well as PAM, Highwire, and RPM-1 (PHR) (Stogios et al. 2005). The BTB-MATH and BTB-kelch proteins are best known as substrate adaptors for Cul3 (Chen and Chen 2016; Dubiel et al. 2018). Interestingly, recent evidence indicates that certain BTB-ZF proteins can be assembled to Cul3 complex for substrate ubiquitination (Mathew et al. 2012), even though many BTB-ZF proteins are mainly involved in transcriptional regulation and chromatin remodeling. A systemic proteomics study identified 53 BTB proteins that interact with Cul3 (Bennett et al. 2010), but it remains unclear precisely how many BTB proteins can be assembled into Cul3 complex in vivo. Structural analysis of the Cul3-SPOP complex revealed that, besides the BTB domain, a 3-box motif located at the C-terminus to the BTB domain is also involved in Cul3 binding (Zhuang et al. 2009). This explains why certain BTB proteins lacking the 3-box region, such as BTBD12, are not copurified with Cul3. Furthermore, KLHL39, which possesses several atypical residues in the BTB domain, cannot bind Cul3 either in cells (Chen et al. 2015).

Cul3 ubiquitin ligases are very versatile in writing the ubiquitin codes. In addition to the conventional proteolytic ubiquitination (usually via the assembly of K48-linked polyubiquitin chain), non-proteolytic ubiquitination, such as monoubiquitination, multi-monoubiquitination, K63-linked ubiquitination, and K33-linked ubiquitination have all been reported (Jerabkova and Sumara 2018). Most of these non-proteolytic ubiquitination events lead to the alteration of subcellular localization of the substrates. Accumulating evidence revealed that different Cullin family ligases seem to be evolved in executing certain specialized cellular functions. For instance, Cul1 family is important in controlling cell cycle progression (Teixeira and Reed 2013; Zheng et al. 2016), whereas chromatin-associated Cul4 ligases are mainly involved in DNA metabolism (Hannah and Zhou 2009, 2015). Recent studies have discovered the role of Cul2 ligases in protein quality control (Koren et al. 2018; Lin et al. 2018). Through catalyzing numerous substrates, the Cul3 ligases are widely involved in many physiological and pathological processes, including cell differentiation, cell cycle regulation, protein trafficking, stress responses, and apoptosis (Dubiel et al. 2018; Jerabkova and Sumara 2018). Currently, the regulation of tumorigenesis and tumor progression has emerged as a prominent function of Cul3 family ligases (Chen and Chen 2016; Cheng et al. 2018). Two substrate adaptors, Keap1 and SPOP, are the most representative cancer-related Cul3 adaptors, and both elicit dual and context-dependent roles in cancer. Additionally, several Cul3 adaptors have been reported to target oncoproteins or tumor suppressor proteins to influence on various types of neoplasms. Genetic alterations and/or aberrant expression of Cul3 adaptors have also been observed. This chapter will overview the functions of Cul3 family ligases in human malignancies and discuss the potential to target this family of ligases for cancer therapy.

12.2 Keap1

12.2.1 The Keap1-Nrf2 Pathway and Its Chemoprevention and Anti-Inflammatory Functions

Keap1 is the prototype of BTB-kelch protein (Cullinan et al. 2004; Kobayashi et al. 2004; Zhang et al. 2004) and was originally discovered as an inhibitor of NF-E2-related factor 2 (Nrf2) (Dinkova-Kostova et al. 2002; Itoh et al. 1999). Nrf2 functions as a transcription factor, which binds to the antioxidant response element to activate the transcription of numerous genes to control antioxidant responses, detoxification of xenobiotics and drugs, autophagy, and metabolism (Hayes and Dinkova-Kostova 2014). Under basal conditions, the Cul3-Keap1 E3 ligase targets Nrf2 for ubiquitination, which facilitates its persistent degradation through the 26S proteasome (Cullinan et al. 2004, Kobayashi et al. 2004, Zhang et al. 2004). However, several cysteine residues in Keap1, such as C151, can be targeted by electrophiles and reactive oxygen species (ROS), resulting in its conformational change to interfere with Nrf2 ubiquitination (Baird et al. 2013; McMahon et al. 2006; Zhang and Hannink 2003). Under these circumstances, Nrf2 is stabilized, thus enabling its translocation to the nucleus to execute its transcription function. Once redox homeostasis is restored, Keap1 brings Nrf2 back into cytosol to turn off its signaling (Sun et al. 2007, 2011). Through this mechanism, the Keap1-Nrf2 pathway is crucial in cellular antioxidation and defense responses.

Oxidative stress and the resulting genome instability play a crucial role in tumorigenesis. Thus, the protective role of Nrf2 could prevent the initiation of tumorigenesis. In support of this idea, *Nrf2* knockout increases the susceptibility to carcinogen-induced tumorigenesis in the many organs/tissues, compared to the wild-type littermates (Bauer et al. 2011; Fahey et al. 2002; Khor et al. 2008; Osburn et al. 2007; Ramos-Gomez et al. 2001; Xu et al. 2006). Conversely, administration of chemicals that activate Nrf2 can reduce tumor incidence in rodent tumor models (Kandoth et al. 2013; Shen et al. 2015; Tao et al. 2013, 2015). The cancer prevention function of Nrf2 is likely attributed to the rapid detoxification of carcinogens, quenching ROS and repairing oxidative damages. Interestingly, certain phytochemicals from food and herbal medicine contain natural electrophiles that can increase Nrf2 via targeting Keap1 cysteine residues. These compounds, along with other type of Nrf2 activators, may be exploited for cancer prevention (Harder et al. 2015; Leinonen et al. 2014).

12.2.2 The Tumor-Promoting Functions of the Keap1-Nrf2 Pathway

Despite the role of Keap1-Nrf2 pathway in cancer prevention, when tumor is formed, tumor cells exploit this pathway for gaining growth and survival advantages to deal with the adverse conditions. Mounting evidence has indicated that Nrf2 activation in cancer promotes tumor progression, metastatic spreading, and resistance to therapy (Rojo de la Vega et al. 2018). Due to these advantages, cancer cells with persistent Nrf2 activation often develop "Nrf2 addiction," and high Nrf2 expression in tumor is often associated with tumor aggressiveness and poor prognosis (Kitamura and Motohashi 2018). In addition, Nrf2 promotes many aspects of cancer hallmarks, either through a direct activation of target genes or an indirect effect on redox regulation. These tumor-promoting functions are summarized below.

12.2.2.1 Proliferation and Metabolic Reprogramming

Multiple lines of evidence have demonstrated a role of the Keap1-Nrf2 pathway in promoting proliferation. For instance, *Nrf2*$^{-/-}$ cells proliferate slower than the wild-type counterparts, whereas *Keap1*$^{-/-}$ cells grow faster (Homma et al. 2009; Lister et al. 2011; Zhang et al. 2015a, 2016). It has been shown that Nrf2 expression is induced by several oncogenes that promote proliferation, such as K-Ras^{G12D}, BRAF^{V619E}, and MYC. This elevation of Nrf2 suppresses ROS production induced by oncogene activation, thus contributing to oncogene-induced proliferation (DeNicola et al. 2011). Nrf2 also regulates growth factor signaling by maintaining the metalloprotease ADAM10 in reduced state, which allows the shedding of epidermal growth factor for its autocrine growth signaling (Chio et al. 2016). In addition, Nrf2 directly activates the expression of a set of genes known to promote proliferation (Malhotra et al. 2010; Wakabayashi et al. 2010).

12.2.2.2 Metabolic Reprogramming

The proliferative effect of the Keap1-Nrf2 pathway can also be attributed to the function of Nrf2 in metabolic reprogramming, thereby providing cancer cells with anabolic precursors, reducing equivalents, energy, and nucleotides for growth. Nrf2 affects a number of metabolic pathways. First, several genes in the pentose phosphate pathways (PPP) are directly or indirectly activated by Nrf2, which, in conjunction with the increased glucose uptake by Nrf2, directs glucose to PPP for purine synthesis (Malhotra et al. 2010; Mitsuishi et al. 2012). Nrf2 also activates the expression of glutaminase (GLS), which converts glutamine to glutamate for providing cancer cells with nitrogen source. The increased expression of malic enzyme 1 (ME1) and several enzymes for glutathione synthesis further converts glutamate to α-ketoglutarate and glutathione (Mitsuishi et al. 2012). In line with these Nrf2 functions, triple deletions of *K-Ras*, *Tp53*, and *Keap1* in lung epithelium result in the generation of lung tumors that are highly dependent on glutaminolysis, showing a high sensitivity to agent that inhibits glutamine transporter SLC1A5 (Romero et al. 2017). Nrf2 also controls amino acid anabolism by activating a number of genes involving in serine/glycine biosynthetic pathways (DeNicola et al. 2015) and stimulates mitochondrial fatty acid oxidation (Ludtmann et al. 2014). Besides regulating key metabolic enzymes, Nrf2 impacts on mitochondria physiology to increase oxidative phosphorylation and activates the expression of several complex IV cytochrome c oxidase subunits (Holmstrom et al. 2013).

12.2.2.3 Invasion and Metastasis

Constitutive activation of Nrf2 in cancer cells has been shown to increase invasion and metastasis (Arfmann-Knubel et al. 2015; Shen et al. 2014; Zhang et al. 2012; Zhao et al. 2017). These functions are associated with several mechanisms. Nrf2 promotes epithelial mesenchymal transition by downregulating E-cadherin and upregulating N-cadherin (Arfmann-Knubel et al. 2015; Shen et al. 2014; Wakabayashi et al. 2010; Zhao et al. 2017). Nrf2 also positively regulates the activity of MMP2 and MMP9 (Zhang et al. 2012) and Rho/RACK pathway (Zhang et al. 2016). On the contrary, Nrf2 expression in tumor microenvironment (TME) has a negative impact on metastasis, as whole-body and myeloid-specific deletion of *Nrf2* increases metastatic incidence through persistent inflammation and redox alterations in immune cells (Hiramoto et al. 2014; Satoh et al. 2010).

12.2.2.4 Resistance to Chemotherapy and Radiotherapy

In response to radiotherapy and certain chemotherapeutic agents that increase ROS, Nrf2 is stabilized. Mounting evidence indicates that this Nrf2 activation renders cancer cells resistant to therapy, whereas Nrf2 knockdown sensitizes them to therapy (Shibata et al. 2008a; Solis et al. 2010; Wang et al. 2008b; Zhang et al. 2010). Cancer cells with an elevated Nrf2 level resulted from Keap1 downregulation similarly show resistance to chemotherapy (Shibata et al. 2008a). Besides the contribution of antioxidant and detoxification to Nrf2-induced survival, Nrf2 can directly inhibit apoptosis by inducing the expression of BCL-2 and BCL-xL, thereby reducing cytochrome C release from mitochondria and diminishing caspase 3/7 activation (Niture and Jaiswal 2012; Tung et al. 2015).

12.2.3 Dysregulation of the Keap1-Nrf2 Pathway in Cancer

Consistent with the tumor-promoting roles, Nrf2 protein level is upregulated in many cancer types. This aberrant upregulation of Nrf2 can occur through multiple mechanisms (Fig. 12.1), in which somatic mutations in proteins functioning in Keap1-Nrf2 pathway represent the main cause for Nrf2 elevation in cancer. *Keap1* gene loss-of-function mutation was originally identified in lung cancer cell lines, leading to decreased binding to Nrf2 (Padmanabhan et al. 2006). Additionally, mutations in the *Keap1* gene were observed in several other cancer types, such as gallbladder cancer, ovarian cancer, liver cancer, and non-small cell lung cancer (Konstantinopoulos et al. 2011; Ohta et al. 2008; Shibata et al. 2008b; Singh et al. 2006; Yoo et al. 2012). Furthermore, a large-scale omics project also identified genetic alterations in *Keap1* and *Nrf2* genes in certain cancer types, especially in lung cancer (Kandoth et al. 2013). In generally, *Keap1* and *Nrf2* mutations are mutually exclusive, consistent with their involvement in the same pathway. Although *Keap1* mutations occur at various places of the coding region, *Nrf2* mutations are found mostly within or surrounded two small regions, the DLG and ETGE motifs (Shibata et al. 2008a, 2011), which are both required for binding Keap1 (Shibata et al. 2008a, 2011; Tong et al. 2006). As a result, these Nrf2 mutants are resistant to Keap1. Additionally, *Nrf2* mutations resulting in the loss of exon 2 have been found in several cancer types, such as lung, head and neck squamous cell carcinoma, and hepatocellular carcinoma, and the exon 2 skipping mutants are also defective in binding Keap1 (Goldstein et al. 2016). Finally, somatic mutations resulting in Nrf2 upregulation in cancers have been detected *RBX1* and *Cul3* as well (Martinez et al. 2014, 2015; Ooi et al. 2013).

As mentioned above, Nrf2 transcription is activated by several oncogenes, and therefore transcriptional activation represents a mechanism for Nrf2 elevation in cancers. In addition, post-transcriptional mechanisms are also reported. For instance, Keap1 promoter hypermethylation has been detected in prostate cancer, lung cancer, glioma, and colon cancer, which results in Nrf2 accumulation (Hanada et al. 2012; Kang et al.

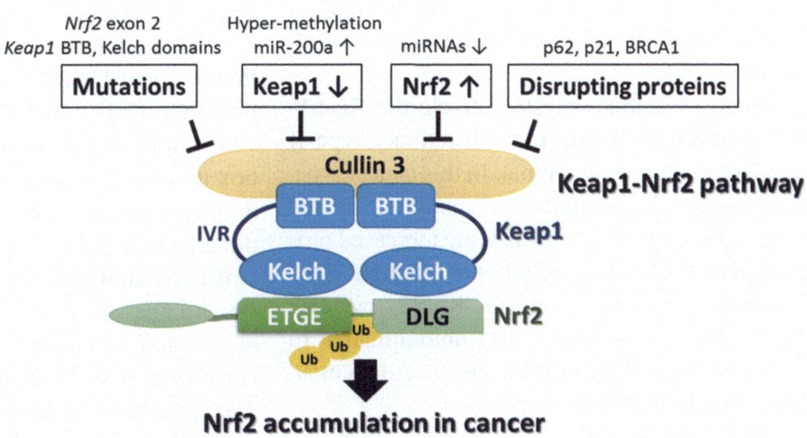

Fig. 12.1 Dysregulation of the Keap1-Nrf2 pathway in cancer

2014; Wang et al. 2008a). Additionally, miR-200a, which is frequently downregulated in tumors, can target Keap1 to indirectly cause Nrf2 upregulation (Eades et al. 2011). In another study, downregulation of miRNAs that target Nrf2 is associated with adverse prognosis of the esophageal cancer (Yamamoto et al. 2014).

Regulation of the Keap1-Nrf2 pathway in tumor can be caused by dysregulation of proteins that disrupt the interaction between Keap1 and Nrf2, which represents a non-canonical mechanism for Nrf2 activation. One such protein is p62, which contains an STGE motif resembling the ETGE motif in Nrf2. Through this binding, p62 acts as a Keap1 pseudosubstrate by competing with Nrf2 for Keap1 binding (Copple et al. 2010; Jain et al. 2010; Komatsu et al. 2010; Lau et al. 2010). Interestingly, p62 is itself a direct target of Nrf2, thus establishing a feedback regulation to reinforce the Nrf2 upregulation (Jain et al. 2010). An additional layer of regulation comes from phosphorylation of the serine residue in the STGE motif by mTOR, leading to an increased binding to Keap1 (Ichimura et al. 2013). Thus, elevated expression of p62, which is frequently observed in hepatocellular carcinoma (Inami et al. 2011; Komatsu et al. 2010), or aberrant activation of mTOR, which occurs in many cancer types (Sabatini 2006), can cause persistent Nrf2 activation. While p62 disrupts Keap1-Nrf2 binding by competing with Nrf2, the p53 downstream target p21 and tumor suppressor BRCA1 bind to ETGE and/or DLG motifs of Nrf2, thus preventing the binding of Nrf2 to Keap1 (Chen et al. 2009; Gorrini et al. 2013).

Finally, a unique way to activate the Keap1-Nrf2 pathway is found in the hereditary type II papillary renal cell carcinoma. In this cancer type, homozygous loss-of-function mutation in the fumarate hydratase gene causes the increased production of Krebs cycle metabolite fumarate. This increased fumarate leads to Keap1 cysteine succination, which blocks Nrf2 ubiquitination to confer highly malignant phenotypes (Adam et al. 2011; Ooi et al. 2011).

12.3 SPOP

12.3.1 SPOP Structure and Biophysical Properties in Relation to Ubiquitination

SPOP (speckle-type BTB/POZ protein) was originally identified as a 374-amino-acid protein and was named based on its presence of the BTB/POZ domain and localization in the nuclear speckles (Nagai et al. 1997). This protein consists of an N-terminal MATH domain, a BTB domain, a BACK domain, and a C-terminal nuclear translocation sequence. In analogous to many other BTB proteins, SPOP functions as a substrate binding subunit of the Cul3 ubiquitin ligase (Kwon et al. 2006). The MATH domain is responsible for substrate recruitment with a substrate binding cleft formed by Y87, F102, Y123, W131, and F133 residues (Zhuang et al. 2009). Correspondingly, SPOP substrate contains an SBC (SPOP binding consensus) motif with the sequence φ-π-S-S/T-S/T (φ = nonpolar; π = polar), which adopts an extended conformation to fit the central shallow groove of the substrate binding cleft of MATH.

Structural and biophysical analyses indicate that SPOP and Cul3 form a 2:2 asymmetric dimer, with the two substrate-binding MATH domains exhibiting a difference in the orientation by 55∘with respect to the BTB/3-box dimer. This increases the structural flexibility between BTB/3-box domain and MATH domain, which allows a SPOP dimer to bind multiple SBCs within a single substrate (Zhuang et al. 2009). Consistent with this notion, several SPOP substrates, such as Puc, Ci, and DAXX, contain multiple SBCs. In the SPOP-Cul3 dimer, the 3-box domain contacts Cul3 with a feature in analog to F box and SOCS box in other Cullin complexes, whereas the BTB domain is responsible for both Cul3 binding and dimerization. However, using a longer SPOP construct containing the entire BACK domain, SPOP forms large oligomers, which is due to the self-association property of the BACK domain (Errington et al. 2012). Importantly, SPOP oligomer possesses a higher ubiquitination ability compared to SPOP dimer, likely due to an increase in substrate binding avidity. The formation of large SPOP oligomer is consistent with the finding that SPOP is recruited to nuclear speckles and other nuclear bodies such as DNA damage foci and PML nuclear bodies (Boysen et al. 2015; Gan et al. 2015; Kwon et al. 2006; Marzahn et al. 2016; Nagai et al. 1997). Recent studies indicate that formation of these intracellular membraneless bodies requires a liquid-liquid phase separation (LLPS) process (Boeynaems et al. 2018). Although SPOP oligomerization is required for LLPS and its localization to the nuclear bodies, it cannot independently undergo LLPS (Marzahn et al. 2016). Subsequent study indicates that substrate interaction together with SPOP oligomerization drives LLPS to form an active ubiquitination compartment (Bouchard et al. 2018). Interestingly, the cancer-associated SPOP mutant (see below) disrupts LLPS, correlating with its functional deficit.

12.3.2 The Tumor-Suppressive Functions of SPOP

12.3.2.1 Prostate Cancer

SPOP elicits context-dependent roles in carcinogenesis (Fig. 12.2). The tumor-suppressive functions of SPOP have been best demonstrated in the prostate cancer setting. Several large-scale genetic analyses revealed that heterozygous missense mutations in *SPOP* gene are observed in ~10% of primary and metastatic prostate cancer, making *SPOP* as one of the most frequently mutated genes in this cancer type (Baca et al. 2013; Barbieri et al. 2012; Berger et al. 2011; Grasso et al. 2012; Kan et al. 2010). Furthermore, SPOP low expression in prostate cancer correlates with poor prognosis (Garcia-Flores et al. 2014). Consistent with a tumor-suppressive role, prostate-specific *SPOP* biallelic deletion in mice leads to the development of prostatic intraepithelial neoplasia (PIN) (Geng et al. 2017). Additionally, knock-in a cancer-associated SPOP mutant ($SPOP^{F133V}$) induces high-grade PIN in conditional heterozygous *PTEN* knockout mice and invasive prostate tumor in conditional homozygous PTEN knockout mice (Blattner et al. 2017). Since SPOP mutation promotes prostate tumorigenesis, it is considered as a driver of prostate cancer.

Unlike Keap1, whose tumor-promoting or suppressing function is mainly mediated by a single substrate, i.e., Nrf2, a number of SPOP substrates can contribute to its tumor-suppressive functions. Among them, androgen receptor (AR) is a highly relevant substrate to prostate cancer, as AR is critical for the initiation and progression of prostate cancer and the development of resistance to anti-androgen therapy (Chen et al. 2004). Intriguingly, SPOP-dependent AR

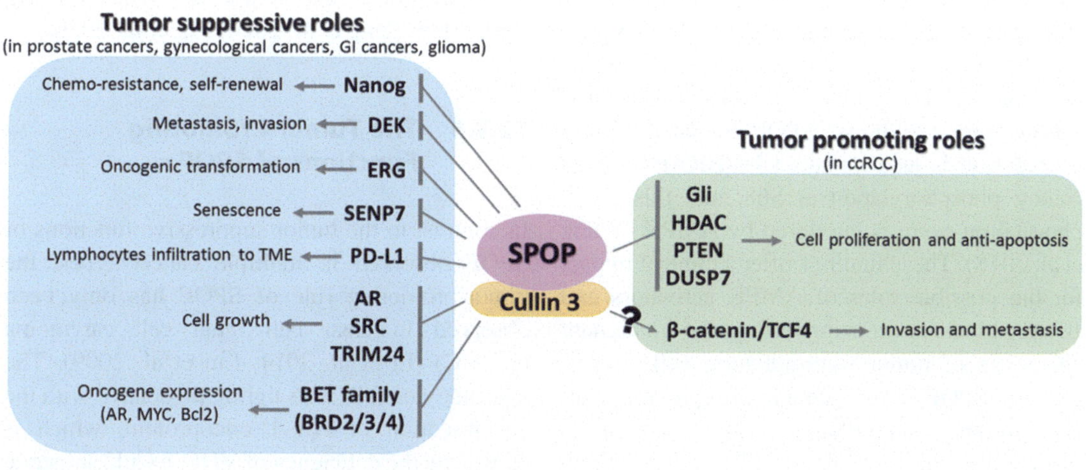

Fig. 12.2 Context-dependent roles of SPOP in cancer

degradation is inhibited by androgens and potentiated by anti-androgens, implying that a ligand-induced conformational change could affect AR binding to SPOP (An et al. 2014). Besides AR, SPOP also targets AR coactivators SRC and TRIM24 for ubiquitination (Geng et al. 2013; Theurillat et al. 2014). Importantly, SPOP mutant fails to act on AR, SRC, and TRIM24 (An et al. 2014; Geng et al. 2013; Groner et al. 2016), which allows the maintenance of a high level of AR signaling for cell proliferation even in the setting of low androgen. Other transcriptional coactivators targeted by SPOP are BET family proteins BRD2/3/4 (Dai et al. 2017; Janouskova et al. 2017; Zhang et al. 2017), which upregulate many oncogenes, including MYC, Bcl-2, and AR to promote oncogenesis. BET inhibitors such as JQ1 and I-BET are emerged as promising anti-cancer therapies (Belkina and Denis 2012). Importantly, SPOP mutants found in prostate cancer cannot bind BRD2/3/4 and therefore confer resistance to BET inhibitors. Other oncogenic targets of SPOP include DEK, ERG, and SENP7 (An et al. 2015; Gan et al. 2015; Theurillat et al. 2014; Zhu et al. 2015), which function in invasion, oncogenic transformation, and suppression of senescence, respectively.

In addition to bulk tumor cells, SPOP is recently shown to act on cancer stem/initiating cells by targeting the pluripotency-determining factor Nanog for ubiquitination and degradation, thereby inhibiting the self-renewal and chemoresistance of cancer stem/initiating cells (Wang et al. 2018; Zhang et al. 2018b). Again, this function is impaired by prostate cancer-associated SPOP mutants, leading to elevated cancer stem cell traits. SPOP-mediated Nanog degradation is also impaired by BRAF-induced Nanog phosphorylation at S68, and this phosphorylation event is interfered by AMPK (Wang et al. 2018). These findings offer a clinical impact for the possible roles of AMPK activators and BRAF blockers in suppressing the malignant characters of tumor stem/initiating cells. Additionally, SPOP is also capable of regulating cancer immune surveillance. This function is mediated by the degradation of immune checkpoint protein PD-L1. SPOP loss-of-function mutation compromises this effect, leading to increased PD-L1 level and reduced infiltration of lymphocytes into TME. Importantly, SPOP level is negatively regulated by CDK4/6, making the combination of CDK4/6 inhibitor with immune checkpoint blockade as a rationale antitumor strategy (Zhang et al. 2018a).

12.3.2.2 Other Cancer Types

SPOP loss-of-function mutations have been frequently detected in gynecological malignancies, including ovarian cancer, breast cancer, and endometrial cancer (Hu et al. 2016; Le Gallo et al. 2012; Li et al. 2011). In certain cancer types, SPOP expression correlates inversely with tumor grade. SPOP targets estrogen receptor-α and progesterone receptor for ubiquitination and degradation in endometrial and breast cancer, respectively (Gao et al. 2015; Zhang et al. 2015b). Similar to prostate cancer, the cancer-associated mutations lose these functions to lead to aberrant elevation of hormone signaling.

SPOP tumor suppressor functions have been implicated in the tumors of GI tracts and glioma. Distinct from prostate and gynecological cancers, few SPOP mutations are discovered from GI cancers (Kim et al. 2014). Rather, reduced expression of SPOP has been reported for liver, gastric, colon cancers, and glioma (Ding et al. 2015; Huang et al. 2015; Kim et al. 2013; Xu et al. 2015). In these cancer types, SPOP substrates such as Gli, HDAC, and ZEB2 may contribute to the tumor-suppressive effect (Huang et al. 2015; Zeng et al. 2014; Zhi et al. 2016).

12.3.3 The Tumor-Promoting Functions of SPOP

In contrast to the tumor-suppressive functions of SPOP observed in multiple cancer types, the tumor-promoting role of SPOP has only been observed in clear cell renal cell carcinoma (ccRCC) (Li et al. 2014; Liu et al. 2009). The pathology of ccRCC is tightly associated with the accumulation of HIF-1 oncoprotein, which is caused by the deficiency of VHL, a Cul2 substrate adaptor targeting HIF-1α (Gossage and Eisen

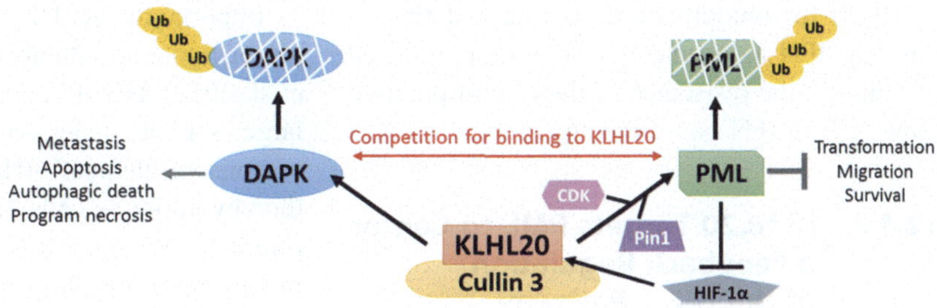

Fig. 12.3 The tumor-promoting functions of KLHL20

2010). Importantly, the pseudohypoxia condition in ccRCC promotes SPOP cytoplasmic retention (Li et al. 2014). This accumulation of cytoplasmic SPOP confers tumor-promoting activities and is associated with unfavorable prognosis of ccRCC (Zhao et al. 2016), which is opposite to the function of nuclear SPOP. Mechanistically, the cytoplasmic SPOP controls the ubiquitination and subsequent degradation of multiple regulators of cell proliferation and apoptosis, such as PTEN, DUSP7, Daxx, and Gli2 (Li et al. 2014). Furthermore, SPOP enhances the activation of the β-catenin/TCF4 axis through unknown mechanism to promote ccRCC invasion and metastasis (Zhao et al. 2016). Small molecular inhibitors that disrupt the binding of SPOP-substrate interaction induce apoptosis of ccRCC cells (Guo et al. 2016). Thus, SPOP targeting may be an attractive therapeutic strategy for ccRCC.

12.4 KLHL20

12.4.1 KLHL20 Targets Tumor Suppressor DAPK in Regulating Interferon Responses

Similar to other KLHL proteins, KLHL20 carries a BTB domain, a BACK domain, and six kelch-repeats and functions as a substrate binding subunit of Cul3 (Lee et al. 2010). KLHL20 was identified as an interacting protein of death-associated protein kinase (DAPK), which elicits pleiotropic tumor-suppressive functions, including the induction of apoptosis, autophagic death, and program necrosis as well as the suppression of tumor cell invasion and metastatic dissemination (Chen et al. 2014a; Ivanovska et al. 2014; Lee et al. 2010). Furthermore, DAPK activity or expression is frequently downregulated in tumors by posttranscriptional, posttranslational, or epigenetic mechanism (Bialik and Kimchi 2004; Chen et al. 2012; Raval et al. 2007; Wang et al. 2007). DAPK interacts with KLHL20 through its kelch domain, which promotes DAPK ubiquitination and subsequent degradation (Lee et al. 2010). In line with this finding, a global analysis identifies DAPK as a substrate candidate of Cul3 (Emanuele et al. 2011). Thus, DAPK is a bona fide substrate of Cul3-KLHL20. Through degrading DAPK, KLHL20 antagonizes the functions of DAPK in suppressing cell survival and tumor cell migration and invasion (Lee et al. 2010) (Fig. 12.3).

Importantly, the KLHL20-mediated DAPK ubiquitination is regulated by interferon (IFN). IFN-α or IFN-γ transcriptionally induces the expression of PML (Lavau et al. 1995; Stadler et al. 1995), the key factor for the assembly of PML nuclear bodies (Bernardi et al. 2006). With the increase of PML expression, KLHL20 is relocated to PML nuclear body, which is mediated in part by a competition between DAPK and PML for binding KLHL20 (Lee et al. 2010). Consequently, DAPK binding to KLHL20 is decreased, and therefore DAPK is stabilized. This DAPK stabilization contributes to apoptosis and autophagic death induced by IFN. In certain multiple myeloma cells where IFN fails to efficiently upregulate PML expression and induce PML nuclear body formation, this KLHL20 relocation is abrogated, and therefore DAPK is persistently associated with

KLHL20 for ubiquitination and degradation. In this way, IFN-induced cell death is compromised, resulting in the resistance of these multiple myeloma cells to IFN-based therapy.

12.4.2 KLHL20 Targets PML to Confer a Feedback Regulation of the HIF-1 Pathway

The localization of KLHL20 in the PML nuclear bodies and the ability of PML to compete with DAPK for binding KLHL20 raise a possibility for PML as a KLHL20 substrate. The *PML* gene was identified based on a t(15:17) chromosome translocation occurred in acute promyelocytic leukemia, and this translocation leads to the generation of the oncogenic PML-RARα fusion protein (de The et al. 1990). The PML protein elicits pleiotropic antitumor effects, such as suppression of proliferation, angiogenesis, cell migration, and metastasis and promotion of apoptosis and senescence (Bernardi et al. 2006; Bernardi and Pandolfi 2007; Reineke et al. 2010; Salomoni et al. 2008). Furthermore, PML suppresses cancer stem cell maintenance and regulates cancer cell metabolism (Carracedo et al. 2012; Ito et al. 2008). In accordance to these tumor-suppressive functions, PML protein, but not its mRNA, is downregulated in many cancer types, including prostate, lung, colon, and breast cancers, brain tumors, non-Hodgkin's lymphoma, and germ cell tumors (Gurrieri et al. 2004). The Cul3-KLHL20 E3 ligase complex is one of the ubiquitin ligases that is responsible for PML ubiquitination and its proteasomal degradation in tumors. However, targeting PML to KLHL20 requires two consecutive posttranslational modifications, that is, phosphorylation at S518 by CDK1/2 kinase followed by prolyl-isomerization of the pS518-P519 peptide bond by Pin1 (Yuan et al. 2011). This mechanism explains the previous observation for a gradual decline of the PML protein abundance with the progression of cell cycle (Dellaire et al. 2006), which correlates with the gradual increase of CDK1/2 kinase activity.

Importantly, KLHL20 is a transcriptional target of hypoxia-inducible factor-1 (HIF-1) (Yuan et al. 2011). HIF-1 activates the transcription of a large set of genes containing the "hypoxia-responsive element" (HRE) on their promoters, thereby allowing cells to adapt the hypoxia environment (Wenger et al. 2005). This transcriptional program plays key roles in promoting many cancer hallmarks, including immortalization, metabolic reprogramming, angiogenesis, growth, migration, invasion, metastasis, cancer stemness maintenance, and treatment resistance (Semenza 2010). The promoter of *KLHL20* gene contains two HREs, and both of them are involved in HIF-1 binding and hypoxia-induced transactivation (Yuan et al. 2011). The induction of KLHL20 by HIF-1 leads to the enhancement of PML ubiquitination and degradation under hypoxia conditions. Interestingly, PML negatively regulates the translation of HIF-1 through a mechanism involving mTOR repression (Bernardi et al. 2006). This PML-induced, mTOR-mediated HIF-1α downregulation, together with the HIF-1-induced, KLHL20-mediated PML degradation, forms a double-negative feedback loop to result in a robust induction of HIF-1α under hypoxia (Yuan et al. 2011). Through this feedback mechanism, KLHL20-dependent PML ubiquitination causes the blockage of various PML tumor-suppressive functions and a robust induction of many tumor hypoxia responses, and these two arms could act in concert to greatly potentiate cancer aggressiveness (Fig. 12.3).

12.4.3 Dysregulation of KLHL20 in Cancer

Consistent with the frequent upregulation of HIF-1α expression in tumors through hypoxia-dependent or independent mechanism (Semenza 2010), KLHL20 expression is higher in prostate cancers than benign prostatic hyperplasia. Furthermore, KLHL20 high expression is associated with the upregulation of HIF-1α, the upregulation of Pin1, but the downregulation of PML (Yuan et al. 2011). More importantly, prostate cancer patients carrying the signature expression pattern of high HIF-1α, high KLHL20, high Pin1, and low PML are increased with the increase of tumor grade. These observations thus highlight a key role of the KLHL20-PML pathway in prostate cancer progression and suggest a promise for targeting this pathway in treating the aggressive prostate cancers.

Besides the upregulation of KLHL20 expression, the Cul3-KLHL20 E3 ligase can be regulated in cancer at the level of its activity. Interestingly, another BTB-kelch protein, KLHL39, functions as an inhibitor of the Cul3-KLHL20 E3 ligase (Chen et al. 2015). Although KLHL39 shares a similar domain architecture with KLHL20, it does not form a complex with Cul3. Instead, KLHL39 binds KLHL20, and this binding is mediated by their kelch domains. In this way, KLHL39 competes with DAPK and PML for binding KLHL20 and functions as a pseudosubstrate of the KLHL20-based ubiquitin ligase. KLHL39 also inhibits the interaction of KLHL20 with Cul3 through a yet unknown mechanism. With these dual inhibitory mechanisms, KLHL39 suppresses KLHL20-mediated ubiquitination and degradation of tumor suppressor proteins DAPK and PML. In human colon cancer patients, KLHL39 low expression is associated with low expression of DAPK and PML. KLHL39 low expression also correlates with higher tumor grade and lymph node and distant metastasis in this cancer type. In accordance with the clinical observations, KLHL39 diminishes the migration, invasion, and metastasis effects on colon cancer cells, and these tumor-suppressive effects are dependent on DAPK and PML upregulation. Thus, KLHL39 functions as a tumor suppressor, and its tumor-suppressive function is mediated in part by blocking KLHL20-dependent ubiquitination of DAPK and PML.

12.5 LZTR1

LZTR1 is an unusual BTB-kelch protein with six-kelch repeats present in the N-terminus followed by two BTB-BACK domains. LTZR1 was originally identified as a Golgi-localized protein with unknown function (Nacak et al. 2006). However, somatic mutations in *LZTR1* have been identified in glioblastoma (Frattini et al. 2013), whereas heterozygous germ line mutations are associated with schwannomatosis, a rare cancer-prone disorder predisposing to late-onset schwannomas and meningiomas (Piotrowski et al. 2014). Subsequently, mutations in *LZTR1* were found in a population of patients with clinical diagnosis as Noonan syndrome (NS), a type of RASopathies with a wide spectrum of developmental disorders and predisposition to certain cancers (Chen et al. 2014b; Yamamoto et al. 2015). Moreover, comprehensive genomic analysis of hepatocellular carcinoma patient samples revealed *LZTR1* as one of the frequently mutated genes in this cancer type (Cancer Genome Atlas Research Network. Electronic address and Cancer Genome Atlas Research 2017). Comprehensive genomic analysis of genes involving in the ubiquitin pathway from 9125 tumor samples across 33 cancer types also found that *LZTR1* is among the frequently mutated genes (Ge et al. 2018). All of these findings support a tight link of LZTR1 to cancer.

Recently, two groups identified the molecular mechanism for the tumor-suppressive function of LZTR1. One group generated LZTR1 heterozygous mice which exhibit several features resembling NS (Steklov et al. 2018). Subsequent studies identified H-Ras as a LZTR1 interacting protein. Another group used a genetic screen to identify genes whose disruption results in resistance to tyrosine kinase inhibitors and recovered LZTR1 (Bigenzahn et al. 2018). Both groups reported that Cul3-LZTR1 targets Ras for monoubiquitination (likely multi-mono-ubiquitination). However, the disease-associated LTZR1 cannot promote Ras ubiquitination due to the inability to bind Ras (for kelch domain mutants) or mislocalization in the cytoplasm (for BTB/BACK domain mutants). Ubiquitination of the K170 residue of H-Ras impairs its membrane localization. Structural simulation suggests that ubiquitination at this residue impedes the insertion of palmitoylated and farnesylated Ras into membrane. Thus, these findings identify LZTR1 as a suppressor of Ras signaling, which explains the association of LZTR1 loss-of-function mutations with several cancer-prone diseases. Furthermore, LZTR mutations could contribute to TKI resistance of cancer cells.

Besides Ras, LZTR1 is found to target the oncoprotein RIT1 (Castel et al. 2019), a Ras-related small GTPase known to regulate cell survival (Shi et al. 2013). Different from Ras, LZTR1 mediates a K48-linked ubiquitination on RIT1, leading to its degradation by the 26S proteasome. RIT1 mutations have been identified in a subset of NS patients (Aoki et al. 2013; Yaoita et al. 2016). Importantly, these disease-associated mutants are defective in interacting with LZTR1, and therefore are resistant to LZTR1-mediated ubiquitination and degradation (Castel et al. 2019). Mouse knocking-in one of such mutations results in phenotypes resembling of NS, accumulation of RIT1 protein, and hyperactivation of MAPK signaling. Cancer-associated LTZR1 mutants are also defective in promoting RIT1 degradation. Thus, promoting proteolytic turnover of Ras and Ras-related GTPases accounts for the tumor-suppressive function of LZTR1.

12.6 Other Cul3 Adaptors with Functions in Cancer

KLHL22 was identified for its role in mitosis, at which it promotes PLK1 monoubiquitination at K492 (Schmucker and Sumara 2014). This modification leads to the dissociation of PLK1 from kinetochore at metaphase to anaphase transition to facilitate a faithful chromosome segregation. Recently, KLHL22 was identified as a regulator of mTOR signaling (Chen et al. 2018), a key cell growth pathway frequently activated in many cancer types (Kim and Guan 2019). In response to amino acid stimulation, KLHL20 targets DEPDC5, an essential component of GATOR1, for K48-linked ubiquitination (Chen et al. 2018). Consequently, GATOR1 is impaired in its GAP activity toward mTOR, leading to mTOR reactivation. Through this mTOR reactivation mechanism, KLHL22 promotes tumorigenesis in nude mice. Furthermore, elevated KLHL22 expression is observed in breast cancer patients, further highlighting a tumor-promoting function of KLHL22.

KLHL6 is a Cul3 adaptor specifically expressed in lymphoid tissues, especially in sheep Peyer's patch and the germinal center B cells in mice and human (Gupta-Rossi et al. 2003). Knockout mice studies indicate an important role of KLHL6 in B-cell differentiation at the level of immature to mature transition in the bone marrow (Bertocci et al. 2017; Kroll et al. 2005). Consistently, the role of KLHL6 in neoplasm is observed in B-cell lymphoma. Recurrent mutations in KLHL6 have been reported for several mature B-cell malignancies, including chronic lymphocytic leukemia, diffuse large B-cell lymphoma (DLBCL), follicular lymphoma, and marginal zone lymphoma (Ganapathi et al. 2016; Kunder et al. 2017; Lohr et al. 2012; Mareschal et al. 2017; Morin et al. 2011; Pasqualucci et al. 2001; Puente et al. 2011; Reddy et al. 2017; Sutton et al. 2015; Weigert et al. 2012). Among them, KLHL6 mutations are most frequently found in DLBCL, the most common type of lymphoma derived from germinal/post-germinal center B cells (Teras et al. 2016). A subtype of DLCBL called ABC-DLCBL relies on the B-cell receptor-mediated NFκB activation for their proliferation and survival (Young et al. 2015). A recent study revealed an intricate link between KLHL6 and NFκB signaling. This group identified the RNA-binding protein Roquin 2 as an ubiquitin substrate of Cul3-KLHL6 (Choi et al. 2018). Roquin 2 binds to the 3′UTR region of a set of RNAs that contain the constitutive-decay element to promote their degradation (Leppek et al. 2013). Interestingly, A20, which participates in the negative feedback regulation of NFκB signaling, is among the targets of Roquin 2 (Choi et al. 2018; Murakawa et al.

2015). Thus, KLHL6-mediated Roquin 2 degradation elevates *A20* mRNA level, which inactivates the NFκB pathway by targeting key regulators of this pathway, such as MALT1 and IKKγ, for degradation (Choi et al. 2018). Through this mechanism, KLHL6 negatively regulates the NFκB pathway. Importantly, both KLHL6 and A20 are the targets of BCR-NFkB signaling, indicating the existence of a double-negative feedback loops to turn off NFκB signaling. The cancer-associated KLHL6 mutants fail to bind Roquin 2, resulting in a sustained NFκB activation to promote lymphoma formation.

KCTD proteins are a subfamily of BTB-kelch proteins. In contrast to KLHL subfamily, KCTD proteins do not contain BACK domain with the 3-box motif essential for Cul3 binding. However, a number of KCTD proteins do bind Cul3 to form oligomeric complexes (Ji et al. 2016). KCTD11 (also known as REN) was identified as a tumor suppressor in medulloblastoma. KCTD11 is deleted or downregulated in medulloblastoma and inhibits medulloblastoma growth by inhibiting Hedgehog pathway (Di Marcotullio et al. 2004), a major oncogenic pathway in this cancer type (Pak and Segal 2016). Subsequent studies identified HDAC1 as a substrate of KCTD11, and KCTD11-mediated HDAC1 degradation potentiates the acetylation of Gli, an effector of the Hedgehog pathway, to increase its transcriptional activity (Canettieri et al. 2010). Interestingly, two other KCTD proteins, namely, KCTD6 and KCTD21, also target HDAC1 ubiquitination, and their epigenetic silencing and deletions are also detected in medulloblastoma (De Smaele et al. 2011), suggesting their collective functions as natural HDAC inhibitors.

12.7 Therapeutic Implications

12.7.1 Clinical Applications of the Keap1-Nrf2 Pathway Modulators

Due to the dual functions of the Keap1-Nrf2 pathway in cancer, modulation of this pathway could offer therapeutic benefits. For example, agents that activate Nrf2 may be exploited for cancer prevention, while Nrf2 inhibitory agents may be used as adjuvants in chemotherapy and radiotherapy to combat the treatment resistance. A number of phytochemicals and small molecular compounds activate Nrf2 by directly modifying the cysteine residues of KEAP1 to inactivate its ubiquitination function. The prototype and most studied agent is sulforaphane A, which has been shown to exert preventive effects against many cancer types (Zhang and Tang 2007). Several other chemicals activating Nrf2 through this mechanism, such as DMF (dimethyl fumarate) and CDDO-me, have been tested in clinical trials (Lu et al. 2016). Compared with these electrophilic inducers, blockage of protein-protein interaction between Keap1 and Nrf2 may be an alternative and more specific approach to activate Nrf2. This area is currently under intensive investigation, and several peptide and small molecular inhibitors, such as tetrahydroisoquinolines and naphthalene sulfonamides, have been developed (Wells 2015). Regarding Nrf2 inhibitors, a set of small molecules have been developed to suppress Nrf2 activity or expression, including brusatol and IM3829 (Lee et al. 2012; Ren et al. 2011). Finally, autophagy pathway that crosstalks with the Keap1-Nrf2 pathway through p62-dependent degradation may be exploited for modulating Nrf2 activity.

12.7.2 Clinical Applications of SPOP Modulators

Based on the context-dependent functions of SPOP in cancer, agents that block SPOP function may offer beneficial effects in ccRCC, whereas SPOP activators may be exploited to combat prostate cancer. To date, only the former has been developed. Using a structure-based design followed by hit optimization, a small molecular compound called compound 6b is generated to interrupt the binding of SPOP to its substrate (Guo et al. 2016). As mentioned above, this compound suppresses the oncogenic pathway of SPOP in ccRCC, induces ccRCC cell death in vitro, and suppresses ccRCC growth in vivo.

Interestingly, compound 6b elicits a minimal effect in cells showing a nuclear localization of SPOP, suggesting its selectivity for targeting ccRCC.

12.8 Conclusions and Future Perspectives

Based on the literatures, a number of Cul3 containing E3 ligase complexes have exerted functions in different aspects of tumor biology and are mutated or dysregulated in various types of human cancer. Furthermore, certain Cul3 complexes, such as Cul3-Keap1 and Cul3-SPOP, show dual roles in tumor, depending on cancer types and tumor stages. This wealth of information makes the regulation of neoplasms as one prominent function of Cul3 family ubiquitin ligases and highlights the potential for modulating the activity of various Cul3 ubiquitin ligases as preventive or therapeutic strategies of this devastating disease. However, what we have known about the role of Cul3 family ligases in cancer is likely only the tip of the iceberg. Among the ~180 BTB proteins identified in human genome, most of them have not been characterized in terms of the capability to be recruited to Cul3 E3 ligase complex and the function or expression/genetic alterations in cancer. In the future, it would be interesting to see how many other BTB proteins can form Cul3 complex and can regulate various cancer hallmarks. For Cul3 adaptors with important functions in cancer, strategies for modulating their activities are needed to be explored. In this aspect, structure-based inhibitor design could be a powerful tool for identifying specific agents that alter the binding between individual adaptor and substrate, as in the case of the development of SPOP inhibitor. In this regard, identifying the specific degron sequences for each Cul3 complex and analyzing the structural basis for the interaction would be an important direction for future studies.

Acknowledgments We thank Ms. Min-Yu Huang for drawing figures. This work was supported by MOST Frontier Grant 107-2321-B001-005 and intramural fund from Institute of Biological Chemistry, Academia Sinica.

References

Adam J, Hatipoglu E, O'Flaherty L, Ternette N, Sahgal N, Lockstone H, Baban D, Nye E, Stamp GW, Wolhuter K, Stevens M, Fischer R, Carmeliet P, Maxwell PH, Pugh CW, Frizzell N, Soga T, Kessler BM, El-Bahrawy M, Ratcliffe PJ, Pollard PJ (2011) Renal cyst formation in Fh1-deficient mice is independent of the Hif/Phd pathway: roles for fumarate in KEAP1 succination and NRF2 signaling. Cancer Cell 20:524–537

An J, Wang C, Deng Y, Yu L, Huang H (2014) Destruction of full-length androgen receptor by wild-type SPOP, but not prostate-cancer-associated mutants. Cell Rep 6:657–669

An J, Ren S, Murphy SJ, Dalangood S, Chang C, Pang X, Cui Y, Wang L, Pan Y, Zhang X, Zhu Y, Wang C, Halling GC, Cheng L, Sukov WR, Karnes RJ, Vasmatzis G, Zhang Q, Zhang J, Cheville JC, Yan J, Sun Y, Huang H (2015) Truncated ERG oncoproteins from TMPRSS2-ERG fusions are resistant to SPOP-mediated proteasome degradation. Mol Cell 59:904–916

Aoki Y, Niihori T, Banjo T, Okamoto N, Mizuno S, Kurosawa K, Ogata T, Takada F, Yano M, Ando T, Hoshika T, Barnett C, Ohashi H, Kawame H, Hasegawa T, Okutani T, Nagashima T, Hasegawa S, Funayama R, Nagashima T, Nakayama K, Inoue S, Watanabe Y, Ogura T, Matsubara Y (2013) Gain-of-function mutations in RIT1 cause Noonan syndrome, a RAS/MAPK pathway syndrome. Am J Hum Genet 93:173–180

Arfmann-Knubel S, Struck B, Genrich G, Helm O, Sipos B, Sebens S, Schafer H (2015) The crosstalk between NRF2 and TGF-beta1 in the epithelial-mesenchymal transition of pancreatic duct epithelial cells. PLoS One 10:e0132978

Baca SC, Prandi D, Lawrence MS, Mosquera JM, Romanel A, Drier Y, Park K, Kitabayashi N, Macdonald TY, Ghandi M, Van Allen E, Kryukov GV, Sboner A, Theurillat JP, Soong TD, Nickerson E, Auclair D, Tewari A, Beltran H, Onofrio RC, Boysen G, Guiducci C, Barbieri CE, Cibulskis K, Sivachenko A, Carter SL, Saksena G, Voet D, Ramos AH, Winckler W, Cipicchio M, Ardlie K, Kantoff PW, Berger MF, Gabriel SB, Golub TR, Meyerson M, Lander ES, Elemento O, Getz G, Demichelis F, Rubin MA, Garraway LA (2013) Punctuated evolution of prostate cancer genomes. Cell 153:666–677

Baird L, Lleres D, Swift S, Dinkova-Kostova AT (2013) Regulatory flexibility in the NRF2-mediated stress response is conferred by conformational cycling of the KEAP1-NRF2 protein complex. Proc Natl Acad Sci U S A 110:15259–15264

Barbieri CE, Baca SC, Lawrence MS, Demichelis F, Blattner M, Theurillat JP, White TA, Stojanov P, Van Allen E, Stransky N, Nickerson E, Chae SS, Boysen G, Auclair D, Onofrio RC, Park K, Kitabayashi N, Macdonald TY, Sheikh K, Vuong T, Guiducci C, Cibulskis K, Sivachenko A, Carter SL, Saksena G,

Voet D, Hussain WM, Ramos AH, Winckler W, Redman MC, Ardlie K, Tewari AK, Mosquera JM, Rupp N, Wild PJ, Moch H, Morrissey C, Nelson PS, Kantoff PW, Gabriel SB, Golub TR, Meyerson M, Lander ES, Getz G, Rubin MA, Garraway LA (2012) Exome sequencing identifies recurrent SPOP, FOXA1 and MED12 mutations in prostate cancer. Nat Genet 44:685–689

Bauer AK, Cho HY, Miller-Degraff L, Walker C, Helms K, Fostel J, Yamamoto M, Kleeberger SR (2011) Targeted deletion of NRF2 reduces urethane-induced lung tumor development in mice. PLoS One 6: e26590

Belkina AC, Denis GV (2012) Bet domain co-regulators in obesity, inflammation and cancer. Nat Rev Cancer 12:465–477

Bennett EJ, Rush J, Gygi SP, Harper JW (2010) Dynamics of cullin-RING ubiquitin ligase network revealed by systematic quantitative proteomics. Cell 143:951–965

Berger MF, Lawrence MS, Demichelis F, Drier Y, Cibulskis K, Sivachenko AY, Sboner A, Esgueva R, Pflueger D, Sougnez C, Onofrio R, Carter SL, Park K, Habegger L, Ambrogio L, Fennell T, Parkin M, Saksena G, Voet D, Ramos AH, Pugh TJ, Wilkinson J, Fisher S, Winckler W, Mahan S, Ardlie K, Baldwin J, Simons JW, Kitabayashi N, Macdonald TY, Kantoff PW, Chin L, Gabriel SB, Gerstein MB, Golub TR, Meyerson M, Tewari A, Lander ES, Getz G, Rubin MA, Garraway LA (2011) The genomic complexity of primary human prostate cancer. Nature 470:214–220

Bernardi R, Pandolfi PP (2007) Structure, dynamics and functions of promyelocytic leukaemia nuclear bodies. Nat Rev Mol Cell Biol 8:1006–1016

Bernardi R, Guernah I, Jin D, Grisendi S, Alimonti A, Teruya-Feldstein J, Cordon-Cardo C, Simon MC, Rafii S, Pandolfi PP (2006) PML inhibits HIF-1alpha translation and neoangiogenesis through repression of mtor. Nature 442:779–785

Bertocci B, Lecoeuche D, Sterlin D, Kuhn J, Gaillard B, De Smet A, LEMBO F, Bole-Feysot C, Cagnard N, Fadeev T, Dahan A, Weill JC, Reynaud CA (2017) KLHL6 deficiency impairs transitional B cell survival and differentiation. J Immunol 199:2408–2420

Bialik S, Kimchi A (2004) DAP-kinase as a target for drug design in cancer and diseases associated with accelerated cell death. Semin Cancer Biol 14:283–294

Bigenzahn JW, Collu GM, Kartnig F, Pieraks M, Vladimer GI, Heinz LX, Sedlyarov V, Schischlik F, Fauster A, Rebsamen M, Parapatics K, Blomen VA, Muller AC, Winter GE, Kralovics R, Brummelkamp TR, Mlodzik M, Superti-Furga G (2018) LZTR1 is a regulator of RAS ubiquitination and signaling. Science 362:1171–1177

Blattner M, Liu D, Robinson BD, Huang D, Poliakov A, Gao D, Nataraj S, Deonarine LD, Augello MA, Sailer V, Ponnala L, Ittmann M, Chinnaiyan AM, Sboner A, Chen Y, Rubin MA, Barbieri CE (2017) SPOP mutation drives prostate tumorigenesis in vivo through coordinate regulation of PI3K/MTOR and AR signaling. Cancer Cell 31:436–451

Boeynaems S, Alberti S, Fawzi NL, Mittag T, Polymenidou M, Rousseau F, Schymkowitz J, Shorter J, Wolozin B, Van Den Bosch L, Tompa P, Fuxreiter M (2018) Protein phase separation: a new phase in cell biology. Trends Cell Biol 28:420–435

Bouchard JJ, Otero JH, Scott DC, Szulc E, Martin EW, Sabri N, Granata D, Marzahn MR, Lindorff-Larsen K, Salvatella X, Schulman BA, Mittag T (2018) Cancer mutations of the tumor suppressor SPOP disrupt the formation of active, phase-separated compartments. Mol Cell 72:19–36 e8

Boysen G, Barbieri CE, Prandi D, Blattner M, Chae SS, Dahija A, Nataraj S, Huang D, Marotz C, Xu L, Huang J, Lecca P, Chhangawala S, Liu D, Zhou P, Sboner A, De Bono JS, Demichelis F, Houvras Y, Rubin MA (2015) SPOP mutation leads to genomic instability in prostate cancer. Elife 4:e09207

Cancer Genome Atlas Research Network, Electronic Address WBE, Cancer Genome Atlas Research N (2017) Comprehensive and integrative genomic characterization of hepatocellular carcinoma. Cell 169:1327–1341 e23

Canettieri G, Di Marcotullio L, Greco A, Coni S, Antonucci L, Infante P, Pietrosanti L, De Smaele E, Ferretti E, Miele E, Pelloni M, De Simone G, Pedone EM, Gallinari P, Giorgi A, Steinkuhler C, Vitagliano L, Pedone C, Schinin ME, Screpanti I, Gulino A (2010) Histone deacetylase and Cullin3-REN(KCTD11) ubiquitin ligase interplay regulates Hedgehog signalling through Gli acetylation. Nat Cell Biol 12:132–142

Canning P, Cooper CD, Krojer T, Murray JW, Pike AC, Chaikuad A, Keates T, Thangaratnarajah C, Hojzan V, Ayinampudi V, Marsden BD, Gileadi O, Knapp S, Von Delft F, Bullock AN (2013) Structural basis for Cul3 protein assembly with the BTB-Kelch family of E3 ubiquitin ligases. J Biol Chem 288:7803–7814

Carracedo A, Weiss D, Leliaert AK, Bhasin M, De Boer VC, Laurent G, Adams AC, Sundvall M, Song SJ, Ito K, Finley LS, Egia A, Libermann T, Gerhart-Hines-Z, Puigserver P, Haigis MC, Maratos-Flier E, Richardson AL, Schafer ZT, Pandolfi PP (2012) A metabolic prosurvival role for PML in breast cancer. J Clin Invest 122:3088–3100

Castel P, Cheng A, Cuevas-Navarro A, Everman DB, Papageorge AG, Simanshu DK, Tankka A, Galeas J, Urisman A, Mccormick F (2019) RIT1 oncoproteins escape LZTR1-mediated proteolysis. Science 363:1226–1230

Chen HY, Chen RH (2016) Cullin 3 ubiquitin ligases in cancer biology: functions and therapeutic implications. Front Oncol 6:113

Chen CD, Welsbie DS, Tran C, Baek SH, Chen R, Vessella R, Rosenfeld MG, Sawyers CL (2004) Molecular determinants of resistance to antiandrogen therapy. Nat Med 10:33–39

Chen W, Sun Z, Wang XJ, Jiang T, Huang Z, Fang D, Zhang DD (2009) Direct interaction between NRF2 and p21(Cip1/WAF1) upregulates the NRF2-mediated antioxidant response. Mol Cell 34:663–673

Chen HY, Lin YM, Chung HC, Lang YD, Lin CJ, Huang J, Wang WC, Lin FM, Chen Z, Huang HD, Shyy JY, Liang JT, Chen RH (2012) miR-103/107 promote metastasis of colorectal cancer by targeting the metastasis suppressors DAPK and KLF4. Cancer Res 72:3631–3641

Chen HY, Lee YR, Chen RH (2014a) The functions and regulations of DAPK in cancer metastasis. Apoptosis 19:364–370

Chen PC, Yin J, Yu HW, Yuan T, Fernandez M, Yung CK, Trinh QM, Peltekova VD, Reid JG, Tworog-Dube E, Morgan MB, Muzny DM, Stein L, Mcpherson JD, Roberts AE, Gibbs RA, Neel BG, Kucherlapati R (2014b) Next-generation sequencing identifies rare variants associated with Noonan syndrome. Proc Natl Acad Sci U S A 111:11473–11478

Chen HY, Hu JY, Chen TH, Lin YC, Liu X, Lin MY, Lang YD, Yen Y, Chen RH (2015) KLHL39 suppresses colon cancer metastasis by blocking KLHL20-mediated PML and DAPK ubiquitination. Oncogene 34:5141–5151

Chen J, Ou Y, Yang Y, Li W, Xu Y, Xie Y, Liu Y (2018) KLHL22 activates amino-acid-dependent mTORC1 signalling to promote tumorigenesis and ageing. Nature 557:585–589

Cheng J, Guo J, Wang Z, North BJ, Tao K, Dai X, Wei W (2018) Functional analysis of Cullin 3 E3 ligases in tumorigenesis. Biochim Biophys Acta Rev Cancer 1869:11–28

Chio IIC, Jafarnejad SM, Ponz-Sarvise M, Park Y, Rivera K, Palm W, Wilson J, Sangar V, Hao Y, Ohlund D, Wright K, Filippini D, Lee EJ, Da Silva B, Schoepfer C, Wilkinson JE, Buscaglia JM, Denicola GM, Tiriac H, Hammell M, Crawford HC, Schmidt EE, Thompson CB, Pappin DJ, Sonenberg N, Tuveson DA (2016) NRF2 promotes tumor maintenance by modulating mRNA translation in pancreatic cancer. Cell 166:963–976

Choi J, Lee K, Ingvarsdottir K, Bonasio R, Saraf A, Florens L, Washburn MP, Tadros S, Green MR, Busino L (2018) Loss of KLHL6 promotes diffuse large B-cell lymphoma growth and survival by stabilizing the mRNA decay factor roquin2. Nat Cell Biol 20:586–596

Copple IM, Lister A, Obeng AD, KitteRINGham NR, Jenkins RE, Layfield R, Foster BJ, GoldRING CE, Park BK (2010) Physical and functional interaction of sequestosome 1 with Keap1 regulates the Keap1-NRF2 cell defense pathway. J Biol Chem 285:16782–16788

Cullinan SB, Gordan JD, Jin J, Harper JW, Diehl JA (2004) The Keap1-BTB protein is an adaptor that bridges NRF2 to a Cul3-based E3 ligase: oxidative stress sensing by a Cul3-Keap1 ligase. Mol Cell Biol 24:8477–8486

Dai X, Gan W, Li X, Wang S, Zhang W, Huang L, Liu S, Zhong Q, Guo J, Zhang J, Chen T, Shimizu K, Beca F, Blattner M, Vasudevan D, Buckley DL, Qi J, Buser L, Liu P, Inuzuka H, Beck AH, Wang L, Wild PJ, Garraway LA, Rubin MA, Barbieri CE, Wong KK, Muthuswamy SK, Huang J, Chen Y, Bradner JE, Wei W (2017) Prostate cancer-associated SPOP mutations confer resistance to BET inhibitors through stabilization of BRD4. Nat Med 23:1063–1071

De Smaele E, Di Marcotullio L, Moretti M, Pelloni M, Occhione MA, Infante P, Cucchi D, Greco A, Pietrosanti L, Todorovic J, Coni S, Canettieri G, Ferretti E, Bei R, Maroder M, Screpanti I, Gulino A (2011) Identification and characterization of KCASH2 and KCASH3, 2 novel Cullin3 adaptors suppressing histone deacetylase and Hedgehog activity in medulloblastoma. Neoplasia 13:374–385

De The H, Chomienne C, Lanotte M, Degos L, Dejean A (1990) The t(15;17) translocation of acute promyelocytic leukaemia fuses the retinoic acid receptor alpha gene to a novel transcribed locus. Nature 347:558–561

Dellaire G, Ching RW, Dehghani H, Ren Y, Bazett-Jones DP (2006) The number of PML nuclear bodies increases in early S phase by a fission mechanism. J Cell Sci 119:1026–1033

Denicola GM, Karreth FA, Humpton TJ, Gopinathan A, Wei C, Frese K, Mangal D, Yu KH, Yeo CJ, Calhoun ES, Scrimieri F, Winter JM, Hruban RH, Iacobuzio-Donahue C, Kern SE, Blair IA, Tuveson DA (2011) Oncogene-induced NRF2 transcription promotes ROS detoxification and tumorigenesis. Nature 475:106–109

Denicola GM, Chen PH, Mullarky E, Sudderth JA, Hu Z, Wu D, Tang H, Xie Y, Asara JM, Huffman KE, Wistuba I, Minna JD, Deberardinis RJ, Cantley LC (2015) NRF2 regulates serine biosynthesis in non-small cell lung cancer. Nat Genet 47:1475–1481

Di Marcotullio L, Ferretti E, De Smaele E, Argenti B, Mincione C, Zazzeroni F, Gallo R, Masuelli L, Napolitano M, Maroder M, Modesti A, Giangaspero F, Screpanti I, Alesse E, Gulino A (2004) REN(KCTD11) is a suppressor of Hedgehog signaling and is deleted in human medulloblastoma. Proc Natl Acad Sci U S A 101:10833–10838

Ding D, Song T, Jun W, Tan Z, Fang J (2015) Decreased expression of the SPOP gene is associated with poor prognosis in glioma. Int J Oncol 46:333–341

Dinkova-Kostova AT, Holtzclaw WD, Cole RN, Itoh K, Wakabayashi N, Katoh Y, Yamamoto M, Talalay P (2002) Direct evidence that sulfhydryl groups of Keap1 are the sensors regulating induction of phase 2 enzymes that protect against carcinogens and oxidants. Proc Natl Acad Sci U S A 99:11908–11913

Dubiel W, Dubiel D, Wolf DA, Naumann M (2018) Cullin 3-based ubiquitin ligases as master regulators of mammalian cell differentiation. Trends Biochem Sci 43:95–107

Eades G, Yang M, Yao Y, Zhang Y, Zhou Q (2011) miR-200a regulates NRF2 activation by targeting Keap1 mRNA in breast cancer cells. J Biol Chem 286:40725–40733

Emanuele MJ, Elia AE, Xu Q, Thoma CR, Izhar L, Leng Y, Guo A, Chen YN, Rush J, Hsu PW, Yen

HC, Elledge SJ (2011) Global identification of modular cullin-RING ligase substrates. Cell 147:459–474

ErRINGton WJ, Khan MQ, Bueler SA, Rubinstein JL, Chakrabartty A, Prive GG (2012) Adaptor protein self-assembly drives the control of a cullin-RING ubiquitin ligase. Structure 20:1141–1153

Fahey JW, Haristoy X, Dolan PM, Kensler TW, Scholtus I, Stephenson KK, Talalay P, Lozniewski A (2002) Sulforaphane inhibits extracellular, intracellular, and antibiotic-resistant strains of Helicobacter pylori and prevents benzo[a]pyrene-induced stomach tumors. Proc Natl Acad Sci U S A 99:7610–7615

Frattini V, Trifonov V, Chan JM, Castano A, Lia M, Abate F, Keir ST, Ji AX, Zoppoli P, Niola F, Danussi C, Dolgalev I, Porrati P, Pellegatta S, Heguy A, Gupta G, Pisapia DJ, Canoll P, Bruce JN, Mclendon RE, Yan H, Aldape K, Finocchiaro G, Mikkelsen T, Prive GG, Bigner DD, Lasorella A, Rabadan R, Iavarone A (2013) The integrated landscape of driver genomic alterations in glioblastoma. Nat Genet 45:1141–1149

Gan W, Dai X, Lunardi A, Li Z, Inuzuka H, Liu P, Varmeh S, Zhang J, Cheng L, Sun Y, Asara JM, Beck AH, Huang J, Pandolfi PP, Wei W (2015) SPOP promotes ubiquitination and degradation of the ERG oncoprotein to suppress prostate cancer progression. Mol Cell 59:917–930

Ganapathi KA, Jobanputra V, Iwamoto F, Jain P, Chen J, Cascione L, Nahum O, Levy B, Xie Y, Khattar P, Hoehn D, Bertoni F, Murty VV, Pittaluga S, Jaffe ES, Alobeid B, Mansukhani MM, Bhagat G (2016) The genetic landscape of dural marginal zone lymphomas. Oncotarget 7:43052–43061

Gao K, Jin X, Tang Y, Ma J, Peng J, Yu L, Zhang P, Wang C (2015) Tumor suppressor SPOP mediates the proteasomal degradation of progesterone receptors (PRs) in breast cancer cells. Am J Cancer Res 5:3210–3220

Garcia-Flores M, Casanova-Salas I, Rubio-Briones J, Calatrava A, Dominguez-Escrig J, Rubio L, Ramirez-Backhaus M, Fernandez-Serra A, Garcia-Casado Z, Lopez-Guerrero JA (2014) Clinico-pathological significance of the molecular alterations of the SPOP gene in prostate cancer. Eur J Cancer 50:2994–3002

Ge Z, Leighton JS, Wang Y, Peng X, Chen Z, Chen H, Sun Y, Yao F, Li J, Zhang H, Liu J, Shriver CD, Hu H, Cancer Genome Atlas Research N, Piwnica-Worms H, Ma L, Liang H (2018) Integrated genomic analysis of the ubiquitin pathway across cancer types. Cell Rep 23:213–226 e3

Geng C, He B, Xu L, Barbieri CE, Eedunuri VK, Chew SA, Zimmermann M, Bond R, Shou J, Li C, Blattner M, Lonard DM, Demichelis F, Coarfa C, Rubin MA, Zhou P, O'Malley BW, Mitsiades N (2013) Prostate cancer-associated mutations in speckle-type POZ protein (SPOP) regulate steroid receptor coactivator 3 protein turnover. Proc Natl Acad Sci U S A 110:6997–7002

Geng C, Kaochar S, Li M, Rajapakshe K, Fiskus W, Dong J, Foley C, Dong B, Zhang L, Kwon OJ, Shah SS, Bolaki M, Xin L, Ittmann M, O'Malley BW, Coarfa C, Mitsiades N (2017) SPOP regulates prostate epithelial cell proliferation and promotes ubiquitination and turnover of c-MYC oncoprotein. Oncogene 36:4767–4777

Genschik P, Sumara I, Lechner E (2013) The emerging family of Cullin3-RING ubiquitin ligases (CRL3s): cellular functions and disease implications. EMBO J 32:2307–2320

Goldstein LD, Lee J, Gnad F, Klijn C, Schaub A, Reeder J, Daemen A, Bakalarski CE, Holcomb T, Shames DS, Hartmaier RJ, Chmielecki J, Seshagiri S, Gentleman R, Stokoe D (2016) Recurrent loss of NFE2L2 Exon 2 is a mechanism for NRF2 pathway activation in human cancers. Cell Rep 16:2605–2617

Gorrini C, Baniasadi PS, Harris IS, Silvester J, Inoue S, Snow B, Joshi PA, Wakeham A, Molyneux SD, Martin B, Bouwman P, Cescon DW, Elia AJ, Winterton-Perks Z, Cruickshank J, Brenner D, Tseng A, Musgrave M, Berman HK, Khokha R, Jonkers J, Mak TW, Gauthier ML (2013) BRCA1 interacts with NRF2 to regulate antioxidant signaling and cell survival. J Exp Med 210:1529–1544

Gossage L, Eisen T (2010) Alterations in VHL as potential biomarkers in renal-cell carcinoma. Nat Rev Clin Oncol 7:277–288

Grasso CS, Wu YM, Robinson DR, Cao X, Dhanasekaran SM, Khan AP, Quist MJ, Jing X, Lonigro RJ, Brenner JC, Asangani IA, Ateeq B, Chun SY, Siddiqui J, Sam L, Anstett M, Mehra R, Prensner JR, Palanisamy N, Ryslik GA, Vandin F, Raphael BJ, Kunju LP, Rhodes DR, Pienta KJ, Chinnaiyan AM, Tomlins SA (2012) The mutational landscape of lethal castration-resistant prostate cancer. Nature 487:239–243

Groner AC, Cato L, De Tribolet-Hardy J, Bernasocchi T, Janouskova H, Melchers D, Houtman R, Cato ACB, Tschopp P, Gu L, Corsinotti A, Zhong Q, Fankhauser C, Fritz C, Poyet C, Wagner U, Guo T, Aebersold R, Garraway LA, Wild PJ, Theurillat JP, Brown M (2016) TRIM24 is an oncogenic transcriptional activator in prostate cancer. Cancer Cell 29:846–858

Guo ZQ, Zheng T, Chen B, Luo C, Ouyang S, Gong S, Li J, Mao LL, Lian F, Yang Y, Huang Y, Li L, Lu J, Zhang B, Zhou L, Ding H, Gao Z, Zhou L, Li G, Zhou R, Chen K, Liu J, Wen Y, Gong L, Ke Y, Yang SD, Qiu XB, Zhang N, Ren J, Zhong D, Yang CG, Liu J, Jiang H (2016) Small-molecule targeting of E3 ligase adaptor SPOP in kidney cancer. Cancer Cell 30:474–484

Gupta-Rossi N, Storck S, Griebel PJ, Reynaud CA, Weill JC, Dahan A (2003) Specific over-expression of deltex and a new Kelch-like protein in human germinal center B cells. Mol Immunol 39:791–799

Gurrieri C, Capodieci P, Bernardi R, Scaglioni PP, Nafa K, Rush LJ, Verbel DA, Cordon-Cardo C, Pandolfi PP (2004) Loss of the tumor suppressor PML in human cancers of multiple histologic origins. J Natl Cancer Inst 96:269–279

Hanada N, Takahata T, Zhou Q, Ye X, Sun R, Itoh J, Ishiguro A, Kijima H, Mimura J, Itoh K, Fukuda S,

Saijo Y (2012) Methylation of the KEAP1 gene promoter region in human colorectal cancer. BMC Cancer 12:66

Hannah J, Zhou P (2009) Regulation of DNA damage response pathways by the cullin-RING ubiquitin ligases. DNA Repair (Amst) 8:536–543

Hannah J, Zhou P (2015) Distinct and overlapping functions of the cullin E3 ligase scaffolding proteins CUL4A and CUL4B. Gene 573:33–45

Harder B, Jiang T, Wu T, Tao S, Rojo De La Vega M, Tian W, Chapman E, Zhang DD (2015) Molecular mechanisms of NRF2 regulation and how these influence chemical modulation for disease intervention. Biochem Soc Trans 43:680–686

Hayes JD, Dinkova-Kostova AT (2014) The NRF2 regulatory network provides an interface between redox and intermediary metabolism. Trends Biochem Sci 39:199–218

Hiramoto K, Satoh H, Suzuki T, Moriguchi T, Pi J, Shimosegawa T, Yamamoto M (2014) Myeloid lineage-specific deletion of antioxidant system enhances tumor metastasis. Cancer Prev Res (Phila) 7:835–844

Holmstrom KM, Baird L, Zhang Y, Hargreaves I, Chalasani A, Land JM, Stanyer L, Yamamoto M, Dinkova-Kostova AT, Abramov AY (2013) NRF2 impacts cellular bioenergetics by controlling substrate availability for mitochondrial respiration. Biol Open 2:761–770

Homma S, Ishii Y, Morishima Y, Yamadori T, Matsuno Y, Haraguchi N, Kikuchi N, Satoh H, Sakamoto T, Hizawa N, Itoh K, Yamamoto M (2009) NRF2 enhances cell proliferation and resistance to anticancer drugs in human lung cancer. Clin Cancer Res 15:3423–3432

Hu X, Yang Z, Zeng M, Liu YI, Yang X, Li Y, Li XU, Yu Q (2016) Speckle-type POZ (pox virus and zinc finger protein) protein gene deletion in ovarian cancer: fluorescence in situ hybridization analysis of a tissue microarray. Oncol Lett 12:658–662

Huang Y, Tan N, Jia D, Jing Y, Wang Q, Li Z, Zhang J, Liu L, Li J, Chen Z, He X (2015) Speckle-type Poz protein is negatively associated with malignancies and inhibits cell proliferation and migration in liver cancer. Tumour Biol 36:9753–9761

Ichimura Y, Waguri S, Sou YS, Kageyama S, Hasegawa J, Ishimura R, Saito T, Yang Y, Kouno T, Fukutomi T, Hoshii T, Hirao A, Takagi K, Mizushima T, Motohashi H, Lee MS, Yoshimori T, Tanaka K, Yamamoto M, Komatsu M (2013) Phosphorylation of p62 activates the Keap1-NRF2 pathway during selective autophagy. Mol Cell 51:618–631

Inami Y, Waguri S, Sakamoto A, Kouno T, Nakada K, Hino O, Watanabe S, Ando J, Iwadate M, Yamamoto M, Lee MS, Tanaka K, Komatsu M (2011) Persistent activation of NRF2 through p62 in hepatocellular carcinoma cells. J Cell Biol 193:275–284

Ito K, Bernardi R, Morotti A, Matsuoka S, Saglio G, Ikeda Y, Rosenblatt J, Avigan DE, Teruya-Feldstein J, Pandolfi PP (2008) PML targeting eradicates quiescent leukaemia-initiating cells. Nature 453:1072–1078

Itoh K, Wakabayashi N, Katoh Y, Ishii T, Igarashi K, Engel JD, Yamamoto M (1999) Keap1 represses nuclear activation of antioxidant responsive elements by NRF2 through binding to the amino-terminal Neh2 domain. Genes Dev 13:76–86

Ivanovska J, Mahadevan V, Schneider-Stock R (2014) DAPK and cytoskeleton-associated functions. Apoptosis 19:329–338

Jain A, Lamark T, Sjottem E, Larsen KB, Awuh JA, Overvatn A, Mcmahon M, Hayes JD, Johansen T (2010) p62/Sqstm1 is a target gene for transcription factor NRF2 and creates a positive feedback loop by inducing antioxidant response element-driven gene transcription. J Biol Chem 285:22576–22591

Janouskova H, El Tekle G, Bellini E, Udeshi ND, Rinaldi A, Ulbricht A, Bernasocchi T, Civenni G, Losa M, Svinkina T, Bielski CM, Kryukov GV, Cascione L, Napoli S, Enchev RI, Mutch DG, Carney ME, Berchuck A, Winterhoff BJN, Broaddus RR, Schraml P, Moch H, Bertoni F, Catapano CV, Peter M, Carr SA, Garraway LA, Wild PJ, Theurillat JP (2017) Opposing effects of cancer-type-specific SPOP mutants on BET protein degradation and sensitivity to BET inhibitors. Nat Med 23:1046–1054

Jerabkova K, Sumara I (2018) Cullin 3, a cellular scripter of the non-proteolytic ubiquitin code. Semin Cell Dev Biol 93:100–110

Ji AX, Chu A, Nielsen TK, Benlekbir S, Rubinstein JL, Prive GG (2016) Structural insights into KCTD protein assembly and Cullin3 recognition. J Mol Biol 428:92–107

Kan Z, Jaiswal BS, Stinson J, Janakiraman V, Bhatt D, Stern HM, Yue P, Haverty PM, Bourgon R, Zheng J, Moorhead M, Chaudhuri S, Tomsho LP, Peters BA, Pujara K, Cordes S, Davis DP, Carlton VE, Yuan W, Li L, Wang W, Eigenbrot C, Kaminker JS, Eberhard DA, WaRING P, Schuster SC, Modrusan Z, Zhang Z, Stokoe D, De Sauvage FJ, Faham M, Seshagiri S (2010) Diverse somatic mutation patterns and pathway alterations in human cancers. Nature 466:869–873

Kandoth C, Mclellan MD, Vandin F, Ye K, Niu B, Lu C, Xie M, Zhang Q, Mcmichael JF, Wyczalkowski MA, Leiserson MD, Miller CA, Welch JS, Walter MJ, Wendl MC, Ley TJ, Wilson RK, Raphael BJ, Ding L (2013) Mutational landscape and significance across 12 major cancer types. Nature 502:333–339

Kang KA, Piao MJ, Kim KC, Kang HK, Chang WY, Park IC, Keum YS, Surh YJ, Hyun JW (2014) Epigenetic modification of NRF2 in 5-fluorouracil-resistant colon cancer cells: involvement of TET-dependent DNA demethylation. Cell Death Dis 5:e1183

Khor TO, Huang MT, Prawan A, Liu Y, Hao X, Yu S, Cheung WK, Chan JY, Reddy BS, Yang CS, Kong AN (2008) Increased susceptibility of NRF2 knockout mice to colitis-associated colorectal cancer. Cancer Prev Res (Phila) 1:187–191

Kim J, Guan KL (2019) mtor as a central hub of nutrient signalling and cell growth. Nat Cell Biol 21:63–71

Kim MS, Je EM, Oh JE, Yoo NJ, Lee SH (2013) Mutational and expressional analyses of SPOP, a candidate tumor suppressor gene, in prostate, gastric and colorectal cancers. APMIS 121:626–633

Kim MS, Kim MS, Yoo NJ, Lee SH (2014) Somatic mutation of SPOP tumor suppressor gene is rare in breast, lung, liver cancers, and acute leukemias. APMIS 122:164–166

Kitamura H, Motohashi H (2018) NRF2 addiction in cancer cells. Cancer Sci 109:900–911

Kobayashi A, Kang MI, Okawa H, Ohtsuji M, Zenke Y, Chiba T, Igarashi K, Yamamoto M (2004) Oxidative stress sensor Keap1 functions as an adaptor for Cul3-based E3 ligase to regulate proteasomal degradation of NRF2. Mol Cell Biol 24:7130–7139

Komatsu M, Kurokawa H, Waguri S, Taguchi K, Kobayashi A, Ichimura Y, Sou YS, Ueno I, Sakamoto A, Tong KI, Kim M, Nishito Y, Iemura S, Natsume T, Ueno T, Kominami E, Motohashi H, Tanaka K, Yamamoto M (2010) The selective autophagy substrate p62 activates the stress responsive transcription factor NRF2 through inactivation of Keap1. Nat Cell Biol 12:213–223

Konstantinopoulos PA, Spentzos D, Fountzilas E, Francoeur N, Sanisetty S, Grammatikos AP, Hecht JL, Cannistra SA (2011) Keap1 mutations and NRF2 pathway activation in epithelial ovarian cancer. Cancer Res 71:5081–5089

Koren I, Timms RT, Kula T, Xu Q, Li MZ, Elledge SJ (2018) The eukaryotic proteome is shaped by E3 ubiquitin ligases targeting C-terminal degrons. Cell 173:1622–1635 e14

Kroll J, Shi X, Caprioli A, Liu HH, Waskow C, Lin KM, Miyazaki T, Rodewald HR, Sato TN (2005) The BTB-kelch protein KLHL6 is involved in B-lymphocyte antigen receptor signaling and germinal center formation. Mol Cell Biol 25:8531–8540

Kunder CA, Roncador G, Advani RH, Gualco G, Bacchi CE, Sabile JM, Lossos IS, Nie K, Tibshirani RJ, Green MR, Alizadeh AA, Natkunam Y (2017) KLHL6 is preferentially expressed in germinal center-derived b-cell lymphomas. Am J Clin Pathol 148:465–476

Kwon JE, La M, Oh KH, Oh YM, Kim GR, Seol JH, Baek SH, Chiba T, Tanaka K, Bang OS, Joe CO, Chung CH (2006) BTB domain-containing speckle-type POZ protein (SPOP) serves as an adaptor of Daxx for ubiquitination by Cul3-based ubiquitin ligase. J Biol Chem 281:12664–12672

Lau A, Wang XJ, Zhao F, Villeneuve NF, Wu T, Jiang T, Sun Z, White E, Zhang DD (2010) A noncanonical mechanism of NRF2 activation by autophagy deficiency: direct interaction between Keap1 and p62. Mol Cell Biol 30:3275–3285

Lavau C, Marchio A, Fagioli M, Jansen J, Falini B, Lebon P, Grosveld F, Pandolfi PP, Pelicci PG, Dejean A (1995) The acute promyelocytic leukaemia-associated PML gene is induced by interferon. Oncogene 11:871–876

Le Gallo M, O'Hara AJ, Rudd ML, Urick ME, Hansen NF, O'Neil NJ, Price JC, Zhang S, England BM, Godwin AK, Sgroi DC, Hieter P, Mullikin JC, Merino MJ, Bell DW (2012) Exome sequencing of serous endometrial tumors identifies recurrent somatic mutations in chromatin-remodeling and ubiquitin ligase complex genes. Nat Genet 44:1310–1315

Lee YR, Yuan WC, Ho HC, Chen CH, Shih HM, Chen RH (2010) The Cullin 3 substrate adaptor KLHL20 mediates DAPK ubiquitination to control interferon responses. EMBO J 29:1748–1761

Lee S, Lim MJ, Kim MH, Yu CH, Yun YS, Ahn J, Song JY (2012) An effective strategy for increasing the radiosensitivity of Human lung Cancer cells by blocking NRF2-dependent antioxidant responses. Free Radic Biol Med 53:807–816

Leinonen HM, Kansanen E, Polonen P, Heinaniemi M, Levonen AL (2014) Role of the Keap1-NRF2 pathway in cancer. Adv Cancer Res 122:281–320

Leppek K, Schott J, Reitter S, Poetz F, Hammond MC, Stoecklin G (2013) Roquin promotes constitutive mRNA decay via a conserved class of stem-loop recognition motifs. Cell 153:869–881

Li C, Ao J, Fu J, Lee DF, Xu J, Lonard D, O'Malley BW (2011) Tumor-suppressor role for the SPOP ubiquitin ligase in signal-dependent proteolysis of the oncogenic co-activator SRC-3/AIB1. Oncogene 30:4350–4364

Li G, Ci W, Karmakar S, Chen K, Dhar R, Fan Z, Guo Z, Zhang J, Ke Y, Wang L, Zhuang M, Hu S, Li X, Zhou L, Calabrese MF, Watson ER, Prasad SM, Rinker-Schaeffer C, Eggener SE, Stricker T, Tian Y, Schulman BA, Liu J, White KP (2014) SPOP promotes tumorigenesis by acting as a key regulatory hub in kidney cancer. Cancer Cell 25:455–468

Lin HC, Yeh CW, Chen YF, Lee TT, Hsieh PY, Rusnac DV, Lin SY, Elledge SJ, Zheng N, Yen HS (2018) C-terminal end-directed protein elimination by CRL2 ubiquitin ligases. Mol Cell 70:602–613 e3

Lister A, Nedjadi T, KitteRINGham NR, Campbell F, Costello E, Lloyd B, Copple IM, Williams S, Owen A, Neoptolemos JP, GoldRING CE, Park BK (2011) NRF2 is overexpressed in pancreatic cancer: implications for cell proliferation and therapy. Mol Cancer 10:37

Liu J, Ghanim M, Xue L, Brown CD, Iossifov I, Angeletti C, Hua S, Negre N, Ludwig M, Stricker T, Al-Ahmadie HA, Tretiakova M, Camp RL, Perera-Alberto M, Rimm DL, Xu T, Rzhetsky A, White KP (2009) Analysis of Drosophila segmentation network identifies a JNK pathway factor overexpressed in kidney cancer. Science 323:1218–1222

Lohr JG, Stojanov P, Lawrence MS, Auclair D, Chapuy B, Sougnez C, Cruz-Gordillo P, Knoechel B, Asmann YW, Slager SL, Novak AJ, Dogan A, Ansell SM, Link BK, Zou L, Gould J, Saksena G, Stransky N, Rangel-Escareno C, Fernandez-Lopez JC, Hidalgo-Miranda A, Melendez-Zajgla J, Hernandez-Lemus E, Schwarz-Cruz Y, Celis A, Imaz-Rosshandler I, Ojesina AI, Jung J, Pedamallu CS, Lander ES, Habermann TM, Cerhan JR, Shipp MA, Getz G, Golub TR (2012) Discovery and prioritization of somatic mutations in diffuse large B-cell lymphoma (DLBCL) by whole-exome sequencing. Proc Natl Acad Sci U S A 109:3879–3884

Lu MC, Ji JA, Jiang ZY, You QD (2016) The Keap1-NRF2-ARE pathway as a potential preventive and therapeutic target: an update. Med Res Rev 36:924–963

Ludtmann MH, Angelova PR, Zhang Y, Abramov AY, Dinkova-Kostova AT (2014) NRF2 affects the efficiency of mitochondrial fatty acid oxidation. Biochem J 457:415–424

Lydeard JR, Schulman BA, Harper JW (2013) Building and remodelling Cullin-RING E3 ubiquitin ligases. EMBO Rep 14:1050–1061

Malhotra D, Portales-Casamar E, Singh A, Srivastava S, Arenillas D, Happel C, Shyr C, Wakabayashi N, Kensler TW, Wasserman WW, Biswal S (2010) Global mapping of binding sites for NRF2 identifies novel targets in cell survival response through Chip-Seq profiling and network analysis. Nucleic Acids Res 38:5718–5734

Mareschal S, Pham-Ledard A, Viailly PJ, Dubois S, Bertrand P, Maingonnat C, Fontanilles M, Bohers E, Ruminy P, Tournier I, Courville P, Lenormand B, Duval AB, Andrieu E, Verneuil L, Vergier B, Tilly H, Joly P, Frebourg T, Beylot-Barry M, Merlio JP, Jardin F (2017) Identification of somatic mutations in primary cutaneous diffuse large B-cell lymphoma, leg type by massive parallel sequencing. J Invest Dermatol 137:1984–1994

Martinez VD, Vucic EA, Thu KL, Pikor LA, Hubaux R, Lam WL (2014) Unique pattern of component gene disruption in the NRF2 inhibitor KEAP1/CUL3/RBX1 E3-ubiquitin ligase complex in serous ovarian cancer. Biomed Res Int 2014:159459

Martinez VD, Vucic EA, Thu KL, Pikor LA, Lam S, Lam WL (2015) Disruption of KEAP1/CUL3/RBX1 E3-ubiquitin ligase complex components by multiple genetic mechanisms: Association with poor prognosis in head and neck cancer. Head Neck 37:727–734

Marzahn MR, Marada S, Lee J, Nourse A, Kenrick S, Zhao H, Ben-Nissan G, Kolaitis RM, Peters JL, Pounds S, ErRINGton WJ, Prive GG, Taylor JP, Sharon M, Schuck P, Ogden SK, Mittag T (2016) Higher-order oligomerization promotes localization of SPOP to liquid nuclear speckles. EMBO J 35:1254–1275

Mathew R, Seiler MP, Scanlon ST, Mao AP, Constantinides MG, Bertozzi-Villa C, Singer JD, Bendelac A (2012) BTB-ZF factors recruit the E3 ligase Cullin 3 to regulate lymphoid effector programs. Nature 491:618–621

Mcmahon M, Thomas N, Itoh K, Yamamoto M, Hayes JD (2006) Dimerization of substrate adaptors can facilitate cullin-mediated ubiquitylation of proteins by a "tetheRING" mechanism: a two-site interaction model for the NRF2-Keap1 complex. J Biol Chem 281:24756–24768

Mitsuishi Y, Taguchi K, Kawatani Y, Shibata T, Nukiwa T, Aburatani H, Yamamoto M, Motohashi H (2012) NRF2 redirects glucose and glutamine into anabolic pathways in metabolic reprogramming. Cancer Cell 22:66–79

Morin RD, Mendez-Lago M, Mungall AJ, Goya R, Mungall KL, Corbett RD, Johnson NA, Severson TM, Chiu R, Field M, Jackman S, Krzywinski M, Scott DW, Trinh DL, Tamura-Wells J, Li S, Firme MR, Rogic S, Griffith M, Chan S, Yakovenko O, Meyer IM, Zhao EY, Smailus D, Moksa M, Chittaranjan S, Rimsza L, Brooks-Wilson A, Spinelli JJ, Ben-Neriah S, Meissner B, Woolcock B, Boyle M, Mcdonald H, Tam A, Zhao Y, Delaney A, Zeng T, Tse K, Butterfield Y, Birol I, Holt R, Schein J, Horsman DE, Moore R, Jones SJ, Connors JM, Hirst M, Gascoyne RD, Marra MA (2011) Frequent mutation of histone-modifying genes in non-Hodgkin lymphoma. Nature 476:298–303

Murakawa Y, Hinz M, Mothes J, Schuetz A, Uhl M, Wyler E, Yasuda T, Mastrobuoni G, Friedel CC, Dolken L, Kempa S, Schmidt-Supprian M, Bluthgen N, Backofen R, Heinemann U, Wolf J, Scheidereit C, Landthaler M (2015) RC3H1 post-transcriptionally regulates A20 mRNA and modulates the activity of the IKK/NF-kappaB pathway. Nat Commun 6:7367

Nacak TG, Leptien K, Fellner D, Augustin HG, Kroll J (2006) The BTB-kelch protein LZTR-1 is a novel Golgi protein that is degraded upon induction of apoptosis. J Biol Chem 281:5065–5071

Nagai Y, Kojima T, Muro Y, Hachiya T, Nishizawa Y, Wakabayashi T, Hagiwara M (1997) Identification of a novel nuclear speckle-type protein, SPOP. FEBS Lett 418:23–26

Niture SK, Jaiswal AK (2012) NRF2 protein up-regulates antiapoptotic protein Bcl-2 and prevents cellular apoptosis. J Biol Chem 287:9873–9886

Ohta T, Iijima K, Miyamoto M, Nakahara I, Tanaka H, Ohtsuji M, Suzuki T, Kobayashi A, Yokota J, Sakiyama T, Shibata T, Yamamoto M, Hirohashi S (2008) Loss of Keap1 function activates NRF2 and provides advantages for lung cancer cell growth. Cancer Res 68:1303–1309

Ooi A, Wong JC, Petillo D, Roossien D, Perrier-Trudova-V, Whitten D, Min BW, Tan MH, Zhang Z, Yang XJ, Zhou M, Gardie B, Molinie V, Richard S, Tan PH, Teh BT, Furge KA (2011) An antioxidant response phenotype shared between hereditary and sporadic type 2 papillary renal cell carcinoma. Cancer Cell 20:511–523

Ooi A, Dykema K, Ansari A, Petillo D, Snider J, Kahnoski R, Anema J, Craig D, Carpten J, Teh BT, Furge KA (2013) CUL3 and NRF2 mutations confer an NRF2 activation phenotype in a sporadic form of papillary renal cell carcinoma. Cancer Res 73:2044–2051

Osburn WO, Karim B, Dolan PM, Liu G, Yamamoto M, Huso DL, Kensler TW (2007) Increased colonic inflammatory injury and formation of aberrant crypt foci in NRF2-deficient mice upon dextran sulfate treatment. Int J Cancer 121:1883–1891

Padmanabhan B, Tong KI, Ohta T, Nakamura Y, Scharlock M, Ohtsuji M, Kang MI, Kobayashi A, Yokoyama S, Yamamoto M (2006) Structural basis

for defects of Keap1 activity provoked by its point mutations in lung cancer. Mol Cell 21:689–700
Pak E, Segal RA (2016) Hedgehog signal transduction: key players, oncogenic drivers, and cancer therapy. Dev Cell 38:333–344
Pasqualucci L, Neumeister P, Goossens T, Nanjangud G, Chaganti RS, Kuppers R, Dalla-Favera R (2001) Hypermutation of multiple proto-oncogenes in B-cell diffuse large-cell lymphomas. Nature 412:341–346
Pintard L, Willems A, Peter M (2004) Cullin-based ubiquitin ligases. Cul3-BTB complexes join the family. EMBO J 23:1681–1687
Piotrowski A, Xie J, Liu YF, Poplawski AB, Gomes AR, Madanecki P, Fu C, Crowley MR, Crossman DK, Armstrong L, Babovic-Vuksanovic D, Bergner A, Blakeley JO, Blumenthal AL, Daniels MS, Feit H, Gardner K, Hurst S, Kobelka C, Lee C, Nagy R, Rauen KA, Slopis JM, Suwannarat P, Westman JA, Zanko A, Korf BR, Messiaen LM (2014) Germline loss-of-function mutations in LZTR1 predispose to an inherited disorder of multiple schwannomas. Nat Genet 46:182–187
Puente XS, Pinyol M, Quesada V, Conde L, Ordonez GR, Villamor N, Escaramis G, Jares P, Bea S, Gonzalez-Diaz M, Bassaganyas L, Baumann T, Juan M, Lopez-Guerra M, Colomer D, Tubio JM, Lopez C, Navarro A, Tornador C, Aymerich M, Rozman M, Hernandez JM, Puente DA, Freije JM, Velasco G, Gutierrez-Fernandez A, Costa D, Carrio A, Guijarro S, Enjuanes A, Hernandez L, Yague J, Nicolas P, Romeo-Casabona CM, Himmelbauer H, Castillo E, Dohm JC, De Sanjose S, Piris MA, De Alava E, San Miguel J, Royo R, Gelpi JL, Torrents D, Orozco M, Pisano DG, Valencia A, Guigo R, Bayes M, Heath S, Gut M, Klatt P, Marshall J, Raine K, Stebbings LA, Futreal PA, Stratton MR, Campbell PJ, Gut I, Lopez-Guillermo A, Estivill X, Montserrat E, Lopez-Otin C, Campo E (2011) Whole-genome sequencing identifies recurrent mutations in chronic lymphocytic leukaemia. Nature 475:101–105
Ramos-Gomez M, Kwak MK, Dolan PM, Itoh K, Yamamoto M, Talalay P, Kensler TW (2001) Sensitivity to carcinogenesis is increased and chemoprotective efficacy of enzyme inducers is lost in NRF2 transcription factor-deficient mice. Proc Natl Acad Sci U S A 98:3410–3415
Raval A, Tanner SM, Byrd JC, Angerman EB, Perko JD, Chen SS, Hackanson B, Grever MR, Lucas DM, Matkovic JJ, Lin TS, Kipps TJ, Murray F, Weisenburger D, Sanger W, Lynch J, Watson P, Jansen M, Yoshinaga Y, Rosenquist R, De Jong PJ, Coggill P, Beck S, Lynch H, De La Chapelle A, Plass C (2007) Downregulation of death-associated protein kinase 1 (DAPK1) in chronic lymphocytic leukemia. Cell 129:879–890
Reddy A, Zhang J, Davis NS, Moffitt AB, Love CL, Waldrop A, Leppa S, Pasanen A, Meriranta L, Karjalainen-Lindsberg ML, Norgaard P, Pedersen M, Gang AO, Hogdall E, Heavican TB, Lone W, Iqbal J, Qin Q, Li G, Kim SY, Healy J, Richards KL, Fedoriw Y, Bernal-Mizrachi L, Koff JL, Staton AD, Flowers CR, Paltiel O, Goldschmidt N, Calaminici M, Clear A, Gribben J, Nguyen E, Czader MB, Ondrejka SL, Collie A, Hsi ED, Tse E, Au-Yeung RKH, Kwong YL, Srivastava G, Choi WWL, Evens AM, Pilichowska M, Sengar M, Reddy N, Li S, Chadburn A, Gordon LI, Jaffe ES, Levy S, Rempel R, Tzeng T, Happ LE, Dave T, Rajagopalan D, Datta J, Dunson DB, Dave SS (2017) Genetic and functional drivers of diffuse large B cell lymphoma. Cell 171:481–494 e15
Reineke EL, Liu Y, Kao HY (2010) Promyelocytic leukemia protein controls cell migration in response to hydrogen peroxide and insulin-like growth factor-1. J Biol Chem 285:9485–9492
Ren D, Villeneuve NF, Jiang T, Wu T, Lau A, Toppin HA, Zhang DD (2011) Brusatol enhances the efficacy of chemotherapy by inhibiting the NRF2-mediated defense mechanism. Proc Natl Acad Sci U S A 108:1433–1438
Rojo De La Vega M, Chapman E, Zhang DD (2018) NRF2 and the hallmarks of cancer. Cancer Cell 34:21–43
Romero R, Sayin VI, Davidson SM, Bauer MR, Singh SX, Leboeuf SE, Karakousi TR, Ellis DC, Bhutkar A, Sanchez-Rivera FJ, Subbaraj L, Martinez B, Bronson RT, Prigge JR, Schmidt EE, Thomas CJ, Goparaju C, Davies A, Dolgalev I, Heguy A, Allaj V, Poirier JT, Moreira AL, Rudin CM, Pass HI, Vander Heiden MG, Jacks T, Papagiannakopoulos T (2017) Keap1 loss promotes Kras-driven lung cancer and results in dependence on glutaminolysis. Nat Med 23:1362–1368
Sabatini DM (2006) mTOR and cancer: insights into a complex relationship. Nat Rev Cancer 6:729–734
Salomoni P, Ferguson BJ, Wyllie AH, Rich T (2008) New insights into the role of PML in tumour suppression. Cell Res 18:622–640
Satoh H, Moriguchi T, Taguchi K, Takai J, Maher JM, Suzuki T, Winnard PT Jr, Raman V, Ebina M, Nukiwa T, Yamamoto M (2010) NRF2-deficiency creates a responsive microenvironment for metastasis to the lung. Carcinogenesis 31:1833–1843
Schmucker S, Sumara I (2014) Molecular dynamics of PLK1 during mitosis. Mol Cell Oncol 1:e954507
Semenza GL (2010) Defining the role of hypoxia-inducible factor 1 in cancer biology and therapeutics. Oncogene 29:625–634
Shen H, Yang Y, Xia S, Rao B, Zhang J, Wang J (2014) Blockage of NRF2 suppresses the migration and invasion of esophageal squamous cell carcinoma cells in hypoxic microenvironment. Dis Esophagus 27:685–692
Shen T, Jiang T, Long M, Chen J, Ren DM, Wong PK, Chapman E, Zhou B, Zhang DD (2015) A curcumin derivative that inhibits vinyl carbamate-induced lung carcinogenesis via activation of the NRF2 protective response. Antioxid Redox Signal 23:651–664
Shi GX, Cai W, Andres DA (2013) Rit subfamily small GTPases: regulators in neuronal differentiation and survival. Cell Signal 25:2060–2068
Shibata T, Kokubu A, Gotoh M, Ojima H, Ohta T, Yamamoto M, Hirohashi S (2008a) Genetic alteration

of Keap1 confers constitutive NRF2 activation and resistance to chemotherapy in gallbladder cancer. Gastroenterology 135:1358–1368 e1-4

Shibata T, Ohta T, Tong KI, Kokubu A, Odogawa R, Tsuta K, Asamura H, Yamamoto M, Hirohashi S (2008b) Cancer related mutations in NRF2 impair its recognition by Keap1-Cul3 E3 ligase and promote malignancy. Proc Natl Acad Sci U S A 105:13568–13573

Shibata T, Kokubu A, Saito S, Narisawa-Saito M, Sasaki H, Aoyagi K, Yoshimatsu Y, Tachimori Y, Kushima R, Kiyono T, Yamamoto M (2011) NRF2 mutation confers malignant potential and resistance to chemoradiation therapy in advanced esophageal squamous cancer. Neoplasia 13:864–873

Singh A, Misra V, Thimmulappa RK, Lee H, Ames S, Hoque MO, Herman JG, Baylin SB, Sidransky D, Gabrielson E, Brock MV, Biswal S (2006) Dysfunctional KEAP1-NRF2 interaction in non-small-cell lung cancer. PLoS Med 3:e420

Solis LM, Behrens C, Dong W, Suraokar M, Ozburn NC, Moran CA, Corvalan AH, Biswal S, Swisher SG, Bekele BN, Minna JD, Stewart DJ, Wistuba II (2010) Nrf2 and Keap1 abnormalities in non-small cell lung carcinoma and association with clinicopathologic features. Clin Cancer Res 16:3743–3753

Stadler M, Chelbi-Alix MK, Koken MH, Venturini L, Lee C, Saib A, Quignon F, Pelicano L, Guillemin MC, Schindler C, Et A (1995) Transcriptional induction of the PML growth suppressor gene by interferons is mediated through an ISRE and a GAS element. Oncogene 11:2565–2573

Steklov M, Pandolfi S, Baietti MF, Batiuk A, Carai P, Najm P, Zhang M, Jang H, Renzi F, Cai Y, Abbasi Asbagh L, Pastor T, De Troyer M, Simicek M, Radaelli E, Brems H, Legius E, Tavernier J, Gevaert K, Impens F, Messiaen L, Nussinov R, Heymans S, Eyckerman S, Sablina AA (2018) Mutations in LZTR1 drive human disease by dysregulating RAS ubiquitination. Science 362:1177–1182

Stogios PJ, Downs GS, Jauhal JJ, Nandra SK, Prive GG (2005) Sequence and structural analysis of BTB domain proteins. Genome Biol 6:R82

Sun Z, Zhang S, Chan JY, Zhang DD (2007) Keap1 controls postinduction repression of the NRF2-mediated antioxidant response by escorting nuclear export of NRF2. Mol Cell Biol 27:6334–6349

Sun Z, Wu T, Zhao F, Lau A, Birch CM, Zhang DD (2011) Kpna6 (Importin {alpha}7)-mediated nuclear import of Keap1 represses the Nrf2-dependent antioxidant response. Mol Cell Biol 31:1800–1811

Sutton LA, Ljungstrom V, Mansouri L, Young E, Cortese D, Navrkalova V, Malcikova J, Muggen AF, Trbusek M, Panagiotidis P, Davi F, Belessi C, Langerak AW, Ghia P, Pospisilova S, Stamatopoulos K, Rosenquist R (2015) Targeted next-generation sequencing in chronic lymphocytic leukemia: a high-throughput yet tailored approach will facilitate implementation in a clinical setting. Haematologica 100:370–376

Tao S, Justiniano R, Zhang DD, Wondrak GT (2013) The NRF2-inducers tanshinone I and dihydrotanshinone protect human skin cells and reconstructed human skin against solar simulated UV. Redox Biol 1:532–541

Tao S, Park SL, Rojo De La Vega M, Zhang DD, Wondrak GT (2015) Systemic administration of the apocarotenoid bixin protects skin against solar UV-induced damage through activation of NRF2. Free Radic Biol Med 89:690–700

Teixeira LK, Reed SI (2013) Ubiquitin ligases and cell cycle control. Annu Rev Biochem 82:387–414

Teras LR, Desantis CE, Cerhan JR, Morton LM, Jemal A, Flowers CR (2016) 2016 US lymphoid malignancy statistics by World Health Organization subtypes. CA Cancer J Clin 66(6):443–459

Theurillat JP, Udeshi ND, ErRINGton WJ, Svinkina T, Baca SC, Pop M, Wild PJ, Blattner M, Groner AC, Rubin MA, Moch H, Prive GG, Carr SA, Garraway LA (2014) Prostate cancer. Ubiquitylome analysis identifies dysregulation of effector substrates in SPOP-mutant prostate cancer. Science 346:85–89

Tong KI, Kobayashi A, Katsuoka F, Yamamoto M (2006) Two-site substrate recognition model for the Keap1-NRF2 system: a hinge and latch mechanism. Biol Chem 387:1311–1320

Tung MC, Lin PL, Wang YC, He TY, Lee MC, Yeh SD, Chen CY, Lee H (2015) Mutant p53 confers chemoresistance in non-small cell lung cancer by upregulating NRF2. Oncotarget 6:41692–41705

Wakabayashi N, Shin S, Slocum SL, Agoston ES, Wakabayashi J, Kwak MK, Misra V, Biswal S, Yamamoto M, Kensler TW (2010) Regulation of notch1 signaling by NRF2: implications for tissue regeneration. Sci Signal 3:ra52

Wang WJ, Kuo JC, Ku W, Lee YR, Lin FC, Chang YL, Lin YM, Chen CH, Huang YP, Chiang MJ, Yeh SW, Wu PR, Shen CH, Wu CT, Chen RH (2007) The tumor suppressor DAPK is reciprocally regulated by tyrosine kinase Src and phosphatase LAR. Mol Cell 27:701–716

Wang R, An J, Ji F, Jiao H, Sun H, Zhou D (2008a) Hypermethylation of the Keap1 gene in human lung cancer cell lines and lung cancer tissues. Biochem Biophys Res Commun 373:151–154

Wang XJ, Sun Z, Villeneuve NF, Zhang S, Zhao F, Li Y, Chen W, Yi X, Zheng W, Wondrak GT, Wong PK, Zhang DD (2008b) NRF2 enhances resistance of cancer cells to chemotherapeutic drugs, the dark side of NRF2. Carcinogenesis 29:1235–1243

Wang X, Jin J, Wan F, Zhao L, Chu H, Chen C, Liao G, Liu J, Yu Y, Teng H, Fang L, Jiang C, Pan W, Xie X, Li J, Lu X, Jiang X, Ge X, Ye D, Wang P (2018) AMPK promotes SPOP-mediated NANOG degradation to regulate prostate cancer cell stemness. Dev Cell 48(3):345–360

Weigert O, Kopp N, Lane AA, Yoda A, Dahlberg SE, Neuberg D, Bahar AY, Chapuy B, Kutok JL, Longtine JA, Kuo FC, Haley T, Salois M, Sullivan TJ, Fisher DC, Fox EA, Rodig SJ, Antin JH, Weinstock DM (2012) Molecular ontogeny of donor-derived follicular lymphomas occurring after hematopoietic cell transplantation. Cancer Discov 2:47–55

Wells G (2015) Peptide and small molecule inhibitors of the Keap1-NRF2 protein-protein interaction. Biochem Soc Trans 43:674–679

Wenger RH, Stiehl DP, Camenisch G (2005) Integration of oxygen signaling at the consensus HRE. Sci STKE 2005:re12

Xu C, Huang MT, Shen G, Yuan X, Lin W, Khor TO, Conney AH, Kong AN (2006) Inhibition of 7,12-dimethylbenz(a)anthracene-induced skin tumorigenesis in C57bl/6 mice by sulforaphane is mediated by nuclear factor E2-related factor 2. Cancer Res 66:8293–8296

Xu J, Wang F, Jiang H, Jiang Y, Chen J, Qin J (2015) Properties and clinical relevance of speckle-type POZ protein in human colorectal cancer. J Gastrointest Surg 19:1484–1496

Yamamoto S, Inoue J, Kawano T, Kozaki K, Omura K, Inazawa J (2014) The impact of miRNA-based molecular diagnostics and treatment of NRF2-stabilized tumors. Mol Cancer Res 12:58–68

Yamamoto GL, Aguena M, Gos M, Hung C, Pilch J, Fahiminiya S, Abramowicz A, Cristian I, Buscarilli M, Naslavsky MS, Malaquias AC, Zatz M, Bodamer O, Majewski J, Jorge AA, Pereira AC, Kim CA, Passos-Bueno MR, Bertola DR (2015) Rare variants in SOS2 and LZTR1 are associated with Noonan syndrome. J Med Genet 52:413–421

Yaoita M, Niihori T, Mizuno S, Okamoto N, Hayashi S, Watanabe A, Yokozawa M, Suzumura H, Nakahara A, Nakano Y, Hokosaki T, Ohmori A, Sawada H, Migita O, Mima A, Lapunzina P, Santos-Simarro F, Garcia-Minaur S, Ogata T, Kawame H, Kurosawa K, Ohashi H, Inoue S, Matsubara Y, Kure S, Aoki Y (2016) Spectrum of mutations and genotype-phenotype analysis in Noonan syndrome patients with RIT1 mutations. Hum Genet 135:209–222

Yoo NJ, Kim HR, Kim YR, An CH, Lee SH (2012) Somatic mutations of the KEAP1 gene in common solid cancers. Histopathology 60:943–952

Young RM, Shaffer AL 3rd, Phelan JD, Staudt LM (2015) B-cell receptor signaling in diffuse large B-cell lymphoma. Semin Hematol 52:77–85

Yuan WC, Lee YR, Huang SF, Lin YM, Chen TY, Chung HC, Tsai CH, Chen HY, Chiang CT, Lai CK, Lu LT, Chen CH, Gu DL, Pu YS, Jou YS, Lu KP, Hsiao PW, Shih HM, Chen RH (2011) A Cullin3-KLHL20 Ubiquitin ligase-dependent pathway targets PML to potentiate HIF-1 signaling and prostate cancer progression. Cancer Cell 20:214–228

Zeng C, Wang Y, Lu Q, Chen J, Zhang J, Liu T, Lv N, Luo S (2014) SPOP suppresses tumorigenesis by regulating Hedgehog/Gli2 signaling pathway in gastric cancer. J Exp Clin Cancer Res 33:75

Zhang DD, Hannink M (2003) Distinct cysteine residues in Keap1 are required for Keap1-dependent ubiquitination of NRF2 and for stabilization of NRF2 by chemopreventive agents and oxidative stress. Mol Cell Biol 23:8137–8151

Zhang Y, Tang L (2007) Discovery and development of sulforaphane as a cancer chemopreventive phytochemical. Acta Pharmacol Sin 28:1343–1354

Zhang DD, Lo SC, Cross JV, Templeton DJ, Hannink M (2004) Keap1 is a redox-regulated substrate adaptor protein for a Cul3-dependent ubiquitin ligase complex. Mol Cell Biol 24:10941–10953

Zhang P, Singh A, Yegnasubramanian S, Esopi D, Kombairaju P, Bodas M, Wu H, Bova SG, Biswal S (2010) Loss of Kelch-like ECH-associated protein 1 function in prostate cancer cells causes chemoresistance and radioresistance and promotes tumor growth. Mol Cancer Ther 9:336–346

Zhang L, Wang N, Zhou S, Ye W, Jing G, Zhang M (2012) Propofol induces proliferation and invasion of gallbladder cancer cells through activation of NRF2. J Exp Clin Cancer Res 31:66

Zhang M, Zhang C, Zhang L, Yang Q, Zhou S, Wen Q, Wang J (2015a) NRF2 is a potential prognostic marker and promotes proliferation and invasion in human hepatocellular carcinoma. BMC Cancer 15:531

Zhang P, Gao K, Jin X, Ma J, Peng J, Wumaier R, Tang Y, Zhang Y, An J, Yan Q, Dong Y, Huang H, Yu L, Wang C (2015b) Endometrial cancer-associated mutants of SPOP are defective in regulating estrogen receptor-alpha protein turnover. Cell Death Dis 6:e1687

Zhang C, Wang HJ, Bao QC, Wang L, Guo TK, Chen WL, Xu LL, Zhou HS, Bian JL, Yang YR, Sun HP, Xu XL, You QD (2016) NRF2 promotes breast cancer cell proliferation and metastasis by increasing RhoA/ROCK pathway signal transduction. Oncotarget 7:73593–73606

Zhang P, Wang D, Zhao Y, Ren S, Gao K, Ye Z, Wang S, Pan CW, Zhu Y, Yan Y, Yang Y, Wu D, He Y, Zhang J, Lu D, Liu X, Yu L, Zhao S, Li Y, Lin D, Wang Y, Wang L, Chen Y, Sun Y, Wang C, Huang H (2017) Intrinsic BET inhibitor resistance in SPOP-mutated prostate cancer is mediated by BET protein stabilization and AKT-mTORC1 activation. Nat Med 23:1055–1062

Zhang J, Bu X, Wang H, Zhu Y, Geng Y, Nihira NT, Tan Y, Ci Y, Wu F, Dai X, Guo J, Huang YH, Fan C, Ren S, Sun Y, Freeman GJ, Sicinski P, Wei W (2018a) Cyclin D-CDK4 kinase destabilizes PD-L1 via cullin 3-SPOP to control cancer immune surveillance. Nature 553:91–95

Zhang J, Chen M, Zhu Y, Dai X, Dang F, Ren J, Ren S, Shulga YV, Beca F, Gan W, Wu F, Lin YM, Zhou X, Decaprio JA, Beck AH, Lu KP, Huang J, Zhao C, Sun Y, Gao X, Pandolfi PP, Wei W (2018b) SPOP promotes Nanog destruction to suppress stem cell traits and prostate cancer progression. Dev Cell 48 (3):329–344.e5

Zhao W, Zhou J, Deng Z, Gao Y, Cheng Y (2016) SPOP promotes tumor progression via activation of beta-catenin/TCF4 complex in clear cell renal cell carcinoma. Int J Oncol 49:1001–1008

Zhao Q, Mao A, Guo R, Zhang L, Yan J, Sun C, Tang J, Ye Y, Zhang Y, Zhang H (2017) Suppression of radiation-induced migration of non-small cell lung

cancer through inhibition of NRF2-Notch Axis. Oncotarget 8:36603–36613
Zheng N, Wang Z, Wei W (2016) Ubiquitination-mediated degradation of cell cycle-related proteins by F-box proteins. Int J Biochem Cell Biol 73:99–110
Zhi X, Tao J, Zhang L, Tao R, Ma L, Qin J (2016) Silencing speckle-type POZ protein by promoter hypermethylation decreases cell apoptosis through upregulating Hedgehog signaling pathway in colorectal cancer. Cell Death Dis 7:e2569
Zhu H, Ren S, Bitler BG, Aird KM, Tu Z, Skordalakes E, Zhu Y, Yan J, Sun Y, Zhang R (2015) SPOP E3 ubiquitin ligase adaptor promotes cellular senescence by degrading the SENP7 deSUMOylase. Cell Rep 13:1183–1193
Zhuang M, Calabrese MF, Liu J, Waddell MB, Nourse A, Hammel M, Miller DJ, Walden H, Duda DM, Seyedin SN, Hoggard T, Harper JW, White KP, Schulman BA (2009) Structures of SPOP-substrate complexes: insights into molecular architectures of BTB-Cul3 ubiquitin ligases. Mol Cell 36:39–50
Zimmerman ES, Schulman BA, Zheng N (2010) Structural assembly of cullin-RING ubiquitin ligase complexes. Curr Opin Struct Biol 20:714–721

CRL3s: The BTB-CUL3-RING E3 Ubiquitin Ligases

13

Pu Wang, Junbin Song, and Dan Ye

Abstract

The ubiquitin proteasome pathway is one of the major regulatory tools used by eukaryotic cells. The evolutionarily conserved cullin family proteins can assemble as many as >600 distinct E3 ubiquitin ligase complexes that regulate diverse cellular pathways. In most of Cullin-RING ubiquitin ligase (CRL) complexes, separate linker and adaptor proteins build the substrate recognition module. Differently, a single BTB-containing adaptor molecule utilizing two protein interaction sites can link the CUL3 scaffold to the substrate, forming as many as 188 CUL3-BTB E3 ligase complexes in mammals. Here, we review the most recent studies on CRL3 complexes, with a focus on the model for CUL3 assembly with its BTB-containing substrate receptors. Also, we summarize the current knowledge of CRL3 substrates and their relevant biological functions. Next, we discuss the mutual exclusivity of somatic mutations in KEAP1, NRF2, and CUL3 in human lung cancer. Finally, we highlight new strategies to expand CUL3 substrates and discuss outstanding questions remaining in the field.

P. Wang · J. Song · D. Ye (✉)
Molecular and Cell Biology Lab, Institutes of Biomedical Sciences, Shanghai Medical College, Fudan University, Shanghai, China
e-mail: yedan@fudan.edu.cn

Keywords

CRL3 · BTB · NRF2-KEAP1-CUL3 · Lung cancer

Abbreviations

AID	Auxin-induced degradation
ANKFY1	Ankyrin repeat and FYVE domain-containing 1
APC	Anaphase-promoting complex
ARE	The antioxidant response element
ATF2	Activating transcription factor 2
BCL2	B-cell CLL/lymphoma 2
BTB	Bric-à-brac, tramtrack and broad
BTB-ZF	BTB-zinc finger
bTRCP	Beta-transducin repeat-containing E3 ubiquitin protein ligase
CENP-A	Centromere protein A
CRL	Cullin-RING ubiquitin ligase
CRN7	Coronin 7
CUL	Cullin
DDB1	Damage-specific DNA binding protein 1
DWD	DDB1-binding WD40
eEF1A1	Eukaryotic translation elongation factor 1 alpha 1
EPS15	Epidermal growth factor receptor pathway substrate 15
ESCRT	Endosomal sorting complexes required for transport
FHIT	Fragile histidine triad gene

Y. Sun et al. (eds.), *Cullin-RING Ligases and Protein Neddylation*, Advances in Experimental Medicine and Biology 1217, https://doi.org/10.1007/978-981-15-1025-0_13

GSTA2	Glutathione S-transferase A2
HECT	Homologous to the E6-AP Carboxyl Terminus
IVR	Intervening region
Keap1	Kelch ECH associating protein 1
KLHDC	Kelch domain-containing protein
KLHL	The Kelch-like
LUSC	Lung squamous cell carcinomas
MEI-1	Meiosis inhibitor protein 1
MVBs	Multivesicular bodies
NEDD8	Neural precursor cell expressed, developmentally downregulated 8
NFE2	Nuclear factor, erythroid 2
NGS	Next-generation sequencing
NQO1	NADPH:quinone oxidoreductase 1
Nrf1	Nuclear respiratory factor 1
NRF2	Nuclear factor erythroid 2-related factor 2
NSCLC	Non-small cell lung cancer
p16 INK4a	Cyclin-dependent kinase inhibitor 2A
PLK1	Polo-like kinases 1
Rb1	Retinoblastoma 1
RBR	RING-In between-RING
RDX	Radixin
RING	Really Interesting New Gene
ROS	Reactive oxygen species
SCF	SKP1-CUL1-F-box
SEC31	Secretory 31
SKP1	S-Phase kinase associated protein 1
SOCS	Suppressor of cytokine signaling
SPOP	Speckle type BTB/POZ protein
SPOPL	Speckle type BTB/POZ protein like
TGN	Trans-Golgi network
ULK1	Unc-51 like autophagy activating kinase 1
WES	Whole-exome sequencing
WNK4	WNK lysine deficient protein kinase 4

13.1 Introduction

Ubiquitin is a protein highly conserved and ubiquitously expressed in all eukaryotes. As a protein modifier, ubiquitin is covalently conjugated to a lysine residue of a substrate protein, resulting in the formation of an isopeptide bond between the amino group of the lysine side chain and the carboxyl group at C-terminus of ubiquitin (Pickart and Eddins 2004). The process is called ubiquitination or ubiquitylation, which regulates numerous biochemical and cellular processes, such as protein degradation, cell cycle, DNA damage repair, vesicular trafficking, epigenetics, and inflammation.

Protein ubiquitylation is catalyzed by sequentially three enzymes, usually called E1, E2, and E3 (Hershko et al. 1983). E1 ubiquitin-activating enzyme is a magnesium-dependent enzyme catalyzing ubiquitin C-terminal acyl adenylation and then transfers the ubiquitin to a catalytic cysteine on E1, forming a thioester bond (Haas et al. 1982). Afterward, ubiquitin is covalently conjugated to the catalytic cysteine of E2 ubiquitin-conjugating enzyme (E2~Ub), which is a small ubiquitin carrier protein (Pickart and Rose 1985). E3 ubiquitin ligases recognize both E2~Ub and the substrate and catalyze the transfer of ubiquitin from E2~Ub complex to a lysine residue of the substrate. Lysine residue on either the substrate or the ubiquitin attached to substrate can be ubiquitylated by E3 with the latter one resulting in a polyubiquitin chain, which can be recognized by proteasome and usually causes substrate degradation (Zheng and Shabek 2017).

Based on the catalytic domain, E3 ubiquitin ligase is classified into RING (*R*eally *I*nteresting *N*ew *G*ene) finger, HECT (*H*omologous to the *E*6-AP *C*arboxyl *T*erminus), U box, PHD finger, and RBR (RING-In between-RING) (Metzger et al. 2014; Zheng and Shabek 2017). *C*ullin-*R*ING ubiquitin *L*igases (CRL) belong to a subfamily of RING finger E3 ligase and comprise the largest subfamily of all ubiquitin ligases (Petroski and Deshaies 2005). Unlike most E3 ligases, in which the catalytic domain and the substrate recognition domain are present in one polypeptide, CRL E3 ligase adopts a complex architecture containing multiple subunits, and each subunit is encoded by an individual gene (Fig. 13.1). In this complex, Cullin is a scaffold protein and its C-terminus binds to a small RING finger protein ROC1 (also known as RBX1), which is the catalytic subunit of CRL (Ohta et al. 1999). ROC1 binds to E2~Ub by physical

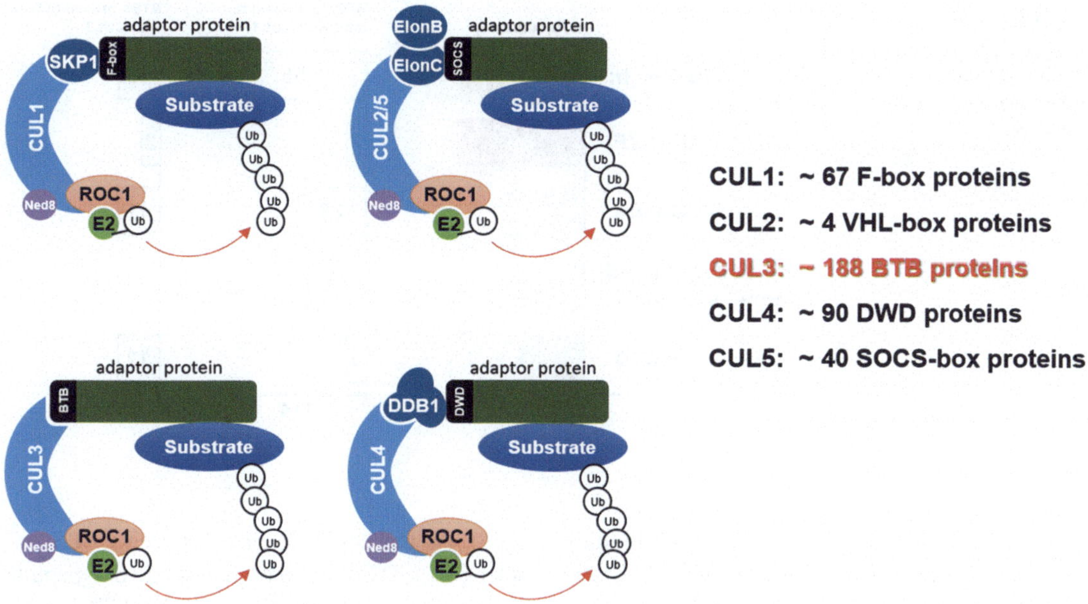

Fig. 13.1 Assembly of Cullin-RING E3 ubiquitin ligases (CRLs)

interacting with both E2 and Ub and locks E2~Ub in an "open" conformation, which exposes the thioester bond between E2 and Ub and facilitates Ub transfer to the lysine residue of the substrate (Buetow and Huang 2016). The N-terminus of Cullin interacts with adaptor proteins which bridge the binding between substrate and Cullin. Most if not all adaptors bind with Cullin through well-characterized domains which allow "multiplexing" the CRL scaffolds to recognize different substrates through different adaptors. For examples, CUL1 binds to Skp1 which in turn binds to F-box domain in substrate adaptor proteins (Bai et al. 1996), CUL2 and CUL5 recruit the substrates through elongin B/C and SOCS domain-containing adaptors (Kamura et al. 2004), and CUL4 recognizes the substrates through DDB1 and DWD domain-containing adaptors (He et al. 2006). Considering the existence of dozens of genes encoding adaptor proteins for each Cullin, and that a single Cullin-adaptor complex could have multiple substrates (e.g., CUL1-SKP1-bTRCP), the substrate spectrum of Cullin family is broad and diverse. There is also a neddylation system which regulates the switch among different adaptors (Merlet et al. 2009) and has been reviewed in details in other chapters of this book.

13.2 The Discovery of BTB-CUL3 Interaction and the Assembly of CRL3

Anaphase-promoting complex (APC) (King et al. 1995) and SKP1-CUL1-F-box (SCF) (Bai et al. 1996; Lyapina et al. 1998; Michel and Xiong 1998) complex were the first two Cullin-RING ubiquitin ligases discovered from studies on the mechanism of eukaryotic cell cycle. Both of them are E3 ligases responsible for coordinating cell cycle progression through degrading essential cell cycle regulators (Glotzer et al. 1991) and are composed a Cullin protein (CUL1 in SCF complex and APC2 in APC complex) as the scaffold and a RING finger protein (ROC1 in SCF complex and APC11 in APC complex) as the catalytic subunit. The shared architecture has defined a new family of RING finger E3 ligases later known as CRLs.

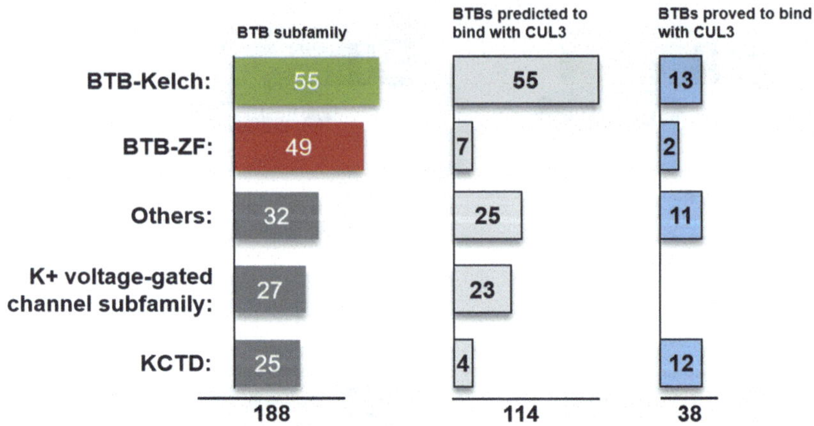

Fig. 13.2 CRL3 recognizes its substrates through different BTB-containing adaptor proteins

CUL3 was first identified as a cyclin E binding protein by yeast two-hybrid screen in 1999 (Singer et al. 1999). Biochemical study has shown CUL3, like CUL1, can bind to ROC1 and catalyze Ub transferring as an E3 ligase (Maeda et al. 2001). In *C. elegans*, CUL3 activity is required for mitotic spindle assembly through degrading MEI-1 (Clark-Maguire and Mains 1994a; Kurz et al. 2002; Pintard et al. 2003a), an oocyte-specific protein essential for meiotic spindle formation (Clark-Maguire and Mains 1994b), but is not required for mitotic spindle function. MEI-26, a BTB (bric-à-brac, tramtrack and broad) domain containing proteins, also known as MEI-1 negative regulator (Dow and Mains 1998), was identified as an adaptor for CUL3-mediated MEI-1 ubiquitylation (Pintard et al. 2003b; Xu et al. 2003). Moreover, CUL3 can be assembled in multiple complexes containing BTB proteins through direct interaction (Furukawa et al. 2003; Geyer et al. 2003; Xu et al. 2003), suggesting that BTB proteins serve as a general adaptor for CRL3.

BTB domain (also known as POZ domain) is a ~120 residues motif originally identified in bric-à-brac, tramtrack and broad complex transcription regulators in *Drosophila* (Bardwell and Treisman 1994; Zollman et al. 1994). BTB domain, whose structure is similar to the adaptor protein of CUL1 (Skp1) and CUL2 (elongin C) interacts with CUL3 through a structure basis conserved in CUL1-Skp1 and CUL2-elongin C interactions (Chaharbakhshi and Jemc 2016; Stogios et al. 2005). As the key difference between CRL3 and other CRLs, BTB proteins are a single polypeptide containing multiple domains, with the BTB domain binding to CUL3 and an additional domain interacting with the substrate. Differently, the substrate recruiting adaptors for other Cullins include more than one polypeptides encoded by multiple independent genes. The BTB domain-containing protein family is divided into several subfamilies according to the additional domain, e.g., BTB-ZF (zinc finger) or BTB-Kelch. By the year of 2018, HUGO (Human Genome Organisation) Gene Nomenclature Committee and Conserved Domain Database in NCBI (National Center for Biotechnology Information) have recorded 188 genes encoding BTB domain-containing proteins in human genome. According to a biochemical and structural study of BTB proteins in 2009, 114 of 188 BTB proteins have the structural basis to bind CUL3 (Zhuang et al. 2009), and 38 of 188 BTB proteins have been reported to form a complex with CUL3 in the literatures (Fig. 13.2). This suggests that CUL3-BTB E3 ligases have a potentially large number of substrates and many of them are still needed to be studied.

13.3 CRL3 Targets Many Substrates and Regulates Diverse Cellular Processes

During the past years, researches have applied different methods to identify potential CRL3 substrates. For example, siRNA targeting BTB-containing SPOPL led to the accumulation of EPS15 protein, which was then identified as a substrate of CRL3-SPOPL (Gschweitl et al. 2016). Moreover, performing yeast two-hybrid screen in a cDNA library by using BTB proteins as "baits," researchers have discovered multiple BTB-interacting CRL3 substrates, such as cEF1A1 (Koiwai et al. 2008), ULK1 (Liu et al. 2016), and ATF2 (Ma et al. 2018). Multiple CRL3 substrates have also been co-purified with ectopically expressed BTB proteins from cells treated with NEDD8-activating enzyme inhibitor MLN4924 or proteasome inhibitor MG132 and then identified by mass spectrometry, such as WNK4 (Shibata et al. 2013) and SEC31 (Jin et al. 2012). CRL3 has constituted a potentially large number of BTB-CUL3-ROC1 E3 ubiquitin ligases and are involved in diverse cellular processes, such as response to oxidative stress, regulation of cell cycle, mitosis, cytokinesis, apoptosis, autophagy, vesicle trafficking, neoplastic transformation, etc. (Table 13.1).

13.3.1 CRL3 Functions in Mitosis and Cytokinesis

Mitosis is the cell division process of eukaryotic cells which separates replicated chromosome into two new daughter cells (McIntosh 2016). By forming a bipolar microtubule (MT) structure attaching kinetochores of chromosomes to centrosomes, mitosis ensures equal distribution of genetic information and dysregulation of mitosis causes genome instability, a hallmark of cancer (Negrini et al. 2010). Mitosis is followed by cytokinesis, the final step of cell division which partitions the cytoplasm of the mother cell into two daughter cells (Green et al. 2012).

CRL3 ubiquitylates several key regulators of mitosis and cytokinesis. For example, polo-like kinases 1 (PLK1) is a polo-like family of serine/threonine kinase localized at multiple subcellular structures during mitosis (Combes et al. 2017). PLK1 can be ubiquitylated by CRL3 through the adaptor protein KLHL22, and this ubiquitylation is required for stable kinetochore-microtubule attachment and faithful chromosome alignment (Beck et al. 2013; Metzger et al. 2013). Aurora-A, a crucial kinase regulating G2 to M transition and mitotic entry (Nikonova et al. 2013), can be ubiquitylated by CRL3-KLHL18, which activates Aurora-A at the centrosomes. Knockdown of CUL3 or KLHL18 was found to delay mitotic entry (Moghe et al. 2012). Aurora-B kinase can also be ubiquitylated by CUL3, utilizing KLHL21 (Huang et al. 2017) or KLHL9/KLHL13 (Sumara et al. 2007) as adaptors, which is indispensable for Aurora-B localization on chromosome, correct chromosome alignment in metaphase, proper midzone, midbody formation, and completion of cytokinesis. In *Drosophila*, CRL3-RDX can stabilize centromere by ubiquitylating centromere histone variant CENP-A, and loss of RDX results in massive chromosome segregation defects during development (Bade et al. 2014). In *C. elegans*, degradation of microtubule severing protein MEI-1 though CRL3-MEL26 mediated poly-ubiquitylation is required for the transition from meiosis to mitosis (Pintard et al. 2003b; Xu et al. 2003), which also leads to the discovery of CUL3-BTB domain interaction. Similarly, in cultured mammalian cell line, MEI-1-like protein p60/Katanin can be degraded by CRL3-KLHDC5, which is required for normal mitosis in mammalian cells (Cummings et al. 2009).

13.3.2 CRL3 Regulates Vesicle Transport

Vesicle transport is a cellular process for the exchange of proteins and lipid through membrane-enclosed vesicles among the plasma membrane and membrane-bound organelles, including Golgi apparatus, endoplasmic reticulum, and endosome (Bonifacino and Glick 2004). CRL3 was first reported to regulate the size of COPII vesicles (Jin et al. 2012) by CRL3-KLHL12-mediated monoubiquitylation of SEC31, a protein coated on COPII vesicles. The monoubiquitylation catalyzed by CRL3-KLHL12 can enlarge COPII vesicles from 60–80 to 200–500 nm in diameter and is crucial for

Table 13.1 Substrates of BTB-CUL3-RING ligases

BTB-containing adaptor	CUL3 substrates	Organism (if not mammals)	Function	Reference (PMID)
ANKFY1	?		Regulate surface level of integrin β1	29038302
BACURDs	RhoA		Control cytoskeleton structure	19782033
BPMs	DREB2A	*Arabidopsis*		28923951
BPOZ-2	eEF1A1		Prevent translation	18459963
GCL	Torso	*Drosophila*	Control the switch between cell lineages	28743001
HIB/SPOP	Ci/Gli	*Drosophila*	Eye morphology	19955409
IBTK	IBTK			25882842
	Pdcd4		Modulate translation	
KBTBD2	p85α		Regulate insulin sensitivity	27708159
KBTBD6/ KBTBD7	TIAM		Restrict TIAM1-RAC1 signaling	25684205
KBTBD8	NOLC1	*Xenopus*	Cell fate determination	26399832
	TCOF1			
KCTD13	RhoA		Synaptic transmission	29088697
KCTD17	PHLPP2		Promote hepatic steatosis	28859855
	Trichoplein		Control ciliogenesis	25270598
KEAP1	NRF2		Response to oxidative stress	15572695, 15601839
	PGAM5		Response to oxidative stress	17046835
	IKKβ		Increase inflammatory responses	26195781
	Bcl-2		Regulate cell apoptosis	26041886
	MCM3		Control cell cycle	27621311
KEL-8	GLR-1	*C. elegans*	Regulate synaptic signaling and plasticity	16394099
KLHDC5	p60/katanin		Regulate cellular microtubule structure and mitosis	19261606
KLHL3	WNK1		Regulate electrolyte homeostasis	23387299
	WNK4		Regulate electrolyte homeostasis	23576762
KLHL9/ KLHL13	Aurora B		Regulate mitotic progression and completion of cytokinesis	17543862
Klhl10	dBruce	*Drosophila*	Regulates caspase activation	17880263
KLHL12	SEC31		Collagen export	22358839
	PLK1		Regulate kinetochore localization during mitosis	24067371, 23455478
	Dopamine D4 receptor		Regulate neurotransmitter receptor function	18303015
KLHL15	PPP2R2B		Promote exchange of protein phosphatase 2A subunit	23135275
KLHL17	GluR6		Regulate neurotransmitter receptor function	17062563
KLHL18	Aurora A		Regulate mitotic entry	23213400
KLHL20	ULK1		Control autophagy termination	26687681
	ATG13		Control autophagy termination	
	VPS34		Control autophagy termination	
	Beclin-1		Control autophagy termination	
	ATG14		Control autophagy termination	
	Coronin 7		Regulate protein trafficking	24768539
	DAPK		Control interferon responses	20389280
	PML		Potentiate HIF-1 signaling and prostate cancer progression	21840486

(continued)

Table 13.1 (continued)

BTB-containing adaptor	CUL3 substrates	Organism (if not mammals)	Function	Reference (PMID)
KLHL21	Aurora B		Promote midbody localization	28620047
LZTR1	RAS		Inhibit RAS signaling	30442762
MEL-26	MEI-1	*C. elegans*	Control mitotic spindle positioning	13679921, 14528312
MATH-BTBs	ATHB6	*Arabidopsis*	Modulate abscisic acid signaling	22172674
RHOBTB2	MSI2		Suppress breast cancer progression	27941885
RhoBTB	LRRC41			22709582
RDX	CENP-A	*Drosophila*	Control centromere maintenance	24636256
SPOP	BRMS1		Suppress breast cancer progression	22085717
	DDIT3/CHOP		Suppress prostate cancer progression	24990631
	Cyclin E1		Suppress prostate cancer progression	30237511
	Erα		Suppress development of endometrial cancer	25766326
	Progesterone receptors		Suppress breast cancer progression	26693071
	ATF2		Suppress prostate cancer progression	29996942
	DAXX		Regulate angiogenesis	28216678
	BRD2/3/4		Related to endometrial cancer and prostate cancer	28805821
	INF2		Suppress mitochondrial fission	28448495
SPOPL	EPS15		Regulate endocytic trafficking and MVBs formation	27008177
ZBTB16	ATG14L		Regulate autophagy	25821988
?	RhoBTB2			15107402
?	Cyclin D1		Regulate G1 cell cycle	11311237

secreting pro-collagen fibers, which is 300–400 nm long and cannot accommodate into smaller COPII vesicles without CRL3-KLHL12. CRL3 can also ubiquitylate CRN7 to form K33-linked polyubiquitin chain (Yuan et al. 2014). Although most poly-ubiquitylation results in proteasome-mediated degradation of substrate protein, K33 polyubiquitin chain attached to CRN7 does not affect its stability but facilitates CRN7 localizing to trans-Golgi network (TGN) through ubiquitin-mediated interaction with EPS15, which facilitates post-Golgi trafficking by promoting TGN F-actin assembly and generating transport carriers from TGN. CRL3 is also important for endocytosis pathway. RNAi-mediated depletion of CUL3 results in trafficking defect of influenza A virus and epidermal growth factor receptor and causes severe morphological defects in late endosomes (Huotari et al. 2012). CUL3 interacting with BTB domain-containing protein ANKFY1 is required for the early endosomal localization of ANKFY1 and can regulate subcellular localization of integrin β1 (Maekawa et al. 2017). ESCRT (endosomal sorting complexes required for transport) machinery is a large cytosolic protein complex and sorts ubiquitin modified membrane protein into intraluminal vesicles of the endosome to form multivesicular bodies (MVBs) (Christ et al. 2017). CRL3-SPOPL can ubiquitylate and degrade EPS15, an adaptor that is associated with ESCRT and endosome. As a result, SPOPL-depleted cells failed to form intraluminal vesicles and to degrade EGFR or other MVB cargos (Gschweitl et al. 2016).

13.4 Alterations of the NRF2-KEAP1-CUL3 Complex in Lung Cancer

The transcription factor nuclear factor erythroid 2-related factor 2 (NRF2) is regarded as one of the main orchestrators of the cellular antioxidant response and is the first discovered mammalian substrate of CRL3. NRF2 together with p45-NFE2, and Nrf1 belongs to the CNC (cap "n" collar) family of basic region-leucine zipper transcription factors (Motohashi and Yamamoto 2004). All of these factors interact with a specific DNA sequence, called the antioxidant response element (ARE), initially identified in the promoters of genes encoding detoxication enzymes GSTA2 (glutathione S-transferase A2) and NQO1 (NADPH:quinone oxidoreductase 1) (Rushmore et al. 1991). NRF2 controls the basal and inducible expression of over 200 genes that contain AREs in their regulatory regions by heterodimerizing with small MAF proteins (Zhu and Fahl 2001). NRF2 target genes regulate redox homeostasis, drug metabolism and excretion, energetic metabolism, iron metabolism, amino acid metabolism, survival, proliferation, autophagy, proteasomal degradation, DNA repair, and mitochondrial physiology (Rojo de la Vega et al. 2018).

The cytoplasmic protein Keap1 (Kelch ECH associating protein 1) is the first reported mammalian adaptor of the Cul3-based E3 ligase system, representing a prime example of how CRL3 regulates their substrates. Germline mutagenesis of the mouse NRF2 gene and phenotypic analyses have revealed that NRF2 and its target genes are indispensable components of defense mechanisms against oxidative and xenobiotic stresses (Motohashi and Yamamoto 2004). Notably, Keap1-deficient mice die by the third week after birth as a result of the abnormal hyperkeratosis of the esophagus and forestomach and consequent feeding problems, and the simultaneous knockout of the NRF2 gene completely reversed the apparent phenotypes and lethality observed in Keap1-null mutant mice, clearly indicating that NRF2 is the major target protein of Keap1 (Wakabayashi et al. 2003). Keap1 is composed of three major characteristic domains, including a BTB dimerization domain, an intervening region (IVR) enriched with cysteine residues, and a Kelch-repeat domain (DGR) that represents a protein docking site. Keap1 associates with the N-terminal region of Cul3 through the IVR domain, thereby acting as an adaptor of Cul3-based E3 ligase (Kobayashi et al. 2004). Cullinan SB et al. reported that NRF2 associates with a Cul3 complex in a Keap1-dependent manner and that Cul3-Keap1 complex regulates NRF2 polyubiquitylation (Cullinan et al. 2004). Later, Furukawa M et al. provided solid data showing that human KEAP1 binds to CUL3 and NRF2 via its BTB and DGR domains, respectively (Furukawa and Xiong 2005). Moreover, endogenous CUL3, KEAP1, and NRF2 form a complex, and this complex exerts its inhibitory effect on NRF2 by promoting its ubiquitylation and subsequent degradation by the 26S proteasome (Furukawa and Xiong 2005).

How does the NRF2-KEAP1-CUL3 complex sense oxidative stress? Numerous studies indicate that KEAP1 acts as a critical intracellular sensor of oxidative stress. Under unstressed conditions, cytosolic Keap1 exists as a homodimer and associates with Cul3. NRF2 is efficiently ubiquitinated by the Keap1-Cul3 ubiquitin ligase complex and rapidly degraded to maintain low basal levels (Itoh et al. 1999; Kobayashi et al. 2004). In the presence of electrophiles or ROS, one or more of the reactive cysteine residues in Keap1 are oxidized, especially cysteine 151 (C151), resulting in a conformational change that impairs NRF2 degradation (McMahon et al. 2010; Zhang and Hannink 2003). As a result, the ubiquitin ligase activity of Keap1 is inactivated and NRF2 ubiquitylation ceases, leading to NRF2 stabilization and subsequent translocation into the nucleus (Zhang et al. 2004). Nuclear NRF2 heterodimerizes with one of the small Maf proteins (transcriptional cofactors), binds to AREs, and initiates the transcription of a variety of cytoprotective genes (Katsuoka et al. 2005; Nguyen et al. 2003).

Due to its essential role in mitigating oxidative stress, a delicate balance of KEAP1-NRF2 signaling is required to maintain cellular homeostasis. Controlled activation of NRF2 in normal cells via the canonical mechanism prevents cancer initiation and is suitable for cancer chemoprevention strategies. However, prolonged (noncanonical) or constitutive (loss of regulatory mechanisms) activation of NRF2 participates in cancer promotion,

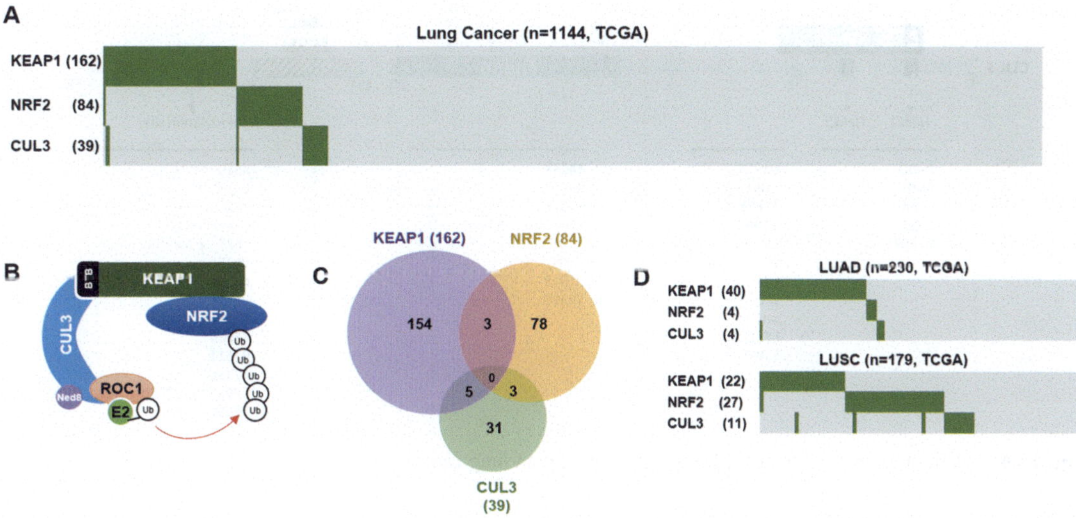

Fig. 13.3 CUL3 mutations tend to be mutually exclusive with KEAP1-NRF2 mutations in lung cancer

progression, and metastasis (Rojo de la Vega et al. 2018). Lung cancer is the most common cause of global cancer-related mortality, leading to over a million deaths each year. In 2012, as part of The Cancer Genome Atlas (TCGA), scientists profiled 178 lung squamous cell carcinomas (LUSC) and discovered mutations in the KEAP1-NRF2 pathway in 34% of LUSC cases, underscoring the significance of this pathway in lung tumorigenesis (Cancer Genome Atlas Research 2012). In 2014, whole-exome sequencing (WES) in 230 previously untreated lung adenocarcinoma samples and matched normal material revealed that KEAP1 gene is mutated in 19% of cases, NRF2 mutations show a lower frequency of 3%, and CUL3 mutations are even less than 1% in these samples (Cancer Genome Atlas Research 2014). In 2018, tumor tissue of 1391 patients with non-small cell lung cancer (NSCLC) was analyzed using next-generation sequencing (NGS), and genomic characterization revealed that the KEAP1 mutations occur with a frequency of 11.3% ($n = 157$), and NRF2 is mutated with a frequency of 3.5% ($n = 49$) (Frank et al. 2018).

According to TCGA database, a catalog of somatic mutations in 1144 lung cancer samples, including both adenocarcinoma and squamous cell carcinoma, revealed 162 and 84 samples carrying mutations in the coding region of KEAP1 and NRF2 genes, respectively, and mutations of both genes were seldom found in this cohort of lung cancer patients (Fig. 13.3a, b). On a molecular level, nearly all somatic mutations in NRF2 occur in specific hotspot regions, that is, the DLG and ETGE motifs responsible for binding to KEAP1 (Cancer Genome Atlas Research 2012; Hast et al. 2014). As such, these mutations liberate NRF2 from KEAP1-mediated ubiquitylation, and high levels of nuclear NRF2 facilitate cancer cell growth and survival as a result of the transactivation of cytoprotective genes. Differently, KEAP1 mutations were not found in specific hotspot regions but heterogeneously spread over the whole protein (Hast et al. 2014). Although not all tumor-derived mutations in KEAP1 directly affect Keap1: NRF2 interface, somatic mutations can reduce KEAP1-mediated repression of NRF2 in lung cancer patients and cell lines (Cancer Genome Atlas Research 2012; Lo et al. 2006; Singh et al. 2006). Loss of KEAP1 function causes constitutive activation of NRF2-mediated cytoprotective gene expression, potentially leading to a pathologic condition.

CUL3 exhibits a higher mutation frequency than the other *CUL* genes in human lung cancer, and the distribution of CUL3 mutations reiterates

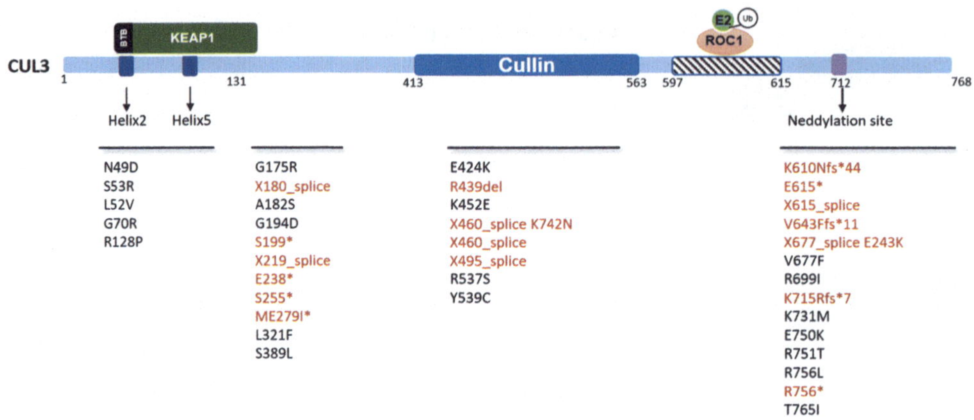

Fig. 13.4 Distribution and types of CUL3 mutation in human lung cancer

the lack of a "mutation cluster region" (Fig. 13.4). Of the 38 mutations found in lung cancer, 17 mutations result in a truncated protein product (highlighted in red). Two mutations, X460_splice and R756L/∗, occur twice in separate tumors, and interestingly, some mutated residues are near the neddylation site of CUL3 at the C-terminus. CUL3 neddylation was first described in the context of CRL3-MEL26-dependent MEI-1 in *C. elegans* (Pintard et al. 2003b), with both neddylation and deneddylation being required for MEI-1 degradation. Likewise, CUL3 neddylation is a prerequisite for Keap1-dependent in vivo ubiquitylation of NRF2 (Lo and Hannink 2006). Whether these tumor-derived mutations near the neddylation site of CUL3 would affect NEDD8 attachment, and subsequent recruitment of ubiquitin-charged E2 enzymes and positioning of the E2 active site for ubiquitin transfer onto substrates, remains unknown and needs further investigation.

It also needs to elucidate whether mutations in CUL3 gene occur during the initiation, promotion, or progression stages of tumor development. Based on the available data in TCGA, the most common mutations occur at a frequency of >50% in NSCLC, including TP53 (the gene encoding p53), Rb1 (retinoblastoma 1), B-cell CLL/lymphoma 2 (BCL2), fragile histidine triad gene (FHIT), and cyclin-dependent kinase inhibitor 2A (which encodes p16 INK4a). However, KEAP1-NRF2 mutations occur at a frequency of ~20% ($n = 246$ of 1144) in lung cancers, and CUL3 mutations show an even lower frequency of ~3% ($n = 39$ of 1144). These low frequencies indicate that the alteration in the NRF2-KEAP1-CUL3 complex is likely to occur during promotion and/or progression, rather than the initiation stage of tumorigenesis. Very interestingly, mutations in KEAP1, NRF2, and CUL3 tend to be mutually exclusive in lung cancers, especially in squamous cell carcinoma (Fig. 13.3c, d). What remains uncertain is which somatic mutations within CUL3 affect its function, to what degree they impact the NRF2-KEAP1 pathway, and mechanistically how its function is compromised.

13.5 Concluding Remarks

During last two decades, large efforts have been taken in studying CUL3 which has led to the discovery of the assembly of ROC1-CUL3-BTB domain E3 ligases, the CUL3 regulation by neddylation and deneddylation, and the biological functions of multiple BTB domain-containing proteins as CUL3 adaptors involved in diverse cellular processes. These exciting discoveries highlight the extraordinary possibility of CRL3 to target a very large number of substrates for ubiquitin-dependent degradation. However, more than half of BTB proteins are predicted to bind with CUL3, and further investigation is needed to elucidate which BTB proteins are able to assemble into the functional CUL3-based complexes and to identify their in vivo substrates. Since BTB-containing proteins are mutated in some human diseases, such as KEAP1 mutations in human lung cancer, reinvestigating their potential roles in ubiquitin-dependent degradation pathways may be rewarding. Moreover, the recently developed "inducible protein degradation system" can achieve a full and rapid deletion of target protein by small molecule perturbations (reviewed by Yesbolatova et al. 2019), including the AID (auxin-induced degradation) and the dTAG systems. Application of these new strategies to immediately target CUL3 degradation coupled with proteomic detection of rapidly accumulated proteins could be a promising strategy to expand CRL3 substrate identification and better understanding of their biological functions.

Acknowledgments This work was supported by the National Key R&D Program of China (No.2016YFA0501800 to D.Y.) and the NSFC grant (No. 31871431, No. 81522033, No. 31821002 to D. Y.).

References

Bade D, Pauleau AL, Wendler A, Erhardt S (2014) The E3 ligase CUL3/RDX controls centromere maintenance by ubiquitylating and stabilizing CENP-A in a CAL1-dependent manner. Dev Cell 28:508–519

Bai C, Sen P, Hofmann K, Ma L, Goebl M, Harper JW, Elledge SJ (1996) SKP1 connects cell cycle regulators to the ubiquitin proteolysis machinery through a novel motif, the F-box. Cell 86:263–274

Bardwell VJ, Treisman R (1994) The POZ domain: a conserved protein-protein interaction motif. Genes Dev 8:1664–1677

Beck J, Maerki S, Posch M, Metzger T, Persaud A, Scheel H, Hofmann K, Rotin D, Pedrioli P, Swedlow JR et al (2013) Ubiquitylation-dependent localization of PLK1 in mitosis. Nat Cell Biol 15:430–439

Bonifacino JS, Glick BS (2004) The mechanisms of vesicle budding and fusion. Cell 116:153–166

Buetow L, Huang DT (2016) Structural insights into the catalysis and regulation of E3 ubiquitin ligases. Nat Rev Mol Cell Biol 17:626–642

Cancer Genome Atlas Research N (2012) Comprehensive genomic characterization of squamous cell lung cancers. Nature 489:519–525

Cancer Genome Atlas Research N (2014) Comprehensive molecular profiling of lung adenocarcinoma. Nature 511:543–550

Chaharbakhshi E, Jemc JC (2016) Broad-complex, tramtrack, and bric-a-brac (BTB) proteins: critical regulators of development. Genesis 54:505–518

Christ L, Raiborg C, Wenzel EM, Campsteijn C, Stenmark H (2017) Cellular functions and molecular mechanisms of the ESCRT membrane-scission machinery. Trends Biochem Sci 42:42–56

Clark-Maguire S, Mains PE (1994a) Localization of the mei-1 gene product of Caenorhabditis elegans, a meiotic-specific spindle component. J Cell Biol 126:199–209

Clark-Maguire S, Mains PE (1994b) mei-1, a gene required for meiotic spindle formation in Caenorhabditis elegans, is a member of a family of ATPases. Genetics 136:533–546

Combes G, Alharbi I, Braga LG, Elowe S (2017) Playing polo during mitosis: PLK1 takes the lead. Oncogene 36:4819–4827

Cullinan SB, Gordan JD, Jin J, Harper JW, Diehl JA (2004) The Keap1-BTB protein is an adaptor that bridges NRF2 to a Cul3-based E3 ligase: oxidative stress sensing by a Cul3-Keap1 ligase. Mol Cell Biol 24:8477–8486

Cummings CM, Bentley CA, Perdue SA, Baas PW, Singer JD (2009) The Cul3/Klhdc5 E3 ligase regulates p60/katanin and is required for normal mitosis in mammalian cells. J Biol Chem 284:11663–11675

Dow MR, Mains PE (1998) Genetic and molecular characterization of the caenorhabditis elegans gene, mel-26, a postmeiotic negative regulator of mei-1, a meiotic-specific spindle component. Genetics 150:119–128

Frank R, Scheffler M, Merkelbach-Bruse S, Ihle MA, Kron A, Rauer M, Ueckeroth F, Konig K, Michels S, Fischer R et al (2018) Clinical and pathological characteristics of KEAP1- and NFE2L2-mutated non-small cell lung carcinoma (NSCLC). Clin Cancer Res 24:3087–3096

Furukawa M, Xiong Y (2005) BTB protein Keap1 targets antioxidant transcription factor NRF2 for ubiquitination by the Cullin 3-Roc1 ligase. Mol Cell Biol 25:162–171

Furukawa M, He YJ, Borchers C, Xiong Y (2003) Targeting of protein ubiquitination by BTB-Cullin 3-Roc1 ubiquitin ligases. Nat Cell Biol 5:1001–1007

Geyer R, Wee S, Anderson S, Yates J, Wolf DA (2003) BTB/POZ domain proteins are putative substrate adaptors for cullin 3 ubiquitin ligases. Mol Cell 12:783–790

Glotzer M, Murray AW, Kirschner MW (1991) Cyclin is degraded by the ubiquitin pathway. Nature 349:132–138

Green RA, Paluch E, Oegema K (2012) Cytokinesis in animal cells. Annu Rev Cell Dev Biol 28:29–58

Gschweitl M, Ulbricht A, Barnes CA, Enchev RI, Stoffel-Studer I, Meyer-Schaller N, Huotari J, Yamauchi Y, Greber UF, Helenius A et al (2016) A SPOPL/Cullin-3 ubiquitin ligase complex regulates endocytic trafficking by targeting EPS15 at endosomes. Elife 5:e13841

Haas AL, Warms JV, Hershko A, Rose IA (1982) Ubiquitin-activating enzyme. Mechanism and role in protein-ubiquitin conjugation. J Biol Chem 257:2543–2548

Hast BE, Cloer EW, Goldfarb D, Li H, Siesser PF, Yan F, Walter V, Zheng N, Hayes DN, Major MB (2014)

Cancer-derived mutations in KEAP1 impair NRF2 degradation but not ubiquitination. Cancer Res 74:808–817
He YJ, McCall CM, Hu J, Zeng Y, Xiong Y (2006) DDB1 functions as a linker to recruit receptor WD40 proteins to CUL4-ROC1 ubiquitin ligases. Genes Dev 20:2949–2954
Hershko A, Heller H, Elias S, Ciechanover A (1983) Components of ubiquitin-protein ligase system. Resolution, affinity purification, and role in protein breakdown. J Biol Chem 258:8206–8214
Huang G, Kaufman AJ, Xu K, Manova K, Singh B (2017) Squamous cell carcinoma-related oncogene (SCCRO) neddylates Cul3 protein to selectively promote midbody localization and activity of Cul3(KLHL21) protein complex during abscission. J Biol Chem 292:15254–15265
Huotari J, Meyer-Schaller N, Hubner M, Stauffer S, Katheder N, Horvath P, Mancini R, Helenius A, Peter M (2012) Cullin-3 regulates late endosome maturation. Proc Natl Acad Sci U S A 109:823–828
Itoh K, Wakabayashi N, Katoh Y, Ishii T, Igarashi K, Engel JD, Yamamoto M (1999) Keap1 represses nuclear activation of antioxidant responsive elements by NRF2 through binding to the amino-terminal Neh2 domain. Genes Dev 13:76–86
Jin L, Pahuja KB, Wickliffe KE, Gorur A, Baumgartel C, Schekman R, Rape M (2012) Ubiquitin-dependent regulation of COPII coat size and function. Nature 482:495–500
Kamura T, Maenaka K, Kotoshiba S, Matsumoto M, Kohda D, Conaway RC, Conaway JW, Nakayama KI (2004) VHL-box and SOCS-box domains determine binding specificity for Cul2-Rbx1 and Cul5-Rbx2 modules of ubiquitin ligases. Genes Dev 18:3055–3065
Katsuoka F, Motohashi H, Ishii T, Aburatani H, Engel JD, Yamamoto M (2005) Genetic evidence that small maf proteins are essential for the activation of antioxidant response element-dependent genes. Mol Cell Biol 25:8044–8051
King RW, Peters JM, Tugendreich S, Rolfe M, Hieter P, Kirschner MW (1995) A 20S complex containing CDC27 and CDC16 catalyzes the mitosis-specific conjugation of ubiquitin to cyclin B. Cell 81:279–288
Kobayashi A, Kang MI, Okawa H, Ohtsuji M, Zenke Y, Chiba T, Igarashi K, Yamamoto M (2004) Oxidative stress sensor Keap1 functions as an adaptor for Cul3-based E3 ligase to regulate proteasomal degradation of NRF2. Mol Cell Biol 24:7130–7139
Koiwai K, Maezawa S, Hayano T, Iitsuka M, Koiwai O (2008) BPOZ-2 directly binds to eEF1A1 to promote eEF1A1 ubiquitylation and degradation and prevent translation. Genes Cells 13:593–607
Kurz T, Pintard L, Willis JH, Hamill DR, Gonczy P, Peter M, Bowerman B (2002) Cytoskeletal regulation by the Nedd8 ubiquitin-like protein modification pathway. Science 295:1294–1298
Liu CC, Lin YC, Chen YH, Chen CM, Pang LY, Chen HA, Wu PR, Lin MY, Jiang ST, Tsai TF et al (2016) Cul3-KLHL20 ubiquitin ligase governs the turnover of ULK1 and VPS34 complexes to control autophagy termination. Mol Cell 61:84–97
Lo SC, Hannink M (2006) CAND1-mediated substrate adaptor recycling is required for efficient repression of NRF2 by Keap1. Mol Cell Biol 26:1235–1244
Lo SC, Li X, Henzl MT, Beamer LJ, Hannink M (2006) Structure of the Keap1:NRF2 interface provides mechanistic insight into NRF2 signaling. EMBO J 25:3605–3617
Lyapina SA, Correll CC, Kipreos ET, Deshaies RJ (1998) Human CUL1 forms an evolutionarily conserved ubiquitin ligase complex (SCF) with SKP1 and an F-box protein. Proc Natl Acad Sci U S A 95:7451–7456
Ma J, Chang K, Peng J, Shi Q, Gan H, Gao K, Feng K, Xu F, Zhang H, Dai B et al (2018) SPOP promotes ATF2 ubiquitination and degradation to suppress prostate cancer progression. J Exp Clin Cancer Res 37:145
Maeda I, Ohta T, Koizumi H, Fukuda M (2001) In vitro ubiquitination of cyclin D1 by ROC1-CUL1 and ROC1-CUL3. FEBS Lett 494:181–185
Maekawa M, Tanigawa K, Sakaue T, Hiyoshi H, Kubota E, Joh T, Watanabe Y, Taguchi T, Higashiyama S (2017) Cullin-3 and its adaptor protein ANKFY1 determine the surface level of integrin beta1 in endothelial cells. Biol Open 6:1707–1719
McIntosh JR (2016) Mitosis. Cold Spring Harb Perspect Biol 8
McMahon M, Lamont DJ, Beattie KA, Hayes JD (2010) Keap1 perceives stress via three sensors for the endogenous signaling molecules nitric oxide, zinc, and alkenals. Proc Natl Acad Sci U S A 107:18838–18843
Merlet J, Burger J, Gomes JE, Pintard L (2009) Regulation of cullin-RING E3 ubiquitin-ligases by neddylation and dimerization. Cell Mol Life Sci 66:1924–1938
Metzger MB, Pruneda JN, Klevit RE, Weissman AM (2014) RING-type E3 ligases: master manipulators of E2 ubiquitin-conjugating enzymes and ubiquitination. Biochim Biophys Acta 1843:47–60
Metzger T, Kleiss C, Sumara I (2013) CUL3 and protein kinases: insights from PLK1/KLHL22 interaction. Cell Cycle 12:2291–2296
Michel JJ, Xiong Y (1998) Human CUL-1, but not other cullin family members, selectively interacts with SKP1 to form a complex with SKP2 and cyclin A. Cell Growth Differ 9:435–449
Moghe S, Jiang F, Miura Y, Cerny RL, Tsai MY, Furukawa M (2012) The CUL3-KLHL18 ligase regulates mitotic entry and ubiquitylates Aurora-A. Biol Open 1:82–91
Motohashi H, Yamamoto M (2004) NRF2-Keap1 defines a physiologically important stress response mechanism. Trends Mol Med 10:549–557
Negrini S, Gorgoulis VG, Halazonetis TD (2010) Genomic instability--an evolving hallmark of cancer. Nat Rev Mol Cell Biol 11:220–228
Nguyen T, Sherratt PJ, Huang HC, Yang CS, Pickett CB (2003) Increased protein stability as a mechanism that enhances NRF2-mediated transcriptional activation of the antioxidant response element. Degradation of NRF2 by the 26 S proteasome. J Biol Chem 278:4536–4541

Nikonova AS, Astsaturov I, Serebriiskii IG, Dunbrack RL Jr, Golemis EA (2013) Aurora A kinase (AURKA) in normal and pathological cell division. Cell Mol Life Sci 70:661–687

Ohta T, Michel JJ, Schottelius AJ, Xiong Y (1999) ROC1, a homolog of APC11, represents a family of cullin partners with an associated ubiquitin ligase activity. Mol Cell 3:535–541

Petroski MD, Deshaies RJ (2005) Function and regulation of cullin-RING ubiquitin ligases. Nat Rev Mol Cell Biol 6:9–20

Pickart CM, Eddins MJ (2004) Ubiquitin: structures, functions, mechanisms. Biochim Biophys Acta 1695:55–72

Pickart CM, Rose IA (1985) Functional heterogeneity of ubiquitin carrier proteins. J Biol Chem 260:1573–1581

Pintard L, Kurz T, Glaser S, Willis JH, Peter M, Bowerman B (2003a) Neddylation and deneddylation of CUL-3 is required to target MEI-1/Katanin for degradation at the meiosis-to-mitosis transition in C. elegans. Curr Biol 13:911–921

Pintard L, Willis JH, Willems A, Johnson JL, Srayko M, Kurz T, Glaser S, Mains PE, Tyers M, Bowerman B et al (2003b) The BTB protein MEL-26 is a substrate-specific adaptor of the CUL-3 ubiquitin-ligase. Nature 425:311–316

Rojo de la Vega M, Chapman E, Zhang DD (2018) NRF2 and the hallmarks of Cancer. Cancer Cell 34:21–43

Rushmore TH, Morton MR, Pickett CB (1991) The antioxidant responsive element. Activation by oxidative stress and identification of the DNA consensus sequence required for functional activity. J Biol Chem 266:11632–11639

Shibata S, Zhang J, Puthumana J, Stone KL, Lifton RP (2013) Kelch-like 3 and Cullin 3 regulate electrolyte homeostasis via ubiquitination and degradation of WNK4. Proc Natl Acad Sci USA 110:7838–7843

Singer JD, Gurian-West M, Clurman B, Roberts JM (1999) Cullin-3 targets cyclin E for ubiquitination and controls S phase in mammalian cells. Genes Dev 13:2375–2387

Singh A, Misra V, Thimmulappa RK, Lee H, Ames S, Hoque MO, Herman JG, Baylin SB, Sidransky D, Gabrielson E et al (2006) Dysfunctional KEAP1-NRF2 interaction in non-small-cell lung cancer. PLoS Med 3:e420

Stogios PJ, Downs GS, Jauhal JJ, Nandra SK, Prive GG (2005) Sequence and structural analysis of BTB domain proteins. Genome Biol 6:R82

Sumara I, Quadroni M, Frei C, Olma MH, Sumara G, Ricci R, Peter M (2007) A Cul3-based E3 ligase removes Aurora B from mitotic chromosomes, regulating mitotic progression and completion of cytokinesis in human cells. Dev Cell 12:887–900

Wakabayashi N, Itoh K, Wakabayashi J, Motohashi H, Noda S, Takahashi S, Imakado S, Kotsuji T, Otsuka F, Roop DR et al (2003) Keap1-null mutation leads to postnatal lethality due to constitutive NRF2 activation. Nat Genet 35:238–245

Xu L, Wei Y, Reboul J, Vaglio P, Shin TH, Vidal M, Elledge SJ, Harper JW (2003) BTB proteins are substrate-specific adaptors in an SCF-like modular ubiquitin ligase containing CUL-3. Nature 425:316–321

Yesbolatova A, Tominari Y, Kanemaki MT (2019) Ligand-induced genetic degradation as a tool for target validation. Drug Discov Today Technol 31:91–98

Yuan WC, Lee YR, Lin SY, Chang LY, Tan YP, Hung CC, Kuo JC, Liu CH, Lin MY, Xu M et al (2014) K33-linked Polyubiquitination of Coronin 7 by Cul3-KLHL20 ubiquitin E3 ligase regulates protein trafficking. Mol Cell 54:586–600

Zhang DD, Hannink M (2003) Distinct cysteine residues in Keap1 are required for Keap1-dependent ubiquitination of NRF2 and for stabilization of NRF2 by chemopreventive agents and oxidative stress. Mol Cell Biol 23:8137–8151

Zhang DD, Lo SC, Cross JV, Templeton DJ, Hannink M (2004) Keap1 is a redox-regulated substrate adaptor protein for a Cul3-dependent ubiquitin ligase complex. Mol Cell Biol 24:10941–10953

Zheng N, Shabek N (2017) Ubiquitin ligases: structure, function, and regulation. Annu Rev Biochem 86:129–157

Zhu M, Fahl WE (2001) Functional characterization of transcription regulators that interact with the electrophile response element. Biochem Biophys Res Commun 289:212–219

Zhuang M, Calabrese MF, Liu J, Waddell MB, Nourse A, Hammel M, Miller DJ, Walden H, Duda DM, Seyedin SN et al (2009) Structures of SPOP-substrate complexes: insights into molecular architectures of BTB-Cul3 ubiquitin ligases. Mol Cell 36:39–50

Zollman S, Godt D, Prive GG, Couderc JL, Laski FA (1994) The BTB domain, found primarily in zinc finger proteins, defines an evolutionarily conserved family that includes several developmentally regulated genes in Drosophila. Proc Natl Acad Sci U S A 91:10717–10721

CRL4 Ubiquitin Pathway and DNA Damage Response

14

Pengbo Zhou and Fan Yan

Abstract

DNA damage occurs in a human cell at an average frequency of 10,000 incidences per day by means of external and internal culprits, damage that triggers sequential cellular responses and stalls the cell cycle while activating specific DNA repair pathways. Failure to remove DNA lesions would compromise genomic integrity, leading to human diseases such as cancer and premature aging. If DNA damage is extensive and cannot be repaired, cells undergo apoptosis. *D*NA *d*amage *r*esponse (DDR) often entails posttranslational modifications of key DNA repair and DNA damage checkpoint proteins, including phosphorylation and ubiquitination. *C*ullin-*R*ING *l*igase *4* (CRL4) enzyme has been found to target multiple DDR proteins for ubiquitination. In this chapter, we will discuss key repair and checkpoint proteins that are subject to ubiquitin-dependent regulation by members of the CRL4 family during ultraviolet light (UV)-induced DNA damage.

Keywords

CRL4 · DCAF · Ubiquitin proteasome system · DNA damage · DNA repair

P. Zhou (✉) · F. Yan
Department of Pathology and Laboratory Medicine, Weill Cornell Medical College, New York, NY, USA
e-mail: pez2001@med.cornell.edu

Abbreviations

6,4-PPs	6,4-Photoproducts
APC/C	The anaphase-promoting complex or cyclosome
ATL	Alkyltransferase-like
BER	Base excision repair
BTB	*B*ric-a-brac, *t*ram-track, and *b*road
CDK	Cyclin-dependent kinase
CPDs	Cyclobutene pyrimidine dimers
CPT	Camptothecin
CR	*C*ullin *r*epeats
CRL4	*C*ullin-*R*ING *l*igase *4*
CS	Cockayne syndrome
CSN	The COP9 signalosome
DCAF	DDB1-cullin 4-associated factor
DDB1	UV-damaged DNA-binding protein 1
DDR	*D*NA *d*amage *r*esponse
E1	Ubiquitin-activating enzyme
E2	Ubiquitin-conjugating enzyme
E3	Ubiquitin ligase
GG-NER	Global genome NER
MMR	Mismatch repair
NER	Nucleotide excision repair
PCNA	Proliferating cell nuclear antigen
PIP	PCNA-interacting protein
RFC	Replication factor C
RING	Really interesting new gene
RPA	Replication protein A
SCF	Skp1-CUL1-F-box

Y. Sun et al. (eds.), *Cullin-RING Ligases and Protein Neddylation*, Advances in Experimental Medicine and Biology 1217, https://doi.org/10.1007/978-981-15-1025-0_14

ssDNA	Single-stranded DNA
TC-NER	Transcription-coupled NER
TFIIH	Transcription factor II H
TLS	Translesion DNA synthesis
TOP1	Topoisomerase 1
Ub	Ubiquitin
UV	Ultraviolet light
XP	Xeroderma pigmentosum

14.1 The CRL4 Family

14.1.1 Architecture of CRL Ubiquitin Ligases

Protein ubiquitination entails covalent attachment of ubiquitin moietie(s) to target proteins via a cascade of enzymatic reactions that involve an E1 ubiquitin-activating enzyme, an E2 ubiquitin-conjugating enzyme, and an E3 ubiquitin ligase. Posttranslational modifications by ubiquitin typically result in the functional alteration of the substrate or proteasomal degradation. While E1 and E2 are required for activation and transfer of ubiquitin, the substrate specificity of any given ubiquitin pathway is conferred exclusively by the E3 ubiquitin ligase.

Elongated modular proteins called cullins serve as scaffolds for assembly of the multi-subunit CRL family of E3 ubiquitin ligase complexes. There are seven cullins in vertebrates, CUL1, 2, 3, 4A, 4B, 5, and 7, that assemble E3 complexes that are structurally similar but functionally distinct (Petroski and Deshaies 2005). The CRL complexes feature a cullin scaffold that brings together two functional subcomplexes (as shown in Fig. 14.1): at its carboxyl (C-) terminus, the RING domain protein Rbx1 (ROC1 or Hrt1) binding the E2 ubiquitin-conjugating enzyme and, at its amino (N-) terminus, specific adaptor and substrate receptor proteins (Fig. 14.1). The substrate-binding domains of cullins, through their adaptors, connect to substrate receptors, which in turn recruit a large array of substrates, some of which are involved in DNA repair and DNA damage checkpoint pathways. For all cullins, the N-terminus folds into three α-helical repeat bundles designated as *c*ullin *r*epeats (CRs), and each CR bundle contains five α-helices (Zheng et al. 2002b). The second and fifth α-helices of CR1 form direct contacts with cullin-specific adaptors, and the specificities of the adaptor interactions are determined by the amino acid residues at the respective binding interfaces, which differ among cullins (Petroski and Deshaies 2005; Zheng et al. 2002b).

The cullin adaptors further select different classes of substrate receptors through specific protein-protein interaction motifs. The various compositions of these adaptor subcomplexes allow CRLs to bind numerous substrates with distinct specificities. For example, in the

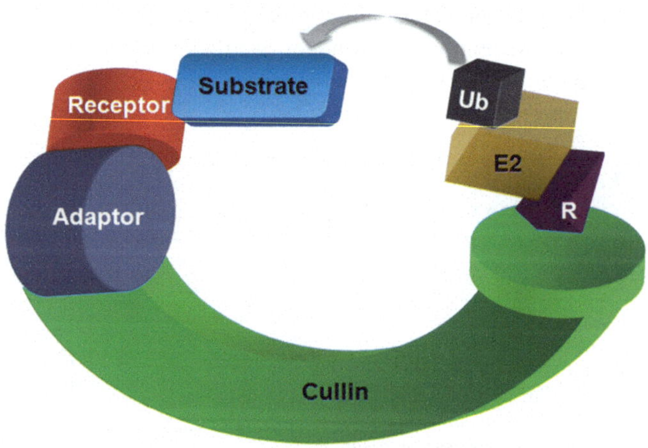

Fig. 14.1 Schematic diagram of the cullin-based ubiquitin ligase complexes. Ub ubiquitin, R Rbx1, E2 ubiquitin-conjugating enzyme

substrate-recruiting module of the Skp1-CUL1-F-box protein (SCF) complex, the Skp1 adaptor binds to CUL1 at its N-terminal CR1 region (Zheng et al. 2002b). Skp1 also interacts with one of the ~70 F-box proteins that function as substrate receptors (Bai et al. 1996; Schulman et al. 2000). These receptors contain additional protein-protein interaction modules (e.g., WD40 or leucine-rich repeats) responsible for substrate recruitment. Other CRLs feature different arrangements of adaptors and substrate receptors: CUL2 and CUL5 bind to the elongin B/C adaptor that interacts with SOCS-box-bearing substrate receptors (Kamura et al. 1998), while CUL3 binds substrates via *b*ric-a-brac, *t*ram-track, and *b*road (BTB)-bearing proteins, which are structurally similar to elongin C (Furukawa et al. 2003; Geyer et al. 2003; Xu et al. 2003). Unlike other CRLs, both the adaptor and substrate receptor functions are conferred by a single BTB protein. These cullin 3 adaptors share a common BTB fold and a second protein-protein interaction module in their tertiary structure (Perez-Torrado et al. 2006).

One adaptor that structurally deviates from those of other CRLs is the DDB1 (UV-damaged DNA-binding protein 1) adaptor of the CUL4A- and CUL4B-based CRL4. DDB1 comprises three consecutive WD40 β-propeller domains (designated BPA, BPB, and BPC) instead of the BTB fold (Li et al. 2006b). It recruits a subclass of WD40 domain-containing substrate receptors that share one or two WDXR signature motifs located within the short linkers, in some cases along with an additional α-helical domain at the N-terminus (e.g., DDB2), connecting two β-strands of a WD40 β-propeller blade (Angers et al. 2006; Jin et al. 2006; Higa et al. 2006; He et al. 2006; Lee and Zhou 2007).

14.1.2 Regulation of CRLs

Overexpression or amplification of several cullins has been observed in human cancers, indicating the need for precise control of CRL activities (Chen et al. 1998). Just as CRLs regulate proteins by covalent modification, at the posttranslational level, they too undergo modifications, which in turn regulate CRL E3 activity. One such modification involves NEDD8, an ubiquitin-like protein, which is covalently attached to a conserved lysine residue in the C-terminal cullin homology domain. NEDD8 modification, or neddylation, activates CRL complexes through multiple mechanisms. Duda et al. showed that neddylation induces a conformational change at the Rbx1-cullin interface that enables the Rbx1-associated E2 ubiquitin-conjugating enzyme to move closer to the adjacent substrate, thus facilitating ubiquitin transfer (Duda et al. 2008).

Neddylation also promotes the initial recruitment of ubiquitin-charged E2 to the Rbx1-cullin complex, thereby enhancing poly-ubiquitination of the substrate (Saha and Deshaies 2008). Furthermore, Nedd8 conjugation to CUL1 prevents recognition by the cullin inhibitor CAND1, which sequesters cullins to prevent them from assembling with Rbx1 and cullin-specific adaptors (Zheng et al. 2002a; Goldenberg et al. 2004; Liu et al. 2002). As such, cullin neddylation effectively excludes CRL association with CAND1 and facilitates assembly of productive CRLs. Thus, neddylation of cullins serves as a critical "on-switch" that facilitates ubiquitin transfer to substrates and prevents inhibition of CRLs. Conversely, the "off-switch" of cullin deneddylation is executed by the JAMM metalloprotease activity of the CSN5 subunit of the COP9 signalosome or CSN (Cope et al. 2002; Lyapina et al. 2001; Cope and Deshaies 2003; Ambroggio et al. 2004). In addition, the CSN complex recruits the deubiquitinating enzyme UBP12, which reverses the autoubiquitination of CRL components by disassembling polyubiquitin chains to prevent proteasomal degradation (Zhou et al. 2003). Therefore, CSN deactivates CRLs but can also prevent the self-destruction of active CRLs. As discussed in the next section, the CSN regulation of CUL4A-DDB1 plays an important part in the nucleotide excision repair process.

14.2 DNA Damage Response to UV or Chemical Carcinogens

Environmental culprits such as radiation or chemical mutagens constantly cause DNA damage. In general three kinds of excision repair are induced to repair single-stranded DNA lesions: base excision repair (BER), DNA mismatch repair (MMR), and nucleotide excision repair (NER) (Griffiths 2008; Morita et al. 2010). BER specializes in non-bulky lesions in DNA and removes only damaged bases by specific glycosylases, while the MMR pathway targets mismatched Watson-Crick base pairs. Bulky DNA lesions induced by UV irradiation, environmental mutagens, and certain chemotherapeutic agents are resolved by NER (Morita et al. 2010).

14.2.1 Nuclear Excision Repair Pathways

As mentioned above, NER is in charge of the removal of bulky DNA lesions induced by UV irradiation, environmental mutagens, and certain chemotherapeutic agents. There are two distinct NER pathways: global genome NER (GG-NER) and transcription-coupled NER (TC-NER) (Gillet and Scharer 2006; Hanawalt and Spivak 2008). TC-NER specifically repairs lesions on the transcribed strand of active genes, while GG-NER identifies and removes DNA adducts elsewhere in the genome, including the untranscribed strand of a transcribed gene and tightly packed heterochromatin. For GG-NER, the DDBs (UV-damaged DNA-binding proteins 1 and 2) and xeroderma pigmentosum complementation group C (XPC)-RAD23B survey the genome and identify UV-induced cyclobutene pyrimidine dimers (CPDs), 6,4-photoproducts (6,4-PPs), and chemical carcinogen-induced bulky DNA adducts. TC-NER is initiated when RNA polymerase stalls at a lesion and involves the TC-NER-specific factors CSA, CSB, and XAB2. Subsequently, both GG-NER and TC-NER enlist core NER factors (XPA, RPA, TFIIH, XPB, XPD, XPF, and XPG) to complete the excision and repair process. In bacteria and lower eukaryotes, the alkyltransferase-like (ATL) proteins participate in NER by binding to bulky alkyl lesions and facilitating their removal (Tubbs et al. 2009; Latypov et al. 2012). An analogous pathway could exist in mammals as well. Alterations of NER pathways are causative or closely associated with genetic disorders, such as the sun-sensitive xeroderma pigmentosum and the hereditary Cockayne syndrome, and patients are predisposed to cancer or accelerated aging and neurodegenerative disorders (Friedberg 2006).

14.2.2 Steps

Sensing DNA Lesions GG-NER and TC-NER utilize distinct "damage-sensing" proteins. DDB (UV-damaged DNA-binding protein) and the XPC-Rad23B complexes are employed by GG-NER to constantly scan the genome for DNA distortions. XPC-Rad23B complex plays an essential role in DNA damage recognition (Shell et al. 2013; Puumalainen et al. 2016), while some types of UV light-induced damage are specifically detected by the heterodimeric DDB1 and DDB2 (XPE) complex (Iovine et al. 2011; Petruseva et al. 2014). Furthermore, XPA has been suggested to participate in lesion recognition (Stout et al. 2005; Takedachi et al. 2010). Li et al. demonstrated that XPC collaborates with XPA and TFIIH to facilitate lesion recognition and verification (Li et al. 2015).

Lesions located at transcriptionally active genes can be recognized and repaired faster than nontranscribed strands or transcriptionally silent DNA. Within the actively transcribed genome segments, the moving RNA polymerase multitasks as a DNA damage sensor when it encounters bulky DNA adducts that stall the transcriptional complex. In this regard, RNA Pol II assumes the role of DNA strand distortion recognition during TC-NER, as the XPC-RAD23B and DDB complexes do in GG-NER (Iyama and Wilson 2013; Wienholz et al. 2019). Following damage sensing by RNA POLII, the Cockayne syndrome protein CSB is recruited to form a stable POLII-CSB-DNA-RNA complex at DNA damage sites (Tantin et al. 1997; Saijo 2013; Velez-Cruz and Egly 2013). Upon initial identification of a damaged site, replication protein A (RPA) and XPA join the NER repair complex and verify the DNA lesion (Iyama and Wilson 2013; Lagerwerf et al. 2011). Other repair proteins are subsequently recruited to the damaged region, as elaborated in later paragraphs.

Dual Incision After the first stage of damage recognition, transcription factor II H (TFIIH) and XPG are recruited to the site of DNA damage. The two subunits of the TFIIH complex, XPD and XPB, act as 5′-3′ and 3′-5′ helicases to unwind DNA to facilitate the access of the NER repair proteins (Mu et al. 2018; Galande et al. 2018). During this phase, XPG not only stabilizes TFIIH but also provides endonuclease activity 3′ to the DNA damage site (Iyer et al. 1996; Sekelsky et al. 2000), whereas the heterodimeric XPF-ERCC1 cleaves DNA at the 5′ end (Faridounnia et al. 2018). To reduce the exposure of single-stranded DNA, the 5′ side incision occurs first and is followed by DNA repair, which begins before the 3′ side incision. The complete dual incision usually removes a single-stranded DNA segment of 25–30 nucleotides that contains the damaged DNA, which is dissociated in an ATP-dependent manner. The complementary ssDNA region is bound to RPA, which serves as a protector of ssDNA along with XPA (Iyama and Wilson 2013; Lagerwerf et al. 2011).

Repair and Ligation Proliferating cell nuclear antigen (PCNA) is introduced to the DNA strand by replication factor C (RFC) (Shiomi and Nishitani 2017), which allows DNA polymerases (δ, ε, and/or κ) to copy from the undamaged strand via translocation. DNA ligase I, flap endonuclease 1, and the ligase-III-XRCC1 complex connect the nicks to complete NER (Moser et al. 2007).

14.3 NER Proteins Targeted by the CRL4 Ubiquitin Ligase

As an E3 ubiquitin ligase, CRL4 is capable of facilitating ubiquitination on numerous substrates, thereby altering the protein structure and function, or sending the protein to proteasome-dependent degradation. Multiple components involved in NER are targeted by CRL4 via distinct DDB1-cullin 4-associated factors (DCAFs).

14.3.1 DNA Damage Sensors

DDB2 and XPC The physical interaction between DDB and the CUL4A scaffold was independently identified by the Raychaudhuri group (Shiyanov et al. 1999; Nag et al. 2001) and the Zhou group (Chen et al. 2001). These findings opened up new avenues of NER: ubiquitin-dependent posttranslational modifications of key NER factors. CUL4A promoted ubiquitination and degradation of DDB2 (Chen et al. 2001; Nag et al. 2001), resulting in suppression of DDB activity for binding to UV-damaged DNA by the DDB1/DDB2 heterodimer (Chen et al. 2001). These findings suggest that NER is compromised in those with *CUL4A* amplification or overexpression, which predisposes these individuals to genomic instability and tumorigenesis. After UV exposure, the DDB heterodimer is recruited to CPDs and 6,4-PPs on chromatin and assembles the CRL4 complex on the damaged DNA. The UV light also induces neddylation of CUL4A and therefore activates the CRL4A ubiquitin ligase, resulting in the rapid ubiquitination and destruction of CRL4A substrates, including DDB2 itself (Chen et al. 2001; Groisman et al. 2003; Matsuda et al. 2005; Nag et al. 2001). In *Cul4a*$^{-/-}$ mouse embryonic fibroblasts, DDB2 is accumulated and is no longer degraded post-UV, indicating that CRL4A is primarily responsible for ubiquitination and proteasomal degradation of DDB2 (Liu et al. 2009).

Although DDB2 is initially identified as a bona fide substrate of the CRL4A complex, it can also function as a DCAF substrate receptor of CRL4^{DDB2} and promote the ubiquitination of other NER proteins. Another DNA damage sensor, XPC, is also a rate-limiting factor in the damage recognition step of NER. DDB2 directly binds UV-damaged DNA and assembles the CRL4A complex, which in turn recruits and ubiquitinates XPC, thereby facilitating NER at inter-nucleosomal sites (Li et al. 2006a; Luijsterburg et al. 2007; Fei et al. 2011; Sugasawa et al. 2005). XPC ubiquitination occurs preferentially at nuclease-“hypersensitive” sites located at "open" chromatin. In cell-free systems, XPC ubiquitination promotes NER reactions when DDB is present (Sugasawa et al. 2005). Moreover, XPC may even protect DDB2 from CRL4A-mediated destruction following UV exposure, thereby allowing DDB2 to recognize DNA lesions at multiple sites and initiate NER (Matsuda et al. 2005). *Cul4a*$^{-/-}$ mice express higher levels of DDB2 and XPC, demonstrate higher rates of NER activity, and exhibit striking resistance against UV-induced skin carcinogenesis (Liu et al. 2009). Despite these advances, the dynamics among CUL4A, DDB2, and XPC during the initiation of NER remains to be further defined.

CSB In addition to regulation of the genome-wide NER factors DDB2 and XPC, the CRL4 complex also associates with Cockayne syndrome (CS) protein A, which is involved in transcription-coupled NER (TC-NER). Patients with CS exhibit sensitivity to light (albeit less severe than XP patients), but the disease is categorized mainly by developmental defects (see Rapin et al. 2006 for a review of CS). When DNA lesions occur on actively transcribed portions of the genome and cause RNA polymerase stalling, the damage is resolved by the TC-NER pathway, which does not require recognition by DDB2 or XPC. Instead, the stalled polymerase acts as a damage sensor—recruiting CSA, CSB, and XAB2 to the damage site and initiating the repair process. After this point, the global genome- (GG-) and TC-NER pathways converge, and excision and repair of CPDs and 6,4-PPs are executed by the same set of repair factors (Nouspikel 2009).

The CUL4A complex can associate with CSA post-UV, and that purified CRL4A^{CSA} complex possesses E3 ubiquitin ligase activity in vitro (Groisman et al. 2003). DDB2 and CSA have structural similarities, and they both interact with DDB1 via an N-terminal helix-loop-helix motif (Fischer et al. 2011). CSA is required for ubiquitin-proteasomal degradation of the TC-NER damage recognition protein CSB. Indeed, the purified CRL4A^{CSA} complex can ubiquitinate CSB in vitro, which is thought to facilitate resumption of transcription post TC-NER (Fischer et al. 2011; Groisman et al. 2006). Whether CRL4^{CSA}-mediated degradation of CSB directly impacts TC-NER has yet to be determined. Additionally, the CRL4-CSA-CSB ubiquitin ligase complex targets the ATF3 transcriptional repressor for ubiquitination and degradation, thereby alleviating transcription arrest induced by genotoxic stress (Epanchintsev et al. 2017). Collectively, CRL4A regulates NER factors that are crucial to the early stages of GG-NER and later stage of transcriptional recovery post TC-NER.

14.3.2 DNA Damage Response Proteins

PCNA CRL4 often employs the DCAF protein Cdt2 to target proliferating cell nuclear antigen (PCNA)-bound proteins. PCNA is a homotrimeric ring complex that encircles double-stranded DNA and recruits a variety of factors involved in DNA replication and repair such as polymerases, chromatin-remodeling proteins, minichromosome maintenance, and GINS complex (Strzalka and Ziemienowicz 2011). Over 50 proteins have been shown to interact with PCNA—many of which contain a PIP (PCNA-interacting protein) box motif (De Biasio and Blanco 2013). PCNA plays an essential role as a docking platform in DNA synthesis and chromatin remodeling, and, as such, it is an ideal contact point for regulatory proteins.

Cdt1 One of the first proteins to be discovered as a bona fide CRL4 substrate was the Cdt1 DNA replication licensing factor, which promotes the formation of the pre-replication complex. Cdt1 collaborates with Cdc6 to load MCM2–MCM7 DNA helicase onto origin recognition complex-bound origins, and these replication origins are licensed for replication (Nishitani et al. 2000; Maiorano et al. 2000; O'Donnell et al. 2013). Following origin firing, Cdt1 is either removed by ubiquitin-proteasomal degradation or sequestered by geminin in order to prevent repeated origin firing and DNA re-replication in each cell cycle.

Following up on Zhong et al.'s finding that CRL4 silencing stabilizes Cdt1 in *Caenorhabditis elegans*, several groups reported that PCNA is required for CRL4-mediated degradation of Cdt1. PIP box mutations disrupt Cdt1-PCNA interactions, promoting DNA re-replication during normal cell cycle progression (Arias and Walter 2006; Higa et al. 2006; Nishitani et al. 2006; Senga et al. 2006; Zhong et al. 2003). Notably, Cdt1 is also degraded by CRL4 during the meiotic cell cycle (Yin et al. 2011). Upon exposure to UV or ionizing radiation, Cdt1 is rapidly ubiquitinated and degraded by the $CRL4^{Cdt2}$-PCNA ubiquitin ligase complex to halt DNA replication origin firing, effectively shutting down DNA replication. Cdt2 (encoded by the DTL gene in humans) serves as the DCAF for CRL4- and PCNA-dependent Cdt1 degradation (Higa et al. 2006; Jin et al. 2006; Sansam et al. 2006). Moreover, following UV irradiation, NER (XPA) and mismatch repair (MMR) proteins were reported to facilitate rapid Cdt1 turnover by activating $CRL4^{cdt2}$ in the G1 phase of the cell cycle (Raman et al. 2011; Shiomi et al. 2012; Tanaka et al. 2017). Importantly, Cdt1 is stabilized only when both CUL4A and CUL4B are silenced by RNAi, indicative of overlapping (redundant) functions for the two $CRL4^{cdt2}$ complexes in Cdt1 turnover following UV or IR.

p21 p21 and the other members of the Cip/Kip family of cyclin-dependent kinase inhibitors suppress Cdk2 and Cdk1 kinase complexes during cell cycle and influence other cellular processes such as senescence, apoptosis, transcription, and cytoskeletal organization (Starostina and Kipreos 2012). During the G1/S transition and G2/M checkpoint, p21 is targeted for ubiquitination and degraded by SCF^{Skp2} and APC/C^{Cdc20}, respectively. In 2008, several groups demonstrated that p21 shares a similar pattern of proteolysis as Cdt1 and is also degraded in the S phase by $CRL4^{Cdt2}$ complexes in a PCNA-dependent manner (Abbas et al. 2008; Kim et al. 2008; Nishitani et al. 2008). Although both CRL4 complexes appear to target p21 for ubiquitination, loss of CUL4A alone is sufficient to stabilize p21 in *CRL4a*$^{-/-}$ primary mouse skin cells and delayed cell cycle progression in mouse embryonic fibroblasts (Liu et al. 2009). RNAi-mediated knockdown of Cul4b results in a massive accumulation of p21 in XEN extra-embryonic endoderm mouse stem cells, which express extremely low endogenous levels of CUL4A (Liu et al. 2012). As such, either CRL4A or CRL4B controls threshold levels of p21, dependent of certain cellular contexts.

In order for DNA repair to occur, cells must be given adequate time to resolve lesions and breakage events. Induction of DNA damage causes transcriptional upregulation of the cyclin-dependent kinase (CDK) inhibitor p21, which blocks entry to the S phase. P21 is a substrate of $CRL4^{Cdt2}$-PCNA. In *Cul4a*$^{-/-}$ mouse-derived embryonic fibroblasts, p21 accumulates to higher levels to halt cell cycle progression, thereby providing cells with an extra 4–6 h to resolve UV-induced DNA damage (Liu et al. 2009). Depletion of p21 reverses the observed reinforcement of DNA damage checkpoint effects in *Cul4a*$^{-/-}$ cells and reduces their DNA repair capacity (Liu et al. 2009). Upon UV irradiation, CRL4A is activated, which then promotes the degradation of p21 to disengage the G1/S checkpoint. Accordingly, cells with excess CUL4A are likely compromised in their ability to launch an effective DNA damage checkpoint response.

p12 The heterotetrameric DNA polymerase δ (Pol δ), which participates in DNA repair as well as DNA replication, is also subjected to CRL4-dependent proteolytic control (Zhang et al. 2013). The p12 subunit of Pol δ contains a PIP box and is rapidly degraded after UV irradiation. Zhang et al. showed that CUL4A and Cdt2 are responsible for UV-induced degradation of p12. Moreover, the PIP box mutations of p12 allow its escape of CRL4 during the S phase (Zhang et al. 2013). These data further reinforce CRL4's role as a key regulator of DNA replication during cell cycle and under genotoxic stress.

Set8 The histone H4 methyltransferase SET8 (Pr-Set7) is required for chromosome compaction in mitosis and maintenance of genome integrity. $CRL4^{CDT2}$ can target SET8 for degradation during the S phase in a PCNA-dependent manner

(Jorgensen et al. 2011; Centore et al. 2010). SET8 degradation requires a conserved PIP-box degron responsible for its interaction with PCNA and recruitment to chromatin where ubiquitination occurs. Efficient degradation of SET8 at the onset of the S phase is required for the regulation of chromatin compaction status and cell cycle progression. During the meiotic cell cycle, Cul4a$^{-/-}$ spermatocytes display highly condensed chromatin and nuclear condensation at the metaphase II stage (Yin et al. 2011; Lin et al. 2016). Moreover, after UV irradiation, the turnover of SET8 is accelerated, mediated by the CRL4^{CDT2} ubiquitin ligase and PCNA. Removal of SET8 supports the modulation of chromatin structure after DNA damage (Havens and Walter 2011).

Chk1 CRL4 ubiquitinates repair proteins (e.g., DDB2, XPC) as well as DNA damage checkpoint factors that arrest the cell cycle. Following UV irradiation, the ATR-Chk1 DNA damage checkpoint pathway is activated, and Chk1 kinase targets numerous proteins. Zhang et al. reported that stress caused by camptothecin treatment led to Chk1 destruction, likely caused by the CUL1-based SCF (*S*kp1-*C*UL1-and *F*-box-containing substrate receptor) and CRL4A ubiquitin ligases (Zhang et al. 2005). Subsequently, Leung-Pineda et al. reported that CRL4A targets Chk1 for ubiquitination and degradation under normal conditions and during replicative stress (Leung-Pineda et al. 2009).

Chromatin Remodeling: H2A, H3, and H4 Damaged DNA is found throughout the genome. When damage occurs in compact chromosomal region or nontranscribed chromatin, relaxation of chromatin is prerequisite for recruitment of factors needed for repair processes. Remodeling of chromatin requires the concerted action of damage recognition proteins, histone acetyltransferases, helicases, and polymerases. In a series of reports, CRL4B^{DDB2} was shown to monoubiquitinate histone H2A on lysines 119 and 120 at damage loci post-UV—relaxing the nucleosome—while the CRL4A complex is less efficient (Guerrero-Santoro et al. 2008; Kapetanaki et al. 2006; Lan et al. 2012). However, loss of CUL4B does not affect global genomic repair of CPDs and 6-4PPs (Liu et al. 2009). Future work should clarify whether CRL4A steps up to assume the function of CRL4B in histone modification and GG-NER during chromatin remodeling when the CRL4B is absent. To this end, DDB2 appears to direct CRL4 toward DNA lesions. CRL4s can also monoubiquitinate histones H3 and H4 post-UV—in turn, destabilizing nucleosomes, stripping damaged DNA, and making lesions more accessible to repair factors such as XPC-RAD23B (Wang et al. 2006).

DNA Unwinding: TOP1 DNA helicase functions to unwind DNA double strands during replication and repair. The torsional strain that results from this unwinding must then be relieved by DNA topoisomerases, which are particularly vital during DNA replication. Cancer cells and bacteria proliferate rapidly, thus highly sensitive to agents that inhibit topoisomerase function. Camptothecins (CPTs) inhibit topoisomerase 1 (TOP1) by stabilizing a TOP1-DNA intermediate structure of cleavage and preventing the religation of nicked DNA strands (reviewed in Pommier 2006 and references therein). These TOP1-DNA-CPT complexes are often resolved by the ubiquitin-proteasome system (Desai et al. 2001). Kerzendorfer et al. showed that lymphoblastoid cell lines derived from patients with CUL4B mutations had increased sensitivity to CPT and that silencing of CUL4A or CUL4B would attenuate TOP1 destruction following CPT treatment (Kerzendorfer et al. 2010).

14.4 CRL4's Role in Other DNA Repair Pathways

Interstrand DNA Cross-Link Repair It is well known that the Vpr proteins of HIV-1 and related viruses are able to block the cell cycle at mitotic entry and therefore essential for the viral life cycle. Recent studies have shown that the Fanconi anemia DNA repair pathway is affected by HIV-1 Vpr. Vpr interacts with the SLX4/MUS81/EME1 endonuclease complex that processes interstrand DNA cross-links. Recombinant Vpr interacts with the C terminus of human SLX4. CRL4-DCAF1 is found to be essential in this process. It can bind to Vpr through DCAF1 (also known as Vpr-binding protein) (Hakata et al. 2014). With the help of CUL4/DDB1, Vpr activates MUS81-EME1's nuclease activity instead of promoting polyubiquitination and degradation of MUS81 (Zhou et al. 2016). Silencing of any component of the SLX4 complex can block CUL4/DDB1/DCAF1/Vpr-mediated cell cycle arrest (Berger et al. 2015). Upon viral infection of cultured cells, SLX4 is recruited to proviral HIV-1 DNA in the presence of Vpr dependent on CUL4/DDB1/DCAF1.

Translesion Synthesis Translesion DNA synthesis (TLS) is a highly conserved mechanism of DNA damage tolerance in eukaryotes. Replication blocking lesions are not removed but instead bypassed to avoid the damaging consequences of a blocked replication fork (Yang and Woodgate 2007). TLS is mainly performed by specialized Y-family TLS polymerases, including Polη, Polκ, Polι, and Rev1 (Lehmann et al. 2007). Y-family polymerases are involved in DNA synthesis and are believed to fall off the replication fork after adding one or two nucleotides across from the lesion. The B-family polymerase comes in to extend the DNA strand and brings TLS to completion (Johnson et al. 2000). Polymerase II in *Escherichia coli* and Polζ in eukaryotes are thought to be TLS polymerases (Wang and Yang 2009; Gan et al. 2008). Polζ is composed of Rev3 (a catalytic subunit) and Rev7 (a regulatory subunit), and it can synthesize across a CPD lesion very efficiently (Nelson et al. 1996) after Y-family polymerases insert nucleotides across from the lesion (Johnson et al. 2000; Neal et al. 2010; Takezawa et al. 2010; Hashimoto et al. 2012; Lee et al. 2014). UV irradiation can cause recruitment of Polη and Rev1 independently to the damage site, and only Rev1 is required for the recruitment of Rev3 (Andersen et al. 2011). Rev7 forms a dimer where one unit binds to the Rev1 C-terminus (Guo et al. 2003) and another to Rev3 (Murakumo et al. 2001), therefore playing a critical role in TLS in response to UV damage (McNally et al. 2008).

In mammalian cells, studies show that Rev7 undergoes ubiquitin-proteasome-mediated degradation upon UV irradiation. In this process, a destruction box in the Rev7 N-terminal is identified as the degron and CRL4A/B as putative E3 ligases (Bhat et al. 2017). Depletion of CUL4B or CUL4A/B result in the partial stabilization of Rev7 after UV damage, suggesting that CRL4 is the cognate E3 ligase of Rev7 and that CRL4A and CRL4B play redundant roles.

DNA Double-Strand Break Repair *Cul4a* gene knockout in mice results in infertility in males but not females (Yin et al. 2011). Decreased spermatozoa number, reduced sperm motility, and defective acrosome formation in the mutant germ cells are accompanied by increased cell death in pachytene/diplotene cells with markedly elevated levels of phospho-p53 and CDT1. Synaptonemal complex assembly and DNA double-strand breakage appear normal; however, *Cul4a*$^{-/-}$ spermatocytes fail to resolve the late recombination nodules, as marked by the persistent MLH1 foci at the diplotene stage. This suggests that CUL4A plays a role in resolving holiday junctions during DNA double-strand break repair. Interestingly, the role of CUL4A in DNA DSB repair is revealed only in male meiotic cells, not somatic cells, because the CUL4A analog CUL4B is not expressed at the pachytene to diplotene transition stage. In somatic cells, the redundant function of CUL4A and 4B effectively masks the role of CRL4 ubiquitin ligase in the homologous recombination repair of DNA double-strand breaks (Yin et al. 2011). The target(s) of CRL4A responsible for meiotic cell cycle arrest at the pachytene-diplotene transition remain to be identified.

14.5 Concluding Remarks

As an E3 ubiquitin ligase, CRL4 is capable of facilitating ubiquitination on numerous substrates specified by distinct DDB1-cullin-associated factors (DCAFs) as a substrate adaptor. Studies have shown that multiple components involved in NER, TLS, and other DNA repair pathways are targeted by CRL4. CRL4 can not only cause substrate degradation but also lead to substrate activation in some cases, highlighting the critical roles played by the CRL4 ubiquitin ligases in DNA damage response.

References

Abbas T, Sivaprasad U, Terai K, Amador V, Pagano M, Dutta A (2008) PCNA-dependent regulation of p21 ubiquitylation and degradation via the CRL4Cdt2 ubiquitin ligase complex. Genes Dev 22 (18):2496–2506. https://doi.org/10.1101/gad.1676108

Ambroggio XI, Rees DC, Deshaies RJ (2004) JAMM: a metalloprotease-like zinc site in the proteasome and signalosome. PLoS Biol 2(1):E2. https://doi.org/10.1371/journal.pbio.0020002

Andersen PL, Xu F, Ziola B, McGregor WG, Xiao W (2011) Sequential assembly of translesion DNA polymerases at UV-induced DNA damage sites. Mol Biol Cell 22(13):2373–2383. https://doi.org/10.1091/mbc.E10-12-0938

Angers S, Li T, Yi X, MacCoss MJ, Moon RT, Zheng N (2006) Molecular architecture and assembly of the DDB1-CUL4A ubiquitin ligase machinery. Nature 443 (7111):590–593. https://doi.org/10.1038/nature05175

Arias EE, Walter JC (2006) PCNA functions as a molecular platform to trigger Cdt1 destruction and prevent re-replication. Nat Cell Biol 8(1):84–90. https://doi.org/10.1038/ncb1346

Bai C, Sen P, Hofmann K, Ma L, Goebl M, Harper JW, Elledge SJ (1996) SKP1 connects cell cycle regulators to the ubiquitin proteolysis machinery through a novel motif, the F-box. Cell 86(2):263–274

Berger G, Lawrence M, Hue S, Neil SJ (2015) G2/M cell cycle arrest correlates with primate lentiviral Vpr interaction with the SLX4 complex. J Virol 89(1):230–240. https://doi.org/10.1128/JVI.02307-14

Bhat A, Qin Z, Wang G, Chen W, Xiao W (2017) Rev7, the regulatory subunit of Polzeta, undergoes UV-induced and Cul4-dependent degradation. FEBS J 284 (12):1790–1803. https://doi.org/10.1111/febs.14088

Centore RC, Havens CG, Manning AL, Li JM, Flynn RL, Tse A, Jin J, Dyson NJ, Walter JC, Zou L (2010) CRL4 (Cdt2)-mediated destruction of the histone methyltransferase Set8 prevents premature chromatin compaction in S phase. Mol Cell 40(1):22–33. https://doi.org/10.1016/j.molcel.2010.09.015

Chen LC, Manjeshwar S, Lu Y, Moore D, Ljung BM, Kuo WL, Dairkee SH, Wernick M, Collins C, Smith HS (1998) The human homologue for the Caenorhabditis elegans cul-4 gene is amplified and overexpressed in primary breast cancers. Cancer Res 58(16):3677–3683

Chen X, Zhang Y, Douglas L, Zhou P (2001) UV-damaged DNA-binding proteins are targets of CUL-4A-mediated ubiquitination and degradation. J Biol Chem 276(51):48175–48182

Cope GA, Deshaies RJ (2003) COP9 signalosome: a multifunctional regulator of SCF and other cullin-based ubiquitin ligases. Cell 114(6):663–671

Cope GA, Suh GS, Aravind L, Schwarz SE, Zipursky SL, Koonin EV, Deshaies RJ (2002) Role of predicted metalloprotease motif of Jab1/Csn5 in cleavage of Nedd8 from Cul1. Science 298(5593):608–611. https://doi.org/10.1126/science.1075901

De Biasio A, Blanco FJ (2013) Proliferating cell nuclear antigen structure and interactions: too many partners for one dancer? Adv Protein Chem Struct Biol 91:1–36. https://doi.org/10.1016/B978-0-12-411637-5.00001-9

Desai SD, Li TK, Rodriguez-Bauman A, Rubin EH, Liu LF (2001) Ubiquitin/26S proteasome-mediated degradation of topoisomerase I as a resistance mechanism to camptothecin in tumor cells. Cancer Res 61(15): 5926–5932

Duda DM, Borg LA, Scott DC, Hunt HW, Hammel M, Schulman BA (2008) Structural insights into NEDD8 activation of cullin-RING ligases: conformational control of conjugation. Cell 134(6):995–1006. https://doi.org/10.1016/j.cell.2008.07.022

Epanchintsev A, Costanzo F, Rauschendorf MA, Caputo M, Ye T, Donnio LM, Proietti-de-Santis L, Coin F, Laugel V, Egly JM (2017) Cockayne's syndrome A and B proteins regulate transcription arrest after genotoxic stress by promoting ATF3 degradation. Mol Cell 68(6):1054–1066. e1056. https://doi.org/10.1016/j.molcel.2017.11.009

Faridounnia M, Folkers GE, Boelens R (2018) Function and interactions of ERCC1-XPF in DNA damage response. Molecules 23(12). https://doi.org/10.3390/molecules23123205

Fei J, Kaczmarek N, Luch A, Glas A, Carell T, Naegeli H (2011) Regulation of nucleotide excision repair by UV-DDB: prioritization of damage recognition to internucleosomal DNA. PLoS Biol 9(10):e1001183. https://doi.org/10.1371/journal.pbio.1001183

Fischer ES, Scrima A, Bohm K, Matsumoto S, Lingaraju GM, Faty M, Yasuda T, Cavadini S, Wakasugi M, Hanaoka F, Iwai S, Gut H, Sugasawa K, Thoma NH (2011) The molecular basis of CRL4DDB2/CSA ubiquitin ligase architecture, targeting, and activation. Cell 147(5):1024–1039. https://doi.org/10.1016/j.cell.2011.10.035

Friedberg EC (2006) DNA repair and mutagenesis, 2nd edn. ASM Press, Washington, D.C.

Furukawa M, He YJ, Borchers C, Xiong Y (2003) Targeting of protein ubiquitination by BTB-Cullin 3-Roc1 ubiquitin ligases. Nat Cell Biol 5(11): 1001–1007. https://doi.org/10.1038/ncb1056

Galande AA, Perween N, Saijo M, Ghaskadbi SS, Ghaskadbi S (2018) Analysis of the conserved NER

helicases (XPB and XPD) and UV-induced DNA damage in hydra. Biochim Biophys Acta Gen Subj 1862 (9):2031–2042. https://doi.org/10.1016/j.bbagen.2018.06.017
Gan GN, Wittschieben JP, Wittschieben BO, Wood RD (2008) DNA polymerase zeta (pol zeta) in higher eukaryotes. Cell Res 18(1):174–183. https://doi.org/10.1038/cr.2007.117
Geyer R, Wee S, Anderson S, Yates J, Wolf DA (2003) BTB/POZ domain proteins are putative substrate adaptors for cullin 3 ubiquitin ligases. Mol Cell 12 (3):783–790
Gillet LC, Scharer OD (2006) Molecular mechanisms of mammalian global genome nucleotide excision repair. Chem Rev 106(2):253–276. https://doi.org/10.1021/cr040483f
Goldenberg SJ, Cascio TC, Shumway SD, Garbutt KC, Liu J, Xiong Y, Zheng N (2004) Structure of the Cand1-Cul1-Roc1 complex reveals regulatory mechanisms for the assembly of the multisubunit cullin-dependent ubiquitin ligases. Cell 119(4): 517–528. https://doi.org/10.1016/j.cell.2004.10.019
Griffiths AJF (2008) Introduction to genetic analysis, 9th edn. W.H. Freeman and Co., New York
Groisman R, Polanowska J, Kuraoka I, Sawada J, Saijo M, Drapkin R, Kisselev AF, Tanaka K, Nakatani Y (2003) The ubiquitin ligase activity in the DDB2 and CSA complexes is differentially regulated by the COP9 signalosome in response to DNA damage. Cell 113 (3):357–367
Groisman R, Kuraoka I, Chevallier O, Gaye N, Magnaldo T, Tanaka K, Kisselev AF, Harel-Bellan A, Nakatani Y (2006) CSA-dependent degradation of CSB by the ubiquitin-proteasome pathway establishes a link between complementation factors of the Cockayne syndrome. Genes Dev 20(11):1429–1434. https://doi.org/10.1101/gad.378206
Guerrero-Santoro J, Kapetanaki MG, Hsieh CL, Gorbachinsky I, Levine AS, Rapic-Otrin V (2008) The cullin 4B-based UV-damaged DNA-binding protein ligase binds to UV-damaged chromatin and ubiquitinates histone H2A. Cancer Res 68(13): 5014–5022. https://doi.org/10.1158/0008-5472.CAN-07-6162
Guo C, Fischhaber PL, Luk-Paszyc MJ, Masuda Y, Zhou J, Kamiya K, Kisker C, Friedberg EC (2003) Mouse Rev1 protein interacts with multiple DNA polymerases involved in translesion DNA synthesis. EMBO J 22(24):6621–6630. https://doi.org/10.1093/emboj/cdg626
Hakata Y, Miyazawa M, Landau NR (2014) Interactions with DCAF1 and DDB1 in the CRL4 E3 ubiquitin ligase are required for Vpr-mediated G2 arrest. Virol J 11:108. https://doi.org/10.1186/1743-422X-11-108
Hanawalt PC, Spivak G (2008) Transcription-coupled DNA repair: two decades of progress and surprises. Nat Rev Mol Cell Biol 9(12):958–970. https://doi.org/10.1038/nrm2549
Hashimoto K, Cho Y, Yang IY, Akagi J, Ohashi E, Tateishi S, de Wind N, Hanaoka F, Ohmori H, Moriya M (2012) The vital role of polymerase zeta and REV1 in mutagenic, but not correct, DNA synthesis across benzo[a]pyrene-dG and recruitment of polymerase zeta by REV1 to replication-stalled site. J Biol Chem 287 (12):9613–9622. https://doi.org/10.1074/jbc.M111.331728
Havens CG, Walter JC (2011) Mechanism of CRL4 (Cdt2), a PCNA-dependent E3 ubiquitin ligase. Genes Dev 25(15):1568–1582. https://doi.org/10.1101/gad.2068611
He YJ, McCall CM, Hu J, Zeng Y, Xiong Y (2006) DDB1 functions as a linker to recruit receptor WD40 proteins to CUL4-ROC1 ubiquitin ligases. Genes Dev 20 (21):2949–2954. https://doi.org/10.1101/gad.1483206
Higa LA, Wu M, Ye T, Kobayashi R, Sun H, Zhang H (2006) CUL4-DDB1 ubiquitin ligase interacts with multiple WD40-repeat proteins and regulates histone methylation. Nat Cell Biol 8(11):1277–1283. https://doi.org/10.1038/ncb1490
Iovine B, Iannella ML, Bevilacqua MA (2011) Damage-specific DNA binding protein 1 (DDB1): a protein with a wide range of functions. Int J Biochem Cell Biol 43 (12):1664–1667. https://doi.org/10.1016/j.biocel.2011.09.001
Iyama T, Wilson DM 3rd (2013) DNA repair mechanisms in dividing and non-dividing cells. DNA Repair (Amst) 12(8):620–636. https://doi.org/10.1016/j.dnarep.2013.04.015
Iyer N, Reagan MS, Wu KJ, Canagarajah B, Friedberg EC (1996) Interactions involving the human RNA polymerase II transcription/nucleotide excision repair complex TFIIH, the nucleotide excision repair protein XPG, and Cockayne syndrome group B (CSB) protein. Biochemistry 35(7):2157–2167. https://doi.org/10.1021/bi9524124
Jin J, Arias EE, Chen J, Harper JW, Walter JC (2006) A family of diverse Cul4-Ddb1-interacting proteins includes Cdt2, which is required for S phase destruction of the replication factor Cdt1. Mol Cell 23(5): 709–721. https://doi.org/10.1016/j.molcel.2006.08.010
Johnson RE, Washington MT, Haracska L, Prakash S, Prakash L (2000) Eukaryotic polymerases iota and zeta act sequentially to bypass DNA lesions. Nature 406(6799):1015–1019. https://doi.org/10.1038/35023030
Jorgensen S, Eskildsen M, Fugger K, Hansen L, Larsen MS, Kousholt AN, Syljuasen RG, Trelle MB, Jensen ON, Helin K, Sorensen CS (2011) SET8 is degraded via PCNA-coupled CRL4(CDT2) ubiquitylation in S phase and after UV irradiation. J Cell Biol 192 (1):43–54. https://doi.org/10.1083/jcb.201009076
Kamura T, Sato S, Haque D, Liu L, Kaelin WGJ, Conaway RC, Conaway JW (1998) The Elongin BC complex interacts with the conserved SOCS-box motif present in members of the SOCS, ras, WD-40 repeat, and

ankyrin repeat families. Genes Dev 12(24):3872–3881. https://doi.org/10.1101/gad.12.24.3872

Kapetanaki MG, Guerrero-Santoro J, Bisi DC, Hsieh CL, Rapic-Otrin V, Levine AS (2006) The DDB1-CUL4ADDB2 ubiquitin ligase is deficient in xeroderma pigmentosum group E and targets histone H2A at UV-damaged DNA sites. Proc Natl Acad Sci U S A 103(8):2588–2593. https://doi.org/10.1073/pnas.0511160103

Kerzendorfer C, Whibley A, Carpenter G, Outwin E, Chiang SC, Turner G, Schwartz C, El-Khamisy S, Raymond FL, O'Driscoll M (2010) Mutations in Cullin 4B result in a human syndrome associated with increased camptothecin-induced topoisomerase I-dependent DNA breaks. Hum Mol Genet 19 (7):1324–1334. https://doi.org/10.1093/hmg/ddq008

Kim Y, Starostina NG, Kipreos ET (2008) The CRL4Cdt2 ubiquitin ligase targets the degradation of p21Cip1 to control replication licensing. Genes Dev 22 (18):2507–2519. https://doi.org/10.1101/gad.1703708

Lagerwerf S, Vrouwe MG, Overmeer RM, Fousteri MI, Mullenders LH (2011) DNA damage response and transcription. DNA Repair (Amst) 10(7):743–750. https://doi.org/10.1016/j.dnarep.2011.04.024

Lan L, Nakajima S, Kapetanaki MG, Hsieh CL, Fagerburg M, Thickman K, Rodriguez-Collazo P, Leuba SH, Levine AS, Rapic-Otrin V (2012) Monoubiquitinated histone H2A destabilizes photolesion-containing nucleosomes with concomitant release of UV-damaged DNA-binding protein E3 ligase. J Biol Chem 287(15):12036–12049. https://doi.org/10.1074/jbc.M111.307058

Latypov VF, Tubbs JL, Watson AJ, Marriott AS, McGown G, Thorncroft M, Wilkinson OJ, Senthong P, Butt A, Arvai AS, Millington CL, Povey AC, Williams DM, Santibanez-Koref MF, Tainer JA, Margison GP (2012) Atl1 regulates choice between global genome and transcription-coupled repair of O (6)-alkylguanines. Mol Cell 47(1):50–60. https://doi.org/10.1016/j.molcel.2012.04.028

Lee J, Zhou P (2007) DCAFs, the missing link of the CUL4-DDB1 ubiquitin ligase. Mol Cell 26 (6):775–780. https://doi.org/10.1016/j.molcel.2007.06.001

Lee YS, Gregory MT, Yang W (2014) Human Pol zeta purified with accessory subunits is active in translesion DNA synthesis and complements Pol eta in cisplatin bypass. Proc Natl Acad Sci U S A 111(8):2954–2959. https://doi.org/10.1073/pnas.1324001111

Lehmann AR, Niimi A, Ogi T, Brown S, Sabbioneda S, Wing JF, Kannouche PL, Green CM (2007) Translesion synthesis: Y-family polymerases and the polymerase switch. DNA Repair (Amst) 6(7):891–899. https://doi.org/10.1016/j.dnarep.2007.02.003

Leung-Pineda V, Huh J, Piwnica-Worms H (2009) DDB1 targets Chk1 to the Cul4 E3 ligase complex in normal cycling cells and in cells experiencing replication stress. Cancer Res 69(6):2630–2637. https://doi.org/10.1158/0008-5472.CAN-08-3382

Li J, Wang QE, Zhu Q, El-Mahdy MA, Wani G, Praetorius-Ibba M, Wani AA (2006a) DNA damage binding protein component DDB1 participates in nucleotide excision repair through DDB2 DNA-binding and cullin 4A ubiquitin ligase activity. Cancer Res 66(17):8590–8597. https://doi.org/10.1158/0008-5472.CAN-06-1115

Li T, Chen X, Garbutt KC, Zhou P, Zheng N (2006b) Structure of DDB1 in complex with a paramyxovirus V protein: viral hijack of a propeller cluster in ubiquitin ligase. Cell 124(1):105–117. https://doi.org/10.1016/j.cell.2005.10.033

Li CL, Golebiowski FM, Onishi Y, Samara NL, Sugasawa K, Yang W (2015) Tripartite DNA lesion recognition and verification by XPC, TFIIH, and XPA in nucleotide excision repair. Mol Cell 59(6):1025–1034. https://doi.org/10.1016/j.molcel.2015.08.012

Lin CY, Chen CY, Yu CH, Yu IS, Lin SR, Wu JT, Lin YH, Kuo PL, Wu JC, Lin SW (2016) Human X-linked intellectual disability factor CUL4B is required for post-meiotic sperm development and male fertility. Sci Rep 6:20227. https://doi.org/10.1038/srep20227

Liu J, Furukawa M, Matsumoto T, Xiong Y (2002) NEDD8 modification of CUL1 dissociates p120 (CAND1), an inhibitor of CUL1-SKP1 binding and SCF ligases. Mol Cell 10(6):1511–1518

Liu L, Lee S, Zhang J, Peters SB, Hannah J, Zhang Y, Yin Y, Koff A, Ma L, Zhou P (2009) CUL4A abrogation augments DNA damage response and protection against skin carcinogenesis. Mol Cell 34(4):451–460. https://doi.org/10.1016/j.molcel.2009.04.020

Liu L, Yin Y, Li Y, Prevedel L, Lacy EH, Ma L, Zhou P (2012) Essential role of the CUL4B ubiquitin ligase in extra-embryonic tissue development during mouse embryogenesis. Cell Res 22(8):1258–1269. https://doi.org/10.1038/cr.2012.48

Luijsterburg MS, Goedhart J, Moser J, Kool H, Geverts B, Houtsmuller AB, Mullenders LH, Vermeulen W, van Driel R (2007) Dynamic in vivo interaction of DDB2 E3 ubiquitin ligase with UV-damaged DNA is independent of damage-recognition protein XPC. J Cell Sci 120 (Pt 15):2706–2716. doi:https://doi.org/10.1242/jcs.008367

Lyapina S, Cope G, Shevchenko A, Serino G, Tsuge T, Zhou C, Wolf DA, Wei N, Deshaies RJ (2001) Promotion of NEDD-CUL1 conjugate cleavage by COP9 signalosome. Science 292(5520):1382–1385. https://doi.org/10.1126/science.1059780

Maiorano D, Moreau J, Mechali M (2000) XCDT1 is required for the assembly of pre-replicative complexes in Xenopus laevis. Nature 404(6778):622–625. https://doi.org/10.1038/35007104

Matsuda N, Azuma K, Saijo M, Iemura S, Hioki Y, Natsume T, Chiba T, Tanaka K (2005) DDB2, the xeroderma pigmentosum group E gene product, is directly ubiquitylated by Cullin 4A-based ubiquitin ligase complex. DNA Repair (Amst) 4(5):537–545. https://doi.org/10.1016/j.dnarep.2004.12.012

McNally K, Neal JA, McManus TP, McCormick JJ, Maher VM (2008) hRev7, putative subunit of hPolzeta, plays a critical role in survival, induction of mutations, and progression through S-phase, of UV ((254nm))-irradiated human fibroblasts. DNA Repair (Amst) 7(4):597–604. https://doi.org/10.1016/j.dnarep.2007.12.013

Morita R, Nakane S, Shimada A, Inoue M, Iino H, Wakamatsu T, Fukui K, Nakagawa N, Masui R, Kuramitsu S (2010) Molecular mechanisms of the whole DNA repair system: a comparison of bacterial and eukaryotic systems. J Nucleic Acids 2010:179594. https://doi.org/10.4061/2010/179594

Moser J, Kool H, Giakzidis I, Caldecott K, Mullenders LH, Fousteri MI (2007) Sealing of chromosomal DNA nicks during nucleotide excision repair requires XRCC1 and DNA ligase III alpha in a cell-cycle-specific manner. Mol Cell 27(2):311–323. https://doi.org/10.1016/j.molcel.2007.06.014

Mu H, Geacintov NE, Broyde S, Yeo JE, Scharer OD (2018) Molecular basis for damage recognition and verification by XPC-RAD23B and TFIIH in nucleotide excision repair. DNA Repair (Amst) 71:33–42. https://doi.org/10.1016/j.dnarep.2018.08.005

Murakumo Y, Ogura Y, Ishii H, Numata S, Ichihara M, Croce CM, Fishel R, Takahashi M (2001) Interactions in the error-prone postreplication repair proteins hREV1, hREV3, and hREV7. J Biol Chem 276(38):35644–35651. https://doi.org/10.1074/jbc.M102051200

Nag A, Bondar T, Shiv S, Raychaudhuri P (2001) The xeroderma pigmentosum group E gene product DDB2 is a specific target of cullin 4A in mammalian cells. Mol Cell Biol 21(20):6738–6747. https://doi.org/10.1128/MCB.21.20.6738-6747.2001

Neal JA, Fletcher KL, McCormick JJ, Maher VM (2010) The role of hRev7, the accessory subunit of hPolzeta, in translesion synthesis past DNA damage induced by benzo[a]pyrene diol epoxide (BPDE). BMC Cell Biol 11:97. https://doi.org/10.1186/1471-2121-11-97

Nelson JR, Lawrence CW, Hinkle DC (1996) Thymine-thymine dimer bypass by yeast DNA polymerase zeta. Science 272(5268):1646–1649

Nishitani H, Lygerou Z, Nishimoto T, Nurse P (2000) The Cdt1 protein is required to license DNA for replication in fission yeast. Nature 404(6778):625–628. https://doi.org/10.1038/35007110

Nishitani H, Sugimoto N, Roukos V, Nakanishi Y, Saijo M, Obuse C, Tsurimoto T, Nakayama KI, Nakayama K, Fujita M, Lygerou Z, Nishimoto T (2006) Two E3 ubiquitin ligases, SCF-Skp2 and DDB1-Cul4, target human Cdt1 for proteolysis. EMBO J 25(5):1126–1136. https://doi.org/10.1038/sj.emboj.7601002

Nishitani H, Shiomi Y, Iida H, Michishita M, Takami T, Tsurimoto T (2008) CDK inhibitor p21 is degraded by a proliferating cell nuclear antigen-coupled Cul4-DDB1Cdt2 pathway during S phase and after UV irradiation. J Biol Chem 283(43):29045–29052. https://doi.org/10.1074/jbc.M806045200

Nouspikel T (2009) DNA repair in mammalian cells: nucleotide excision repair: variations on versatility. Cell Mol Life Sci 66(6):994–1009. https://doi.org/10.1007/s00018-009-8737-y

O'Donnell M, Langston L, Stillman B (2013) Principles and concepts of DNA replication in bacteria, archaea, and eukarya. Cold Spring Harb Perspect Biol 5(7). https://doi.org/10.1101/cshperspect.a010108

Perez-Torrado R, Yamada D, Defossez PA (2006) Born to bind: the BTB protein-protein interaction domain. BioEssays 28(12):1194–1202. https://doi.org/10.1002/bies.20500

Petroski MD, Deshaies RJ (2005) Function and regulation of cullin-RING ubiquitin ligases. Nat Rev Mol Cell Biol 6(1):9–20. https://doi.org/10.1038/nrm1547

Petruseva IO, Evdokimov AN, Lavrik OI (2014) Molecular mechanism of global genome nucleotide excision repair. Acta Nat 6(1):23–34

Pommier Y (2006) Topoisomerase I inhibitors: camptothecins and beyond. Nat Rev Cancer 6 (10):789–802. https://doi.org/10.1038/nrc1977

Puumalainen MR, Ruthemann P, Min JH, Naegeli H (2016) Xeroderma pigmentosum group C sensor: unprecedented recognition strategy and tight spatiotemporal regulation. Cell Mol Life Sci 73(3):547–566. https://doi.org/10.1007/s00018-015-2075-z

Raman M, Havens CG, Walter JC, Harper JW (2011) A genome-wide screen identifies p97 as an essential regulator of DNA damage-dependent CDT1 destruction. Mol Cell 44(1):72–84. https://doi.org/10.1016/j.molcel.2011.06.036

Rapin I, Weidenheim K, Lindenbaum Y, Rosenbaum P, Merchant SN, Krishna S, Dickson DW (2006) Cockayne syndrome in adults: review with clinical and pathologic study of a new case. J Child Neurol 21(11):991–1006. https://doi.org/10.1177/08830738060210110101

Saha A, Deshaies RJ (2008) Multimodal activation of the ubiquitin ligase SCF by Nedd8 conjugation. Mol Cell 32 (1):21–31. https://doi.org/10.1016/j.molcel.2008.08.021

Saijo M (2013) The role of Cockayne syndrome group A (CSA) protein in transcription-coupled nucleotide excision repair. Mech Ageing Dev 134(5–6):196–201. https://doi.org/10.1016/j.mad.2013.03.008

Sansam CL, Shepard JL, Lai K, Ianari A, Danielian PS, Amsterdam A, Hopkins N, Lees JA (2006) DTL/CDT2 is essential for both CDT1 regulation and the early G2/M checkpoint. Genes Dev 20(22):3117–3129. https://doi.org/10.1101/gad.1482106

Schulman BA, Carrano AC, Jeffrey PD, Bowen Z, Kinnucan ER, Finnin MS, Elledge SJ, Harper JW, Pagano M, Pavletich NP (2000) Insights into SCF ubiquitin ligases from the structure of the Skp1-Skp2 complex. Nature 408(6810):381–386. https://doi.org/10.1038/35042620

Sekelsky JJ, Hollis KJ, Eimerl AI, Burtis KC, Hawley RS (2000) Nucleotide excision repair endonuclease genes in Drosophila melanogaster. Mutat Res 459 (3):219–228

Senga T, Sivaprasad U, Zhu W, Park JH, Arias EE, Walter JC, Dutta A (2006) PCNA is a cofactor for Cdt1 degradation by CUL4/DDB1-mediated N-terminal ubiquitination. J Biol Chem 281(10):6246–6252. https://doi.org/10.1074/jbc.M512705200

Shell SM, Hawkins EK, Tsai MS, Hlaing AS, Rizzo CJ, Chazin WJ (2013) Xeroderma pigmentosum complementation group C protein (XPC) serves as a general sensor of damaged DNA. DNA Repair (Amst) 12 (11):947–953. https://doi.org/10.1016/j.dnarep.2013.08.013

Shiomi Y, Nishitani H (2017) Control of genome integrity by RFC complexes; conductors of PCNA loading onto and unloading from chromatin during DNA replication. Genes (Basel) 8(2). https://doi.org/10.3390/genes8020052

Shiomi Y, Hayashi A, Ishii T, Shinmyozu K, Nakayama J, Sugasawa K, Nishitani H (2012) Two different replication factor C proteins, Ctf18 and RFC1, separately control PCNA-CRL4Cdt2-mediated Cdt1 proteolysis during S phase and following UV irradiation. Mol Cell Biol 32(12):2279–2288. https://doi.org/10.1128/MCB.06506-11

Shiyanov P, Nag A, Raychaudhuri P (1999) Cullin 4A associates with the UV-damaged DNA-binding protein DDB. J Biol Chem 274(50):35309–35312. https://doi.org/10.1074/jbc.274.50.35309

Starostina NG, Kipreos ET (2012) Multiple degradation pathways regulate versatile CIP/KIP CDK inhibitors. Trends Cell Biol 22(1):33–41. https://doi.org/10.1016/j.tcb.2011.10.004

Stout GJ, Oosten M, Acherrat FZ, Wit J, Vermeij WP, Mullenders LH, Gruijl FR, Backendorf C (2005) Selective DNA damage responses in murine Xpa−/−, Xpc−/− and Csb−/− keratinocyte cultures. DNA Repair (Amst) 4(11):1337–1344. https://doi.org/10.1016/j.dnarep.2005.07.012

Strzalka W, Ziemienowicz A (2011) Proliferating cell nuclear antigen (PCNA): a key factor in DNA replication and cell cycle regulation. Ann Bot 107 (7):1127–1140. https://doi.org/10.1093/aob/mcq243

Sugasawa K, Okuda Y, Saijo M, Nishi R, Matsuda N, Chu G, Mori T, Iwai S, Tanaka K, Hanaoka F (2005) UV-induced ubiquitylation of XPC protein mediated by UV-DDB-ubiquitin ligase complex. Cell 121 (3):387–400. https://doi.org/10.1016/j.cell.2005.02.035

Takedachi A, Saijo M, Tanaka K (2010) DDB2 complex-mediated ubiquitylation around DNA damage is oppositely regulated by XPC and Ku and contributes to the recruitment of XPA. Mol Cell Biol 30(11):2708–2723. https://doi.org/10.1128/MCB.01460-09

Takezawa J, Ishimi Y, Aiba N, Yamada K (2010) Rev1, Rev3, or Rev7 siRNA abolishes ultraviolet light-induced translesion replication in HeLa cells: a comprehensive study using alkaline sucrose density gradient sedimentation. J Nucleic Acids 2010:750296. https://doi.org/10.4061/2010/750296

Tanaka M, Takahara M, Nukina K, Hayashi A, Sakai W, Sugasawa K, Shiomi Y, Nishitani H (2017) Mismatch repair proteins recruited to ultraviolet light-damaged sites lead to degradation of licensing factor Cdt1 in the G1 phase. Cell Cycle 16(7):673–684. https://doi.org/10.1080/15384101.2017.1295179

Tantin D, Kansal A, Carey M (1997) Recruitment of the putative transcription-repair coupling factor CSB/-ERCC6 to RNA polymerase II elongation complexes. Mol Cell Biol 17(12):6803–6814. https://doi.org/10.1128/mcb.17.12.6803

Tubbs JL, Latypov V, Kanugula S, Butt A, Melikishvili M, Kraehenbuehl R, Fleck O, Marriott A, Watson AJ, Verbeek B, McGown G, Thorncroft M, Santibanez-Koref MF, Millington C, Arvai AS, Kroeger MD, Peterson LA, Williams DM, Fried MG, Margison GP, Pegg AE, Tainer JA (2009) Flipping of alkylated DNA damage bridges base and nucleotide excision repair. Nature 459(7248):808–813. https://doi.org/10.1038/nature08076

Velez-Cruz R, Egly JM (2013) Cockayne syndrome group B (CSB) protein: at the crossroads of transcriptional networks. Mech Ageing Dev 134(5–6):234–242. https://doi.org/10.1016/j.mad.2013.03.004

Wang F, Yang W (2009) Structural insight into translesion synthesis by DNA Pol II. Cell 139(7):1279–1289. https://doi.org/10.1016/j.cell.2009.11.043

Wang H, Zhai L, Xu J, Joo HY, Jackson S, Erdjument-Bromage H, Tempst P, Xiong Y, Zhang Y (2006) Histone H3 and H4 ubiquitylation by the CUL4-DDB-ROC1 ubiquitin ligase facilitates cellular response to DNA damage. Mol Cell 22(3):383–394. https://doi.org/10.1016/j.molcel.2006.03.035

Wienholz F, Zhou D, Turkyilmaz Y, Schwertman P, Tresini M, Pines A, van Toorn M, Bezstarosti K, Demmers JAA, Marteijn JA (2019) FACT subunit Spt16 controls UVSSA recruitment to lesion-stalled RNA Pol II and stimulates TC-NER. Nucleic Acids Res 47:4011. https://doi.org/10.1093/nar/gkz055

Xu L, Wei Y, Reboul J, Vaglio P, Shin TH, Vidal M, Elledge SJ, Harper JW (2003) BTB proteins are substrate-specific adaptors in an SCF-like modular ubiquitin ligase containing CUL-3. Nature 425 (6955):316–321. https://doi.org/10.1038/nature01985

Yang W, Woodgate R (2007) What a difference a decade makes: insights into translesion DNA synthesis. Proc Natl Acad Sci U S A 104(40):15591–15598. https://doi.org/10.1073/pnas.0704219104

Yin Y, Lin C, Kim ST, Roig I, Chen H, Liu L, Veith GM, Jin RU, Keeney S, Jasin M, Moley K, Zhou P, Ma L (2011) The E3 ubiquitin ligase Cullin 4A regulates meiotic progression in mouse spermatogenesis. Dev Biol 356(1):51–62. https://doi.org/10.1016/j.ydbio.2011.05.661

Zhang YW, Otterness DM, Chiang GG, Xie W, Liu YC, Mercurio F, Abraham RT (2005) Genotoxic stress targets human Chk1 for degradation by the ubiquitin-proteasome pathway. Mol Cell 19(5):607–618. https://doi.org/10.1016/j.molcel.2005.07.019

Zhang S, Zhao H, Darzynkiewicz Z, Zhou P, Zhang Z, Lee EY, Lee MY (2013) A novel function of CRL4(Cdt2): regulation of the subunit structure of DNA polymerase delta in response to DNA damage and during the S phase. J Biol Chem 288(41):29550–29561. https://doi.org/10.1074/jbc.M113.490466

Zheng J, Yang X, Harrell JM, Ryzhikov S, Shim EH, Lykke-Andersen K, Wei N, Sun H, Kobayashi R, Zhang H (2002a) CAND1 binds to unneddylated CUL1 and regulates the formation of SCF ubiquitin E3 ligase complex. Mol Cell 10(6):1519–1526

Zheng N, Schulman BA, Song L, Miller JJ, Jeffrey PD, Wang P, Chu C, Koepp DM, Elledge SJ, Pagano M, Conaway RC, Conaway JW, Harper JW, Pavletich NP (2002b) Structure of the Cul1-Rbx1-Skp1-F boxSkp2 SCF ubiquitin ligase complex. Nature 416(6882): 703–709. https://doi.org/10.1038/416703a

Zhong W, Feng H, Santiago FE, Kipreos ET (2003) CUL-4 ubiquitin ligase maintains genome stability by restraining DNA-replication licensing. Nature 423(6942):885–889. https://doi.org/10.1038/nature01747

Zhou C, Wee S, Rhee E, Naumann M, Dubiel W, Wolf DA (2003) Fission yeast COP9/signalosome suppresses cullin activity through recruitment of the deubiquitylating enzyme Ubp12p. Mol Cell 11(4):927–938

Zhou X, DeLucia M, Ahn J (2016) SLX4-SLX1 protein-independent down-regulation of MUS81-EME1 protein by HIV-1 viral protein R (Vpr). J Biol Chem 291 (33):16936–16947. https://doi.org/10.1074/jbc.M116.721183

Cullin 4-DCAF Proteins in Tumorigenesis 15

Zhuan Zhou, Xinxin Song, Cindy M. Wavelet, and Yong Wan

Abstract

Cullin-RING ligase 4 (CRL4), a member of the cullin-RING ligase family, orchestrates a variety of critical cellular processes and pathophysiological events. Recent results from mouse genetics, clinical analyses, and biochemical studies have revealed the impact of CRL4 in development and cancer etiology and elucidated its in-depth mechanism on catalysis of ubiquitination as a ubiquitin E3 ligase. Here, we summarize the versatile roles of the CRL4 E3 ligase complexes in tumorigenesis dependent on the evidence obtained from knockout and transgenic mouse models as well as biochemical and pathological studies.

Keywords

Cullin 4 · DCAF · CRL4 ligase and tumorigenesis · Mouse models · Cancer etiology

Abbreviations

6-4PP	6-4 photoproduct
APC	Anaphase-promoting complex
BRCA1	Breast cancer 1
CAND1	Cullin-associated NEDD8-dissociated protein 1
Cbl	Casitas B-lineage lymphoma
CDC25A	Cell division cycle 25A
CDT1	Chromatin licensing and DNA replication factor 1
CHIP	Carboxyl terminus of Hsc70-interacting protein
CK1	Casein kinase 1α
COP1	Constitutive photomorphogenesis protein 1
CPD	Cyclobutane pyrimidine dimer
CRBN	Cereblon
CRL4	Cullin-RING ligase 4
CSN5	COP9 signalosome subunit 5
DCAFs	DDB1 CUL4-associated factors
DDB1	UV-damaged DNA-binding protein 1
DDB2	UV-damaged DNA-binding protein 2
EloB/C	Elongin B/C
ETV5	E26 transformation-specific variant 5
GNB	G protein subunit beta
Gnb3	G protein subunit beta 3
GRK2	G protein-coupled receptor kinase 2
GS	Glutamine synthetase
GSPT1	G1-to-S phase transition 1
HBO1	Histone acetyltransferase
HBx	Hepatitis B virus X protein
HECT	Homologous to the E6AP carboxyl terminus

Z. Zhou · X. Song · C. M. Wavelet · Y. Wan (✉)
Department of Obstetrics and Gynecology and Pharmacology, Northwestern University Feinberg School of Medicine, Chicago, IL, USA

Robert H. Lurie Comprehensive Cancer Center, Lurie Medical Research Center, Chicago, IL, USA
e-mail: yong.wan@northwestern.edu

Y. Sun et al. (eds.), *Cullin-RING Ligases and Protein Neddylation*, Advances in Experimental Medicine and Biology 1217, https://doi.org/10.1007/978-981-15-1025-0_15

HMSN	Axonal hereditary motor and sensory neuropathy
HOX	Homeodomain transcription factors
HSC	Hematopoietic stem cells
IAPs	Inhibitor of apoptosis protein
IKZF1/3	Ikaros family zinc finger protein 1/3
KMT5A	Lysine methyltransferase 5A
MAGE	Melanoma antigen gene
MM	Multiple myeloma
NEDD8	Neural precursor cell-expressed, developmentally downregulated 8
NF2	Neurofibromin 2
NLS	Nuclear localization signal
NRF2	NFE2-related factor 2
PARC	p53-associated Parkin-like cytoplasmic protein
PML	Promyelocytic leukemia
PPARγ	Peroxisome proliferator-activated receptor γ
RBR	RING-between RING-RING
RBX1	RING box protein 1
RING	Really interesting new gene
ROC1	Regulator of cullins 1
SET8	Set domain histone methyltransferase-8
SMC	Structural maintenance of chromosomes
ßTrCP	Beta-transducin repeat containing
UPS	Ubiquitin-proteasome system
VHL	von Hippel-Lindau tumor-suppressor protein
VprBP	Vpr (HIV-1) binding protein
WDR	WD repeat-containing protein
WDR4	WD repeat domain 4
WSS	Woodhouse-Sakati syndrome
XLMR	X-linked mental retardation
XPC	Xeroderma pigmentosum group C protein

15.1 Introduction

Genomic studies have revealed the presence of over 600 putative E3 ligases in the human genome (Fajner et al. 2017; Zheng and Shabek 2017; Morreale and Walden 2016). According to the presence of characteristic domains and mechanisms of ubiquitin moiety transfer, E3 ligases can be classified into three major categories, including the HECT (homologous to the E6AP carboxyl terminus) family, RBR (RING-between RING-RING), and the RING (really interesting new gene) family (Morreale and Walden 2016; Liu et al. 2018). Particularly, to catalyze transfer of ubiquitin onto the substrates, RING E3 ligases bind to ubiquitin-E2 thiolester complex and recognize targets in a simultaneous way, while HECT and RBR E3 ligases perform two processes: catalysis of a thiolester bond with ubiquitin on the E3, followed by a transfer to a catalytic cysteine from the E3 to the substrate (Zheng and Shabek 2017; Morreale and Walden 2016; Natarajan and Takeda 2017). Furthermore, the RING ligases harboring a RING catalytic domain can be divided into several subfamilies, including multi-subunit complex cullin-RING ligases (CRLs) and anaphase-promoting complex (APC), monomeric RING c-Cbl (casitas B-lineage lymphoma), homodimeric RING IAPs (inhibitor of apoptosis protein), heterodimeric RING BRCA1 (breast cancer 1), monomeric U-box-containing E4B (ubiquitin conjugation factor E4 B), and homodimeric U-box E3 CHIP (carboxyl terminus of Hsc70-interacting protein) (Zheng and Shabek 2017; Morreale and Walden 2016).

The cullin-RING ligases (CRLs), an evolutionarily conserved family first discovered almost two decades ago, belong to a superfamily of RING E3 ubiquitin ligases (Nguyen et al. 2017; Petroski and Deshaies 2005). CRLs structurally contain three principal components: an organized cullin scaffold, a RING-finger protein, and a substrate-targeting unit (Petroski and Deshaies 2005). The cullin scaffold is a key organizer of complex assembly, serving the platform for the RING-finger component as well as substrate-targeting unit. About eight cullin scaffold proteins have been studied in *Homo sapiens*, including CUL1, CUL2, CUL3, CUL4a, CUL4b, CUL5, CUL7, and CUL9/PARC (p53-associated Parkin-like cytoplasmic protein) (Sarikas et al. 2011) (Fig. 15.1). The RING-finger component works with the E2-ubiquitin thiolester, resulting in the transfer of ubiquitin to the target protein that is bound to the substrate recognition factor.

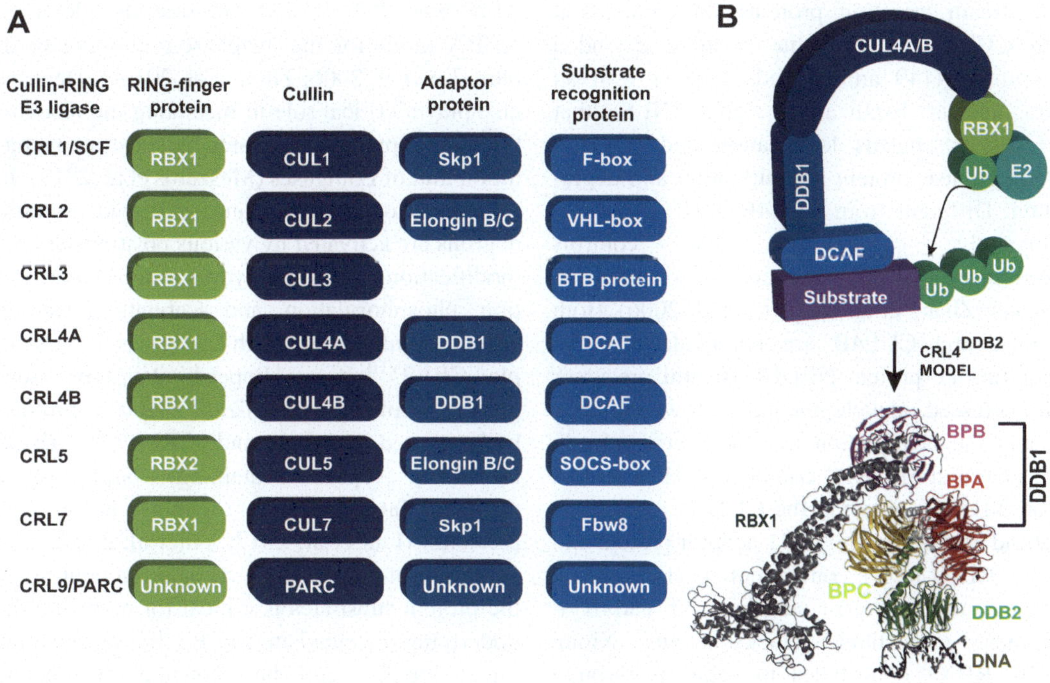

Fig. 15.1 Overview of Cullin-RING E3 ligase and CRL4 components. (**a**) The modularity of Cullin-RING E3 ligases. Eight types of Cullin (CUL1, CUL2, CUL3, CUL4A, CUL4B, CUL5, CUL7, and CUL9/PARC) exist in the human genome. The composition of the Cullin-RING E3 ligases typically contains Cullin protein, adaptor protein, and substrate recognition protein. (**b**) General CRL4 composition. Up panel shows the scaffold protein CUL4A/B tethers both substrate-recognition subunit DDB1/DCAF and RING-finger proteins RBX1. RBX1 recruits E2-ubiquitin complex and transfers the ubiquitin moiety. The down panel shows the modular architecture of CRL4-DDB1^{DDB2} DNA complexes. Green, yellow, and magenta indicate the DDB2 BPA, BPB, and BPC motif, respectively; gray indicates DNA; cyan indicates RBX1

Two RING-finger components have been revealed, including RBX1 (RING box protein 1) and RBX2 (RING box protein 2), which are also called ROC1 (regulator of cullins 1) and ROC2 (regulator of cullins 2), respectively (Petroski and Deshaies 2005). The substrate-targeting unit has both the adaptor activity and substrate receptor function (Lydeard et al. 2013; Cui et al. 2016). In CRL3, the substrate-targeting unit is a single molecule containing a bric-a-brac/tramtrack/broad (BTB) adaptor domain that binds to CUL3. CRL1 and CRL4 utilize an adaptor subunit, Skp1 or DDB1 (UV-damaged DNA-binding protein 1), respectively, to link the substrate receptor to CUL1 or CUL4. In CRL2 and CRL5, a heterodimeric elongin B (EloB)/elongin C (EloC) adaptor is used for connecting the substrate receptor to CUL2. The substrate receptor determines the specificity of a CRL for targeted protein degradation (Natarajan and Takeda 2017; Bulatov and Ciulli 2015). More than 400 substrate recognition proteins have also been deciphered (Lydeard et al. 2013; Cui et al. 2016). All the above subunits can be further assembled into distinct cullin-RING ligase complexes, comprising almost 20% of ubiquitination events in the ubiquitin-proteasome system (UPS) allowing for the assembly of distinct substrate recognition components that share a basic catalytic core, generating flexibility in the CRL specificity (Chen et al. 2015a; Zhao and Sun 2013).

The cullin 4-RING ligases (CRL4) consist of two scaffold proteins, including CUL4A (cullin 4A) and CUL4B (cullin 4B). Although CUL4A and CUL4B share 82% identity of genomic sequences, they target a different spectrum of downstream targeting proteins. In comparison with CUL4A, CUL4B has a more extended N-terminus (149 amino acids longer) with an extra nuclear localization signal (NLS) that governs its nucleus localization and essential role in nuclear protein ubiquitination and degradation. Different from CUL4B, CUL4A is predominantly cytoplasmic and controls ubiquitination and degradation of cytoplasmic proteins (Zhang et al. 2003; Li et al. 2006). Both CUL4A and CUL4B are neddylated by the ubiquitin-like protein NEDD8 (neural precursor cell-expressed, developmentally downregulated 8), and the neddylation modification results in the activation of CRL4 complexes (Hannah and Zhou 2015). In general, the CUL4A or CUL4B scaffold interacts with DDB1 adaptor protein and RBX1 RING-finger component to stabilize the complex through their amino- and carboxyl-terminus, respectively (Jackson and Xiong 2009). RBX1 contributes to catalysis through recruitment of E2-ubiquitin thiolester complex as well as promoting the transfer of the ubiquitin moiety to the substrate. Either Cul4A or Cul4B can be bridged to the substrate recognition component by DDB1. DDB1 contains three WD40 propeller domains (BPA, BPB, and BPC) and acts as the connection element between CUL4A/CUL4B scaffold and DCAFs (Hannah and Zhou 2015; Sang et al. 2015). The DCAFs (DDB1-CUL4-associated factors) are substrate recognition factors for CRL4, which contain over 100 members and have been identified in various species including *Drosophila melanogaster* and *Arabidopsis thaliana* (Angers et al. 2006; Jin et al. 2006; Zhang et al. 2008; Tamori et al. 2010). While the E3 ligase activity of CRL4 is determined by CUL4-RBX1, the substrate specificity of CRL4 is controlled by various DCAFs (Chen et al. 2017a) (Fig. 15.1).

15.2 Structure of CRL4 and Their Regulation by Neddylation

Degrons confer protein instability and are characterized by specific short amino acids that stretch within a protein substrate, such as the KEN box, D-box, and the recently identified ABBA motif for the anaphase-promoting complex (Lo et al. 2006; Zhou et al. 2016a). Degrons also play a critical role in mediating the recognition of the protein substrates by substrate recognition unit of E3 ligases (Meszaros et al. 2017). In response to intrinsic and extrinsic stimuli, degrons are activated by various posttranslational modifications such as hydroxylation, methylation, phosphorylation, and acetylation, leading to affinity interactions with E3 ligases. For example, $CRL2^{VHL}$ (von Hippel-Lindau tumor-suppressor protein) recognizes HIF-α's proline-hydroxylated degron, and $CRL1^{\beta TrCP}$ (beta-transducin repeat containing) binds to a phosphorylated degron within IκBα and β-catenin (Lau et al. 2012; Yu et al. 2001). The definition of the substrate recognition degrons in the protein substrates is critical for revealing the underlying mechanism for E3 ligase substrate direct binding and has potential therapeutic values for pathological protein-specific targeting degradation (Meszaros et al. 2017).

Several degrons for CRL4 substrate recognition factors that recognize their targeting proteins have been revealed recently, for example, the consensus (D/E)-(D/E)-(x)-x-x-V-P-(D/E) degron motif in the substrates such as ETV1, ETV5, JUN, ACC, and Trib1 recognized by DCAF protein COP1 (constitutive photomorphogenesis protein 1) and the consensus Q-x-x-(L/V/I/M)-x-x-(F/Y)-(F/Y) motif in the substrates such as PCNA, p21, SET, and Cdt1 recognized by DCAF protein CDT2/DCAF2 (Fig. 15.2a) (Uljon et al. 2016; Havens et al. 2012). The DCAF protein DCAF1/VprBP (Vpr (HIV-1) binding protein) has been reported to recognize a "mono-methyl degron" on substrates RORα via its putative chromo domain (Lee et al. 2012). The DCAF protein cereblon (CRBN) can recognize an "acetylated degron" on the substrate glutamine synthetase (GS) and leads to GS protein ubiquitination and degradation in response to high glutamine (Van Nguyen et al. 2016) (Fig. 15.2a). Nevertheless, more effort is needed to identify the degrons of CRL4 substrates.

CRLs can be conjugated with NEDD8, which is essential for their function of ubiquitin ligases.

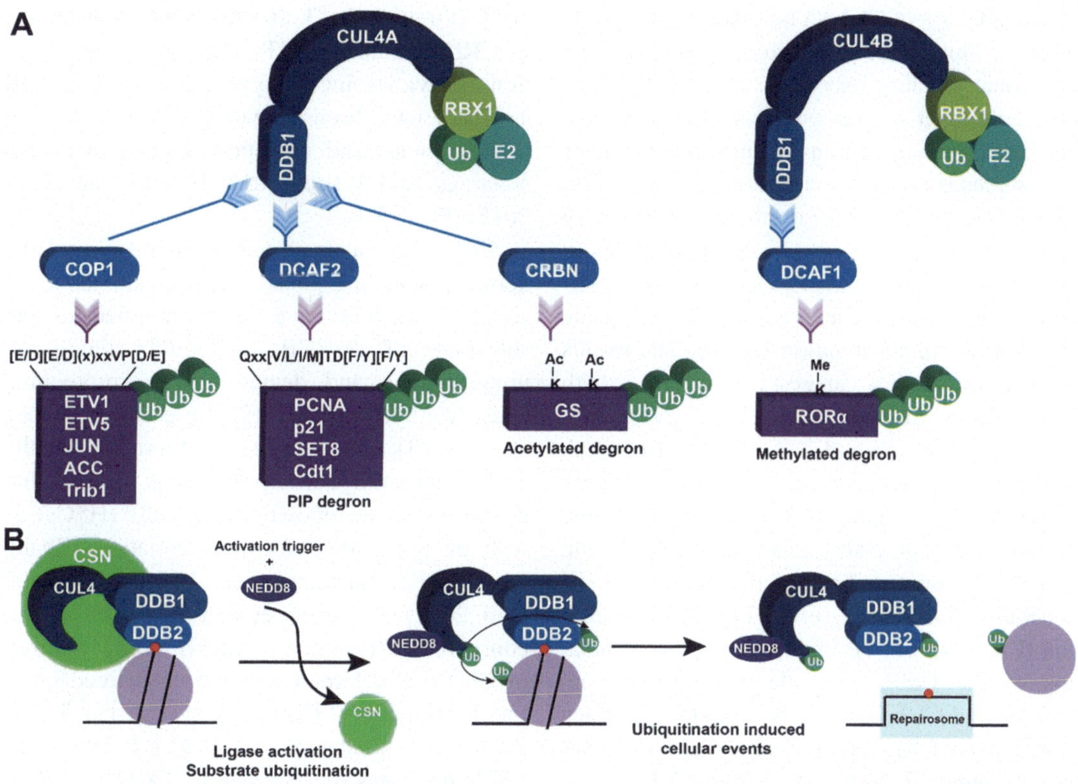

Fig. 15.2 Substrate recognition and neddylation regulation of CRL4. (**a**) The typical substrates motif recognized by (1) CUL4A-DDB1 substrate recognition protein COP1, DCAF2, CRBN and (2) CUL4B-DDB1 substrate recognition protein DCAF1. (**b**) Neddylation regulation of CRL4. While CSN inhibits Cul4-DDB1^{DDB2} activity, CUL4 neddylation by NEDD leads to release of CSN and ligase activation that, in turn, activates ligase activity

The covalent modification of NEDD8 to CUL4A and CUL4B is catalyzed through E1-activating enzyme, E2-conjugating enzyme UBC12, Rbx1 as an E3, and DCN1 as a co-E3 (Duda et al. 2011). Removal of neddylation is completed by the CSN5 (COP9 signalosome subunit 5), which cleaves the isopeptidic bond between cullin and NEDD8 (Cope and Deshaies 2003). The exchange factor CAND1 (cullin-associated NEDD8-dissociated protein 1) can also bind to cullins without neddylation and thus inhibits NEDD8 conjugation and cullin neddylation and CRL function as well (Duda et al. 2011). For example, during the regulation of CRL4-DCAF, the inactive CSN5-CRL4-DDB2 was recruited into the nucleosome embedded lesion; then CSN5 was released, and CUL4 was neddylated to activate the ubiquitin ligase for ubiquitination of substrates such as DDB2 (UV-damaged DNA-binding protein 2), histones, and XPC (xeroderma pigmentosum group C protein), thereby leading to coordinated DNA damage response that includes proteasomal degradation of the targeted proteins and recruitment of the binding partners (Scrima et al. 2011) (Fig. 15.2b).

15.3 Physiological Roles of Cullin 4 E3 Ligases

15.3.1 Physiological Relevance of Cullin 4 E3 Ligases Based on Animal Model

Results based on knockout mouse models have shown that CRL4 plays an important role in early embryonic development as well as tissue or organogenesis. Earlier studies indicated that genetic ablation of CUL4A has shown embryonic lethality, but subsequent studies suggest the embryonic lethality may be caused by an unintended deletion of the essential *Pcid2* gene, a conserved subunit of translation initiation factor 3 complexes, COP9 signalosome and 26S proteasome which resides on the complementary strand of DNA adjacent to Cul4a (Liu et al. 2009; Li et al. 2002). In mammals, CUL4A and CUl4B genes are broadly coexpressed and assemble structurally similar ubiquitin ligases. The results of multiple studies suggest no overt growth abnormalities in germline *Cul4a or Cul4b* knockouts (Jackson and Xiong 2009; Liu et al. 2009, 2012; Chen et al. 2012), which is likely caused by the redundancy between CUL4A and CUL4B (Liu et al. 2009). Deletion of *CUL4A* and *CUL4B* in mouse embryonic fibroblasts, as well as cultured tumor cells, results in growth retardation (Liu et al. 2009). Importantly, genetic ablation of the Ddb1, the sole adaptor for both CUL4A and CUL4B, is embryonic lethal, which indicates that CRL4 is essential for mice development (Cang et al. 2006, 2007) (Fig. 15.3).

Genetic ablation of CUL4A and CUL4B in mice models suggested that CRL4 plays an essential role in mouse reproduction. Previous genetic studies showed distinct expression profiles of CUL4A and CUL4B in male meiosis (Yin et al. 2011, 2016; Kopanja et al. 2011). It was demonstrated that $Cul4b^{(\Delta)/Y}$ male mice are sterile, exhibiting a progressive loss of germ cells and sperm deformation. $Cul4a^{-/-}$ male mice are infertile due to the primary spermatocytes being arrested and deficient in progression through late prophase I (Yin et al. 2011, 2016; Kopanja et al. 2011). Though $Cul4a^{-/-}$ male mice are sterile, these mice are still capable of bearing and delivering live pups, highlighting the critical roles of CRL4A in the male reproductive system but not that of females (Yin et al. 2011, 2016; Kopanja et al. 2011). Interestingly, disruption of the whole CRL4 by genetic ablation of Ddb1 results in female infertility (Yu et al. 2015a). Conditional ablation of adaptor protein *Ddb1* or its critical substrate recognition protein DCAF1 or CDT2 all led to infertility in females due to loss of oocytes and defection of ovulation (Xu et al. 2017; Yu et al. 2013, 2015b). Several substrates of CRL4 including TET1 (ten-eleven translocation methylcytosine dioxygenase 1), p53, as well as CDT1 have been suggested to result in germ cell apoptosis and defection (Liu et al. 2018; Natarajan and Takeda 2017; Bulatov and Ciulli 2015) (Fig. 15.3).

CRL4 regulates protein homeostasis in the hematopoietic lineage. HOX (homeodomain transcription factors) have been identified as the substrates of the CRL4 ubiquitin ligase for ubiquitination and degradation by proteomic/yeast two-hybrid assays (Zhang et al. 2003; Lee et al. 2013). HOX genes are degraded via the CRL4 ubiquitin ligase-proteasome system that determines hematopoietic stem cells (HSCs) as well as progenitors for differentiation (Zhang et al. 2003; Sauvageau et al. 1994; Pineault et al. 2002; Giampaolo et al. 1995). Moreover, a conserved CRL4 degron LxCxE motif, localized in HOX helix 1 region, has been defined in all family members of HOX (Lee et al. 2013). Transduction of nondegradable HOXB4 into adult HSCs promotes the expansion of CD34+ HSCs and multipotent progenitors, leading to enhanced long-term bone marrow engraftment. It is also reported that abnormal degradation of HOXA9 facilitated by both CUL4A and CUL4B enhances proliferation of HSCs in vitro and maintains the primitive state of HSCs (Zhang et al. 2003; Lee et al. 2013). CRL4 is also involved in nervous system regulation. Murine brain-specific conditional deletion of adaptor protein *Ddb1* leads to the neonatal mortality due to an increase of p53 (Cang et al. 2006). Furthermore, murine hippocampus- and cerebral cortex-specific conditional deletion of *Ddb1* generates results in epilepsy through the changes of the activity of the BK (Ca^{2+} and voltage-activated K^{+}) channel, highlighting the potential role of CRL4 in disorders with neuroelectrophysiology (Liu et al. 2014). CRL4 is also involved in regulation of other organ systems such as the liver, skin, and pancreas. CRL4 appears to play differential roles in hepatic cells in a context-dependent manner. Conditional knockout *Ddb1* in murine hepatocytes results in disrupted gluconeogenesis which takes place in the liver (Tong et al. 2017).

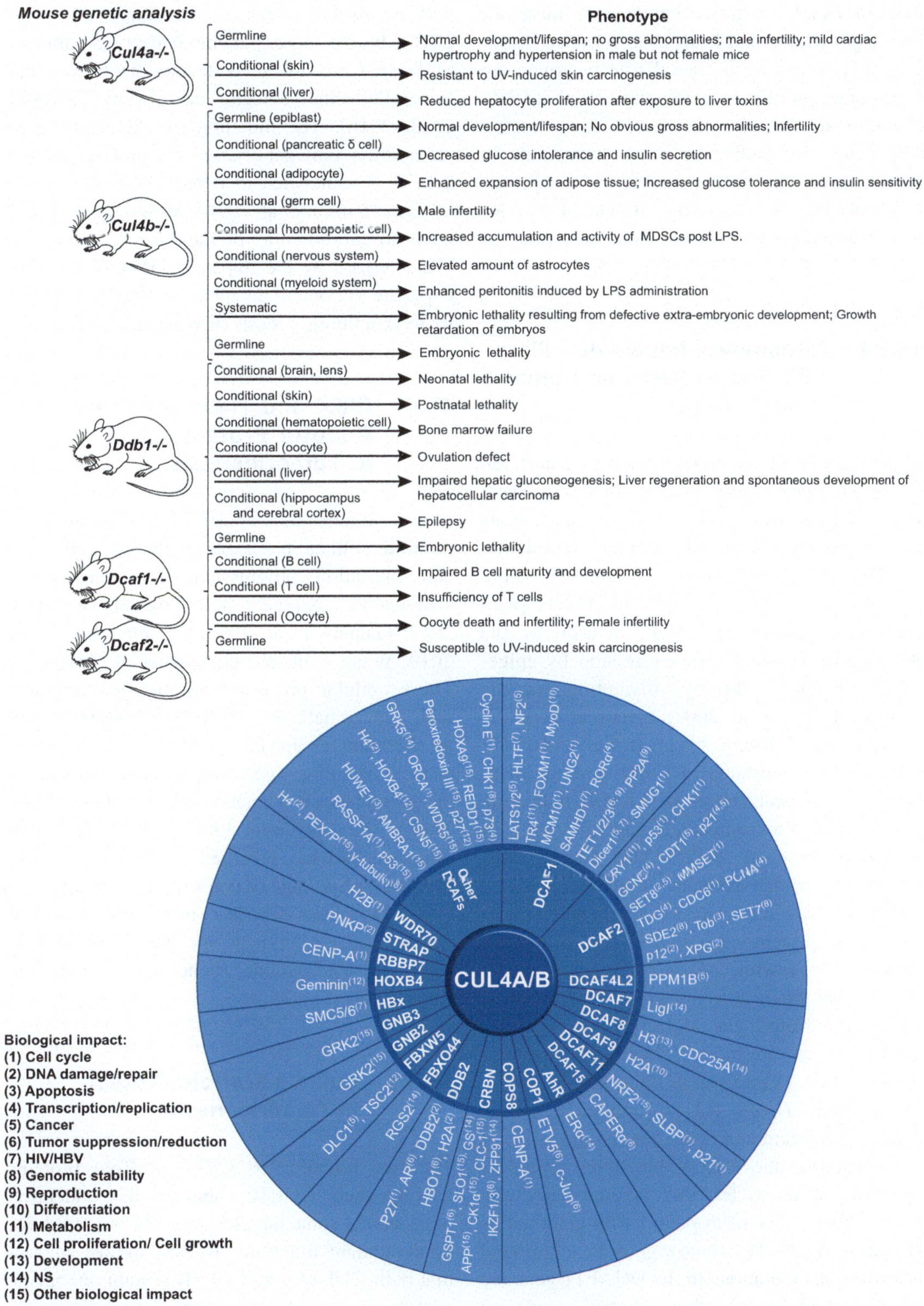

Fig. 15.3 Physiological roles of Cullin 4 E3 ligases. Upper panel summaries the phenotypes of knockout mouse models of *Cul4a*, *Cul4b*, *Ddb1*, *Dcaf1*, *Dcaf2*. Down panels summarize CRL4 substrate recognition proteins, substrates, and their involved biological impact

Loss of *Ddb1* in murine hepatocytes increases liver regeneration and leads to liver cancer spontaneously, implying a critical role in governing hepatocyte proliferation (Yamaji et al. 2010). Conditional knockout *Cul4b* in murine adipocytes and pancreatic delta cells leads to insulin sensitivity enhancement, which appears to phenocopy the beneficial effects of PPARγ (peroxisome proliferator-activated receptor γ) agonists (Li et al. 2017a, b) (Fig. 15.3).

15.3.2 Pathological Impact of Cullin 4 E3 Ligases Based on Human Clinical Studies

Mutations of CRL4 subunits have been connected to various human diseases. Recent studies have shown that the mutations in *Cul4b* gene accounts for triggering X-linked mental retardation (XLMR, also called Cabezas syndrome), possibly caused by the stabilization of WDR5 (WD repeat-containing protein 5) which orchestrates the decrease in neuronal gene expression by epigenetic silencing, thereby disrupting neurite outgrowth (Zou et al. 2007; Tarpey et al. 2007; Nakagawa and Xiong 2011). In the *Cul4b*$^{(\Delta)/Y}$ mice model, restoration of *Cul4b* gene dramatically reduces occurrences of XLMR, indicating *Cul4b* genes as a promising therapeutic target for XLMR (Chen et al. 2012). Other than XLMR, among the human patients of microcephaly, a CRL4 substrate recognition subunit WD repeat-containing protein 62 (WDR62) is frequently detected. Meanwhile, *Cul4b*$^{(\Delta)/Y}$ mice model also shows brain arteriovenous malformation, which implicated the defection of CUL4b-WDR62 as a causation of brain arteriovenous malformation (Vulto-van Silfhout et al. 2015). In addition, *Dcaf8* R317C mutation, which defects the interaction with the DDB1 adaptor and therefore the CRL4 functional complex assembly, leads to HMSN2 (axonal hereditary motor and sensory neuropathy) with giant axons (Klein et al. 2014). Homozygous or compound heterozygous mutations in the DCAF17 gene are the causation of Woodhouse-Sakati syndrome (WSS), a rare, multisystem genetic condition with autosomal recessive inheritance (Ali et al. 2016). Furthermore, patients harboring a mutant of *Dcaf14* are linked to developmental retardation, intellectual defects, and obesity (Webster et al. 2016). The mutation of CRL4 substrate recognition unit gene *Gnb3* (G protein subunit beta 3) is connected to hereditary hypertension (El Din Hemimi et al. 2016). Mutations on *Ddb2* lead to xeroderma pigmentosum, which is characterized by the impaired DNA repair after exposure to UV and potent increasing susceptibility to skin tumorigenesis (Kapetanaki et al. 2006).

15.4 CRL4 and Their Substrate Adaptor Proteins in Tumorigenesis

It was demonstrated that CRL4 is involved in multiple cellular processes, including cell cycle and migration progression, determination of senescence, autophagy, and apoptosis (Xu et al. 2017; Galanos et al. 2016; Ravichandran et al. 2019; Wang et al. 2017a; Song et al. 2015a, b). These cellular processes are tightly correlated with cancer hallmarks such as maintaining cell proliferation, escape of cell death, override senescence, promoting migration, invasion, as well as distant metastasis (Hanahan and Weinberg 2011). Thus, the potential role for CRL4 in tumorigenesis has emerged in the past decade. The oncogenic and tumor suppressing roles for CRL4 have been described in studies from mice genetic models, clinical specimens, as well as molecular and cellular benchworks (Lee and Zhou 2012) (Fig. 15.3).

15.4.1 Cullin 4 Scaffold Protein in Tumorigenesis

The two homologous CRL4 scaffold proteins CUL4A and CUL4B, sharing 82% identity, have shared some targeted proteins but still have nonredundant functions. Recent studies suggest that both CUL4A and CUL4B sustain oncogenic activity in a wide spectrum of tumorigenesis, implicating that the whole CRL4 complex is potentially involved in malignancies, since these two scaffold proteins are essential to the entire complex formation and function (Hannah and Zhou 2015; Yang et al. 2014; Wang et al. 2015a, b; Yuan et al. 2015a; Jia et al. 2017).

Previous studies implicated the crucial roles of CUL4A following DNA damage. Upon cellular exposure to genotoxic stress, CUL4A influences DNA replication through modulating the protein stability of the DNA replication licensing factor CDT1 (chromatin licensing and DNA replication factor 1), histone methyltransferase KMT5A (lysine methyltransferase 5A), and cyclin-dependent kinase inhibitor p21 (Jin et al. 2006; Higa et al. 2003; Hu et al. 2004; Zhong et al. 2003; Kim et al. 2008; Abbas et al. 2008; Centore et al. 2010). CUL4A can limit the DNA repair capacity in response to UV. *Cul4a*$^{-/-}$ cells showed increased checkpoint function for DNA repair and DNA damage response. Genetic ablation of *Cul4a* in the skin led to elevated resistance to UV-induced skin carcinogenesis (Liu et al. 2009). Upon cellular exposure to UV, cells conduct the nucleotide excision repair (NER) pathway to remove the UV-induced cyclobutane pyrimidine dimers (CPDs) and 6-4 photoproducts (6-4PPs) and activate the G1/S checkpoint to arrest the cell cycle progression till the DNA damage resolved. It is reported that CUL4A is an important regulator for DNA damage response through the ubiquitin-proteasomal pathway to degrade DDB2, XPC, as well as p21, crucial factors involved in DNA damage recognition or the checkpoint pathway (Liu et al. 2009). Therefore, suppressing the function of CUL4A constitutes a potential new strategy in cancer prevention and treatment.

In malignancies, CUL4A protein is usually accumulated, caused by frequent gene amplification and mRNA level upregulation through the canonical Wnt signaling pathway (Miranda-Carboni et al. 2008). Upregulation of CUL4A induces lung tumorigenesis in mice (Yang et al. 2014; Wang et al. 2015a, b). Abnormal CUL4A accumulation has been detected in multiple human cancer specimens, such as lung carcinoma, including lung cancer (Jia et al. 2017; Hrecka et al. 2016; Wang et al. 2014a), gastric carcinoma (Gong et al. 2017; Deng et al. 2016; Wang et al. 2014b), colon cancer (Wang et al. 2014b; Sui et al. 2017; Ren et al. 2016a, b), as well as breast cancer (Wang et al. 2014b; Chen et al. 1998; Wang et al. 2014c), negatively correlated with the prognostic survival (Song et al. 2015b; Jia et al. 2017; Hrecka et al. 2016; Wang et al. 2014a; Birner et al. 2012; Huang et al. 2017; Zhang et al. 2017a; Jiang et al. 2013). The oncogenic impact of CUL4A, the core component of the CRL4 complex, largely depends on its downstream substrates which contribute to tumor cell proliferation, migration, invasion, metastasis, as well as drug resistance (Sui et al. 2017; Wang et al. 2014c; Pan et al. 2015).

Similarly, CUL4B is also identified as an oncoprotein among multiple malignant tumors. Murine *Cul4b* hepatocyte transgenic model results in liver carcinogenesis (Yuan et al. 2015a). It is reported that CUL4B is aberrantly expressed in a broad set of malignancies such as colorectal cancer (Song et al. 2015b; Jiang et al. 2013), lung cancer (Jia et al. 2017; Mi et al. 2017), and pancreatic cancer (Zhang et al. 2018), negatively correlating with patient survival (Jiang et al. 2013; Li et al. 2017c). Aberrant association between CUL4B and its substrates were also connected to the activation of cell growth signals and epigenetic silencing, resulting in tumor expansion (Yang et al. 2015; Yuan et al. 2015b). Deregulation of some miRNAs (miR-194/300) may modulate CUL4B expression that in turn affects tumorigenesis, although the mechanisms remain unclear (Mi et al. 2017; Zhang et al. 2018).

15.4.2 RBX1, a RING-Finger Protein, in Tumorigenesis

RING-finger RBX1, the catalytic component, is thought to be a tumorigenic protein in multiple cancers including lung cancer (Xing et al. 2016), ovarian cancer (Pan et al. 2013), bladder cancer (Migita et al. 2014), and stomach cancer (Xing et al. 2016). However, there are limited mouse models and investigations to address the physiological relevance of RBX1 in carcinogenesis.

15.4.3 DDB1, an Adapter Protein, in Tumorigenesis

The adaptor DDB1 is critical in tumorigenesis in a context-dependent manner. It was shown that hepatocyte-specific deletion of *Ddb1* knockout mouse models renders spontaneous liver cancer development (Yamaji et al. 2010). DDB1 accumulation is observed in ovarian cancer, which is consistent with other subunits in the CRL4 complex (Pan et al. 2013). This differential role of DDB1 may contribute to the neoplastic diversity of the CLR4 complexes or may reflect other CLR4-independent function in other regulatory networks.

15.4.4 Cullin 4 Substrate Recognition Proteins Involved in Tumorigenesis

15.4.4.1 Cullin 4 Substrate Recognition Proteins Act as Tumor Suppressors

Cereblon (CRBN)

CRBN is a pivotal substrate recognition receptor for CRL4 that is adversely associated with tumorigenesis. The CRL4-CRBN complex has many impacts on neurodegeneration, homeostasis of membrane excitability, anti-epileptogenesis, and so on, depending on the downstream substrates (Liu et al. 2014; Del Prete et al. 2016; Chen et al. 2015b). The expression levels of CRBN have significantly dropped in multiple myeloma cells as a critical tumor suppressive gene (Broyl et al. 2013). CK1 (casein kinase 1α), GSPT1 (G1-to-S phase transition 1), as well as IKZF1/3 (Ikaros family zinc finger protein 1/3) might not be the typical targeting proteins of CRBN; however, the degradation of CK1α, GSPT1, and IKZF1/3 by CRL4-CRBN can be induced by thalidomide and its derivatives such as pomalidomide and lenalidomide in several treatments such as multiple myeloma, myelodysplastic syndrome, and leukemia, respectively (Kroenke et al. 2015; Matyskiela et al. 2016; Kroenke et al. 2014). The underlying mechanisms are thought to be associated with the inhibition of cell cycle progression (Figs. 15.3 and 15.4).

DDB2

In human cancer, DDB2 has been thought to be one of the most important tumor-suppressor genes. CRL4-DDB2-substrate interactions regulate cell cycle progression and genome stability (Kapetanaki et al. 2006; Yan et al. 2011). Mice with *Ddb2* knockout exhibit increased sensitivity to the skin tumorigenesis triggered by UV in rodent models (Yoon et al. 2005). Furthermore, DDB2 downregulation is often observed in skin cancer (Kapetanaki et al. 2006), colon cancer (Roy et al. 2013), and human prostate cancer (Chen et al. 2017b). Furthermore, destabilization of oncogenic targeting proteins such as HBO1 (histone acetyltransferase, also known as MYST2, KAT7) (Matsunuma et al. 2016) may be associated with DDB2-mediated tumor-suppressive effects (Figs. 15.3 and 15.4).

COP1 (Constitutive Photomorphogenesis 1)

Results from the mouse genetic analysis indicated that *Cop1*$^{-/-}$ mice are embryonic lethal while *Cop1*$^{+/-}$ mice (hypomorphic alleles, partial loss of function) are viable and fertile (Migliorini et al. 2011). The Cop1 protein level dropped 90% in the *Cop1*$^{+/-}$ mice and reduced 15–20% in body weight, resulting in decreased organ sizes (Migliorini et al. 2011). In renal cell carcinoma, COP1 protein shows reduced expression (Ta et al. 2016). In addition, results from biochemical studies demonstrated that CRL4^{COP1} inhibits several oncogenic transcription factors by degrading the substrates such as c-Jun, ETV5 (E26 transformation-specific variant 5) in lung carcinogenesis (Zhang et al. 2017b; Wertz et al. 2004) (Fig. 15.3).

AhR (Arylhydrocarbon Receptor)

AhR, a ligand-activated transcription factor known to be involved with detoxification and metabolic pathways, mediates the adverse effects of dioxins, including modulation of sex steroid hormone signaling (Andersson et al. 2002). The

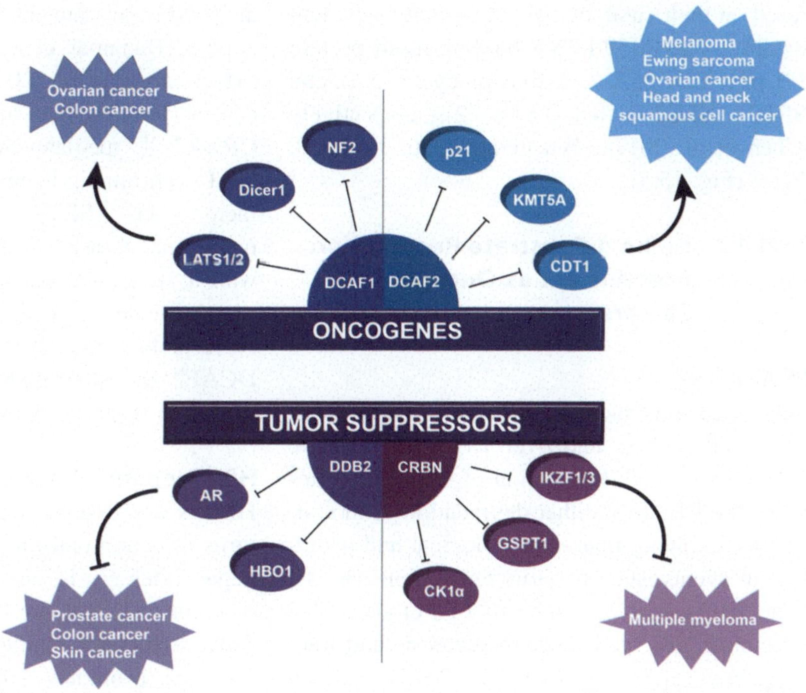

Fig. 15.4 Major tumor suppressors and oncogenic substrate recognition proteins and substrates of CRL4. Upper panel summarizes the oncogenic substrate recognition proteins DCAF1 and DCAF2 and their corresponding tumor suppressive substrates in cancer. Down panel shows the tumor suppressive substrate recognition proteins DDB2 and CRBN and their corresponding tumor oncogenic substrates in cancer

ligand-activated AhR directly associates with estrogen or androgen receptors (ERα or AR) as a substrate recognition subunit to recruit ERα/AR (Wormke et al. 2003; Ohtake et al. 2009). AhR subsequently induces the proteolysis of ERα/AR through assembling a ubiquitin ligase complex, CUL4B-AhR. In the CUL4B-AhR complex, AhR acts as a substrate recognition subunit to recruit ERα/AR. The tumor-suppressive effect of AhR was observed in many types of cancers such as glioblastoma, colon, liver, and skin cancers (Wang et al. 2013; Khanal et al. 2017; Spink et al. 2015; Gabriely et al. 2017; Sun et al. 2015) (Fig. 15.3).

DCAF12

In *Drosophila*, DCAF12 promotes evoked neurotransmitter release and homeostatic plasticity via regulating glutamate receptor subunits GluRIIA, GluRIIC, and GluRIID in postsynaptic (Patron et al. 2019) and maintains tissue homeostasis via regulating Diapl cleavage in response to pro-apoptotic signals (Hwangbo et al. 2016). Melanoma antigen genes (MAGEs) are emerging as important oncogenic drivers that are normally restricted to expression in male germ cells but are aberrantly expressed in cancers and promote tumorigenesis (Lee and Potts 2017). The MAGE-A3/6 proteins are repressors of autophagy and are downregulated in response to nutrient deprivation. Short-term cellular starvation promotes rapid MAGE-A3/6 degradation in a proteasome-dependent manner. The CRL4^{DCAF12} E3 ubiquitin ligase specific degrades MAGE-A3/6 and this degradation is required for starvation-induced autophagy and tumor growth in response to nutrient status (Ravichandran et al. 2019) (Fig. 15.3).

Others

WDR70 gene mutations that lead to loss of function were documented in studies of carcinogenesis of ovarian cancer (Guo et al. 2016). WDR70 enhances histone H2B stability through mono-ubiquitination of H2B during cell cycle and thus suppressing tumorigenesis (Guo et al. 2016). DCAF15, DCAF11, and DCAF8 have been demonstrated as potential tumor suppressors based on their catalytic role in regulating protein stability of RBM39 (RNA binding motif protein 39), p21, CDC25A (cell division cycle 25A), and NRF2 (NFE2-related factor 2), respectively (Chen et al. 2017a; Wu et al. 2016; Lo et al. 2017) (Fig. 15.3).

15.4.4.2 Cullin 4 Substrate Recognition Proteins Act as Oncogenic Factors

DCAF1

The CRL4 substrate recognition unit DCAF1 has been well-studied, and over 15 substrates have been documented. Regulation of these substrates by DCAF1 leads to either degradation or modification, involving many physiological and pathological events, such as virus replication (Hrecka et al. 2016), germ cell meiosis (Yu et al. 2015a; Jung et al. 2015), skeletal myogenesis and transcriptional repression (Lee et al. 2012), and cell cycle progression (Wang et al. 2017a). DCAF1 was demonstrated as a critical oncoprotein in the CRL4 complex in ovarian cancer as well as colorectal cancer patient samples (Wang et al. 2017a; Ren et al. 2016b). The identified substrates include LATS1 (large tumor suppressor 1) (Li et al. 2014; Ni et al. 2017), NF2 (neurofibromin 2), and Dicer1 through protein degradation (Ren et al. 2016b; Huang and Chen 2008). These substrates normally function as the tumor suppressors to hamper various subsequent oncogenic signaling pathways, such as Hippo/YAP (Li et al. 2014; Ni et al. 2017) and JAK/STAT3 (Ren et al. 2016b). Nevertheless, genetic mouse models are unfortunately not available to decipher the role DCAF1 in carcinogenesis (Figs. 15.3 and 15.4).

DCAF2

DCAF2 has been recognized as a major oncogenic receptor of CRL4. It is involved in multiple biological events, such as genomic stability (Oda et al. 2010; Tardat et al. 2010; Abbas et al. 2010), gluconeogenesis (Tong et al. 2015, 2017), transcriptional suppression (Li et al. 2011), and cell cycle progression (Huh and Piwnica-Worms 2013) either by degrading or activating specific substrates. Elevated expression of DCAF2 is detected in ovarian cancer (Pan et al. 2013), melanoma (Benamar et al. 2016), Ewing sarcoma (Mackintosh et al. 2012), and head and neck cancer (Vanderdys et al. 2018). Mechanistically, $CRL4^{DCAF2}$ mediates the targeted degradation of CDT1 (chromatin licensing and DNA replication factor 1), SET8 (set domain histone methyltransferase-8, also known as KMT5A), as well as p21 to boost the proliferation of melanoma (Benamar et al. 2016; Higa et al. 2006). Nevertheless, the detailed biochemical studies of DCAF2 in other malignancies remain largely unknown (Figs. 15.3 and 15.4).

HBx (Hepatitis B Virus X Protein)

HBx is expressed depending on the hepatitis B virus infection and the replication within human hepatocytes. It is also a substrate recognition receptor of CRL4 resulting in the degradation of SMC (structural maintenance of chromosomes) 5/6 that promotes HBV replication (Murphy et al. 2016). HBx transgenic mice developed hepatocyte tumors (Wang et al. 2012; Koike et al. 1994). HBx was detected to be accumulated in human hepatocyte cancer (Ding et al. 2010), adenoid cystic carcinoma (Xie et al. 2014), and intrahepatic cholangiocarcinoma samples (Zhou et al. 2012), indicating a broad oncogenic role for HBx. Other than the HBV replication, SMC5/6 also regulated cell division and proliferation. It was also demonstrated that HBx could modulate the apoptotic machinery and activate a series of downstream oncogenes (Hwang et al. 2017; Jiang et al. 2016) (Fig. 15.3).

WDR4

Currently, WDR4 (WD repeat domain 4) has been identified as another oncogenic substrate recognition unit of CRL4. It caused the degradation of the PML (promyelocytic leukemia), a well-known tumor suppressor that controls cell proliferation, apoptosis, and senescence, through the ubiquitination modification in lung cancer (Wang et al. 2017b) (Fig. 15.3).

FBXW5

F-box protein FBXW5 was the targeting subunit of $CRL4^{DDB1}$ in ubiquitinating HsSAS-6 (Puklowski et al. 2011) and DLC1 (DLC1 Rho

GTPase-activating protein) (Kim et al. 2013) followed by degradation. DLC1 is a tumor suppressor which is frequently lost in cancer. Depletion of FBXW5, CUL4A, or DDB1 led to upregulated DLC1 protein levels and growth inhibition (Kim et al. 2013) (Fig. 15.3).

DCAF4

Functional variants in DCAF4 gene loci rs2535913 have been reported to be associated with leucocyte telomere length (Mangino et al. 2015) and lung cancer risk by possibly upregulating mRNA expression and decreasing methylation status (Yan et al. 2017; Liu et al. 2017). ST7 gene, a candidate tumor-suppressor gene identified recently at human chromosome 7q31.1, was also detected because LOH at this site has also been widely reported in stomach cancer, prostate cancer, and colon cancer (Wang et al. 2015a; Zhang et al. 2017a; Zenklusen et al. 2001). The CRL4^{DCAF4} E3 ligase specifically directs degradation of ST7. Inflammation-dependent overexpression of c-Myc enhances CRL4^{DCAF4} E3 ligase activity and promotes ubiquitination of ST7 in colitis-associated cancer (Liu et al. 2019) (Fig. 15.3).

Others

It was reported that additional CRL4 substrate recognition receptors are responsible for triggering the onset and progression of malignancies, including DCAF13 in hepatocellular carcinoma (Cao et al. 2017) and DCAF6 (also known as Nuclear Receptor Interaction Protein, NRIP) in prostate cancer (Chen et al. 2017b). The mechanism for DCAF13 and DCAF6 involved in tumorigenesis are mediated through genomic instability, apoptosis, and cell cycle regulation (Fig. 15.3).

15.4.4.3 Physiological Context-Dependent Substrate Recognition Proteins

Physiologically, GNB (G protein subunit beta) 2/3 are reported as context-dependent substrate recognition unit for CRL4. CRL4-GNB2/3 degrades of GRK2 (G protein-coupled receptor kinase 2) leading to cardiovascular protective effects. Nevertheless, in mammary carcinogenesis, the roles of GRK2 were linked to tumorigenesis (Nogues et al. 2016), but its tumor-suppressing role was reported in hepatocellular carcinoma (Ma et al. 2016). Unfortunately, none of mouse genetic models are available to decipher their physiological relevance yet (Ma et al. 2016; Zhou et al. 2016b; Rivas et al. 2013) (Fig. 15.3).

15.5 Perspective

In general, cullin proteins play essential roles in carcinogenesis through multiple mechanisms, including but not limited to DNA damage response and DNA repair, oxidative stress sensation and relief, cell cycle regulation, autophagy regulation, senescence and apoptosis regulation, chromatin remodeling, cytoskeleton regulation, or oncogenic signal transduction. Recent studies underscore the role of CRL4 in tumorigenesis and its potential in clinical translation. Previous results based on genetic mice models and detailed biochemical characterization as well as clinic-related studies have well dissected the molecular mechanistic and pathological features of scaffold protein CUL4A or CUL4B, RING-finger protein RBX1, and substrate recognition units DDB1 and DCAFs. Furthermore, the increased number of CRL4 substrates could link its function to more aspects. Besides, some CRL4-targeted agents such as thalidomide (targeting CRBN) and the sulfonamides (targeting DCAF15) have shed light on treatment of certain types of cancer such as multiple myeloma (MM) and colon cancer, while their mechanisms of action remain unknown. More future in-depth animal and therapeutic studies will lead to a better understanding and clinical application of cullin-DCAF proteins in tumorigenesis and cancer treatments.

References

Abbas T, Sivaprasad U, Terai K, Amador V, Pagano M, Dutta A (2008) PCNA-dependent regulation of p21 ubiquitylation and degradation via the CRL4(Cdt2) ubiquitin ligase complex. Genes Dev 22(18): 2496–2506

Abbas T, Shibata E, Park J, Jha S, Karnani N, Dutta A (2010) CRL4(Cdt2) regulates cell proliferation and histone gene expression by targeting PR-Set7/Set8 for degradation. Mol Cell 40(1):9–21

Ali RH, Shah K, Nasir A, Steyaert W, Coucke PJ, Ahmad W (2016) Exome sequencing revealed a novel biallelic deletion in the DCAF17 gene underlying woodhouse Sakati syndrome. Clin Genet 90(3):263–269

Andersson P, McGuire J, Rubio C, Gradin K, Whitelaw ML, Pettersson S et al (2002) A constitutively active dioxin/aryl hydrocarbon receptor induces stomach tumors. Proc Natl Acad Sci U S A 99(15):9990–9995

Angers S, Li T, Yi X, MacCoss MJ, Moon RT, Zheng N (2006) Molecular architecture and assembly of the DDB1-CUL4A ubiquitin ligase machinery. Nature 443(7111):590–593

Benamar M, Guessous F, Du K, Corbett P, Obeid J, Gioeli D et al (2016) Inactivation of the CRL4-CDT2-SET8/ p21 ubiquitylation and degradation axis underlies the therapeutic efficacy of pevonedistat in melanoma. EBioMedicine 10:85–100

Birner P, Schoppmann A, Schindl M, Dinhof C, Jesch B, Berghoff AS et al (2012) Human homologue for Caenorhabditis elegans CUL-4 protein overexpression is associated with malignant potential of epithelial ovarian tumours and poor outcome in carcinoma. J Clin Pathol 65(6):507–511

Broyl A, Kuiper R, van Duin M, van der Holt B, el Jarari L, Bertsch U et al (2013) High cereblon expression is associated with better survival in patients with newly diagnosed multiple myeloma treated with thalidomide maintenance. Blood 121(4):624–627

Bulatov E, Ciulli A (2015) Targeting Cullin-RING E3 ubiquitin ligases for drug discovery: structure, assembly and small-molecule modulation. Biochem J 467: 365–386

Cang Y, Zhang J, Nicholas SA, Bastien J, Li B, Zhou P et al (2006) Deletion of DDB1 in mouse brain and lens leads to p53-dependent elimination of proliferating cells. Cell 127(5):929–940

Cang Y, Zhang J, Nicholas SA, Kim AL, Zhou P, Goff SP (2007) DDB1 is essential for genomic stability in developing epidermis. Proc Natl Acad Sci USA 104 (8):2733–2737

Cao J, Hou P, Chen J, Wang P, Wang W, Liu W et al (2017) The overexpression and prognostic role of DCAF13 in hepatocellular carcinoma. Tumor Biol 39 (6)

Centore RC, Havens CG, Manning AL, Li J-M, Flynn RL, Tse A et al (2010) CRL4(Cdt2)-mediated destruction of the histone methyltransferase Set8 prevents premature chromatin compaction in S phase. Mol Cell 40 (1):22–33

Chen C-Y, Tsai M-S, Lin C-Y, Yu IS, Chen Y-T, Lin S-R et al (2012) Rescue of the genetically engineered Cul4b mutant mouse as a potential model for human X-linked mental retardation. Hum Mol Genet 21(19):4270–4285

Chen H-H, Fan P, Chang S-W, Tsao Y-P, Huang H-P, Chen S-L (2017b) NRIP/DCAF6 stabilizes the androgen receptor protein by displacing DDB2 from the CUL4A-DDB1 E3 ligase complex in prostate cancer. Oncotarget 8(13):21501–21515

Chen LC, Manjeshwar S, Lu Y, Moore D, Ljung BM, Kuo WL et al (1998) The human homologue for the Caenorhabditis elegans cul-4 gene is amplified and overexpressed in primary breast cancers. Cancer Res 58(16):3677–3683

Chen Y-A, Peng Y-J, Hu M-C, Huang J-J, Chien Y-C, Wu J-T et al (2015b) The Cullin 4A/B-DDB1-Cereblon E3 ubiquitin ligase complex mediates the degradation of CLC-1 chloride channels. Sci Rep 5

Chen Z, Sui J, Zhang F, Zhang C (2015a) Cullin family proteins and tumorigenesis: genetic association and molecular mechanisms. J Cancer 6(3):233–242

Chen Z, Wang K, Hou C, Jiang K, Chen B, Chen J et al (2017a) CRL4B(DCAF11) E3 ligase targets p21 for degradation to control cell cycle progression in human osteosarcoma cells. Sci Rep 7

Cope GA, Deshaies RJ (2003) COP9 signalosome: a multifunctional regulator of SCF and other cullin-based ubiquitin ligases. Cell 114(6):663–671

Cui D, Xiong X, Zhao Y (2016) Cullin-RING ligases in regulation of autophagy. Cell Div 11

Del Prete D, Rice RC, Rajadhyaksha AM, D'Adamio L (2016) Amyloid precursor protein (APP) May act as a substrate and a recognition unit for CRL4(CRBN) and Stub1 E3 Ligases facilitating ubiquitination of proteins involved in presynaptic functions and neurodegeneration. J Biol Chem 291(33):17209–17227

Deng J, Lei W, Xiang X, Zhang L, Lei J, Gong Y et al (2016) Cullin 4A (CUL4A), a direct target of miR-9 and miR-137, promotes gastric cancer proliferation and invasion by regulating the hippo signaling pathway. Oncotarget 7(9):10037–10050

Ding R, Han S, Lu Y, Guo C, Xie H, Zhang N et al (2010) Overexpressed Id-1 is associated with patient prognosis and HBx expression in hepatitis B virus-related hepatocellular carcinoma. Cancer Biol Ther 10(3): 299–307

Duda DM, Scott DC, Calabrese MF, Zimmerman ES, Zheng N, Schulman BA (2011) Structural regulation of cullin-RING ubiquitin ligase complexes. Curr Opin Struct Biol 21(2):257–264

El Din Hemimi NS, Mansour AA, Abdelsalam MM (2016) Prediction of the risk for essential hypertension among carriers of C825T genetic polymorphism of G protein beta3 (GNB3) gene. Biomark Insights 11:69–75

Fajner V, Maspero E, Polo S (2017) Targeting HECT-type E3 ligases - insights from catalysis, regulation and inhibitors. Febs Lett 591(17):2636–2647

Gabriely G, Wheeler MA, Takenaka MC, Quintana FJ (2017) Role of AHR and HIF-1alpha in glioblastoma metabolism. Trends Endocrinol Metab 28(6):428–436

Galanos P, Vougas K, Walter D, Polyzos A, Maya-Mendoza A, Haagensen EJ et al (2016) Chronic p53-independent p21 expression causes genomic

instability by deregulating replication licensing. Nat Cell Biol 18(7):777–789
Giampaolo A, Pelosi E, Valtieri M, Montesoro E, Sterpetti P, Samoggia P et al (1995) HOXB gene expression and function in differentiating purified hematopoietic progenitors. Stem Cells 13(Suppl 1): 90–105
Gong Y, Xiang X-J, Feng M, Chen J, Fang Z-L, Xiong J-P (2017) CUL4A promotes cell invasion in gastric cancer by activating the NF-kappa B signaling pathway. Biol Targets Therapy 11:45–53
Guo L-D, Wang D, Yang F, Liang Y-J, Yang X-Q, Qin Y-Y et al (2016) Functional analysis of DNA damage repair factor WDR70 and Its mutation in ovarian cancer. Sichuan da xue xue bao Yi xue ban 47(4):501–506
Hanahan D, Weinberg RA (2011) Hallmarks of cancer: the next generation. Cell 144(5):646–674
Hannah J, Zhou P (2015) Distinct and overlapping functions of the cullin E3 ligase scaffolding proteins CUL4A and CUL4B. Gene 573(1):33–45
Havens CG, Shobnam N, Guarino E, Centore RC, Zou L, Kearsey SE et al (2012) Direct role for proliferating cell nuclear antigen in substrate recognition by the E3 ubiquitin ligase CRL4(Cdt2). J Biol Chem 287(14): 11410–11421
Higa LA, Wu M, Ye T, Kobayashi R, Sun H, Zhang H (2006) CUL4-DDB1 ubiquitin ligase interacts with multiple WD40-repeat proteins and regulates histone methylation. Nat Cell Biol 8(11):1277–1U48
Higa LAA, Mihaylov IS, Banks DP, Zheng JY, Zhang H (2003) Radiation-mediated proteolysis of CDT1 by CUL4-ROC1 and CSN complexes constitutes a new checkpoint. Nat Cell Biol 5(11):1008–1015
Hrecka K, Hao C, Shun M-C, Kaur S, Swanson SK, Florens L et al (2016) HIV-1 and HIV-2 exhibit divergent interactions with HLTF and UNG2 DNA repair proteins. Proc Natl Acad Sci U S A 113(27):E3921–E3E30
Hu J, McCall CM, Ohta T, Xiong Y (2004) Targeted ubiquitination of CDT1 by the DDB1-CUL4A-ROC1 ligase in response to DNA damage. Nat Cell Biol 6 (10):1003–1009
Huang G-K, Liu T-T, Weng S-W, You H-L, Wei Y-C, Chen C-H et al (2017) CUL4A overexpression as an independent adverse prognosticator in intrahepatic cholangiocarcinoma. BMC Cancer 17
Huang J, Chen J (2008) VprBP targets Merlin to the Roc1-Cul4A-DDB1 E3 ligase complex for degradation. Oncogene 27(29):4056–4064
Huh J, Piwnica-Worms H (2013) CRL4(CDT2) targets CHK1 for PCNA-independent destruction. Mol Cell Biol 33(2):213–226
Hwang G, Sun F, O'Brien M, Eppig JJ, Handel MA, Jordan PW (2017) SMC5/6 is required for the formation of segregation-competent bivalent chromosomes during meiosis I in mouse oocytes. Development 144 (9):1648–1660
Hwangbo DS, Biteau B, Rath S, Kim J, Jasper H (2016) Control of apoptosis by drosophila DCAF12. Dev Biol 413(1):50–59
Jackson S, Xiong Y (2009) CRL4s: the CUL4-RING E3 ubiquitin ligases. Trends Biochem Sci 34(11):562–570
Jia L, Yan F, Cao W, Chen Z, Zheng H, Li H et al (2017) Dysregulation of CUL4A and CUL4B ubiquitin Ligases in lung cancer. J Biol Chem 292(7):2966–2978
Jiang T, Tang H-M, Wu Z-H, Chen J, Lu S, Zhou C-Z et al (2013) Cullin 4B is a novel prognostic marker that correlates with colon cancer progression and pathogenesis. Med Oncol 30(2)
Jiang T, Liu M, Wu J, Shi Y (2016) Structural and bio chemical analysis of Bcl-2 interaction with the hepatitis B virus protein HBx. Proc Natl Acad Sci U S A 113 (8):2074–2079
Jin J, Arias EE, Chen J, Harper JW, Walter JC (2006) A family of diverse Cul4-Ddb1-interacting proteins includes Cdt2, which is required for S phase destruction of the replication factor Cdt1. Mol Cell 23(5): 709–721
Jung E-S, Sim YJ, Jeong H-S, Kim S-J, Yun Y-J, Song J-H et al (2015) Jmjd2C increases MyoD transcriptional activity through inhibiting G9a-dependent MyoD degradation. Biochim Biophys Acta 1849(8):1081–1094
Kapetanaki MG, Guerrero-Santoro J, Bisi DC, Hsieh CL, Rapic-Otrin V, Levine AS (2006) The DDB1-CUL4A (DDB2) ubiquitin ligase is deficient in xeroderma pigmentosum group E and targets histone H2A at UV-damaged DNA sites. Proc Natl Acad Sci USA 103(8):2588–2593
Khanal T, Choi K, Leung YK, Wang J, Kim D, Janakiram V et al (2017) Loss of NR2E3 represses AHR by LSD1 reprogramming, is associated with poor prognosis in liver cancer. Sci Rep 7(1):10662
Kim TY, Jackson S, Xiong Y, Whitsett TG, Lobello JR, Weiss GJ et al (2013) CRL4A-FBXW5-mediated degradation of DLC1 Rho GTPase-activating protein tumor suppressor promotes non-small cell lung cancer cell growth. Proc Natl Acad Sci U S A 110(42): 16868–16873
Kim Y, Starostina NG, Kipreos ET (2008) The CRL4 (Cdt2) ubiquitin ligase targets the degradation of p21 (Cip1) to control replication licensing. Genes Dev 22 (18):2507–2519
Klein CJ, Wu Y, Vogel P, Goebel HH, Boennemann C, Zukosky K et al (2014) Ubiquitin ligase defect by DCAF8 mutation causes HMSN2 with giant axons. Neurology 82(10):873–878
Koike K, Moriya K, Iino S, Yotsuyanagi H, Endo Y, Miyamura T et al (1994) High-level expression of hepatitis B virus HBx gene and hepatocarcinogenesis in transgenic mice. Hepatology 19(4):810–819
Kopanja D, Roy N, Stoyanova T, Hess RA, Bagchi S, Raychaudhuri P (2011) Cul4A is essential for spermatogenesis and male fertility. Dev Biol 352(2): 278–287
Kroenke J, Udeshi ND, Narla A, Grauman P, Hurst SN, McConkey M et al (2014) Lenalidomide causes selective degradation of IKZF1 and IKZF3 in multiple myeloma cells. Science 343(6168):301–305
Kroenke J, Fink EC, Hollenbach PW, MacBeth KJ, Hurst SN, Udeshi ND et al (2015) Lenalidomide induces

ubiquitination and degradation of CK1a in del (5q) MDS. Nature 523(7559):183–U02
Lau AW, Fukushima H, Wei W (2012) The Fbw7 and Beta-TRCP E3 ubiquitin ligases and their roles in tumorigenesis. Front Biosci 17:2197–2212
Lee AK, Potts PR (2017) A comprehensive guide to the MAGE family of ubiquitin ligases. J Mol Biol 429 (8):1114–1142
Lee J, Zhou P (2012) Pathogenic role of the CRL4 ubiquitin ligase in human disease. Front Oncol 2:21
Lee J, Shieh JH, Zhang J, Liu L, Zhang Y, Eom JY et al (2013) Improved ex vivo expansion of adult hematopoietic stem cells by overcoming CUL4-mediated degradation of HOXB4. Blood 121(20): 4082–4089
Lee JM, Lee JS, Kim H, Kim K, Park H, Kim J-Y et al (2012) EZH2 generates a methyl Degron that is recognized by the DCAF1/DDB1/CUL4 E3 ubiquitin ligase complex. Mol Cell 48(4):572–586
Li B, Jia N, Kapur R, Chun KT (2006) Cul4A targets p27 for degradation and regulates proliferation, cell cycle exit, and differentiation during erythropoiesis. Blood 107(11):4291–4299
Li BH, Ruiz JC, Chun KT (2002) CUL-4A is critical for early embryonic development. Mol Cell Biol 22 (14):4997–5005
Li P, Song Y, Zan W, Qin L, Han S, Jiang B et al (2017a) Lack of CUL4B in adipocytes promotes PPAR gamma-mediated adipose tissue expansion and insulin sensitivity. Diabetes 66(2):300–313
Li P, Zhang L, Yang M, Qi M, Jin X, Han B (2017c) Cul4B is a novel prognostic marker in cholangiocarcinoma. Oncol Lett 14(2):1265–1274
Li Q, Cui M, Yang F, Li N, Jiang B, Yu Z et al (2017b) A cullin 4B-RING E3 ligase complex fine-tunes pancreatic delta cell paracrine interactions. J Clin Investig 127 (7):2631–2646
Li W, Cooper J, Zhou L, Yang C, Erdjument-Bromage H, Zagzag D et al (2014) Merlin/NF2 loss-driven tumorigenesis linked to CRL4(DcAF1)-mediated inhibition of the hippo pathway kinases Latsl and 2 in the nucleus. Cancer Cell 26(1):48–60
Li Y, Jaramillo-Lambert A, Hao J, Yang Y, Zhu W (2011) The stability of histone acetyltransferase general control non-derepressible (Gcn) 5 is regulated by Cullin4-RING E3 ubiquitin ligase. J Biol Chem 286 (48):41344–41352
Liu H, Liu Z, Wang Y, Stinchcombe TE, Owzar K, Han Y et al (2017) Functional variants in DCAF4 associated with lung cancer risk in European populations. Carcinogenesis 38(5):541–551
Liu H, Lu W, He H, Wu J, Zhang C, Gong H et al (2019) Inflammation-dependent overexpression of c-Myc enhances CRL4(DCAF4) E3 ligase activity and promotes ubiquitination of ST7 in colitis-associated cancer. J Pathol 248(4):464–475
Liu J, Ye J, Zou X, Xu Z, Feng Y, Zou X et al (2014) CRL4A(CRBN) E3 ubiquitin ligase restricts BK channel activity and prevents epileptogenesis. Nat Commun 5
Liu L, Lee S, Zhang J, Peters SB, Hannah J, Zhang Y et al (2009) CUL4A abrogation augments DNA damage response and protection against skin carcinogenesis. Mol Cell 34(4):451–460
Liu L, Yin Y, Li Y, Prevedel L, Lacy EH, Ma L et al (2012) Essential role of the CUL4B ubiquitin ligase in extra-embryonic tissue development during mouse embryogenesis. Cell Res 22(8):1258–1269
Liu L, Wong CC, Gong B, Yu J (2018) Functional significance and therapeutic implication of ring-type E3 ligases in colorectal cancer. Oncogene 37(2):148–159
Lo JY, Spatola BN, Curran SP (2017) WDR23 regulates NRF2 independently of KEAP1. Plos Genet 13(4)
Lo S-C, Li X, Henzl MT, Beamer LJ, Hannink M (2006) Structure of the Keap1: Nrf2 interface provides mechanistic insight into Nrf2 signaling. EMBO J 25(15): 3605–3617
Lydeard JR, Schulman BA, Harper JW (2013) Building and remodelling Cullin-RING E3 ubiquitin ligases – 'Ubiquitylation: mechanism and functions' review series. EMBO Rep 14(12):1050–1061
Ma Y, Han C-C, Huang Q, Sun W-Y, Wei W (2016) GRK2 overexpression inhibits IGF1-induced proliferation and migration of human hepatocellular carcinoma cells by downregulating EGR1. Oncol Rep 35 (5):3068–3074
Mackintosh C, Ordonez JL, Garcia-Dominguez DJ, Sevillano V, Llombart-Bosch A, Szuhai K et al (2012) 1q gain and CDT2 overexpression underlie an aggressive and highly proliferative form of Ewing sarcoma. Oncogene 31(10):1287–1298
Mangino M, Christiansen L, Stone R, Hunt SC, Horvath K, Eisenberg DT et al (2015) DCAF4, a novel gene associated with leucocyte telomere length. J Med Genet 52(3):157–162
Matsunuma R, Niida H, Ohhata T, Kitagawa K, Sakai S, Uchida C et al (2016) UV damage-induced phosphorylation of HBO1 triggers CRL4(DDB2)-mediated degradation to regulate cell proliferation. Mol Cell Biol 36 (3):394–406
Matyskiela ME, Lu G, Ito T, Pagarigan B, Lu C-C, Miller K et al (2016) A novel cereblon modulator recruits GSPT1 to the CRL4(CRBN) ubiquitin ligase. Nature 535(7611):252–257
Meszaros B, Kumar M, Gibson TJ, Uyar B, Dosztanyi Z (2017) Degrons in cancer. Sci Signal 10(470)
Mi J, Zou Y, Lin X, Lu J, Liu X, Zhao H et al (2017) Dysregulation of the miR-194-CUL4B negative feedback loop drives tumorigenesis in non-small-cell lung carcinoma. Mol Oncol 11(3):305–319
Migita K, Takayama T, Matsumoto S, Wakatsuki K, Tanaka T, Ito M et al (2014) Prognostic impact of RING box protein-1 (RBX1) expression in gastric cancer. Gastric Cancer 17(4):601–609
Migliorini D, Bogaerts S, Defever D, Vyas R, Denecker G, Radaelli E et al (2011) Cop1 constitutively regulates

c-Jun protein stability and functions as a tumor suppressor in mice. J Clin Investig 121(4):1329–1343
Miranda-Carboni GA, Krum SA, Yee K, Nava M, Deng QE, Pervin S et al (2008) A functional link between Wnt signaling and SKP2-independent p27 turnover in mammary tumors. Genes Dev 22(22):3121–3134
Morreale FE, Walden H (2016) Types of ubiquitin ligases. Cell 165(1):248–2e1
Murphy CM, Xu Y, Li F, Nio K, Reszka-Blanco N, Li X et al (2016) Hepatitis B virus X protein promotes degradation of SMC5/6 to enhance HBV replication. Cell Rep 16(11):2846–2854
Nakagawa T, Xiong Y (2011) X-linked mental retardation gene CUL4B targets ubiquitylation of H3K4 methyltransferase component WDR5 and regulates neuronal gene expression. Mol Cell 43(3):381–391
Natarajan C, Takeda K (2017) Regulation of various DNA repair pathways by E3 ubiquitin ligases. J Cancer Res Ther 13(2):157–169
Nguyen HC, Wang W, Xiong Y. Cullin-RING E3 ubiquitin ligases: bridges to destruction. Subcell Biochem 2017;83:323–347
Ni W, Zhang Y, Zhan Z, Ye F, Liang Y, Huang J et al (2017) A novel lncRNA uc.134 represses hepatocellular carcinoma progression by inhibiting CUL4A-mediated ubiquitination of LATS1. J Hematol Oncol 10
Nogues L, Reglero C, Rivas V, Salcedo A, Lafarga V, Neves M et al (2016) G protein-coupled receptor kinase 2 (GRK2) promotes breast tumorigenesis through a HDAC6-Pin1 Axis. EBioMedicine 13: 132–145
Oda H, Huebner MR, Beck DB, Vermeulen M, Hurwitz J, Spector DL et al (2010) Regulation of the histone H4 monomethylase PR-Set7 by CRL4(cdt2)-mediated PCNA-dependent degradation during DNA damage. Mol Cell 40(3):364–376
Ohtake F, Fujii-Kuriyama Y, Kato S (2009) AhR acts as an E3 ubiquitin ligase to modulate steroid receptor functions. Biochem Pharmacol 77(4):474–484
Pan W-W, Zhou J-J, Yu C, Xu Y, Guo L-J, Zhang H-Y et al (2013) Ubiquitin E3 ligase CRL4(CDT2/DCAF2) as a potential chemotherapeutic target for ovarian surface epithelial cancer. J Biol Chem 288(41): 29680–29691
Pan Y, Wang B, Yang X, Bai F, Xu Q, Li X et al (2015) CUL4A facilitates hepatocarcinogenesis by promoting cell cycle progression and epithelial-mesenchymal transition. Sci Rep 5
Patron LA, Nagatomo K, Eves DT, Imad M, Young K, Torvund M et al (2019) Cul4 ubiquitin ligase cofactor DCAF12 promotes neurotransmitter release and homeostatic plasticity. J Cell Biol 218(3):993–1010
Petroski MD, Deshaies RJ (2005) Function and regulation of cullin-RING ubiquitin ligases. Nat Rev Mol Cell Biol 6(1):9–20
Pineault N, Helgason CD, Lawrence HJ, Humphries RK (2002) Differential expression of Hox, Meis1, and Pbx1 genes in primitive cells throughout murine hematopoietic ontogeny. Exp Hematol 30(1):49–57
Puklowski A, Homsi Y, Keller D, May M, Chauhan S, Kossatz U et al (2011) The SCF-FBXW5 E3-ubiquitin ligase is regulated by PLK4 and targets HsSAS-6 to control centrosome duplication. Nat Cell Biol 13 (8):1004–1009
Ravichandran R, Kodali K, Peng J, Potts PR (2019) Regulation of MAGE-A3/6 by the CRL4-DCAF12 ubiquitin ligase and nutrient availability. EMBO Rep 20(7):e47352
Ren W, Sun Z, Zeng Q, Han S, Zhang Q, Jiang L (2016a) Aberrant expression of CUL4A is associated with IL-6/ STAT3 activation in colorectal cancer progression. Arch Med Res 47(3):214–222
Ren W, Shen S, Sun Z, Shu P, Shen X, Bu C et al (2016b) Jak-STAT3 pathway triggers DICER1 for proteasomal degradation by ubiquitin ligase complex of CUL4A (DCAF1) to promote colon cancer development. Cancer Lett 375(2):209–220
Rivas V, Carmona R, Munoz-Chapuli R, Mendiola M, Nogues L, Reglero C et al (2013) Developmental and tumoral vascularization is regulated by G protein-coupled receptor kinase 2. J Clin Investig 123(11): 4714–4730
Roy N, Bommi PV, Bhat UG, Bhattacharjee S, Elangovan I, Li J et al (2013) DDB2 suppresses epithelial-to-mesenchymal transition in colon cancer. Cancer Res 73(12):3771–3782
Sang Y, Yan F, Ren X (2015) The role and mechanism of CRL4 E3 ubiquitin ligase in cancer and its potential therapy implications. Oncotarget 6(40):42590–42602
Sarikas A, Hartmann T, Pan ZQ (2011) The cullin protein family. Genome Biol 12(4):220
Sauvageau G, Lansdorp PM, Eaves CJ, Hogge DE, Dragowska WH, Reid DS et al (1994) Differential expression of homeobox genes in functionally distinct CD34(+) subpopulations of human bone-marrow cells. Proc Natl Acad Sci USA 91(25):12223–12227
Scrima A, Fischer ES, Lingaraju GM, Bohm K, Cavadini S, Thoma NH (2011) Detecting UV-lesions in the genome: the modular CRL4 ubiquitin ligase does it best. FEBS Lett 585(18):2818–2825
Song B, Zhan H, Bian Q, Li J (2015b) Knockdown of CUL4B inhibits proliferation and promotes apoptosis of colorectal cancer cells through suppressing the Wnt/ beta-catenin signaling pathway. Int J Clin Exp Pathol 8 (9):10394–10402
Song J, Zhang J, Shao J (2015a) Knockdown of CUL4A inhibits invasion and induces apoptosis in osteosarcoma cells. Int J Immunopathol Pharmacol 28(2): 263–269
Spink BC, Bloom MS, Wu S, Sell S, Schneider E, Ding X et al (2015) Analysis of the AHR gene proximal promoter GGGGC-repeat polymorphism in lung, breast, and colon cancer. Toxicol Appl Pharmacol 282(1): 30–41
Sui X, Zhou H, Zhu L, Wang D, Fan S, Zhao W (2017) CUL4A promotes proliferation and metastasis of colorectal cancer cells by regulating H3K4 trimethylation in epithelial-mesenchymal transition. Oncotargets Therapy 10:735–743

Sun F, Indran IR, Zhang ZW, Tan MH, Li Y, Lim ZL et al (2015) A novel prostate cancer therapeutic strategy using icaritin-activated arylhydrocarbon-receptor to co-target androgen receptor and its splice variants. Carcinogenesis 36(7):757–768

Ta L, Xuan C, Xing N, Zhu X (2016) COP1 is downregulated in renal cell carcinoma (RCC) and inhibits the migration of RCC ACHN cells in vitro. Mol Med Rep 14(2):1371–1378

Tamori Y, Bialucha CU, Tian AG, Kajita M, Huang YC, Norman M et al (2010) Involvement of Lgl and Mahjong/VprBP in cell competition. PLoS Biol 8(7): e1000422

Tardat M, Brustel J, Kirsh O, Lefevbre C, Callanan M, Sardet C et al (2010) The histone H4 Lys 20 methyltransferase PR-Set7 regulates replication origins in mammalian cells. Nat Cell Biol 12(11): 1086–1U82

Tarpey PS, Raymond FL, O'Meara S, Edkins S, Teague J, Butler A et al (2007) Mutations in CUL4B, which encodes a ubiquitin E3 ligase subunit, cause an X-linked mental retardation syndrome associated with aggressive outbursts, seizures, relative macrocephaly, central obesity, hypogonadism, pes cavus, and tremor. Am J Hum Genet 80(2):345–352

Tong X, Zhang D, Guha A, Arthurs B, Cazares V, Gupta N et al (2015) CUL4-DDB1-CDT2 E3 ligase regulates the molecular clock activity by promoting ubiquitination-dependent degradation of the mammalian CRY1. PLoS One 10(10):e0139725

Tong X, Zhang D, Charney N, Jin E, VanDommelon K, Stamper K et al (2017) DDB1-mediated CRY1 degradation promotes FOXO1-driven gluconeogenesis in liver. Diabetes 66(10):2571–2582

Uljon S, Xu X, Durzynska I, Stein S, Adelmant G, Marto JA et al (2016) Structural basis for substrate selectivity of the E3 ligase COP1. Structure 24(5):687–696

Van Nguyen T, Lee JE, Sweredoski MJ, Yang S-J, Jeon S-J, Harrison JS et al (2016) Glutamine triggers acetylation-dependent degradation of glutamine synthetase via the thalidomide receptor cereblon. Mol Cell 61(6):809–820

Vanderdys V, Allak A, Guessous F, Benamar M, Read PW, Jameson MJ et al (2018) The neddylation inhibitor pevonedistat (MLN4924) suppresses and radiosensitizes head and neck squamous carcinoma cells and tumors. Mol Cancer Ther 17(2):368–380

Vulto-van Silfhout AT, Nakagawa T, Bahi-Buisson N, Haas SA, Hu H, Bienek M et al (2015) Variants in CUL4B are associated with cerebral malformations. Hum Mutat 36(1):106–117

Wang C, Yang W, Yan H-X, Luo T, Zhang J, Tang L et al (2012) Hepatitis B virus X (HBx) induces tumorigenicity of hepatic progenitor cells in 3,5-diethoxycarbonyl-1,4-dihydrocollidine-treated HBx transgenic mice. Hepatology 55(1):108–120

Wang K, Li Y, Jiang YZ, Dai CF, Patankar MS, Song JS et al (2013) An endogenous aryl hydrocarbon receptor ligand inhibits proliferation and migration of human ovarian cancer cells. Cancer Lett 340(1):63–71

Wang X, Arceci A, Bird K, Mills CA, Choudhury R, Kernan JL et al (2017a) VprBP/DCAF1 regulates the degradation and nonproteolytic activation of the cell cycle transcription factor FoxM1. Mol Cell Biol 37(13)

Wang Y, Zhang P, Liu Z, Wang Q, Wen M, Wang Y et al (2014a) CUL4A overexpression enhances lung tumor growth and sensitizes lung cancer cells to Erlotinib via transcriptional regulation of EGFR. Mol Cancer 13

Wang Y, Wen M, Kwon Y, Xu Y, Liu Y, Zhang P et al (2014b) CUL4A induces epithelial-mesenchymal transition and promotes cancer metastasis by regulating ZEB1 expression. Cancer Res 74(2):520–531

Wang Y, Ma G, Wang Q, Wen M, Xu Y, He X et al (2014c) Involvement of CUL4A in regulation of multidrug resistance to P-gp substrate drugs in breast cancer cells. Molecules 19(1):159–176

Wang Y, Xu Z, Mao J-H, Hsieh D, Au A, Jablons DM et al (2015a) PR-Set7 is degraded in a conditional Cul4A transgenic mouse model of lung cancer. Zhongguo Fei Ai Za Zhi 18(6):345–350

Wang Y, Xu Z, Mao J-H, Hung M-S, Hsieh D, Au A et al (2015b) Analysis of lung tumor initiation and progression in transgenic mice for Cre-inducible overexpression of Cul4A gene. Thoracic Cancer 6(4):480–487

Wang Y-T, Chen J, Chang C-W, Jen J, Huang T-Y, Chen C-M et al (2017b) Ubiquitination of tumor suppressor PML regulates prometastatic and immunosuppressive tumor microenvironment. J Clin Investig 127(8): 2982–2997

Webster E, Cho MT, Alexander N, Desai S, Naidu S, Bekheirnia MR et al (2016) De novo PHIP-predicted deleterious variants are associated with developmental delay, intellectual disability, obesity, and dysmorphic features. Cold Spring Harb Mol Case Stud 2(6): a001172

Wertz IE, O'Rourke KM, Zhang ZM, Dornan D, Arnott D, Deshaies RJ et al (2004) Human De-etiolated-1 regulates c-Jun by assembling a CUL4A ubiquitin ligase. Science 303(5662):1371–1374

Wormke M, Stoner M, Saville B, Walker K, Abdelrahim M, Burghardt R et al (2003) The aryl hydrocarbon receptor mediates degradation of estrogen receptor alpha through activation of proteasomes. Mol Cell Biol 23(6):1843–1855

Wu Y, Zhou L, Wang X, Lu J, Zhang R, Liang X et al (2016) A genome-scale CRISPR-Cas9 screening method for protein stability reveals novel regulators of Cdc25A. Cell Discov 2

Xie L, Wang W, Xu B, Liu Y (2014) Experimental study on hepatitis B-virus X gene expression in adenoid cystic carcinoma. Hua Xi Kou Qiang Yi Xue Za Zhi 32(4):328–330

Xing R, Chen K-B, Xuan Y, Feng C, Xue M, Zeng Y-C (2016) RBX1 expression is an unfavorable prognostic factor in patients with non-small cell lung cancer. Surg Oncol 25(3):147–151

Xu Y-W, Cao L-R, Wang M, Xu Y, Wu X, Liu J et al (2017) Maternal DCAF2 is crucial for maintenance of genome stability during the first cell cycle in mice. J Cell Sci 130(19):3297–3307

Yamaji S, Zhang M, Zhang J, Endo Y, Bibikova E, Goff SP et al (2010) Hepatocyte-specific deletion of DDB1 induces liver regeneration and tumorigenesis. Proc Natl Acad Sci USA 107(51):22237–22242

Yan H, Bi L, Wang Y, Zhang X, Hou Z, Wang Q et al (2017) Integrative analysis of multi-omics data reveals distinct impacts of DDB1-CUL4 associated factors in human lung adenocarcinomas. Sci Rep 7(1):333

Yan Y, Zhang X, Legerski RJ (2011) Artemis interacts with the Cul4A-DDB1(DDB2) ubiquitin E3 ligase and regulates degradation of the CDK inhibitor p27. Cell Cycle, 4109 10(23):4098

Yang Y, Liu R, Qiu R, Zheng Y, Huang W, Hu H et al (2015) CRL4B promotes tumorigenesis by coordinating with SUV39H1/HP1/DNMT3 A in DNA methylation-based epigenetic silencing. Oncogene 34(1):104–118

Yang Y-L, Hung M-S, Wang Y, Ni J, Mao J-H, Hsieh D et al (2014) Lung tumourigenesis in a conditional Cul4A transgenic mouse model. J Pathol 233(2):113–123

Yin Y, Lin C, Kim ST, Roig I, Chen H, Liu L et al (2011) The E3 ubiquitin ligase Cullin 4A regulates meiotic progression in mouse spermatogenesis. Dev Biol 356 (1):51–62

Yin Y, Liu L, Yang C, Lin C, Veith GM, Wang C et al (2016) Cell autonomous and nonautonomous function of CUL4B in mouse spermatogenesis. J Biol Chem 291(13):6923–6935

Yoon T, Chakrabortty A, Franks R, Valli T, Kiyokawa H, Raychaudhuri P (2005) Tumor-prone phenotype of the DDB2-deficient mice. Oncogene 24(3):469–478

Yu C, Zhang Y-L, Pan W-W, Li X-M, Wang Z-W, Ge Z-J et al (2013) CRL4 complex regulates mammalian oocyte survival and reprogramming by activation of TET proteins. Science 342(6165):1518–1521

Yu C, Ji S-Y, Sha Q-Q, Sun Q-Y, Fan H-Y (2015a) CRL4-DCAF1 ubiquitin E3 ligase directs protein phosphatase 2A degradation to control oocyte meiotic maturation. Nat Commun 6

Yu C, Xu Y-W, Sha Q-Q, Fan H-Y (2015b) CRL4 (DCAF1) is required in activated oocytes for follicle maintenance and ovulation. Mol Hum Reprod 21(2): 195–205

Yu F, White SB, Zhao Q, Lee FS (2001) HIF-1alpha binding to VHL is regulated by stimulus-sensitive proline hydroxylation. Proc Natl Acad Sci USA 98(17): 9630–9635

Yuan J, Jiang B, Zhang A, Qian Y, Tan H, Gao J et al (2015a) Accelerated hepatocellular carcinoma development in CUL4B transgenic mice. Oncotarget 6(17): 15209–15221

Yuan J, Han B, Hu H, Qian Y, Liu Z, Wei Z et al (2015b) CUL4B activates Wnt/beta-catenin signalling in hepatocellular carcinoma by repressing Wnt antagonists. J Pathol 235(5):784–795

Zenklusen JC, Conti CJ, Green ED (2001) Mutational and functional analyses reveal that ST7 is a highly conserved tumor-suppressor gene on human chromosome 7q31. Nat Genet 27(4):392–398

Zhang J-Q, Chen S, Gu J-N, Zhu Y, Zhan Q, Cheng D-F et al (2018) MicroRNA-300 promotes apoptosis and inhibits proliferation, migration, invasion and epithelial-mesenchymal transition via the Wnt/−catenin signaling pathway by targeting CUL4B in pancreatic cancer cells. J Cell Biochem 119(1): 1027–1040

Zhang T-J, Xue D, Zhang C-D, Zhang Z-D, Liu Q-R, Wang J-Q (2017a) Cullin 4A is associated with epithelial to mesenchymal transition and poor prognosis in perihilar cholangiocarcinoma. World J Gastroenterol 23(13):2318–2329

Zhang Y, Morrone G, Zhang JX, Chen XA, Lu XL, Ma L et al (2003) CUL-4A stimulates ubiquitylation and degradation of the HOXA9 homeodomain protein. EMBO J 22(22):6057–6067

Zhang Y, Feng S, Chen F, Chen H, Wang J, McCall C et al (2008) Arabidopsis DDB1-CUL4 ASSOCIATED FACTOR1 forms a nuclear E3 ubiquitin ligase with DDB1 and CUL4 that is involved in multiple plant developmental processes. Plant Cell 20(6):1437–1455

Zhang Z, Newton K, Kummerfeld SK, Webster J, Kirkpatrick DS, Phu L et al (2017b) Transcription factor Etv5 is essential for the maintenance of alveolar type II cells. Proc Natl Acad Sci U S A 114(15): 3903–3908

Zhao Y, Sun Y (2013) Cullin-RING ligases as attractive anti-cancer targets. Curr Pharm Des 19(18):3215–3225

Zheng N, Shabek N (2017) Ubiquitin ligases: structure, function, and regulation. In: Kornberg RD (eds) Annual review of biochemistry, vol 86, pp 129–157

Zhong WW, Feng H, Santiago FE, Kipreos ET (2003) CUL-4 ubiquitin ligase maintains genome stability by restraining DNA-replication licensing. Nature 423 (6942):885–889

Zhou L, Wang M-Y, Liang Z-Y, Zhou W-X, You L, Pan B-J et al (2016b) G-protein-coupled receptor kinase 2 in pancreatic cancer: clinicopathologic and prognostic significance. Hum Pathol 56:171–177

Zhou YM, Cao L, Li B, Zhang XZ, Yin ZF (2012) Expression of HBx protein in hepatitis B virus-infected intrahepatic cholangiocarcinoma. Hepatobiliary Pancreat Dis Int 11(5):532–535

Zhou Z, He M, Shah AA, Wan Y (2016a) Insights into APC/C: from cellular function to diseases and therapeutics. Cell Div 11:9

Zou Y, Liu Q, Chen B, Zhang X, Guo C, Zhou H et al (2007) Mutation in CUL4B, which encodes a member of cullin-RING ubiquitin ligase complex, causes X-linked mental retardation. Am J Hum Genet 80 (3):561–566

Cullin RING Ligase 5 (CRL-5): Neddylation Activation and Biological Functions

16

Shizhen Zhang and Yi Sun

Abstract

Cullin-5 (Cul-5) was originally identified as an arginine vasopressin (AVP) receptor due to its homology to a vasopressin-activated calcium-mobilizing protein 1 (VACM-1). Cul-5 has subsequently gained much attention after being identified as the key component of CRL-5 (Cullin-RING ligase-5) that mediates ubiquitylation and degradation of several key cellular proteins associated with human cancers and viral infections. Structurally, Cul-5 interacts with the Elongin B/C complex, a RING finger protein (RBX2/SAG), and a SOCS protein to form a CRL-5 E3 ubiquitin ligase protein complex. CRL-5, by controlling turnover of a variety of substrates, is implicated in several biological processes and human diseases. Activation of CRL-5 requires Cul-5 neddylation, catalyzed by a neddylation enzyme cascade, consisting of the E1 NEDD8-activating enzyme (NAE), the E2 neddylation conjugating enzyme (UBE2F), and E3 neddylation ligase (RBX2/SAG). RBX2/SAG, therefore, serves as both Cul-5 neddylation E3 and CRL-5 ubiquitylation E3. Here, we review the current knowledge on CRL-5, its activation by the UBE2F-SAG, its regulation of various signaling pathways via substrate degradation, and its implications in human cancers.

S. Zhang
Institute of Translational Medicine, Zhejiang University School of Medicine, Hangzhou, Zhejiang, China

Y. Sun (✉)
Institute of Translational Medicine, Zhejiang University School of Medicine, Hangzhou, Zhejiang, China

Division of Radiation and Cancer Biology, Department of Radiation Oncology, University of Michigan, Ann Arbor, MI, USA
e-mail: sunyi@umich.edu

Keywords

CRL-5 · SAG/RBX2 · Tumorigenesis · UBE2F · Viral proteins

Abbreviations

APOBEC3G	Apolipoprotein B editing complex 3G
APS	Adapter protein with a pleckstrin homology and Src homology 2 domain
AVP	Arginine vasopressin
CAND1	Cullin-associated neddylation-dissociated 1
CH	Cullin homology
CKB	Cytosolic creatine kinase
CR	Cullin repeats
CRL-5	Cullin RING ligase 5
CRLs	Cullin-RING ligases
CSN	COP9 signalosome complex
CTD	Carboxyl-terminal domain
Cul-5	Cullin-5

Y. Sun et al. (eds.), *Cullin-RING Ligases and Protein Neddylation*, Advances in Experimental Medicine and Biology 1217, https://doi.org/10.1007/978-981-15-1025-0_16

DCNL	The defective in cullin neddylation protein-like proteins
DSBs	Double-stranded DNA breaks
EBV	Epstein-Barr virus
ER	Endoplasmic reticulum
FAK	Focal adhesion kinase
HAdV	Human adenoviruses
HIF-α	Hypoxia-inducible factor-α
HIPK2	Homeodomain-interacting protein kinase 2
HIV-1	Human immunodeficiency virus-1
HSP90	Heat shock protein 90
ID2	Inhibitor of DNA binding 2
iNOS	Inducible nitric oxide synthase
IRS4	Insulin receptor substrate 4
JAK2	Janus kinase 2
KSHV	Kaposi's sarcoma-associated herpesvirus
LANA	Latency-associated nuclear antigen
LPS	Lipopolysaccharide
LRRK2	Leucine-rich repeat kinase 2
MCK	Mitochondrial creatine kinase
MLL	Mixed-lineage leukemia
MuHV-4	Murid herpesvirus-4
NAE	NEDD8-activating enzyme
NHEJ	Nonhomologous end-joining DNA repair system
NSCLC	Non-small lung cancers
NTD	Amino-terminal domain
PTHrP	Parathyroid hormone-related peptide
RBX2	RING-box protein 2
ROS	Reactive oxygen species
Rpb1	RNA polymerase II B1
SAG	Sensitive to apoptosis gene
SCLC	Small cell lung cancers
SH2	Src homology 2
SOCS	Suppressor of cytokine signaling
TCGA	The Cancer Genome Atlas
TNF-R2	Tumor necrosis factor receptor 2
TRIAD1	Two RING finger and double RING finger linked 1
VACM-1	Vasopressin-activated calcium-mobilizing protein 1

16.1 Introduction

Cul-5 is the least conserved member of the cullin family, but highly homologous among various species (Byrd et al. 1997). In kidney cells, Cul-5 localizes to the cell membrane where it binds to AVP to control body fluid and blood pressure, thus retaining the homeostasis (Burnatowska-Hledin et al. 1995). Otherwise, Cul-5, as a scaffold component, complexes with adaptor proteins Elongin B/C, a RING finger protein RBX2/SAG, and a substrate receptor SOCS protein to form a CRL-5 E3 ligase complex. Structurally, Cul-5 possesses a long stalklike amino-terminal domain (NTD), which contains three cullin repeats (CR1, CR2, CR3), and a globular carboxyl-terminal domain (CTD), harboring a highly conserved signature cullin homology (CH) domain (Sarikas et al. 2011), along with the lysine residue (Lys^{724}) for covalent NEDD8 attachment, a process known as neddylation for CRL-5 activation (Fig. 16.1). The amino-terminal helices H2 and H5 of CR1 are used to anchor the cognate adaptor, Elongin B/C. The CTD of Cul-5 binds to SAG, which recruits the ubiquitin-loaded E2 conjugating enzymes for catalysis of ubiquitylation reaction (Petroski and Deshaies 2005). The fourth component of CRL-5 is the suppressor of cytokine signaling (SOCS) protein, responsible for the recognition of cellular substrates, involved in various cellular functions. Interestingly, some viral proteins, including the viral infectivity factor (Vif) in human immunodeficiency virus-1 (HIV-1), adenovirus proteins E4orf6 and E1B55K, and latency-associated nuclear antigen (LANA) in Kaposi's sarcoma-associated herpesvirus (KSHV), can hijack host CRL-5 complex to trigger the ubiquitylation and degradation of host defensive proteins (Yu et al. 2003; Querido et al. 2001; Cai et al. 2006). Thus, CRL-5 is involved in regulation of both cellular functions and viral infections.

16.2 The Family of Substrate Receptors and Their Substrates

In mammalian cells, four families of substrate receptors were identified in CRL-5 E3 ligase, and every family member contains a C-terminal SOCS box, consisting of a BC box for Elongin B/C binding and a Cul-5 box for Cul-5 binding (Fig. 16.1). CRL-5 ligases target a wide range of proteins for ubiquitylation and degradation with substrate specificity determined by substrate receptors.

16.2.1 Cytokine-Inducible SH2 (Src Homology 2) Domain-Containing SOCS Box Proteins

The SH2 domain-containing SOCS box proteins (SOCS1-7) have a central SH2 domain and a C-terminally located SOCS box, which consists of a BC box and a Cul-5 box with an approximately 40-amino acid motif (Fig. 16.1). The SOCS box interacts with Cul-5 via its amino acid sequence LPΦP (Φ represents a hydrophobic residue) within the Cul5 box (Mahrour et al. 2008; Okumura et al. 2012), and the BC box is responsible for binding to Elongin B/C (Endo et al. 1997) (Fig. 16.1). SOCS1 and SOCS3 have been extensively studied in leukocytes with the activity to inhibit JAK family tyrosine kinase signaling, which are mediated by both Cul5-independent and Cul5-dependent mechanisms (Mahrour et al. 2008; Kazi et al. 2014; Linossi and Nicholson 2015). As the substrate-recognizing subunits, the family of SOCS proteins is involved in ubiquitylation and subsequent proteasomal degradation of a variety of cellular proteins by CRL-5. Specifically, SOCS1 suppresses the signal transduction via targeting for degradation of a variety of cellular proteins, including Vav (De Sepulveda et al. 2000), focal adhesion kinase (FAK) (Liu et al. 2003), the NF-κB family member p65/RelA (Ryo et al. 2003), myeloid differentiation primary-response gene 88 adaptor-like protein (MAL) (Mansell et al. 2006), the Janus kinase 2 (JAK2) (Ungureanu et al. 2002), the TEL-JAK2 onco-fusion protein (Kamizono et al. 2001; Frantsve et al. 2001), Cdh1 (Parrillas et al. 2013), HPV E7 (Kamio et al. 2004), and insulin receptor substrates IRS1 and IRS2 (Rui et al. 2002). SOCS3 specifically binds to the phosphorylated immunoreceptor tyrosine-based inhibitory motif

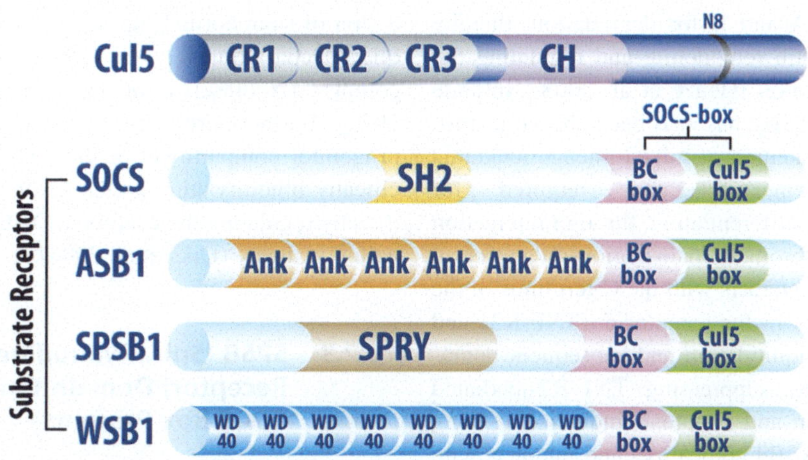

Fig. 16.1 Domain structures of Cul-5 and SOCS box containing proteins. Cullin repeat 1 (CR1) anchors the cognate adaptor proteins, and the cullin homology (CH) domain at the carboxyl-terminus is critical for the binding of the RING finger protein. The N8 site indicates the position of the neddylation site. The SOCS box consists of a BC box and a Cul5 box in the order indicated. *SH2* Src homology 2 phosphotyrosine-binding domain, *WD40* WD40 repeats, *SPRY* Sp1A/ryanodine receptor domain, *Ank* ankyrin repeats

of CD33, resulting in accelerated CD33 degradation (Orr et al. 2007). SOCS6, on the other hand, is involved in the degradation of Cas and other unidentified Src substrates to inhibit Src-dependent cell transformation (Teckchandani et al. 2014). SOCS6 also targets p56^{lck} and c-KIT for degradation (Choi et al. 2010; Lamsoul et al. 2016), whereas SOCS7 degrades Dab1 (Simo and Cooper 2013), an essential protein for neuron migration and positioning (Tissir and Goffinet 2003), thus playing an important role in normal neuron positioning during cerebral development.

16.2.2 Ankyrin Repeat-Containing SOCS Box (ASB) Family

The ASB family includes 18 members from ASB1 to ASB18, all of which contain two functional domains, the C-terminal SOCS box domain and an upstream ankyrin repeat region (Kile et al. 2002). Several members of the ASB family are able to interact with Cul5-SAG to form ubiquitin ligase complexes (Kohroki et al. 2005). ASB1 is expressed widely in a variety of organs, and *ASB1* knockout mice show no significantly phenotypes, but with a diminution of spermatogenesis (Kile et al. 2001). On the other hand, ASB2 promotes the polyubiquitylation of the actin-binding protein filamins A and B for degradation, thereby modulating actin remodeling and regulating the cell differentiation (Heuze et al. 2008; Burande et al. 2009). ASB2 has also been shown to promote the degradation of mixed-lineage leukemia (MLL) protein, a factor required for hematopoietic differentiation, through interaction with its PHD/bromodomain region (Wang et al. 2012). ASB3 interacts with the C-terminus of the tumor necrosis factor receptor 2 (TNF-R2) and triggers its ubiquitylation and subsequent degradation, thereby suppressing TNF-R2-mediated JNK activation and apoptosis induction (Chung et al. 2005). ASB4 colocalizes and interacts with the insulin receptor substrate 4 (IRS4) in neurons of the hypothalamus to promote IRS4 ubiquitylation and degradation, thus modulating neuron sensitivity to circulating insulin levels (Li et al. 2011). ASB4 also binds to and promotes ubiquitylation and degradation of inhibitor of DNA binding 2 (ID2), thus mediating vascular differentiation in the placenta (Townley-Tilson et al. 2014). ASB6 is restrictedly expressed in adipose tissue. In 3T3-L1 adipocytes, ASB6 regulated the insulin signaling pathway by targeted ubiquitylation and degradation of the adapter protein with a pleckstrin homology and Src homology 2 domain (APS) upon activation of the insulin receptor (Wilcox et al. 2004). ASB7 is involved in the regulation of cell division by promoting the degradation of DDA3, a critical factor that controls chromosome compression and segregation via modulating the dynamics of the mitotic spindle (Uematsu et al. 2016). ASB9 is predominantly expressed in the kidney and testes, where it promotes the ubiquitination and degradation of brain-type cytosolic creatine kinase (CKB) (Debrincat et al. 2007) and ubiquitous mitochondrial creatine kinase (uMtCK) (Kwon et al. 2010). ASB10 is induced by inflammation cytokines and involved in protein degradation pathways in glaucoma (Keller and Wirtz 2017). The observation that ASB10 forms a complex with Cul-5, SAG, and Elongin B/C (Andresen et al. 2014) suggests a possible role of CRL-5 in the process. Finally, ASB11 is an endoplasmic reticulum (ER)-related ubiquitin ligase, which promotes ubiquitylation and degradation of ribophorin 1, an integral protein of the oligosaccharyltransferase (OST) glycosylation complex (Andresen et al. 2014; Kelleher et al. 1992). Furthermore, ASB11 affects the neural progenitor compartment of the embryos by specifically ubiquitylating Delta A for degradation, thereby regulating the canonical Delta-Notch signaling pathway (Diks et al. 2008).

16.2.3 SPSB (SplA/Ryanodine Receptor) Domain-Containing SOCS Box Proteins

The SPSB family is characterized by a central SPRY (*SplA/ryanodine receptor*)/domain and a C-terminal SOCS box with four members (Perfetto et al. 2013; Hilton et al. 1998). SPSB1, SPSB2, and SPSB4 were reported to promote ubiquitylation and degradation of inducible nitric oxide synthase (iNOS/NOS2) (Nishiya et al. 2011; Kuang et al. 2010; Lewis et al. 2011). Given that iNOS is responsible for sustained production of NO upon stimulation by microbes or cytokines (Lowenstein and Padalko 2004), SPSB1 and SPSB4, therefore, play important roles in preventing the overproduction of NO by triggering iNOS degradation (Lewis et al. 2011; Matsumoto et al. 2011). At the physiological aspect, *SPSB2* knockout in macrophages results in excessive production of iNOS and NO to kill more *Leishmania* major parasites (Kuang et al. 2010). Furthermore, SPSB1 was reported to negatively regulate the TGF-β signaling pathway through an interaction with type II TGFβ receptor (TβRII) via its SPRY domain, leading to enhanced ubiquitylation and degradation of TβRII (Liu et al. 2015), whereas SPSB3 overexpression significantly inhibits tumor metastasis by promoting polyubiquitylation and degradation of SNAIL upon phosphorylation mediated by GSK-3β (Liu et al. 2018).

16.2.4 WD Repeat-Containing SOCS Box Protein 1 (WSB1)

WSB1 is another member of SOCS protein responsible for ubiquitylation and degradation of several key regulatory proteins. First, WSB1 promotes ubiquitylation and degradation of homeodomain-interacting protein kinase 2 (HIPK2) (Choi et al. 2008), a nuclear protein kinase that triggers apoptosis in part by activation of p53 (Puca et al. 2009). Thus, approaches that block WSB1-mediated HIPK2 degradation, such as treatment with adriamycin or cisplatin, enhance the DNA damage-induced apoptosis (Choi et al. 2008). Second, WSB1 was reported to promote ubiquitylation and degradation of von Hippel-Lindau tumor suppressor (pVHL), thus stabilizing hypoxia-inducible factor-α (HIF-α) under both normoxic and hypoxic conditions. The highly level of HIF-α maintained by WSB1 is responsible for enhanced cancer metastasis (Kim et al. 2015). Third, WSB1 promotes degradation of Rho-binding protein RhoGDI2 under tumor hypoxic environment, thereby inducing Rac1 activation to stimulate osteosarcoma cell migration and invasion (Cao et al. 2015). Fourth, a recent study reported that WSB1 triggers the ubiquitylation and degradation of ATM to bypass the oncogene-induced senescence, contributing to abnormal cell proliferation and cellular transformation (Kim et al. 2017). Finally, WSB1 was reported to promote ubiquitylation of yet two additional proteins, not for degradation, but for functional modulation. The first protein is the mutant of leucine-rich repeat kinase 2 protein (LRRK2), whose expression in neurons causes abnormal neurite process and nuclear condensation, indicating neuronal toxicity, frequently seen in Parkinson's disease (Lim et al. 2019). WSB1 promotes ubiquitylation of LRRK2 through the K27 and K29 linkages, which contributes to the formation of LRRK2 aggregation for neuronal protection (Nucifora et al. 2016). Therefore, *WSB1* knockdown exhibits enhanced neuronal toxicity with decreased protein aggregation in LRRK2 mutant *Drosophila* model (Nucifora et al. 2016). The second protein is thyroid hormone-activating enzyme type II iodothyronine deiodinase (D2). The WD-40 propeller of WSB-1 is capable of interacting with an 18-amino-acid loop in D2 to cause D2 polyubiquitylation. Such ubiquitylated D2 subsequently induces parathyroid hormone-related peptide (PTHrP) to regulate chondrocyte differentiation (Dentice et al. 2005).

16.2.5 Other SOCS Box Proteins

Rab40 was reported to interact with Elongin B/C and Cul-5 at the Golgi apparatus of *Xenopus*, forming a ubiquitin ligase complex to regulate the ubiquitylation and localization of the Rap2 GTPase, thereby playing an essential role in the non-canonical Wnt pathway (Kamura et al. 2001). MUF1 was shown to have a ubiquitin ligase activity after complexing with the Cul-5/Elongin BC complex. However, its specific substrate has not been identified (Kamura et al. 2001). Finally, in response to UV irradiation, Elongin A was reported to bind with Elongin B/C to form an Elongin ABC complex and then assembles with the Cul-5 and SAG module to promote ubiquitylation and degradation of the large subunit of RNA polymerase II B1 (Rpb1) (Yasukawa et al. 2008). VHL, a SOCS box-like protein, but lacking the C-terminal sequence of the SOCS box, is well-known to interact with endogenous Elongin B/C, Cul-2, and RBX1 to form an active E3 ubiquitin ligase for HIF-1α degradation (Kamura et al. 2004). Our earlier study showed that VHL is also associated with SAG/Cul-5, particularly under hypoxic conditions, to facilitate HIF-1α degradation to keep HIF-1α levels under control (Tan et al. 2008). A summary of the CRL-5 receptor components and their corresponding substrates is listed in Table 16.1.

16.3 Cul-5 Neddylation and CRL-5 Activation

It is well-known that activation of CRL-5 requires the attachment of NEDD8, a ubiquitin-like protein onto the Lys^{724} residue located at the C-terminus of Cul-5, in a process known as neddylation (Duda et al. 2008). More specifically, NEDD8 is first activated by NAE E1 and then transferred from the active site Cys of NAE onto the active site Cys of UBE2F. NEDD8-loaded UBE2F is then recognized by SAG E3 on the same surface recognized by NAE, to catalyze the transfer of the NEDD8 molecule from UBE2F onto the Lys^{724} residue at the winged-helix B motif of Cul-5. Such modulation of Cul-5 leads to a conformation change of the cullin-RING interface and results in the catalytically active CRL state (Rabut and Peter 2008) (Fig. 16.2). Conversely, inactivation of cullins occurs through removal of NEDD8 from cullins by a process known as deneddylation via the COP9 signalosome complex (CSN) (Lyapina et al. 2001). Cullin neddylation activates Cullin-E3 ligase activity via several mechanisms. First, it prevents the inhibitory binding of cullin-associated neddylation-dissociated 1 (CAND1) to cullins (Duda et al. 2008); second, it induces the conformational change at the cullin-RBX interface, allowing the ubiquitin-loaded E2s to move closer to the acceptor lysine residue of substrate proteins; third, NEDD8 is capable of promoting the formation of higher-order cullin-RBX complexes to increase the catalytic efficiency of some Cullin-E3 ligases (Soucy et al. 2009). Mammalian cells contain a single neddylation E1, a heterodimer of catalytic subunit UBA3/NAEβ and regulatory subunit APPBP1/NAE1, two neddylation E2s, UBE2M (also known as UBC12) and UBE2F, and several neddylation E3s, mainly consisting of a few RING domain-containing proteins, such as RBX1 and SAG (Zhou et al. 2018a).

16.3.1 Neddylation E2: UBE2F and UBE2M

Two neddylation E2s have distinct features by structural comparison between UBE2F and UBE2M. Both E2s bind to ubiquitin-fold domain and UBA3 hydrophobic groove of E1 through its core domain and N-terminal motif, respectively (Huang et al. 2004). Biochemically, these two E2s have a certain degree of binding selectivity; UBE2M couples with RBX1 to neddylate Cul1-Cul4, whereas UBE2F is relatively specific for SAG to promote Cul-5 neddylation (Huang et al. 2009). In both E2s, the N-terminal methionine is acetylated, which facilitates their respective binding to the PONY domain pocket of DCNL, another neddylation E3, thus enhancing the efficiency of cullin neddylation (Huang et al. 2009; Monda et al. 2013). Biologically, in NIH3T3 cells with knockdown of *UBE2M*, but not of *UBE2F*, suppress cell growth (Huang et al. 2009). While both E2s are recruited to DNA damage sites in response to IR or other DNA-damaging agents, only depletion of *UBE2M*, but not *UBE2F*, sensitizes cells to DNA damaging agents (Brown et al. 2015; Cukras et al. 2014).

We recently found that UBE2F is subjected to negative regulation by its family member, UBE2M (Zhou et al. 2018b). Specifically, UBE2M is a stress-inducible protein. Upon transcriptional induction by hypoxia or mitogen, UBE2M acts as a dual E2 for both neddylation

Table 16.1 CRL-5 E3 ligase and their substrates

Substrate receptors	Substrates	Involving pathways	Potential links to diseases	Refs.
SH2 domain-containing proteins with a SOCS box (SOCS)				
SOCS1	p65/RelA	Transcription		Ryo et al. (2003)
	FAK	Cell signaling		Liu et al. (2003)
	IRS1/2	Cell signaling	Insulin resistance syndromes, diabetes	Rui et al. (2002)
	JAK2	Cell signaling		Ungureanu et al. (2002)
	MAL	Cell signaling		Mansell et al. (2006)
	TEL-JAK2	Cell signaling	Leukemia	Kamizono et al. (2001)
	Vav	Cell signaling		De Sepulveda et al. (2000)
	Cdh1	Cell cycle		Parrillas et al. (2013)
	HPV E7	Cell signaling	Cervical cancer	Kamio et al. (2004)
SOCS3	CD33 receptor	Cell signaling		Orr et al. (2007)
SOCS6	cKit receptor	Cell signaling		Lamsoul et al. (2016)
	Cas	Epithelial cell transformation		Teckchandani et al. (2014)
	p56lck	T cell activation		Choi et al. (2010)
SOCS7	Dab1	Cell signaling	Cerebral development	Simo and Cooper (2013)
Ankyrin repeat-containing proteins with a SOCS box (ASB)				
ASB2	Filamin A/B	Cell motility, cell differentiation	Leukemia	Heuze et al. (2008), Burande et al. (2009)
	MLL	Epigenetic regulation		Wang et al. (2012)
ASB3	TNF-R2	Cell signaling		Chung et al. (2005)
ASB4	IRS4	Cell signaling		Li et al. (2011)
	ID2	Transcription, cell differentiation	Preeclampsia	Townley-Tilson et al. (2014)
ASB6	APS	Cell signaling		Wilcox et al. (2004)
ASB7	DDA3	Cell signaling		Uematsu et al. (2016)
ASB9	Creatine kinase B	Metabolism		Debrincat et al. (2007)
	uMtCK	Metabolism		Kwon et al. (2010)
ASB11	Dela A	Cell signaling		Diks et al. (2008)
	Ribophorin 1	Cell signaling		Andresen et al. (2014)
SPRY domain-containing proteins with a SOCS box (SPSB)				
SPSB1	iNOS	Innate host response	Chronic infections	Lewis et al. (2011)
	TGF-beta II receptor	Cell signaling		Liu et al. (2015)
SPSB2	iNOS	Innate host response	Chronic infections	Kuang et al. (2010)
SPSB3	SNAIL	Cell signaling		Liu et al. (2018)
SPSB4	iNOS	Innate host response	Chronic infections	Nishiya et al. (2011)
WD repeat-containing proteins with a SOCS box (WSB)				
WSB1	HIPK2	Cell signaling		Choi et al. (2008)
	D2	Metabolism		Dentice et al. (2005)
	pVHL	Ubiquitylation	Cancer metastasis	Kim et al. (2015)
	RhoGDI2	Cell signaling	Cancer metastasis	Cao et al. (2015)
	ATM	Senescence		Kim et al. (2017)
	LRRK2	Protein aggregation	Parkinson's disease	Nucifora et al. (2016)

(continued)

Table 16.1 (continued)

Substrate receptors	Substrates	Involving pathways	Potential links to diseases	Refs.
Other SOCS box proteins				
Rad40	Rap2	Cell signaling		Kamura et al. (2001)
Elongin A	Rab1	Transcription		Yasukawa et al. (2008)
Viral proteins with a SOCS box				
HIV-1 Vif	APOBEC3F	Antiviral activity	Viral infections	Liu et al. (2005)
	APOBEC3G	Antiviral activity	Viral infections	Sheehy et al. (2003)
E4rf6/E1B55K				
	p53	Cell cycle, apoptosis	Viral infections	Steegenga et al. (1998)
	DNA ligase IV	DNA repair		Baker et al. (2007)
	Mre11	DNA repair		Stracker et al. (2002)
	Integrin a3	Cell motility		Dallaire et al. (2009)
	Rep52	Viral DNA replication		Nayak et al. (2008)
BLZF1	p53	Cell cycle, apoptosis	Viral infections	Sato et al. (2009)
LANA	p53	Cell cycle, apoptosis	Viral infections	Cai et al. (2006)
	pVHL	Ubiquitylation		Cai et al. (2006)
MuHV-4 ORF73	p65/RelA	Transcription	Viral infections	Rodrigues et al. (2009)
Substrates of Cul5/E3 with unknown receptors				
Unknown	DEPTOR	Autophagy		Tan et al. (2016)
Unknown	NOXA	Apoptosis	Lung cancer	Zhou et al. (2017)
Unknown	TRAF6	Inflammation		Zhu et al. (2016)
Unknown	TRIAD1	Ubiquitylation	Hematopoiesis	Kelsall et al. (2013)

and ubiquitylation. It promotes Cul-3 neddylation to activate CRL3^{Keap1} E3 ligase; it also serves as E2 for ubiquitylation E3 Parkin/DJ-1, leading to ubiquitylation and degradation of UBE2F (Zhou et al. 2018b).

16.3.2 Neddylation and Ubiquitylation E3: SAG

SAG, also known as RBX2, ROC2, or RNF7, was first cloned by us as a redox-inducible protein (Duan et al. 1999), and it was later found to be the second family member of RBX RING protein with ligase activity when complexed with other CRL components (Swaroop et al. 2000). Human *SAG* gene consists of four exons and three introns, which is mapped onto chromosome 3q22-24 with three splicing variants and two family pseudogenes (Swaroop et al. 2001). Structurally, *SAG* encodes a protein of 113 amino acids, with 12 cysteine residues and a zinc-binding C3H2C3 RING finger domain at the C-terminus (Duan et al. 1999). This characteristic of SAG structure confers its dual functionalities. First, when acting alone, SAG has a nonenzymatic antioxidant activity to scavenge ROS at the expense of self-oligomerization via formation of intra- and intermolecule disulfide bonds (Swaroop et al. 2001). Second, when complex with other components of CRL, SAG possesses an intrinsic ubiquitin E3 ligase activity (Tan et al. 2010). Furthermore, when cooperating with UBE2F, SAG acts as a neddylation E3 to neddylate Cul-5 and activates CRL-5 (Huang et al. 2009).

16.3.2.1 Antioxidant Role of SAG

SAG has well-defined antioxidant activity by generating thiol/disulfide redox buffer and chelating metals with enriched cysteine residues. SAG forms oligomers by hydrogen peroxide, reversible by antioxidant dithiothreitol (DTT), or by the alkylating agent NEM, indicating that SAG oligomerization is induced by the formation of inter- or intramolecular disulfide bonds (Swaroop et al. 2001). SAG also binds to metal ions, such as zinc, iron, or copper, to inhibit the copper-induced lipid peroxidation in vitro (Duan et al. 1999; Swaroop et al. 2001). Importantly, the antioxidant protective role of SAG can also be expanded to the in vivo mouse models. Injection of SAG-expression adenovirus or purified cell-penetrable SAG protein (Tat-SAG) attenuates ischemia/oxidative stress-induced damages in the mouse brain (Kim et al. 2010; Yang et al. 2001).

16.3.2.2 SAG as a RING Component of SCF E3 Ubiquitin Ligase

Our earlier study found that purified SAG binds to Cul-1, and SAG-Cul1 complex promotes the formation of high molecular weight smears in a ligase reaction mixture containing ubiquitin, E1, E2, and ATP, indicative of polyubiquitylation in the E1- and E2-dependent manner (Swaroop et al. 2000). The polyubiquitylation activity of SAG-Cul1 is dependent on the RING structure of SAG, since SAG RING mutants completely abrogate this reaction. Furthermore, like the RBX1-Cul1 complex, the SAG-Cul1 complex promotes polyubiquitylation of phosphorylated IκB in an in vitro ubiquitylation assay (Tan et al. 2010). Under overexpressed conditions, both RBX1 and SAG are capable of binding to six members of the cullin proteins (Cul1-3, Cul-4A-B, and Cul-5) (Ohta et al. 1999), whereas under physiological conditions, RBX1 is preferentially associated with Cullin 1–4, whereas SAG is selectively to interact with Cul-5 (Kamura et al. 2004; Huang et al. 2009). These two RBX family members, along with seven cullins, many adaptors, and substrate receptors, assemble into numerous ubiquitin ligase complexes (Cardozo and Pagano 2004; Jia and Sun 2009), which are responsible for ubiquitylation of 20% cellular proteins doomed for proteasome degradation (Soucy et al. 2009).

Most recently, we found that SAG is a dual ubiquitin E3, capable of promoting substrate polyubiquitylation via both K48 and K11 linkages. Specifically, on one hand, SAG binds to UBCH5C E2 and couples with Cul-1 to promote polyubiquitylation of the substrates via the K48 linkage. On the other hand, SAG binds to UBE2C/2S E2 and couples with Cul-5 to promote polyubiquitylation of the substrates via the K11 linkage (Kuang et al. 2016) (Fig. 16.2), even though both are for targeted degradation via proteasome system. In contrast, RBX1, another SAG family member, is only capable of binding with various K48-linked E2s, but not K11-linked UBE2C/2S, for substrate polyubiquitylation via the K48 linkage (Kuang et al. 2016). This distinct difference in biochemical feature may explain why SAG and RBX1 are functionally non-redundant during mouse development (Tan et al. 2009, 2011).

16.3.2.3 SAG as a Neddylation E3 Ligase

NEDD8 E3 ligases catalyze the process of transferring the NEDD8 from the E2 conjugating enzyme onto the target substrate. Until now, all reported NEDD8 E3 ligases are capable of functioning as ubiquitin E3s (Enchev et al. 2015), with a majority of belonging to the RING domain-containing subclass (Deshaies and Joazeiro 2009). Two well-known NEDD8 E3 ligases are RBX1 and SAG. As mentioned before, RBX1 specifically couples with UBE2M E2 to promote neddylation of cullins 1–4, while SAG couples with UBE2F E2 to promote Cul-5 neddylation (Huang et al. 2009). Both the E3 RING domains and the UFD of NAE bind to the same surface on the NEDD8 E2 enzymes, resulting in toggling of relative affinities to ensure the unidirectionality of the neddylation process (Eletr et al. 2005). The interaction between the E3 RING domain and E2-bound NEDD8 is required for catalyzing the transfer of the NEDD8 molecule into a Lys residue or the N terminus of the target substrate. Furthermore, cullin neddylation by SAG or RBX1 is aided by E3 DCNLs (Enchev et al. 2015). A summary of SAG acting as a dual E3 for ubiquitylation and neddylation is shown in Fig. 16.2.

16.4 Substrates of CRL-5 E3 Ligase with Unknown Substrate Receptors

An array of CRL-5 substrates with corresponding receptor proteins are listed in Table 16.1. Several CRL-5 substrates with significant biological

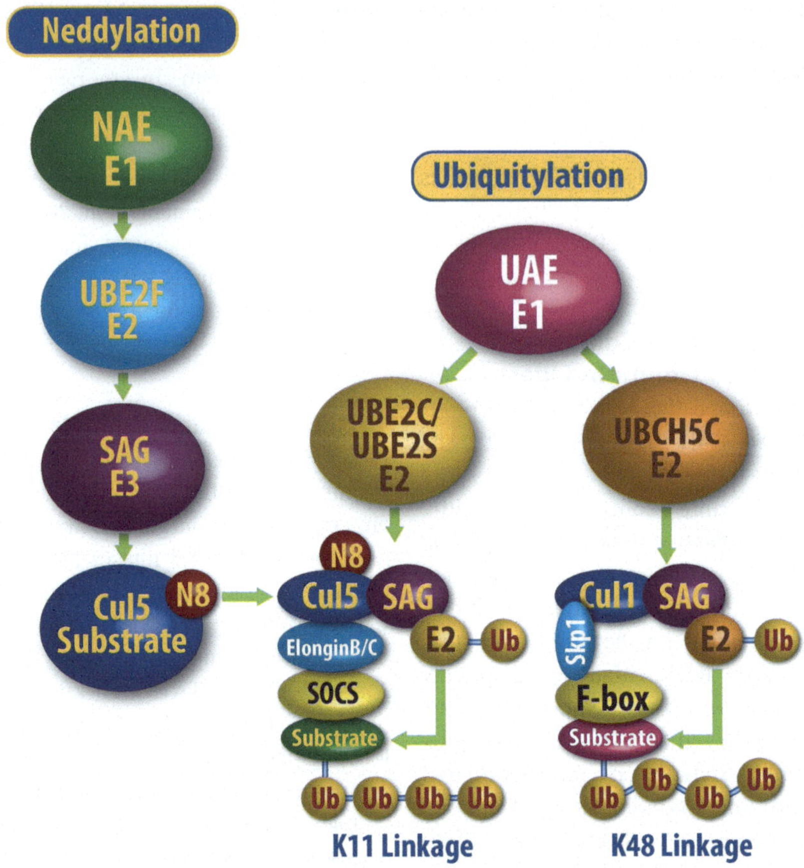

Fig. 16.2 Neddylation activation of Cul-5 and SAG-Cul5 E3 ligase activity. CRL-5 activation requires Cul-5 neddylation, which is catalyzed by neddylation E1 NAE, neddylation E2 UBE2F, and neddylation E3 SAG with Cul-5 as the substrate (left panel). Upon Cul-5 neddylation, SAG binds to Cul-5 via its N-terminus and ubiquitylation E2, UBE2C and UBE2S, via its RING domain, and acts as a ubiquitylation E3 to catalyze the ubiquitin transfer from E2 to a substrate and form polyubiquitylation chain via the K11 linkage. On the other hand, SAG complexes with Cul-1 to bind with E2 UBCH5C to catalyze polyubiquitylation of substrate via the K48 linkage (right panel). Thus, SAG is a dual E3 for both neddylation and ubiquitylation; and SAG binds to two types of E2 responsible for polyubiquitylation chains via the K11 and K48 linkage, respectively

functions, but unknown corresponding receptors, are reviewed in the following.

16.4.1 DEPTOR

DEPTOR is a naturally occurring inhibitor of mTORC1 and mTORC2 through a direct binding to mTOR and acts as a tumor suppressor in a context-dependent manner (Peterson et al. 2009). Our earlier study, along with two other groups, showed that DEPTOR accumulates in starvation conditions and contributes to autophagy induction, whereas upon stimulation by serum or growth factors, DEPTOR is phosphorylated and recognized by βTrCP, followed by ubiquitylation and degradation by SCF E3 ligase (also known as CRL1) to ensure mTOR activation (Zhao et al. 2011; Duan et al. 2011; Gao et al. 2011). Our later study showed that SAG can complex with either Cul-1 or Cul-5 to promote ubiquitylation and degradation of DEPTOR (Tan et al. 2016). Importantly, negative regulation of DEPTOR by SAG has biological consequences. In cell culture model, *SAG* knock-down suppresses growth, survival, and migration of human prostate cancer cells via inactivation of the PI3K/AKT/mTOR signaling axis through

DEPTOR accumulation; whereas in a mouse prostate cancer model, *Sag* deletion significantly inhibits prostate tumorigenesis triggered by *Pten* loss as a result of suppressed proliferation due to DEPTOR accumulation (Tan et al. 2016). Furthermore, another group has shown that the Cul5/Elongin B complex promotes ubiquitylation and degradation of DEPTOR to negatively regulate autophagy under nutrient-rich conditions (Antonioli et al. 2014).

16.4.2 Heat Shock Protein 90 (Hsp90) Client Proteins

Hsp90, a molecular chaperone with approximately 350 client proteins, is responsible for the correct folding of proteins, which facilitates proteins to attain their proper stabilization and activity (Taipale et al. 2012). Two studies showed that upon treatment of human cancer cells with the clinical HSP90 inhibitor 17-AAG, Cul5/SAG E3 are actively involved or required for degradation of several HSP90 clients, including ErbB2, HIF-1α, $BRAF^{V600E}$, AKT, and CDK4 (Ehrlich et al. 2009; Samant et al. 2014). Thus, it appears that Cul-5 E3 plays a role in regulation of the cellular response to HSP90 inhibition.

16.4.3 NOXA

NOXA is a pro-apoptotic member of Bcl-2 protein family, which can form hetero- or homodimers and act as pro-apoptotic regulator (Oda et al. 2000). Our recent studies showed that Cul5/SAG E3 targets NOXA for degradation to protect cancer cells from apoptosis, ensuring an apoptosis-escaping mechanism in lung cancer cells. Specifically, UBE2F NEDD8-E2 incorporates with SAG E3 to induce Cul-5 neddylation, leading to activation of CRL-5 E3 to promote NOXA polyubiquitylation via K11 linkage for proteasomal degradation (Jia et al. 2010; Zhou et al. 2017). Most recently, we further found that upon stress stimuli (e.g., hypoxia), UBE2M was induced and then formed a complex with Parkin/DJ-1 to promote UBE2F ubiquitylation and degradation, leading to Cul5/E3 inactivation and subsequent NOXA accumulation for apoptosis induction (Zhou et al. 2018b).

16.4.4 TRAF6

A recent study showed that Cul-5 directly binds to TRAF6 via the C-terminal domain of Cul-5 and the TRAF-C domain of TRAF6 and promotes TRAF6 polyubiquitylation via the K63 linkage in response to lipopolysaccharide (LPS) stimulation (Zhu et al. 2016). While homozygous deletion of *Cul-5* is embryonic lethal, heterozygous *Cul-5* deletion improves mouse survival and reduces proinflammatory cytokine production in response to LPS challenge due to reduced activation of NF-κB and MAPK signals (Zhu et al. 2016). Given that TRAF6 is an intrinsic E3 ligase, capable of self-poly-ubiquitylation, it remains to be determined whether TRAF6 is indeed a true Cul-5 substrate without involving a substrate receptor protein (Zhu et al. 2016).

Finally, the neddylated Cul5/SAG complex was shown to interact with and surprisingly enhances rather than inhibits the E3 ligase activity of TRIAD1 (two RING finger and double RING finger linked) (Kelsall et al. 2013), a distinct class of E3 ubiquitin ligases implicated in the process of hematopoiesis, mainly inhibiting myeloid colony formation (Marteijn et al. 2005).

16.5 Virus-Mediated Hijacking of CRL-5

Several strains of human virus were reported to hijack CRL-5 to promote ubiquitylation and degradation of a variety of host antiviral proteins. This unique feature makes CRL-5 E3 as a promising antivirus target for drug discovery efforts.

16.5.1 Human Immunodeficiency Virus-1 (HIV-1)

Apolipoprotein B editing complex 3G (APOBEC3G/A3G) is a potent anti-retroviral cytidine deaminase with a broad antiviral activity by inducing C to U mutations in the viral minus

DNA strand during reverse transcription, which causes the deleterious G to A mutations in the coding strand (Suspene et al. 2004). To overcome host antiviral protective system, the HIV Vif protein hijacks CRL-5 to promote ubiquitylation and degradation of A3G (Sheehy et al. 2003). The BC box and SOCS box of Vif are required for the interaction with Elongin B/C and Cul-5, respectively. The knockdown of *SAG*, but not its family member *RBX1*, impairs Vif-induced A3G degradation (Wang et al. 2015). Moreover, neddylation of Cul-5 by UBE2F/SAG is required for Vif-mediated degradation of A3G, since pharmacological inhibition of the NEDD8 E1 with MLN4924 or knockdown of *UBE2F* bypasses the effect of Vif, restoring the restriction potential of A3G (Stanley et al. 2012). A non-NEDD8ylatable mutant Cul-5(m) was also shown to inhibit Vif-induced ubiquitination and degradation of A3G (Yu et al. 2003). Notably, A3G without ubiquitylation is still degraded in a Vif-dependent manner, suggesting that the polyubiquitylation of Vif, rather than polyubiquitylation of AG3, serves as a vehicle to transport A3G into proteasomes for degradation (Dang et al. 2008). Another antiviral factor, APOBEC3F/A3F, was also degraded by HIV-1 Vif via hijacking CRL-5 (Liu et al. 2005).

16.5.2 Human Adenoviruses (HAdV)

E4orf6 is 34 kDa product from open reading frame 6 of human adenovirus early region 4 (E4) with three BC boxes. E4orf6 cooperates with the viral E1B55K protein product to form an E3 ubiquitin ligase with Cul-5 to reduce the level of the p53 via the proteasome pathway (Querido et al. 2001; Steegenga et al. 1998), by triggering Cul-5 localization from the cytoplasm to the nucleus and CRL-5 activation via facilitating neddylation (Guo et al. 2019). Furthermore, as a consequence of adenovirus infection, E4orf6/E1B55K-Cul5 complex was shown to promote the degradation of DNA ligase IV, an enzyme that plays a pivotal role in repairing of double-stranded DNA breaks (DSBs) by performing the joining step of the nonhomologous end-joining DNA repair system (NHEJ) (Baker et al. 2007). The E4orf6/E1B55K-Cul5 E3 ligase also promotes the degradation of Mre11, a member of the MRN DNA repair complex (Stracker et al. 2002). Furthermore, the de novo-expressed, preassembled capsid proteins and Rep52 are also degraded by E4orf6/E1B55K-Cul5 E3 (Nayak et al. 2008), and this degradative activity of E4Orf6 can be overcome by virus-associated RNA, thereby increasing the capsid proteins and Rep52 to the levels necessary for efficient virus production (Nayak and Pintel 2007). Finally, the E4orf6/E1B55K-Cul5 E3 ligase complex is involved in the degradation of $\alpha3$, a component of integrin $\alpha3\beta1$, which plays an important role in the regulation of cellular adhesion through the binding with a variety of extracellular matrix substrates, including bronectin, collagen, vitronectin, and laminins, thereby playing an important role in virus spread (Dallaire et al. 2009).

16.5.3 Epstein-Barr Virus (EBV)

EBV is a human γ-herpesvirus that is associated with several B cell and epithelial cell malignancies. BZLF1 (known as Zta, EB1, or ZEBRA) is a transcriptional transactivator that promotes an EBV lytic cycle cascade by inducing EBV early gene expression (Chevallier-Greco et al. 1986). Importantly, BZLF1 couples with Cul-5 to form an active ubiquitin ligase to promote ubiquitylation and degradation of p53, a required process for efficient viral propagation in the lytic replication stage (Sato et al. 2009).

16.5.4 Kaposi's Sarcoma-Associated Herpesvirus (KSHV)

The KSHV-encoded latency-associated nuclear antigen (LANA) complex was initially identified as a DNA binding, nuclear transcription factor that contributes to KSHV latent replication and regulates virus latency. LANA contains a putative SOCS box and can form a complex with Elongin B/C and Cul-5 for ubiquitination and degradation of pVHL and p53 tumor suppressor proteins (Cai et al. 2006).

16.5.5 Murid Herpesvirus-4 (MuHV-4)

MuHV-4 is a gamma herpesvirus that is genetically related to the human pathogens EBV and KSHV (Simas and Efstathiou 1998). The latency-associated protein ORF73 encoded by MuHV-4 is able to interact with Elongin C and Cul-5 to reconstitute an active E3 ubiquitin ligase to target the NF-κB family member p65/RelA for polyubiquitylation and subsequent proteasomal degradation. Such viral inhibition of NF-κB activity is critical for the establishment of a propitious environment for the maintenance of latent infection and progression of KSHV-associated tumors (Rodrigues et al. 2009).

16.6 The Physiological Role of Cul-5 in Human Cancers

The role of CRL-5 components in human cancers has been extensively reviewed. For example, the SOCS family members are extensively involved in inflammation and cancer, largely acting in suppression of proliferation (Inagaki-Ohara et al. 2013; Jiang et al. 2017). The RING component SAG is largely oncogenic (Sun and Li 2013), required for lung tumorigenesis triggered by *Kras* activation (Li et al. 2014) or for prostate tumorigenesis triggered by *Pten* loss (Tan et al. 2016), but it is tumor suppressive in skin tumorigenesis, triggered by *Kras* activation (Xie et al. 2015). This book chapter will only focus on the potential role of Cul-5 in human cancers.

Most studies imply that Cul-5 exerts a tumor suppressive role, as evidenced mainly by frequently downregulation of Cul-5 in various human cancers (Kalla et al. 2007; Fay et al. 2003; Xu et al. 2012; Devor et al. 2016; Tapia-Laliena et al. 2019). Specifically, in a cancer profiling array study, Cul-5 expression was approximately 2.2-fold lower in the breast cancer tissues versus the matched normal tissues (Fay et al. 2003). Ectopic Cul-5 overexpression significantly suppressed the proliferation of breast cancer cells (Burnatowska-Hledin et al. 2004), by inhibiting MAPK phosphorylation to reduce nuclear localization of estrogen receptor ER, eventually leading to suppression of growth of estrogen-dependent cells (Johnson et al. 2007). Conversely, inhibition of Cul-5 by microRNA-19a and microRNA-19b significantly promoted proliferation and invasion of cervical carcinoma cells and gastric cancer cells (Xu et al. 2012; Zhu et al. 2019). The siRNA-based *Cul-5* knockdown caused centriole overduplication and mitotic errors and also induced structural chromosomal damage in renal cell carcinoma (Tapia-Laliena et al. 2019). Furthermore, an immunohistochemistry staining study showed that Cul-5 expression is frequently lower in renal cell carcinoma and a reduced *Cul-5* expression or *Cul-5* deletion is associated with a significantly worse overall patient survival by the analysis of The Cancer Genome Atlas (TCGA) database (https://tcga-data.nci.nih.gov/tcga) (Tapia-Laliena et al. 2019). In small cell lung cancers (SCLC), genetic deletion of *Cul-5* or *SOCS3* by CRISPR/Ccas9 impaired CRL-5-mediated degradation of integrin β1, leading to stabilization of integrin β1 to activate the downstream focal adhesion kinase/SRC (FAK/SRC) signaling, eventually driving growth and metastasis of SCLC (Zhao et al. 2019). Moreover, low expression of Cul-5 and SOCS3 and relatively high expression of integrin β1 are significantly associated with worse patient survival, suggesting Cul-5 is a tumor suppressor (Zhao et al. 2019). However, in non-small lung cancers (NSCLC), CRL-5 specifically targets NOXA for polyubiquitylation via the K11 linkage and proteasomal degradation to inhibit apoptosis and increase the survival of lung cancer cells (Zhou et al. 2017). Furthermore, elevated expression of Cul-5, coupled with low expression of NOXA, was found to predict poor patient survival in NSCLC patients, suggesting Cul-5 has oncogenic activity (Zhou et al. 2017).

To better understand the expression pattern and prognostic value of Cul-5 in human cancers, we examined Cul-5 expression in 32 types of human cancers using the UALCAN, an easy-to-

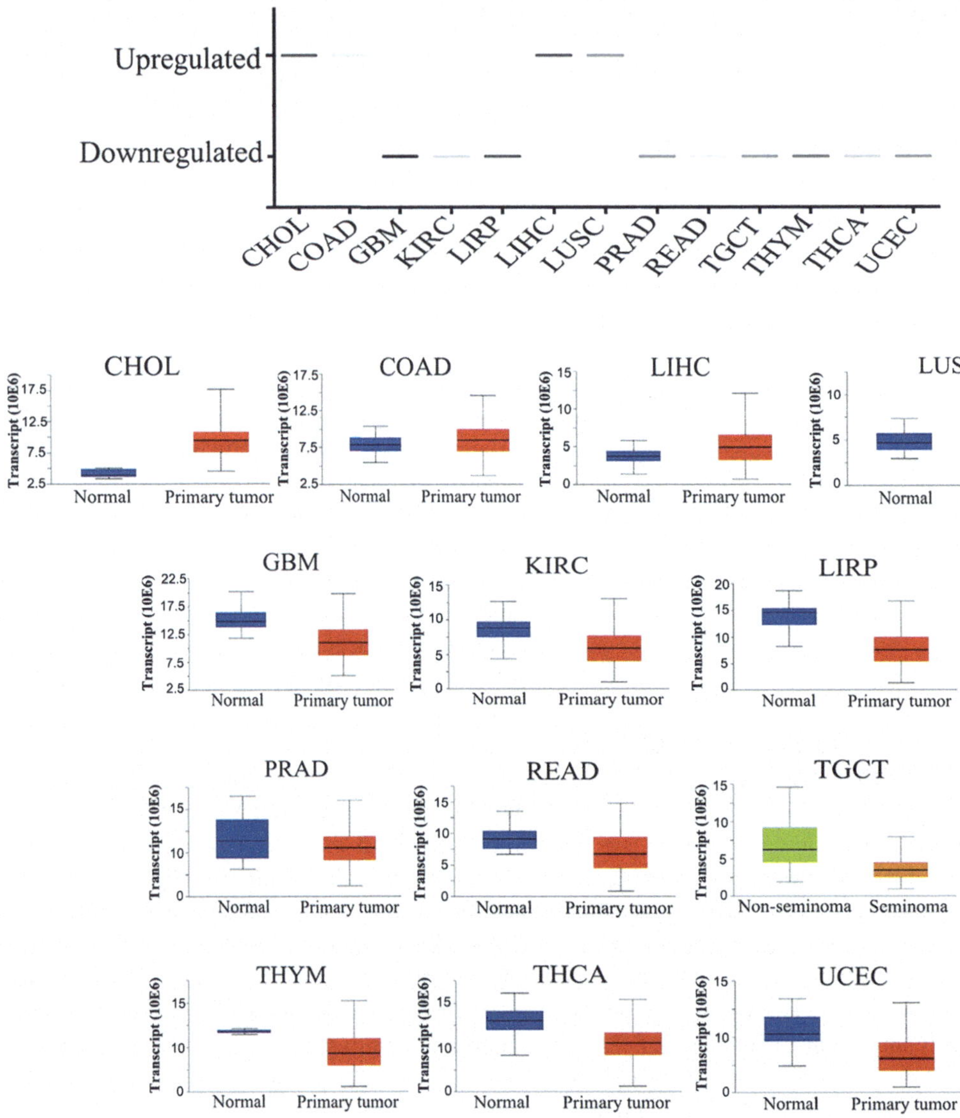

Fig. 16.3 Expression of Cul-5 mRNA in human cancer tissues. The search of the UALCAN database (http://ualcan.path.uab.edu) revealed that Cul-5 mRNA is upregulated in four types of human cancers and downregulated in nine types of human cancers. The abbreviations are as follows: *CHOL* cholangiocarcinoma, *COAD* colon adenocarcinoma, *LIHC* liver hepatocellular carcinoma, *LUSC* lung squamous cell carcinoma, *GBM* glioblastoma multiforme, *KIRC* kidney renal clear cell carcinoma, *KTRP* kidney renal papillary cell carcinoma, *PRAD* prostate adenocarcinoma, *READ* rectum adenocarcinoma, *TGCT* testicular germ cell tumors, *THYM* thymoma, *THCA* thyroid carcinoma, *UCEC* uterine corpus endometrial carcinoma

use, interactive web portal to perform in-depth analyses of TCGA gene expression data (http://ualcan.path.uab.edu/) (Chandrashekar et al. 2017). Notably, compared to normal tissues, Cul-5 mRNA is upregulated in four types, but downregulated in nine types of human cancers (Fig. 16.3). Kaplan-Meier analysis revealed that higher levels of Cul-5 mRNA are associated with a worse patient survival for liver hepatocellular carcinoma and kidney chromophobe, but a better patient survival for kidney renal clear cell carcinoma, prostate adenocarcinoma, and rectum

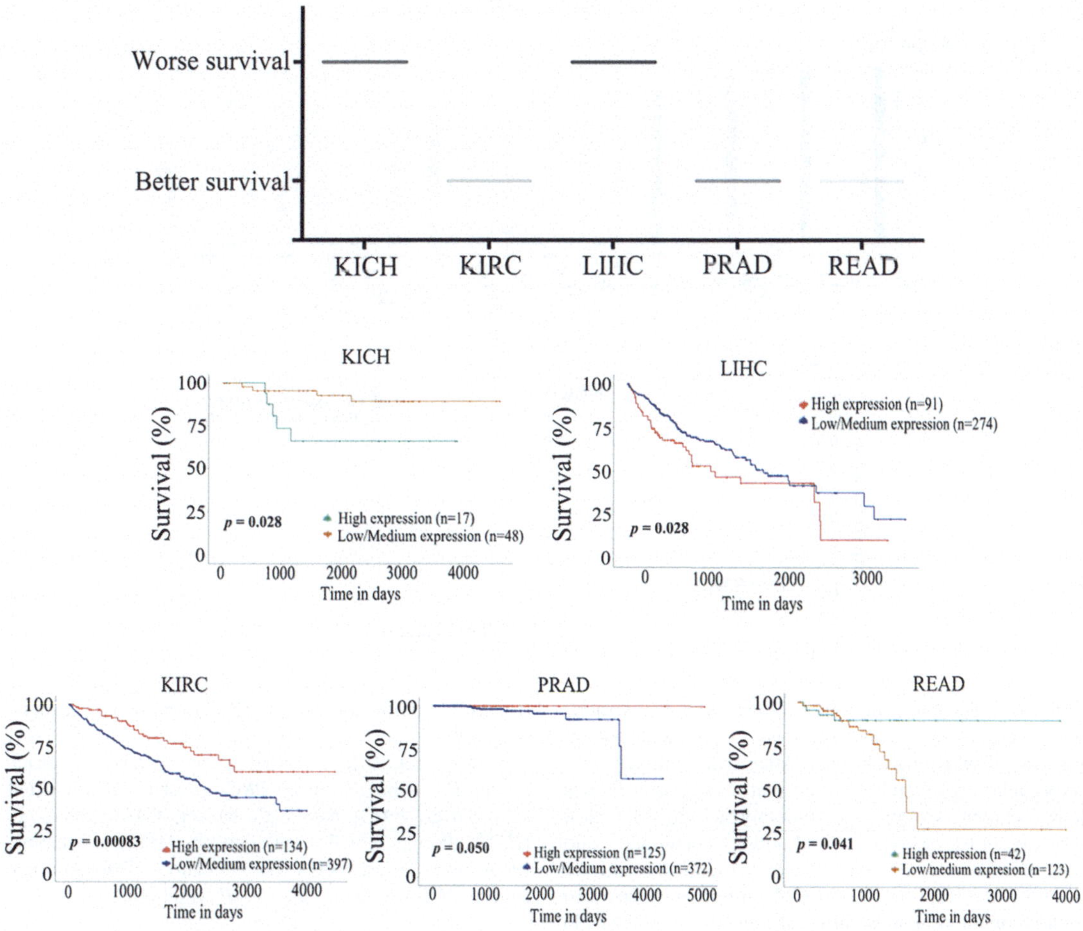

Fig. 16.4 Cul-5 mRNA levels in various human cancer tissues associated with patient survival. The UALCAN (http://ualcan.path.uab.edu) search also revealed that elevated levels of Cul-5 mRNA are associated with the worse survival for kidney chromophobe and liver hepatocellular carcinoma, but the better survival for kidney renal clear cell carcinoma, prostate adenocarcinoma, and rectum adenocarcinoma. *KICH* kidney chromophobe, *LIHC* liver hepatocellular carcinoma, *KIRC* kidney renal clear cell carcinoma, *PRAD* prostate adenocarcinoma, *READ* rectum adenocarcinoma

adenocarcinoma (Fig. 16.4). We further analyzed the significance of Cul-5 protein levels in human cancers using the database of the Human Protein Atlas (https://www.proteinatlas.org). Notably, Cul-5 staining is mainly detected at the moderate levels in the moderate cytoplasm in most of cancer tissues, and in more than 50% cases of lymphomas, melanomas, and lung, ovarian, and cervical cancers, the Cul-5 staining is weak or negative (Fig. 16.5a). At the protein levels, only one type of human cancer showed a statistically significant correlation between the expression and the patient survival. That is, the higher Cul-5 staining predicts a better prognosis in renal cancer patients (Fig. 16.5b).

16.7 Future Perspectives

The UBE2F-SAG-Cul-5 axis is exclusively presented in metazoans by phylogenetic analyses, whereas the UBE2M-RBX1-Cul1-4 axis appears in all eukaryotes; the UBE2F-SAG-Cul5 is, therefore, regarded as a distinct pathway in regulation

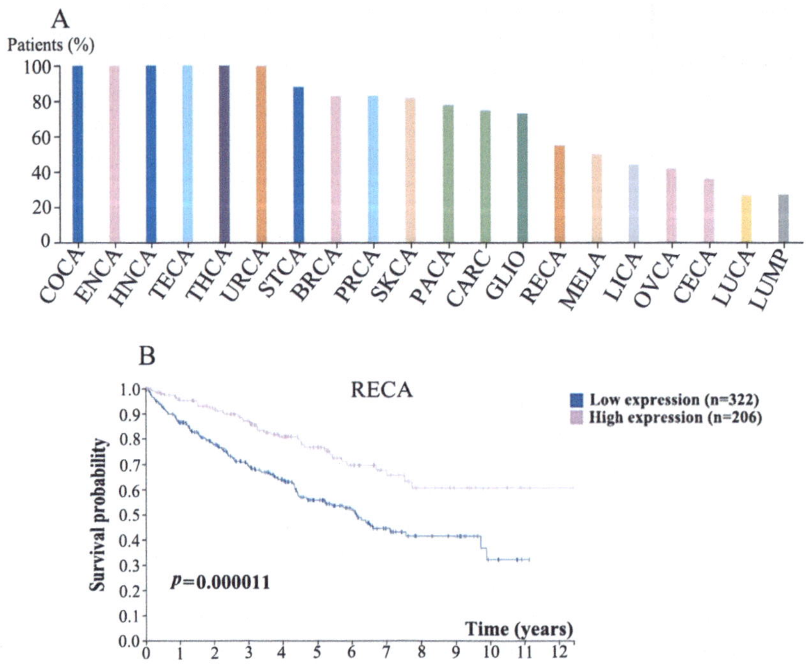

Fig. 16.5 The Cul-5 protein immunostaining in various human cancer tissues and their association with patient survival. The Human Protein Atlas data (https://www.proteinatlas.org) search revealed that most cancer tissues have moderate cytoplasmic immunoreactivity of Cul-5, whereas most cases of lymphomas, melanomas, and lung, ovarian, and cervical cancers are stained weakly or negative (**a**). High levels of Cul-5 protein are associated with favorable prognosis for renal cancer ($p = 0.000011$) (**b**). *COCA* colorectal cancer, *ENCA* endometrial cancer, *HNCA* head and neck cancer, *TECA* testis cancer, *THCA* thyroid cancer, *URCA* urothelial cancer, *STCA* stomach cancer, *BRCA* breast cancer, *PRCA* prostate cancer, *SKCA* skin cancer, *PACA* pancreatic cancer, *CARC* carcinoid, *GLIO* glioma, *RECA* renal cancer, *MELA* melanoma, *LICA* liver cancer, *OVCA* ovarian cancer, *CECA* cervical cancer, *LUCA* lung cancer, *LYMP* lymphoma

of CRL-5 E3 ligase activity (Huang et al. 2009). CRL-5 is likely involved in regulation of many biological processes, given that a variety of important signal molecules are its substrates (Table 16.1) for proteasomal degradation. However, many of reported studies were conducted in the cell culture setting under overexpressed conditions; the physiological relevance or significance is questionable. Furthermore, total *Cul-5* knockout is embryonic lethal (Zhu et al. 2016), whereas no study on tissue specific *Cul-5* knockout has been reported. The study of Cul-5 involvement in tumorigenesis under physiological setting is, therefore, lacking. The future studies focusing on the activity and functions of CRL-5 E3 ligase should be directed in the following aspects.

16.7.1 Generation of Conditional Mouse Models to Study the Role of Cul-5 in Tumorigenesis

It is urgent to generate and characterize conditional *Cul-5* knockout mice model to study its role in tumorigenesis in various organs triggered by oncogene activation or tumor suppressor inactivation induced by chemical carcinogens or radiation. Given all the data collected from human clinical tumor tissues with regard to Cul-5 expression levels and prognosis correlation only provide an association with the particular type(s) of cancer without elucidation of any cause-consequence relationship, these mouse studies will provide under the physiological settings whether *Cul-5* is a cooperative oncogene or tumor suppressor in a tissue-/context-dependent manner.

16.7.2 Generation of Conditional Mouse Models to Study the Roles of Each Receptor Subunit in Tumorigenesis

Among the substrate receptor families, there are many members (Fig. 16.1), and each member is capable of complexing with the rest three components to constitute a variety of CRL-5 E3s. This large array of E3s must play a variety of functions under certain physiological or pathological conditions or in response to environmental stresses, which has not been addressed in many current studies. One feasible approach is to generate conditional knockout mouse models for each receptor family member to fully understand their functions under the physiological settings or during tumorigenesis, triggered by either oncogene activation or tumor suppressor inactivation or by environmental insults.

16.7.3 Identification of Additional Substrates of CRL-5 E3s

Many members of CRL-5 receptors are the orphan receptor without corresponding substrates identified and characterized, which limited our full understanding of CRL-5 functions. Thus, the use of various current technologies and methodologies to identify and characterize additional downstream ubiquitin substrates, especially under biological significant and physiological relevant settings, is much needed.

16.7.4 Dynamic Regulation of CRL-5 E3 Ligase Activity

The enzymatic activity of each of CRLs can be regulated at the level of the subunit assembly (see earlier chapters of this book). The regulation of CRL assembly and activity is a very dynamic and precise process. The expression levels of some subunits fluctuate at particular cellular or developmental stages (e.g., at the different phases of cell cycle), which dictate the assembly of a given CRL (Bennett et al. 2010). Also the assembly of CRL is dependent upon CAND1 that can act as a subunit exchange factor (Bennett et al. 2010). Furthermore, the kinetics of CSN association with CRLs and the subsequent deneddylation are also subject to precise regulation. The thorough elucidation of these mechanism regulations, particularly for CRL-5, would lead to a better understanding of its biochemical activity and the consequent biological functions under the physiological or pathological conditions.

16.7.5 CRL-5 as an Attractive Antivirus Drug Target

Several infectious viruses are capable of hijacking CRL-5 to degrade host antiviral proteins, eventually obtaining anti-host property to propagate in the host cells. Inhibition of CRL-5 by MLN4924, a small molecular inhibitor of cullin neddylation, leading to general inactivation of all CRLs (Soucy et al. 2009), indeed showed potent antivirus activity (Stanley et al. 2012; Guo et al. 2019; Becker et al. 2019; Hughes et al. 2015; Kraus et al. 2017; Le-Trilling et al. 2016). Since a general inhibition of all CRLs by MLN4924 likely has some cytotoxic effect, the discovery of selective CRL-5 inhibitors as a novel class of antivirus therapeutic drugs would be an ideal approach.

References

Andresen CA, Smedegaard S, Sylvestersen KB, Svensson C, Iglesias-Gato D, Cazzamali G, Nielsen TK, Nielsen ML, Flores-Morales A (2014) Protein interaction screening for the ankyrin repeats and suppressor of cytokine signaling (SOCS) box (ASB) family identify Asb11 as a novel endoplasmic reticulum resident ubiquitin ligase. J Biol Chem 289(4): 2043–2054

Antonioli M, Albiero F, Nazio F, Vescovo T, Perdomo AB, Corazzari M, Marsella C, Piselli P, Gretzmeier C, Dengjel J, Cecconi F, Piacentini M, Fimia GM (2014) AMBRA1 interplay with cullin E3 ubiquitin ligases

regulates autophagy dynamics. Dev Cell 31(6): 734–746
Baker A, Rohleder KJ, Hanakahi LA, Ketner G (2007) Adenovirus E4 34k and E1b 55k oncoproteins target host DNA ligase IV for proteasomal degradation. J Virol 81(13):7034–7040
Becker T, Le-Trilling VTK, Trilling M (2019) Cellular cullin RING ubiquitin ligases: druggable host dependency factors of cytomegaloviruses. Int J Mol Sci 20 (7)
Bennett EJ, Rush J, Gygi SP, Harper JW (2010) Dynamics of cullin-RING ubiquitin ligase network revealed by systematic quantitative proteomics. Cell 143(6): 951–965
Brown JS, Lukashchuk N, Sczaniecka-Clift M, Britton S, le Sage C, Calsou P, Beli P, Galanty Y, Jackson SP (2015) Neddylation promotes ubiquitylation and release of Ku from DNA-damage sites. Cell Rep 11 (5):704–714
Burande CF, Heuze ML, Lamsoul I, Monsarrat B, Uttenweiler-Joseph S, Lutz PG (2009) A label-free quantitative proteomics strategy to identify E3 ubiquitin ligase substrates targeted to proteasome degradation. Mol Cell Proteomics 8(7):1719–1727
Burnatowska-Hledin MA, Spielman WS, Smith WL, Shi P, Meyer JM, Dewitt DL (1995) Expression cloning of an AVP-activated, calcium-mobilizing receptor from rabbit kidney medulla. Am J Phys 268(6 Pt 2): F1198–F1210
Burnatowska-Hledin MA, Kossoris JB, Van Dort CJ, Shearer RL, Zhao P, Murrey DA, Abbott JL, Kan CE, Barney CC (2004) T47D breast cancer cell growth is inhibited by expression of VACM-1, a cul-5 gene. Biochem Biophys Res Commun 319(3):817–825
Byrd PJ, Stankovic T, McConville CM, Smith AD, Cooper PR, Taylor AM (1997) Identification and analysis of expression of human VACM-1, a cullin gene family member located on chromosome 11q22-23. Genome Res 7(1):71–75
Cai QL, Knight JS, Verma SC, Zald P, Robertson ES (2006) EC5S ubiquitin complex is recruited by KSHV latent antigen LANA for degradation of the VHL and p53 tumor suppressors. PLoS Pathog 2(10): e116
Cao J, Wang Y, Dong R, Lin G, Zhang N, Wang J, Lin N, Gu Y, Ding L, Ying M, He Q, Yang B (2015) Hypoxia-induced WSB1 promotes the metastatic potential of osteosarcoma cells. Cancer Res 75(22): 4839–4851
Cardozo T, Pagano M (2004) The SCF ubiquitin ligase: insights into a molecular machine. Nat Rev Mol Cell Biol 5(9):739–751
Chandrashekar DS, Bashel B, Balasubramanya SAH, Creighton CJ, Ponce-Rodriguez I, Chakravarthi B, Varambally S (2017) UALCAN: a portal for facilitating tumor subgroup gene expression and survival analyses. Neoplasia 19(8):649–658
Chevallier-Greco A, Manet E, Chavrier P, Mosnier C, Daillie J, Sergeant A (1986) Both Epstein-Barr virus (EBV)-encoded trans-acting factors, EB1 and EB2, are required to activate transcription from an EBV early promoter. EMBO J 5(12):3243–3249
Choi DW, Seo YM, Kim EA, Sung KS, Ahn JW, Park SJ, Lee SR, Choi CY (2008) Ubiquitination and degradation of homeodomain-interacting protein kinase 2 by WD40 repeat/SOCS box protein WSB-1. J Biol Chem 283(8):4682–4689
Choi YB, Son M, Park M, Shin J, Yun Y (2010) SOCS-6 negatively regulates T cell activation through targeting p56lck to proteasomal degradation. J Biol Chem 285 (10):7271–7280
Chung AS, Guan YJ, Yuan ZL, Albina JE, Chin YE (2005) Ankyrin repeat and SOCS box 3 (ASB3) mediates ubiquitination and degradation of tumor necrosis factor receptor II. Mol Cell Biol 25(11):4716–4726
Cukras S, Morffy N, Ohn T, Kee Y (2014) Inactivating UBE2M impacts the DNA damage response and genome integrity involving multiple cullin ligases. PLoS One 9(7):e101844
Dallaire F, Blanchette P, Groitl P, Dobner T, Branton PE (2009) Identification of integrin alpha3 as a new substrate of the adenovirus E4orf6/E1B 55-kilodalton E3 ubiquitin ligase complex. J Virol 83(11):5329–5338
Dang Y, Siew LM, Zheng YH (2008) APOBEC3G is degraded by the proteasomal pathway in a Vif-dependent manner without being polyubiquitylated. J Biol Chem 283(19):13124–13131
De Sepulveda P, Ilangumaran S, Rottapel R (2000) Suppressor of cytokine signaling-1 inhibits VAV function through protein degradation. J Biol Chem 275 (19):14005–14008
Debrincat MA, Zhang JG, Willson TA, Silke J, Connolly LM, Simpson RJ, Alexander WS, Nicola NA, Kile BT, Hilton DJ (2007) Ankyrin repeat and suppressors of cytokine signaling box protein asb-9 targets creatine kinase B for degradation. J Biol Chem 282(7): 4728–4737
Dentice M, Bandyopadhyay A, Gereben B, Callebaut I, Christoffolete MA, Kim BW, Nissim S, Mornon JP, Zavacki AM, Zeold A, Capelo LP, Curcio-Morelli C, Ribeiro R, Harney JW, Tabin CJ, Bianco AC (2005) The Hedgehog-inducible ubiquitin ligase subunit WSB-1 modulates thyroid hormone activation and PTHrP secretion in the developing growth plate. Nat Cell Biol 7(7):698–705
Deshaies RJ, Joazeiro CA (2009) RING domain E3 ubiquitin ligases. Annu Rev Biochem 78:399–434
Devor EJ, Schickling BM, Reyes HD, Warrier A, Lindsay B, Goodheart MJ, Santillan DA, Leslie KK (2016) Cullin-5, a ubiquitin ligase scaffold protein, is significantly underexpressed in endometrial adenocarcinomas and is a target of miR-182. Oncol Rep 35(4):2461–2465
Diks SH, Sartori MA, da Silva JL, Hillebrands RJ, Bink HHV, van Rooijen C, Brouwers A, Chitnis AB, Peppelenbosch MP, Zivkovic D (2008) d-Asb11 is an essential mediator of canonical Delta-Notch signalling. Nat Cell Biol 10(10):1190–1198

Duan H, Wang Y, Aviram M, Swaroop M, Loo JA, Bian J, Tian Y, Mueller T, Bisgaier CL, Sun Y (1999) SAG, a novel zinc RING finger protein that protects cells from apoptosis induced by redox agents. Mol Cell Biol 19 (4):3145–3155

Duan S, Skaar JR, Kuchay S, Toschi A, Kanarek N, Ben-Neriah Y, Pagano M (2011) mTOR generates an auto-amplification loop by triggering the betaTrCP- and CK1alpha-dependent degradation of DEPTOR. Mol Cell 44(2):317–324

Duda DM, Borg LA, Scott DC, Hunt HW, Hammel M, Schulman BA (2008) Structural insights into NEDD8 activation of cullin-RING ligases: conformational control of conjugation. Cell 134(6):995–1006

Ehrlich ES, Wang T, Luo K, Xiao Z, Niewiadomska AM, Martinez T, Xu W, Neckers L, Yu XF (2009) Regulation of Hsp90 client proteins by a Cullin5-RING E3 ubiquitin ligase. Proc Natl Acad Sci U S A 106(48): 20330–20335

Eletr ZM, Huang DT, Duda DM, Schulman BA, Kuhlman B (2005) E2 conjugating enzymes must disengage from their E1 enzymes before E3-dependent ubiquitin and ubiquitin-like transfer. Nat Struct Mol Biol 12(10): 933–934

Enchev RI, Schulman BA, Peter M (2015) Protein neddylation: beyond cullin-RING ligases. Nat Rev Mol Cell Biol 16(1):30–44

Endo TA, Masuhara M, Yokouchi M, Suzuki R, Sakamoto H, Mitsui K, Matsumoto A, Tanimura S, Ohtsubo M, Misawa H, Miyazaki T, Leonor N, Taniguchi T, Fujita T, Kanakura Y, Komiya S, Yoshimura A (1997) A new protein containing an SH2 domain that inhibits JAK kinases. Nature 387 (6636):921–924

Fay MJ, Longo KA, Karathanasis GA, Shope DM, Mandernach CJ, Leong JR, Hicks A, Pherson K, Husain A (2003) Analysis of CUL-5 expression in breast epithelial cells, breast cancer cell lines, normal tissues and tumor tissues. Mol Cancer 2:40

Frantsve J, Schwaller J, Sternberg DW, Kutok J, Gilliland DG (2001) Socs-1 inhibits TEL-JAK2-mediated transformation of hematopoietic cells through inhibition of JAK2 kinase activity and induction of proteasome-mediated degradation. Mol Cell Biol 21(10):3547–3557

Gao D, Inuzuka H, Tan MK, Fukushima H, Locasale JW, Liu P, Wan L, Zhai B, Chin YR, Shaik S, Lyssiotis CA, Gygi SP, Toker A, Cantley LC, Asara JM, Harper JW, Wei W (2011) mTOR drives its own activation via SCF(betaTrCP)-dependent degradation of the mTOR inhibitor DEPTOR. Mol Cell 44(2):290–303

Guo H, Shen S, Li Y, Bi R, Zhang N, Zheng W, Deng Y, Yang Y, Yu XF, Wang C, Wei W (2019) Adenovirus oncoprotein E4orf6 triggers Cullin5 neddylation to activate the CLR5 E3 ligase for p53 degradation. Biochem Biophys Res Commun 516(4):1242–1247

Heuze ML, Lamsoul I, Baldassarre M, Lad Y, Leveque S, Razinia Z, Moog-Lutz C, Calderwood DA, Lutz PG (2008) ASB2 targets filamins A and B to proteasomal degradation. Blood 112(13):5130–5140

Hilton DJ, Richardson RT, Alexander WS, Viney EM, Willson TA, Sprigg NS, Starr R, Nicholson SE, Metcalf D, Nicola NA (1998) Twenty proteins containing a C-terminal SOCS box form five structural classes. Proc Natl Acad Sci U S A 95(1):114–119

Huang DT, Miller DW, Mathew R, Cassell R, Holton JM, Roussel MF, Schulman BA (2004) A unique E1-E2 interaction required for optimal conjugation of the ubiquitin-like protein NEDD8. Nat Struct Mol Biol 11(10):927–935

Huang DT, Ayrault O, Hunt HW, Taherbhoy AM, Duda DM, Scott DC, Borg LA, Neale G, Murray PJ, Roussel MF, Schulman BA (2009) E2-RING expansion of the NEDD8 cascade confers specificity to cullin modification. Mol Cell 33(4):483–495

Hughes DJ, Wood JJ, Jackson BR, Baquero-Perez B, Whitehouse A (2015) NEDDylation is essential for Kaposi's sarcoma-associated herpesvirus latency and lytic reactivation and represents a novel anti-KSHV target. PLoS Pathog 11(3):e1004771

Inagaki-Ohara K, Kondo T, Ito M, Yoshimura A (2013) SOCS, inflammation, and cancer. JAKSTAT 2(3): e24053

Jia L, Sun Y (2009) RBX1/ROC1-SCF E3 ubiquitin ligase is required for mouse embryogenesis and cancer cell survival. Cell Div 4:16

Jia L, Yang J, Hao X, Zheng M, He H, Xiong X, Xu L, Sun Y (2010) Validation of SAG/RBX2/ROC2 E3 ubiquitin ligase as an anticancer and radiosensitizing target. Clin Cancer Res 16(3):814–824

Jiang M, Zhang WW, Liu P, Yu W, Liu T, Yu J (2017) Dysregulation of SOCS-mediated negative feedback of cytokine signaling in carcinogenesis and its significance in cancer treatment. Front Immunol 8:70

Johnson AE, Le IP, Buchwalter A, Burnatowska-Hledin MA (2007) Estrogen-dependent growth and estrogen receptor (ER)-alpha concentration in T47D breast cancer cells are inhibited by VACM-1, a cul 5 gene. Mol Cell Biochem 301(1–2):13–20

Kalla C, Scheuermann MO, Kube I, Schlotter M, Mertens D, Dohner H, Stilgenbauer S, Lichter P (2007) Analysis of 11q22-q23 deletion target genes in B-cell chronic lymphocytic leukaemia: evidence for a pathogenic role of NPAT, CUL5, and PPP2R1B. Eur J Cancer 43(8):1328–1335

Kamio M, Yoshida T, Ogata H, Douchi T, Nagata Y, Inoue M, Hasegawa M, Yonemitsu Y, Yoshimura A (2004) SOCS1 [corrected] inhibits HPV-E7-mediated transformation by inducing degradation of E7 protein. Oncogene 23(17):3107–3115

Kamizono S, Hanada T, Yasukawa H, Minoguchi S, Kato R, Minoguchi M, Hattori K, Hatakeyama S, Yada M, Morita S, Kitamura T, Kato H, Nakayama K, Yoshimura A (2001) The SOCS box of SOCS-1 accelerates ubiquitin-dependent proteolysis of TEL-JAK2. J Biol Chem 276(16):12530–12538

Kamura T, Burian D, Yan Q, Schmidt SL, Lane WS, Querido E, Branton PE, Shilatifard A, Conaway RC, Conaway JW (2001) Muf1, a novel Elongin

BC-interacting leucine-rich repeat protein that can assemble with Cul5 and Rbx1 to reconstitute a ubiquitin ligase. J Biol Chem 276(32):29748–29753

Kamura T, Maenaka K, Kotoshiba S, Matsumoto M, Kohda D, Conaway RC, Conaway JW, Nakayama KI (2004) VHL-box and SOCS-box domains determine binding specificity for Cul2-Rbx1 and Cul5-Rbx2 modules of ubiquitin ligases. Genes Dev 18(24): 3055–3065

Kazi JU, Kabir NN, Flores-Morales A, Ronnstrand L (2014) SOCS proteins in regulation of receptor tyrosine kinase signaling. Cell Mol Life Sci 71(17): 3297–3310

Kelleher DJ, Kreibich G, Gilmore R (1992) Oligosaccharyltransferase activity is associated with a protein complex composed of ribophorins I and II and a 48 kd protein. Cell 69(1):55–65

Keller KE, Wirtz MK (2017) Working your SOCS off: the role of ASB10 and protein degradation pathways in glaucoma. Exp Eye Res 158:154–160

Kelsall IR, Duda DM, Olszewski JL, Hofmann K, Knebel A, Langevin F, Wood N, Wightman M, Schulman BA, Alpi AF (2013) TRIAD1 and HHARI bind to and are activated by distinct neddylated Cullin-RING ligase complexes. EMBO J 32(21):2848–2860

Kile BT, Metcalf D, Mifsud S, DiRago L, Nicola NA, Hilton DJ, Alexander WS (2001) Functional analysis of Asb-1 using genetic modification in mice. Mol Cell Biol 21(18):6189–6197

Kile BT, Schulman BA, Alexander WS, Nicola NA, Martin HM, Hilton DJ (2002) The SOCS box: a tale of destruction and degradation. Trends Biochem Sci 27 (5):235–241

Kim DW, Lee SH, Jeong MS, Sohn EJ, Kim MJ, Jeong HJ, An JJ, Jang SH, Won MH, Hwang IK, Cho SW, Kang TC, Lee KS, Park J, Yoo KY, Eum WS, Choi SY (2010) Transduced Tat-SAG fusion protein protects against oxidative stress and brain ischemic insult. Free Radic Biol Med 48(7):969–977

Kim JJ, Lee SB, Jang J, Yi SY, Kim SH, Han SA, Lee JM, Tong SY, Vincelette ND, Gao B, Yin P, Evans D, Choi DW, Qin B, Liu T, Zhang H, Deng M, Jen J, Zhang J, Wang L, Lou Z (2015) WSB1 promotes tumor metastasis by inducing pVHL degradation. Genes Dev 29 (21):2244–2257

Kim JJ, Lee SB, Yi SY, Han SA, Kim SH, Lee JM, Tong SY, Yin P, Gao B, Zhang J, Lou Z (2017) WSB1 overcomes oncogene-induced senescence by targeting ATM for degradation. Cell Res 27(2):274–293

Kohroki J, Nishiyama T, Nakamura T, Masuho Y (2005) ASB proteins interact with Cullin5 and Rbx2 to form E3 ubiquitin ligase complexes. FEBS Lett 579(30): 6796–6802

Kraus RJ, Yu X, Cordes BA, Sathiamoorthi S, Iempridee T, Nawandar DM, Ma S, Romero-Masters JC, McChesney KG, Lin Z, Makielski KR, Lee DL, Lambert PF, Johannsen EC, Kenney SC, Mertz JE (2017) Hypoxia-inducible factor-1alpha plays roles in Epstein-Barr virus's natural life cycle and tumorigenesis by inducing lytic infection through direct binding to the immediate-early BZLF1 gene promoter. PLoS Pathog 13(6):e1006404

Kuang P, Tan M, Zhou W, Zhang Q, Sun Y (2016) SAG/RBX2 E3 ligase complexes with UBCH10 and UBE2S E2s to ubiquitylate beta-TrCP1 via K11-linkage for degradation. Sci Rep 6:37441

Kuang Z, Lewis RS, Curtis JM, Zhan Y, Saunders BM, Babon JJ, Kolesnik TB, Low A, Masters SL, Willson TA, Kedzierski L, Yao S, Handman E, Norton RS, Nicholson SE (2010) The SPRY domain-containing SOCS box protein SPSB2 targets iNOS for proteasomal degradation. J Cell Biol 190(1):129–141

Kwon S, Kim D, Rhee JW, Park JA, Kim DW, Kim DS, Lee Y, Kwon HJ (2010) ASB9 interacts with ubiquitous mitochondrial creatine kinase and inhibits mitochondrial function. BMC Biol 8:23

Lamsoul I, Uttenweiler-Joseph S, Moog-Lutz C, Lutz PG (2016) Cullin 5-RING E3 ubiquitin ligases, new therapeutic targets? Biochimie 122:339–347

Le-Trilling VT, Megger DA, Katschinski B, Landsberg CD, Ruckborn MU, Tao S, Krawczyk A, Bayer W, Drexler I, Tenbusch M, Sitek B, Trilling M (2016) Broad and potent antiviral activity of the NAE inhibitor MLN4924. Sci Rep 6:19977

Lewis RS, Kolesnik TB, Kuang Z, D'Cruz AA, Blewitt ME, Masters SL, Low A, Willson T, Norton RS, Nicholson SE (2011) TLR regulation of SPSB1 controls inducible nitric oxide synthase induction. J Immunol 187(7):3798–3805

Li H, Tan M, Jia L, Wei D, Zhao Y, Chen G, Xu J, Zhao L, Thomas D, Beer DG, Sun Y (2014) Inactivation of SAG/RBX2 E3 ubiquitin ligase suppresses KrasG12D-driven lung tumorigenesis. J Clin Invest 124(2):835–846

Li JY, Chai B, Zhang W, Wu X, Zhang C, Fritze D, Xia Z, Patterson C, Mulholland MW (2011) Ankyrin repeat and SOCS box containing protein 4 (Asb-4) colocalizes with insulin receptor substrate 4 (IRS4) in the hypothalamic neurons and mediates IRS4 degradation. BMC Neurosci 12:95

Lim SY, Tan AH, Ahmad-Annuar A, Klein C, Tan LCS, Rosales RL, Bhidayasiri R, Wu YR, Shang HF, Evans AH, Pal PK, Hattori N, Tan CT, Jeon B, Tan EK, Lang AE (2019) Parkinson's disease in the Western Pacific region. Lancet Neurol

Linossi EM, Nicholson SE (2015) Kinase inhibition, competitive binding and proteasomal degradation: resolving the molecular function of the suppressor of cytokine signaling (SOCS) proteins. Immunol Rev 266(1):123–133

Liu B, Sarkis PT, Luo K, Yu Y, Yu XF (2005) Regulation of Apobec3F and human immunodeficiency virus type 1 Vif by Vif-Cul5-ElonB/C E3 ubiquitin ligase. J Virol 79(15):9579–9587

Liu E, Cote JF, Vuori K (2003) Negative regulation of FAK signaling by SOCS proteins. EMBO J 22(19): 5036–5046

Liu S, Nheu T, Luwor R, Nicholson SE, Zhu HJ (2015) SPSB1, a novel negative regulator of the transforming

growth factor-beta signaling pathway targeting the type II receptor. J Biol Chem 290(29):17894–17908
Liu Y, Zhou H, Zhu R, Ding F, Li Y, Cao X, Liu Z (2018) SPSB3 targets SNAIL for degradation in GSK-3beta phosphorylation-dependent manner and regulates metastasis. Oncogene 37(6):768–776
Lowenstein CJ, Padalko E (2004) iNOS (NOS2) at a glance. J Cell Sci 117(Pt 14):2865–2867
Lyapina S, Cope G, Shevchenko A, Serino G, Tsuge T, Zhou C, Wolf DA, Wei N, Shevchenko A, Deshaies RJ (2001) Promotion of NEDD-CUL1 conjugate cleavage by COP9 signalosome. Science 292(5520):1382–1385
Mahrour N, Redwine WB, Florens L, Swanson SK, Martin-Brown S, Bradford WD, Staehling-Hampton K, Washburn MP, Conaway RC, Conaway JW (2008) Characterization of Cullin-box sequences that direct recruitment of Cul2-Rbx1 and Cul5-Rbx2 modules to Elongin BC-based ubiquitin ligases. J Biol Chem 283 (12):8005–8013
Mansell A, Smith R, Doyle SL, Gray P, Fenner JE, Crack PJ, Nicholson SE, Hilton DJ, O'Neill LA, Hertzog PJ (2006) Suppressor of cytokine signaling 1 negatively regulates Toll-like receptor signaling by mediating Mal degradation. Nat Immunol 7(2):148–155
Marteijn JA, van Emst L, Erpelinck-Verschueren CA, Nikoloski G, Menke A, de Witte T, Lowenberg B, Jansen JH, van der Reijden BA (2005) The E3 ubiquitin-protein ligase Triad1 inhibits clonogenic growth of primary myeloid progenitor cells. Blood 106(13):4114–4123
Matsumoto K, Nishiya T, Maekawa S, Horinouchi T, Ogasawara K, Uehara T, Miwa S (2011) The ECS (SPSB) E3 ubiquitin ligase is the master regulator of the lifetime of inducible nitric-oxide synthase. Biochem Biophys Res Commun 409(1):46–51
Monda JK, Scott DC, Miller DJ, Lydeard J, King D, Harper JW, Bennett EJ, Schulman BA (2013) Structural conservation of distinctive N-terminal acetylation-dependent interactions across a family of mammalian NEDD8 ligation enzymes. Structure 21 (1):42–53
Nayak R, Pintel DJ (2007) Positive and negative effects of adenovirus type 5 helper functions on adeno-associated virus type 5 (AAV5) protein accumulation govern AAV5 virus production. J Virol 81(5):2205–2212
Nayak R, Farris KD, Pintel DJ (2008) E4Orf6-E1B-55k-dependent degradation of de novo-generated adeno-associated virus type 5 Rep52 and capsid proteins employs a cullin 5-containing E3 ligase complex. J Virol 82(7):3803–3808
Nishiya T, Matsumoto K, Maekawa S, Kajita E, Horinouchi T, Fujimuro M, Ogasawara K, Uehara T, Miwa S (2011) Regulation of inducible nitric-oxide synthase by the SPRY domain- and SOCS box-containing proteins. J Biol Chem 286(11): 9009–9019
Nucifora FC Jr, Nucifora LG, Ng CH, Arbez N, Guo Y, Roby E, Shani V, Engelender S, Wei D, Wang XF, Li T, Moore DJ, Pletnikova O, Troncoso JC, Sawa A, Dawson TM, Smith W, Lim KL, Ross CA (2016) Ubiquitination via K27 and K29 chains signals aggregation and neuronal protection of LRRK2 by WSB1. Nat Commun 7:11792
Oda E, Ohki R, Murasawa H, Nemoto J, Shibue T, Yamashita T, Tokino T, Taniguchi T, Tanaka N (2000) Noxa, a BH3-only member of the Bcl-2 family and candidate mediator of p53-induced apoptosis. Science 288(5468):1053–1058
Ohta T, Michel JJ, Schottelius AJ, Xiong Y (1999) ROC1, a homolog of APC11, represents a family of cullin partners with an associated ubiquitin ligase activity. Mol Cell 3(4):535–541
Okumura F, Matsuzaki M, Nakatsukasa K, Kamura T (2012) The role of Elongin BC-containing ubiquitin ligases. Front Oncol 2:10
Orr SJ, Morgan NM, Elliott J, Burrows JF, Scott CJ, McVicar DW, Johnston JA (2007) CD33 responses are blocked by SOCS3 through accelerated proteasomal-mediated turnover. Blood 109(3): 1061–1068
Parrillas V, Martinez-Munoz L, Holgado BL, Kumar A, Cascio G, Lucas P, Rodriguez-Frade JM, Malumbres M, Carrera AC, van Wely KH, Mellado M (2013) Suppressor of cytokine signaling 1 blocks mitosis in human melanoma cells. Cell Mol Life Sci 70 (3):545–558
Perfetto L, Gherardini PF, Davey NE, Diella F, Helmer-Citterich M, Cesareni G (2013) Exploring the diversity of SPRY/B30.2-mediated interactions. Trends Biochem Sci 38(1):38–46
Peterson TR, Laplante M, Thoreen CC, Sancak Y, Kang SA, Kuehl WM, Gray NS, Sabatini DM (2009) DEPTOR is an mTOR inhibitor frequently overexpressed in multiple myeloma cells and required for their survival. Cell 137(5):873–886
Petroski MD, Deshaies RJ (2005) Function and regulation of cullin-RING ubiquitin ligases. Nat Rev Mol Cell Biol 6(1):9–20
Puca R, Nardinocchi L, Sacchi A, Rechavi G, Givol D, D'Orazi G (2009) HIPK2 modulates p53 activity towards pro-apoptotic transcription. Mol Cancer 8:85
Querido E, Blanchette P, Yan Q, Kamura T, Morrison M, Boivin D, Kaelin WG, Conaway RC, Conaway JW, Branton PE (2001) Degradation of p53 by adenovirus E4orf6 and E1B55K proteins occurs via a novel mechanism involving a Cullin-containing complex. Genes Dev 15(23):3104–3117
Rabut G, Peter M (2008) Function and regulation of protein neddylation. 'Protein modifications: beyond the usual suspects' review series. EMBO Rep 9 (10):969–976
Rodrigues L, Filipe J, Seldon MP, Fonseca L, Anrather J, Soares MP, Simas JP (2009) Termination of NF-kappaB activity through a gammaherpesvirus protein that assembles an EC5S ubiquitin-ligase. EMBO J 28(9):1283–1295
Rui L, Yuan M, Frantz D, Shoelson S, White MF (2002) SOCS-1 and SOCS-3 block insulin signaling by

ubiquitin-mediated degradation of IRS1 and IRS2. J Biol Chem 277(44):42394–42398
Ryo A, Suizu F, Yoshida Y, Perrem K, Liou YC, Wulf G, Rottapel R, Yamaoka S, Lu KP (2003) Regulation of NF-kappaB signaling by Pin1-dependent prolyl isomerization and ubiquitin-mediated proteolysis of p65/RelA. Mol Cell 12(6):1413–1426
Samant RS, Clarke PA, Workman P (2014) E3 ubiquitin ligase Cullin-5 modulates multiple molecular and cellular responses to heat shock protein 90 inhibition in human cancer cells. Proc Natl Acad Sci U S A 111(18): 6834–6839
Sarikas A, Hartmann T, Pan ZQ (2011) The cullin protein family. Genome Biol 12(4):220
Sato Y, Shirata N, Kudoh A, Iwahori S, Nakayama S, Murata T, Isomura H, Nishiyama Y, Tsurumi T (2009) Expression of Epstein-Barr virus BZLF1 immediate-early protein induces p53 degradation independent of MDM2, leading to repression of p53-mediated transcription. Virology 388(1):204–211
Sheehy AM, Gaddis NC, Malim MH (2003) The antiretroviral enzyme APOBEC3G is degraded by the proteasome in response to HIV-1 Vif. Nat Med 9(11): 1404–1407
Simas JP, Efstathiou S (1998) Murine gammaherpesvirus 68: a model for the study of gammaherpesvirus pathogenesis. Trends Microbiol 6(7):276–282
Simo S, Cooper JA (2013) Rbx2 regulates neuronal migration through different cullin 5-RING ligase adaptors. Dev Cell 27(4):399–411
Soucy TA, Smith PG, Milhollen MA, Berger AJ, Gavin JM, Adhikari S, Brownell JE, Burke KE, Cardin DP, Critchley S, Cullis CA, Doucette A, Garnsey JJ, Gaulin JL, Gershman RE, Lublinsky AR, McDonald A, Mizutani H, Narayanan U, Olhava EJ, Peluso S, Rezaei M, Sintchak MD, Talreja T, Thomas MP, Traore T, Vyskocil S, Weatherhead GS, Yu J, Zhang J, Dick LR, Claiborne CF, Rolfe M, Bolen JB, Langston SP (2009) An inhibitor of NEDD8-activating enzyme as a new approach to treat cancer. Nature 458 (7239):732–736
Stanley DJ, Bartholomeeusen K, Crosby DC, Kim DY, Kwon E, Yen L, Cartozo NC, Li M, Jager S, Mason-Herr J, Hayashi F, Yokoyama S, Krogan NJ, Harris RS, Peterlin BM, Gross JD (2012) Inhibition of a NEDD8 cascade restores restriction of HIV by APOBEC3G. PLoS Pathog 8(12):e1003085
Steegenga WT, Riteco N, Jochemsen AG, Fallaux FJ, Bos JL (1998) The large E1B protein together with the E4orf6 protein target p53 for active degradation in adenovirus infected cells. Oncogene 16(3):349–357
Stracker TH, Carson CT, Weitzman MD (2002) Adenovirus oncoproteins inactivate the Mre11-Rad50-NBS1 DNA repair complex. Nature 418(6895):348–352
Sun Y, Li H (2013) Functional characterization of SAG/RBX2/ROC2/RNF7, an antioxidant protein and an E3 ubiquitin ligase. Protein Cell 4(2):103–116
Suspene R, Sommer P, Henry M, Ferris S, Guetard D, Pochet S, Chester A, Navaratnam N, Wain-Hobson S, Vartanian JP (2004) APOBEC3G is a single-stranded DNA cytidine deaminase and functions independently of HIV reverse transcriptase. Nucleic Acids Res 32(8): 2421–2429
Swaroop M, Wang Y, Miller P, Duan H, Jatkoe T, Madore S, Sun Y (2000) Yeast homolog of human SAG/ROC2/Rbx2/Hrt2 is essential for cell growth, but not for germination: chip profiling implicates its role in cell cycle regulation. Oncogene 19:2855–2866
Swaroop M, Gosink M, Sun Y (2001) SAG/ROC2/Rbx2/Hrt2, a component of SCF E3 ubiquitin ligase: genomic structure, a splicing variant, and two family pseudogenes. DNA Cell Biol 20(7):425–434
Taipale M, Krykbaeva I, Koeva M, Kayatekin C, Westover KD, Karras GI, Lindquist S (2012) Quantitative analysis of HSP90-client interactions reveals principles of substrate recognition. Cell 150(5):987–1001
Tan M, Gu Q, He H, Pamarthy D, Semenza GL, Sun Y (2008) SAG/ROC2/RBX2 is a HIF-1 target gene that promotes HIF-1 alpha ubiquitination and degradation. Oncogene 27(10):1404–1411
Tan M, Davis SW, Saunders TL, Zhu Y, Sun Y (2009) RBX1/ROC1 disruption results in early embryonic lethality due to proliferation failure, partially rescued by simultaneous loss of p27. Proc Natl Acad Sci U S A 106(15):6203–6208
Tan M, Zhu Y, Kovacev J, Zhao Y, Pan ZQ, Spitz DR, Sun Y (2010) Disruption of Sag/Rbx2/Roc2 induces radiosensitization by increasing ROS levels and blocking NF-kappaB activation in mouse embryonic stem cells. Free Radic Biol Med 49(6):976–983
Tan M, Zhao Y, Kim SJ, Liu M, Jia L, Saunders TL, Zhu Y, Sun Y (2011) SAG/RBX2/ROC2 E3 ubiquitin ligase is essential for vascular and neural development by targeting NF1 for degradation. Dev Cell 21(6): 1062–1076
Tan M, Xu J, Siddiqui J, Feng F, Sun Y (2016) Depletion of SAG/RBX2 E3 ubiquitin ligase suppresses prostate tumorigenesis via inactivation of the PI3K/AKT/ mTOR axis. Mol Cancer 15(1):81
Tapia-Laliena MA, Korzeniewski N, Pena-Llopis S, Scholl C, Frohling S, Hohenfellner M, Duensing A, Duensing S (2019) Cullin 5 is a novel candidate tumor suppressor in renal cell carcinoma involved in the maintenance of genome stability. Oncogene 8(1):4
Teckchandani A, Laszlo GS, Simo S, Shah K, Pilling C, Strait AA, Cooper JA (2014) Cullin 5 destabilizes Cas to inhibit Src-dependent cell transformation. J Cell Sci 127(Pt 3):509–520
Tissir F, Goffinet AM (2003) Reelin and brain development. Nat Rev Neurosci 4(6):496–505
Townley-Tilson WH, Wu Y, Ferguson JE 3rd, Patterson C (2014) The ubiquitin ligase ASB4 promotes trophoblast differentiation through the degradation of ID2. PLoS One 9(2):e89451
Uematsu K, Okumura F, Tonogai S, Joo-Okumura A, Alemayehu DH, Nishikimi A, Fukui Y, Nakatsukasa K, Kamura T (2016) ASB7 regulates spindle dynamics and genome integrity by targeting

DDA3 for proteasomal degradation. J Cell Biol 215 (1):95–106

Ungureanu D, Saharinen P, Junttila I, Hilton DJ, Silvennoinen O (2002) Regulation of Jak2 through the ubiquitin-proteasome pathway involves phosphorylation of Jak2 on Y1007 and interaction with SOCS-1. Mol Cell Biol 22(10):3316–3326

Wang J, Muntean AG, Hess JL (2012) ECSASB2 mediates MLL degradation during hematopoietic differentiation. Blood 119(5):1151–1161

Wang X, Wang X, Wang W, Zhang J, Wang J, Wang C, Lv M, Zuo T, Liu D, Zhang H, Wu J, Yu B, Kong W, Wu H, Yu X (2015) Both Rbx1 and Rbx2 exhibit a functional role in the HIV-1 Vif-Cullin5 E3 ligase complex in vitro. Biochem Biophys Res Commun 461(4):624–629

Wilcox A, Katsanakis KD, Bheda F, Pillay TS (2004) Asb6, an adipocyte-specific ankyrin and SOCS box protein, interacts with APS to enable recruitment of elongins B and C to the insulin receptor signaling complex. J Biol Chem 279(37):38881–38888

Xie CM, Wei D, Zhao L, Marchetto S, Mei L, Borg JP, Sun Y (2015) Erbin is a novel substrate of the Sag-betaTrCP E3 ligase that regulates KrasG12D-induced skin tumorigenesis. J Cell Biol 209(5):721–737

Xu XM, Wang XB, Chen MM, Liu T, Li YX, Jia WH, Liu M, Li X, Tang H (2012) MicroRNA-19a and -19b regulate cervical carcinoma cell proliferation and invasion by targeting CUL5. Cancer Lett 322(2):148–158

Yang GY, Pang L, Ge HL, Tan M, Ye W, Liu XH, Huang FP, Wu DC, Che XM, Song Y, Wen R, Sun Y (2001) Attenuation of ischemia-induced mouse brain injury by SAG, a redox- inducible antioxidant protein. J Cereb Blood Flow Metab 21(6):722–733

Yasukawa T, Kamura T, Kitajima S, Conaway RC, Conaway JW, Aso T (2008) Mammalian Elongin A complex mediates DNA-damage-induced ubiquitylation and degradation of Rpb1. EMBO J 27 (24):3256–3266

Yu X, Yu Y, Liu B, Luo K, Kong W, Mao P, Yu XF (2003) Induction of APOBEC3G ubiquitination and degradation by an HIV-1 Vif-Cul5-SCF complex. Science 302(5647):1056–1060

Zhao G, Gong L, Su D, Jin Y, Guo C, Yue M, Yao S, Qin Z, Ye Y, Tang Y, Wu Q, Zhang J, Cui B, Ding Q, Huang H, Hu L, Chen Y, Zhang P, Hu G, Chen L, Wong KK, Gao D, Ji H (2019) Cullin5 deficiency promotes small-cell lung cancer metastasis by stabilizing integrin beta1. J Clin Invest 129(3):972–987

Zhao Y, Xiong X, Sun Y (2011) DEPTOR, an mTOR inhibitor, is a physiological substrate of SCF (betaTrCP) E3 ubiquitin ligase and regulates survival and autophagy. Mol Cell 44(2):304–316

Zhou L, Zhang W, Sun Y, Jia L (2018a) Protein neddylation and its alterations in human cancers for targeted therapy. Cell Signal 44:92–102

Zhou W, Xu J, Li H, Xu M, Chen ZJ, Wei W, Pan Z, Sun Y (2017) Neddylation E2 UBE2F promotes the survival of lung cancer cells by activating CRL5 to degrade NOXA via the K11 linkage. Clin Cancer Res 23(4):1104–1116

Zhou W, Xu J, Tan M, Li H, Li H, Wei W, Sun Y (2018b) UBE2M is a stress-inducible dual E2 for neddylation and ubiquitylation that promotes targeted degradation of UBE2F. Mol Cell 70(6):1008–1024.e6

Zhu Y, Li L, Hou D, Ouyang Y, Guo X, Wang Y, Li J, Gong K (2019) MicroRNA-19a regulates the proliferation, migration and invasion of human gastric cancer cells by targeting CUL5. Arch Biochem Biophys 662:93–100

Zhu Z, Wang L, Hao R, Zhao B, Sun L, Ye RD (2016) Cutting edge: a Cullin-5-TRAF6 interaction promotes TRAF6 polyubiquitination and lipopolysaccharide signaling. J Immunol 197(1):21–26

Cullin-RING E3 Ubiquitin Ligase 7 in Growth Control and Cancer

17

Zhen-Qiang Pan

Abstract

CRL7^{Fbxw8} is an E3 ubiquitin ligase complex, containing cullin7 (CUL7) as a scaffold, the F-box protein Fbxw8 as a substrate receptor, the Skp1 adaptor, and the ROC1/Rbx1 RING finger protein for working with E2 enzyme to facilitate ubiquitin transfer. This chapter provides an update on studies linking CRL7^{Fbxw8} to hereditary human growth retardation disease, as at least 64 *cul7* germ line mutations were found in patients with autosomal recessive 3-M syndrome. CRL7^{Fbxw8} interacts with two additional 3-M associated proteins OBSL1 and CCDC8, leading to subcellular localization of the E3 complex to regions including plasma membrane, centrosome, and Golgi. At least ten mammalian cellular proteins were identified or implicated as CRL7^{Fbxw8} substrates. Discussion focuses on the possible impact of CRL7^{Fbxw8}-mediated proteolytic or non-proteolytic pathways in growth control and cancer.

Keywords

E3 ubiquitin ligase · Cullin 7 · 3-M disease · Growth signaling

Z.-Q. Pan (✉)
Department of Oncological Sciences, Icahn School of Medicine at Mount Sinai, New York, NY, USA
e-mail: zhen-qiang.pan@mssm.edu

Abbreviation

3-M	An autosomal recessive disorder characterized by pre- and postnatal growth retardation and named after the initials of three researchers (Miller, McKusick, and Malvaux) who first identified the disease
Akt	Protein kinase B
CCDC8	Coiled-coil domain containing 8
CDK	Cyclin-dependent kinase
CRL	Cullin-RING E3 ubiquitin ligase
CRL7^{Fbxw8}	A member of CRL family that contains four proteins known as cullin7, Fbxw8, Skp1, and ROC1/Rbx1
CUL	Cullin
Grb2	Growth factor receptor-bound protein 2
HPK1	Hematopoietic progenitor kinase 1
IGF	Insulin-like growth factor
IRS	Insulin receptor substrate
IUGR	Intrauterine growth retardation
MAPK/ERK	Mitogen-activated protein kinase
MEK	Mitogen-activated protein kinase
mTOR	Mechanistic target of rapamycin
OBSL1	Obscurin-like 1
PI3-K	Phosphoinositide 3-kinase
Rheb	Ras homologue enriched in brain
S/Ser	Serine

Y. Sun et al. (eds.), *Cullin-RING Ligases and Protein Neddylation*, Advances in Experimental Medicine and Biology 1217, https://doi.org/10.1007/978-981-15-1025-0_17

S6K1	s6 kinase 1
SH2	Src homology 2
TSC1/2	Tuberous sclerosis 1/2 complex

17.1 The CRL7^{Fbxw8} Complex

The CRL7^{Fbxw8} complex is a member of Cullin-RING E3 ubiquitin ligase (CRL) family (Petroski and Deshaies 2005; Sarikas et al. 2011). CRL7^{Fbxw8} was originally isolated and identified by Dias and colleagues using biochemical affinity purification and mass spectrometry (Dias et al. 2002). It contains four subunits (Fig. 17.1a) including cullin 7 (CUL7, also known as KIAA0076, p185, or p193), the WD40 repeat-containing F-box protein Fbxw8 (also named Fbx29, Fbw6, or Fbw8), the adapter protein Skp1 (S-phase kinase-associated protein 1), and the RING (for Really Interesting New Gene) finger protein ROC1 (also termed Rbx1 or Hrt1). The core CRL7^{Fbxw8} composition (CUL7, Fbxw8, Skp1, and ROC1) was independently identified and reported by Arai et al. (2003).

CRL7^{Fbxw8} has two unique biochemical properties. First, CUL7 is an atypical cullin family protein (Fig. 17.1b). The primary function of CUL7 is to provide a molecular scaffold that organizes an E3 CRL complex. However, human CUL7 contains 1698 amino acids, a size more than double that of a canonical cullin molecule (CUL1-5). As elaborated below, CUL7 appears to comprise multiple protein-protein interaction domains, enabling a range of biochemical functions for both proteolytic and non-proteolytic activities. Secondly, while CUL7 assembles an SCF/CRL1-like complex, it exhibits a remarkable selectivity by interacting with Skp1-Fbxw8 predominantly. Since the initial report of the CRL7^{Fbxw8} complex by Dias et al. (2002) and Arai et al. (2003), many independent reports have confirmed the selective CUL7-Fbxw8 association (Bae et al. 2017; Kim et al. 2014; Kong et al. 2012; Li et al. 2017; Litterman et al. 2011; Okabe et al. 2006; Tsunematsu et al. 2006; Wang et al. 2014; Yan et al. 2014). Currently there are no high-resolution structural models of the CUL7-Fbxw8 interactions to help understand the molecular basis for the selectivity.

Despite abundant evidence for CRL7^{Fbxw8} as a dominant E3 form, it may not be the only type of subunit organization. There are independent reports demonstrating interactions between CUL7 and CUL1 (Tsunematsu et al. 2006) or CUL9 (also named PARC; Skaar et al. 2007; Li et al. 2014). Additional CUL7-based E3 complexes have been reported as well (Kong et al. 2019; Luo et al. 2019; Shah and Maddika 2018).

17.2 CRL7^{Fbxw8}, 3-M Syndrome, and E3 Subcellular Distribution

3-M is a human autosomal recessive growth retardation syndrome (reviewed by Clayton et al. 2012; Huber et al. 2011). It is characterized by small birth size and postnatal growth restriction associated with a range of minor anomalies (including a triangular-shaped face, flat cheeks, full lips, short chest, and prominent fleshy heels). In 2005 a landmark report from Huber and colleagues (Huber et al. 2005) first linked *cul7* germ line mutations to 3-M syndrome. This connection has since been confirmed and extended by many independent groups concerning patients worldwide (Al-Dosari et al. 2012; Dauber et al. 2013; Hasegawa et al. 2016; Hu et al. 2017; Lugli et al. 2016; Meazza et al. 2013; Simsek-Kiper et al. 2019). To date, 64 3-M-linked *cul7* mutations have been reported and can be readily accessed in the database of HGMD. These mutations span the entire CUL7 coding sequence. Many mutations are expected to disable CUL7 activity by mechanisms of mRNA decay or significant protein truncations. There are, however, substitution mutations that could help CUL7 structural activity relationship analysis. For instance, 3-M-derived missense mutation H1464P resides in the CUL7 cullin domain (Fig. 17.1b) and was shown to cause reduction of the E3 ligase activity (Huber et al. 2005). In addition, CUL7 mutation may be responsible for the 3-M-like, Yakuts short stature syndrome (Maksimova et al. 2007).

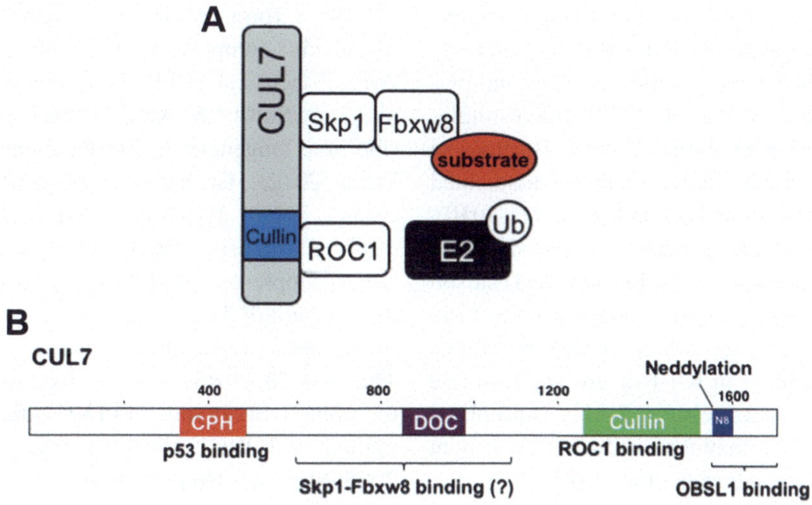

Fig. 17.1 (**a**) Organization of CRL7^{Fbxw8}. While Fbxw8 is a substrate receptor, ROC1 works with an E2 ubiquitin-conjugating enzyme for transferring ubiquitin to the bound substrate. (**b**) Domain organization of CUL7. The Fbxw8 binding site is derived from an early study (Huber et al. 2005). Future work is needed to precisely locate the binding interface. The requirement for the CUL7 C-terminus for binding to OBSL1 was shown by Litterman et al. (2011)

Subsequent studies have discovered that mutations in obscurin-like 1 (OBSL1) (Hanson et al. 2009; Huber et al. 2010) and coiled-coil domain containing 8 (CCDC8) (Hanson et al. 2011) contribute to 3-M syndrome as well. To date, 22 and 3 3-M-linked *obsl1* and *ccdc8* mutations, respectively, have been reported (HGMD). CUL7 appears to be the major gene responsible for 3-M syndrome. The prevalence of 3-M mutations was around 70% in CUL7, 20% in OBSL1, and below 10% in CCDC8 (Huber et al. 2011; Hanson et al. 2012).

CUL7, OBSL1, and CCDC8 appear to be associated physically as well. Early co-immunoprecipitation studies revealed interaction between CCDC8 and OBSL1 (Hanson et al. 2011). It was subsequently found that OBSL1 binds to CUL7 (Litterman et al. 2011). Yan et al. (2014) then provided evidence that CUL7, OBSL1, and CCDC8 are in a complex designated as 3-M complex that also include Fbxw8. OBSL1 is a cytoskeletal adaptor protein linking the internal cytoskeleton of cells to the cell membrane. CCDC8 contains multiple protein-protein interaction domains capable of interacting with OBSL1, then CRL7^{Fbxw8}, as well as additional proteins through its C-terminally located PxLPxL motif (Nie et al. 2015).

In a more recent study (Wang et al. 2019), Xiong and colleagues have shown that CCDC8 was localized on the plasma membrane exclusively. Phosphorylation of CCDC8 by CK2 and GSK3 enabled binding to OBSL1 and then CUL7, resulting in assembly of the membrane-associated 3-M complex. These authors further identified the plasma membrane protein LL5β as a substrate of 3-M complex. Inhibition of the CCDC8 phosphorylation by Wnt signaling caused disruption of membrane localization of the 3-M complex and accumulation of LL5β. Such defects were also observed in cells expressing CUL7 or OBSL1 carrying 3-M-derived mutations. Deletion of *Ccdc8* in mice caused defects in trophoblast migration and placental development and exhibited intrauterine growth restriction and perinatal lethality.

17.3 Role of CRL7^{Fbxw8} in Growth Control

The link of *cul7* mutations to human hereditary syndromes 3-M and Yakuts (Huber et al. 2005; Maksimova et al. 2007) strongly suggests a role for CRL7^{Fbxw8} in growth control. Consistent with human genetics evidence, targeted disruption of the *cul7* gene in mice resulted in severe intrauterine growth retardation (IUGR) with significantly smaller fetuses at later gestational stages and placenta anomalies (Arai et al. 2003). Interestingly, the CUL7 gene is upregulated up to 10 times in IUGR and 15 times in preeclampsia associated with IUGR (Gascoin-Lachambre et al. 2010). Dysregulation of the growth hormone signaling appears to be a feature of 3-M syndrome (Hanson et al. 2012). For example, fibroblast cells from 3-M patients carrying *cul7* or *ccdc8* mutations showed impairment in IGF1 or growth hormone signaling, respectively. 3-M fibroblasts containing *obsl1* mutations exhibited impairment in both pathways (Hanson et al. 2012).

Disruption of the *fbxw8* gene resulted in a less severe phenotype with abnormalities mainly restricted to the placenta and growth (Tsunematsu et al. 2006; Tsutsumi et al. 2008). Approximately 30% of the homozygous *fbxw8−/−* offspring reached adulthood, even though their body sizes were smaller than wild-type littermates throughout postnatal development. Thus, CUL7 and Fbxw8 have overlapping function in growth control, consistent with a hypothesis that CUL7 employs Fbxw8 to mediate proliferative activity. On the other hand, the more severe phenotype of the *cul7−/−* mice implicates Fbxw8-independent functions.

A few proteolytic mechanisms have been proposed to explain the role for CRL7$^{\mathrm{Fbxw8}}$ in growth control as summarized below.

IRS1 and mTORC1 Negative Feedback Loop Insulin, or insulin-like growth factor (IGF), stimulates growth by initiating binding to their receptors. The ligand-bound receptor tyrosine kinases then phosphorylate the insulin receptor substrate (IRS) such as IRS1 at multiple tyrosine residues. The resulting phospho-tyrosines provide docking sites capable of recruiting SH2 (Src homology 2)-containing signaling proteins that include PI3-K (phosphoinositide 3-kinase) and Grb2 (growth factor receptor-bound protein 2), thus activating the downstream Akt (protein kinase B; via *PI3-K*) and RAS (through Grb2) pathways, respectively. Activated Akt inhibits the TSC1/2 (tuberous sclerosis 1/2) complex, thereby liberating the small G-protein Rheb (Ras homologue enriched in brain). This leads to activation of mTORC1 (protein kinase mechanistic target of rapamycin complex 1; Laplante and Sabatini 2012; Zoncu et al. 2011) and its downstream effector kinase S6K1 (s6 kinase 1), resulting in elevated ribosome biogenesis and cell growth (Copps and White 2012; Harrington et al. 2005; Shah and Hunter 2005). Hyper-activated mTORC1/S6K1 catalyze multisite IRS1 seryl-phosphorylation, which suppresses IRS1's ability to interact with the insulin/IGF-1 receptors and promotes proteasomal degradation (Zhande et al. 2002). This mTORC1/IRS1 negative feedback attenuates the strength or duration of PI3-K activity to ensure optimal mTORC1 signaling (Harrington et al. 2005; Shah and Hunter 2005).

Several lines of evidence suggest a role for CRL7$^{\mathrm{Fbxw8}}$ in the mTORC1/ IRS1 negative feedback control by targeting IRS1 for ubiquitin-dependent degradation. Xu et al. (2008) have shown that Fbxw8 binds to IRS1 and promotes its ubiquitination and proteasomal degradation; inactivation/deletion of Fbxw8 and CUL7, respectively, accumulates IRS1. Importantly, Fbxw8-induced degradation of IRS1 depends on mTORC1 activity. In a support of these observations, embryonic fibroblasts of *cul7−/−* mice were found to accumulate IRS1 and exhibit increased activation of IRS1 downstream pathways Akt and MEK/ERK. It was proposed that hyper-activated mTORC1/S6K1 spark multisite seryl-phosphorylation of IRS1, triggering the binding of IRS1 to CRL7$^{\mathrm{Fbxw8}}$, resulting in IRS1 ubiquitination and degradation, and in turn causing attenuation of the PI3-K/Akt activities.

Additional biochemical (Xu et al. 2012) and physiological (Scheufele et al. 2014) evidences were provided in follow-up studies. IRS1 degradation signal sequence was mapped to its N-terminal 574 amino acid residues. Within this segment, Ser-307/Ser-312 and Ser-527 constitute S6K1 phosphorylation consensus sites, which were found indispensable for supporting CRL7$^{\mathrm{Fbxw8}}$-mediated degradation (Xu et al. 2012). Using in vitro reconstitution system, the ubiquitination of bacterially expressed IRS1 N-terminal fragment by CRL7$^{\mathrm{Fbxw8}}$ was stimulated by S6K1 albeit at low levels. In contrast, CRL7$^{\mathrm{Fbxw8}}$ supported efficient ubiquitination of IRS1 N-terminal fragment in hyper-phosphorylated form, which was isolated from infected insect cells. These data suggest requirement of additional phosphorylation by kinases yet to be identified. It was proposed that the requirement of multisite phosphorylation in the N terminus of IRS1 for its turnover might ensure that complete IRS1 degradation occurs only when mTORC1 and S6K1 reach exceedingly high levels. In addition, enhanced AKT and MAP kinase phosphorylation were observed in *cul7−/−* mouse embryonic fibroblasts upon insulin stimulation (Scheufele et al. 2014). Consistent with this, CUL7 knockdown by RNA interference in C2C12 myotubes led to elevated levels of insulin signaling pathways and cellular glucose uptake. The CUL7 depletion decreased the capacity of these cells to mediate insulin-induced degradation of IRS1. In mouse models, heterozygosity of either *cul7* or *fbxw8* elevated PI3-K/AKT activation in skeletal muscle tissue upon insulin stimulation when compared to the wild-type controls. Finally, enhanced insulin sensitivity and plasma glucose clearance were observed in *cul7+/−* or *fbxw8+/−* mice.

An independent investigation has revealed an mTORC2-dependent feedback inhibition of IRS1 by directly phosphorylating Fbxw8, resulting in enhanced stability of this F-box protein that promotes IRS1 degradation (Kim et al. 2012). Collectively, these studies have implicated roles for CRL7^{Fbxw8} in impacting both mTORC1 and mTORC2 signaling.

However, conflicting reports have appeared. Ponyeam and Hagen (2012) failed to observe accumulation of IRS1 in cells depleted of CUL7 although the phosphorylation status of IRS1 was not examined. More recently, Yoneyama et al. (2018) have identified human IRS1 S422 as a residue critical for phosphorylation by mTORC1 that appears to trigger interactions with SCF/CRL1^{ßTrCP} for degradation. Future work is needed, however, to provide evidence for direct binding of ßTrCP to the IRS1 S422 degron peptide, which differs significantly from the well-defined ßTrCP substrate-binding consensus motif.

TBC1D3 and Growth Factor Signaling Hominoid-specific TBC1D3 oncoprotein enhances growth factor receptor signaling and subsequently promotes cellular proliferation and survival. TBC1D3 is degraded in response to growth factor signaling, thereby constituting a growth factor-driven negative feedback loop (Kong et al. 2012). Multiple lines of evidence suggest that CRL7^{Fbxw8} targets TBC1D3 for ubiquitination and degradation in response to serum and growth factor stimulation.

Hippo Signaling and Cardiomyocyte Proliferation Using the cardiomyocyte model, it was revealed that inhibition of cardiomyocyte proliferation may be related to the accumulation of the Hippo kinases Mst1 and Lats1/2, suggesting a role for Hippo-YAP signaling in cardiac development. CUL7 was shown to be involved in controlling the abundance of Mst1 and therefore participates in Hippo-Yap signaling and cardiomyocyte proliferation (Zou et al. 2018).

17.4 Role of CRL7^{Fbxw8} in Cancer

p53 Kasper et al. (2006) and Andrews et al. (2006) reported the CUL7-p53 interactions that were mapped to the CUL7 CPH domain (Fig. 17.1b) and p53's tetramerization domain. A follow-up NMR study by Kaustov et al. (2007) provided high-resolution structural model for the interactions between CUL7's CPH and p53's tetramerization domains. Based on available evidence, the consequence of the CUL7-p53 interactions appears to antagonize p53's tumor suppressor activity. Cell culture studies have shown that CUL7 expression resulted in decrease of p53 transcription activity (Andrews et al. 2006), increase of the rate of cell proliferation in a manner that requires intact p53 (Andrews et al. 2006), and inhibition of p53 activation in response to DNA damage (Jung et al. 2007). Additional evidence includes the effects of CUL7 in suppressing Myc-induced apoptosis, although whether such an effect depends on the CUL7-p53 interactions has not been addressed (Kim et al. 2007). Thus far there is no evidence that CRL7^{Fbxw8} plays a role in modifying p53 by ubiquitin that leads to changes in p53 stability (Andrews et al. 2006; Jung et al. 2007).

SV40 T Antigen and Transformation CUL7 was originally identified by immunoprecipitation studies as a host cell protein p185 (Kohrman and Imperiale 1992) or p193 (Daud et al. 1993) that was associated with simian virus 40 large T antigen in early 1990s, long before its eventual recognition as a component of an E3 ubiquitin ligase complex (Arai et al. 2003; Dias et al. 2002). Studies by Decaprio and colleagues have mapped the CUL7 binding site to T antigen amino acids 69–83 (Kasper et al. 2005). Intriguingly, T antigen mutant defective in binding to CUL7, while still capable of interacting with p53 and pRb, was unable to induce proliferation in mouse embryo fibroblasts. These data suggest that the ability of T antigen to transform requires not only p53 and pRB but also inactivation of CUL7 activity. These results imply a role for CUL7 as a tumor suppressor, at least in the presence of the potent oncoprotein T antigen.

In an effort to substantiate these studies, Hartmann et al. (2014) have shown that wild-type T antigen, but not the mutant (Δ69–83) deficient in binding to CUL7, inhibited the degradation of the $CRL7^{Fbxw8}$ substrate IRS1 by the 26S proteasome. Accumulation and prolonged half-life of IRS1 were observed in cells expressing T antigen. Consistent with this, $CRL7^{Fbxw8}$-dependent IRS1 ubiquitination in vitro was inhibited by purified T antigen. Moreover, cells expressing T antigen, or depleted of CUL7 by RNA interference, showed enhanced activation of IRS1 downstream signaling pathways PI3-K/AKT and Erk mitogen-activated pathway kinase, as well as upregulation of the downstream target gene c-fos. Finally, elevated IRS1 protein levels and activation of downstream signaling were detected in T antigen-positive carcinoma of carcinoembryonic antigen 424/SV40 LT transgenic mice. Altogether, these results suggest a role for T antigen in protecting IRS1 from degradation by $CRL7^{Fbxw8}$. Such viral activity may play a role in sustaining high levels of pro-mitogenic IRS1 downstream signaling pathways.

Collectively, these studies may reconcile the CUL7 oncogene/tumor suppressor paradox (Sarikas et al. 2008). In normal cells, the mTORC1/IRS1 feedback functions to ensure proper mTOR signaling (Harrington et al. 2005; Shah and Hunter 2005). Loss of CUL7 results in sustained mTOR signaling, leading to senescence (Xu et al. 2008). This is in keeping with a role for $CRL7^{Fbxw8}$ in growth control. However, the potent oncoprotein T antigen commands high levels of cell proliferation. Breaking the mTORC1/IRS1 negative feedback loop by T antigen-mediated inhibition of IRS1 degradation may be necessary to sustain pro-mitogenic signaling, thereby meeting proliferative demands.

Cyclin D1 and Cell Cycle Cell cycle progression into S phase requires removing cyclin D1 through re-localization and degradation. Okabe et al. (2006) have provided evidence suggesting that sustained MAPK signaling, a feature unique to cancer cells, resulted in cyclin D1 phosphorylation at T286, which triggered interactions with $CRL7^{Fbxw8}$ leading to ubiquitin-dependent proteasomal degradation. Fbxw8 knockdown caused a significant accumulation of cyclin D1, as well as cytoplasmic sequestration of CDK1, leading to a severe reduction of cell proliferation. Constitutive nuclear expression of cyclin D1-T286A reversed these effects. These findings support a role for $CRL7^{Fbxw8}$ in cancer cell proliferation through proteolysis of cyclin D1. However, mouse embryonic fibroblasts (MEFs) from *fbxw8−/−* mice or the wild type showed similar rate of cyclin D1 degradation. These genetic analyses raised questions on a significant role for Fbxw8 in cyclin D1 degradation during normal cell cycle progression (Kanie et al. 2012).

HPK1, MAPK, and Pancreatic Cancer Hematopoietic progenitor kinase 1 (HPK1) inhibits MEK1/2-mediated ERK activation and is lost in >95% pancreatic cancer through proteasome-mediated degradation. HPK1 may function as a novel tumor suppressor, and loss of HPK1 plays a critical role in the development of pancreatic cancer. $CRL7^{Fbxw8}$ targets HPK1 for degradation in a manner that requires HPK1 autophosphorylation (Wang et al. 2014). Knockdown of Fbxw8 restores endogenous HPK1 protein expression and inhibits cell proliferation of pancreatic cancer cells. These findings suggest a role for CRL7^{Fbxw8} in constituting a negative feedback loop to restrain the growth-inhibitory activity of HPK1 and that CRL7^{Fbxw8} promotes pancreatic cancer cell proliferation.

CUL7, CUL9, and Microtubule Dynamics Yan et al. (2014) have linked CUL7, OBSL1, and CCDC8 to the control of microtubule dynamics. It was observed that CUL7 depletion results in altered microtubule dynamics, prometaphase arrest, tetraploidy, and mitotic cell death. Importantly, these defects were observed in CUL7 mutated 3-M cells as well and were rescued by expression of the wild-type CUL7, but not by 3-M-derived mutants. Similar defects were observed in cells depleted of OBSL1 or CCDC8. It was proposed that CUL7, OBSL1, and CCDC8 proteins form a 3-M complex that functions in maintaining microtubule, genome integrity, and normal development.

The CUL7/microtubule dynamics appears to be connected with cullin 9 (CUL9) (Li et al. 2014). *Cul9* null mice develop spontaneous tumors in multiple organs. It was observed that the microtubule and mitosis defects caused by knockdown of CUL7 or OBSL1 were rescued by depletion of CUL9. It was shown that CUL7 inhibits the CUL9-mediated ubiquitination and degradation of survivin. It was proposed that a 3M-CUL9-survivin pathway is critical for maintaining microtubule and genome integrity, normal development, and tumor suppression.

17.5 Role of CRL7^{Fbxw8} in Stem Cell Self Renewal

Nanog Nanog regulates human and mouse embryonic stem (ES) cell self-renewal activity. Activation of ERK signaling inhibits ES cell self-renewal and induces differentiation. It was shown that this inhibition is mediated by the ability of ERK1 to phosphorylate Nanog, which leads to binding to Fbxw8 and ubiquitination-mediated degradation (Kim et al. 2014).

OCT4 The POU transcription factor OCT4 is critical for maintaining the undifferentiated state of embryonic stem cells (ESCs) and generating induced pluripotent stem cells (iPSCs). It was observed that c-Jun N-terminal kinases (JNKs) directly phosphorylated OCT4 at serine 347, which triggered the binding of Fbxw8, leading to increased OCT4 proteasomal degradation (Bae et al. 2017).

17.6 Role of CRL7^{Fbxw8} in Neurons

Golgi Litterman et al. (2011) reported that CRL7^{Fbxw8} is Golgi associated as a result of CUL7-OBSL1 interactions. Inactivation of CRL7^{Fbxw8} through depletion of Fbxw8 impairs Golgi structure and function and dramatically inhibits the elaboration and growth of dendrites in primary neurons and in the developing rat cerebellum in vivo. CRL7^{Fbxw8} targets the Golgi protein Grasp65 for ubiquitination and degradation, thereby critically regulating the structural integrity and function of the Golgi apparatus and dendrite development in neurons.

Eag1 Potassium Channels Hsu et al. (2017) have observed interactions between CUL7 and rat Eag1, both of which appear to co-localize at synaptic regions in neurons. CUL7 appears to target endoplasmic reticulum- and plasma membrane-localized rat Eag1 to the proteasome and the lysosome, respectively, for protein degradation. These findings suggest a role for CUL7 in quality control of Eag1 channels.

17.7 Concluding Remarks

Since its discovery in 2002, CRL7^{Fbxw8} has been shown or implicated in control of the stability of more than ten protein substrates. Table 17.1 provides a summary of these substrates with key biological role(s) and identified/implicated kinase (s) involved in the proteolytic signaling. Note that six CRL7^{Fbxw8} substrates have played roles in growth control (Fig. 17.2). The discovery of two additional 3-M-linked proteins OBSL1 and

Table 17.1 CRL7^{Fbxw8} substrates

Substrates	Biological role	Kinase(s)	Reference
• IRS1	Insulin/IGF-1 signaling	S6K, mTORC1	Xu et al. (2008, 2012)
• Cyclin D1	Cell cycle progression	MAPK	Okabe et al. (2006)
• HPK1	Inhibition of MEK1/2-mediated activation of ERK	HPK1, autophosphorylation	Wang et al. (2014)
• TBC1D3	Growth factor receptor signaling, proliferation, and survival	Unknown	Kong et al. (2012)
• Mst1[a]	Hippo signaling		Zou et al. (2018)
• Nanog	Embryonic stem cell self-renewal	ERK1	Kim et al. (2014)
• OCT4	Embryonic stem cell self-renewal, cell differentiation	JNK	Bae et al. (2017)
• Grasp65	Neuronal morphogenesis, dendrite growth and patterning, Golgi function	Unknown	Litterman et al. (2011)
• Eag1[a]	Neuron, potassium (K+) channels		Hsu et al. (2017)
• HIV-1 Gag	HIV-1 assembly		Wei et al. (2015)
• MRFAP1	Anaphase-telophase transition, genomic stability	Unknown	Li et al. (2017)

[a]The abundance of these proteins was regulated by CUL7 although the role of Fbxw8 was not reported

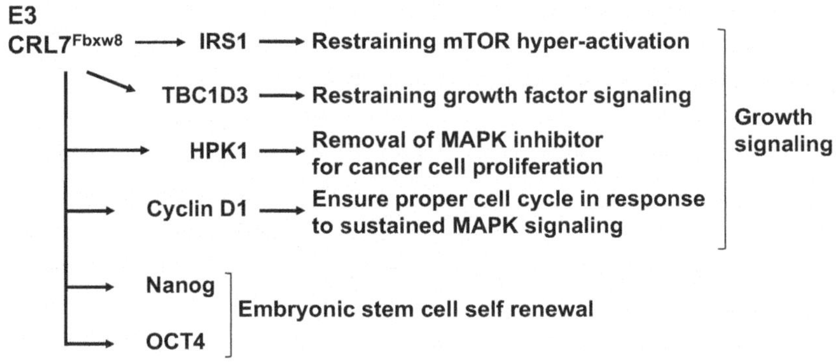

Fig. 17.2 Proteolytic pathways of CRL7^{Fbxw8} in growth signaling. Shown are six CRL7^{Fbxw8} substrates in cell proliferation and stem cell self-renewal

CCDC8 and their physical association with CRL7^{Fbxw8} underscore the significance of distinct subcellular locations of the E3 complex (Fig. 17.3) in its biological function and role in growth retardation disease.

It remains unclear whether the biological defects observed in mouse *cul7* knockout (Arai et al. 2003) and/or human 3-M syndrome bearing *cul7* mutations (Huber et al. 2005) can be attributed to aberrant accumulation of any of the CRL7^{Fbxw8} substrates discovered to date (Table 17.1). It is possible that we have not yet identified the CRL7^{Fbxw8} substrate that plays a predominant growth regulatory role and that, when dysregulated, leads to 3-M growth retardation. Alternatively, 3-M may be a disease caused by dysregulation of multiple proteolytic pathways affected by CRL7^{Fbxw8}. It should also be mindful that CUL7 has non-proteolytic functions (such as binding to p53, Fig. 17.1b) that may play significant role in cell proliferation and 3-M syndrome.

It is hoped that continuing efforts using genetic and biochemical approaches will lead to better understanding of the role of CRL7^{Fbxw8} in growth

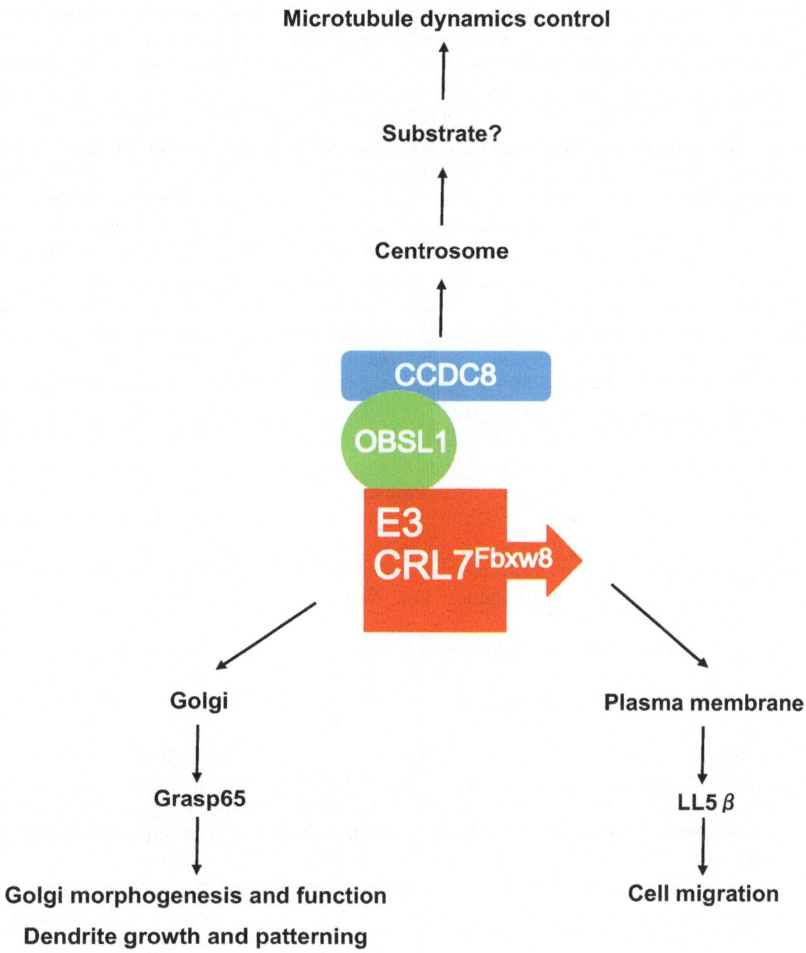

Fig. 17.3 Subcellular localization of CRL7^{Fbxw8} as a result of interactions with OBSL1 and CCDC8. Shown are a few distinct subcellular locations of CRL7^{Fbxw8} due to recruitment through interactions with OBSL1 and CCDC8

control and cancer. For example, advanced mouse models may be created to resemble the human 3-M syndrome and to more precisely define the role of CRL7^{Fbxw8} in cell proliferation signaling pathways. In-depth characterization of the proteolytic and non-proteolytic functions of CRL7^{Fbxw8} may yield new mechanistic insights. It is hopeful that such discoveries would enable the birth of innovative therapeutic approaches for the treatment of 3-M and related growth retardation syndromes.

Acknowledgment This work was supported by NIH grant GM122751 to Z-Q P.

References

Al-Dosari MS, Al-Shammari M, Shaheen R, Faqeih E, Alghofely MA, Boukai A, Alkuraya FS (2012) 3M syndrome: an easily recognizable yet underdiagnosed cause of proportionate short stature. J Pediatr 161 (1):139-45.e1

Andrews P, He YJ, Xiong Y (2006) Cytoplasmic localized ubiquitin ligase cullin 7 binds to p53 and promotes cell growth by antagonizing p53 function. Oncogene 25:4534–4548

Arai T, Kasper JS, Skaar JR, Ali SH, Takahashi C, DeCaprio JA (2003) Targeted disruption of p185/Cul7 gene results in abnormal vascular morphogenesis. Proc Natl Acad Sci U S A 100(17):9855–9860

Bae KB, Yu DH, Lee KY, Yao K, Ryu J, Lim DY, Zykova TA, Kim MO, Bode AM, Dong Z (2017) Serine

347 phosphorylation by JNKs negatively regulates OCT4 protein stability in mouse embryonic stem cells. Stem Cell Rep 9(6):2050–2064

Clayton PE, Hanson D, Magee L, Murray PG, Saunders E, Abu-Amero SN, Moore GE, Black GC (2012) Exploring the spectrum of 3-M syndrome, a primordial short stature disorder of disrupted ubiquitination. Clin Endocrinol 77(3):335–342

Copps KD, White MF (2012) Regulation of insulin sensitivity by serine/threonine phosphorylation of insulin receptor substrate proteins IRS1 and IRS2. Diabetologia 55(10):2565–2582

Dauber A, Stoler J, Hechter E, Safer J, Hirschhorn JN (2013) Whole exome sequencing reveals a novel mutation in CUL7 in a patient with an undiagnosed growth disorder. J Pediatr 162(1):202–4.e1

Daud AI, Lanson NA Jr, Claycomb WC, Field LJ (1993) Identification of SV40 large T-antigen-associated proteins in cardiomyocytes from transgenic mice. Am J Phys 264:H1693–H1700

Dias DC, Dolios G, Wang R, Pan ZQ (2002) CUL7: A DOC domain-containing cullin selectively binds Skp1. Fbx29 to form an SCF-like complex. Proc Natl Acad Sci U S A 99:16601–16606

Gascoin-Lachambre G, Buffat C, Rebourcet R, Chelbi ST, Rigourd V, Mondon F, Mignot TM, Legras E, Simeoni U, Vaiman D, Barbaux S (2010) Cullins in human intra-uterine growth restriction: expressional and epigenetic alterations. Placenta 31:151–157

Hanson D, Murray PG, Sud A, Temtamy SA, Aglan M, Superti-Furga A, Holder SE, Urquhart J, Hilton E, Manson FD, Scambler P, Black GC, Clayton PE (2009) The primordial growth disorder 3-M syndrome connects ubiquitination to the cytoskeletal adaptor OBSL1. Am J Hum Genet 84(6):801–806

Hanson D, Murray PG, O'Sullivan J, Urquhart J, Daly S, Bhaskar SS, Biesecker LG, Skae M, Smith C, Cole T, Kirk J, Chandler K, Kingston H, Donnai D, Clayton PE, Black GC (2011) Exome sequencing identifies CCDC8 mutations in 3-M syndrome, suggesting that CCDC8 contributes in a pathway with CUL7 and OBSL1 to control human growth. Am J Hum Genet 89(1):148–153

Hanson D, Murray PG, Coulson T, Sud A, Omokanye A, Stratta E, Sakhinia F, Bonshek C, Wilson LC, Wakeling E, Temtamy SA, Aglan M, Rosser EM, Mansour S, Carcavilla A, Nampoothiri S, Khan WI, Banerjee I, Chandler KE, Black GC, Clayton PE (2012) Mutations in CUL7, OBSL1 and CCDC8 in 3-M syndrome lead to disordered growth factor signaling. J Mol Endocrinol 49(3):267–275

Harrington LS, Findlay GM, Lamb RF (2005) Restraining PI3K: mTOR signaling goes back to the membrane. Trends Biochem Sci 30:35–42

Hartmann T, Xu X, Kronast M, Zimmermann W, Hurwitz J, Pan ZQ, Engelhardt S, Sarikas A (2014) Inhibition of Cullin-RING E3 ubiquitin ligase 7 by simian virus 40 large T antigen. PNAS 111 (9):3371–3376

Hasegawa K, Tanaka H, Higuchi Y, Yamashita M, Tsukahara H (2016) Changes in facial appearance from neonate to adult in 3-M syndrome patient with novel CUL7 gene mutations. J Pediatr Endocrinol Metab 29(2):241–246

Hsu PH, Ma YT, Fang YC, Huang JJ, Gan YL, Chang PT, Jow GM, Tang CY, Jeng CJ (2017) Cullin 7 mediates proteasomal and lysosomal degradations of rat Eag1 potassium channels. Sci Rep 7:40825

Hu X, Li H, Gui B, Xu Y, Wang J, Li N, Su J, Zhang S, Song Y, Wang Y, Luo J, Fan X, Wang J, Chen S, Gong C, Shen Y (2017) Prenatal and early diagnosis of Chinese 3-M syndrome patients with novel pathogenic variants. Clin Chim Acta 474:159–164

Huber C, Dias-Santagata D, Glaser A, O'Sullivan J, Brauner R, Wu K, Pearce K, Wang R, Luisa M, Uzielli G, Dagoneau N, Chemaitilly W, Superti-Furga-A, Dos Santos H, Mégarbané A, Morin G, Gillessen-Kaesbach G, Hennekam R, Brunner H, Graeme Black GCM, Clayton PE, Read A, Le Merrer M, Scambler PJ, Munnich A, Pan ZQ, Winter R, Cormier-Daire V (2005) Identification of CUL7 mutations in the 3-M syndrome. Nat Genet 37:1119–1124

Huber C, Fradin M, Edouard T, Le Merrer M, Alanay Y, Da Silva DB, David A, Hamamy H, van Hest L, Lund AM, Michaud J, Oley C, Patel C, Rajab A, Skidmore DL, Stewart H, Tauber M, Munnich A, Cormier-Daire V (2010) OBSL1 mutations in 3-M syndrome are associated with a modulation of IGFBP2 and IGFBP5 expression levels. Hum Mutat 31(1):20–26

Huber C, Munnich A, Cormier-Daire V (2011) The 3M syndrome. Best Pract Res Clin Endocrinol Metab 25 (1):143–151

Jung P, Verdoodt B, Bailey A, Yates JR 3rd, Menssen A, Hermeking H (2007) Induction of cullin 7 by DNA damage attenuates p53 function. Proc Natl Acad Sci U S A 104:11388–11393

Kanie T, Onoyama I, Matsumoto A, Yamada M, Nakatsumi H, Tateishi Y, Yamamura S, Tsunematsu R, Matsumoto M, Nakayama KI (2012) Genetic reevaluation of the role of F-box proteins in cyclin D1 degradation. Mol Cell Biol 32(3):590–605

Kasper JS, Kuwabara H, Arai T, Ali SH, DeCaprio JA (2005) Simian virus 40 large T antigen's association with the CUL7 SCF complex contributes to cellular transformation. J Virol 79(18):11685–11692

Kasper JS, Arai T, DeCaprio JA (2006) A novel p53-binding domain in CUL7. Biochem Biophys Res Commun 348(1):132–138

Kaustov L, Lukin J, Lemak A, Duan S, Ho M, Doherty R, Penn LZ, Arrowsmith CH (2007) The conserved CPH domains of Cul7 and PARC are protein-protein interaction modules that bind the tetramerization domain of p53. J Biol Chem 282:11300–11307

Kim SH, Kim MO, Cho YY, Yao K, Kim DJ, Jeong CH, Yu DH, Bae KB, Cho EJ, Jung SK, Lee MH, Chen H, Kim JY, Bode AM, Dong Z (2014) ERK1 phosphorylates Nanog to regulate protein stability and stem cell self-renewal. Stem Cell Res 13(1):1–11

Kim SJ, DeStefano MA, Oh WJ, Wu CC, Vega-Cotto NM, Finlan M, Liu D, Su B, Jacinto E (2012) mTOR complex 2 regulates proper turnover of insulin receptor substrate-1 via the ubiquitin ligase subunit Fbw8. Mol Cell 48(6):875–887

Kim SS, Shago M, Kaustov L, Boutros PC, Clendening JW, Sheng Y, Trentin GA, Barsyte-Lovejoy D, Mao DY, Kay R, Jurisica I, Arrowsmith CH, Penn LZ (2007) CUL7 is a novel antiapoptotic oncogene. Cancer Res 67:9616–9622

Kohrman DC, Imperiale MJ (1992) Simian virus 40 large T antigen stably complexes with a 185-kilodalton host protein. J Virol 66:1752–1760

Kong C, Samovski D, Srikanth P, Wainszelbaum MJ, Charron AJ, Liu J, Lange JJ, Chen PI, Pan ZQ, Su X, Stahl PD (2012) Ubiquitination and degradation of the hominoid-specific oncoprotein TBC1D3 is mediated by CUL7 E3 ligase. PLoS One 7(9):e46485

Kong Y, Wang Z, Huang M, Zhou Z, Li Y, Miao H, Wan X, Huang J, Mao X, Chen C (2019) CUL7 promotes cancer cell survival through promoting Caspase-8 ubiquitination. Int J Cancer. https://doi.org/10.1002/ijc.32239

Laplante M, Sabatini DM (2012) mTOR signaling in growth control and disease. Cell 149(2):274–293

Li DZ, Liu SF, Zhu L, Wang YX, Chen YX, Liu J, Hu G, Yu X, Li J, Zhang J, Wu ZX, Lu H, Liu W, Liu B (2017) FBXW8-dependent degradation of MRFAP1 in anaphase controls mitotic cell death. Oncotarget 8 (57):97178–97186

Li Z, Pei XH, Yan J, Yan F, Cappell KM, Whitehurst AW, Xiong Y (2014) CUL9 mediates the functions of the 3M complex and ubiquitylates survivin to maintain genome integrity. Mol Cell 54(5):805–819

Litterman N, Ikeuchi Y, Gallardo G, O'Connell BC, Sowa ME, Gygi SP, Harper JW, Bonni A (2011) An OBSL1-Cul7Fbxw8 ubiquitin ligase signaling mechanism regulates Golgi morphology and dendrite patterning. PLoS Biol 9(5):e1001060

Lugli L, Bertucci E, Mazza V, Elmakky A, Ferrari F, Neuhaus C, Percesepe A (2016) Pre- and post-natal growth in two sisters with 3-M syndrome. Eur J Med Genet 59(4):232–236

Luo Y, Liu Y, Wu L, Ma X, Liu Q, Huang F, Zhang X, Zhang Y, Zhang J, Luo H, Yang Y, Lu G, Tang X, Li L, Zeng Y, Pan T, Zhang H (2019) CUL7 E3 ubiquitin ligase mediates the degradation of activation-induced Cytidine Deaminase and regulates the Ig class switch recombination in B lymphocytes. J Immunol 203(1):269–281

Maksimova N, Hara K, Miyashia A, Nikolaeva I, Shiga A, Nogovicina A, Sukhomyasova A, Argunov V, Shvedova A, Ikeuchi T, Nishizawa M, Kuwano R, Onodera O (2007) Clinical, molecular and histopathological features of short stature syndrome with novel CUL7 mutation in Yakuts: new population isolate in Asia. J Med Genet 44:772–778

Meazza C, Lausch E, Pagani S, Bozzola E, Calcaterra V, Superti-Furga A, Silengo M, Bozzola M (2013) 3-M syndrome associated with growth hormone deficiency: 18 year follow-up of a patient. Ital J Pediatr 39:21

Nie J, Xu C, Jin J, Aka JA, Tempel W, Nguyen V, You L, Weist R, Min J, Pawson T, Yang XJ (2015) Ankyrin repeats of ANKRA2 recognize a PxLPxL motif on the 3M syndrome protein CCDC8. Structure 23 (4):700–712

Okabe H, Lee SH, Phuchareon J, Albertson DG, McCormick F, Tetsu O (2006) A critical role for FBXW8 and MAPK in cyclin D1 degradation and cancer cell proliferation. PLoS One 1:e128

Petroski MD, Deshaies RJ (2005) Function and regulation of cullin-RING ubiquitin ligases. Nat Rev Mol Cell Biol 6:9–20

Ponyeam W, Hagen T (2012) Characterization of the Cullin7 E3 ubiquitin ligase--heterodimerization of cullin substrate receptors as a novel mechanism to regulate cullin E3 ligase activity. Cell Signal 24 (1):290–295

Sarikas A, Xu X, Field LJ, Pan ZQ (2008) The Cullin7 E3 ubiquitin ligase: a novel player in growth control. Cell Cycle 7(20):3154–3161

Sarikas A, Hartmann T, Pan ZQ (2011) The cullin protein family. Genome Biol 12(4):220

Scheufele F, Wolf B, Kruse M, Hartmann T, Lempart J, Muehlich S, Pfeiffer AF, Field LJ, Charron MJ, Pan ZQ, Engelhardt S, Sarikas A (2014) Evidence for a regulatory role of Cullin-RING E3 ubiquitin ligase 7 in insulin signaling. Cell Signal 26(2):233–239

Shah OJ, Hunter T (2005) Tuberous sclerosis and insulin resistance. Unlikely bedfellows reveal a TORrid affair. Cell Cycle 4:46–51

Shah VJ, Maddika S (2018) CRL7^{SMU1} E3 ligase complex-driven H2B ubiquitylation functions in sister chromatid cohesion by regulating SMC1 expression. J Cell Sci 131(8)

Simsek-Kiper PO, Taskiran E, Kosukcu C, Arslan UE, Cormier-Daire V, Gonc N, Ozon A, Alikasifoglu A, Kandemir N, Utine GE, Alanay Y, Alikasifoglu M, Boduroglu K (2019) Further expanding the mutational spectrum and investigation of genotype-phenotype correlation in 3M syndrome. Am J Med Genet A 179 (7):1157–1172

Skaar JR, Florens L, Tsutsumi T, Arai T, Tron A, Swanson SK, Washburn MP, DeCaprio JA (2007) PARC and CUL7 form atypical cullin RING ligase complexes. Cancer Res 67(5):2006–2014

Tsunematsu R, Nishiyama M, Kotoshiba S, Saiga T, Kamura T, Nakayama KI (2006) Fbxw8 is essential for Cul1-Cul7 complex formation and for placental development. Mol Cell Biol 26(16):6157–6169

Tsutsumi T, Kuwabara H, Arai T, Xiao Y, Decaprio JA (2008) Disruption of the Fbxw8 gene results in pre- and postnatal growth retardation in mice. Mol Cell Biol 28:743–751

Wang H, Chen Y, Lin P, Li L, Zhou G, Liu G, Logsdon C, Jin J, Abbruzzese JL, Tan TH, Wang H (2014) The CUL7/F-box and WD repeat domain containing 8 (CUL7/Fbxw8) ubiquitin ligase promotes degradation of hematopoietic progenitor kinase 1. J Biol Chem 289(7):4009–4017

Wang P, Yan F, Li Z, Yu Y, Parnell SE, Xiong Y (2019) Impaired plasma membrane localization of ubiquitin ligase complex underlies 3-M syndrome development. J Clin Invest 130:pii: 129107. https://doi.org/10.1172/JCI129107

Wei M, Zhao X, Liu M, Huang Z, Xiao Y, Niu M, Shao Y, Kleiman L (2015) Inhibition of HIV-1 assembly by coiled-coil domain containing protein 8 in human cells. Sci Rep 5:14724

Xu X, Sarikas A, Dias-Santagata DC, Dolios G, Lafontant PJ, Tsai SC, Zhu W, Nakajima H, Nakajima HO, Field LJ, Wang R, Pan ZQ (2008) The CUL7 E3 ubiquitin ligase targets insulin receptor substrate 1 for ubiquitin-dependent degradation. Mol Cell 30:403–414

Xu X, Keshwani M, Meyer K, Sarikas A, Taylor S, Pan ZQ (2012) Identification of insulin receptor substrate 1's degradation determinants for signaling cullin-RING E3 ubiquitin ligase 7-mediated ubiquitination. J Biol Chem 287(48):40758–662012

Yan J, Yan F, Li Z, Sinnott B, Cappell KM, Yu Y, Mo J, Duncan JA, Chen X, Cormier-Daire V, Whitehurst AW, Xiong Y (2014) The 3M complex maintains microtubule and genome integrity. Mol Cell 54 (5):791–804

Yoneyama Y, Inamitsu T, Chida K, Iemura SI, Natsume T, Maeda T, Hakuno F, Takahashi SI (2018) Serine Phosphorylation by mTORC1 Promotes IRS-1 Degradation through SCFβ-TRCP E3 Ubiquitin Ligase. iScience 5:1–18

Zhande R, Mitchell JJ, Wu J, Sun XJ (2002) Molecular mechanism of insulin-induced degradation of insulin receptor substrate 1. Mol Cell Biol 22:1016–1026

Zoncu R, Efeyan A, Sabatini DM (2011) mTOR: from growth signal integration to cancer, diabetes and ageing. Nat Rev Mol Cell Biol 12(1):21–35

Zou J, Ma W, Li J, Littlejohn R, Zhou H, Kim IM, Fulton DJR, Chen W, Weintraub NL, Zhou J, Su H (2018) Neddylation mediates ventricular chamber maturation through repression of hippo signaling. Proc Natl Acad Sci U S A 115(17):E4101–E4110

Targeting Protein Neddylation for Cancer Therapy

18

Lisha Zhou and Lijun Jia

Abstract

Neddylation is a posttranslational modification that conjugates a ubiquitin-like protein NEDD8 to substrate proteins. The best-characterized substrates of neddylation are the cullin subunits of cullin-RING E3 ubiquitin ligase complexes (CRLs). CRLs as the largest family of E3 ubiquitin ligases control many important biological processes, including tumorigenesis, through promoting ubiquitylation and subsequent degradation of a variety of key regulatory proteins. The process of protein neddylation is overactivated in multiple types of human cancers, providing a sound rationale as an attractive anticancer therapeutic strategy, evidenced by the development of the NEDD8-activating enzyme (NAE) inhibitor MLN4924 (also known as pevonedistat). Recently, increasing evidence strongly indicates that neddylation inhibition by MLN4924 exerts anticancer effects mainly by triggering cell apoptosis, senescence, and autophagy and causing angiogenesis suppression, inflammatory responses, and chemo-/radiosensitization in a context-dependent manner. Here, we briefly summarize the latest progresses in this field, focusing on the preclinical studies to validate neddylation modification as a promising anticancer target.

L. Zhou (✉)
Department of Biochemistry, Medical College, Taizhou University, Taizhou, China
e-mail: lishazhou@tzc.edu.cn

L. Jia (✉)
Cancer Institute, Longhua Hospital, Shanghai University of Traditional Chinese Medicine, Shanghai, China
e-mail: ljjia@shutcm.edu.cn

Keywords

Neddylation · Cancer target · Apoptosis · Senescence · Autophagy · Inflammatory responses

Abbreviations

4E-BP1	Eukaryotic translation initiation factor 4E-binding protein 1
AICD	APP intracellular domain
APPBP1	Amyloid beta precursor protein binding protein 1
ATF4	Activating transcription factor 4
BAX	BCL2-associated X, apoptosis regulator
BCA3	Breast cancer-associated protein 3
BIK	BCL2 Interacting killer
CAM	Chorioallantoic membrane
c-CBL	Casitas B-lineage lymphoma
CDT1/2	Chromatin licensing and DNA replication factor 1/2
c-FLIP	CASP8 and FADD-like apoptosis regulator
CRLs	Cullin-RING ligases
CSN	COP9 signalosome complex
DCN1	Defective in cullin neddylation 1

Y. Sun et al. (eds.), *Cullin-RING Ligases and Protein Neddylation*, Advances in Experimental Medicine and Biology 1217, https://doi.org/10.1007/978-981-15-1025-0_18

DCNL	DCN1-like protein
Deptor	DEP domain-containing mTOR-interacting protein
DR5	Death receptor 5
E2F-1	E2F transcriptional factor 1
EGFR	Epidermal growth factor receptor
FANCD2	FA complementation group D2
FBXO11	F-box protein 11
HIF1α/2α	Hypoxia-inducible factor 1α/2α
HUR	Hu antigen R
HUVECs	Human umbilical vein endothelial cells
IAPs	Inhibitors of apoptosis
IL-6	Interleukin 6
MAPK	Mitogen-activated protein kinase
MDM2	Murine double minute 2
mTOR	Mammalian target of rapamycin
NAC	*N*-acetyl cysteine
NAE1	NEDD8-activating enzyme E1 subunit 1
NAEβ	NEDD8-activating enzyme E1 subunit β
NEDD8	Neural precursor cell expressed, developmentally downregulated 8
NEDP1	NEDD8-specific protease 1
NF-κB	Nuclear factor-κB
ORC1	Origin recognition complex subunit 1
PAMPs	Pathogen-associated molecular patterns
PfUCH54	54-kDa plasmodium falciparum ubiquitin C-terminal hydrolase
PINK1	PTEN-induced putative kinase 11
pIκB	p-inhibitor of nuclear factor kappa B
PONY	Potentiating neddylation
pVHL	p-von Hippel-Lindau
RBX1/2	RING box protein 1/2
RhoA	Ras homolog family member A
RING	Really interesting new gene
RNF111	Ring finger protein 111
ROS	Reactive oxygen species
S6K1	Ribosomal protein S6 kinase B1
SKP1/2	S-phase kinase-associated protein 1/2
SMURF1	SMAD-specific E3 ubiquitin protein ligase 1
TGFβ-RII	Transforming growth factor β type II receptor
TNFα	Tumor necrosis factor α
TRAIL	Tumor necrosis factor-related apoptosis-inducing ligand
TRIM40	Tripartite motif containing 40
TSC1	TSC complex subunit 1
UBA3	Ubiquitin-like modifier activating enzyme 3
UBC12	Ubiquitin-conjugating enzyme 12
UBE2F	Ubiquitin-conjugating enzyme E2F
UBE2M	Ubiquitin-conjugating enzyme E2M
UCH-L1/3	Ubiquitin carboxyl-terminal esterase L1/3
USP21	Ubiquitin-specific peptidase 21
VEGF	Vascular endothelial growth factor
WEE1	WEE1 G2 checkpoint kinase 1
β-Trcp	β-Transducin repeat containing E3 ubiquitin protein ligase

18.1 Introduction

Protein neddylation is a process of conjugating a ubiquitin-like molecule NEDD8 (neuronal precursor cell-expressed developmentally downregulated protein 8) to a lysine residue of targeted substrate proteins (Kamitani et al. 1997; Rabut and Peter 2008; Xirodimas 2008). NEDD8 is one of the most studied ubiquitin-like molecules, with 60% identity and 80% homology to ubiquitin (Kamitani et al. 1997). Similar to ubiquitin, NEDD8 is attached to the substrates by forming an isopeptide chain between its C-terminal glycine residue (Gly76) and a lysine residue of the targeted protein (Zhao et al. 2014). However, NEDD8 is first synthesized as a precursor with five additional residues downstream from Gly76, which need to be cleaved by C-terminal hydrolases (Rabut and Peter 2008; Zhao et al. 2014). After that, the mature NEDD8 is activated in an ATP-dependent manner by the NEDD8-activating enzyme E1 (Walden et al. 2003). The activated NEDD8 is then transferred to the NEDD8-conjugating enzymes through a trans-thiolation reaction (Gong and Yeh 1999; Huang et al. 2005). Ultimately, a substrate-specific E3 ligase transfers NEDD8 from the

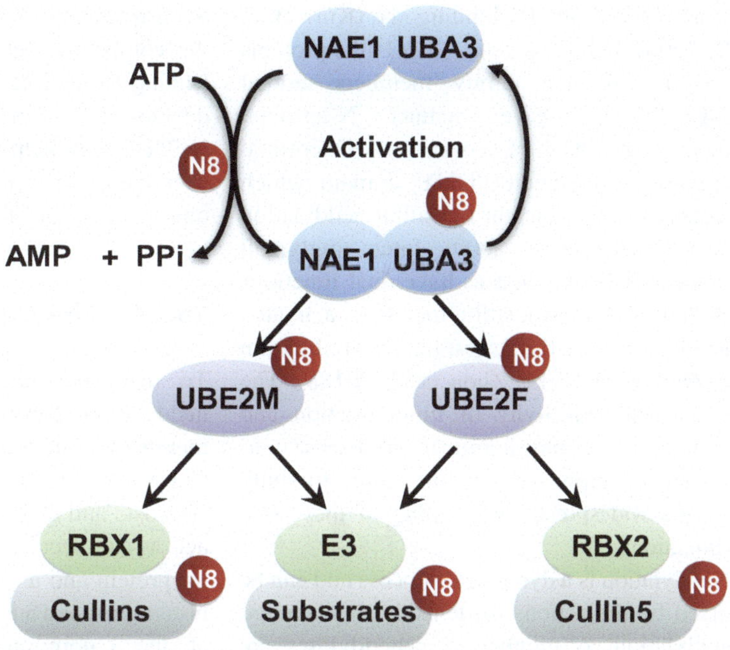

Cullins	Non-Cullins		E3	
Cul1	AICD	p53	c-CBL	RBX1
Cul2	BCA3	MDM2	DCNL1-5	RBX2
Cul3	E2F-1	Parkin	FBXO11	RNF111
Cul4A	EGFR	pVHL	IAPs	TFB3
Cul4B	HIF1/2α	RPs	MDM2	TRIM40
	HuR	SMURF1		
	p53	TGF-βII		

Fig. 18.1 The three-step enzymatic reaction for protein neddylation. Neddylation is a posttranslational modification that conjugates a ubiquitin-like protein NEDD8 to substrate proteins. The involving enzymes, reported neddylation E1/E2s/E3s, and substrates (Non-Cullins) are listed. N8: NEDD8

charged E2 to a lysine residue in its target protein via the covalent attachment (Fig. 18.1).

The NEDD8 cascade consists of a single E1 (NAE), two E2s (UBE2M/UBC12 and UBE2F) and several E3s. NAE is a heterodimer consisting of NAE1/APPBP1 and UBA3/NAEβ (Bohnsack and Haas 2003). UBE2M and UBE2F are two independent E2s with distinct biological functions. UBE2M pairs with RBX1 to regulate neddylation of cullin-1, cullin-2, cullin-3, cullin-4A, and cullin-4B, whereas UBE2F is highly specific to the neddylation of RBX2-associated cullin-5 (Huang et al. 2009). Interestingly, glycyl-tRNA synthetase, an essential enzyme for protein synthesis, binds with NAE1 subunit of E1 to capture and protect activated UBE2M E2 before reaching downstream target (Mo et al. 2016). There are at least ten different NEDD8 E3 ligases, most of which contain the really interesting new gene (RING) domain. Besides the best-characterized RBX1/2 (Huang et al. 2009; Duan et al. 1999), reported NEDD8 E3 ligases include MDM2 (Xirodimas et al. 2004), c-CBL (Oved et al. 2006; Zuo et al. 2013), FBXO11 (Abida et al. 2007), RNF111 (Ma et al. 2013), IAPs (Broemer et al. 2010), TFB3 (Rabut et al. 2011), and TRIM40 (Noguchi et al. 2011). In *yeast* and *Caenorhabditis elegans*, defective in cullin neddylation 1 (DCN1) has been identified as a NEDD8 E3 ligase, which does not contain a RING domain for its catalytic activity and directly interacts with the UBE2M on a surface overlapping with the E1-binding site (Kurz et al. 2005, 2008). In human cells, DCN1-like proteins (DCNLs) with five family members termed DCNL1–DCNL5 have distinct N-terminal domains but share a conserved C-terminal potentiating neddylation (PONY) domain, which is necessary and sufficient for cullin neddylation (Meyer-Schaller et al. 2009). Interestingly, all reported NEDD8 E3 ligases have dual functions of ubiquitin E3 ligases; cullin members, activated by neddylation, are the largest family of ubiquitin E3s (Zhao et al. 2014; Zhou et al. 2018a). The crosstalk and selectivity of ubiquitination and neddylation under physiological and disease condition are therefore a fascinating topic to study given the widespread overlapping of these two modifications.

Neddylation is a dynamical process and can be reversed by the actions of deneddylases, termed deneddylation. A number of deneddylases are able to remove NEDD8 from the neddylated substrate, including the COP9 signalosome (CSN), NEDP1, USP21, Ataxin-3, PfUCH54, UCH-L1, and UCH-L3 (Watson et al. 2011). Of these enzymes, only CSN and NEDP1 are NEDD8 specific, while the rest are also considered to regulate deubiquitination (Rabut and Peter 2008; Xirodimas 2008). CSN, a zinc metalloprotease consisting of eight subunits, is the best-characterized NEDD8 deneddylase (Wei et al. 2008). Loss of CSN function in mice accumulates neddylated cullins and various neddylated proteins with unknown identifier (Su et al. 2011, 2013). The cysteine protease NEDP1 is another specific deneddylase with the ability to process NEDD8 precursor and remove NEDD8 from neddylated substrates (Hemelaar et al. 2004; Reverter et al. 2005; Shen et al. 2005).

18.2 Substrates of Neddylation

In contrast with a wide spectrum of ubiquitylated proteins, much less is known about the identities of NEDD8 targets. Owing to the relative low steady-state abundance, except for cullins, neddylated proteins are hardly detectable under physiological condition. Moreover, NEDD8 overexpression likely causes controversial targets via competing the lysine residues of ubiquitin, leading to aberrant effects of cells. Recently, a developed deconjugation-resistant form of NEDD8 may help to address this challenge, via stabilizing the neddylated form of cullins and other non-cullin substrates (Coleman et al. 2017).

18.2.1 The Cullin Family

To date, the best-characterized physiological neddylation substrates are the cullin family members, including Cul-1, Cul-2, Cul-3, Cul-4A, Cul-4B, Cul-5, Cul-7, and Cul-9 (Petroski and Deshaies 2005). Each cullin protein acts as a molecular scaffold that binds to an adaptor protein and a substrate receptor protein at the N-terminus and a RING protein, RBX1 or RBX2, at the C-terminus to assemble cullin-RING ligases (CRLs) (Petroski and Deshaies 2005; Deshaies 1999). By pairing individual cullin with different F-box proteins, CRLs control the degradation of about 20% of all cellular proteins and, not surprisingly, regulate many aspects of biological processes including cell cycle progression, DNA repair, signal transduction, and tumorigenesis (Petroski and Deshaies 2005; Soucy et al. 2009). Generally, neddylation of cullins at their conserved C-terminal lysine residue causes conformational change, which disrupts the inhibitory binding by CAND1 and activates CRLs to facilitate the transfer of ubiquitin from charged E2 to the substrates (Merlet et al. 2009; Sakata et al. 2007; Duda et al. 2008; Zheng et al. 2002). After completion of ubiquitination, the deneddylase CSN binds to neddylated cullins and enables deneddylation of cullins, leading to the dissociation of the CRL complex and release of E2 and the substrate for the next round of ubiquitination (Lyapina et al. 2001). CAND1-mediated dynamic neddylation of cullins is required for the recycling of CRLs, enabling it to assemble with different substrate receptors to catalyze the ubiquitination of many different substrates to maintain cellular homeostasis (Lo and Hannink 2006). CSN deficiency causes sustained cullin neddylation, followed by autoubiquitination of CRL components and subsequent self-destruction, eventually leading to defects in cell cycle progression, signal transduction, and development (Wei et al. 2008; Lee et al. 2011). Thus, it is important to understand mechanistically how neddylation and deneddylation are precisely regulated under physiological conditions or in response to both internal and external stimuli.

18.2.2 Non-cullin Substrates of Neddylation

Besides cullins, several non-cullin proteins have been identified as the substrates of neddylation (Table 18.1). For instance, emerged evidence suggests that the stability and function of the tumor suppressor p53 are regulated by neddylation at different levels. Neddylation of p53 prevents its nuclear translocation and suppresses its transcriptional activity, whereas neddylation-resistant p53 mutants retain the transcriptional activity (Xirodimas et al. 2004). Consistently, overexpression or downregulation of NAE1 inhibits or enhances p53 activity, respectively (Guihard et al. 2012). The well-known E3 NEDD8 ligases for p53 neddylation are MDM2 and FBXO11 (Xirodimas et al. 2004; Abida et al. 2007), the stability of which is directly or indirectly controlled by neddylation themselves. Specifically, neddylation of MDM2 increases its stability, whereas NEDP1-induced deneddylation results in its destabilization (Watson et al. 2010). Moreover, ribosome proteins L11 and S14 could bind to and subsequently inhibit MDM2, leading to p53 activation. Interestingly, the stability and function of these two ribosome proteins are also regulated by neddylation (Mahata et al. 2012; Sundqvist et al. 2009; Zhang et al. 2014). The activity of p53 is also affected by neddylation since cullin-1 neddylation is required for the regulation of autoubiquitination of FBXO11, thereby promoting p53 activity (Cope and Deshaies 2006). Other reported non-cullin substrates of neddylation include tumor suppressor pVHL (Stickle et al. 2004), oncoproteins Hu antigen R (HuR) (Abdelmohsen and Gorospe 2010; Embade et al. 2012), receptor proteins such as EGFR (Oved et al. 2006) and TGF-β type II receptor (Zuo et al. 2013), transcriptional regulators such as HIF1α/HIF2α (Ryu et al. 2011), breast cancer-associated protein 3 (BCA3) (Gao et al. 2006), APP intracellular

Table 18.1 Substrates of neddylation

Substrates	E3 ligases	Functions of neddylation
Cullins	RBX1/2, DCN1	Increases CRLs activity
p53	MDM2/ FBXO11	Inhibits p53 transcriptional activity
HuR	MDM2	Increases protein stability and promotes nuclear localization
MDM2	MDM2	Increases protein stability
BCA3	–	Recruits histone deacetylase SIRT1 to inhibit the transcriptional activity of NF-κB
EGFR	c-CBL	Enhances ubiquitination of EGFR and subsequent lysosomal degradation
Ribosomal proteins	MDM2	Increases protein stability and promotes nuclear localization
pVHL	–	Promotes the interaction between pVHL and fibronectin
TGF-βII	c-CBL	Increases protein stability
Histone H4	RNF111	Activates DNA damage-induced ubiquitination
E2F-1		Reduces stability and inhibit transcriptional activity
AICD	–	Inhibits the transcriptional activity of AICD by disrupting the interaction between AICD and transcriptional co-activators (Fe65 and Tip60)
Parkin/PINK1		Increases its E3 ligase activity and stabilizes PINK 1 55 kDa fragment
SMURF1		Increases E3 ubiquitin ligase activity
HIF1α/HIF2α		Increases protein stability

domain (AICD) (Lee et al. 2008), and E2F-1 (Loftus et al. 2012), HECT-domain ubiquitin E3 ligase SMURF1, and RBR ubiquitin E3 ligase Parkin (Enchev et al. 2015; Xie et al. 2014). Genome-wide proteomic analysis has identified numerous proteins with abundance changes after MLN4924 treatment, and a subset of them could be neddylation substrates, awaiting experimental validation. It is worth noting that except for neddylated cullins, regulatory mechanisms and physiological functions of non-cullin substrates are still poorly understood.

18.3 Neddylation Modification as an Attractive Anticancer Target

Recently, a series of studies have clearly shown that the levels of catalytic enzyme (e.g., NEDD8 E1, NAE1/UBA3, and NEDD8 E2, UBE2M/UBE2F) in the neddylation pathway are significantly increased in several human cancers, including lung cancer, liver cancer, glioblastoma, breast cancer, etc., when compared to adjacent normal tissues (Xie et al. 2014, 2017; Li et al. 2014; Gao et al. 2014; Barbier-Torres et al. 2015; Hua et al. 2015; Chen et al. 2016; Jia et al. 2019; Tian et al. 2019). Moreover, overexpression of these neddylation modifying enzymes confers worse overall patient survival, suggesting that elevated neddylation modification may represent an oncogenic event during carcinogenesis (Li et al. 2014; Gao et al. 2014; Barbier-Torres et al. 2015; Xie et al. 2017; Chen et al. 2016; Jia et al. 2019; Tian et al. 2019). Higher expressions of neddylation enzymes may remarkably increase the modification levels of CRLs, leading to ubiquitylation and degradation of many tumor suppressor substrates of CRLs (e.g., p21 and p27). In contrast, the blockage of overactivated neddylation-CRLs pathway causes the accumulation of tumor suppressors via inhibiting the degradation, thus an evident antitumor effect (Zhao et al. 2014; Zhou et al. 2018a) (Fig. 18.2). Therefore, in normal tissues, the neddylation modification and subsequent CRL activation is tightly regulated and controlled to maintain the homeostasis (Fig. 18.2). Taken together, overactivation of the neddylation pathway in tumors provides multiple potential antitumor molecular targets, such as catalytic enzymes E1, E2, and E3, laying a solid scientific basis for developing cancer drugs targeting this pathway.

Indeed, MLN4924 (also known as pevonedistat), a potent and highly selective small molecular inhibitor of NAE, discovered in 2009 (Soucy et al. 2009), has been a promising anticancer agent currently in clinical development. As an adenosine sulfamate derivative, MLN4924 forms a steady-state covalent adduct

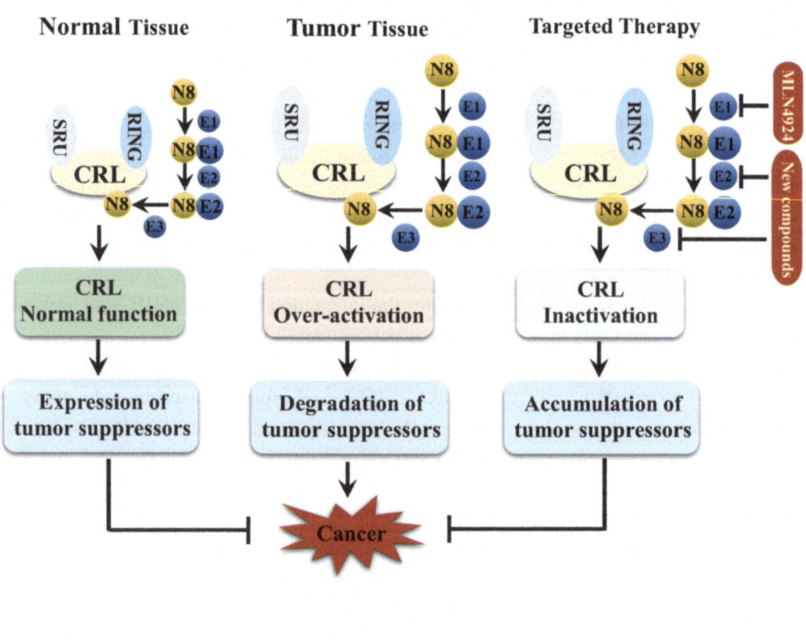

Fig. 18.2 Neddylation activates CRLs to regulate tumorigenesis. In normal tissues, the neddylation modification and subsequent CRL activation is tightly regulated and controlled to maintain the homeostasis. Blockage of overactivated neddylation-CRLs pathway causes the accumulation of tumor suppressors via inhibiting the degradation, thus an evident antitumor effect

with NEDD8, catalyzed by NAE. The MLN4924-NEDD8 adduct resembles the adenylated NEDD8 at the active site of NAE, blocks NAE enzymatic activity, and terminates neddylation cascade (Brownell et al. 2010). Given that NAE is the unique enzyme responsible for the first step of neddylation cascade, its inhibitor MLN4924 could completely block neddylation pathway. By doing so, MLN4924 inactivates the entire CRLs by effectively blocking cullin neddylation, resulting in the accumulation of various CRLs substrates and consequently suppressing cancer cells growth via multiple mechanisms (Soucy et al. 2009; Brownell et al. 2010; Luo et al. 2012a; Jia et al. 2011; Pan et al. 2012).

As a first-in-class agent with potent antitumor activity and well-tolerated toxicity in a range of preclinical studies, MLN4924 has been employed in a series of phase I/II/III clinical trials alone or in combinations with chemo-/radiotherapy. So far, 30 clinical trials have been registered on the clinical trials.gov website (https://www.clinicaltrials.gov/) (Table 18.2). The first five phase I clinical trials, which are now completed, demonstrated the validity of MLN4924 as a therapeutic target in the clinical application, as evidenced by the beneficial effects in terms of partial response (PR), completed responses (CR), and prolonged stable disease (SD) (Sarantopoulos et al. 2016; Shah et al. 2016; Swords et al. 2015, 2018; Bhatia et al. 2016). Given those promising clinical effects, several phase II clinical trials, in combination with other chemotherapies, are currently recruiting patients. Notably, a phase III trial of a combination of MLN4924 with azacytidine has been launched in patients with acute myelogenous leukemia (AML), myelodysplastic syndrome (MS), and chronic myelomonocytic leukemia (CMML). Implementation of these clinical trials will demonstrate the safety and tolerability of MLN4924 in combination with standard chemotherapies and the potential efficacy of MLN4924 in the treatment of fatal human cancers.

Despite the considerable success in treating cancer with MLN4924, some cancer cells have developed MLN4924 resistance due to the emergence of heterozygous mutations in UBA3, the catalytic subunit of NAE, and the direct target of MLN4924 (Toth et al. 2012; Milhollen et al. 2012). Thus, it is desirable to develop more specific small molecule inhibitors selectively targeting neddylation E2s (UBE2M/UBC12 or UBE2F) or E3s (such as RBX1/2, DCN1). Indeed, several small molecule inhibitors targeting UBC2M-DCN1 (E2-E3) binding have been discovered (Scott et al. 2017; Zhou et al. 2017a, 2018b; Wang et al. 2019; Hammill et al. 2018), based upon their crystal structures (Scott et al. 2011, 2014). The specific inhibitors DI-404 and DI-571 selectively inhibit cullin-3 neddylation to cause accumulation of CRL3 substrate NRF2 (Zhou et al. 2017a, 2018b). Another inhibitor, WS-383, targets DCN1 and inhibits cullin-3/1 neddylation selectively over other cullins and also induces the accumulation of p21, p27, and NRF2 (Wang et al. 2019). However, biological activity of these compounds needs further investigation.

18.4 Cellular Responses to Neddylation Inhibition

Neddylation inhibition by MLN4924 exerts significant anticancer effects mainly by the inactivation of CRLs, which leads to the accumulation of a large number of critical CRLs substrates, thus suppressing tumor cell growth via multiple mechanisms in a cell-type-dependent manner (Fig. 18.3).

18.4.1 Induction of Apoptosis

Accumulating evidence supports that neddylation inhibition by MLN4924 effectively induces cell apoptosis in multiple cancer cells, through mechanisms involving the accumulation of CRLs substrates (Fig. 18.3). Early reports showed that MLN4924 causes the inactivation of $CRL1^{SKP2}$ and $CRL4^{CDT2}$ and the subsequent stabilization of CDT1, which triggers DNA re-replication and S phase arrest to induce apoptosis (Milhollen et al. 2011; Hu et al. 2004; Li

Table 18.2 Clinical trials of MLN4924

Clinical trials. gov identifier	Cancer	Treatment	Phase	Status	Time initiated	Patient number	Outcome	Ref.
NCT00722488	HM, MM, lymphoma, HL	Alone	I	Completed	2008,07	56	1PR	Shah et al. 2016
NCT01011530	Metastatic melanoma	Alone	I	Completed	2009,11	37	1 PR; 15 SDs	Bhatia et al. 2016
NCT00911066	AML, MS, and ALL	Azacitidine	I	Completed	2009,03	72	4CRs, 2PRs	Swords et al. 2015
NCT00677170	Solid tumor	Alone	I	Completed	2008,03	62	9SDs	Sarantopoulos et al. 2016
NCT01814826	AML	Azacitidine	I	Completed	2013,03	64	20CRs, 7PRs	Swords et al. 2018
NCT01862328	Solid tumor	Docetaxel, gemcitabine, carboplatin, paclitaxel	I	Completed	2013,05	64	NR	NR
NCT02122770	Solid tumor	Fluconazole, itraconazole, docetaxel, carboplatin, paclitaxel	I	Completed	2014,04	52	NR	NR
NCT03057366	Solid tumors, neoplasms	Docetaxel, carboplatin, paclitaxel	I	Completed	2017,02	8	NR	NR
NCT03323034	Neoplasm and lymphoma	Temozolomide, irinotecan	I	Recruiting	2017,11	76	NR	NR
NCT03330106	Advanced solid neoplasm	Docetaxel, carboplatin, paclitaxel	I	Recruiting	2017,11	45	NR	NR
NCT03486314	Advanced solid neoplasm	Rifampin, docetaxel, carboplatin, paclitaxel	I	Recruiting	2018,8	20	NR	NR
NCT03459859	AML and MS	Cytarabine	I	Recruiting	2018,03	18	NR	NR
NCT03386214	Myelofibrosis	Ruxolitinib	I	Recruiting	2018,04	18	NR	NR
NCT03479268	CLL or Non-HL	Ibrutinib	I	Recruiting	2018,03	30	NR	NR
NCT03009240	AML	Decitabine	I	Recruiting	2017,08	30	NR	NR
NCT03330821	AML	Cytarabine; idarubicin	I/II	Recruiting	2018,04	53	NR	NR
NCT03709576	AML	Azacitidine	II	Recruiting	2018,07	30	NR	NR
NCT03268954	MS, CMML, and AML	Azacitidine	III	Recruiting	2017,11	450	NR	NR
NCT03319537	Mesothelioma	Pemetrexed, cisplatin	I/II	Recruiting	2017,10	42	NR	NR
NCT03349281	ALL	Vincristine, dexamethasone PEG-asparaginase, doxorubicin	I	Recruiting	2019,03	18	NR	NR
NCT03228186	Non-small cell lung cancer	Docetaxel	I	Recruiting	2018,01	37	NR	NR
NCT03238248	MS, MN	Azacitidine	I	Recruiting	2017,08	71	NR	NR
NCT03013998	AML	Azacitidine	I/II	Recruiting	2016,11	500	NR	NR

(continued)

Table 18.2 (continued)

Clinical trials. gov identifier	Cancer	Treatment	Phase	Status	Time initiated	Patient number	Outcome	Ref.
NCT02610777	MS, CMML, and AML	Azacitidine	II	Not yet recruiting	2016,04	120	NR	NR
NCT02782468	Leukemia, myeloid, AML	Azacitidine	I	Not yet recruiting	2016,03	37	NR	NR
NCT03772925	AML, MS	Belinostat	I	Not yet recruiting	2019,03	45	NR	NR
NCT03814005	MS, CMML, and AML	Azacitidine	I	Not yet recruiting	2019,02	60	NR	NR
NCT03770260	Multiple myeloma	Ixazomib citrate	I	Not yet recruiting	2019,07	54	NR	NR
NCT03745352	AML	Azacitidine	I	Not yet recruiting	2019,03	72	NR	NR
NCT03813147	AML, MS	Azacitidine, fludarabine phosphate, cytarabine	I	Not yet recruiting	2019,05	23	NR	NR

CR complete response, *PR* partial response, *SD* stable disease, *NR* not reported
AML acute myelogenous leukemia, *MS* myelodysplastic syndromes, *ALL* acute lymphoblastic leukemia, *CMML* chronic myelomonocytic leukemia, *CLL* chronic lymphocytic leukemia, *MN* myeloproliferative neoplasm, *HL* Hodgkin lymphoma, *HM* hematologic malignancies, *MM* multiple myeloma

et al. 2003). Subsequently, many studies showed that MLN4924 causes a time-dependent stabilization of CRL1$^{\beta TRCP}$ substrate IκB, which can inactivate NF-κB, thus inducing apoptosis (Milhollen et al. 2010; Swords et al. 2010; Godbersen et al. 2014). MLN4924 also inhibits NF-κB-mediated miR-155 expression, resulting in the accumulation of apoptosis inducers including SHIP1, an inhibitor of the PI3K/Akt pathway, and PU.1, a transcription factor important for myeloid differentiation (Khalife et al. 2015). The later studies showed that MLN4924 triggers apoptosis by upregulating classical pro-apoptotic proteins (e.g., NOXA) and downregulating anti-apoptotic proteins (e.g., c-IAP2) (Li et al. 2014; Chen et al. 2016; Godbersen et al. 2014; Yao et al. 2014; Dengler et al. 2014; Czuczman et al. 2016; Knorr et al. 2015; Tong et al. 2017; Leclerc et al. 2016; Wang et al. 2017; Zhang et al. 2018). Specifically, MLN4924-induced NOXA accumulation is mediated by the inactivation of two CRLs: (a) RBX2-CRL5, which facilitates NOXA ubiquitylation via K11 linkage for proteasomal degradation (Jia et al. 2010; Zhou et al. 2017b, 2018c), and (b) RBX1-CRL1, which is involved in the degradation of NOXA transcription activator (e.g., c-Myc and ATF4) (Chen et al. 2016; Knorr et al. 2015; Zhang et al. 2018).

In addition to inducing mitochondrial-mediated endogenous cell apoptosis, MLN4924 also effectively triggers the activation of death receptor-mediated exogenous apoptosis pathway. Specifically, MLN4924 causes the accumulation of ATF4, a well-characterized substrate of CRLs, subsequently transactivating CHOP. As a transcription factor, the activated CHOP further induces the expression of DR5, thus activating Caspase 8 to trigger extrinsic apoptosis (Chen et al. 2016). Moreover, activation of Caspase 8 also produces apoptotic promoter tBid in part via the cleavage of Bid, which then translocates into the mitochondria to activate endogenous cell apoptosis (Chen et al. 2016). Downregulation of ATF4 and CHOP expression significantly inhibits the transcription of DR5 and attenuates MLN4924-induced apoptosis (Chen et al. 2016). These findings suggest that induction of ATF4-

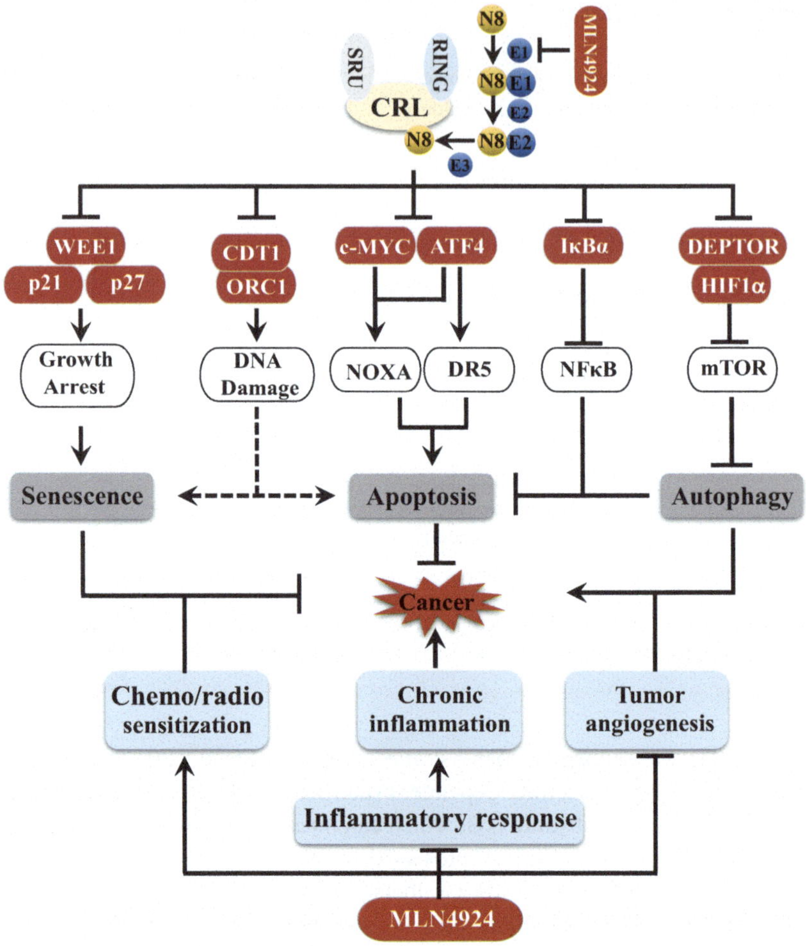

Fig. 18.3 MLN4924 triggers multiple anticancer mechanisms. Neddylation inhibition by MLN4924 exerts anticancer effects mainly by triggering cell apoptosis, senescence, and autophagy and causing angiogenesis suppression, inflammatory responses, and chemo-/radiosensitization

CHOP-DR5 axis and subsequent exogenous apoptosis is an important mechanism of MLN4924-triggered apoptosis.

18.4.2 Induction of Senescence

Senescence is a state of irreversible cell cycle arrest, characterized by the enlarged and flattened cellular morphology and positive staining of senescence-associated β-galactosidase. Neddylation inhibition by MLN4924 induces characteristic senescence phenotypes and further suppresses malignant phenotypes of cancer cells (Jia et al. 2011; Lin et al. 2010a). It is noteworthy that MLN4924-induced senescence appears to be a universal phenotype in a wide range of cancer types, including lung cancer, colon cancer, glioblastoma, lymphoma, etc. (Li et al. 2014; Gao et al. 2014; Hua et al. 2015; Lin et al. 2010a; Wang et al. 2015; Zhang et al. 2016; Lan et al. 2016; Huang et al. 2015). Notably, given that MLN4924 could induce irreversible senescence of cancer cells at a minimal dose and that normal fibroblasts are relatively insensitive to MLN4924, it is possible to use low doses of drugs to achieve a higher therapeutic index (Jia et al. 2011).

Mechanistic studies revealed that MLN4924-induced senescence is associated with DNA damage response, likely triggered by DNA re-replication due to the accumulation of DNA-licensing proteins (e.g., CDT1 and ORC1), resulting from inactivation of CRLs (Jia et al. 2011; Lin et al. 2010a; Wang et al. 2015)

(Fig. 18.3). Furthermore, MLN4924-induced senescence is irreversible and coupled with persistent accumulation of $CRL1^{SKP2}$ and $CRL4^{CDT2}$ substrate p21, but not pRB/p16 and p53, as evidenced by the fact that simultaneous knockdown of *p21* remarkably abrogates MLN4924-induced senescence (Jia et al. 2011; Wang et al. 2015; Lin et al. 2010b) (Fig. 18.3). Besides p21, another $CRL1^{SKP2}$ substrate p27 is also involved in the MLN4924-induced senescence (Jia et al. 2011; Wang et al. 2015) (Fig. 18.3). Consistently, cellular senescence, induced by SKP2 inactivation, depends on p21 and p27, but instead p53 (Lin et al. 2010b). Given that MLN4924-induced senescence is largely independent of p53, a great deal of human cancers, regardless of the state of p53, can be treated with MLN4924.

18.4.3 Induction of Autophagy

MLN4924 not only induces apoptosis and senescence to mediate the killing of cancer cells but also induces protective autophagy of cancer cells. To this end, MLN4924 was shown to effectively induce autophagy with a time-dependent conversion of LC3-I to LC3-II in multiple human cancer lines, indicating a general phenomenon (Zhao et al. 2011a, 2012, 2014; Luo et al. 2012a, b; Yang et al. 2013; Chen et al. 2015; Lv et al. 2018) (Fig. 18.3). Mechanistically, MLN4924-induced autophagy is mainly mediated by the inactivation of mTOR, as evidenced by the reduced phosphorylation of mTOR itself and two mTORC1 substrates, S6K1 and 4E-BP1. MLN4924-induced mTORC1 inactivation is attributed to the accumulation of (Zhao et al. 2012) (i) DEPTOR, a substrate of $CRL1^{\beta\text{-TrCP}}$ and a natural inhibitor of mTORCs, and (ii) HIF1α, a widely known substrate of $CRL2^{VHL}$, which causes the activation of the HIF1-α-REDD1-TSC1 axis. Meanwhile, ROS stress seems to participate in the induction of MLN4924-triggered autophagy, as ROS scavenger NAC could significantly block autophagy response upon MLN4924 treatment (Luo et al. 2012a).

Functionally, neddylation inactivation by MLN4924 induces pro-survival autophagy, and blockage of autophagy with chloroquine, a clinically available autophagy inhibitor, significantly enhances the anticancer effect of MLN4924 by inducing apoptosis (Zhao et al. 2012; Yang et al. 2012a, 2013). The enhanced apoptosis by dual inhibition of autophagy and neddylation pathway is attributed to the upregulation of pro-apoptotic proteins (e.g., NOXA) and downregulation of anti-apoptotic proteins (e.g., c-IAP). Specifically, the induction of NOXA is partially triggered by DNA damage stress, as evidenced by MLN4924 causing the accumulation of DNA replication licensing proteins CDT1 and ORC1 (Chen et al. 2015) (Fig. 18.4). Besides, oxidative stress induced by the dual inhibition is also involved in the NOXA induction, based on that MLN4924-chloroquine combination causes mitochondrial depolarization and results in ROS production (Chen et al. 2015). These findings provide important preclinical evidence for clinical testing of combinatorial inhibition of neddylation and autophagy.

18.4.4 Suppression of Angiogenesis

Aberrant angiogenesis, a major feature of malignant tumors, promotes tumorigenesis, tumor invasion, and metastasis and serves as a promising anticancer drug target. Recent studies showed that all the components of neddylation modification, including E1, E2, and E3 enzymes, are expressed and functional in human umbilical vein endothelial cells (HUVECs) (Li et al. 2014; Yao et al. 2014). Besides, MLN4924-treated HUVECs showed significant decrease in the levels of total NEDD8-conjugated proteins and cullin neddylation, thus suppressing the formation of capillary-like tube networks, transwell migration, and migrated distance in a dose-dependent manner. MLN4924 also exerts suppressive effect on angiogenesis in several classical angiogenic assays (e.g., in vitro aortic ring, in vivo chick embryo chorioallantoic membrane, and matrigel plug). Furthermore, MLN4924 also strongly inhibits tumor angiogenesis in mouse model of pancreatic cancer and melanoma, leading to suppression of tumor growth and metastasis (Yao et al. 2014; Tan et al. 2014; Jin et al. 2018).

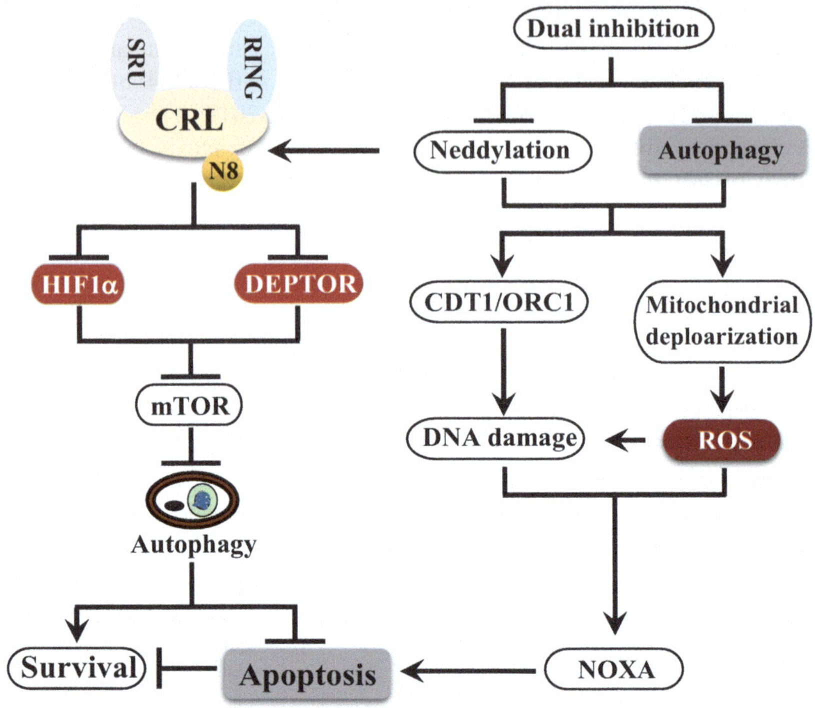

Fig. 18.4 Combination of MLN4924 with inhibitors of autophagy (e.g., chloroquine) promotes the apoptosis of cancer cells

Mechanistically, the activity of MLN4924 against endothelial cells activation is largely mediated by inactivating CRLs and subsequently accumulating different sets of CRLs substrates. At the early stages post MLN4924 treatment, when cell viability is not obviously disturbed, the suppressive effect of MLN4924 on HUVECs is attributed to the accumulation of CRLs substrate RhoA, a member of the Rho GTPase family, which inhibits cell migration and capillary tube formation (Yao et al. 2014; Hill et al. 1995; Chen et al. 2009). Importantly, knockdown of accumulated *RhoA* remarkably rescues MLN4924-mediated angiogenesis inhibition, suggesting its causal roles (Yao et al. 2014). With prolonged exposure time, MLN4924 induces DNA damage response, cell cycle arrest, and apoptosis of endothelial cells due to the accumulation of cell cycle-related CRLs substrates (e.g., p21, p27, and WEE1) and pro-apoptotic protein (e.g., NOXA, which is transactivated by CRLs substrate ATF4) (Yao et al. 2014). Collectively, these findings reveal that neddylation pathway plays a crucial role in the regulation of tumor angiogenesis and provides evidence for the future development of neddylation inhibitors (e.g., MLN4924) as a novel class of anti-angiogenic drugs.

18.4.5 Regulation of Infiltrated Immune Cells

Different types of infiltrated immune cells are involved in tumor microenvironment (TME) and play critical roles in all stages of tumor development from initiation, promotion, and progression to metastasis. Thus, targeting these immune cells is likely to be a promising anticancer strategy. Recently, a number of studies have implicated a potential role of neddylation modification in the regulation of functions of several immune cells, including macrophages, T cells, and dendritic cells (DCs). For macrophages, blocking neddylation, either pharmacologically (e.g., MLN4924) or genetically (e.g., siRNA), represses lipopolysaccharides (LPS)-induced production of proinflammatory cytokines (e.g., TNF-α and IL-6), through inhibiting CRL1-mediated IκB degradation to block NF-κB translocation and transcriptional activation (Chang et al. 2012; Li et al. 2013). Apart from promoting the functions of macrophages, neddylation pathway is also essential for their proliferation and survival, through facilitating cell cycle progression and preventing apoptosis (Li et al. 2013). Consistently, manipulation of RBX2 was found to regulate macrophage survival/death and immune response when challenged by pathogen-associated molecular patterns (PAMPs), with accumulation of pro-apoptotic proteins (e.g., BAX) to induce apoptosis (Chang and Ding 2014).

Similar to macrophages, neddylation pathway in DCs is associated with the activity of DCs and its immune regulation. Researchers found that MLN4924 remarkably suppresses the production of cytokines TNF-α and IL-6, through the accumulation of CRL substrates Deptor and IκB (Cheng et al. 2016; Mathewson et al. 2013). Moreover, secretion of IL-12p70, a key cytokine produced by DCs for Th1 differentiation, and the expression of co-stimulatory molecules are significantly suppressed with MLN4924 treatment, suggesting the restricted capacity in T-cell activation and immune responses (Cheng et al. 2016). In addition, MLN4924 treatment or *NEDD8* knockdown could also trigger apoptosis or necroptosis of DCs in a caspase-dependent manner, resulting in the reduction of functional DCs (Cheng et al. 2016; El-Mesery et al. 2015). For T cells, blockade of neddylation pathway either by MLN4924 treatment or siRNA-mediated depletion of UBE2M induces $CD4^+$ T cells G0/G1 phase arrest, leading to much slower cell division than control T cells (Jin et al. 2013). Moreover, neddylation inhibition leads to impaired antigen-driven cytokine production (e.g., IFN-γ, IL-2 and IL-4), which is required for efficient Th1 and Th2 differentiation, demonstrating a potent positive function of neddylation pathway in T-cell activation (Jin et al. 2013; Cheng et al. 2018). Taken together, these findings highlight a critical role of the neddylation pathway in regulation of several immune cell types and thus a promising role in antitumor immune responses.

18.4.6 MLN4924 Serves as a Chemo-/Radiosensitizer

In addition to the anticancer effect via abovementioned mechanisms as a single agent, recent studies have shown that MLN4924 can also sensitize several types of cancer cells to chemo- or radiotherapy. As for chemotherapy, there are several possible mechanisms triggered by MLN4924 as a potential sensitizer (Table 18.3): (i) MLN4924 sensitizes acute myeloid leukemia cells to azacitidine via inducing the accumulation of pro-apoptotic proteins such as CDT1 and WEE1 (Visconte et al. 2016), and a phase 1b study of pevonedistat with azacitidine based on synergistic activity was seen preclinically (Swords et al. 2018). MLN4924 sensitizes AML cells to HDAC inhibitor belinostat via triggering the expression of pro-apoptotic protein Bim (Zhou et al. 2016) and increases cellular sensitivity to cytarabine by disrupting nucleotide metabolism (Nawrocki et al. 2015).

(ii) The combination of MLN4924 and retinoic acid, a classical anti-leukemia chemotherapeutic drug, can significantly enhance apoptosis by inducing the expression of c-Jun and NOXA (Tan et al. 2011).

(iii) MLN4924 increases cellular sensitivity to bortezomib in multiple myeloma cells, by disrupting AKT and mTOR signaling with the accumulation of REDD1 (Gu et al. 2014).

(iv) MLN4924 sensitizes head and neck cancer cells and neoplastic B cells to TRAIL-induced apoptosis by promoting the degradation of anti-apoptotic protein C-FLIP (Zhao et al. 2011b; Paiva et al. 2017).

(v) MLN4924 significantly enhances the sensitivity of ovarian cancer, lung cancer, cholangiocarcinoma, and urothelial carcinoma to platinum-based chemotherapeutics, by inducing DNA damage and oxidative stress and increasing the expression of pro-apoptotic proteins such as BIK/NOXA (Nawrocki et al. 2013; Lin et al. 2015, 2018; Ho et al. 2015).

(vi) MLN4924 increases the efficacy of gemcitabine, a standard chemotherapy in

Table 18.3 MLN4924 serves as a chemosensitizer

Drug	Cancer	Mechanisms	Ref.
Azacitidine	Acute myeloid leukemia	Apoptosis via inducing accumulation of CDT1 and WEE1	Visconte et al. 2016
Belinostat	Acute myeloid leukemia	Apoptosis via inducing accumulation of Bim	Zhou et al. 2016
Cytarabine	Acute myeloid leukemia	Disrupts nucleotide metabolism	Nawrocki et al. 2015
Retinoic acid	Leukemia	Apoptosis via inducing accumulation of c-Jun and NOXA	Tan et al. 2011
Bortezomib	Myeloma	Disrupts AKT and mTOR signaling with the accumulation of REDD1	Gu et al. 2014
TRAIL	Head and neck cancer	Apoptosis by promoting the degradation of anti-apoptotic protein C-FLIP	Zhao et al. 2011b; Paiva et al. 2017
	Lymphoid malignancies		
Platinum	Ovarian cancer	Induces DNA damage and oxidative stress and increases the expression of pro-apoptotic proteins such as BIK/NOXA	Nawrocki et al. 2013; Lin et al. 2015, 2018; Ho et al. 2015
	Lung cancer		
	Cholangiocarcinoma		
	Urothelial carcinoma		
Gemcitabine	Pancreatic cancer	Apoptosis via inducing accumulation of NOXA and ERBIN	Li et al. 2017
ICL agents	Cervical carcinoma	Suppresses cells growth via inhibiting DNA damage-induced activation of FANCD3 and CHK1	Kee et al. 2012; Garcia et al. 2014
	Colon cancer		
	Osteosarcoma		
	Melanoma		
	Lung cancer		

pancreatic cancer, via inducing accumulation of NOXA and ERBIN, a natural occurring inhibitor of RAS-MAPK pathway (Li et al. 2017).

(vii) Finally, in several types of cancer cells, the combination of MLN4924 with DNA inter-strand cross-link (ICL) agents (e.g., mitomycin and hydroxyurea) suppresses cells growth via inhibiting DNA damage-induced activation of FANCD3 and CHK1 (Kee et al. 2012; Garcia et al. 2014).

Besides chemosensitization, MLN4924 also displays a sensitizing effect on radiotherapy. In pancreatic and colorectal cancer, MLN4924 effectively sensitizes cells to ionizing radiation, which is mainly attributed to the accumulation of CRLs substrates (e.g., CDT1, WEE1, NOXA), in parallel with DNA damage, G2/M phase cell cycle arrest, and apoptosis (Wei et al. 2012; Wan et al. 2016). In breast cancer cells, MLN4924-induced radiosensitization is at least partially mediated by the accumulation of p21, but not CDT1 or WEE1 (Yang et al. 2012b), whereas combination of MLN4924 with two-deoxy-D-glucose (2DG) enhances the efficacy of radiotherapy (Oladghaffari et al. 2017). In prostate cancer cells, MLN4924 significantly increases the efficacy of ionizing radiation against the hormone-resistant prostate cancer cells, by triggering G2 cell cycle arrest, DNA damage, and apoptosis, which is attributed to accumulated WEE1/p21/p27 (Wang et al. 2016). Finally, in head and neck squamous cells, MLN4924 radiosensitization was observed with the accumulation of CDT1 and subsequent induction of DNA re-replication (Vanderdys et al. 2018). Collectively, MLN4924 can serve either as an anticancer agent alone or more effectively in combination with chemo- or radiotherapy for a variety of chemo-/radio-resistant cancers.

In conclusion, MLN4924 exerts anticancer effects mainly by inducing apoptosis, senescence, and autophagy in a cell-type-dependent manner. It's worth investigating what determines the fate of cells when exposed to MLN4924—drug dosage, treatment time, or intrinsic changes within the cancer cells? Thus, identification of effective biomarkers for determination of cell fate will be very helpful to maximize the therapeutic efficacy of MLN4924 as a single or combined therapy. Finally, MLN4924, in addition to acting on tumor cells, also influences the functions of multiple important components of the tumor microenvironment, including immune cells and endothelial cells, all of which are crucial for tumorigenesis, further highlighting the overall anticancer efficacy of neddylation inhibition.

Acknowledgments The Chinese Minister of Science and Technology grant (2016YFA0501800), National Natural Science Foundation of China (Grant Nos. 81820108022, 81625018, 81572340, 81772470, 81602072, 81401893, 81702244, and 81871870), Innovation Program of Shanghai Municipal Education Commission (2019-01-07-00-10-E00056), National Thirteenth Five-Year Science and Technology Major Special Project for New Drug and Development (2017ZX09304001), and Program of Shanghai Academic/Technology Research Leader (18XD1403800) supported this work.

References

Abdelmohsen K, Gorospe M (2010) Posttranscriptional regulation of cancer traits by HuR. Wiley Interdiscip Rev RNA 1(2):214–229

Abida WM et al (2007) FBXO11 promotes the Neddylation of p53 and inhibits its transcriptional activity. J Biol Chem 282(3):1797–1804

Barbier-Torres L et al (2015) Stabilization of LKB1 and Akt by neddylation regulates energy metabolism in liver cancer. Oncotarget 6(4):2509–2523

Bhatia S et al (2016) A phase I study of the investigational NEDD8-activating enzyme inhibitor pevonedistat (TAK-924/MLN4924) in patients with metastatic melanoma. Invest New Drugs 34(4):439–449

Bohnsack RN, Haas AL (2003) Conservation in the mechanism of Nedd8 activation by the human AppBp1-Uba3 heterodimer. J Biol Chem 278(29):26823–26830

Broemer M et al (2010) Systematic in vivo RNAi analysis identifies IAPs as NEDD8-E3 ligases. Mol Cell 40 (5):810–822

Brownell JE et al (2010) Substrate-assisted inhibition of ubiquitin-like protein-activating enzymes: the NEDD8 E1 inhibitor MLN4924 forms a NEDD8-AMP mimetic in situ. Mol Cell 37(1):102–111

Chang SC, Ding JL (2014) Ubiquitination by SAG regulates macrophage survival/death and immune response during infection. Cell Death Differ 21 (9):1388–1398

Chang FM et al (2012) Inhibition of neddylation represses lipopolysaccharide-induced proinflammatory cytokine production in macrophage cells. J Biol Chem 287 (42):35756–35767

Chen Y et al (2009) Cullin mediates degradation of RhoA through evolutionarily conserved BTB adaptors to control actin cytoskeleton structure and cell movement. Mol Cell 35(6):841–855

Chen P et al (2015) Synergistic inhibition of autophagy and neddylation pathways as a novel therapeutic approach for targeting liver cancer. Oncotarget 6 (11):9002–9017

Chen P et al (2016) Neddylation inhibition activates the extrinsic apoptosis pathway through ATF4-CHOP-DR5 axis in human esophageal cancer cells. Clin Cancer Res 22(16):4145–4157

Cheng M et al (2016) Inhibition of neddylation regulates dendritic cell functions via Deptor accumulation driven mTOR inactivation. Oncotarget 7(24):35643–35654

Cheng Q et al (2018) Neddylation contributes to CD4+ T cell-mediated protective immunity against blood-stage Plasmodium infection. PLoS Pathog 14(11):e1007440

Coleman KE et al (2017) SENP8 limits aberrant neddylation of NEDD8 pathway components to promote cullin-RING ubiquitin ligase function. Elife 6

Cope GA, Deshaies RJ (2006) Targeted silencing of Jab1/Csn5 in human cells downregulates SCF activity through reduction of F-box protein levels. BMC Biochem 7:1

Czuczman NM et al (2016) Pevonedistat, a NEDD8-activating enzyme inhibitor, is active in mantle cell lymphoma and enhances rituximab activity in vivo. Blood 127(9):1128–1137

Dengler MA et al (2014) Discrepant NOXA (PMAIP1) transcript and NOXA protein levels: a potential Achilles' heel in mantle cell lymphoma. Cell Death Dis 5: e1013

Deshaies RJ (1999) SCF and Cullin/Ring H2-based ubiquitin ligases. Annu Rev Cell Dev Biol 15:435–467

Duan H et al (1999) SAG, a novel zinc RING finger protein that protects cells from apoptosis induced by redox agents. Mol Cell Biol 19(4):3145–3155

Duda DM et al (2008) Structural insights into NEDD8 activation of cullin-RING ligases: conformational control of conjugation. Cell 134(6):995–1006

El-Mesery M et al (2015) MLN4924 sensitizes monocytes and maturing dendritic cells for TNF-dependent and -independent necroptosis. Br J Pharmacol 172 (5):1222–1236

Embade N et al (2012) Murine double minute 2 regulates Hu antigen R stability in human liver and colon cancer through NEDDylation. Hepatology 55(4):1237–1248

Enchev RI, Schulman BA, Peter M (2015) Protein neddylation: beyond cullin-RING ligases. Nat Rev Mol Cell Biol 16(1):30–44

Gao F et al (2006) Neddylation of a breast cancer-associated protein recruits a class III histone deacetylase that represses NFkappaB-dependent transcription. Nat Cell Biol 8(10):1171–1177

Gao Q et al (2014) Neddylation pathway is up-regulated in human intrahepatic cholangiocarcinoma and serves as a potential therapeutic target. Oncotarget 5 (17):7820–7832

Garcia K et al (2014) Nedd8-activating enzyme inhibitor MLN4924 provides synergy with mitomycin C through interactions with ATR, BRCA1/BRCA2, and chromatin dynamics pathways. Mol Cancer Ther 13 (6):1625–1635

Godbersen JC et al (2014) The Nedd8-activating enzyme inhibitor MLN4924 thwarts microenvironment-driven NF-kappaB activation and induces apoptosis in chronic lymphocytic leukemia B cells. Clin Cancer Res 20(6):1576–1589

Gong L, Yeh ET (1999) Identification of the activating and conjugating enzymes of the NEDD8 conjugation pathway. J Biol Chem 274(17):12036–12042

Gu Y et al (2014) MLN4924, an NAE inhibitor, suppresses AKT and mTOR signaling via upregulation of REDD1 in human myeloma cells. Blood 123 (21):3269–3276

Guihard S et al (2012) The NEDD8 conjugation pathway regulates p53 transcriptional activity and head and neck cancer cell sensitivity to ionizing radiation. Int J Oncol 41(4):1531–1540

Hammill JT et al (2018) Discovery of an orally bioavailable inhibitor of defective in cullin neddylation 1 (DCN1)-mediated cullin neddylation. J Med Chem 61(7):2694–2706

Hemelaar J et al (2004) Specific and covalent targeting of conjugating and deconjugating enzymes of ubiquitin-like proteins. Mol Cell Biol 24(1):84–95

Hill CS, Wynne J, Treisman R (1995) The Rho family GTPases RhoA, Rac1, and CDC42Hs regulate transcriptional activation by SRF. Cell 81(7):1159–1170

Ho IL et al (2015) MLN4924 synergistically enhances cisplatin-induced cytotoxicity via JNK and Bcl-xL pathways in human urothelial carcinoma. Sci Rep 5:16948

Hu J et al (2004) Targeted ubiquitination of CDT1 by the DDB1-CUL4A-ROC1 ligase in response to DNA damage. Nat Cell Biol 6(10):1003–1009

Hua W et al (2015) Suppression of glioblastoma by targeting the overactivated protein neddylation pathway. Neuro Oncol 17(10):1333–1343

Huang DT et al (2005) Structural basis for recruitment of Ubc12 by an E2 binding domain in NEDD8's E1. Mol Cell 17(3):341–350

Huang DT et al (2009) E2-RING expansion of the NEDD8 cascade confers specificity to cullin modification. Mol Cell 33(4):483–495

Huang J et al (2015) NEDD8 inhibition overcomes CKS1B-induced drug resistance by upregulation of p21 in multiple myeloma. Clin Cancer Res 21 (24):5532–5542

Jia L et al (2010) Validation of SAG/RBX2/ROC2 E3 ubiquitin ligase as an anticancer and radiosensitizing target. Clin Cancer Res 16(3):814–824

Jia L, Li H, Sun Y (2011) Induction of p21-dependent senescence by an NAE inhibitor, MLN4924, as a mechanism of growth suppression. Neoplasia 13 (6):561–569

Jia X et al (2019) Neddylation inactivation facilitates FOXO3a nuclear export to suppress estrogen receptor transcription and improve fulvestrant sensitivity. Clin Cancer Res

Jin HS et al (2013) Neddylation pathway regulates T-cell function by targeting an adaptor protein Shc and a protein kinase Erk signaling. Proc Natl Acad Sci U S A 110(2):624–629

Jin Y et al (2018) Neddylation blockade diminishes hepatic metastasis by dampening cancer stem-like cells and angiogenesis in uveal melanoma. Clin Cancer Res 24(15):3741–3754

Kamitani T et al (1997) Characterization of NEDD8, a developmentally down-regulated ubiquitin-like protein. J Biol Chem 272(45):28557–28562

Kee Y et al (2012) Inhibition of the Nedd8 system sensitizes cells to DNA interstrand cross-linking agents. Mol Cancer Res 10(3):369–377

Khalife J et al (2015) Pharmacological targeting of miR-155 via the NEDD8-activating enzyme inhibitor MLN4924 (Pevonedistat) in FLT3-ITD acute myeloid leukemia. Leukemia 29(10):1981–1992

Knorr KL et al (2015) MLN4924 induces Noxa upregulation in acute myelogenous leukemia and synergizes with Bcl-2 inhibitors. Cell Death Differ 22 (12):2133–2142

Kurz T et al (2005) The conserved protein DCN-1/Dcn1p is required for cullin neddylation in C. elegans and S. cerevisiae. Nature 435(7046):1257–1261

Kurz T et al (2008) Dcn1 functions as a scaffold-type E3 ligase for cullin neddylation. Mol Cell 29(1):23–35

Lan H et al (2016) Neddylation inhibitor MLN4924 suppresses growth and migration of human gastric cancer cells. Sci Rep 6:24218

Leclerc GM et al (2016) The NEDD8-activating enzyme inhibitor pevonedistat activates the eIF2alpha and mTOR pathways inducing UPR-mediated cell death in acute lymphoblastic leukemia. Leuk Res 50:1–10

Lee MR et al (2008) Inhibition of APP intracellular domain (AICD) transcriptional activity via covalent conjugation with Nedd8. Biochem Biophys Res Commun 366(4):976–981

Lee MH et al (2011) Roles of COP9 signalosome in cancer. Cell Cycle 10(18):3057–3066

Li X et al (2003) The SCF(Skp2) ubiquitin ligase complex interacts with the human replication licensing factor Cdt1 and regulates Cdt1 degradation. J Biol Chem 278(33):30854–30858

Li L et al (2013) Neddylation pathway regulates the proliferation and survival of macrophages. Biochem Biophys Res Commun 432(3):494–498

Li L et al (2014) Overactivated neddylation pathway as a therapeutic target in lung cancer. J Natl Cancer Inst 106 (6):dju083

Li H et al (2017) Inhibition of neddylation modification sensitizes pancreatic cancer cells to gemcitabine. Neoplasia 19(6):509–518

Lin JJ et al (2010a) NEDD8-targeting drug MLN4924 elicits DNA rereplication by stabilizing Cdt1 in S phase, triggering checkpoint activation, apoptosis, and senescence in cancer cells. Cancer Res 70 (24):10310–10320

Lin HK et al (2010b) Skp2 targeting suppresses tumorigenesis by Arf-p53-independent cellular senescence. Nature 464(7287):374–379

Lin WC et al (2015) MLN4924, a Novel NEDD8-activating enzyme inhibitor, exhibits antitumor activity and enhances cisplatin-induced cytotoxicity in human cervical carcinoma: in vitro and in vivo study. Am J Cancer Res 5(11):3350–3362

Lin S et al (2018) Neddylation inhibitor MLN4924 induces G2 cell cycle arrest, DNA damage and sensitizes esophageal squamous cell carcinoma cells to cisplatin. Oncol Lett 15(2):2583–2589

Lo SC, Hannink M (2006) CAND1-mediated substrate adaptor recycling is required for efficient repression of Nrf2 by Keap1. Mol Cell Biol 26(4):1235–1244

Loftus SJ et al (2012) NEDDylation regulates E2F-1-dependent transcription. EMBO Rep 13(9):811–818

Luo Z et al (2012a) The Nedd8-activating enzyme inhibitor MLN4924 induces autophagy and apoptosis to suppress liver cancer cell growth. Cancer Res 72 (13):3360–3371

Luo Z et al (2012b) Inactivation of the Cullin (CUL)-RING E3 ligase by the NEDD8-activating enzyme inhibitor MLN4924 triggers protective autophagy in cancer cells. Autophagy 8(11):1677–1679

Lv Y et al (2018) The Nedd8-activating enzyme inhibitor MLN4924 suppresses colon cancer cell growth via triggering autophagy. Korean J Physiol Pharmacol 22 (6):617–625

Lyapina S et al (2001) Promotion of NEDD-CUL1 conjugate cleavage by COP9 signalosome. Science 292 (5520):1382–1385

Ma T et al (2013) RNF111-dependent neddylation activates DNA damage-induced ubiquitination. Mol Cell 49(5):897–907

Mahata B, Sundqvist A, Xirodimas DP (2012) Recruitment of RPL11 at promoter sites of p53-regulated genes upon nucleolar stress through NEDD8 and in an Mdm2-dependent manner. Oncogene 31 (25):3060–3071

Mathewson N et al (2013) Neddylation plays an important role in the regulation of murine and human dendritic cell function. Blood 122(12):2062–2073

Merlet J et al (2009) Regulation of cullin-RING E3 ubiquitin-ligases by neddylation and dimerization. Cell Mol Life Sci 66(11–12):1924–1938

Meyer-Schaller N et al (2009) The human Dcn1-like protein DCNL3 promotes Cul3 neddylation at membranes. Proc Natl Acad Sci U S A 106 (30):12365–12370

Milhollen MA et al (2010) MLN4924, a NEDD8-activating enzyme inhibitor, is active in diffuse large B-cell lymphoma models: rationale for treatment of NF-{kappa}B-dependent lymphoma. Blood 116 (9):1515–1523

Milhollen MA et al (2011) Inhibition of NEDD8-activating enzyme induces rereplication and apoptosis in human tumor cells consistent with deregulating CDT1 turnover. Cancer Res 71(8):3042–3051

Milhollen MA et al (2012) Treatment-emergent mutations in NAEbeta confer resistance to the NEDD8-activating enzyme inhibitor MLN4924. Cancer Cell 21 (3):388–401

Mo Z et al (2016) Neddylation requires glycyl-tRNA synthetase to protect activated E2. Nat Struct Mol Biol 23(8):730–737

Nawrocki ST et al (2013) Disrupting protein NEDDylation with MLN4924 is a novel strategy to target cisplatin resistance in ovarian cancer. Clin Cancer Res 19 (13):3577–3590

Nawrocki ST et al (2015) The NEDD8-activating enzyme inhibitor MLN4924 disrupts nucleotide metabolism and augments the efficacy of cytarabine. Clin Cancer Res 21(2):439–447

Noguchi K et al (2011) TRIM40 promotes neddylation of IKKgamma and is downregulated in gastrointestinal cancers. Carcinogenesis 32(7):995–1004

Oladghaffari M et al (2017) MLN4924 and 2DG combined treatment enhances the efficiency of radiotherapy in breast cancer cells. Int J Radiat Biol 93(6):590–599

Oved S et al (2006) Conjugation to Nedd8 instigates ubiquitylation and down-regulation of activated receptor tyrosine kinases. J Biol Chem 281 (31):21640–21651

Paiva C et al (2017) Pevonedistat, a Nedd8-activating enzyme inhibitor, sensitizes neoplastic B-cells to death receptor-mediated apoptosis. Oncotarget 8 (13):21128–21139

Pan Y et al (2012) Induction of cell senescence by targeting to Cullin-RING Ligases (CRLs) for effective cancer therapy. Int J Biochem Mol Biol 3(3):273–281

Petroski MD, Deshaies RJ (2005) Function and regulation of cullin-RING ubiquitin ligases. Nat Rev Mol Cell Biol 6(1):9–20

Rabut G, Peter M (2008) Function and regulation of protein neddylation. 'Protein modifications: beyond the usual suspects' review series. EMBO Rep 9 (10):969–976

Rabut G et al (2011) The TFIIH subunit Tfb3 regulates cullin neddylation. Mol Cell 43(3):488–495

Reverter D et al (2005) Structure of a complex between Nedd8 and the Ulp/Senp protease family member Den1. J Mol Biol 345(1):141–151

Ryu JH et al (2011) Hypoxia-inducible factor alpha subunit stabilization by NEDD8 conjugation is reactive oxygen species-dependent. J Biol Chem 286 (9):6963–6970

Sakata E et al (2007) Direct interactions between NEDD8 and ubiquitin E2 conjugating enzymes upregulate cullin-based E3 ligase activity. Nat Struct Mol Biol 14(2):167–168

Sarantopoulos J et al (2016) Phase I study of the investigational NEDD8-activating enzyme inhibitor pevonedistat (TAK-924/MLN4924) in patients with advanced solid tumors. Clin Cancer Res 22(4):847–857

Scott DC et al (2011) N-terminal acetylation acts as an avidity enhancer within an interconnected multiprotein complex. Science 334(6056):674–678

Scott DC et al (2014) Structure of a RING E3 trapped in action reveals ligation mechanism for the ubiquitin-like protein NEDD8. Cell 157(7):1671–1684
Scott DC et al (2017) Blocking an N-terminal acetylation-dependent protein interaction inhibits an E3 ligase. Nat Chem Biol 13(8):850–857
Shah JJ et al (2016) Phase I study of the novel investigational NEDD8-activating enzyme inhibitor pevonedistat (MLN4924) in patients with relapsed/ refractory multiple myeloma or lymphoma. Clin Cancer Res 22(1):34–43
Shen LN et al (2005) Structural basis of NEDD8 ubiquitin discrimination by the deNEDDylating enzyme NEDP1. EMBO J 24(7):1341–1351
Soucy TA et al (2009) An inhibitor of NEDD8-activating enzyme as a new approach to treat cancer. Nature 458 (7239):732–736
Stickle NH et al (2004) pVHL modification by NEDD8 is required for fibronectin matrix assembly and suppression of tumor development. Mol Cell Biol 24 (8):3251–3261
Su H et al (2011) Perturbation of cullin deneddylation via conditional Csn8 ablation impairs the ubiquitin-proteasome system and causes cardiomyocyte necrosis and dilated cardiomyopathy in mice. Circ Res 108 (1):40–50
Su H et al (2013) The COP9 signalosome is required for autophagy, proteasome-mediated proteolysis, and cardiomyocyte survival in adult mice. Circ Heart Fail 6(5):1049–1057
Sundqvist A et al (2009) Regulation of nucleolar signalling to p53 through NEDDylation of L11. EMBO Rep 10(10):1132–1139
Swords RT et al (2010) Inhibition of NEDD8-activating enzyme: a novel approach for the treatment of acute myeloid leukemia. Blood 115(18):3796–3800
Swords RT et al (2015) Pevonedistat (MLN4924), a first-in-class NEDD8-activating enzyme inhibitor, in patients with acute myeloid leukaemia and myelodysplastic syndromes: a phase 1 study. Br J Haematol 169(4):534–543
Swords RT et al (2018) Pevonedistat, a first-in-class NEDD8-activating enzyme inhibitor, combined with azacitidine in patients with AML. Blood 131 (13):1415–1424
Tan M et al (2011) Inactivation of SAG E3 ubiquitin ligase blocks embryonic stem cell differentiation and sensitizes leukemia cells to retinoid acid. PLoS One 6 (11):e27726
Tan M, Li H, Sun Y (2014) Endothelial deletion of Sag/Rbx2/Roc2 E3 ubiquitin ligase causes embryonic lethality and blocks tumor angiogenesis. Oncogene 33 (44):5211–5220
Tian DW et al (2019) Neural precursor cell expressed, developmentally downregulated 8 promotes tumor progression and predicts poor prognosis of patients with bladder cancer. Cancer Sci 110(1):458–467
Tong S et al (2017) MLN4924 (Pevonedistat), a protein neddylation inhibitor, suppresses proliferation and migration of human clear cell renal cell carcinoma. Sci Rep 7(1):5599
Toth JI et al (2012) A gatekeeper residue for NEDD8-activating enzyme inhibition by MLN4924. Cell Rep 1 (4):309–316
Vanderdys V et al (2018) The neddylation inhibitor pevonedistat (MLN4924) suppresses and radiosensitizes head and neck squamous carcinoma cells and tumors. Mol Cancer Ther 17(2):368–380
Visconte V et al (2016) Comprehensive quantitative proteomic profiling of the pharmacodynamic changes induced by MLN4924 in acute myeloid leukemia cells establishes rationale for its combination with azacitidine. Leukemia 30(5):1190–1194
Walden H et al (2003) The structure of the APPBP1-UBA3-NEDD8-ATP complex reveals the basis for selective ubiquitin-like protein activation by an E1. Mol Cell 12(6):1427–1437
Wan J et al (2016) Radiosensitization of human colorectal cancer cells by MLN4924: an inhibitor of NEDD8-activating enzyme. Technol Cancer Res Treat 15 (4):527–534
Wang Y et al (2015) Targeting protein neddylation with an NEDD8-activating enzyme inhibitor MLN4924 induced apoptosis or senescence in human lymphoma cells. Cancer Biol Ther 16(3):420–429
Wang X et al (2016) Radiosensitization by the investigational NEDD8-activating enzyme inhibitor MLN4924 (pevonedistat) in hormone-resistant prostate cancer cells. Oncotarget 7(25):38380–38391
Wang J et al (2017) Targeting neddylation pathway with MLN4924 (Pevonedistat) induces NOXA-dependent apoptosis in renal cell carcinoma. Biochem Biophys Res Commun 490(4):1183–1188
Wang S et al (2019) Development of highly potent, selective, and cellular active triazolo[1,5-a]pyrimidine-based inhibitors targeting the DCN1-UBC12 protein-protein interaction. J Med Chem 62(5):2772–2797
Watson IR et al (2010) Chemotherapy induces NEDP1-mediated destabilization of MDM2. Oncogene 29 (2):297–304
Watson IR, Irwin MS, Ohh M (2011) NEDD8 pathways in cancer, Sine Quibus Non. Cancer Cell 19(2):168–176
Wei N, Serino G, Deng XW (2008) The COP9 signalosome: more than a protease. Trends Biochem Sci 33(12):592–600
Wei D et al (2012) Radiosensitization of human pancreatic cancer cells by MLN4924, an investigational NEDD8-activating enzyme inhibitor. Cancer Res 72 (1):282–293
Xie P et al (2014) The covalent modifier Nedd8 is critical for the activation of Smurf1 ubiquitin ligase in tumorigenesis. Nat Commun 5:3733
Xie P et al (2017) Promoting tumorigenesis in nasopharyngeal carcinoma, NEDD8 serves as a potential theranostic target. Cell Death Dis 8(6):e2834
Xirodimas DP (2008) Novel substrates and functions for the ubiquitin-like molecule NEDD8. Biochem Soc Trans 36(Pt 5):802–806

Xirodimas DP et al (2004) Mdm2-mediated NEDD8 conjugation of p53 inhibits its transcriptional activity. Cell 118(1):83–97

Yang D et al (2012a) Protective autophagy induced by RBX1/ROC1 knockdown or CRL inactivation via modulating the DEPTOR-MTOR axis. Autophagy 8 (12):1856–1858

Yang D et al (2012b) The p21-dependent radiosensitization of human breast cancer cells by MLN4924, an investigational inhibitor of NEDD8 activating enzyme. PLoS One 7(3):e34079

Yang D et al (2013) Induction of autophagy and senescence by knockdown of ROC1 E3 ubiquitin ligase to suppress the growth of liver cancer cells. Cell Death Differ 20(2):235–247

Yao WT et al (2014) Suppression of tumor angiogenesis by targeting the protein neddylation pathway. Cell Death Dis 5:e1059

Zhang J et al (2014) hCINAP is a novel regulator of ribosomal protein-HDM2-p53 pathway by controlling NEDDylation of ribosomal protein S14. Oncogene 33 (2):246–254

Zhang Y et al (2016) MLN4924 suppresses neddylation and induces cell cycle arrest, senescence, and apoptosis in human osteosarcoma. Oncotarget 7(29):45263–45274

Zhang W et al (2018) The Nedd8-activating enzyme inhibitor MLN4924 (TAK-924/Pevonedistat) induces apoptosis via c-Myc-Noxa axis in head and neck squamous cell carcinoma. Cell Prolif:e12536

Zhao Y, Xiong X, Sun Y (2011a) DEPTOR, an mTOR inhibitor, is a physiological substrate of SCF (betaTrCP) E3 ubiquitin ligase and regulates survival and autophagy. Mol Cell 44(2):304–316

Zhao L et al (2011b) The NEDD8-activating enzyme inhibitor, MLN4924, cooperates with TRAIL to augment apoptosis through facilitating c-FLIP degradation in head and neck cancer cells. Mol Cancer Ther 10 (12):2415–2425

Zhao Y et al (2012) Targeting Cullin-RING ligases by MLN4924 induces autophagy via modulating the HIF1-REDD1-TSC1-mTORC1-DEPTOR axis. Cell Death Dis 3:e386

Zhao Y, Morgan MA, Sun Y (2014) Targeting Neddylation pathways to inactivate cullin-RING ligases for anticancer therapy. Antioxid Redox Signal 21(17):2383–2400

Zheng N et al (2002) Structure of the Cul1-Rbx1-Skp1-F boxSkp2 SCF ubiquitin ligase complex. Nature 416 (6882):703–709

Zhou L et al (2016) The NAE inhibitor pevonedistat interacts with the HDAC inhibitor belinostat to target AML cells by disrupting the DDR. Blood 127 (18):2219–2230

Zhou H, Lu J, Liu L, Bernard D, Yang C-Y, Fernandez-Salas E, Chinnaswamy K, Layton S, Stuckey J, Yu Q, Zhou W, Pan Z-Q, Sun Y, Wang S (2017a) A potent small-molecule inhibitor of the DCN1-UBC12 interaction that selectively blocks cullin 3 neddylation. Nat Commun 8(1):1150

Zhou W et al (2017b) Neddylation E2 UBE2F promotes the survival of lung cancer cells by activating CRL5 to degrade NOXA via the K11 linkage. Clin Cancer Res 23(4):1104–1116

Zhou L et al (2018a) Protein neddylation and its alterations in human cancers for targeted therapy. Cell Signal 44:92–102

Zhou H et al (2018b) High-affinity peptidomimetic inhibitors of the DCN1-UBC12 protein-protein interaction. J Med Chem 61(5):1934–1950

Zhou W et al (2018c) UBE2M is a stress-inducible dual E2 for neddylation and ubiquitylation that promotes targeted degradation of UBE2F. Mol Cell 70 (6):1008–1024 e6

Zuo W et al (2013) c-Cbl-mediated neddylation antagonizes ubiquitination and degradation of the TGF-beta type II receptor. Mol Cell 49(3):499–510

Targeting Cullin-RING Ubiquitin Ligases and the Applications in PROTACs

19

Longyuan Gong, Danrui Cui, Xiufang Xiong, and Yongchao Zhao

Abstract

Cullin-RING ligases (CRLs), the largest family of E3 ubiquitin ligases, have become an attractive target for drug discovery, primarily due to their ability to regulate the degradation of numerous functionally and structurally diverse proteins, thereby controlling a myriad of biological processes. As the abnormal expressions of CRLs and their substrate proteins are associated with human diseases, elucidating their roles in these physiological and pathological processes will facilitate CRL-targeting drug development for the treatment of these diseases. Notably, these studies are also providing new concepts for the design of potential small-molecule therapeutics targeting CRLs and for the use of CRLs to degrade "undruggable" proteins. In this chapter, we systematically review the development of small molecules that target CRLs and especially emphasize the applications of CRLs in a chemical chimera for protein degradation, termed proteolysis-targeting chimeras (PROTACs).

Keywords

Cullin-RING ligase · Small-molecule inhibitors · Proteolysis targeting chimera · PROTAC

Abbreviations

A3G	APOBEC3G
AD	Alzheimer's disease
AML	Acute myeloid leukemia
AR	Androgen receptor
BTK	Bruton's tyrosine kinase
BTZ	Bortezomib
CAND1	Cullin-associated and neddylation-dissociated-1
CDK9	Cyclin-dependent kinase 9
COI1	Coronatine-insensitive protein 1
CRBN	Cereblon
CRL	Cullin-RING ligase
CRPC	Castration-resistant prostate cancer
DHT	Dihydrotestosterone
E2	Ubiquitin-conjugating enzyme
E3	Ubiquitin ligase
ER	Estrogen receptor

L. Gong · D. Cui · Y. Zhao (✉)
Key Laboratory of Combined Multi-Organ Transplantation, Ministry of Public Health, First Affiliated Hospital, Zhejiang University School of Medicine, Hangzhou, China

Institute of Translational Medicine, Zhejiang University School of Medicine, Hangzhou, China
e-mail: yongchao@zju.edu.cn

X. Xiong (✉)
Cancer Institute of the Second Affiliated Hospital, Zhejiang University School of Medicine, Hangzhou, China

Institute of Translational Medicine, Zhejiang University School of Medicine, Hangzhou, China
e-mail: xiufang@zju.edu.cn

Y. Sun et al. (eds.), *Cullin-RING Ligases and Protein Neddylation*, Advances in Experimental Medicine and Biology 1217, https://doi.org/10.1007/978-981-15-1025-0_19

ERRα	Estrogen-related receptor alpha
GCN5	General control nonderepressible 5
HBV	Hepatitis B virus
HBX	X protein of HBV
HDAC	Histone deacetylase
HIF1α	Hypoxia-inducible factor 1α
KEAP1	Kelch-like ECH-associated protein 1
NAE	NEDD8 activating enzyme
NEDD8	Neural precursor cell expressed, developmentally downregulated 8
NHL	Non-Hodgkin's lymphoma
PCAF	P300/CBP-associated factor
POI	Protein of interest
PPI	Protein-protein interaction
PROTAC	Proteolysis-targeting chimeric molecule
SAR	Structure-activity relationship
SCF	Skp1-CUL1-F-box
SMER	Small-molecule enhancer of rapamycin
Ub	Ubiquitin
UPS	Ubiquitin proteasome system
VHL	von Hippel-Lindau protein

19.1 Introduction

Due to mounting reports on the abnormal expressions of CRLs, the largest family of E3 ubiquitin ligases, and their substrate proteins in human diseases, including diabetes, inflammatory diseases, and cancer (Cohen and Tcherpakov 2010; Huang and Dixit 2016), there is growing enthusiasm for developing small molecules targeting CRLs for clinical applications (Zhao and Sun 2013; Zhao et al. 2014; Bulatov and Ciulli 2015). Small molecules can act directly by disrupting receptor-substrate interactions, interfering with CRL assembly, and inducing conformational changes, resulting in the inhibition of CRL activity or overall kinetic changes (Zhao and Sun 2013). The specificity of the inhibitors can be improved by inhibiting the components of specific subgroups of molecules involved in protein degradation in the ubiquitin proteasome system (UPS), and researchers are making efforts to develop more therapeutic methods in this way (Fig. 19.1). What's more, there is a new drug development strategy utilizing the UPS to remove unwanted or damaged proteins, termed proteolysis-targeting chimeras (PROTACs) (Sakamoto et al. 2001), whose bioavailability, stability, and efficacy are being proved by emerging studies, and a variety of PROTACs with good efficiency in recruiting CRLs are being optimized to enter clinical trials (Lu et al. 2015).

19.2 Inhibitors Targeting CRL

The proteasome inhibitor bortezomib and the second-generation proteasome inhibitors carfilzomib and ixazomib have been successfully applied to the treatment of relapsed/refractory multiple myeloma and mantle cell lymphoma (Richardson et al. 2017). Thus, the UPS, the central protein degradation system in mammalian cells, has proven its potential as valuable drug target for the treatment of human cancer. However, in recent clinical trials, proteasome inhibitors, mainly bortezomib, have been found to cause side effects, such as neuropathy, that interfere with its application in the treatment of relapsed/refractory lymphoma or multiple myeloma (Manasanch and Orlowski 2017). It is expected that proteasome inhibitors will inevitably present challenges including broad cellular impacts, potential risks of side effects, and increasing resistance due to the extensive block of protein degradation. Therefore, finding more specific and selective targets upstream of the proteasome is widely recognized as a promising therapeutic approach that may alleviate some of the limitations of proteasome inhibitors. Ubiquitin E3 ligases, including CRLs, have attracted particular attention in this regard because they are responsible for recognizing a number of specific substrates and mediating their ubiquitination and degradation in series of important physiological and pathological processes (Zhao and Sun 2013; Bulatov and Ciulli 2015).

Protein-protein interactions (PPIs) have long been recognized as one of the most desirable

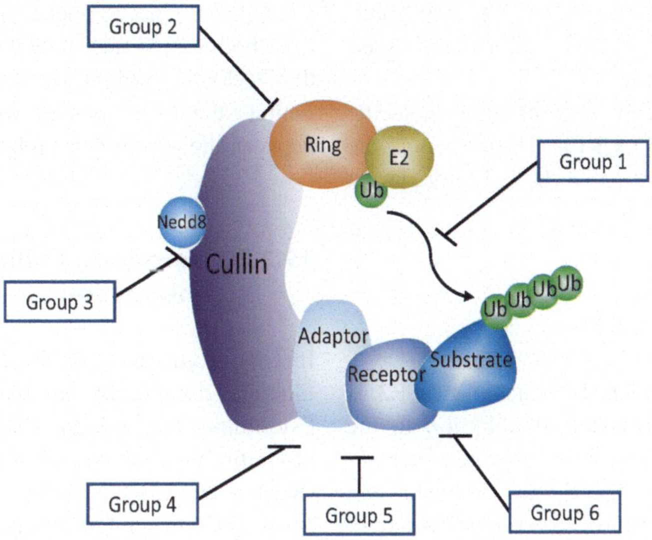

Fig. 19.1 Potential targeting points in CRLs. The inhibitors which are developed to disrupt the interaction between the components of CRL complex can be classified into six groups: (1) group 1, targeting the transfer of ubiquitin from the E2-Ub conjugate to the lysine residues on the substrate; (2) group 2, disrupting the interaction between Cullin and RING protein; (3) group 3, inhibiting the neddylation of Cullins; (4) group 4, disrupting the interaction between Cullin and adaptor; (5) group 5, targeting the interaction between adaptor and receptor; (6) group 6, targeting the interaction between receptor and substrate

types of drug targets, as many PPIs have high specificity and play an important role in biological activities (Wells and McClendon 2007). The use of small molecules to modulate PPIs has many advantages. Specifically, they have higher bioavailability than nucleic acid, polypeptide, and protein therapies and are more likely to be administered orally, resulting in better compliance (Bulatov and Ciulli 2015). However, the development of small-molecule compounds targeting PPIs of CRLs have traditionally been considered challenging because it is difficult to obtain compounds with high binding affinity to large and often uncharacteristic binding interfaces. This property has previously led to difficulties in the development of compounds targeting E3 ligases (Bulatov and Ciulli 2015). Encouragingly, in recent years, with the success of small molecules targeting E3 ligases, such as IAP (inhibitor of apoptotic proteins)/caspase (Wang et al. 2004), MDM2/p53 (Vassilev et al. 2004), and VHL/HIF-1α (Galdeano et al. 2014), researchers have become particularly interested in drug development targeting ubiquitin E3 ligases. These examples provide important and encouraging precedents for the discovery of small-molecule compounds that target ubiquitin E3 ligases.

However, it is important to be clear that (1) CRL E3 ligases, consisting of multiple independent subunits that function together in a complex, have many sites for potential chemical intervention (Fig. 19.1); (2) at present, due to the lack of clear ligand binding sites in most CRL E3 ligases, it is challenging to develop small-molecule inhibitors; and (3) each CRL targets different substrates for degradation; at the same time, a particular substrate may be recognized by different CRLs. Thus, each protein must be treated separately according to its specific structure and functional properties. Based on these factors, the successful development of lead compounds targeting CRL E3 is extremely challenging. It is necessary to combine the techniques of high-throughput screening, structure-based virtual screening and docking analysis, and structural chemistry, to accurately develop small-molecule inhibitors and elucidate their mechanisms of actions.

Next, we will classify current small-molecule inhibitors targeting CRLs, based on the components of the CRL that the inhibitors target (Fig. 19.1 and Table 19.1).

19.2.1 Targeting E2

The interaction of CRL E3 ligase with E2 is responsible for the transfer of ubiquitin to the lysine residues on the substrate protein specifically recognized by E3 ligase (Petroski and Deshaies 2005). Therefore, targeting E2 is attractive for the regulation of substrates. From this perspective, the range of affected substrate proteins is still large. However, if the profile of E2-acting CRL complexes can be accurately defined by various techniques, the range of substrate proteins affected can be narrowed. The E2-E3 interaction is rather weak and unstable with equilibrium dissociation constant (KD) values in the low micromolar range, so even selective ligands can serve as effective modulators of the PPI, but the potential is moderate (Ye and Rape 2009). Additionally, to date, only a few crystal structures of E2-E3 complexes are available, most of which involve non-CRL E3 ligases (Ye and Rape 2009; Pruneda et al. 2012).

CC0651, the first inhibitor targeting E2, was discovered via high-throughput screening of small-molecule inhibitors against the ubiquitination of $p27^{Kip1}$ by the SCF^{SKP2} E3 complex (Ceccarelli et al. 2011). Biochemical studies showed that CC0651 specifically targets the E2 binding enzyme Cdc34 that interacts with CRL1, although Cdc34 bears significant structural resemblance to other E2 enzymes. CC0651 stabilizes the interaction of E2 with ubiquitin and locks the complex in an inactive conformation, thereby inhibiting the transfer of ubiquitin from the E2-Ub conjugate to the lysine residues on the substrate. The crystal structure of the Cdc34-Ub-CC0651 complex reveals that CC0651 is located in a composite binding pocket composed of residues from both Cdc34 and ubiquitin, which determines that CC0651 bridges the Cdc34-Ub chain and inhibits the hydrolysis of weak thioester bonds between the two proteins (Huang et al. 2014).

19.2.2 Targeting Cullin-RING Interaction

In CRL complexes, RING protein binds to the globular domain of the Cullin scaffold at its C-terminus to facilitate the recruitment of E2 ubiquitin-conjugating enzyme to CRL complexes (Sarikas et al. 2011; Tron et al. 2012). Specifically, Cullin and RBX1 are docked mainly by conserved interfacial interactions involving a unique structure motif (VLYRLWLN) of Cullin and the N-terminal β-strand of RBX1 (Shafique et al. 2018). In addition, the intermolecular α/β hydrophobic core essentially makes Cullin-RBX1 inseparable. RBX2 (SAG) binds preferentially to Cullin5 and plays a key role in regulating cancer cell proliferation via apoptosis induction (Li et al. 2014). Thus, targeting RBX1 or RBX2 to inactivate CRLs would produce potential therapeutic value.

Although some of the crystal structures of CRLs have been elucidated (Zheng et al. 2002; Duda et al. 2008; Stanley et al. 2008; Fischer et al. 2011; Cardote et al. 2017), the current study of this structural connection predominantly targets the RBX1-CRL complex, mainly due to the pro-oncoprotein role of RBX1 in human cancers, such as lung cancer and liver cancer (Jia et al. 2009; Xu et al. 2015), and the specificity of RBX2 toward Cullin5. Through drug-like profiling based on virtual structure screening and drug docking analysis, researchers have proposed a set of inhibitors that can perturb the connection interface between CRLs and RBX1. Among them, one compound termed C64 (Shafique et al. 2018) was found to bind to the RBX1-binding groove (VLYRLWLN) of Cullins and showed a wide range of binding characteristics selective for CRLs.

Table 19.1 Inhibitors targeting CRLs

Group	CRL	Target	Compound	Structure	Disease model	Refs.
Group 1		Cdc34	CC0651	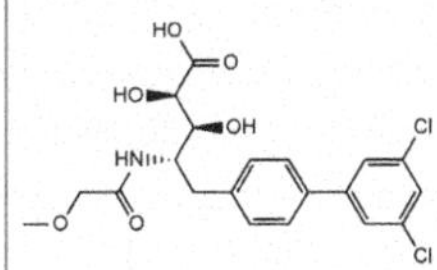	HCT116 colorectal cancer cell lines	Ceccarelli et al. (2011) and Seol et al. (1999)
Group 2		CUL-RBX1	C64	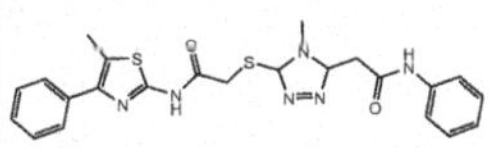	NA	Shafique et al. (2018)
Group 3		NEDD8	MLN4924	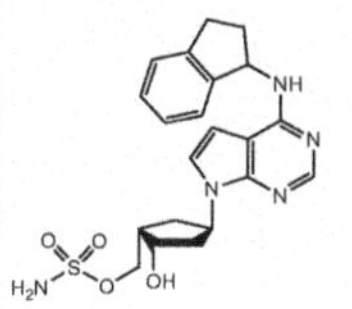	Solid tumor and hematologic malignancies	Zhao et al. (2014), Soucy et al. (2009) and Brownell et al. (2010)
Group 4	CRL1	CUL1-Skp1	Ubvs	NA	NA	Gorelik et al. (2016)
Group 5	CRL1	Skp1-Skp2	CpdA	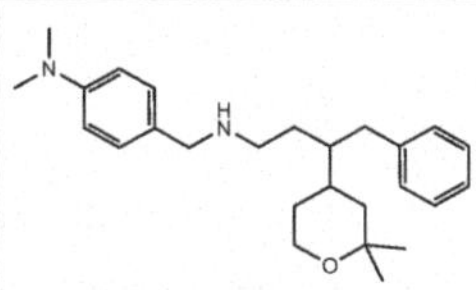	Multiple myeloma	Chen et al. (2008)
			Compound 25	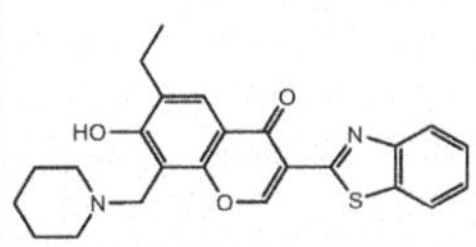	Prostate and lung tumors	Chan et al. (2013)
Group 5			DT204	NA	Myeloma	Malek et al. (2017)
Group 6	CRL1	Skp2-p27	C1, C2, C16, and C20	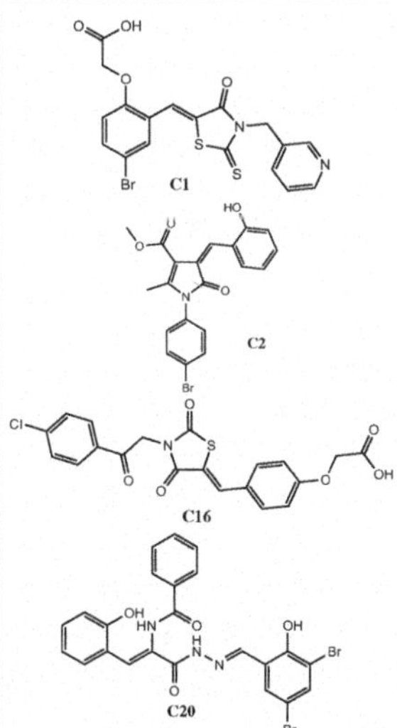	Endometrial cancer	Pavlides et al. (2013) and Wu et al. (2012)

(continued)

Table 19.1 (continued)

Group	CRL	Target	Compound	Structure	Disease model	Refs.
Group 6			SMIP001 and SMIP004	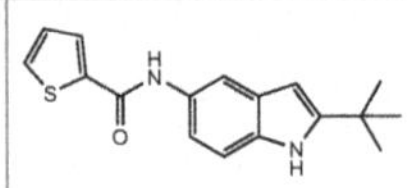 SMIP001 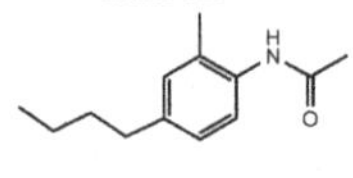 SMIP004	Prostate cancer	Rico-Bautista et al. (2010)
Group 6	CRL1	Fbxo3-Fbxl2	BC-1215	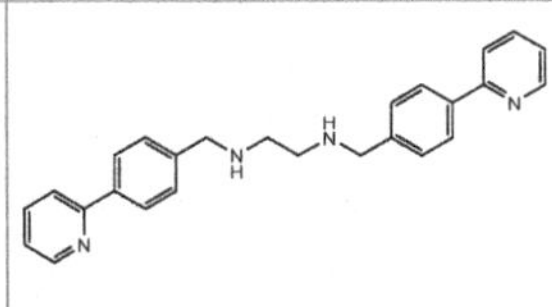	Cecal ligation and puncture (CLP)-induced sepsis	Mallampalli et al. (2013) and Chen et al. (2013)
Group 6	CRL1	Cdc4-Sic1	SCF-I2	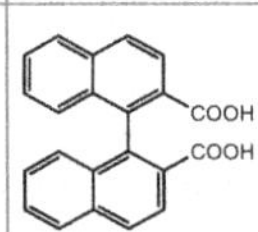	NA	Orlicky et al. (2010)
Group 6	CRL1	Met30-Met4	SMER3	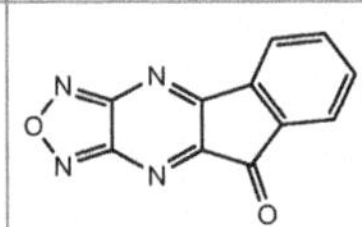	Lung cancer cells	Sarkar et al. (2007) and Aghajan et al. (2010)
Group 6	CRL1	TIR1-Aux/IAA	Probe 8	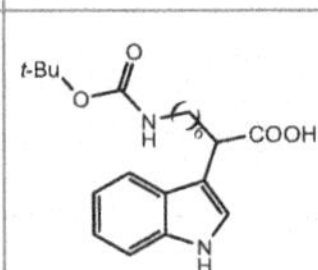	NA	Hayashi et al. (2008)
Group 6	CRL1	COI1-JAZ1	JA-Ile	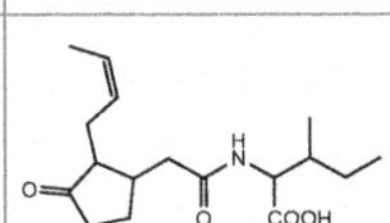	NA	Thines et al. (2007)
Group 6	CRL1	β-Trcp or IκBα	GS143	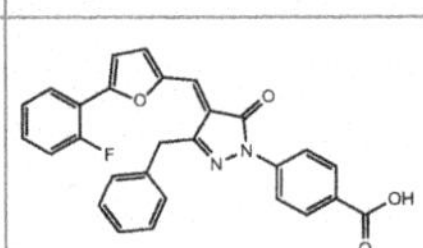	NA	Nakajima et al. (2008)
Group 6	CRL2	VHL-HIF1α	Compound 15, Compound 7, VH298	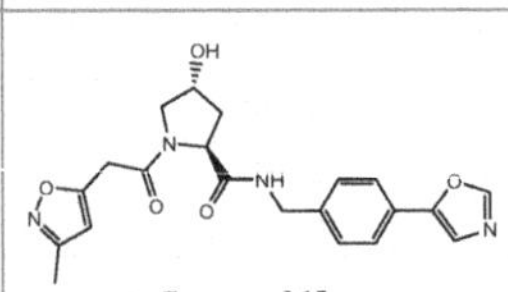 Compound 15 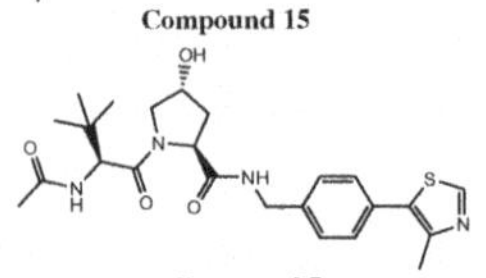 Compound 7 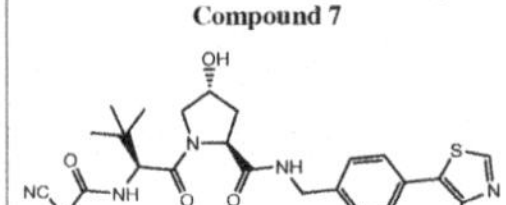 VH298	NA	Galdeano et al. (2014), Buckley et al. (2012b) and Frost et al. (2016)

(continued)

Table 19.1 (continued)

Group	CRL	Target	Compound	Structure	Disease model	Refs.
Group 6	CRL3	KEAP1-Nrf2	SRS1a, SRS-59, Cpd15, and Cpd16	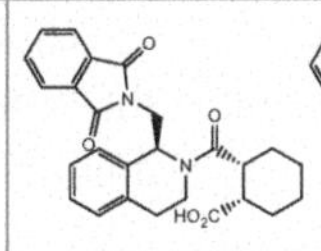 (*S, R, S*)-**1a** 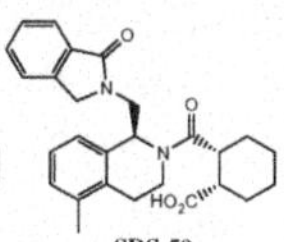 **SRS-59** 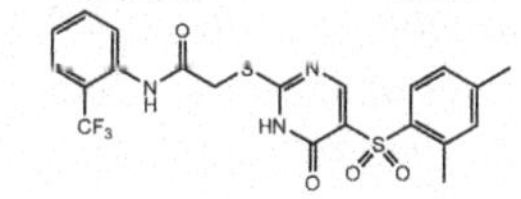 **Cpd15** 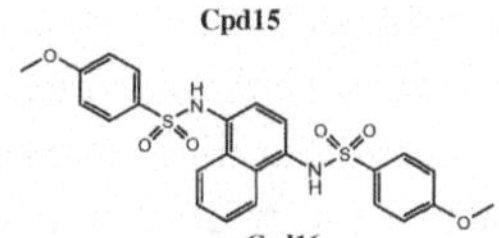 **Cpd16**	NA	Hu et al. (2013), Jnoff et al. (2014), Yasuda et al. (2016) and Marcotte et al. (2013)
Group 6	CRL4A	CRBN	Thalidomide, lenalidomide, pomalidomide	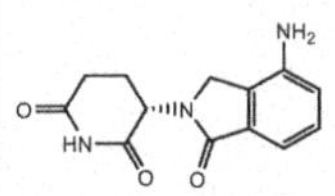 **Lenalidomide** 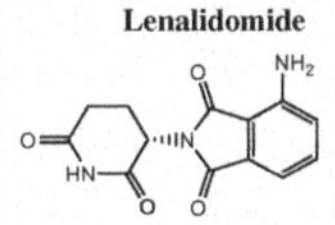 **Pomalidomide** 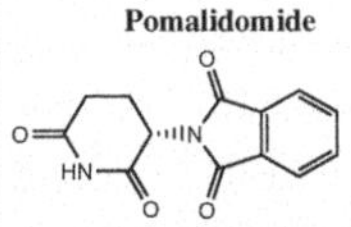 **Thalidomide**	Myeloma	Fischer et al. (2014) and Lopez-Girona et al. (2012)
Group 6	CRL5	EloBC-CBFβ	RN-18, MM-1 and MM-2	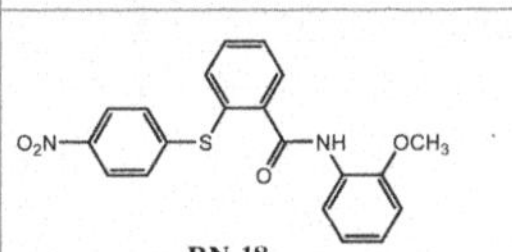 **RN-18** 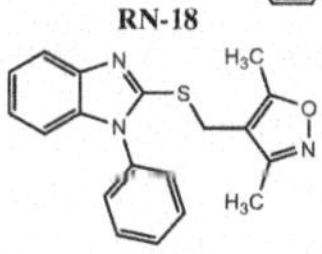 **MM-1** 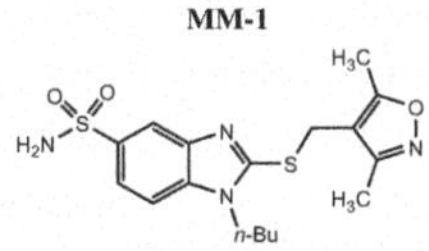 **MM-2**	NA	Nathans et al. (2008) and Matsui et al. (2014)

Given that the Cullin-RING module constitutes the active center of the CRL E3 ligase, inhibitors targeting this site should be more effective than blocking the downstream effects of CRLs. The elucidation of the structure of different Cullins along with particular RING proteins would provide a structural basis for the development of more specific targeting inhibitors against the interaction of Cullin-RING.

19.2.3 Targeting the Neddylation of Cullins

It is generally accepted that Cullin neddylation has a range of functional consequences for the activity of CRL E3 ligases (Bulatov and Ciulli 2015), which include disrupting the inhibitory binding of CAND1 (Cullin-associated and neddylation-dissociated-1), a large sequestration factor that binds to Cullins (Duda et al. 2008; Liu et al. 2002), enhancing the recruitment of E2 to CRLs (Saha and Deshaies 2008), facilitating ubiquitin transfer, improving the access of the activated E2 to the end of the nascent polyubiquitin chain (Reitsma et al. 2017), etc. To date, the most successful small-molecule inhibitor of NEDD8 (neural precursor cell expressed, developmentally downregulated 8) modification of Cullin is MLN4924 (pevonedistat), which selectively and potently targets NEDD8 activating enzyme (NAE) (Soucy et al. 2009). MLN4924 is structurally similar to the adenylate intermediate of the NAE catalytic reaction, and the crystal structure of the NAE-NEDD8-MLN4924 ternary complex revealed the formation of a covalent NEDD8-MLN4924 adduct that blocks the active site of NAE (Nawrocki et al. 2012). Inhibition of NEDD8 transfer by MLN4924 results in the inactivation of CRLs, which further inhibits the ubiquitination and degradation of their substrates. Currently, this compound is in several phase I/II/III clinical trials for a variety of cancers, such as relapsed/refractory multiple myeloma or lymphoma (Zhao et al. 2014; Nawrocki et al. 2012; Shah et al. 2016; Zhou et al. 2018). However, MLN4924 treatment has been found to cause drug resistance through NAEβ mutations (Milhollen et al. 2012). Thus, a new generation of inhibitors is urgently needed. The promising clinical effect and the side effects of MLN4924 inspire the pursuit of inhibitors targeting other downstream enzymes in the neddylation process.

19.2.4 Targeting the Cullin-Adaptor Interaction

In a CRL complex, the adaptor protein links to the N-terminal part of Cullin scaffold protein and mediates attachment of the receptor protein to recognize the substrate proteins (Petroski and Deshaies 2005). The adaptor protein in each CRL complex has been confirmed. However, the studies on drug candidates are limited by a small number of known crystal structure data. Currently, inhibitors targeting this site are mostly against CUL1-Skp1 in SCF (Zheng et al. 2002; Sandoval et al. 2015) and CUL2-ElonginB/C in CRL2 (Cardote et al. 2017; Nguyen et al. 2015).

The Skp1-F-box complex is tightly bound to the N-terminus CUL1, which is highly conserved among CUL1 orthologues from different species (Petroski and Deshaies 2005). The structure of CUL1-Skp1 has been elucidated (Zheng et al. 2002). Through structure-based design and phage display, a team of researchers have found a series of Ub variants (Ubvs) with specificity that prevent CUL1-Skp1 from interfacing with their interface by strengthening weak interactions between Ub and F-box proteins and preventing CUL1 from binding to the same surface (Gorelik et al. 2016). Theoretically, these Ubvs can inhibit the activity of most F-box proteins and can be used as a specific inhibitor targeting SCF E3 ligases for basic experimental research or to promote the development of small-molecule inhibitors with potential tumor therapeutic effects.

Based on the crystal structure of CRL2-VHL (Cardote et al. 2017), two peptides were developed to specifically target the binding site of CUL2 and its adaptor protein ElonginB/C with micromolar dissociation constants through a structure-oriented biophysical method (Cardote and Ciulli 2017). Further studies have revealed that these short linear peptides act as ligands that interact with the EloC binding site, named the EloC site, thereby disrupting CUL2-ElonginB/C. Although small-molecule compounds that effectively target this site have not yet been found, this study has laid the foundation for the development of small molecules with the potential to selectively target CUL2- ElonginB/C PPI in CRLs.

19.2.5 Targeting the Adaptor-Receptor Interaction

Except for CRL9, whose structure is not yet defined, and CRL3, each CRL complex has a specific adaptor protein linked to a receptor protein. However, investigations targeting the adaptor-receptor have mainly focused on the Skp1-F-box interaction in the SCF complex. It has been determined that there are 69 F-box proteins in mammalian cells, all of which interact with Skp1 using the F-box domain. Current studies have found that abnormal F-box protein-mediated proteolysis can cause disorders in human diseases, such as pancreatic cancer (Wang et al. 2014, 2016). It is worth noting that inhibitors targeting F-box proteins have shown great therapeutic potential (Chan et al. 2013). Thus, more genetic and mechanistic studies on F-box proteins would facilitate the rational design of drug candidates targeting F-Box proteins.

Of the 69 F-box proteins, Skp2 was found to be excessively expressed in human cancers, correlating with poor prognosis and promoting tumorigenesis, implying its oncogenic property (Gstaiger et al. 2001; Hao and Huang 2015; Chan et al. 2014). Using high-throughput screening, a compound CpdA was identified to interfere with SCF^{Skp2} enzymatic activity in vitro (Chen et al. 2008). Further studies found that CpdA caused cell cycle arrest at G1/S phase and p27-dependent cell killing by blocking the assembly of Skp2 into the SCF complex. CpdA has been found to be active against patient-derived plasma cells as well as myeloid and lymphoblastic leukemia cells, exhibiting preferential activity against tumor cells, and relatively less against other normal cells. These findings provide a rational framework for the further development of antitumor drugs targeting Skp2. In addition, a small molecule called Compound 25, obtained by a high-throughput in silico screening of large and diverse chemical libraries, has been confirmed to target the F-box motif of Skp2 in an in vitro pulldown assay and dose-dependent binding experiments to prevent Skp1-Skp2 linkage, thereby eliminating the E3 activity of SCF^{Skp2} on its substrate proteins p27 and Akt (Chan et al. 2013). Compound 25 selectively inhibits Skp2 without affecting the activity of other F-box proteins (e.g., Fbw7 and β-TrCP). In terms of biological function, Compound 25 effectively induces cell senescence, inhibits glycolysis, and suppresses the growth of prostate and lung tumors in vivo. Mechanistically, Compound 25 was shown to inhibit SCF^{Skp2}-mediated p27 ubiquitination and degradation and Akt K63-linked ubiquitination and activation. Although Compound 25 prevented the formation of the Skp2-Skp1 complex and inhibited the activity of SCF^{Skp2}, it did not destroy Skp2-Skp1 complex already present in the cells.

Using chemical library screening, a novel compound was identified, designated DT204, which synergistically enhances the induction of apoptosis by the proteasome inhibitor bortezomib (BTZ) apoptosis by reducing the binding of Skp2 to Cullin-1 and Commd1, a Cullin-1-binding protein, therefore decreasing SCF^{Skp2} ubiquitin ligase activity (Malek et al. 2017; Mao et al. 2011). Cotreatment with DT204 and BTZ in a mouse model overcame resistance to BTZ and reduced the growth of myeloma in vivo. The development of DT204 provided proof-of-concept evidence of a rationally designed combination of drugs comprising an inhibitor to treat BTZ-resistant disease (Malek et al. 2017).

19.2.6 Targeting the Receptor-Substrate Interaction

19.2.6.1 SCF^{Skp2}-p27

Many studies have shown that SCF^{Skp2} promotes ubiquitination and degradation of its substrate protein p27, which accelerates disease progression (Haque et al. 2015; Wang et al. 2017), suggesting that targeting the interfaces between p27-Skp2 might be a promising strategy. In view of the relatively large number of published crystal structures of the components of SCF^{Skp2} alone or in combination with other proteins, structure-based drug design methods are particularly attractive for the development of effective inhibitors. A number of potential compounds have been successfully discovered in the field using virtual screening and computer-assisted methods.

By virtual screening of 315,000 compounds based on the published crystal structure of Skp2/Cks1, in which the accessory protein Cks1 is required for interaction with the receptor Skp2 and its substrate p27, four compounds, C1, C2, C16, and C20, were found to stabilize p27 levels in a Skp2-dependent manner through targeting the pockets at the Skp2-Cks1-p27 interface and thereby inhibiting p27 ubiquitination (Pavlides et al. 2013; Wu et al. 2012). It was later determined that C2 and C20 inhibited cancer cell proliferation by increasing the levels of nuclear p27, suggesting that these compounds can be further optimized to treat SCF^{Skp2}-dependent cancers.

Using high-throughput chemical genetic methods in human prostate cancer cell lines stably overexpressing Skp2, a study identified two small molecules, SMIP001 and SMIP004, that increase nuclear p27 levels (Rico-Bautista et al. 2010). Meanwhile, SMIPs also induced G1 phase delay and cell cycle arrest by upregulating p21 levels and inhibiting CDK2 activity. Although the precise targets of these two compounds have not yet been accurately elucidated, the authors have proposed a method for identifying bioactive small molecules with selective anticancer activity by using nuclear p27 as an endpoint.

19.2.6.2 SCF^{Fbxo3}- Fbxl2

Fbxo3, a poorly characterized F-box protein, was recently shown to effectively stimulate the secretion of cytokines in human inflammatory cells by mediating the ubiquitination and degradation of Fbxl2, a panreactive TRAF inhibitor (Mallampalli et al. 2013). TRAF is an important mediator of inflammation, innate and adaptive immune responses, and apoptosis (Inoue et al. 2000). The E3 ligase for Fbxl2, Fbxo3, targets Fbxl2 via a bacterial-like highly conserved ApaG domain at its C-terminus. Using molecular docking analysis and virtual screening, researchers developed the benzathine penicillin derivative BC-1215 as an inhibitor of the Fbxo3 protein (Mallampalli et al. 2013). Through disrupting the interaction of Fbxo3 with Fbxl2 and blocking the SCF^{Fbxo3}-mediated ubiquitination of Fbxl2, BC-1215 exhibited anti-inflammatory properties in a cytokine-driven mouse model of inflammation (Mallampalli et al. 2013).

19.2.6.3 SCF^{Cdc4}-Sic1

A biplanar dicarboxylic acid compound called SCF-I2 was identified as binding to Cdc4 from a fluorescence polarization screen of 50,000 compounds (Orlicky et al. 2010). This compound is a racemic mixture. However, only the (R)-(+) enantiomer binds to Cdc4. Structural studies have shown that SCF-I2 inhibits Cdc4 by inserting into a position 25 Å away from the substrate recognition site in Cdc4 and blocking the binding of the WD40β domain to the substrate. Subsequent mutations in the SCF-I2 binding site render its inhibition ineffective, which explains the specificity of its allosteric inhibition mechanism. However, the in vivo activity of SCF-I2 was not significant, probably because of its poor cell permeability (Orlicky et al. 2010).

19.2.6.4 SCF^{Met30}-Met4

Interestingly, a small-molecule compound, SMER3 (small-molecule enhancer of rapamycin), was found to inhibit SCF^{Met30} effectively and selectively in a phenotypic-based chemical genetics screen in yeast of small-molecule compounds that enhance rapamycin action. Genomic, genetic, and biochemical analyses indicate that SMER3 can affect cellular nutrient responses and proliferation by inhibiting SCF^{Met30} ubiquitin ligase. SMER3 exhibits high specificity for Met30 when analyzed with the effects of other different F-box proteins, including the closely related Cdc4. In addition to SMER3, several other SEMRs discovered by the same team were shown to effectively enhance autophagy in a Huntington's disease model and reduce the toxicity of rapamycin through unknown mechanisms (Sarkar et al. 2007).

19.2.6.5 SCF^{TIR1}-Aux/IAA

Based on the structure of the hormone auxin, a compound called probe 8 was designed to inhibit the ability of SCF^{TIR1} ubiquitin ligase to recognize its substrate protein Aux/IAAs in *Arabidopsis*, thereby mediating an auxin antagonism reaction dependent on TIR1, an F-box protein in SCF^{TIR1} ubiquitin ligase (Hayashi et al. 2008). The crystal structure of the small-molecule probe 8 and TIR1 complex shows that the long alkyl chain of probe 8 blocks the interaction between TIR1 and Aux/IAA. However, no ubiquitination experiments were conducted to confirm the inhibitory activity of probe 8 on

SCF^{TIR1} or the suppression of subsequent substrate ubiquitination (Hayashi et al. 2008).

19.2.6.6 SCF^{COI1}-JAZ1

Interestingly, a similar mechanism was also found for COI1 (coronatine-insensitive protein 1). In a phenotypic study of the plant and yeast two-hybrid system, the jasmonate ZIM-domain (JAZ) protein family was found to play a vital role in jasmonate signaling. JAZ1, a jasmonate signaling repressor, was identified as a substrate of SCF^{COI1}, and the jasmonoyl-isoleucine (JA-Ile) conjugate can promote the interaction of the F-box protein COI1 with its substrate JAZ1, resulting in JAZ1 ubiquitination and degradation by the proteasome pathway in SCF^{COI1}-dependent fashion (Thines et al. 2007).

19.2.6.7 $CRL2^{VHL}$-HIF1α

The von Hippel-Lindau protein (VHL) is a substrate recognition component of the multisubunit $CRL2^{VHL}$ E3 ligase. The current target site of small-molecule inhibitors targeting the CRL2 subunit is VHL-HIF1α. Hypoxia-inducible factor 1α (HIF1α), a signaling hub to coordinate the activities of many signaling molecules and transcription factors that affect tumorigenesis (Balamurugan 2016), is a classical substrate of VHL, and $CRL2^{VHL}$ is responsible for the ubiquitination and degradation of HIF1α. However, under hypoxic conditions, prolyl hydroxylase domain enzymes (PHDs) are unable to hydroxylate HIF1α, leading to the accumulation of HIF1α, which in turn leads to the upregulation of genes involved in hypoxia. Thus, stabilizing HIF1α levels by PHD inhibitors may be a strategy for treating chronic anemia in clinical studies (Buckley et al. 2012a).

Using the minimal hydroxyproline recognition unit and a combination of in silico design and structure-guided medicinal chemistry, a series of small-molecule inhibitors were synthesized to target the surface of p-VHL/HIF1α, and Compound 15 showed great potential as a lead compound (Buckley et al. 2012b). Further fragment-based approaches dissecting binding hot spots allowed these inhibitors to be validated in order to rationalize fragmentary contributions at the PPI (Van Molle et al. 2012). In a parallel structure-driven and metrics-driven design, Galdeano and coworkers optimized the previous generation of small molecules targeting pVHL in binding affinities and lipophilicity and developed the inhibitor with a greater potential, Compound 7 (Galdeano et al. 2014).

Recently, Julianty and colleagues reported a small molecule, designated VH298, as a potent and selective VHL inhibitor that triggers a functional dose-dependent response downstream of HIF1α hydroxylation in the hypoxia-signaling cascade in cells (Frost et al. 2016). VH298 was also found to stimulate the production of endogenous erythropoietin (EPO), indicating its potential to benefit patients with anemia resulting from insufficient EPO synthesis (Haase 2010). However, since the VHL inhibitors may also turn on HIF and oncogenic signaling in cancers, the implication of these implications in cancer remains unclear.

19.2.6.8 $CRL3^{KEAP1}$-Nrf2

The CRL3 complexes are unique in the CRL family in that the CRL3 adaptor BTB protein acts directly as a substrate recognition protein to mediate substrate ubiquitination (Sarikas et al. 2011). The best characterized BTB protein is KEAP1 (Kelch-like ECH-associated protein 1), one of the Kelch-like family members. KEAP1 regulates the oxidative stress response by controlling the levels of the transcription factor Nrf2 (nuclear factor erythroid 2-related factor 2), which correlates with many diseases, including neurodegenerative, cardiovascular, metabolic, and inflammatory diseases, and has attracted much attention from the pharmaceutical field (Taguchi and Yamamoto 2017). The development of inhibitors disrupting KEAP1-Nrf2 protein-protein interactions has become a hot topic in current research, and some effective inhibitors have been found.

From a homogenous fluorescence polarization-based high-throughput screening of the NIH MLPCN (Molecular Libraries Probe Centers Network) small-molecule compound library, a small molecule (hit 1) was identified as a direct inhibitor disrupting the protein-protein interaction of KEAP1-Nrf2. Furthermore, among the eight possible isomers, only one stereoisomer, designated ML334 (LH601A) or SRS1a, was demonstrated to be active. Compared to other stereoisomers, this molecule has at least 100× greater binding affinity for the Kelch domain of KEAP1 (Hu et al. 2013). Using this molecule as a lead compound, a highly effective and efficient KEAP1 inhibitor, SRS-59, was designed (Jnoff et al. 2014).

Cpd15 and Cpd16 are additional successful KEAP1-Nrf2 inhibitors obtained from high-throughput screening of the Evotec Lead Discovery library, which contains 267,551 compounds, for binding to the Kelch domain of KEAP1. To understand the binding mode of Cpd15, a crystal structure of the Kelch-DC/Cpd15 complex was solved. The crystal structure confirmed that these two compounds can indeed bind to the central cavity of the Kelch-DC domain, and the binding sites of the two compounds are similar but not identical. Between them, Cpd16 has stronger inhibition potential due to its deeper insertion into the central cavity (Yasuda et al. 2016).

19.2.6.9 CRL4A^{CRBN}

The crystal structure of CUL4A-DDB1-CRBN in complex with thalidomide, lenalidomide, and pomalidomide demonstrates that CRBN, although lacking the WD40 fold compared with other canonical DCAFs, is similar in size and position to the other substrate receptors of the CRL4A^{CRBN} complex (Fischer et al. 2014). Using a seven α-helical bundle domain (HBD), CRBN interacts with DDB1 at two of the three seven-bladed WD40 β-propellers (BPA and BPC), thus functioning as a substrate recognition subunit for CRL4A. Thalidomide and its derivatives, used as immunomodulatory drugs in hematologic malignancies, block the degradation of endogenous substrates of CRL4A^{CRBN} and promote the ubiquitination of IKAROS family transcription factors IKZF1 and IKZF3 (Fischer et al. 2014; Kronke et al. 2014; Lu et al. 2014) and casein kinase 1α (CK1α) (Petzold et al. 2016) by hijacking CRL4A^{CRBN} E3 ligase. These three ligands bind to the same hydrophobic pocket of CRBN in a similar binding mode and exhibit low KD values (250 nM for thalidomide, 178 nM for lenalidomide, and 157 nM for pomalidomide), thereby showing similar affinities and close interactions for CRBN (Chamberlain et al. 2014).

19.2.6.10 CRL5^{Vif} -CBFβ

The Vif protein in HIV plays an important role in the replication of HIV virus. It has been reported that after invading normal cells, Vif recruits the CRL5 complex to degrade the anti-retroviral factors present in cells, thereby facilitating the replication of the virus in cells (Zhou et al. 2012). A structural study of the Vif hijacking of CRL5 revealed that Vif was incorporated into the E3 ligase by the formation of a Vif-CBFβ-EloBC-Cul5 multimeric complex (Guo et al. 2014). Vif serves as a substrate recognition subunit in the CRL5 complex, and its classical substrates include APOBEC3G (A3G), which is one of the most potent retroviral restriction factors. Therefore, blocking this process is of great significance for antiviral therapy. Several compounds that were found to control viral replication have also been included in the ranks of CRL inhibitors. It has been reported that RN-18, a compound identified by high-throughput screening, increases the levels of A3G in a Vif-dependent manner and increases the binding of A3G to virions without inhibiting proteasome-mediated general protein degradation (Nathans et al. 2008). In addition, MM-1 and MM-2, two structurally similar small-molecule compounds, were found to inhibit the degradation of A3G by HIV-1 Vif. However, the precise mechanism of inhibition remains unclear (Matsui et al. 2014). Structural data would greatly enhance the understanding of virus hijacking and provide an opportunity to rationalize the design of anti-HIV drugs that target HIV-1 Vif.

Given that (1) CRLs with multiple components are the largest family of E3 ubiquitin ligases, consisting of more than 400 ligases, and are responsible for the ubiquitination of ~20% of cellular proteins degraded through UPS; (2) CRLs have been validated as attractive anti-cancer targets by various approaches, such as genetic and small-molecule approaches; (3) in cancer cells, highly expressed CRLs preferentially target the degradation of tumor suppressor proteins to sustain unlimited proliferation, the development of inhibitors targeting CRLs is receiving increasing attention for therapeutic intervention. With more structural and functional information of CRLs being elucidated, targeted drug discovery for CRLs will realize the great potential due to their substrate versatility and wide temporospatial distributions.

19.3 CRLs in PROTACs

Proteolysis-targeting chimeras (PROTACs), synthetic heterobifunctional chimeric molecules, are an emerging targeting technique that utilizes the UPS to remove unwanted or damaged proteins. One ligand at the end of the PROTAC is responsible for the recruitment of a ubiquitin ligase, and the other ligand is designed to identify the target protein for degradation. In addition, the two-terminal ligand is tethered by a suitable linker, which is typically an ethylene-oxy-based aliphatic chain (Sakamoto et al. 2001). The PROTAC molecule brings the ubiquitin ligase and the protein of interest (POI) into a close proximity, where the ubiquitin-tagged reaction can occur effectively. The ubiquitin ligase promotes polyubiquitination and degradation through the 26S proteasome, thereby degrading the protein rather than inhibiting it (Fig. 19.2). Therefore, PROTACs target proteins beyond the scope of the ubiquitin proteasome system (UPS) and seize proteins that other canonical drugs cannot bind. It is thought that PROTACs can be used to target proteins traditionally considered "undruggable," such as the oncogenic protein MYC or the tau protein entangled in Alzheimer's disease (Scudellari 2019). Moreover, PROTACs have been shown to have significantly better biological effects in the case of drug resistance induced by inhibition of the protein function by

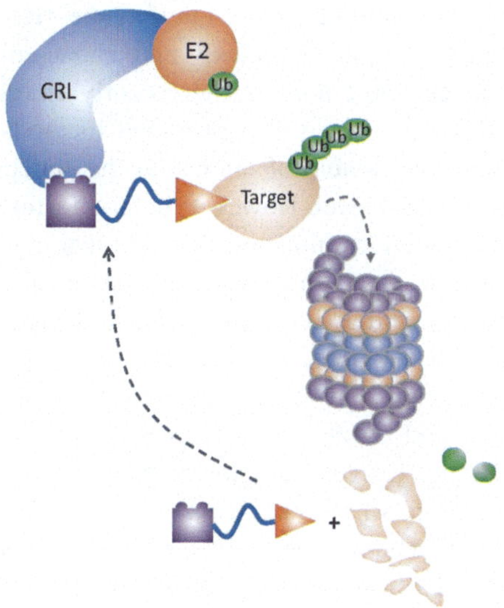

Fig. 19.2 Schematic diagram of PROTAC based on CRL. A PROTAC molecule consists of a ligand for recruiting CRL E3 ligase, a linker, and a ligand binding to the target protein. The PROTAC molecule hijacks the target protein and CRL together, leading to target protein ubiquitination and degradation by the proteasome

small-molecule inhibitors alone (Lu et al. 2015; Abruzzese et al. 2016; Mullard 2019).

Inhibiting abnormal protein expression in vivo has been one of the strategies for drug development. Small-molecule inhibitors have occupied an important position in drug discovery in the past decades through occupying active sites. Small-molecule-mediated "occupancy-driven" inhibition of protein function is the basic principle of an array of clinically used agents, which are currently the main type of therapies targeted toward intracellular abnormal proteins (Bondeson and Crews 2017). However, the traditional occupant form of protein inhibition has the following limitations. (1) The competitive nature (dose-dependent) of small molecular inhibitors may lead to toxicity and off-target effects. Specifically, small-molecule inhibitors require high dose levels to achieve >90% target engagement (Adjei 2006), but pharmacologically relevant inhibition is usually achieved, which may result in off-target effects. (2) The requirement to bind to the active site limits its wider application. (3) High genomic mutations in human cancer result in drug resistance (Lord and Ashworth 2013; Wood et al.

2016). In addition, RNA interference (RNAi), antisense oligonucleotides, and gene-editing techniques, most notably the recently emerging CRISPR/Cas9, have also provided methods for blocking the protein of interest at the genetic or transcriptional level. However, several features of these nucleotides hindered their clinical application: (1) the instability of nucleotides in plasma; (2) nucleic acid delivery issues due to the requirement of extensive efforts to develop nanomaterials and other packaging technologies (Conde and Artzi 2015; Tokatlian and Segura 2010); and (3) ease of capture by the liver and aggregation, which may cause toxicity (Fontana 2014). As a natural protein degradation system in the cell, UPS undoubtedly provides an excellent source for drug discovery.

CRL has been regarded as one of the key components of the degradation of targeted proteins since the advent of small-molecule PROTACs, based on CRL2^{VHL} and CRL4^{CRBN}. CRLs, including CRL1, CRL2, CRL3, CRL4A, CRL4B, CRL5, CRL7, and CRL9, consisting of more than 400 ligases, are distributed in almost all cell types and throughout the whole lifespan of cells. Furthermore, each CRL E3 ligase can be hijacked by different ligands, which target the POI. Thus, theoretically, CRLs can assemble millions of different PROTACs (Fig. 19.3). This space-infinite feature is inspiring to researchers. In particular, PROTACs have great potential to eliminate "undruggable" protein targets, such as transcription factors, which are hardly inhibited using conventional therapeutic modalities. The development of PROTACs is currently in the full swing and, as the first PROTACs (ARV-110, targeted degradation of androgen receptor), applying to prostate cancer, has entered

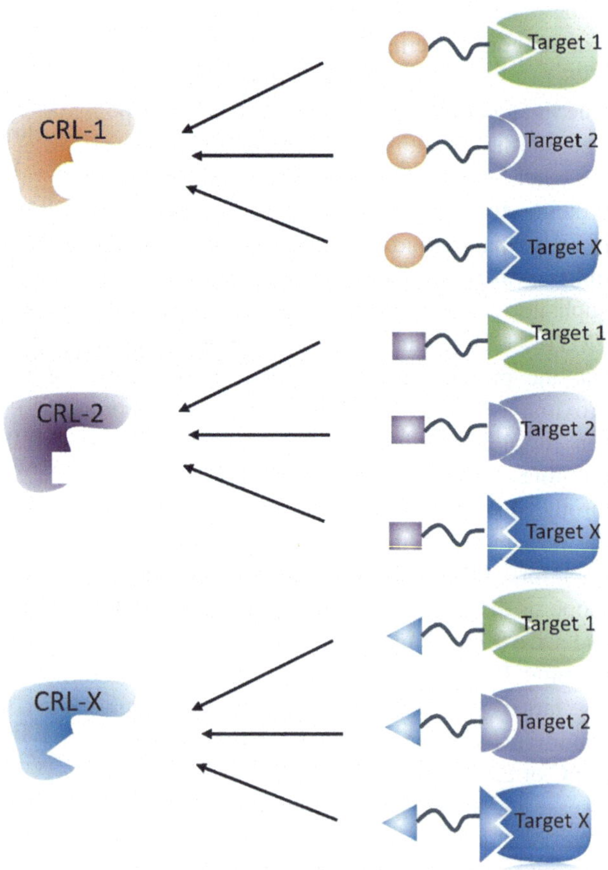

Fig. 19.3 Schematic diagram of the general application of CRLs in PROTAC technology. CRLs, including CRL1, CRL2, CRL3, CRL4A, CRL4B, CRL5, CRL7, and CRL9, consist of a total of more than 400 ligases. Each CRL E3 ligase complex can be hijacked by various ligands, which target multiple proteins (target 1, 2, X). Thus, theoretically, CRLs can be used to assemble millions of PROTACs

clinical trials in the mid of 2019. Other PROTAC products are commercially available for research use. It is widely believed that in the next decade, PROTACs will set off another wave in drug development. In the next sections, we will briefly introduce the reported PROTACs that target CRLs, focusing on the important role of CRLs in PROTACs, and highlight the diversity of CRLs, which presents unlimited possibilities for the development of PROTACs (Table 19.2).

19.3.1 CRL-Based Peptide PROTACs

19.3.1.1 $SCF^{\beta TrCP}$

PROTAC-1

The first peptide PROTAC, PROTAC-1, contains an IκBα phosphorylation peptide that is recognized by $SCF^{\beta TrCP}$, a linker and ovalicin that binds with MetAP-2 (methionine aminopeptidase-2). Thus, Protac-1 can trigger MetAP-2 ubiquitination and degradation (Sakamoto et al. 2001). Importantly, MetAP-2 is not a physiological substrate for $SCF^{\beta TrCP}$ ubiquitin ligase. In the Protac-1 chimeric molecule, one end is a phosphorylated polypeptide of IκBα (DRHDpSGLDpSM) recognized by the ubiquitin ligase $SCF^{\beta TrCP}$, and the other end is a small molecule, ovalicin (OVA), that specifically binds to MetAP-2. Protac-1 acts as an intermediate bridge to recruit MetAP-2 protein to the $SCF^{\beta TrCP}$ complex, which promotes the ubiquitination of MetAP-2. Indeed, Protac-1 binds to MetAP-2 protein in a dose-dependent manner. The endogenous $SCF^{\beta TrCP}$ complex targets the MetAP-2/Protac-1 protein for degradation in a time-dependent manner but has no effect on the free MetAP-2 protein. Moreover, the proteolytic inhibitors (LLnL and Epox) abrogate Protac-1-mediated MetAP-2 degradation, suggesting that the degradation is proteasome specific but not due to alternative pathways, such as those involving lysosomes, or the proteases, such as caspases (Cui et al. 2016). Thus, the Protac-1 chimeric molecule is indeed capable of recruiting the $SCF^{\beta TrCP}$ complex to mediate the degradation of MetAP-2.

PROTAC-2/3

Based on Protac-1, Protac-2 and Protac-3 were then successfully synthesized. In both cases, the ligand responsible for the recruitment of ubiquitin ligase is still a phosphorylated polypeptide of IκBα that is specifically recognized by $SCF^{\beta TrCP}$. The other end of Protac-2 is estradiol, which specifically binds to estrogen receptor (ER), while dihydrotestosterone (DHT) in Protac-3 has a high affinity with androgen receptor (AR). Both PROTAC molecules were then demonstrated to induce the degradation of targeted receptors in a cellular system by microinjecting PROTACs into HEK293 cells (Sakamoto et al. 2003). This work not only demonstrated that PROTACs can increase the turnover of specific target proteins in cells but also, for the first time, extended the PROTAC approach to proteins that play an important role in human disease.

19.3.1.2 $CRL2^{VHL}$

PROTAC-4

Protac-4 is the first cell-penetrating PROTAC with its ubiquitin ligase ligand ALAPIIP for recruiting $CRL2^{VHL}$. ALAPIIP, a seven-amino acid recognition sequence of hypoxia-inducible factor 1α (HIF1α), is the smallest polypeptide sequence for VHL recognition (Hon et al. 2002; Min et al. 2002). The other end of the protein ligand AP21998 specifically binds to the mutated FKBP12 (F36V) but not to the wild-type FKBP12. Under normal oxygen conditions, Pro564 on HIF1α protein is hydroxylated by proline hydroxylase, which is recognized by $CRL2^{VHL}$ and initiates the polyubiquitination and degradation of HIF1α. Therefore, under normal oxygen conditions, HIF1α is continuously degraded and maintained at a low level (Masoud and Li 2015). The carboxy terminus of the ALAPYIP polypeptide is fused to eight dextran-arginine tags, which enables Protac-4 to penetrate cells in a way resembling that of the HIV Tat protein. On the other hand, the tag helps prevent Protac-4 degradation through other nonspecific proteolysis. In HeLa cells stably expressing EGFP-FKBP12 (F36V), Protac-4 treatment

Table 19.2 PROTACs based on CRLs

Category	CRL	Target	Structure	PROTAC	Year	Ref.
Peptide	CRL1$^{\beta\text{-Trcp}}$	MetAP-2	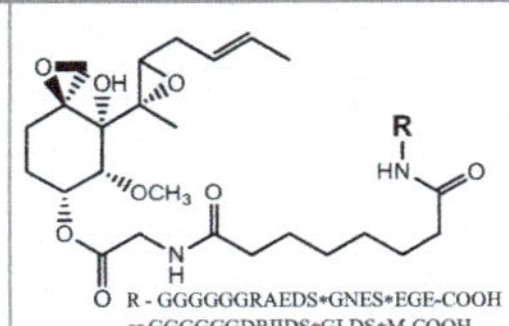	Protac-1	2001	Sakamoto et al. (2001)
Peptide	CRL1$^{\beta\text{-Trcp}}$	Estrogen receptor	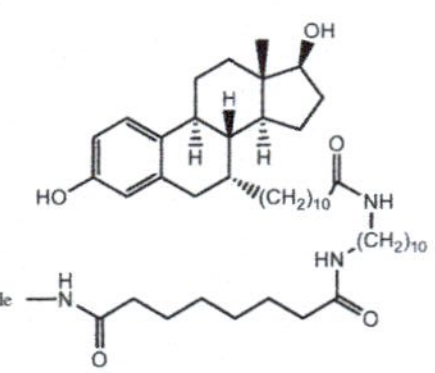	Protac-2	2003	Sakamoto et al. (2003)
Peptide	CRL1$^{\beta\text{-Trcp}}$	Androgen receptor	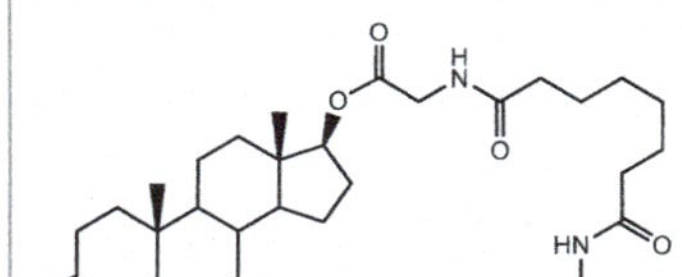	Protac-3	2003	Sakamoto et al. (2003)
Peptide	CRL2^{VHL}	(F36V)FKBP12	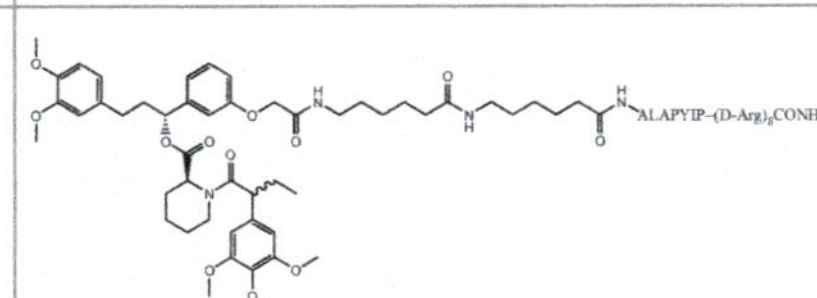	Protac-4	2004	Schneekloth Jr. et al. (2004)
Peptide	CRL2^{VHL}	Androgen receptor	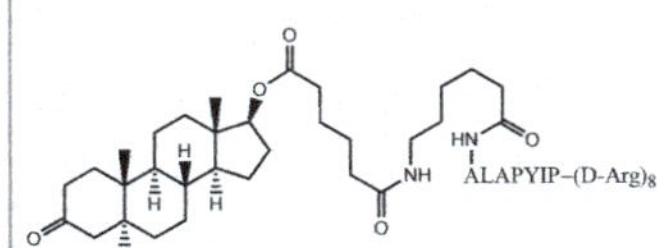	Protac-5	2004	Schneekloth Jr. et al. (2004)
Peptide	CRL2^{VHL}	HBX	CPP Oligomerization domain ODD domain FITC-RRRRRRRRLCLRPVGAESRGRPVSGPFGGMLAPYIPM CPP ODD domain Oligomerization domain FITC-RRRRRRRRMLAPYIPMGGLCLRPVGAESRGRPVSGPFG	X-protein-targeting PROTAC	2014	Montrose and Krissansen (2014)

Peptide	$CRL2^{VHL}$	TrkA		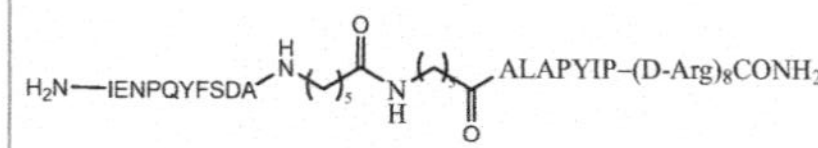	$^{TrkA}PP_{FRS2\alpha}$	2013	Hines et al. (2013)
Peptide	$CRL2^{VHL}$	AKT		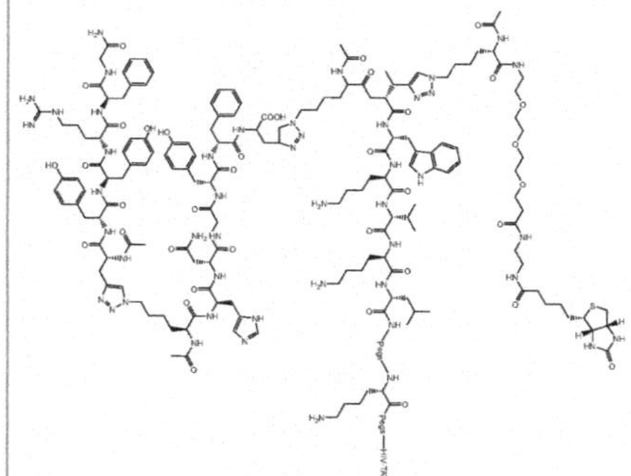	CPP_tri_a-PR	2016	Henning et al. (2016)
Peptide	$CRL2^{VHL}$	Tau		YQQYQDATADEQG GSGS-ALAPYIP-RRRRRRRR-$CONH_2$	TH006	2016	Chu et al. (2016)
Small molecule	$CRL2^{VHL}$	Nuclear hormone receptor	ERRα	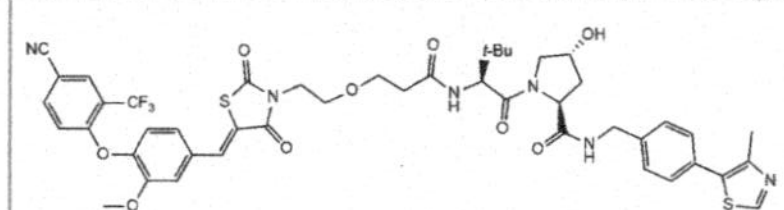	PROTAC_ERRα	2015	Bondeson et al. (2015)
Small molecule	$CRL2^{VHL}$		Androgen receptor	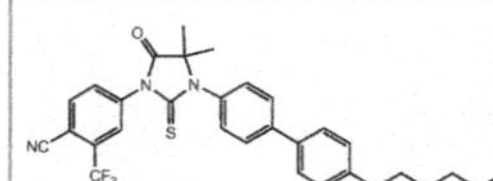	ARCC-4	2018	Salami et al. (2018)
Small molecule	$CRL2^{VHL}$	Kinase	RIPK2	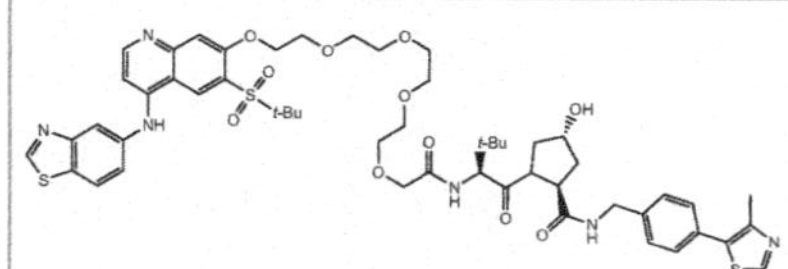	PROTAC_RIPK2	2015	Bondeson et al. (2015)
Small molecule	$CRL2^{VHL}$		cABL	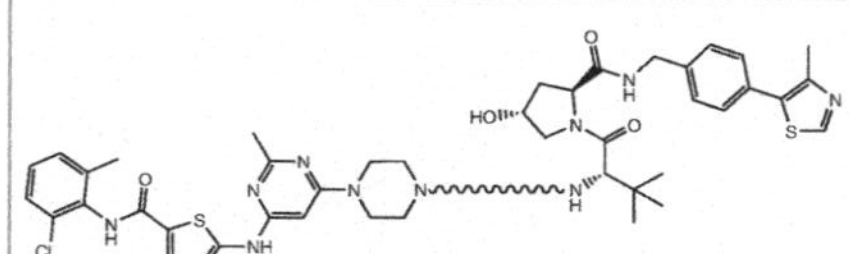	DAS-VHL	2016	Lai et al. (2016)

(continued)

Table 19.2 (continued)

Category	CRL	Target		Structure	PROTAC	Year	Ref.
Small molecule	$CRL2^{VHL}$	Epigenetic targets	BRD4	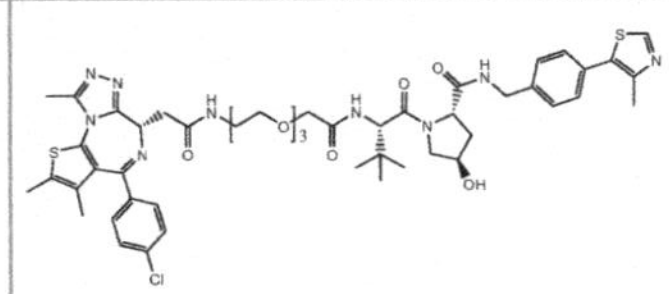	MZ1	2015	Zengerle et al. (2015)
Small molecule	$CRL2^{VHL}$		BRD4	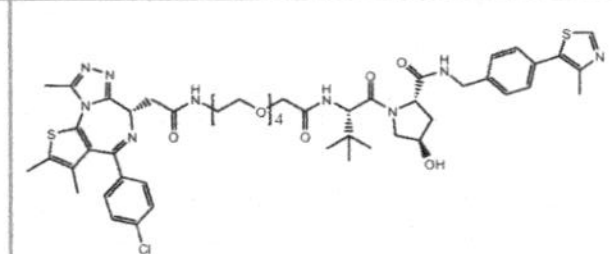	MZ2	2015	Zengerle et al. (2015)
Small molecule	$CRL2^{VHL}$		BRD2/3/4	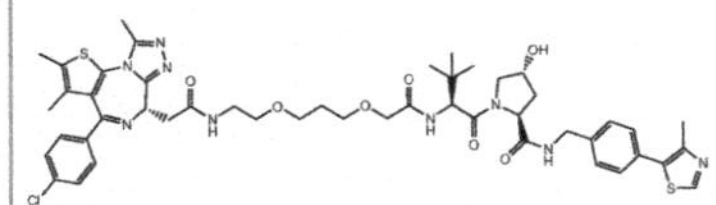	ARV-771	2016	Sun et al. (2018b)
Small molecule	$CRL2^{VHL}$		TRIM24	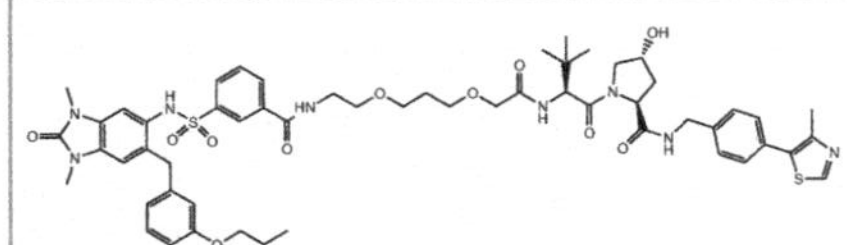	dTRIM24	2018	Gechijian et al. (2018)
Small molecule	$CUL4^{CRBN}$	Kinase	CDK9	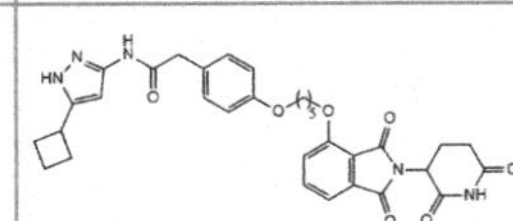	PROTAC 3	2016	Robb et al. (2017)
Small molecule	$CUL4^{CRBN}$		cABL, BCR-ABL	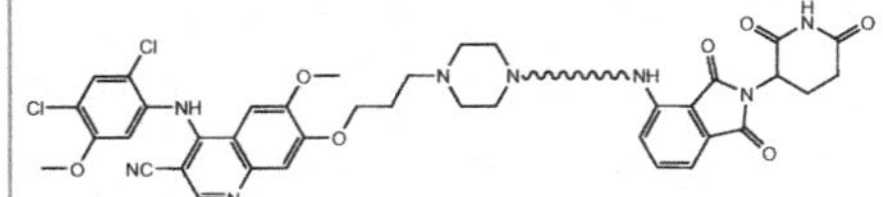	BOS-CRBN	2016	Lai et al. (2016)
Small molecule	$CUL4^{CRBN}$		cABL, BCR-ABL	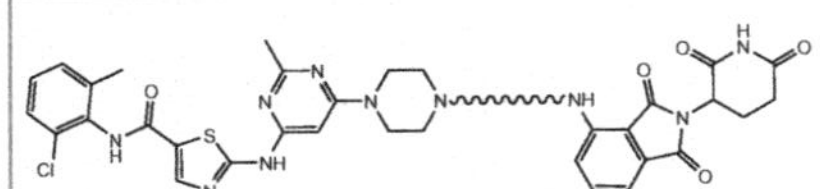	DAS-CRBN	2016	Lai et al. (2016)

Small molecule	$CUL4^{CRBN}$		BTK	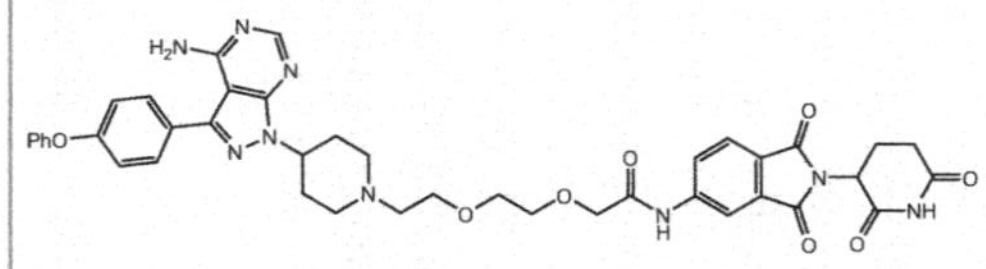	MT-802	2018	Buhimschi et al. (2018)
Small molecule	$CUL4^{CRBN}$		PI3K	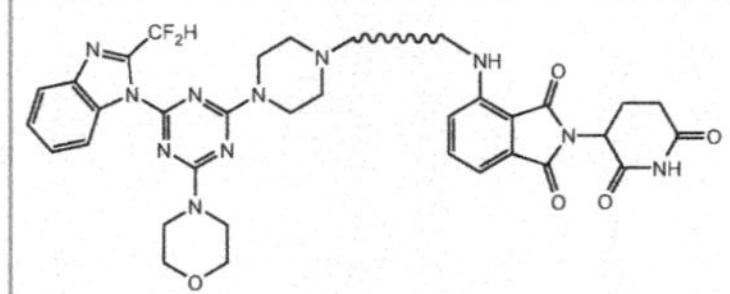	NA	2018	Li et al. (2018)
Small molecule	$CUL4^{CRBN}$	Epigenetic targets	BRD4	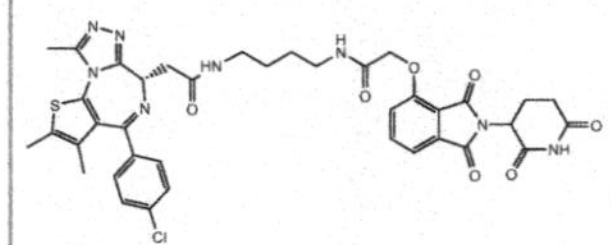	dBET1	2015	Winter et al. (2015)
Small molecule	$CUL4^{CRBN}$		BRD4	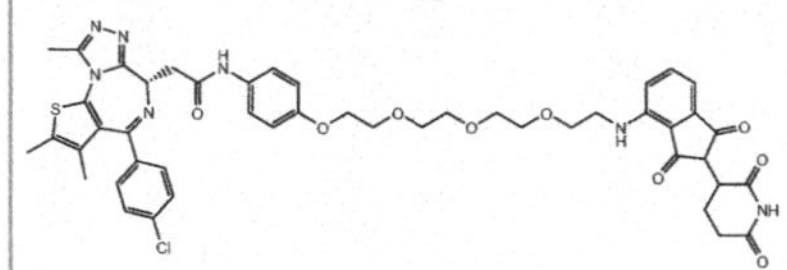	ARV-825	2015	Lu et al. (2015)
Small molecule	$CUL4^{CRBN}$		BRD9	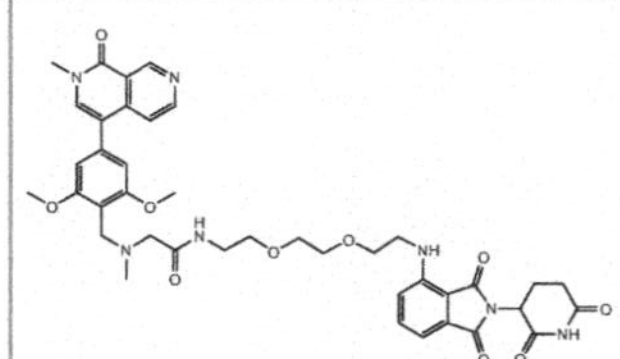	dBRD9	2017	Remillard et al. (2017)
Small molecule	$CUL4^{CRBN}$		Sirt2	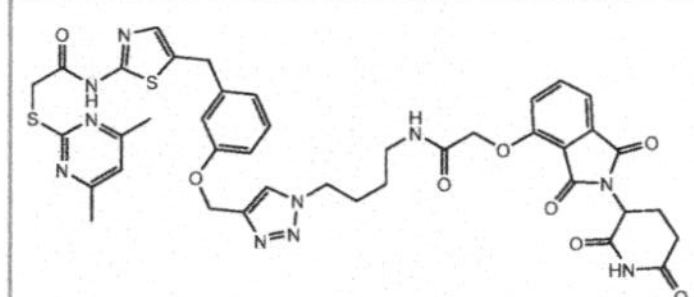	Compound 12	2018	Schiedel et al. (2018)

(continued)

Table 19.2 (continued)

Category	CRL	Target	Structure	PROTAC	Year	Ref.
Small molecule	CUL4^{CRBN}	HDAC6	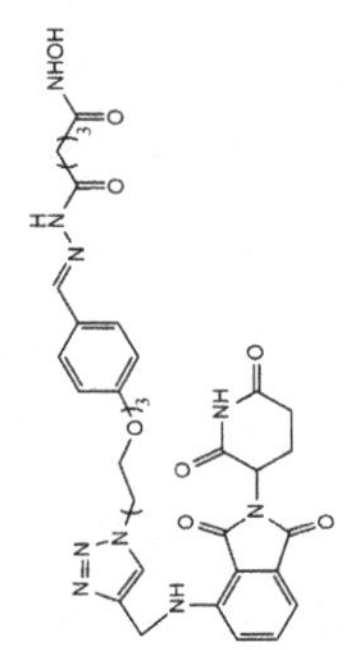	dHDAC6	2018	Yang et al. (2018)
Small molecule	CUL4^{CRBN}	PCAF/ GCN5	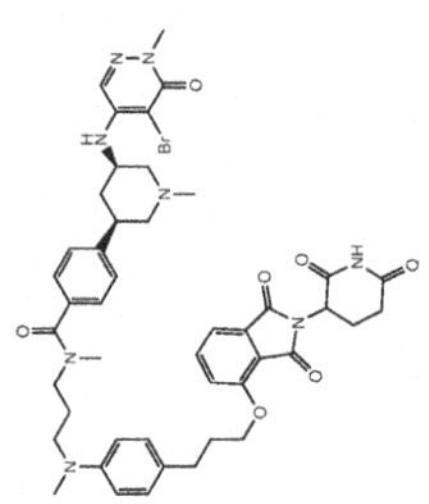	GSK983	2018	Bassi et al. (2018)

significantly reduced the levels of EGFP-FKBP12 (F36V). Importantly, Protac-4 had no effect on the fluorescence in 786-O cells without VHL expression. Thus, Protac-4 was proven to enter cells and act as a bridge to target FKBP12 (F36V) for degradation by CRL2^{VHL} (Schneekloth Jr. et al. 2004).

PROTAC-5

PROTAC-5 was used to verify the robustness of PROTAC technology by using DHT instead of AP21998 to identify androgen receptors with the same ubiquitin ligase ligand (Schneekloth Jr. et al. 2004). Fluorescence microscopy and immunoblotting showed that Protac-5 treatment significantly reduced the level of GFP-AR in 293 cells that stably expressed GFP-AR. Furthermore, pretreatment with proteasome inhibitors abrogated the decrease of GFP-AR caused by Protac-5 treatment. Biologically, Protac-5 targeted AR for degradation by the proteasome, inhibited prostate cancer cell proliferation, and induced cell cycle arrest in an AR-dependent manner (Rodriguez-Gonzalez et al. 2008). Therefore, the PROTAC technology has been proven once again to have practical application value.

PROTAC for AKT

More recently, VHL-recruiting peptidic PROTACs, CPP-tri_a-PR, were utilized to induce the degradation of Akt. In this PROTAC, the protein ligand was constructed using in situ click chemistry (Lewis et al. 2002) to allow a protein of interest to select its own high-affinity binders, named protein-catalyzed capture agents (PCCs). These synthetic peptides have some similarities with monoclonal antibodies but are smaller in size and have higher stability. AKT, a protein kinase that is involved in many human diseases, has been proven to be a challenging target for degradation by ubiquitination. However, CPP-tri_a-PR triggers the rapid degradation of endogenous AKT after 30 min of treatment, which validates that the epitope targeting selectivity of PCCs can be coupled with CRLs to construct PROTACs to eliminate challenging targets (Henning et al. 2016).

PROTAC for HBX Protein of Hepatitis B Virus (HBV)

This PROTAC molecule recruits the X protein of HBV (HBX) by the oligomerization domain of X protein and hijacks CRL2^{VHL} E3 ligase using the oxygen-dependent degradation (ODD) domain of HIF1α. In addition, a polyarginine cell-penetrating peptide (CPP) at its N-terminus assists it in entering cells. Biologically, this PROTAC leads to the inhibition of HBV infection by inducing HBX degradation (Montrose and Krissansen 2014).

TH006

A PROTAC designated as TH006, containing a Tau-recognition moiety and a CRL2^{VHL} E3 ligase-binding moiety, is designed to promote Tau degradation, which decreases the toxicity of amyloid β (Aβ) in Alzheimer's disease (AD) (Chu et al. 2016). The dysregulation of Tau is associated with Alzheimer's disease. Reducing Tau levels with chemical molecules such as PROTACs provides a novel strategy to regulate proteins such as Tau, which are difficult to target for the treatment of human diseases.

Taken together, the evidence shows that cell-penetrating PROTACs associated with CRLs can achieve chemical knockout against specific target proteins. In particular, they can eliminate pathogenic proteins in cells, exhibiting great potential value.

19.3.2 CRL-Based All-Small-Molecule PROTACs

In 2008, a PROTAC molecule using nutlin-3a to recruit E3 ligase MDM2 containing a RING structure was successfully designed to target the degradation of ARs (Schneekloth et al. 2008). Subsequently, the potency of PROTAC was increased by using RG7112 and RG7388 instead of nutlin-3a (Ding et al. 2013; Vu et al. 2013). Although the ability of these PROTACs to degrade the protein of interest is not as strong as that of the pro-peptide PROTACs, the application of these small-molecule PROTACs has opened up a new field for the further investigation of small-molecule PROTACs.

Early PROTACs were mainly designed based on E3 ubiquitin ligase peptide ligands (e.g., IκBα polypeptide, which is recognized by βTrCP); however, peptides have some intrinsic disadvantages, such as poor metabolic stability and difficulty in penetrating cell membranes. Given that small molecules can rapidly diffuse into cells and have a higher affinity for target proteins, the current research on PROTACs is mainly focused on using small molecules as an E3 ubiquitin ligase ligand. More importantly, compared to peptide PROTACs, small-molecule PROTACs have greater metabolic stability and can catalytically degrade target proteins in a superstoichiometric manner. This type of PROTAC, which has been reported to hijack CRLs, can be divided into two categories, VHL and CRBN.

19.3.2.1 CRL2^{VHL}-Based PROTAC

The first small-molecule ligand for VHL E3 ligase was identified by computer simulation and fragment-based screening in 2012 (Buckley et al. 2012b). Subsequent structure-activity relationship (SAR) studies and optimized design have led to the development of a series of small-molecule inhibitors that compete for binding to the HIF1 binding site of VHL (Galdeano et al. 2014; Buckley et al. 2012a). Given that VHL-based peptide PROTACs have shown higher efficiency in protein targeting, the discovery of high-affinity small-molecule ligands for the CRL2^{VHL} ubiquitin ligase complex might greatly facilitate the development of VHL-based PROTAC technology.

VHL-Based PROTAC and Nuclear Hormone Receptor

1. Several VHL-based small-molecule PROTACs targeting nuclear hormone receptors were successfully developed, the first of which was PROTAC_ERRα (Bondeson et al. 2015). PROTAC_ERRα targets the degradation of 50% ERRα, an orphan receptor estrogen-related receptor alpha (ERRα) in the nucleus, at a concentration of 100 nM in MCF-7 breast

cancer cells. Notably, PROTAC_ERRα has also been proven to be equally effective in the xenograft nude mouse tumor model: it reduced ERRα in the heart (44%), kidney (44%), and transplanted breast cancer (39%) of nude mice, respectively (Bondeson et al. 2015).

2. Given that AR mediates gene expression and remains a vital driver during prostate cancer progression, many therapeutic strategies have been developed to focus on regulating AR, such as androgen deprivation therapy combined with AR antagonists (Gustafson et al. 2015). However, this strategy leads in most cases to castration-resistant prostate cancer (CRPC) in the later stage of treatment and causes a poor prognosis. Therefore, targeting AR for degradation via PROTACs might be a better strategy to avoid drug resistance in prostate cancer. A PROTAC named ARCC-4 is able to degrade approximately 95% of cellular AR and its clinically relevant mutants, showing great potential to address the hurdle of drug resistance. Unlike enzalutamide, ARCC-4 retains an antiproliferative effect in an environment with high levels of androgen (Salami et al. 2018).

VHL-Based PROTAC and Protein Kinase

PROTAC_RIPK2, a VHL-based small-molecule PROTAC, targets the degradation of RIPK2, a serine/threonine kinase that is involved in the activation of NFκB and MAPK signaling pathways and plays an important regulatory role in innate immunity (Humphries et al. 2015). The ligand recognizing RIPK2 is used to produce a PROTAC_RIPK2 molecule by a 12-atom linkage to the VHL ligand. PROTAC_RIPK2 degraded 95% of RIPK2 at a concentration of 10 nM and minimized the level of RIPK2 within 4 h, whereas the original level of RIPK2 was quite stable with a half-life of 60 h (Bondeson et al. 2015).

VHL-Based PROTAC and Epigenetic Targets

1. MZ1 and MZ2 are small-molecule PROTACs that target BRD4 for degradation by linking JQ1 and CRL2^{VHL} ligand molecules through different lengths of linkers (Zengerle et al. 2015). Interestingly, MZ1, with a shorter linker chain, has higher degradation efficiency, suggesting that the length and composition of the linker have an important regulatory effect on the activity of the PROTAC molecules. In addition, in contrast to CRBN-based dBET1, the hijacking of CRL2^{VHL} ubiquitin ligase by MZ1 and MZ2 resulted in the specific degradation of BRD4 but not BRD2 and BRD3 of the BET family, suggesting that the type of ubiquitin ligase recruited also provides a certain selectivity for the degradation of the target protein.
2. In addition to CRBN-based PROTACs, which will be introduced in the next section, a picomolar VHL-based pan-BET-targeting PROTAC was also developed to show superior antiproliferative activity compared to BET inhibitors. This pan-BET-targeting PROTAC also exhibited activity against a 22RV-1 CRPC mouse xenograft, thus expanding the potential application of PROTAC technology to the treatment of solid tumor malignancies (Raina et al. 2016).
3. Bromodomain-containing transcriptional regulator TRIM24, also known as TIF1α, has shown dependency in numerous cancers, including acute leukemia. However, the selective ligands targeting the bromodomain of TRIM24 have no effective antiproliferative responses. Thus, the potent TRIM24 inhibitor IACS-9571 was used to form a heterobifunctional degrader by hijacking CRL2^{VHL} to degrade TRIM24, named dTRIM24. Compared with IACS-9571, dTRIM24 is much more effective in regulating genome-wide transcription at TRIM24 target genes, indicating that this chemical knockout technique has greater potential than target inhibitors alone (Gechijian et al. 2018).

19.3.2.2 CRL4A^{CRBN}-Based PROTAC

In recent years, phthalimides (thalidomide, pomalidomide, lenalidomide), a type of potent immunomodulatory drugs, were found to bind to the E3 ubiquitin ligase cereblon (CRBN), leading to the degradation of several proteins (Fischer et al. 2014; Ito et al. 2010). Notably, lenalidomide, a highly effective drug for the treatment of multiple myeloma, can cause the selective ubiquitination and degradation of two essential transcription factors in multiple myeloma, IKZF1 and IKZF3, by the $CRL4A^{CRBN}$ ubiquitin ligase (Kronke et al. 2014; Lu et al. 2014). IKZF1 and IKZF3 are not optimal physiological substrates of $CRL4A^{CRBN}$ ubiquitin ligase. Thus, thalidomide and its derivatives are PROTACs and are widely used and effective in the clinic, which demonstrates that PROTACs will eventually be developed as a novel class of drugs for the treatment of human diseases.

CRBN-Based PROTAC and Kinase

1. Cyclin-dependent kinase 9 (CDK9) is a component of the positive transcription elongation factor b (P-TEFB), which promotes transcription elongation (Liu et al. 2017). CDK9 promotes cancer development through facilitating Pol II pause release, a rate-limiting step in normal transcriptional regulation (Gressel et al. 2017). There have already been several CDK9 inhibitors in clinical treatment. SNS-032, originally developed as a selective CDK2 inhibitor, was later observed to be a potent inhibitor of CDK7 and CDK9 (Olson et al. 2018). Currently, targeting CDK9 selectively among CDKs was achieved in three independent research groups. By hijacking $CRL4A^{CRBN}$, a molecular PROTAC that selectively degrades CDK9, designated PROTAC 3, has been found to selectively degrade CDK9 in a dose-dependent manner in HCT116 cells (Robb et al. 2017). In addition, other small-molecule PROTACs, designed by changing different chemical groups at the para position on the phenyl ring according to different surfaces among the CDKs, have also shown inhibition toward diverse CDKs.
2. Inhibition of PI3K, a component of the PI3K/AKT/mTOR pathway that is crucial for carcinogenesis (Aoki and Fujishita 2017), has been proven to be efficacious in the treatment of many human cancers (Bertacchini et al. 2015). However, the development of PI3K inhibitors for clinical application is hindered by acquired drug resistance (Park et al. 2017). PROTACs could be an alternative strategy to overcome this problem. Small-molecule PROTACs recruit $CRL4A^{CRBN}$ by conjugating pomalidomide and ZSTK474, a pan-class I PI3K inhibitor; four of these PROTACs have been demonstrated to induce the time- and dose-dependent degradation of PI3K and mediate downstream protein expression (Li et al. 2018). A mechanistic study showed that autophagy was induced to inhibit cancer growth, in contrast to the effects of inhibitors acting alone (Li et al. 2018).
3. Inhibitor of Bruton's tyrosine kinase (BTK), a nonreceptor tyrosine kinase essential for B cell maturation, has been proven to be an efficacious way of treating non-Hodgkin's lymphoma (NHL) (Corneth et al. 2016). Ibrutinib is a BTK inhibitor that is able to bind Cys481 of BTK and was approved by the FDA for the treatment of several types of NHL. However, the C481S (cysteine to serine mutation at position 481) mutation has been reported to give rise to ibrutinib resistance and poor prognosis (Wu et al. 2016). PROTAC technology has come to researchers' attention as an approach to overcome drug resistance. In contrast to inhibitors of BTK, PROTACs have been designed to eliminate BTK by inducing the degradation of both wild-type and C481S mutant BTK (Zorba et al. 2018). These PROTACs were designed to recruit $CRL4A^{CRBN}$ ligase to trigger BTK ubiquitination and degradation via UPS (Buhimschi et al. 2018; Sun et al. 2018a).

CRBN-Based PROTAC and Epigenetic Targets

1. BRD4 plays a key role in the proliferation and survival of tumor cells. JQ1, one of the inhibitors of BRD4, was conjugated to a thalidomide derivative to form a PROTAC, termed dBET1, capable of inducing BRD4 degradation at low nanomolar concentrations (Winter et al. 2015).

2. OTX015 is a potent small-molecule drug candidate targeting BRD4 (Henssen et al. 2016). The fully small-molecule PROTAC ARV-825 consists of a short alkyl chain linking pomalidomide and OTX015 (Lu et al. 2015). ARV-825 can target almost all of the BRD4 proteins in the cells within 6 h at a concentration of 10 nM, and the effective time can last for more than 24 h, which reflects the ability of ARV-825 to efficiently degrade BRD4. Studies have shown that the inhibitory effect of degradation of the target protein itself by chemical knockout is much better than the inhibition of small-molecule compounds (Scheepstra et al. 2019). Indeed, the ability of ARV-825 to inhibit c-MYC, a downstream target of RBD4, was significantly higher than that of BRD4 inhibitors JQ1 and OTX015. Biologically, the ability of ARV-825 to induce apoptosis and inhibit tumor growth was also significantly higher than that of JQ1 and OTX015 (Lu et al. 2015).
3. Another bromodomain-containing protein, BRD9, a component of the human ATP-dependent chromatin remodeling BAF (SWI/SNF) complex, has recently been found to play a key role in acute myeloid leukemia (AML) (Hohmann et al. 2016). Using I-BRD9, a bromodomain probe, to capture BRD9, a series of PROTACs were developed with different linkers or ligands to hijack CRBN or VHL E3 ligases. Through comparative biochemical and biological assays, a lead BRD9 PROTAC degrader associated with $CRL4^{CRBN}$, termed dBRD9, was developed. This first-in-class chemical degrader of BRD9 with its rapid and potent activity makes it ideally suited to explore the function of the BAF complex (Remillard et al. 2017).
4. The Sirtuin family of proteins, NAD^+-dependent deacetylases, are involved in a variety of cellular processes, including inflammation, aging, and apoptosis. Sirt2 is one of seven isotypes of Sirtuins in human cells and has been reported to be involved in mammalian metabolism (Gomes et al. 2015) and cancers (Jing et al. 2016). Due to the lack of compounds that have sufficient isotype selectivity and good pharmacokinetic properties, PROTACs selectively degrading Sirt2 were designed (Schiedel et al. 2018). In this case, the Sirtuin-rearranging ligands (SirReals) act as a highly potent and isotype-selective Sirt2-recruiting ligands; thalidomide is responsible for hijacking $CRL4A^{CRBN}$ E3 ligase. SirReal-based PROTAC, Compound 12, triggered Sirt2, not Sirt1 degradation, leading to hyperacetylation of the microtubules (Schiedel et al. 2018).
5. HDAC6, a zinc-dependent histone deacetylase (HDAC), has been found to be highly expressed in malignancy and chronic disease processes (Batchu et al. 2016). PROTACs targeting HDAC were developed by conjugating nonselective HDAC inhibitors with $CRL4A^{CRBN}$ E3 ubiquitin ligase ligands (thalidomide and its derivatives). One PROTAC, named as dHDAC6, could selectively degrade HDAC6 and cause the accumulation of Ac-α-Tubulin (K40) (Yang et al. 2018).
6. P300/CBP-associated factor (PCAF) and general control nonderepressible 5 (GCN5) are highly homologous multidomain proteins that are able to modify histones (acetyltransferase) and recognize modified histones (bromodomain) (Nagy and Tora 2007). The production of cytokines, including TNFα and IL-6, stimulated by LPS, was reduced due to PCAF downregulation, suggesting the vital role of PCAF in immune function (Qiu et al. 2019). A small molecule, PROTAC, designated as GSK983, was developed via GSK4027, a potent and selective inhibitor targeting the bromodomains of PCAF and GCN5, conjugating CRBN ligands. GSK983 is able to degrade PCAF/GCN5 and modulate the expression of multiple inflammatory mediators in LPS-stimulated macrophages and dendritic cells (Bassi et al. 2018).

Taken together, current PROTACs prove that the organic synthesis of small-molecule PROTAC technology is an effective strategy to eliminate endogenous "undruggable" disease-causing proteins. The small-molecule ligands for capturing target proteins can be any specific compound binding with the target protein, not limited to inhibitors. Similarly, the ligands for hijacking CRL E3 ligases can also be any specific compounds that bind the substrate receptors of CRLs. Additionally, the different linkers coupling these two ligands exhibit different efficiencies for target degradation (Crew et al. 2017). Thus, small-molecule PROTAC technology offers great hope for future investigation. However, a number of fundamental questions that limit the future therapeutic application of PROTACs need to be urgently addressed: how to reduce the molecular weight and increase the bioavailability; how to reduce the toxicity and improve the pharmacokinetics; how to precisely control the tissue distribution; etc. Eventually, PROTACs will be developed as a novel class of drugs for the treatment of human diseases by eliminating endogenous disease-causing proteins.

19.4 Conclusion

In recent years, advances in basic research on CRLs, as well as the development of new technologies such as PROTACs, have significantly promoted innovative efforts in drug discovery. Although many studies have validated CRLs as attractive anticancer targets, there are no drugs directly targeting CRLs in the clinic. Thus, extensive basic investigation into the functions of CRLs will accelerate the development of a novel class of anticancer agents for the treatment of human cancers. In addition, with improving PROTAC technology, CRLs have been widely utilized. In particular thalidomide and its derivatives, $CRL4A^{CRBN}$-based PROTACs, are widely used and rather effective in the clinic. Given the nature of CRLs, PROTACs obtained by hijacking CRLs may achieve tissue- or time-specific function due to the spatiotemporal expression of CRLs, which makes PROTAC molecules more flexible for application. Significantly, the first batch PROTACs by Arvinas targeting ERα and AR, ARV-471 and ARV-110, have entered into phase I clinical trials. It is hopeful that small-molecule inhibitors targeting specific CRL E3 and CRL-based PROTACs would be developed as a novel class of anticancer drugs.

Funding This work was supported by the National Key R&D Program of China (2016YFA0501800 to Y.Z. and X.X.), the National Natural Science Foundation of China (81672728, 81972591, and 81721091 to Y.Z. and 81572708 and 81974429 to X.X.), and the Natural Science Foundation of Zhejiang Province (LR16C050001 to Y.Z.).

References

Abruzzese MP, Bilotta MT, Fionda C, Zingoni A, Soriani A, Vulpis E, Borrelli C, Zitti B, Petrucci MT, Ricciardi MR, Molfetta R, Paolini R, Santoni A, Cippitelli M (2016) Inhibition of bromodomain and extra-terminal (BET) proteins increases NKG2D ligand MICA expression and sensitivity to NK cell-mediated cytotoxicity in multiple myeloma cells: role of cMYC-IRF4-miR-125b interplay. J Hematol Oncol 9:134

Adjei AA (2006) What is the right dose? The elusive optimal biologic dose in phase I clinical trials. J Clin Oncol Off J Am Soc Clin Oncol 24:4054–4055

Aghajan M, Jonai N, Flick K, Fu F, Luo M, Cai X, Ouni I, Pierce N, Tang X, Lomenick B, Damoiseaux R, Hao R, Del Moral PM, Verma R, Li Y, Li C, Houk KN, Jung ME, Zheng N, Huang L, Deshaies RJ, Kaiser P, Huang J (2010) Chemical genetics screen for enhancers of rapamycin identifies a specific inhibitor of an SCF family E3 ubiquitin ligase. Nat Biotechnol 28:738–742

Aoki M, Fujishita T (2017) Oncogenic roles of the PI3K/AKT/mTOR Axis. Curr Top Microbiol Immunol 407:153–189

Balamurugan K (2016) HIF-1 at the crossroads of hypoxia, inflammation, and cancer. Int J Cancer 138:1058–1066

Bassi ZI, Fillmore MC, Miah AH, Chapman TD, Maller C, Roberts EJ, Davis LC, Lewis DE, Galwey NW, Waddington KE, Parravicini V, Macmillan-Jones AL, Gongora C, Humphreys PG, Churcher I, Prinjha RK, Tough DF (2018) Modulating PCAF/GCN5 immune cell function through a PROTAC approach. ACS Chem Biol 13:2862–2867

Batchu SN, Brijmohan AS, Advani A (2016) The therapeutic hope for HDAC6 inhibitors in malignancy and chronic disease. Clin Sci 130:987–1003

Bertacchini J, Heidari N, Mediani L, Capitani S, Shahjahani M, Ahmadzadeh A, Saki N (2015) Targeting PI3K/AKT/mTOR network for treatment of leukemia. Cell Mol Life Sci 72:2337–2347

Bondeson DP, Crews CM (2017) Targeted protein degradation by small molecules. Annu Rev Pharmacol Toxicol 57:107–123

Bondeson DP, Mares A, Smith IE, Ko E, Campos S, Miah AH, Mulholland KE, Routly N, Buckley DL, Gustafson JL, Zinn N, Grandi P, Shimamura S, Bergamini G, Faelth-Savitski M, Bantscheff M, Cox C, Gordon DA, Willard RR, Flanagan JJ, Casillas LN, Votta BJ, den Besten W, Famm K, Kruidenier L, Carter PS, Harling JD, Churcher I, Crews CM (2015) Catalytic in vivo protein knockdown by small-molecule PROTACs. Nat Chem Biol 11:611–617

Brownell JE, Sintchak MD, Gavin JM, Liao H, Bruzzese FJ, Bump NJ, Soucy TA, Milhollen MA, Yang X, Burkhardt AL, Ma J, Loke HK, Lingaraj T, Wu D, Hamman KB, Spelman JJ, Cullis CA, Langston SP, Vyskocil S, Sells TB, Mallender WD, Visiers I, Li P, Claiborne CF, Rolfe M, Bolen JB, Dick LR (2010) Substrate-assisted inhibition of ubiquitin-like protein-activating enzymes: the NEDD8 E1 inhibitor MLN4924 forms a NEDD8-AMP mimetic in situ. Mol Cell 37:102–111

Buckley DL, Gustafson JL, Van Molle I, Roth AG, Tae HS, Gareiss PC, Jorgensen WL, Ciulli A, Crews CM (2012a) Small-molecule inhibitors of the interaction between the E3 ligase VHL and HIF1alpha. Angew Chem Int Ed Engl 51:11463–11467

Buckley DL, Van Molle I, Gareiss PC, Tae HS, Michel J, Noblin DJ, Jorgensen WL, Ciulli A, Crews CM (2012b) Targeting the von Hippel-Lindau E3 ubiquitin ligase using small molecules to disrupt the VHL/HIF-1alpha interaction. J Am Chem Soc 134:4465–4468

Buhimschi AD, Armstrong HA, Toure M, Jaime-Figueroa S, Chen TL, Lehman AM, Woyach JA, Johnson AJ, Byrd JC, Crews CM (2018) Targeting the C481S Ibrutinib-resistance mutation in Bruton's tyrosine kinase using PROTAC-mediated degradation. Biochemistry 57:3564–3575

Bulatov E, Ciulli A (2015) Targeting Cullin-RING E3 ubiquitin ligases for drug discovery: structure, assembly and small-molecule modulation. Biochem J 467:365–386

Cardote TAF, Ciulli A (2017) Structure-guided design of peptides as tools to probe the protein-protein interaction between Cullin-2 and Elongin BC substrate adaptor in Cullin RING E3 ubiquitin ligases. ChemMedChem 12:1491–1496

Cardote TAF, Gadd MS, Ciulli A (2017) Crystal structure of the Cul2-Rbx1-EloBC-VHL ubiquitin ligase complex. Structure 25:901–911 e3

Ceccarelli DF, Tang X, Pelletier B, Orlicky S, Xie W, Plantevin V, Neculai D, Chou YC, Ogunjimi A, Al Hakim A, Varelas X, Koszela J, Wasney GA, Vedadi M, Dhe-Paganon S, Cox S, Xu S, Lopez-Girona A, Mercurio F, Wrana J, Durocher D, Meloche S, Webb DR, Tyers M, Sicheri F (2011) An allosteric inhibitor of the human Cdc34 ubiquitin-conjugating enzyme. Cell 145:1075–1087

Chamberlain PP, Lopez-Girona A, Miller K, Carmel G, Pagarigan B, Chie-Leon B, Rychak E, Corral LG, Ren YJ, Wang M, Riley M, Delker SL, Ito T, Ando H, Mori T, Hirano Y, Handa H, Hakoshima T, Daniel TO, Cathers BE (2014) Structure of the human Cereblon-DDB1-lenalidomide complex reveals basis for responsiveness to thalidomide analogs. Nat Struct Mol Biol 21:803–809

Chan CH, Morrow JK, Li CF, Gao Y, Jin G, Moten A, Stagg LJ, Ladbury JE, Cai Z, Xu D, Logothetis CJ, Hung MC, Zhang S, Lin HK (2013) Pharmacological inactivation of Skp2 SCF ubiquitin ligase restricts cancer stem cell traits and cancer progression. Cell 154:556–568

Chan CH, Morrow JK, Zhang S, Lin HK (2014) Skp2: a dream target in the coming age of cancer therapy. Cell Cycle 13:679–680

Chen Q, Xie W, Kuhn DJ, Voorhees PM, Lopez-Girona A, Mendy D, Corral LG, Krenitsky VP, Xu W, Moutouh-de Parseval L, Webb DR, Mercurio F, Nakayama KI, Nakayama K, Orlowski RZ (2008) Targeting the p27 E3 ligase SCF(Skp2) results in p27- and Skp2-mediated cell-cycle arrest and activation of autophagy. Blood 111:4690–4699

Chen BB, Coon TA, Glasser JR, McVerry BJ, Zhao J, Zhao Y, Zou C, Ellis B, Sciurba FC, Zhang Y, Mallampalli RK (2013) A combinatorial F box protein directed pathway controls TRAF adaptor stability to regulate inflammation. Nat Immunol 14:470–479

Chu TT, Gao N, Li QQ, Chen PG, Yang XF, Chen YX, Zhao YF, Li YM (2016) Specific knockdown of endogenous tau protein by peptide-directed ubiquitin-proteasome degradation. Cell Chem Biol 23:453–461

Cohen P, Tcherpakov M (2010) Will the ubiquitin system furnish as many drug targets as protein kinases? Cell 143:686–693

Conde J, Artzi N (2015) Are RNAi and miRNA therapeutics truly dead? Trends Biotechnol 33:141–144

Corneth OBJ, Klein Wolterink RGJ, Hendriks RW (2016) BTK Signaling in B cell differentiation and autoimmunity. Curr Top Microbiol Immunol 393:67–105

Crew AP, Raina K, Dong H, Qian Y, Wang J, Vigil D, Serebrenik YV, Hamman BD, Morgan A, Ferraro C, Siu K, Neklesa TK, Winkler JD, Coleman KG, Crews CM (2017) Identification and characterization of Von Hippel-Lindau-recruiting proteolysis targeting chimeras (PROTACs) of TANK-binding kinase 1. J Med Chem 61(2):583–598

Cui D, Xiong X, Zhao Y (2016) Cullin-RING ligases in regulation of autophagy. Cell Div 11:8

Ding Q, Zhang Z, Liu JJ, Jiang N, Zhang J, Ross TM, Chu XJ, Bartkovitz D, Podlaski F, Janson C, Tovar C, Filipovic ZM, Higgins B, Glenn K, Packman K, Vassilev LT, Graves B (2013) Discovery of RG7388, a potent and selective p53-MDM2 inhibitor in clinical development. J Med Chem 56:5979–5983

Duda DM, Borg LA, Scott DC, Hunt HW, Hammel M, Schulman BA (2008) Structural insights into NEDD8 activation of cullin-RING ligases: conformational control of conjugation. Cell 134:995–1006

Fischer ES, Scrima A, Bohm K, Matsumoto S, Lingaraju GM, Faty M, Yasuda T, Cavadini S, Wakasugi M, Hanaoka F, Iwai S, Gut H, Sugasawa K, Thoma NH (2011) The molecular basis of CRL4DDB2/CSA ubiquitin ligase architecture, targeting, and activation. Cell 147:1024–1039

Fischer ES, Bohm K, Lydeard JR, Yang H, Stadler MB, Cavadini S, Nagel J, Serluca F, Acker V, Lingaraju GM, Tichkule RB, Schebesta M, Forrester WC, Schirle M, Hassiepen U, Ottl J, Hild M, Beckwith RE, Harper JW, Jenkins JL, Thoma NH (2014) Structure of the DDB1-CRBN E3 ubiquitin ligase in complex with thalidomide. Nature 512:49–53

Fontana RJ (2014) Pathogenesis of idiosyncratic drug-induced liver injury and clinical perspectives. Gastroenterology 146:914–928

Frost J, Galdeano C, Soares P, Gadd MS, Grzes KM, Ellis L, Epemolu O, Shimamura S, Bantscheff M, Grandi P, Read KD, Cantrell DA, Rocha S, Ciulli A (2016) Potent and selective chemical probe of hypoxic signalling downstream of HIF-alpha hydroxylation via VHL inhibition. Nat Commun 7:13312

Galdeano C, Gadd MS, Soares P, Scaffidi S, Van Molle I, Birced I, Hewitt S, Dias DM, Ciulli A (2014) Structure-guided design and optimization of small molecules targeting the protein-protein interaction between the von Hippel-Lindau (VHL) E3 ubiquitin ligase and the hypoxia inducible factor (HIF) alpha subunit with in vitro nanomolar affinities. J Med Chem 57:8657–8663

Gechijian LN, Buckley DL, Lawlor MA, Reyes JM, Paulk J, Ott CJ, Winter GE, Erb MA, Scott TG, Xu M, Seo HS, Dhe-Paganon S, Kwiatkowski NP, Perry JA, Qi J, Gray NS, Bradner JE (2018) Functional TRIM24 degrader via conjugation of ineffectual bromodomain and VHL ligands. Nat Chem Biol 14:405–412

Gomes P, Fleming Outeiro T, Cavadas C (2015) Emerging role of Sirtuin 2 in the regulation of mammalian metabolism. Trends Pharmacol Sci 36:756–768

Gorelik M, Orlicky S, Sartori MA, Tang X, Marcon E, Kurinov I, Greenblatt JF, Tyers M, Moffat J, Sicheri F, Sidhu SS (2016) Inhibition of SCF ubiquitin ligases by engineered ubiquitin variants that target the Cul1 binding site on the Skp1-F-box interface. Proc Natl Acad Sci U S A 113:3527–3532

Gressel S, Schwalb B, Decker TM, Qin W, Leonhardt H, Eick D, Cramer P (2017) CDK9-dependent RNA polymerase II pausing controls transcription initiation. elife 6:e29736

Gstaiger M, Jordan R, Lim M, Catzavelos C, Mestan J, Slingerland J, Krek W (2001) Skp2 is oncogenic and overexpressed in human cancers. Proc Natl Acad Sci U S A 98:5043–5048

Guo Y, Dong L, Qiu X, Wang Y, Zhang B, Liu H, Yu Y, Zang Y, Yang M, Huang Z (2014) Structural basis for hijacking CBF-beta and CUL5 E3 ligase complex by HIV-1 Vif. Nature 505:229–233

Gustafson JL, Neklesa TK, Cox CS, Roth AG, Buckley DL, Tae HS, Sundberg TB, Stagg DB, Hines J, McDonnell DP, Norris JD, Crews CM (2015) Small-molecule-mediated degradation of the androgen receptor through hydrophobic tagging. Angew Chem Int Ed Engl 54:9659–9662

Haase VH (2010) Hypoxic regulation of erythropoiesis and iron metabolism. Am J Physiol Renal Physiol 299:F1–F13

Hao Z, Huang S (2015) E3 ubiquitin ligase Skp2 as an attractive target in cancer therapy. Front Biosci 20:474–490

Haque I, Banerjee S, De A, Maity G, Sarkar S, Majumdar M, Jha SS, McGragor D, Banerjee SK (2015) CCN5/WISP-2 promotes growth arrest of triple-negative breast cancer cells through accumulation and trafficking of p27(Kip1) via Skp2 and FOXO3a regulation. Oncogene 34:3152–3163

Hayashi K, Tan X, Zheng N, Hatate T, Kimura Y, Kepinski S, Nozaki H (2008) Small-molecule agonists and antagonists of F-box protein-substrate interactions in auxin perception and signaling. Proc Natl Acad Sci U S A 105:5632–5637

Henning RK, Varghese JO, Das S, Nag A, Tang G, Tang K, Sutherland AM, Heath JR (2016) Degradation of Akt using protein-catalyzed capture agents. J Pept Sci 22:196–200

Henssen A, Althoff K, Odersky A, Beckers A, Koche R, Speleman F, Schafers S, Bell E, Nortmeyer M, Westermann F, De Preter K, Florin A, Heukamp L, Spruessel A, Astrahanseff K, Lindner S, Sadowski N, Schramm A, Astorgues-Xerri L, Riveiro ME, Eggert A, Cvitkovic E, Schulte JH (2016) Targeting MYCN-driven transcription by BET-Bromodomain inhibition. Clin Cancer Res 22:2470–2481

Hines J, Gough JD, Corson TW, Crews CM (2013) Post-translational protein knockdown coupled to receptor tyrosine kinase activation with phosphoPROTACs. Proc Natl Acad Sci U S A 110:8942–8947

Hohmann AF, Martin LJ, Minder JL, Roe JS, Shi J, Steurer S, Bader G, McConnell D, Pearson M, Gerstberger T, Gottschamel T, Thompson D, Suzuki Y, Koegl M, Vakoc CR (2016) Sensitivity and engineered resistance of myeloid leukemia cells to BRD9 inhibition. Nat Chem Biol 12:672–679

Hon WC, Wilson MI, Harlos K, Claridge TD, Schofield CJ, Pugh CW, Maxwell PH, Ratcliffe PJ, Stuart DI, Jones EY (2002) Structural basis for the recognition of hydroxyproline in HIF-1 alpha by pVHL. Nature 417:975–978

Hu L, Magesh S, Chen L, Wang L, Lewis TA, Chen Y, Khodier C, Inoyama D, Beamer LJ, Emge TJ, Shen J, Kerrigan JE, Kong AN, Dandapani S, Palmer M, Schreiber SL, Munoz B (2013) Discovery of a small-molecule inhibitor and cellular probe of Keap1-Nrf2 protein-protein interaction. Bioorg Med Chem Lett 23:3039–3043

Huang X, Dixit VM (2016) Drugging the undruggables: exploring the ubiquitin system for drug development. Cell Res 26:484–498

Huang H, Ceccarelli DF, Orlicky S, St-Cyr DJ, Ziemba A, Garg P, Plamondon S, Auer M, Sidhu S, Marinier A, Kleiger G, Tyers M, Sicheri F (2014) E2 enzyme inhibition by stabilization of a low-affinity interface with ubiquitin. Nat Chem Biol 10:156–163

Humphries F, Yang S, Wang B, Moynagh PN (2015) RIP kinases: key decision makers in cell death and innate immunity. Cell Death Differ 22:225–236

Inoue J, Ishida T, Tsukamoto N, Kobayashi N, Naito A, Azuma S, Yamamoto T (2000) Tumor necrosis factor receptor-associated factor (TRAF) family: adapter proteins that mediate cytokine signaling. Exp Cell Res 254:14–24

Ito T, Ando H, Suzuki T, Ogura T, Hotta K, Imamura Y, Yamaguchi Y, Handa H (2010) Identification of a primary target of thalidomide teratogenicity. Science (New York, NY) 327:1345–1350

Jia L, Soengas MS, Sun Y (2009) ROC1/RBX1 E3 ubiquitin ligase silencing suppresses tumor cell growth via sequential induction of G2-M arrest, apoptosis, and senescence. Cancer Res 69:4974–4982

Jing H, Hu J, He B, Negron Abril YL, Stupinski J, Weiser K, Carbonaro M, Chiang YL, Southard T, Giannakakou P, Weiss RS, Lin H (2016) A SIRT2-selective inhibitor promotes c-Myc Oncoprotein degradation and exhibits broad anticancer activity. Cancer Cell 29:297–310

Jnoff E, Albrecht C, Barker JJ, Barker O, Beaumont E, Bromidge S, Brookfield F, Brooks M, Bubert C, Ceska T, Corden V, Dawson G, Duclos S, Fryatt T, Genicot C, Jigorel E, Kwong J, Maghames R, Mushi I, Pike R, Sands ZA, Smith MA, Stimson CC, Courade JP (2014) Binding mode and structure-activity relationships around direct inhibitors of the Nrf2-Keap1 complex. ChemMedChem 9:699–705

Kronke J, Udeshi ND, Narla A, Grauman P, Hurst SN, McConkey M, Svinkina T, Heckl D, Comer E, Li X, Ciarlo C, Hartman E, Munshi N, Schenone M, Schreiber SL, Carr SA, Ebert BL (2014) Lenalidomide causes selective degradation of IKZF1 and IKZF3 in multiple myeloma cells. Science (New York, NY) 343:301–305

Lai AC, Toure M, Hellerschmied D, Salami J, Jaime-Figueroa S, Ko E, Hines J, Crews CM (2016) Modular PROTAC design for the degradation of oncogenic BCR-ABL. Angew Chem Int Ed Engl 55:807–810

Lewis WG, Green LG, Grynszpan F, Radic Z, Carlier PR, Taylor P, Finn MG, Sharpless KB (2002) Click chemistry in situ: acetylcholinesterase as a reaction vessel for the selective assembly of a femtomolar inhibitor from an array of building blocks. Angew Chem Int Ed Engl 41:1053–1057

Li H, Tan M, Jia L, Wei D, Zhao Y, Chen G, Xu J, Zhao L, Thomas D, Beer DG, Sun Y (2014) Inactivation of SAG/RBX2 E3 ubiquitin ligase suppresses KrasG12D-driven lung tumorigenesis. J Clin Investig 124:835–846

Li W, Gao C, Zhao L, Yuan Z, Chen Y, Jiang Y (2018) Phthalimide conjugations for the degradation of oncogenic PI3K. Eur J Med Chem 151:237–247

Liu J, Furukawa M, Matsumoto T, Xiong Y (2002) NEDD8 modification of CUL1 dissociates p120 (CAND1), an inhibitor of CUL1-SKP1 binding and SCF ligases. Mol Cell 10:1511–1518

Liu X, Gao Y, Ye H, Gerrin S, Ma F, Wu Y, Zhang T, Russo J, Cai C, Yuan X, Liu J, Chen S, Balk SP (2017) Positive feedback loop mediated by protein phosphatase 1alpha mobilization of P-TEFb and basal CDK1 drives androgen receptor in prostate cancer. Nucleic Acids Res 45:3738–3751

Lopez-Girona A, Mendy D, Ito T, Miller K, Gandhi AK, Kang J, Karasawa S, Carmel G, Jackson P, Abbasian M, Mahmoudi A, Cathers B, Rychak E, Gaidarova S, Chen R, Schafer PH, Handa H, Daniel TO, Evans JF, Chopra R (2012) Cereblon is a direct protein target for immunomodulatory and antiproliferative activities of lenalidomide and pomalidomide. Leukemia 26:2326–2335

Lord CJ, Ashworth A (2013) Mechanisms of resistance to therapies targeting BRCA-mutant cancers. Nat Med 19:1381–1388

Lu G, Middleton RE, Sun H, Naniong M, Ott CJ, Mitsiades CS, Wong K-K, Bradner JE, Kaelin WG (2014) The myeloma drug Lenalidomide promotes the Cereblon-dependent destruction of Ikaros proteins. Science 343:305–309

Lu J, Qian Y, Altieri M, Dong H, Wang J, Raina K, Hines J, Winkler JD, Crew AP, Coleman K, Crews CM (2015) Hijacking the E3 ubiquitin ligase Cereblon to efficiently target BRD4. Chem Biol 22:755–763

Malek E, Abdel-Malek MA, Jagannathan S, Vad N, Karns R, Jegga AG, Broyl A, van Duin M, Sonneveld P, Cottini F, Anderson KC, Driscoll JJ (2017) Pharmacogenomics and chemical library screens reveal a novel SCF(SKP2) inhibitor that overcomes Bortezomib resistance in multiple myeloma. Leukemia 31:645–653

Mallampalli RK, Coon TA, Glasser JR, Wang C, Dunn SR, Weathington NM, Zhao J, Zou C, Zhao Y, Chen BB (2013) Targeting F box protein Fbxo3 to control cytokine-driven inflammation. J Immunol 191:5247–5255

Manasanch EE, Orlowski RZ (2017) Proteasome inhibitors in cancer therapy. Nat Rev Clin Oncol 14:417–433

Mao X, Gluck N, Chen B, Starokadomskyy P, Li H, Maine GN, Burstein E (2011) COMMD1 (copper metabolism MURR1 domain-containing protein 1) regulates Cullin RING ligases by preventing CAND1 (Cullin-associated Nedd8-dissociated protein 1) binding. J Biol Chem 286:32355–32365

Marcotte D, Zeng W, Hus JC, McKenzie A, Hession C, Jin P, Bergeron C, Lugovskoy A, Enyedy I, Cuervo H, Wang D, Atmanene C, Roecklin D, Vecchi M, Vivat V, Kraemer J, Winkler D, Hong V, Chao J, Lukashev M, Silvian L (2013) Small molecules inhibit the interaction of Nrf2 and the Keap1 Kelch domain through a non-covalent mechanism. Bioorg Med Chem 21:4011–4019

Masoud GN, Li W (2015) HIF-1alpha pathway: role, regulation and intervention for cancer therapy. Acta Pharm Sin B 5:378–389

Matsui M, Shindo K, Izumi T, Io K, Shinohara M, Komano J, Kobayashi M, Kadowaki N, Harris RS, Takaori-Kondo A (2014) Small molecules that inhibit Vif-induced degradation of APOBEC3G. Virol J 11:122

Milhollen MA, Thomas MP, Narayanan U, Traore T, Riceberg J, Amidon BS, Bence NF, Bolen JB, Brownell J, Dick LR, Loke HK, McDonald AA, Ma J, Manfredi MG, Sells TB, Sintchak MD, Yang X, Xu Q, Koenig EM, Gavin JM, Smith PG (2012) Treatment-emergent mutations in NAEbeta confer resistance to the NEDD8-activating enzyme inhibitor MLN4924. Cancer Cell 21:388–401

Min JH, Yang H, Ivan M, Gertler F, Kaelin WG Jr, Pavletich NP (2002) Structure of an HIF-1alpha -pVHL complex: hydroxyproline recognition in signaling. Science (New York, NY) 296:1886–1889

Montrose K, Krissansen GW (2014) Design of a PROTAC that antagonizes and destroys the cancer-forming X-protein of the hepatitis B virus. Biochem Biophys Res Commun 453:735–740

Mullard A (2019) First targeted protein degrader hits the clinic. Nat Rev Drug Discov

Nagy Z, Tora L (2007) Distinct GCN5/PCAF-containing complexes function as co-activators and are involved in transcription factor and global histone acetylation. Oncogene 26:5341–5357

Nakajima H, Fujiwara H, Furuichi Y, Tanaka K, Shimbara N (2008) A novel small-molecule inhibitor of NF-kappaB signaling. Biochem Biophys Res Commun 368:1007–1013

Nathans R, Cao H, Sharova N, Ali A, Sharkey M, Stranska R, Stevenson M, Rana TM (2008) Small-molecule inhibition of HIV-1 Vif. Nat Biotechnol 26:1187–1192

Nawrocki ST, Griffin P, Kelly KR, Carew JS (2012) MLN4924: a novel first-in-class inhibitor of NEDD8-activating enzyme for cancer therapy. Expert Opin Investig Drugs 21:1563–1573

Nguyen HC, Yang H, Fribourgh JL, Wolfe LS, Xiong Y (2015) Insights into Cullin-RING E3 ubiquitin ligase recruitment: structure of the VHL-EloBC-Cul2 complex. Structure 23:441–449

Olson CM, Jiang B, Erb MA, Liang Y, Doctor ZM, Zhang Z, Zhang T, Kwiatkowski N, Boukhali M, Green JL, Haas W, Nomanbhoy T, Fischer ES, Young RA, Bradner JE, Winter GE, Gray NS (2018) Pharmacological perturbation of CDK9 using selective CDK9 inhibition or degradation. Nat Chem Biol 14:163–170

Orlicky S, Tang X, Neduva V, Elowe N, Brown ED, Sicheri F, Tyers M (2010) An allosteric inhibitor of substrate recognition by the SCF(Cdc4) ubiquitin ligase. Nat Biotechnol 28:733–737

Park KS, Yang H, Choi J, Seo S, Kim D, Lee CH, Jeon H, Kim SW, Lee DH (2017) The HSP90 inhibitor, NVP-AUY922, attenuates intrinsic PI3K inhibitor resistance in KRAS-mutant non-small cell lung cancer. Cancer Lett 406:47–53

Pavlides SC, Huang KT, Reid DA, Wu L, Blank SV, Mittal K, Guo L, Rothenberg E, Rueda B, Cardozo T, Gold LI (2013) Inhibitors of SCF-Skp2/Cks1 E3 ligase block estrogen-induced growth stimulation and degradation of nuclear p27kip1: therapeutic potential for endometrial cancer. Endocrinology 154:4030–4045

Petroski MD, Deshaies RJ (2005) Function and regulation of cullin-RING ubiquitin ligases. Nat Rev Mol Cell Biol 6:9–20

Petzold G, Fischer ES, Thoma NH (2016) Structural basis of lenalidomide-induced CK1alpha degradation by the CRL4(CRBN) ubiquitin ligase. Nature 532:127–130

Pruneda JN, Littlefield PJ, Soss SE, Nordquist KA, Chazin WJ, Brzovic PS, Klevit RE (2012) Structure of an E3: E2~Ub complex reveals an allosteric mechanism shared among RING/U-box ligases. Mol Cell 47:933–942

Qiu L, Xu C, Chen J, Li Q, Jiang H (2019) Downregulation of the transcriptional co-activator PCAF inhibits the proliferation and migration of vascular smooth muscle cells and attenuates NF-kappaB-mediated inflammatory responses. Biochem Biophys Res Commun 513:41–48

Raina K, Lu J, Qian Y, Altieri M, Gordon D, Rossi AM, Wang J, Chen X, Dong H, Siu K, Winkler JD, Crew AP, Crews CM, Coleman KG (2016) PROTAC-induced BET protein degradation as a therapy for castration-resistant prostate cancer. Proc Natl Acad Sci U S A 113:7124–7129

Reitsma JM, Liu X, Reichermeier KM, Moradian A, Sweredoski MJ, Hess S, Deshaies RJ (2017) Composition and regulation of the cellular repertoire of SCF ubiquitin ligases. Cell 171:1326–1339.e14

Remillard D, Buckley DL, Paulk J, Brien GL, Sonnett M, Seo HS, Dastjerdi S, Wuhr M, Dhe-Paganon S, Armstrong SA, Bradner JE (2017) Degradation of the BAF complex factor BRD9 by Heterobifunctional ligands. Angew Chem Int Ed Engl 56:5738–5743

Richardson PG, Hofmeister CC, Raje NS, Siegel DS, Lonial S, Laubach J, Efebera YA, Vesole DH, Nooka AK, Rosenblatt J, Doss D, Zaki MH, Bensmaine A, Herring J, Li Y, Watkins L, Chen MS, Anderson KC (2017) Pomalidomide, bortezomib and low-dose dexamethasone in lenalidomide-refractory and proteasome inhibitor-exposed myeloma. Leukemia 31:2695–2701

Rico-Bautista E, Yang CC, Lu L, Roth GP, Wolf DA (2010) Chemical genetics approach to restoring p27Kip1 reveals novel compounds with antiproliferative activity in prostate cancer cells. BMC Biol 8:153

Robb CM, Contreras JI, Kour S, Taylor MA, Abid M, Sonawane YA, Zahid M, Murry DJ, Natarajan A, Rana S (2017) Chemically induced degradation of CDK9 by a proteolysis targeting chimera (PROTAC). Chem Commun (Camb) 53:7577–7580

Rodriguez-Gonzalez A, Cyrus K, Salcius M, Kim K, Crews CM, Deshaies RJ, Sakamoto KM (2008) Targeting steroid hormone receptors for ubiquitination and degradation in breast and prostate cancer. Oncogene 27:7201–7211

Saha A, Deshaies RJ (2008) Multimodal activation of the ubiquitin ligase SCF by Nedd8 conjugation. Mol Cell 32:21–31

Sakamoto KM, Kim KB, Kumagai A, Mercurio F, Crews CM, Deshaies RJ (2001) Protacs: chimeric molecules that target proteins to the Skp1-Cullin-F box complex for ubiquitination and degradation. Proc Natl Acad Sci U S A 98:8554–8559

Sakamoto KM, Kim KB, Verma R, Ransick A, Stein B, Crews CM, Deshaies RJ (2003) Development of protacs to target cancer-promoting proteins for ubiquitination and degradation. Mol Cell Proteomics 2:1350–1358

Salami J, Alabi S, Willard RR, Vitale NJ, Wang J, Dong H, Jin M, McDonnell DP, Crew AP, Neklesa TK, Crews CM (2018) Androgen receptor degradation by the proteolysis-targeting chimera ARCC-4 outperforms enzalutamide in cellular models of prostate cancer drug resistance. Commun Biol 1:100

Sandoval D, Hill S, Ziemba A, Lewis S, Kuhlman B, Kleiger G (2015) Ubiquitin-conjugating enzyme Cdc34 and ubiquitin ligase Skp1-cullin-F-box ligase (SCF) interact through multiple conformations. J Biol Chem 290:1106–1118

Sarikas A, Hartmann T, Pan ZQ (2011) The cullin protein family. Genome Biol 12:220

Sarkar S, Perlstein EO, Imarisio S, Pineau S, Cordenier A, Maglathlin RL, Webster JA, Lewis TA, O'Kane CJ, Schreiber SL, Rubinsztein DC (2007) Small molecules enhance autophagy and reduce toxicity in Huntington's disease models. Nat Chem Biol 3:331–338

Scheepstra M, Hekking KFW, van Hijfte L, Folmer RHA (2019) Bivalent ligands for protein degradation in drug discovery. Comput Struct Biotechnol J 17:160–176

Schiedel M, Herp D, Hammelmann S, Swyter S, Lehotzky A, Robaa D, Olah J, Ovadi J, Sippl W, Jung M (2018) Chemically induced degradation of Sirtuin 2 (Sirt2) by a proteolysis targeting chimera (PROTAC) based on Sirtuin rearranging ligands (SirReals). J Med Chem 61:482–491

Schneekloth JS Jr, Fonseca FN, Koldobskiy M, Mandal A, Deshaies R, Sakamoto K, Crews CM (2004) Chemical genetic control of protein levels: selective in vivo targeted degradation. J Am Chem Soc 126:3748–3754

Schneekloth AR, Pucheault M, Tae HS, Crews CM (2008) Targeted intracellular protein degradation induced by a small molecule: en route to chemical proteomics. Bioorg Med Chem Lett 18:5904–5908

Scudellari M (2019) Protein-slaying drugs could be the next blockbuster therapies. Nature 567:298–300

Seol JH, Feldman RM, Zachariae W, Shevchenko A, Correll CC, Lyapina S, Chi Y, Galova M, Claypool J, Sandmeyer S, Nasmyth K, Deshaies RJ, Shevchenko A, Deshaies RJ (1999) Cdc53/cullin and the essential Hrt1 RING-H2 subunit of SCF define a ubiquitin ligase module that activates the E2 enzyme Cdc34. Genes Dev 13:1614–1626

Shafique S, Ali W, Kanwal S, Rashid S (2018) Structural basis for Cullins and RING component inhibition: targeting E3 ubiquitin pathway conductors for cancer therapeutics. Int J Biol Macromol 106:532–543

Shah JJ, Jakubowiak AJ, O'Connor OA, Orlowski RZ, Harvey RD, Smith MR, Lebovic D, Diefenbach C, Kelly K, Hua Z, Berger AJ, Mulligan G, Faessel HM, Tirrell S, Dezube BJ, Lonial S (2016) Phase I study of the novel investigational NEDD8-activating enzyme inhibitor Pevonedistat (MLN4924) in patients with relapsed/refractory multiple myeloma or lymphoma. Clin Cancer Res 22:34–43

Soucy TA, Smith PG, Milhollen MA, Berger AJ, Gavin JM, Adhikari S, Brownell JE, Burke KE, Cardin DP, Critchley S, Cullis CA, Doucette A, Garnsey JJ, Gaulin JL, Gershman RE, Lublinsky AR, McDonald A, Mizutani H, Narayanan U, Olhava EJ, Peluso S, Rezaei M, Sintchak MD, Talreja T, Thomas MP, Traore T, Vyskocil S, Weatherhead GS, Yu J, Zhang J, Dick LR, Claiborne CF, Rolfe M, Bolen JB, Langston SP (2009) An inhibitor of NEDD8-activating enzyme as a new approach to treat cancer. Nature 458:732–736

Stanley BJ, Ehrlich ES, Short L, Yu Y, Xiao Z, Yu XF, Xiong Y (2008) Structural insight into the human immunodeficiency virus Vif SOCS box and its role in human E3 ubiquitin ligase assembly. J Virol 82:8656–8663

Sun Y, Zhao X, Ding N, Gao H, Wu Y, Yang Y, Zhao M, Hwang J, Song Y, Liu W, Rao Y (2018a) PROTAC-induced BTK degradation as a novel therapy for mutated BTK C481S induced ibrutinib-resistant B-cell malignancies. Cell Res 28:779–781

Sun B, Fiskus W, Qian Y, Rajapakshe K, Raina K, Coleman KG, Crew AP, Shen A, Saenz DT, Mill CP, Nowak AJ, Jain N, Zhang L, Wang M, Khoury JD, Coarfa C, Crews CM, Bhalla KN (2018b) BET protein proteolysis targeting chimera (PROTAC) exerts potent lethal activity against mantle cell lymphoma cells. Leukemia 32:343–352

Taguchi K, Yamamoto M (2017) The KEAP1-NRF2 system in cancer. Front Oncol 7:85

Thines B, Katsir L, Melotto M, Niu Y, Mandaokar A, Liu G, Nomura K, He SY, Howe GA, Browse J (2007) JAZ repressor proteins are targets of the SCF (COI1) complex during jasmonate signalling. Nature 448:661–665

Tokatlian T, Segura T (2010) siRNA applications in nanomedicine. Wiley Interdiscip Rev Nanomed Nanobiotechnol 2:305–315

Tron AE, Arai T, Duda DM, Kuwabara H, Olszewski JL, Fujiwara Y, Bahamon BN, Signoretti S, Schulman BA, DeCaprio JA (2012) The glomuvenous malformation protein Glomulin binds Rbx1 and regulates cullin RING ligase-mediated turnover of Fbw7. Mol Cell 46:67–78

Van Molle I, Thomann A, Buckley DL, So EC, Lang S, Crews CM, Ciulli A (2012) Dissecting fragment-based lead discovery at the von Hippel-Lindau protein:

hypoxia inducible factor 1alpha protein-protein interface. Chem Biol 19:1300–1312

Vassilev LT, Vu BT, Graves B, Carvajal D, Podlaski F, Filipovic Z, Kong N, Kammlott U, Lukacs C, Klein C, Fotouhi N, Liu EA (2004) In vivo activation of the p53 pathway by small-molecule antagonists of MDM2. Science (New York, NY) 303:844–848

Vu B, Wovkulich P, Pizzolato G, Lovey A, Ding Q, Jiang N, Liu JJ, Zhao C, Glenn K, Wen Y, Tovar C, Packman K, Vassilev L, Graves B (2013) Discovery of RG7112. a small-molecule MDM2 inhibitor in clinical development. ACS Med Chem Lett 4:466–469

Wang Z, Cuddy M, Samuel T, Welsh K, Schimmer A, Hanaii F, Houghten R, Pinilla C, Reed JC (2004) Cellular, biochemical, and genetic analysis of mechanism of small molecule IAP inhibitors. J Biol Chem 279:48168–48176

Wang Z, Liu P, Inuzuka H, Wei W (2014) Roles of F-box proteins in cancer. Nat Rev Cancer 14:233–247

Wang H, Maitra A, Wang H (2016) The emerging roles of F-box proteins in pancreatic tumorigenesis. Semin Cancer Biol 36:88–94

Wang ST, Ho HJ, Lin JT, Shieh JJ, Wu CY (2017) Simvastatin-induced cell cycle arrest through inhibition of STAT3/SKP2 axis and activation of AMPK to promote p27 and p21 accumulation in hepatocellular carcinoma cells. Cell Death Dis 8:e2626

Wells JA, McClendon CL (2007) Reaching for high-hanging fruit in drug discovery at protein-protein interfaces. Nature 450:1001–1009

Winter GE, Buckley DL, Paulk J, Roberts JM, Souza A, Dhe-Paganon S, Bradner JE (2015) Drug Development. Phthalimide conjugation as a strategy for in vivo target protein degradation. Science (New York, NY) 348:1376–1381

Wood K, Hensing T, Malik R, Salgia R (2016) Prognostic and predictive value in KRAS in non-small-cell lung cancer: a review. JAMA Oncol 2:805–812

Wu L, Grigoryan Arsen V, Li Y, Hao B, Pagano M, Cardozo Timothy J (2012) Specific small molecule inhibitors of Skp2-mediated p27 degradation. Chem Biol 19:1515–1524

Wu J, Liu C, Tsui ST, Liu D (2016) Second-generation inhibitors of Bruton tyrosine kinase. J Hematol Oncol 9:80

Xu J, Li L, Yu G, Ying W, Gao Q, Zhang W, Li X, Ding C, Jiang Y, Wei D, Duan S, Lei Q, Li P, Shi T, Qian X, Qin J, Jia L (2015) The neddylation-cullin 2-RBX1 E3 ligase axis targets tumor suppressor RhoB for degradation in liver cancer. Mol Cell Proteomics 14:499–509

Yang K, Song Y, Xie H, Wu H, Wu YT, Leisten ED, Tang W (2018) Development of the first small molecule histone deacetylase 6 (HDAC6) degraders. Bioorg Med Chem Lett 28:2493–2497

Yasuda D, Nakajima M, Yuasa A, Obata R, Takahashi K, Ohe T, Ichimura Y, Komatsu M, Yamamoto M, Imamura R, Kojima H, Okabe T, Nagano T, Mashino T (2016) Synthesis of Keap1-phosphorylated p62 and Keap1-Nrf2 protein-protein interaction inhibitors and their inhibitory activity. Bioorg Med Chem Lett 26:5956–5959

Ye Y, Rape M (2009) Building ubiquitin chains: E2 enzymes at work. Nat Rev Mol Cell Biol 10:755–764

Zengerle M, Chan KH, Ciulli A (2015) Selective small molecule induced degradation of the BET Bromodomain protein BRD4. ACS Chem Biol 10:1770–1777

Zhao Y, Sun Y (2013) Cullin-RING ligases as attractive anti-cancer targets. Curr Pharm Des 19:3215–3225

Zhao Y, Morgan MA, Sun Y (2014) Targeting Neddylation pathways to inactivate cullin-RING ligases for anticancer therapy. Antioxid Redox Signal 21:2383–2400

Zheng N, Schulman BA, Song L, Miller JJ, Jeffrey PD, Wang P, Chu C, Koepp DM, Elledge SJ, Pagano M, Conaway RC, Conaway JW, Harper JW, Pavletich NP (2002) Structure of the Cul1-Rbx1-Skp1-F box Skp2 SCF ubiquitin ligase complex. Nature 416:703–709

Zhou X, Evans SL, Han X, Liu Y, Yu XF (2012) Characterization of the interaction of full-length HIV-1 Vif protein with its key regulator CBFbeta and CRL5 E3 ubiquitin ligase components. PLoS One 7:e33495

Zhou L, Zhang W, Sun Y, Jia L (2018) Protein neddylation and its alterations in human cancers for targeted therapy. Cell Signal 44:92–102

Zorba A, Nguyen C, Xu Y, Starr J, Borzilleri K, Smith J, Zhu H, Farley KA, Ding W, Schiemer J, Feng X, Chang JS, Uccello DP, Young JA, Garcia-Irrizary CN, Czabaniuk L, Schuff B, Oliver R, Montgomery J, Hayward MM, Coe J, Chen J, Niosi M, Luthra S, Shah JC, El-Kattan A, Qiu X, West GM, Noe MC, Shanmugasundaram V, Gilbert AM, Brown MF, Calabrese MF (2018) Delineating the role of cooperativity in the design of potent PROTACs for BTK. Proc Natl Acad Sci U S A 115: E7285–e7292

Targeting DCN1-UBC12 Protein-Protein Interaction for Regulation of Neddylation Pathway

20

Haibin Zhou, Jianfeng Lu, Chao-Yie Yang, Yi Sun, and Shaomeng Wang

Abstract

Protein neddylation is one type of posttranslational modifications that regulates the activity of the substrate proteins. Neddylation modification is catalyzed by NEDD8-activating enzyme (NAE, E1), NEDD8-conjugating enzyme (E2), and NEDD8 ligase (E3) to attach NEDD8, an ubiquitin-like molecule, to a lysine residue of a substrate protein. The best known neddylation substrates are cullin family members, which are scaffold components of cullin-RING ligases (CRLs), and cullin neddylation is required for activation of CRLs. In mammalian cells, there are one E1, two E2s (UBC12/UBE2M and UBE2F), and over a dozen E3s. MLN4924, the first-in-class small-molecule inhibitor of NAE, blocks the entire neddylation modification to inactivate activity of all CRLs. MLN4924 is currently in the Phase I/II clinical trials for anticancer application.

In the last few years, targeting protein-protein interactions of the neddylation complexes has been pursued as a potential strategy to selectively inhibit the activity of individual CRL. Analysis of the co-crystal structures of DCN1, a co-E3 for neddylation, and its binding partners UBC12 (a neddylation E2) suggested that it may be amenable for the design of potent, small-molecule inhibitors. In this chapter, we will review the discovery of small-molecule inhibitors that block the interactions of DCN1 with UBC12 (hereafter called DCN1 inhibitors) from a number of laboratories, including ours, leading to selective inactivation of CRL-1 and/or CRL-3. We will also discuss potential therapeutic applications of these small-molecule inhibitors.

Keywords

CRL3 · UBC12/UBE2M · DCN1 · Small-molecule inhibitor · Protein-protein interaction

H. Zhou · J. Lu
Department of Internal Medicine, University of Michigan, Ann Arbor, MI, USA

C.-Y. Yang
Department of Pharmaceutical Sciences, College of Pharmacy, The University of Tennessee Health Sciences Center, Memphis, TN, USA

Y. Sun
Division of Radiation and Cancer Biology, Department of Radiation Oncology, University of Michigan, Ann Arbor, MI, USA

S. Wang (✉)
Department of Internal Medicine, University of Michigan, Ann Arbor, MI, USA

Department of Pharmacology, University of Michigan, Ann Arbor, MI, USA

Department of Medicinal Chemistry, University of Michigan, Ann Arbor, MI, USA
e-mail: shaomeng@umich.edu

Abbreviations

BLI	Biolayer interferometry
CRLs	Cullin-RING ligases

Y. Sun et al. (eds.), *Cullin-RING Ligases and Protein Neddylation*, Advances in Experimental Medicine and Biology 1217, https://doi.org/10.1007/978-981-15-1025-0_20

Cul	Cullin
DCN1	Defective in cullin neddylation protein 1
FP	Fluorescence polarization
HTS	High-throughput screening
NAE	NEDD8-activating enzyme
NEDD8	Neural precursor cell-expressed developmentally downregulated protein-8
NRF2	Nuclear factor erythroid 2-related factor 2
RBX1	RING-box protein 1
TR-FRET	Time-resolved fluorescence energy transfer
UPS	Ubiquitin-proteasome system

20.1 Introduction

Elimination of misfolded or damaged proteins is an actively regulated cellular process and is critical for cells to maintain homeostasis (Hershko and Ciechanover 1998; Hershko 2005). The regulated destruction of intracellular proteins is initiated by covalent modifications of target proteins with ubiquitin, followed by degradation by the proteasome system (UPS) (Hershko and Ciechanover 1998; Hershko 2005). The cullin-RING ligases (CRLs), a central component of the UPS, regulate the turnover of approximately 20% of cellular proteins and play a critical role in normal cellular physiology and various human diseases (Bedford et al. 2011; Petroski and Deshaies 2005; Bulatov and Ciulli 2015; Nalepa et al. 2006). Each CRL system consists of a number of proteins with an individual cullin protein being the central component. To date, a total of eight subfamilies of mammalian CRLs has been identified based upon the presence of eight individual cullin proteins (cullins 1, 2, 3, 4A, 4B, 5, 7, and 9) (Petroski and Deshaies 2005; Zhao and Sun 2013; Zhao et al. 2014).

The activities of CRLs are controlled by covalent attachment of the *n*eural precursor cell-*e*xpressed *d*evelopmentally *d*ownregulated protein-*8* (NEDD8), a ubiquitin-like protein (Gong and Yeh 1999; Pan et al. 2004; Deshaies et al. 2010; Duda et al. 2008), to substrate proteins, a process known as neddylation. Neddylation is analogous to the process of ubiquitination, wherein the ubiquitin-like protein NEDD8 is conjugated to its target proteins. The neddylation cascade begins with the activation of NEDD8 by the NEDD8-activating enzyme (NAE), an E1 enzyme, followed by transfer of the activated NEDD8 to one of two NEDD8-specific E2 enzymes, UBC12 (also known as UBE2M) and UBE2F. In the final step of this cascade, E3 enzymes catalyze the transfer of NEDD8 from E2 enzymes to target substrates, with cullins being the best-characterized substrates. Neddylation of cullins results in the activation of CRLs. The structural and biochemical mechanisms underlying the E1-E2-E3 cascade reaction in the NEDD8 pathway have been elegantly elucidated by Schulman and colleagues (Huang et al. 2004; Kurz et al. 2005, 2008; Scott et al. 2010, 2011, 2014).

In the last decade, targeting neddylation has been pursued as a potential therapeutic strategy (Zhao and Sun 2013; Zhao et al. 2014; Soucy et al. 2010; Watson et al. 2011). The most advanced compound from these efforts, MLN4924, was developed as an inhibitor of the E1 enzyme NAE. MLN4924 is very effective in achieving broad inhibition of all CRLs and induces accumulation of CRL substrate proteins (Soucy et al. 2009a; Brownell et al. 2010). MLN4924 is currently in clinical development for the treatment of human cancer and has demonstrated clinical efficacy (Soucy et al. 2009b; Abidi and Xirodimas 2015). Because MLN4924 blocks degradation of protein substrates regulated by all CRLs, it is not suitable for many potential therapeutic applications. Therefore, selective inhibition of individual CRLs is highly desirable but has proven to be challenging.

Our laboratory has proposed that targeting protein-protein interactions of the neddylation enzymes within the CRL E3 ligase complex can be a potential strategy to specifically inhibit the activity of individual CRL. Analysis of the co-crystal structures of the cullin 1 E3 ligase complex suggested that the interactions between DCN1, a co-E3 ligase, and its binding partners UBC12 or UBE2F may be amenable for the design of potent, small-molecule inhibitors. In the last few years, a number of laboratories, including ours, have designed small-molecule inhibitors to block the interactions of DCN1 with UBC12 or UBE2F (hereafter called DCN1 inhibitors). Hereafter, we will review the discoveries of DCN1 inhibitors reported in the last few years.

20.2 Structural Basis for the Design of Small-Molecule Inhibitors of UBC12-DCN1 Interaction

We analyzed the X-ray crystallographic structure of the Rbx1-UBC12~NEDD8-Cul1-DCN1 complex (Scott et al. 2011, 2014) and identified a number of protein-protein interaction sites potentially suitable for the design of small-molecule inhibitors to disrupt the complex of the cullin 1 CRL E3 ligase. Among them, the DCN1-UBC12 protein-protein interaction appears to be particularly suited for the design of small-molecule inhibitors (Fig. 20.1).

The DCN1-UBC12 interaction is mediated primarily by a well-defined binding groove in DCN1 and an N-terminal 12-residue peptide of UBC12 protein (Fig. 20.1a) (Scott et al. 2011, 2014). Computational analysis of the DCN1-binding groove using a cosolvent mapping method identified four hydrophobic hotspots, namely P1, P2, P3, and P4, which cover those regions occupied by residues Met1, Ile2, and Leu4 of UBC12, respectively (Fig. 20.1b). The cluster of these hotspots also suggests the feasibility of designing potent inhibitors of the DCN1-UBC12 interaction. Indeed, a number of laboratories have reported the discovery of small-molecule inhibitors of the DCN1-UBC12 protein-protein interaction (hereafter called DCN1 inhibitors) in the last few years (Scott et al. 2017; Zhou et al. 2017, 2018, 2019; Hammill et al. 2018a, b; Wang et al. 2019) as detailed below.

20.3 Design of Peptidomimetic Inhibitors of the UBC12-DCN1 Interaction

Peptidomimetics often serve as useful starting points to target protein-protein interactions. Our laboratories have carried out extensive modifications on the UBC12 peptide, which led to the design of potent peptidomimetics. The key steps are summarized in Fig. 20.2 (Zhou et al. 2017, 2018).

First, to identify a minimal peptide motif with significant affinity to DCN1, a series of peptides were synthesized and tested by sequential truncation of C-terminal residues of the N-terminal acetylated, 12-residue UBC12 peptide **1**, which led to identification of the 4-residue peptide **2** with a K_i value of 130 μM to DCN1. Peptide

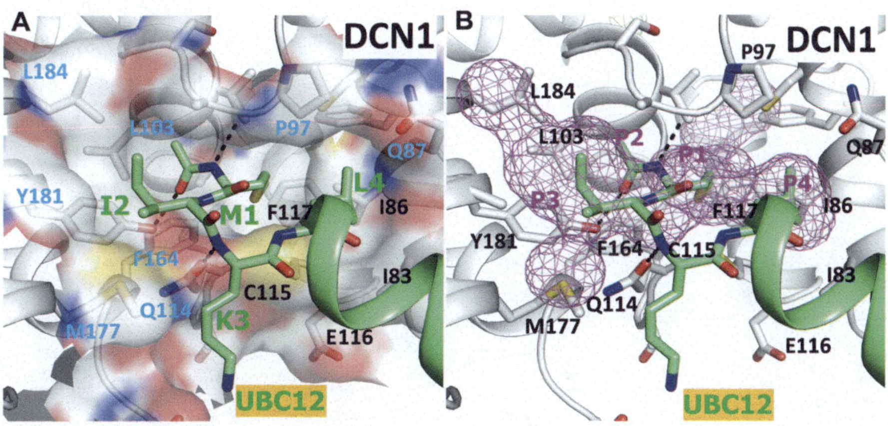

Fig. 20.1 (**a**) Crystal structure of DCN1 complexed with UBC12 peptide (green). (**b**) Hydrophobic hotspots (purple mesh) at the UBC12 peptide binding site identified by cosolvent mapping. Subsites for design and optimization are labeled

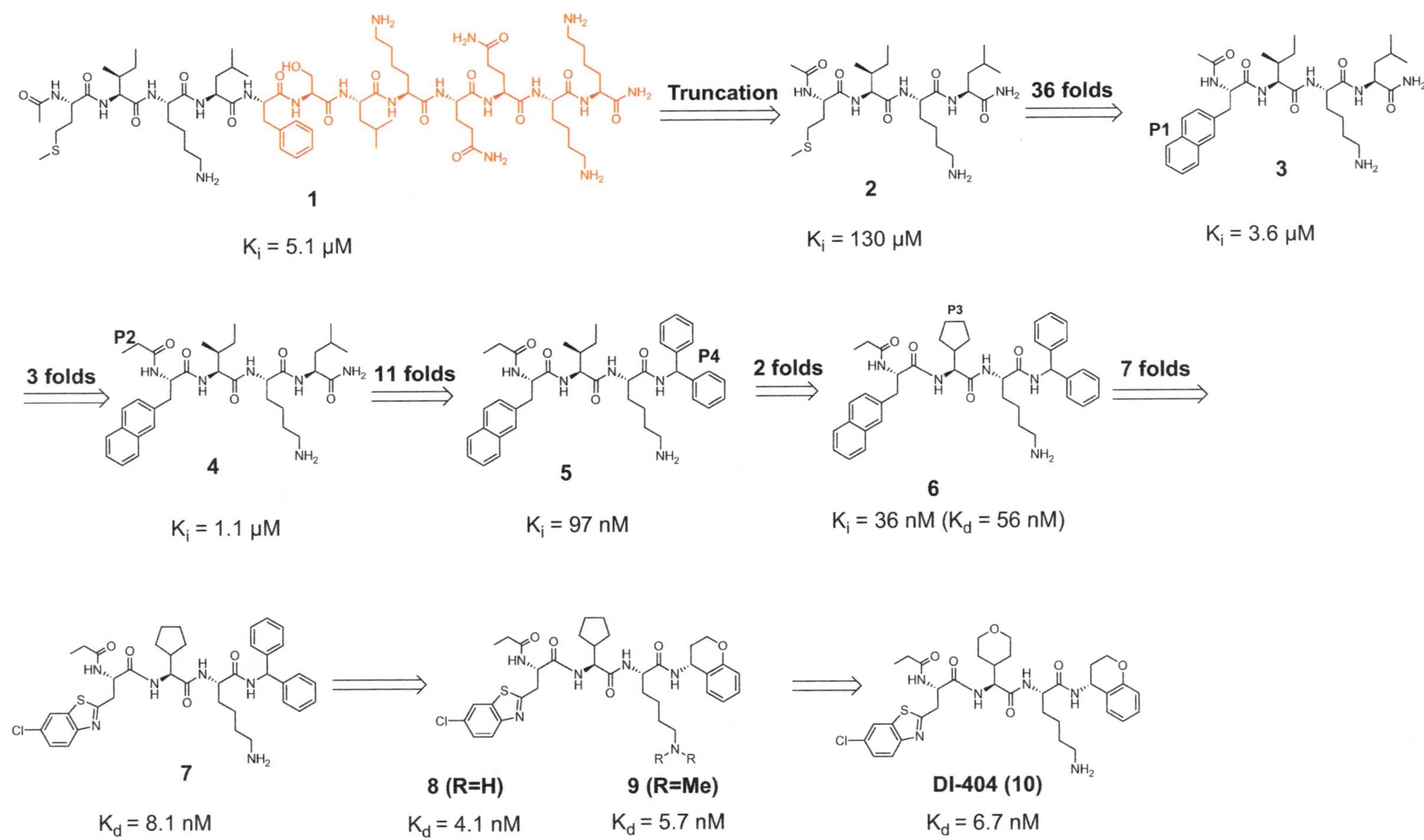

Fig. 20.2 Design of peptidomimetic inhibitors of the DCN1-UBC12 interaction

2 was used as the template molecule for the next round of optimization.

As shown in Fig. 20.1, the side chain of Met1 of UBC12 inserts deeply into a hydrophobic pocket in DCN1. Extensive modifications of the Met1 side chain of the peptide **2** were performed by replacing the Met1 residue with commercially available natural and unnatural amino acids. This effort resulted in compound **3**, which is 36 times more potent than peptide **2**.

The N-terminal acetylation of UBC12 is critical for its binding to DCN1 as the methyl of the N-terminal acetyl group is projected toward the hydrophobic pocket P2, and the carbonyl group of the N-terminal acetyl group forms hydrogen bond with Y181, and the NH group of the N-terminal acetyl group forms a hydrogen bond with P97 of DCN1 (Fig. 20.2) (Kurz et al. 2008). Probing pocket P2 led to the discovery of compound **4** by substitution of the methyl group in **3** with an ethyl group, which improved the binding affinity by threefold.

Next, the optimization of the interactions between the C-terminus of compound **4** and the P4 pocket was performed. As shown in Fig. 20.1, the hydrophobic side chain of L4 is directed to the P4 hydrophobic pocket, while the carbamoyl group of L4 is oriented toward the solvent and lacks any specific interaction with DCN1. Therefore, various hydrophobic moieties were selected to replace L4 in compound **4** in order to improve the potency. The benzhydryl group was found to be the best replacement of L4 in compound **4**, which yielded compound **5** with a K_i value of 97 nM to DCN1, and is thus 11 times more potent than **4**. The increased binding affinity may stem from extensive association of a phenyl ring with the P4 pocket in which the orientation of this phenyl ring was restricted by another phenyl ring in benzhydryl group of compound **5**. Additionally, the peptidic nature of compound **5** was reduced by replacement of the primary amide with a phenyl ring.

The P3 pocket was also explored by modification of the I2 in compound **5**. The P3 pocket shows good tolerability to various hydrophobic groups, and compound **6** with cyclo-pentylglycine achieves the best binding affinity to DCN1.

Because the biggest boost for the binding affinity was obtained from the exploration of the P1 pocket, further optimization was carried out to gain more potency. Replacement of the naphthalen-2-yl group in **6** with a 6-chlorobenzo[d]thiazol-2-yl group led to compound **7**. Compound **7** achieved a K_d value of 8.1 nM in a label-free biolayer interferometry (BLI) method and is eight times more potent than compound **6** in the same assay.

To reduce the hydrophobicity of **7**, the diphenylmethyl group was replaced by chromane, which yielded compound **8** with a K_d value of 4.1 nM to DCN1. Changing the primary amine group in **8** to dimethylamine resulted in **9**, which has a K_d value of 5.7 nM, similar to that of compound **8**. Replacing the cyclopentyl group in **9** with tetrahydro-2H-pyran generated DI-404, which has a calculated *c*Log*P* value of 1.9 and is much more soluble than compound **7**, which has a calculated *c*Log*P* of 5.3. DI-404 has a K_d value of 6.7 nM and is thus a high-affinity probe compound (Fig. 20.2).

The co-crystal structure of DCN1 in complex with compound **9** (Fig. 20.3a; PDB ID, 6B5Q) shows that compound **9** captures all the critical interactions with DCN1 as observed in the UBC12 peptide 1/DCN1 co-crystal structure. The 6-chlorobenzothiazol-2-yl group of **9** penetrated deeply in the hydrophobic pocket P1. Notably, there is induced conformational change of F109 and F117 residues to give more room to accommodate **9** (Fig. 20.3b).

20.4 Structure-Based Discovery of DI-591 as a Potent, Selective, Cell-Permeable, and Drug-Like Small-Molecule Inhibitor of DCN1

Although DI-404 binds to DCN1 with a high affinity, it only shows modest cellular activity, which may be due to its peptiditic nature. To obtain more drug-like small-molecule DCN1 inhibitors, our laboratories have performed structure-based design starting from compound **3**, with key steps shown in Fig. 20.4.

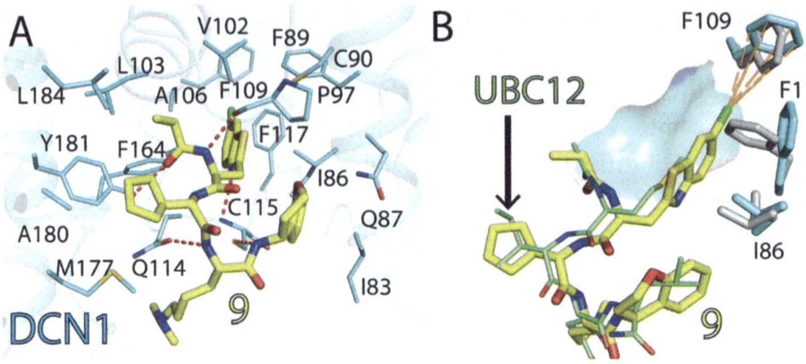

Fig. 20.3 (**a**) Co-crystal structure of DCN1 (cyan) in complex with compound **9** (yellow; PDB ID, 6B5Q). (**b**) Overlay of DCN1-**9** complex (PDB ID, 6B5Q) with DCN1-UBC12 (green) peptide (PDB ID, 3TDU). Conformational changes of I86, F109, and F117 at the pocket P1 to accommodate 6-chlorobenzothiazole in **9** are compared with those with UBC12 peptide. Hydrogen bonds are shown as red dashed lines

First, replacement of the naphthalen-2-yl group in compound **3** with a 6-chlorobenzothiazol-2-yl or 6-isopropylbenzothiazol-2-yl group yielded compounds **12** and **13**, respectively, which are >100 times more potent than compound **3**. Compound **13** binds to DCN1 with K_i value of 4.7 nM, 7 times more potent than **12**.

Despite its high binding affinity to DCN1, compound **13** is not ideal for cell-based studies due to its limited aqueous solubility owing to the combination of the 3 hydrophobic groups and 5 peptide bonds. Therefore, L4 and K3 in compound **13** were subsequently removed to yield compound **14** and **15**, which is 37 and 62 times less potent than compound **13**, respectively. Compound **16** was obtained by converting the C-terminal amide of **15** into a free amino group, and compound **16** has a K_i value of 66.4 nM and is four times more potent than **15**. More importantly, **16** has greatly improved aqueous solubility (>20 mM at pH = 2.0 and 7.4).

Replacement of the methyl group in **16** with an ethyl group generated **17** which has a K_i value of 19 nM, 3 times more potent than **15**. Replacement of the isobutyl group with a cyclohexyl group generated **18**, which has a K_i value of 13 nM to DCN1 and is slightly more potent than **17**. DI-591 was obtained by linking a morpholinyl group to the core structure of compound **18** through an ethylene spacer and an amide bond. DI-591 binds to DCN1 with a K_i value of 12 nM and has excellent aqueous solubility in both acidic and neutral conditions (>20 mM at pH 2.0 and 7.4).

We determined the co-crystal structure of DI-591 and DCN1, which showed similar interaction between amino acid residues of DCN1 with DI-404 (Fig. 20.5). The induced conformational changes of F109 and F117 in DCN1 were observed to provide more space for the bicyclic aromatic group of DI-591. This bicyclic ring of DI-591 makes extensive hydrophobic contacts with this subpocket in DCN1. The propionyl group and cyclohexyl group also form hydrophobic interaction with DCN1. Hydrogen bonds formed between DI-591 and DCN1 critically contribute to the high binding affinity.

20.5 Discovery of Small-Molecule Inhibitors of DCN1-UBC12 Protein-Protein Interaction Through High-Throughput Screening (HTS)

The Guy and Schulman research groups have reported the discovery of potent, non-peptidic, small-molecule DCN1 inhibitors (Scott et al. 2017; Hammill et al. 2018a, b).

A time-resolved fluorescence energy transfer (TR-FRET) assay was developed and employed to screen a library containing 601,194 unique chemicals. One of the hits from the screen, NAcM-HIT, was validated, and importantly, the

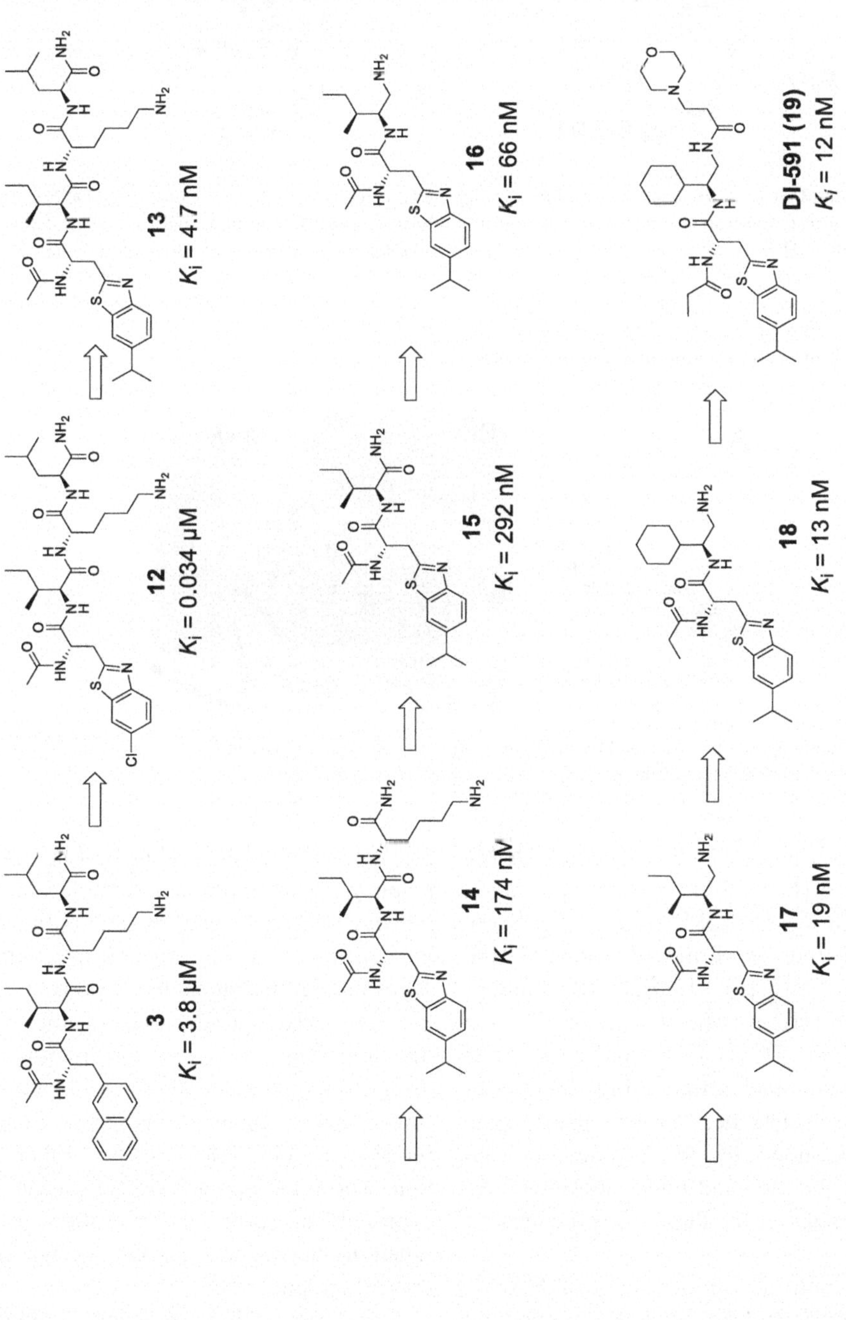

Fig. 20.4 Structure-based design of small-molecule inhibitors of the DCN1-UBC12 interaction

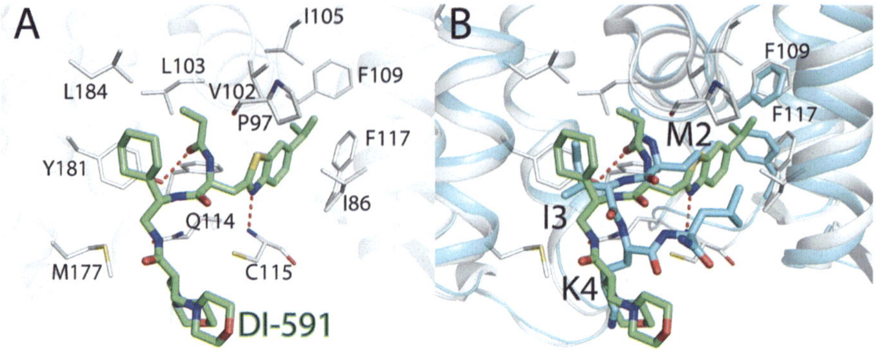

Fig. 20.5 Structural basis of the high affinity binding of DI-591 to DCN1. (**a**) Co-crystal structure of DCN1 (gray) complexed with **DI-591** (green). DCN1 residues interacting with **DI-591** are shown as sticks with carbon atoms in gray, nitrogen atoms in blue, oxygen atoms in red, and sulfur atoms in yellow. Hydrogen bonds are shown as dashed lines. (**b**) Overlay of DCN1 (gray)-**DI-591** (green) complex structure with DCN1 (light cyan)-UBC12 peptide complex structure (dark cyan) (PDB ID: 3TDU). Side chains of residues Phe109 and Phe117 from the DCN1-UBC12 structure are shown as cyan sticks, and these residues change their conformations upon **DI-591** binding

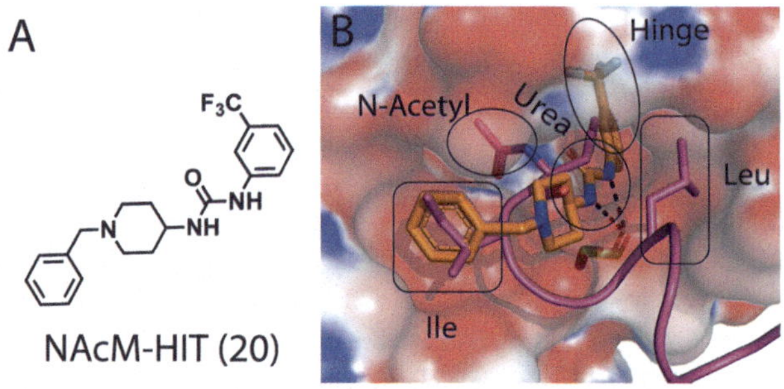

Fig. 20.6 (**a**) Chemical structure of the HTS hit NAcM-HIT. (**b**) Overlay of NAcM-HIT (orange): DCN1 (electrostatic potential surface) and UBE2MNAc (magenta):DCN1 X-ray co-crystal structures

co-crystal structure of NAcM-HIT and DCN1 was determined (Fig. 20.6). Overlay of NAcM-HIT:DCN1 and UBE2MNAc:DCN1 X-ray co-crystal structures confirmed that NAcM-HIT occupies the targeted UBE2MNAc-binding pocket in DCN1, which also revealed five subpockets within DCN1 that could be targeted to improve compound affinity (Fig. 20.6b). In addition to the isoleucine, *N*-acetyl, and leucine subpockets occupied by those moieties from UBE2MNAc, the urea and hinge pockets represent areas populated by those substructures of NAcM-HIT.

Based on both crystal structures of NAcM-HIT:DCN1 and UBE2MNAc:DCN1, compound **22** was derived and is 100 times more potent than the initial hit compound **20** (NAcM-HIT) (Fig. 20.7). Growing the NAcM-HIT to access the Leu pocket resulted in tenfold increase in potency, and additional tenfold increase in potency was achieved by the optimization of interactions in the Ile pocket, as confirmed by X-ray crystal structures. In the process of medicinal chemistry optimizations of the initial hit, determination of X-ray co-structures of the designed inhibitors bound to DCN1 not only confirmed binding mode of this series of compounds but also provided solid basis for further optimization. The unoccupied *N*-acetyl pocket was also explored; however, the modified compounds failed to access this pocket, shown by X-ray crystallography.

Consistent with its potency in inhibition of DCN1-UBE2M protein-protein interaction in biochemical assay, compound **22** is able to bind to DCN1 and selectively reduce steady-state

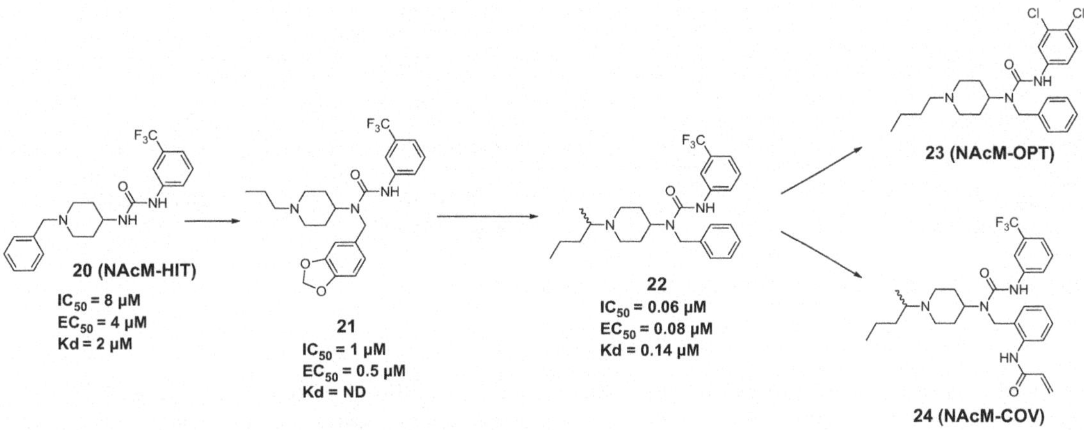

Fig. 20.7 Discovery of potent, non-peptidic, small-molecule DCN1 inhibitors

levels of neddylated cullin 1 and cullin 3 in cells. However, the poor microsomal stability of compound **22** makes it unsuitable for in vivo studies. Therefore, another round of optimization of compound **22** was performed employing a rational, structure-based design as well as an empirical chemical approach, which led to the discovery of an orally available analogue, **23** (NAcM-OPT). Compound **23** has similar biochemical and cellular activities as compound **22**, and **23** indeed reduced the steady-state neddylation of cullin 1 and cullin 3, which led to the modest upregulation of two known CRL substrates, NRF2 and p27 (Hammill et al. 2018b). Importantly, compound **23** is significantly more stable both in vitro and in vivo and shows good oral bioavailability in mice, making it a very attractive compound for in vivo studies.

Based on the obtained X-ray crystal structures, covalently targeting DCN1's Cys115 was attempted, which resulted in the discovery of DCN1 covalent inhibitor **24** (NAcM-COV). Both X-ray co-structure of NAcM-COV bound to DCN1 and MS analysis of DCN1 immunoprecipitates from NAcM-COV-treated cells confirmed the formation of the covalent linkage between NAcM-COV and Cys115. Compared to the non-covalent control compound, NAcM-COV is >30 times more potent in the TR-FRET assay and 2.5-fold in the neddylation experiment (Fig. 20.7).

20.6 Discovery of Triazolo[1,5-α]pyrimidines as DCN1 Inhibitors

The Liu research group has reported the discovery of triazolo[1,5-α]pyrimidine-based inhibitors starting from the hit compound **25** (E1) identified from their molecular library and subsequent structure-based optimizations.

The hit compound **25** (E1) was identified by high-throughput screening and was used as the starting point to perform further structural optimizations (Fig. 20.8). Different heterocycles were explored and found that the triazolopyridine is a privileged scaffold for the binding to DCN1. To improve the binding affinity of **25** to DCN1, a variety of substituents attached to this core scaffold (**26**) were investigated.

As the R3 group is predicted to be exposed to the solvent, soluble group is more favorable than the hydrophobic group, as shown by compound **27** being 40 times more potent than the hit compound **25**. For R1, bulky groups are detrimental to the binding affinity and a methyl group gave the best affinity. Based on the predicted model, the R2 group is presumably inserted into a deep subpocket, and 4-chloro benzyl group was found to be the best substituent on this position. The resulting compound WS-383 (**28**) binds potently to DCN1 (IC_{50} = 11 nM) and selectively inhibits neddylation of Cul3 and Cul1, which lead to the

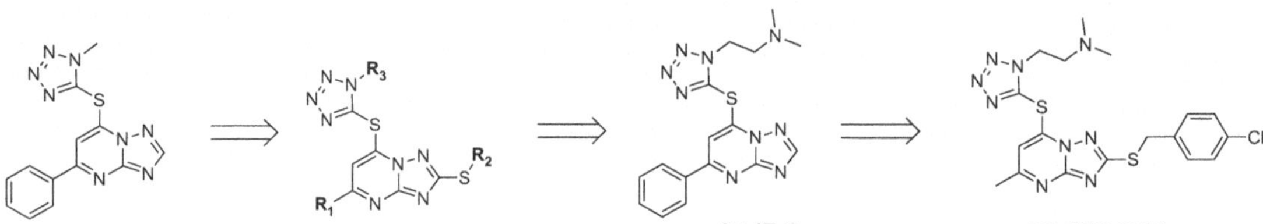

25 (E1)
IC50 = 5.82 μM

26

27 (E4)
IC_{50} = 130 nM

28 (WS-383)
11 nM

Fig. 20.8 Triazolo[1,5-α]pyrimidine-based DCN1 inhibitors

accumulation of p21, p27, and NRF2 (Fig. 20.8) (Wang et al. 2019).

20.7 Pyrimidine-Based DCN1 Inhibitors

The Zhao and Liu laboratories have reported the discovery of pyrimidine-based small molecular DCN1 inhibitors (Fig. 20.9) (Zhou et al. 2019). An in-house molecular library was screened, which led to identify DC-1. A series of chemical optimizations of DC-1 were performed to generate DC-2 which is 80 times more potent than DC-1. Studies showed that DC-2 is able to inhibit the interaction of UBE2M and DCN1 and decrease cullin 3 neddylation, leading to the accumulation of its substrate, NRF2.

20.8 Discovery of Pyrazolopyridone DCN1 Inhibitors

The Guy research group recently reported the optimization of pyrazolopyridone (**31**) as a new class of DCN1 inhibitors (Fig. 20.10) (Kim et al. 2019). Compound **31** was identified as an initial hit from screening. Based on the co-crystal structure of compound **31** in complex with DCN1, a series of pyrazolopyridones have been designed to improve the potency of **31** to DCN1. The optimized compound **32** has additional

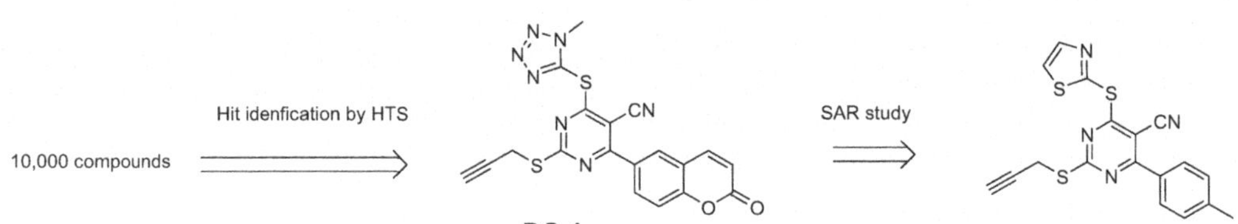

DC-1
IC_{50} = 1.2 μM

DC-2
IC_{50} = 15 nM

Fig. 20.9 Pyrimidine-based small molecular DCN1 inhibitors

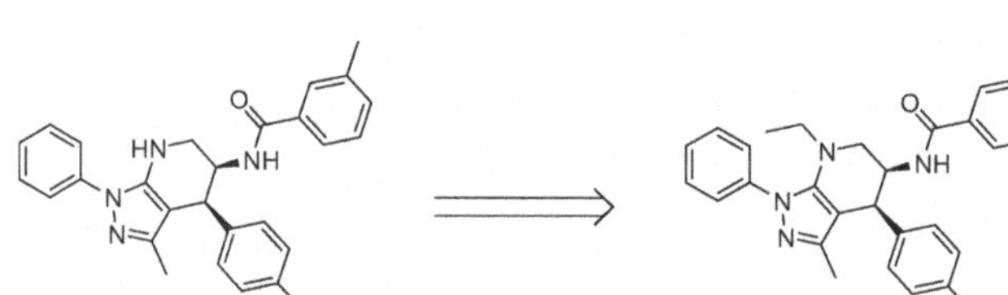

31
IC_{50} = 5.1 μM

32
IC_{50} = 0.2 μM

Fig. 20.10 Pyrazolopyridone-based DCN1 inhibitors

hydrophobic interaction with DCN1 and is 25-fold more potent than **31**. Compound **32** is able to engage cellular DCN1 and selectively inhibit the neddylation of Cul1 and Cul3.

20.9 Biological Effect of Blocking the UBC12-DCN1 Interaction on Neddylation of Individual Cullins

Previous studies have demonstrated that DCN1 acts together with RBX1 as a co-E3 ligase to facilitate neddylation of cullin 1 as well as other cullins. Therefore, the effect on neddylation of individual cullins was investigated using different classes of DCN1 inhibitors.

Our laboratory demonstrated that DI-591, an inhibitor selective for DCN1 and DCN2 over other DCN members (DCN3, DCN4, and DCN5), is effective in inhibition of the neddylation of cullin 3 and only has a minimal or moderate effect on the neddylation of other cullin members examined in a panel of cell lines of different tissue origins. Although DI-591 binds to both DCN1 and DCN2 with similar potency, we further demonstrated that DCN1, but not DCN2, is the underlying molecular target for the cellular effect of DI-591 in the cell lines we investigated. The inhibitory effect of DI-591 on the neddylation of cullin 3 is specific because an inactive, enantiomeric control compound of DI-591 (DI-591DD) has no effect on the neddylation of cullin 3 (Zhou et al. 2017).

It is known that the cullin 3 E3 ligase regulates the degradation of NRF2 (nuclear factor erythroid 2-related factor 2) protein, a transcriptional factor (Kobayashi et al. 2004; Cullinan et al. 2004; Venugopal and Jaiswal 1998; Gorrini et al. 2013). Hence, selective inhibition of the neddylation of the cullin 3 E3 ligase should block the degradation of NRF2 protein and result in the accumulation of NRF2 protein without affecting the transcription of the *NRF2* gene. Indeed, DI-591 was found to induce accumulation of NRF2 protein in a dose-dependent manner in a number of cell lines of different tissue origins (Zhou et al. 2017). Using DI-591DD as the inactive control, the induced accumulation of NRF2 protein by DI-591 was shown to be dependent upon DCN1 (Zhou et al. 2017). The induction of NRF2 protein was further confirmed using different classes of non-peptide, small-molecule inhibitors in subsequent studies (Wang et al. 2019; Zhou et al. 2019).

In the studies published by the Guy and Schulman laboratories, both covalent and non-covalent DCN1 inhibitors effectively inhibit neddylation of both cullin 1 and cullin 3 and induce accumulation of p27 and NRF2 proteins, the substrates of cullin 1 and cullin 3 E3 ligases, respectively. In contrast to the data obtained using DI-591, no selective inhibition of neddylation of cullin 3 over cullin 1 was observed in the reports by the Schulman and Guy laboratories (Hammill et al. 2018a, b). In the subsequent studies published by the groups of Liu and Zhao, they also observed dual inhibition of both neddylation of cullin 3 and cullin 1 using different classes of DCN1 inhibitors. Consistently, the groups of Liu and Zhao also reported induced accumulation of NRF2, a substrate for the cullin 3 E3 ligase, as well as p21 and p27 proteins, two substrate proteins for the cullin 1 E3 ligase (Wang et al. 2019; Zhou et al. 2019). The selective inhibition of Cul3, but not Cul1 by compound DI-591, which is not observed by DCN1 inhibitors reported by the other groups, deserves further investigation.

20.10 Therapeutic Potential of DCN1 Inhibitors

Using DI-591 as a biological tool compound, our laboratory demonstrated that DI-591 is quite effective in inducing accumulation of NRF2 protein (Zhou et al. 2017). Furthermore, DI-591 does not induce upregulation of NRF2 mRNA in cells, indicating that the induced accumulation of NRF2 protein by DI-591 is not due to oxidative cellular stress.

NRF2 is a transcriptional factor and a master regulator of antioxidant responses, regulating numerous detoxifying and antioxidant genes, such as the Phase II detoxification enzymes heme oxygenase (HO1) and NAD(P)H:quinone oxidoreductase-1 (NQO1) (Kensler et al. 2007). Indeed, we demonstrated that DI-591, but not its enantiomeric control DI-591DD, robustly increased the mRNA levels of NQO1 and HO1, indicating specific transcriptional activation of NRF2. Consequently, the HO1 protein level in cells was robustly increased by DI-591 (Zhou et al. 2017).

Activation of NRF2 has been pursued as a therapeutic strategy for the treatment of human diseases, including multiple sclerosis (Bomprezzi 2015). Indeed, dimethyl fumarate, which activates NRF2 by covalently binding to KEAP1, has been approved by the FDA for the treatment of relapsing multiple sclerosis (Bomprezzi 2015). Other activators of NRF2 such as bardoxolone methyl and omaveloxolone, which also activate NRF2 by covalently binding to KEAP1, are being evaluated in the Phase II and/or Phase III clinical trials for the treatment of several chronic diseases in which mitochondrial dysfunction and inflammation are implicated (https://www.reatapharma.com/our-science/pipeline/). Therefore, it can be envisioned that potent DCN1 inhibitors with optimized potencies and pharmacokinetic properties may have a therapeutic potential for the treatment of multiple sclerosis, as well as other chronic diseases in which mitochondrial dysfunction and inflammation are implicated. It is worth noting that DI-591 does not show any cytotoxicity in human cancer cell lines and normal liver cells. It is possible that optimized DI-591-like DCN1 inhibitors may have a potential therapeutic application to protect normal tissues from acute toxicity induced by oxidative stress through upregulation of NRF2.

Because those potent DCN1 inhibitors reported by the groups of the Guy and Schulman laboratories, as well as those from the Liu and Zhao groups, also inhibit neddylation of cullin 1, further optimization of these compounds may have potential therapeutic applications for the treatment of human diseases for which inhibition of cullin 1 neddylation may provide a benefit. For example, p21 and p27 were shown to be upregulated by these DCN1 inhibitors reported by these groups. Because p21 is a cell cycle regulator and p27 is a tumor suppressor, upregulation of these two genes and their gene products may find potential applications for the treatment of human cancers. Of note, the increased levels of p21 and p27 proteins by these selective DCN1 inhibitors are still modest, as compared to MLN4924, suggesting a need to further improve the binding affinities to DCN1 and the potencies for these compounds in cells.

20.11 Concluding Remarks

Selective targeting neddylation of individual cullin E3 ligases using small molecules not only provides new insights into the roles of individual cullin ligases in different biological processes but also has the potential for the development of new therapeutics for the treatment of human diseases. In the last few years, targeting the interactions of DCN1, a co-E3 ligase, with its binding partners UBC12 and UBC2E, has been pursued as a strategy to target the neddylation of individual cullins. These efforts have yielded potent non-covalent and covalent small-molecule inhibitors with nanomolar binding affinities to DCN1 and DCN2. Furthermore, these compounds also show excellent selectivity (>1000 times) over other DCN members (DCN3, 4 and 5). Biological studies using these DCN1 inhibitors demonstrate that they effectively inhibit the neddylation of cullin 1 and/or cullin 3 and have no or minimal effect on other cullin members. As expected, the substrate proteins regulated by cullin 1 or cullin 3 E3 ligase are accumulated by these DCN1 inhibitors, including NRF2 for cullin 3 and p21 and p27 for cullin 1. Of note, the cellular potencies for these DCN1 inhibitors are still moderate, suggesting a need to further improve their cellular potencies. Although no study has been performed in relevant disease models in animals, the discovery of these potent DCN1 inhibitors would pave the way for such studies in the near future. Collectively, these recent studies provide a framework to target other protein-protein interactions within different cullin E3 ligase complexes as a new approach to selectively inhibit the neddylation of individual cullins.

References

Abidi N, Xirodimas DP (2015) Regulation of cancer-related pathways by protein NEDDylation and strategies for the use of NEDD8 inhibitors in the clinic. Endocr Relat Cancer 22(1):T55–T70

Bedford L, Lowe J, Dick LR, Mayer RJ, Brownell JE (2011) Ubiquitin-like protein conjugation and the ubiquitin-proteasome system as drug targets. Nat Rev Drug Discov 10(1):29–46

Bomprezzi R (2015) Dimethyl fumarate in the treatment of relapsing-remitting multiple sclerosis: an overview. Ther Adv Neurol Disord 8(1):20–30

Brownell JE, Sintchak MD, Gavin JM, Liao H, Bruzzese FJ, Bump NJ, Soucy TA, Milhollen MA, Yang XF, Burkhardt AL, Ma JY, Loke HK, Lingaraj T, Wu DY, Hamman KB, Spelman JJ, Cullis CA, Langston SP, Vyskocil S, Sells TB, Mallender WD, Visiers I, Li P, Claiborne CF, Rolfe M, Bolen JB, Dick LR (2010) Substrate-assisted inhibition of ubiquitin-like protein-activating enzymes: the NEDD8 E1 inhibitor MLN4924 forms a NEDD8-AMP mimetic in situ. Mol Cell 37(1):102–111

Bulatov E, Ciulli A (2015) Targeting Cullin-RING E3 ubiquitin ligases for drug discovery: structure, assembly and small-molecule modulation. Biochem J 467 (3):365–386

Cullinan SB, Gordan JD, Jin JO, Harper JW, Diehl JA (2004) The Keap1-BTB protein is an adaptor that bridges Nrf2 to a Cul3-based E3 ligase: oxidative stress sensing by a Cul3-Keap1 ligase. Mol Cell Biol 24 (19):8477–8486

Deshaies RJ, Emberley ED, Saha A (2010) Control of Cullin-Ring ubiquitin ligase activity by Nedd8. Conjug Deconjug Ubiquitin Fam Modifiers 54:41–56

Duda DM, Borg LA, Scott DC, Hunt HW, Hammel M, Schulman BA (2008) Structural insights into NEDD8 activation of cullin-RING ligases: conformational control of conjugation. Cell 134(6):995–1006

Gong LM, Yeh ETH (1999) Identification of the activating and conjugating enzymes of the NEDD8 conjugation pathway. J Biol Chem 274(17):12036–12042

Gorrini C, Harris IS, Mak TW (2013) Modulation of oxidative stress as an anticancer strategy. Nat Rev Drug Discov 12(12):931–947

Hammill JT, Scott DC, Min J, Connelly MC, Holbrook G, Zhu F, Matheny A, Yang L, Singh B, Schulman BA, Guy RK (2018a) Piperidinyl ureas chemically control defective in cullin neddylation 1 (DCN1)-mediated cullin neddylation. J Med Chem 61(7):2680–2693

Hammill JT, Bhasin D, Scott DC, Min J, Chen Y, Lu Y, Yang L, Kim HS, Connelly MC, Hammill C, Holbrook G, Jeffries C, Singh B, Schulman BA, Guy RK (2018b) Discovery of an orally bioavailable inhibitor of defective in cullin neddylation 1 (DCN1)-mediated cullin neddylation. J Med Chem 61 (7):2694–2706

Hershko A (2005) The ubiquitin system for protein degradation and some of its roles in the control of the cell division cycle. Cell Death Differ 12(9):1191–1197

Hershko A, Ciechanover A (1998) The ubiquitin system. Annu Rev Biochem 67:425–479

Huang DT, Miller DW, Mathew R, Cassell R, Holton JM, Roussel MF, Schulman BA (2004) A unique E1-E2 interaction required for optimal conjugation of the ubiquitin-like protein NEDD8. Nat Struct Mol Biol 11(10):927–935

Kensler TW, Wakabayashi N, Biswal S (2007) Cell survival responses to environmental stresses via the Keap1-Nrf2-ARE pathway. Annu Rev Pharmacol Toxicol 47:89–116

Kim HS, Hammill JT, Scott DC, Chen Y, Min J, Rector J, Singh B, Schulman BA, Guy RK (2019) Discovery of novel pyrazolo-pyridone DCN1 inhibitors controlling cullin neddylation. J Med Chem

Kobayashi A, Kang MI, Okawa H, Ohtsuji M, Zenke Y, Chiba T, Igarashi K, Yamamoto M (2004) Oxidative stress sensor Keap1 functions as an adaptor for Cul3-based E3 ligase to regulate for proteasomal degradation of Nrf2. Mol Cell Biol 24(16):7130–7139

Kurz T, Ozlu N, Rudolf F, O'Rourke SM, Luke B, Hofmann K, Hyman AA, Bowerman B, Peter M (2005) The conserved protein DCN-1/Dcn1p is required for cullin neddylation in C-elegans and S-cerevisiae. Nature 435(7046):1257–1261

Kurz T, Chou YC, WillemS AR, Meyer-Schaller N, Hecht ML, Tyers M, Peter M, Sicheri F (2008) Dcn1 functions as a scaffold-type E3 ligase for cullin neddylation. Mol Cell 29(1):23–35

Nalepa G, Rolfe M, Harper JW (2006) Drug discovery in the ubiquitin-proteasome system. Nat Rev Drug Discov 5(7):596–613

Pan ZQ, Kentsis A, Dias DC, Yamoah K, Wu K (2004) Nedd8 on cullin: building an expressway to protein destruction. Oncogene 23(11):1985–1997

Petroski MD, Deshaies RJ (2005) Function and regulation of cullin-RING ubiquitin ligases. Nat Rev Mol Cell Biol 6(1):9–20

Scott DC, Monda JK, Grace CRR, Duda DM, Kriwacki RW, Kurz T, Schulman BA (2010) A dual E3 mechanism for Rub1 ligation to Cdc53. Mol Cell 39 (5):784–796

Scott DC, Monda JK, Bennett EJ, Harper JW, Schulman BA (2011) N-terminal acetylation acts as an avidity enhancer within an interconnected multiprotein complex. Science 334(6056):674–678

Scott DC, Sviderskiy VO, Monda JK, Lydeard JR, Cho SE, Harper JW, Schulman BA (2014) Structure of a RING E3 trapped in action reveals ligation mechanism for the ubiquitin-like protein NEDD8. Cell 157 (7):1671–1684

Scott DC, Hammill JT, Min J, Rhee DY, Connelly M, Sviderskiy VO, Bhasin D, Chen Y, Ong SS, Chai SC, Goktug AN, Huang G, Monda JK, Low J, Kim HS, Paulo JA, Cannon JR, Shelat AA, Chen T, Kelsall IR, Alpi AF, Pagala V, Wang X, Peng J, Singh B, Harper JW, Schulman BA, Guy RK (2017) Blocking an N-terminal acetylation-dependent protein interaction inhibits an E3 ligase. Nat Chem Biol 13(8):850–857

Soucy TA, Smith PG, Milhollen MA, Berger AJ, Gavin JM, Adhikari S, Brownell JE, Burke KE, Cardin DP, Critchley S, Cullis CA, Doucette A, Garnsey JJ, Gaulin JL, Gershman RE, Lublinsky AR, McDonald A, Mizutani H, Narayanan U, Olhava EJ, Peluso S, Rezaei M, Sintchak MD, Talreja T, Thomas MP, Traore T, Vyskocil S, Weatherhead GS, Yu J, Zhang J, Dick LR, Claiborne CF, Rolfe M, Bolen JB, Langston SP (2009a) An inhibitor of NEDD8-activating enzyme as a new approach to treat cancer. Nature 458(7239):732–736

Soucy TA, Smith PG, Rolfe M (2009b) Targeting NEDD8-activated Cullin-RING ligases for the treatment of cancer. Clin Cancer Res 15(12):3912–3916

Soucy TA, Dick LR, Smith PG, Milhollen MA, Brownell JE (2010) The NEDD8 conjugation pathway and its relevance in cancer biology and therapy. Genes Cancer 1(7):708–716

Venugopal R, Jaiswal AK (1998) Nrf2 and Nrf1 in association with Jun proteins regulate antioxidant response element-mediated expression and coordinated induction of genes encoding detoxifying enzymes. Oncogene 17(24):3145–3156

Wang S, Zhao L, Shi XJ, Ding L, Yang L, Wang ZZ, Shen D, Tang K, Li XJ, Mamun M, Li H, Yu B, Zheng YC, Wang S, Liu HM (2019) Development of highly potent, selective, and cellular active triazolo[1,5-a] pyrimidine-based inhibitors targeting the DCN1-UBC12 protein-protein interaction. J Med Chem 62 (5):2772–2797

Watson IR, Irwin MS, Ohh M (2011) NEDD8 pathways in cancer, Sine Quibus Non. Cancer Cell 19(2):168–176

Zhao YC, Sun Y (2013) Cullin-RING ligases as attractive anti-cancer targets. Curr Pharm Des 19(18):3215–3225

Zhao YC, Morgan MA, Sun Y (2014) Targeting neddylation pathways to inactivate Cullin-RING ligases for anticancer therapy. Antioxid Redox Signal 21(17):2383–2400

Zhou H, Lu J, Liu L, Bernard D, Yang C-Y, Fernandez-Salas E, Chinnaswamy K, Layton S, Stuckey J, Yu Q, Zhou W, Pan Z, Sun Y, Wang S (2017) A potent small-molecule inhibitor of the DCN1-UBC12 interaction that selectively blocks cullin 3 neddylation. Nat Commun 8(1):1150

Zhou H, Zhou W, Zhou B, Liu L, Chern TR, Chinnaswamy K, Lu J, Bernard D, Yang CY, Li S, Wang M, Stuckey J, Sun Y, Wang S (2018) High-affinity peptidomimetic inhibitors of the DCN1-UBC12 protein-protein interaction. J Med Chem 61(5):1934–1950

Zhou W, Ma L, Ding L, Guo Q, He Z, Yang J, Qiao H, Li L, Yang J, Yu S, Zhao L, Wang S, Liu HM, Suo Z, Zhao W (2019) Potent 5-Cyano-6-phenyl-pyrimidin-based derivatives targeting DCN1-UBE2M interaction. J Med Chem 62(11):5382–5403

Neddylation-Independent Activities of MLN4924

21

Hongmei Mao and Yi Sun

Abstract

MLN4924, also known as pevonedistat, is a highly selective small-molecule inhibitor of NEDD8 (neuronal precursor cell-expressed developmentally downregulated protein 8)-activating enzyme (NAE) to block the entire neddylation modification cascade, leading to inactivation of cullin-RING ligases (CRLs), since activation of CRLs requires cullin neddylation. MLN4924 showed impressive anticancer activity in many preclinical studies and is currently in several Phase I/II clinical trials for anticancer therapy as a single agent or in combination with chemotherapeutic drugs.

In addition to well-characterized anti-neddylation activity, recent studies showed that MLN4924 has several neddylation-independent activities. First, MLN4924 triggers EGFR dimerization to activate EGFR and its downstream RAS/MAPK and PI3K/AKT1 signals, leading to enhanced tumor sphere formation, accelerated EGF-mediated wound healing, and inhibited ciliogenesis. Second, MLN4924 induces PKM2 tetramerization to promote glycolysis, thus affecting energy metabolism. Third, MLN4924 inhibits the interaction between ACT1 (NF-κB activator 1) and TRAF6 (tumor necrosis factor receptor-associated factor 6) and attenuates IL-17A-mediated activation of NF-κB to reduce pulmonary inflammation. Fourth, MLN4924 inhibits IRF3 binding to the *IFN-β* promoter to inhibit IFN-β production. And finally, MLN4924 activates the JNK signaling pathway to reduce c-FLIP levels, thus enhancing TRAIL-induced apoptosis. This chapter will summarize these neddylation-independent activities of MLN4924 and discuss the underlying mechanisms and potential therapeutic applications.

Keywords

Ciliogenesis · Glycolysis · MLN4924 · PKM2 · Tumor sphere

H. Mao
Institute of Translational Medicine, Zhejiang University School of Medicine, Hangzhou, Zhejiang, China

The Gilbert Family Neurofibromatosis Institute, Centers for Cancer and Immunology Research and Neuroscience Research, The Children's National Hospital, Washington, DC, USA

Y. Sun (✉)
Institute of Translational Medicine, Zhejiang University School of Medicine, Hangzhou, Zhejiang, China

Division of Radiation and Cancer Biology, Department of Radiation Oncology, University of Michigan, Ann Arbor, MI, USA
e-mail: sunyi@umich.edu

Abbreviations

c-FLIP	Cellular FLICE-inhibitory protein
CRL	Cullin-RING ubiquitin ligases

Y. Sun et al. (eds.), *Cullin-RING Ligases and Protein Neddylation*, Advances in Experimental Medicine and Biology 1217, https://doi.org/10.1007/978-981-15-1025-0_21

CXCL1	C-X-C motif chemokine ligand 1
DBD	DNA-binding domain
ECAR	Extracellular acidification rate
EGFR	Epidermal growth factor receptor
ERK	Extracellular-signal-regulated kinase
IL-17A	Interleukin-17A
IRF3	Interferon regulatory factor 3
IRFs	Interferon regulatory factors
JNK	c-Jun-amino-terminal kinase
MAPK	Mitogen-activated protein kinase
NAE	NEDD8-activating enzyme
NDEP1	NEDD8-specific protease 1
NEDD8	Neuronal precursor cell-expressed developmentally downregulated protein 8
PI3K	Phosphoinositide 3-kinase
PK	Pyruvate kinase
PKM2	Pyruvate kinase isozymes M2
PRDs	Positive regulatory domains
RIG-I	Retinoic acid-inducible gene I
RTK	Receptor tyrosine kinases
SAE	SUMO-activating enzyme
SAICAR	Succinyl-5-aminoimidazole-4-carboxamide-1-ribose-5′-phosphate
SEFIR	SEF/IL-17R
TLR3/4	Toll-like receptor 3/4
TNF	Tumor necrosis factor
TNFSF	Tumor necrosis factor (TNF) superfamily
TRAF2/5	Tumor necrosis factor receptor-associated factor protein 2/5
TRAF6	Tumor necrosis factor receptor-associated factor protein 6
TRAFs	Tumor necrosis factor receptor-associated factor proteins
TRAIL	TNF-related apoptosis-inducing ligand
UAE	Ubiquitin-activating enzyme

21.1 Introduction

MLN4924, also known as pevonedistat, is the first-in-class small-molecule inhibitor of neddylation (Soucy et al. 2009), currently in several Phase I/II clinical trials for anticancer applications as a single agent or in combination with chemotherapeutic drugs (Zhou et al. 2018; Zhao et al. 2014; Nawrocki et al. 2012). Numerous preclinical studies using both in vitro cell culture and in vivo xenograft and genetically modified mouse models showed that MLN4924 suppresses growth and survival of many human cancer cells by inducing growth arrest, apoptosis, autophagy, and senescence as a single agent, as well as sensitizes cancer cells to chemotherapeutic drugs and radiation when used in combination. In addition, MLN4924 has been shown to inhibit angiogenesis and inflammation and modulate tumor microenvironment to retard tumorigenesis (Zhou et al. 2018, 2019a). This is achieved by inactivation of cullin neddylation, a process required for activity of CRLs (Fig. 21.1).

Recently, accumulated lines of evidence showed that MLN4924 has several activities independent of its neddylation inhibition, which is summarized in this chapter. Elucidation of these additional mechanisms of MLN4924 action would provide sound strategy to design combinational therapy for improved efficacy as an anticancer agent and at the same time offer new opportunities for additional therapeutic applications.

21.2 MLN4924 Triggers EGFR Dimerization

21.2.1 Stimulation of Tumor Sphere Formation and In Vivo Tumorigenesis

EGFR (epidermal growth factor receptor), also known as ErbB1, belongs to the ErbB family of receptor tyrosine kinases (RTKs) that transduces important growth factor signaling from the extracellular milieu to the cell (Oda et al. 2005). Dimerization of EGFR in response to ligand stimulation results in the activation of EGFR tyrosine kinase activity and receptor trans-autophosphorylation, thereby triggering the downstream signaling cascades, including the RAS/RAF/MEK/ERK (Ras/MAPK) pathway and the PI3K/AKT pathway (Scaltriti and

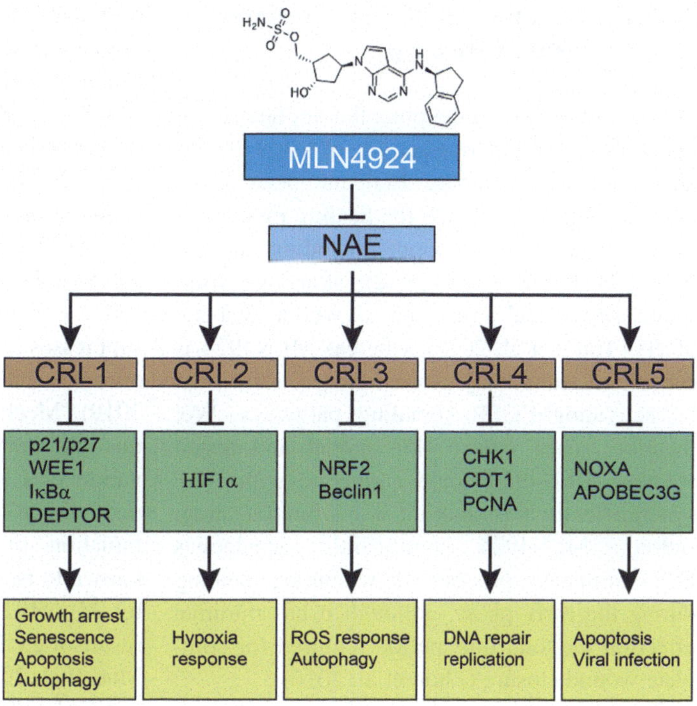

Fig. 21.1 Neddylation-dependent activity of MLN4924. As a potent inhibitor of neddylation E1 NAE, MLN4924 inactivates all CRL E3s by blocking cullin neddylation to cause the accumulation of a variety of key signaling substrates, leading to growth arrest, senescence, apoptosis, or autophagy, as well as regulating DNA repair and replication process and various stress responses, including hypoxia, ROS, and viral infection

Baselga 2006). These two signaling pathways regulate cell survival, cell cycle progression, differentiation, apoptosis, wound healing, and tissue repair. Given the frequent activation of the EGFR/RAS/MAPK pathway and EGFR/PI3K/AKT pathway in human cancers (Engelman 2009; McCubrey et al. 2007), both oncogenic pathways have become important drug targets for cancer therapy.

MLN4924 has shown impressive anticancer activity in preclinical studies using multiple cancer cell lines derived from a variety of human cancers (Zhao et al. 2014). Since cancer stem cells (CSCs) or tumor-initiating cells (TICs) play a key role in cancer initiation and relapse (Nguyen et al. 2012), we determined the potential inhibitory role of MLN4924 in CSC by tumor sphere formation assays. Unexpectedly, we found that at low drug concentrations (30–100 nM), MLN4924 actually stimulates in vitro tumor sphere formation and in vivo tumorigenesis of both cancer cells and embryonic stem cells (Zhou et al. 2016). We followed up this interesting observation mechanistically and confirmed that MLN4924 inactivates SCF^{FBXW7} E3 to cause accumulation of c-Myc, one of the four Yamanaka factors that are capable of reprogramming fibroblasts into induced pluripotent stem cells (Takahashi and Yamanaka 2006). Unexpectedly, we also found that MLN induces EGFR dimerization, as measured by two independent assays: protein cross-linking-based Western blotting and the proximity ligase assay (PLA), also known as proximity ligation immunofluorescence (PLI) assay (Gajadhar and Guha 2010). Dimerization of EGFR triggers EGFR activation as well as activation of its downstream signals, including RAS/RAF/MEK/ERK and PI3K/AKT1/mTOR pathways (Zhou et al. 2016). Using various kinase inhibitors targeting these signaling pathways individually, we found that the tumor sphere stimulating activity of MLN4924 is mainly mediated by the RAS/MAPK pathway (Zhou et al. 2016).

21.2.2 Stimulation of EGF-Mediated Wound Healing

We then investigated the potential application of this stem cell-stimulating activity of MLN4924 in an in vivo animal model. EGF has been previously shown to accelerate the healing process of various types of skin wounds, including massive burns (Wenczak et al. 1992), skin loss from trauma or medical procedures, as well as diabetic ulcers (Tiaka et al. 2012), whereas MLN4924 is capable of stimulating proliferation of stem cells by activating EGFR signaling pathways. We, therefore, used mouse skin model and tested whether MLN4924 could cooperate with EGF synergistically in promoting skin wound healing. Indeed, MLN4924 significantly accelerates EGF-stimulated process of wounding healing during the early phase, although it has minimal effect on the total time period required for complete wound closure (Zhou et al. 2016).

Taken together, our study revealed a neddylation-independent activity of MLN4924 by triggering EGFR dimerization, although the detailed mechanism remains elusive. It is conceivable that the lower dose of MLN4924 promotes stem cell proliferation, whereas the higher dose kills cancer cells. While our study raises a concern in anticancer application of MLN4924, it provides, however, an opportunity for future development of MLN4924 as an agent for stem cell therapy and tissue regeneration at lower dose.

21.2.3 Inhibition of Ciliogenesis

Ciliogenesis is a process of formation of the cilium, including motile cilium and primary cilium. Primary cilium is a microtubule-based organelle projecting from the apical surface of most eukaryotic cells, which is involved in many physiological and developmental processes (Goetz and Anderson 2010). It is a highly dynamic process consisting of assembly (including initiation and elongation) and disassembly (Walz 2017). Defects in ciliogenesis are responsible for many human ciliopathies, including cancer (Eggenschwiler and Anderson 2007; Liu et al. 2018).

Given the biological significance of ciliogenesis in human health and diseases (Gerdes et al. 2009), we investigated potential role of neddylation in the process and found that blockage of neddylation by MLN4924 effectively blocks ciliogenesis at both stages of synthesis and degradation. Specifically, MLN4924 suppresses cilia initiation or assembly and promotes cilia disassembly as well (Mao et al. 2019). Mechanistic study further revealed that this inhibitory activity is not mediated by inactivation of CRLs, whose activity requires cullin neddylation (Deshaies et al. 2010), given no accumulation of several cilia-associated proteins, known to be CRL substrates. Rather, it is caused by MLN4924-mediated AKT1 activation as a result of EGFR dimerization (Zhou et al. 2016; Mao et al. 2019), since small-molecule inhibitors of EGFR, PI3K, or AKT1 all abrogate MLN4924 inhibitory effect on cilia formation (Mao et al. 2019). We then focused on the downstream AKT1 using both siRNA-based genetic and small-molecule inhibitor-based pharmacological approaches. Among three AKT family members (AKT1–AKT3), only AKT1 is involved in this process. Specifically, MLN4924-induced activation of AKT1 at the Ser^{473} residue plays a major role in suppression of cilia formation, as evidenced by complete rescue via *siAKT1*-based knockdown and small molecule inhibitor of AKT1-Ser^{473} (Mao et al. 2019).

We further extended this in vitro cell culture observation to an in vivo animal model, again in mouse skin, for its physiological significance. It is well-established that hair regrowth requires primary cilia (Goetz and Anderson 2010; Lehman et al. 2009; Rishikaysh et al. 2014; Abe and Tanaka 2017). We hypothesized that MLN4924 inhibition of ciliogenesis would suppress hair regrowth after depilation. It is indeed the case. MLN4924 administration significantly inhibited hair regrowth with reduced number of skin follicles as well as reduced number of cilia-positive hair follicles (Mao et al. 2019).

Collectively, our study demonstrated that cilia inhibitory effect of MLN4924 can be extended from in vitro cell culture setting to an in vivo mouse model. MLN4924 may, therefore, have a novel application for the treatment of human cancers which rely on cilia for growth or drug resistance (Jenks et al. 2018). Furthermore, given that AKT1 negatively regulates ciliogenesis, selective AKT1 inhibitor may have an application to promote cilia formation in the human diseases of ciliopathies (e.g., polycystic kidney) (Eggenschwiler and Anderson 2007).

21.3 MLN4924 Triggers PKM2 Tetramerization

Pyruvate kinase (PK) is a rate-limiting glycolytic enzyme that catalyzes the last irreversible step in glycolysis and mediates the conversion of phosphoenolpyruvate (PEP) to pyruvate, thus affecting glycolytic rate (Witney et al. 2015). The M2 isoform of pyruvate kinase (PKM2), one splice variant encoded by *PKM* gene, is the major form of PK in cancer cells and plays an important role in cancer Warburg effect (Chen et al. 2011; Dayton et al. 2016). Indeed, overexpression of PKM2 was found in diverse types of human cancer including breast, prostate, blood, cervix, kidney, bladder, papillary thyroid, and colon cancer to promote the Warburg effect for tumor formation (Wong et al. 2015).

We recently investigated potential effect of neddylation on glycolysis. Using an assay to measure extracellular acidification rate (ECAR), an indicator of glycolysis, we found that MLN4924 causes a significant increase in media acidification in time- and dose-dependent but neddylation-independent manners (Zhou et al. 2019b). Consistent with increased ECAR, MLN4924 also increases cellular utilization of glucose and production of glycolytic end products (pyruvate and lactate) (Zhou et al. 2019b). Mechanistically, we found that MLN4924 significantly increases PKM2 enzymatic activity in both time- and dose-dependent manners in part by triggering PKM2 homotetramer, which is the most enzymatically active form (Dayton et al. 2016). Using bacterially expressed and purified PKM2 protein in an in vitro biochemical assay, we found that MLN4924 increases PKM2 activity even slightly better than SAICAR (succinyl-5-aminoimidazole-4-carboxamide-1-ribose-5′-phosphate), a well-known PKM2 activator (Keller et al. 2012), which is an intermediate of the de novo purine nucleotide synthesis pathway, accumulated under glucose starvation condition to activate PKM2 in cancer cells for generating energy (Li et al. 2018). Structural comparison showed a high similarity between MLN4924 and SAICAR (Fig. 21.2), providing a potential mechanism of MLN4924 action in promoting PKM2 tetramerization.

What is the biological significance of this finding? Given that glycolysis promotes cancer formation, MLN4924-induced glycolysis would in theory counteract its anticancer activity. We, therefore, tested our working hypothesis that the combination of MLN4924 with a PKM2 inhibitor would enhance anticancer activity of the former. Indeed, MLN4924-induced PKM2 activation confers a better survival for breast cancer cells, and the combination of MLN4924 and PKM2 inhibitor shikonin significantly suppresses breast cancer cell growth both in vitro cell culture setting and in two in vivo breast cancer xenograft models (Zhou et al. 2019b). Our study, therefore, provides a sound strategy for mechanism-based combinational targeted therapy to increase the anticancer efficacy of MLN4924.

21.4 MLN4924 Inhibits the Interaction Between ACT1 and TRAFs

ACT1 (NF-κB activator 1), also known as CIKS (connection to IKK and SAPK/JNK), is a cytoplasmic protein that activates NF-κB through its association with the IκB kinase complex, thus regulating inflammatory responses (Chang et al. 2006; Kanamori et al. 2002). ACT1 has a SEF/IL-17R (SEFIR) domain at the carboxyl terminus that mediates its interaction with IL-17 receptor complex and a tumor necrosis factor receptor-associated factor (TRAF) binding domain at the

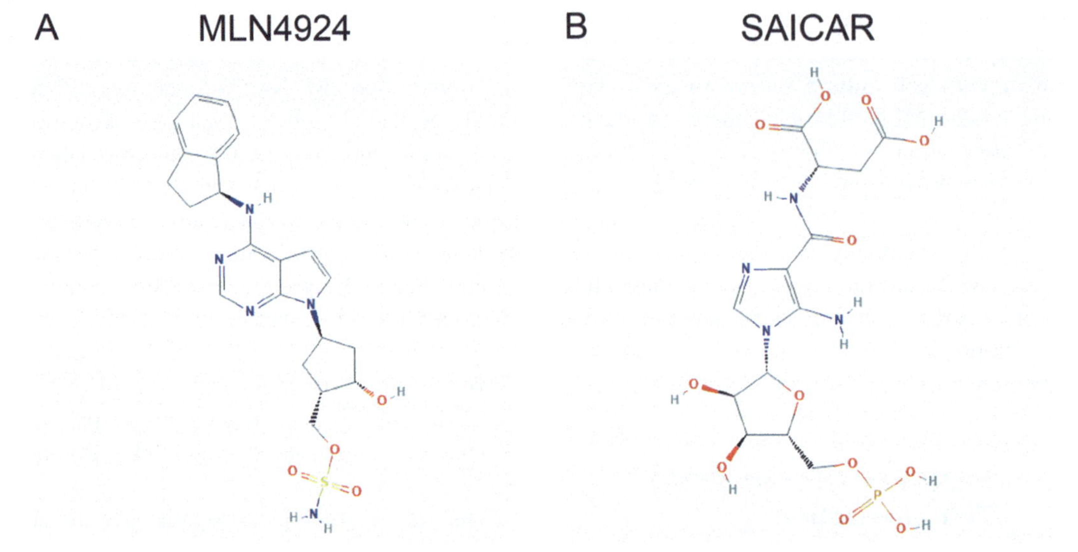

Fig. 21.2 Structural comparison of (**a**) MLN4924 and (**b**) SAICAR. MLN4924 (National Center for Biotechnology Information n.d.-a) is structurally related to SAICAR (National Center for Biotechnology Information n.d.-b). The differences are shown below, which may not be critical in their ability to trigger the PKM2 tetramerization: (1) MLN4924 has a sulfamate, while SAICAR has the phosphate; (2) MLN4924 has a carbocycle, and SAICAR has the ribose sugar. One hydroxyl on the five-member ring of SAICAR is absent in MLN4924; (3) MLN4924 has a deazapurine base substituted with an aminoindane at N6, while SAICAR has an imidazole ring. (4) MLN4924 has a (R)-2,3-dihydro-1H-inden-1-amin, while SAICAR has a L-aspartate

amino terminus that binds to TRAFs (Chang and Dong 2011). Upon ACT1 binding, TRAF6 is ubiquitylated via the K63 linkage to activate several inflammatory processes, including production of proinflammatory cytokines and modulation of multiple receptor families with immunoregulatory functions (Walsh et al. 2015). Deregulation of ACT1 and TRAF6 is responsible for disorganized lymphoid tissues, inflammatory autoimmune disease, and cancer (Walsh et al. 2015; Qian et al. 2007).

Two previous studies have shown that MLN4924 suppresses inflammatory responses, such as a mucosal inflammatory response and liver fibrosis via targeting Cul-2/HIF-1 and Cul-1/c-Jun, respectively (Curtis et al. 2015; Zubiete-Franco et al. 2017). Most recently, MLN4924 was found to inhibit IL-17A-induced pulmonary inflammation in an in vivo mouse model by significantly reducing neutrophilia and the expression of proinflammatory cytokines and chemokines, including IL-1β, IL-6, and CXCL-1 (C-X-C motif chemokine ligand 1) (Hao et al. 2019). Two underlying mechanisms appear to be involved in this process. First, MLN4924 suppresses IL-17A-induced stabilization of mRNAs encoding these proinflammatory cytokines and chemokines, but the underlying mechanism remains elusive. Second, MLN4924 blocks the interaction between ACT1 and TRAF6 to inactivate the NF-κB pathway, since MLN4924 (1) markedly inhibited ACT1 activation of NF-κB activity in a luciferase reporter assay; (2) greatly reduced the interaction of ACT1 with TRAF6 in a co-immunoprecipitation pull-down assay; and (3) attenuated ACT1-mediated polyubiquitylation of TRAF6. These effects collectively lead to NF-κB inactivation and reduced the production of proinflammatory cytokines. Interestingly, MLN4924 also targets the interactions between ACT1 and TRAF2/5 (Hao et al. 2019). MLN4924-induced disruption of the ACT1-TRAFs binding appears to be largely independent of its neddylation inhibition, although it is unknown how MLN4924 targets these protein-protein interactions. On the other hand, the study indeed showed that MLN4924 blocks IL-17A-induced degradation of IκBα, a well-known NFκB inhibitor and a substrate of CRL1/SCF E3 (Tan et al. 1999), suggesting that neddylation-dependent mechanism should also possibly contribute to NFκB inactivation. Correctively, MLN4924 might be a promising drug for the treatment of IL-17A-induced pulmonary inflammatory diseases, regardless of its involving mechanisms.

21.5 MLN4924 Interrupts IRF3 Binding to the Promoter of the *IFN-β*

Interferon regulatory factors (IRFs) are a family of transcription factors involved in many aspects of innate and adaptive immune responses (Taniguchi et al. 2001). IRF3 is one of the nine family members which share significant homology in their N-terminal DNA-binding domain (DBD) that recognizes specific DNA sequences (Jefferies 2019). IFN-β is one of type I IFNs that are widely expressed cytokines with potent antiviral and growth-inhibitory effects. After viral infection, TRF3 transactivates IFN-β expression via binding to its promoter region to trigger the production of numerous antiviral genes (Honda et al. 2006; Kim and Maniatis 1997). On the other hand, overexpression of IFNα/β is involved in several autoimmune diseases, including systemic lupus erythematosus (Meyer 2009).

MLN4924, but not *NEDD8* knockdown, was reported to inhibit IFN-β expression induced by TLR3/TLR4- and retinoic acid-inducible gene-I in various cell lines. MLN4924 also inhibits IFN-β production upon stimulation by LPS or polyinosinic-polycytidylic acid in an in vivo mouse model (Song et al. 2016). Mechanistically, MLN4924 inhibits IFN-β expression by inhibiting IRF3 binding to the *IFN-β* promoter, as demonstrated by the luciferase-based reporter assay and ChIP-based DNA binding assay. This activity appears to be neddylation-independent, since inhibition of neddylation by *NEDD8* knockdown and overexpression of NEDP1, a deneddylase, have no effects on IRF3-induced activation of the *IFN-β* promoter (Song et al. 2016). Again, regardless of underlying mechanisms, the study provides an implication for potential application of MLN4924 in the treatment of the autoimmune diseases induced by IFN-β overproduction.

21.6 MLN4924 Activates the JNK Signal to Trigger c-FLIP Degradation

The Jun N-terminal kinases or JNKs (JNK1, JNK2, and JNK3) are MAPK superfamily involved in a wide array of signaling events to control cell proliferation, motility, metabolism, DNA repair, and programmed cell death and to maintain the cell homeostasis (Dhanasekaran and Reddy 2017). Phosphorylated JNKs either phosphorylate and transactivate c-Jun to upregulate pro-apoptotic genes or directly modulate the activities of mitochondrial pro- and anti-apoptotic proteins (Dhanasekaran and Reddy 2008). Dysregulated JNK signaling is responsible for many diseases including neurodegeneration, chronic inflammation, and cancer (Dhanasekaran and Reddy 2017).

It is well-established that MLN4924 induces apoptosis in human cancer cells by triggering accumulation of pro-apoptotic substrates (Zhou et al. 2018). In a study using head and neck squamous cell carcinoma (HNSCC) cells, MLN4924 alone or in combination with TNF-related apoptosis-inducing ligand (TRAIL) effectively induces apoptosis to inhibit cell growth (Zhao et al. 2011). Interestingly, the mechanistic studies revealed that MLN4924-induced apoptosis is causally related to a rapid activation of JNK signaling pathway, which triggers proteasome-dependent degradation of c-FLIP (cellular FLICE-inhibitory protein), a master anti-apoptotic protein (Safa 2012). The supporting lines of evidence include the following: (1) MLN4924 decreased the c-FLIP stability by promoting c-FLIP ubiquitylation for targeted degradation; (2) MLN4924 caused the increase of both phospho-c-Jun and total c-Jun and the decrease of c-FLIP, which were rescued by the JNK-specific inhibitor, SP600125; and (3) ectopic c-FLIP expression or JNK inhibition protected HNSCC cells from apoptosis induced by the combination of MLN4924 and TRAIL, indicating a causal effect. However, it is yet unknown how MLN4924 activates the JNK signal and what is the physiological E3 ubiquitin ligase responsible for c-FLIP degradation upon MLN4924 activation of the JNK signal, although the involvement of Itch E3 ligase was excluded. Importantly, this MLN4924-induced JNK activation/c-FLIP degradation appears to be neddylation independent, since *NEDD8* knockdown cannot mimic the MLN4924 effects (Zhao et al. 2011). However, it is conceivable that neddylation-dependent mechanism must also be involved in apoptosis induction via the c-FLIP-independent pathways.

21.7 Summary and Perspectives

Although MLN4924 is a highly selective inhibitor of neddylation E1 NAE with little or no off-target activity against ubiquitin E1 UAE or sumoylation E1 SAE (Soucy et al. 2009), increasing number of studies unexpectedly revealed that MLN4924 has additional biochemical activities other than NAE inhibition, leading to a variety of biological consequences. This chapter summarized currently reported neddylation-independent "off-target" effects of MLN4924 with unexpected action mechanisms involving induction of EGFR dimerization, induction of PKM2 tetramerization, or other unknown mechanisms involving the disruption of the protein-protein interaction (ACT1-TRAFs), inhibition of the protein-DNA binding (IRF3-IFNβ), or activation of the JNK signal to trigger c-FLIP degradation (Fig. 21.3).

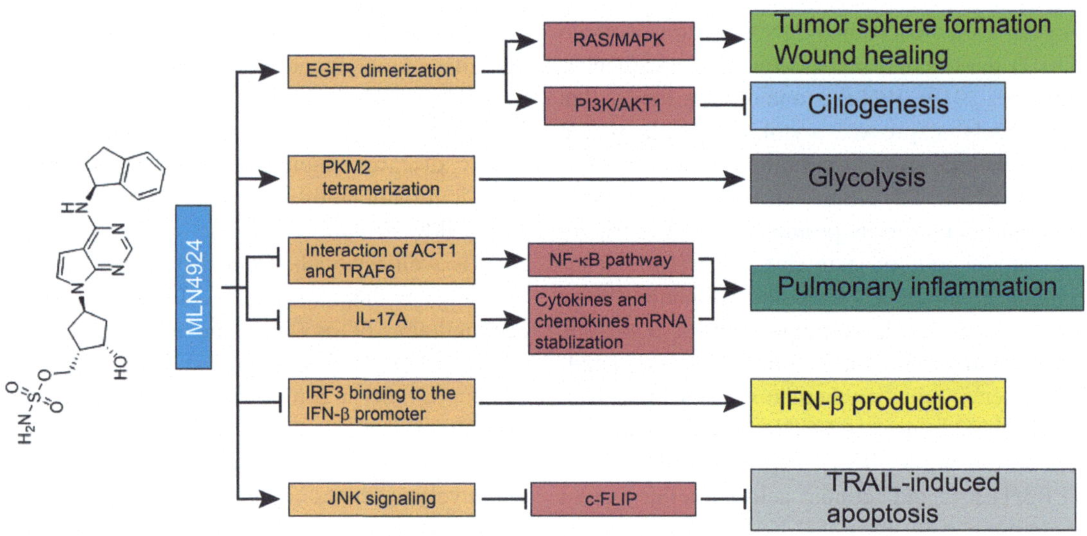

Fig. 21.3 Neddylation-independent activities of MLN4924. (A) Induction of EGFR dimerization: MLN4924 activates EGFR by triggering EGFR dimerization to activate its downstream RAS/MAPK pathway to promote (a) tumor sphere formation and EGF-mediated wound healing and (b) PI3K/AKT1 pathway to inhibit ciliogenesis, respectively. (B) Induction of PKM2 tetramerization: MLN4924 activates PKM2 via promoting its tetramerization to increase glycolysis. (C) Disruption of the ACT1-TRAFs interaction: MLN4924 inhibits (a) the interaction of ACT1 and TRAF6 to inactivate the NF-κB pathway, resulting in reduced secretion of inflammatory cytokines, and (b) IL-17A-induced stabilization of mRNAs encoding the cytokines and chemokines. (D) Inhibition of IRF3 binding to the IFN-β promoter: MLN4924 inhibits IRF3 binding to the IFN-β promoter to suppress the production of IFN-β. (E) Activation of the JNK signaling: MLN4924 activates the JNK signaling to promote c-FLIP degradation and enhances TRAIL-induced apoptosis

As of December 13, 2019, 10 years after initial MLN4924 discovery (Soucy et al. 2009), the PubMed search under keyword "MLN4924" identified 298 publications and 205 publications under keyword "pevonedistat," a designated name for MLN4924 clinical trials. Although a major overlap exists between two searches, it clearly shows the fields of cullin-RING ligases and neddylation modification are greatly promoted by the discovery of MLN4924/pevonedistat. The publication of this book, the first one in the field, is also a good and strong indication. The study with the use of MLN4924 also promotes the drug discovery efforts in identification of small molecule inhibitors of neddylation E2 or E3 with expected higher specificity and lower cytotoxicity.

It is our sincere hope that pevonedistat will be eventually approved for the clinical use as the first-in-class anticancer drug and the discovery of its "off-target" effects would provide sound strategy for effective combinational therapy to eliminate such off-target effects or even for new applications in the treatment of other human diseases with these novel mechanisms of action for MLN4924. Examples include stem cell therapy and/or tissue regeneration, abnormal ciliogenesis, noninfectious pulmonary inflammation, or autoimmune related diseases.

References

Abe Y, Tanaka N (2017) Roles of the hedgehog signaling pathway in epidermal and hair follicle development, homeostasis, and cancer. J Dev Biol 5(4):12

Chang SH, Dong C (2011) Signaling of interleukin-17 family cytokines in immunity and inflammation. Cell Signal 23(7):1069–1075

Chang SH, Park H, Dong C (2006) Act1 adaptor protein is an immediate and essential signaling component of interleukin-17 receptor. J Biol Chem 281 (47):35603–35607

Chen J, Xie J, Jiang Z, Wang B, Wang Y, Hu X (2011) Shikonin and its analogs inhibit cancer cell glycolysis by targeting tumor pyruvate kinase-M2. Oncogene 30 (42):4297–4306

Curtis VF, Ehrentraut SF, Campbell EL, Glover LE, Bayless A, Kelly CJ et al (2015) Stabilization of HIF through inhibition of Cullin-2 neddylation is protective in mucosal inflammatory responses. FASEB J 29 (1):208–215

Dayton TL, Jacks T, Vander Heiden MG (2016) PKM2, cancer metabolism, and the road ahead. EMBO Rep 17 (12):1721–1730

Deshaies RJ, Emberley ED, Saha A (2010) In: Groettrup M (ed) Conjugation and deconjugation of ubiquitin family modifiers: subcellular biochemistry. Springer, Berlin, pp 41–56

Dhanasekaran DN, Reddy EP (2008) JNK signaling in apoptosis. Oncogene 27(48):6245–6251

Dhanasekaran DN, Reddy EP (2017) JNK-signaling: a multiplexing hub in programmed cell death. Genes Cancer 8(9–10):682–694

Eggenschwiler JT, Anderson KV (2007) Cilia and developmental signaling. Annu Rev Cell Dev Biol 23:345–373

Engelman JA (2009) Targeting PI3K signalling in cancer: opportunities, challenges and limitations. Nat Rev Cancer 9(8):550–562

Gajadhar A, Guha A (2010) A proximity ligation assay using transiently transfected, epitope-tagged proteins: application for in situ detection of dimerized receptor tyrosine kinases. BioTechniques 48(2):145–152

Gerdes JM, Davis EE, Katsanis N (2009) The vertebrate primary cilium in development, homeostasis, and disease. Cell 137(1):32–45

Goetz SC, Anderson KV (2010) The primary cilium: a signalling centre during vertebrate development. Nat Rev Genet 11(5):331–344

Hao R, Song Y, Li R, Wu Y, Yang X, Li X et al (2019) MLN4924 protects against interleukin-17A-induced pulmonary inflammation by disrupting ACT1-mediated signaling. Am J Physiol Lung Cell Mol Physiol 316(6):L1070–L1L80

Honda K, Takaoka A, Taniguchi T (2006) Type I interferon [corrected] gene induction by the interferon regulatory factor family of transcription factors. Immunity 25(3):349–360

Jefferies CA (2019) Regulating IRFs in IFN driven disease. Front Immunol 10:325

Jenks AD, Vyse S, Wong JP, Kostaras E, Keller D, Burgoyne T et al (2018) Primary cilia mediate diverse kinase inhibitor resistance mechanisms in cancer. Cell Rep 23(10):3042–3055

Kanamori M, Kai C, Hayashizaki Y, Suzuki H (2002) NF-kappaB activator Act1 associates with IL-1/toll pathway adaptor molecule TRAF6. FEBS Lett 532 (1–2):241–246

Keller KE, Tan IS, Lee YS (2012) SAICAR stimulates pyruvate kinase isoform M2 and promotes cancer cell survival in glucose-limited conditions. Science 338 (6110):1069–1072

Kim TK, Maniatis T (1997) The mechanism of transcriptional synergy of an in vitro assembled interferon-beta enhanceosome. Mol Cell 1(1):119–129

Lehman JM, Laag E, Michaud EJ, Yoder BK (2009) An essential role for dermal primary cilia in hair follicle morphogenesis. J Investig Dermatol 129(2):438–448

Li YH, Li XF, Liu JT, Wang H, Fan LL, Li J et al (2018) PKM2, a potential target for regulating cancer. Gene 668:48–53

Liu H, Kiseleva AA, Golemis EA (2018) Ciliary signalling in cancer. Nat Rev Cancer 18(8):511–524
Mao H, Tang Z, Li H, Sun B, Tan M, Fan S et al (2019) Neddylation inhibitor MLN4924 suppresses cilia formation by modulating AKT1. Protein Cell:1–19. https://doi.org/10.1007/s13238-019-0614-3. Mar 9. pii: 10.1007/s13238-019-0614-3
McCubrey JA, Steelman LS, Chappell WH, Abrams SL, Wong EW, Chang F et al (2007) Roles of the Raf/MEK/ERK pathway in cell growth, malignant transformation and drug resistance. Biochim Biophys Acta 1773(8):1263–1284
Meyer O (2009) Interferons and autoimmune disorders. Joint Bone Spine 76(5):464–473
National Center for Biotechnology Information (n.d.-a) PubChem Database. Pevonedistat, CID=16720766, https://pubchem.ncbi.nlm.nih.gov/compound/Pevonedistat. Accessed on July 21 2019
National Center for Biotechnology Information (n.d.-b) PubChem Database. Saicar, CID=160666, https://pubchem.ncbi.nlm.nih.gov/compound/Saicar. Accessed on July 21 2019
Nawrocki ST, Griffin P, Kelly KR, Carew JS (2012) MLN4924: a novel first-in-class inhibitor of NEDD8-activating enzyme for cancer therapy. Expert Opin Investig Drugs 21(10):1563–1573
Nguyen LV, Vanner R, Dirks P, Eaves CJ (2012) Cancer stem cells: an evolving concept. Nat Rev Cancer 12 (2):133–143
Oda K, Matsuoka Y, Funahashi A, Kitano H (2005) A comprehensive pathway map of epidermal growth factor receptor signaling. Mol Syst Biol 1:2005 0010
Qian Y, Liu C, Hartupee J, Altuntas CZ, Gulen MF, Jane-Wit D et al (2007) The adaptor Act1 is required for interleukin 17-dependent signaling associated with autoimmune and inflammatory disease. Nat Immunol 8(3):247–256
Rishikaysh P, Dev K, Diaz D, Qureshi WM, Filip S, Mokry J (2014) Signaling involved in hair follicle morphogenesis and development. Int J Mol Sci 15 (1):1647–1670
Safa AR (2012) c-FLIP, a master anti-apoptotic regulator. Exp Oncol 34(3):176–184
Scaltriti M, Baselga J (2006) The epidermal growth factor receptor pathway: a model for targeted therapy. Clin Cancer Res 12(18):5268–5272
Song H, Huai W, Yu Z, Wang W, Zhao J, Zhang L et al (2016) MLN4924, a first-in-class NEDD8-activating enzyme inhibitor, attenuates IFN-beta production. J Immunol 196(7):3117–3123
Soucy TA, Smith PG, Milhollen MA, Berger AJ, Gavin JM, Adhikari S et al (2009) An inhibitor of NEDD8-activating enzyme as a new approach to treat cancer. Nature 458(7239):732–736
Takahashi K, Yamanaka S (2006) Induction of pluripotent stem cells from mouse embryonic and adult fibroblast cultures by defined factors. Cell 126(4):663–676
Tan P, Fuchs SY, Chen A, Wu K, Gomez C, Ronai Z et al (1999) Recruitment of a ROC1-CUL1 ubiquitin ligase by Skp1 and HOS to catalyze the ubiquitination of IkBa. Mol Cell 3:527–533
Taniguchi T, Ogasawara K, Takaoka A, Tanaka N (2001) IRF family of transcription factors as regulators of host defense. Annu Rev Immunol 19:623–655
Tiaka EK, Papanas N, Manolakis AC, Georgiadis GS (2012) Epidermal growth factor in the treatment of diabetic foot ulcers: an update. Perspect Vasc Surg Endovasc Ther 24(1):37–44
Walsh MC, Lee J, Choi Y (2015) Tumor necrosis factor receptor- associated factor 6 (TRAF6) regulation of development, function, and homeostasis of the immune system. Immunol Rev 266(1):72–92
Walz G (2017) Role of primary cilia in non-dividing and post-mitotic cells. Cell Tissue Res 369(1):11–25
Wenczak BA, Lynch JB, Nanney LB (1992) Epidermal growth factor receptor distribution in burn wounds. Implications for growth factor-mediated repair. J Clin Investig 90(6):2392–2401
Witney TH, James ML, Shen B, Chang E, Pohling C, Arksey N et al (2015) PET imaging of tumor glycolysis downstream of hexokinase through noninvasive measurement of pyruvate kinase M2. Sci Transl Med 7 (310):310ra169
Wong N, Ojo D, Yan J, Tang D (2015) PKM2 contributes to cancer metabolism. Cancer Lett 356(2 Pt A):184–191
Zhao L, Yue P, Lonial S, Khuri FR, Sun SY (2011) The NEDD8-activating enzyme inhibitor, MLN4924, cooperates with TRAIL to augment apoptosis through facilitating c-FLIP degradation in head and neck cancer cells. Mol Cancer Ther 10(12):2415–2425
Zhao Y, Morgan MA, Sun Y (2014) Targeting neddylation pathways to inactivate Cullin-RING ligases for anticancer therapy. Antioxid Redox Signal 21 (17):2383–2400
Zhou X, Tan M, Nyati MK, Zhao Y, Wang G, Sun Y (2016) Blockage of neddylation modification stimulates tumor sphere formation in vitro and stem cell differentiation and wound healing in vivo. Proc Natl Acad Sci U S A 113(21):E2935–E2944
Zhou L, Zhang W, Sun Y, Jia L (2018) Protein neddylation and its alterations in human cancers for targeted therapy. Cell Signal 44:92–102
Zhou L, Jiang Y, Luo Q, Li L, Jia L (2019a) Neddylation: a novel modulator of the tumor microenvironment. Mol Cancer 18(1):77
Zhou Q, Li H, Li Y, Tan M, Fan S, Cao C et al (2019b) Inhibiting neddylation modification alters mitochondrial morphology and reprograms energy metabolism in cancer cells. JCI Insight 4(4):e121582
Zubiete-Franco I, Fernandez-Tussy P, Barbier-Torres L, Simon J, Fernandez-Ramos D, Lopitz-Otsoa F et al (2017) Deregulated neddylation in liver fibrosis. Hepatology 65(2):694–709

The manufacturer's authorised representative in the EU is Springer
Nature Customer Service Centre GmbH, Europaplatz 3, 69115 Heidelberg,
Germany. If you have any concerns regarding our products, please
contact ProductSafety@springernature.com

Printed and bound by CPI Group (UK) Ltd, Croydon, CR0 4YY
30/09/2026
02225556-0019